Urban Ornithology

Urban Ornithology

150 Years of Birds in New York City

P. A. Buckley, Walter Sedwitz,
William J. Norse, and John Kieran

COMSTOCK PUBLISHING ASSOCIATES
an imprint of
Cornell University Press
Ithaca and London

Copyright © 2018 by Cornell University

All rights reserved. Except for brief quotations in a review, this book, or parts thereof, must not be reproduced in any form without permission in writing from the publisher. For information, address Cornell University Press, Sage House, 512 East State Street, Ithaca, New York 14850. Visit our website at cornellpress.cornell.edu.

First published 2018 by Cornell University Press

Printed in the United States of America

Library of Congress Cataloging-in-Publication Data

Names: Buckley, P. A., author. | Sedwitz, Walter, –1985, author. |
   Norse, William J., –2006, author. | Kieran, John, 1892–1981, author.
Title: Urban ornithology : 150 years of birds in New York City /
   P.A. Buckley, Walter Sedwitz, William J. Norse, and John Kieran.
Description: Ithaca : Comstock Publishing Associates, an imprint of
   Cornell University Press, 2018. | Includes bibliographical references and indexes.
Identifiers: LCCN 2017054478 (print) | LCCN 2017057126 (ebook) |
   ISBN 9781501719639 (epub/mobi) | ISBN 9781501719622 (pdf) |
   ISBN 9781501719615 | ISBN 9781501719615 (cloth ; alk. paper)
Subjects: LCSH: Birds—New York (State)—New York—History. |
   Bird populations—New York (State)—New York—History.
Classification: LCC QL684.N7 (ebook) | LCC QL684.N7 B825 2018 (print) |
   DDC 598.09747—dc23
LC record available at https://lccn.loc.gov/2017054478

To John L. Young and Catherine (Minty) O'Brien

Students and Stewards of
Van Cortlandt Park's Birds
and Their Environment

Contents

List of Figures   ix
List of Tables   xi
Preface   xiii
Acknowledgments   xv

Introduction   1
Avifaunal Overview   27
    Terminology   27
    Subarea Coverage   27
    Historical and Recent Data Sources   28
    Van Cortlandt Contrasted with Central and Prospect Parks   34
    Ecological Connectivity   36
    Breeding Species   36
    Winter Species   48
    Migration   52
    Resource Concerns   59
    The Future   67
Introduction to the Species Accounts   73
Species Accounts   81

Appendixes
    1. Tables   413
    2. Scientific Names of All Organisms Mentioned in This Book Other Than Birds in the Species Accounts   459
    3. Glossary of Symbols, Abbreviations, and Terms Used in This Book   462
    4. Names of All Observers Appearing Anywhere in the Body of This Book   465
    5. Stranger in a Strange Land: An Andean Gull in the Bronx   468
    6. Specimens in Museum Collections from Van Cortlandt Park, Kingsbridge Meadows, Woodlawn Cemetery and Jerome Reservoir, and Riverdale   470

Literature Cited   487
About the Authors   495
Indexes
    English Bird Names   497
    Scientific Bird Names   503
    Subjects   509

Figures

Please note: Except as noted, all views are looking approximately NNE. Also, identical tags are sometimes placed at multiple locations on a single feature.

Plate 1, facing page 1. Largely tidal Kingsbridge Meadows looking southwest in 1906 from modern W. 238th St. and appearing just as it did when Bicknell worked it from 1872 to 1901. At left rear is wooded Marble Hill, to its right in the distance is Inwood Hill, and fringing the west side of Tibbett's Brook is wooded Riverdale. The dominant meadow vegetation is Smooth Cordgrass, which near land is augmented by Saltmarsh Cordgrass, Marsh Spikegrass, and cattails, with nary a trace of *Phragmites*. Today the entire area is residential city streets, having been so since the 1930s. Image from the Edward Wenzel collection in the New-York Historical Society.

1. Northwest Bronx and adjacent Yonkers in 1891, showing all 7 study subareas   2
2. Northwest Bronx and adjacent Yonkers in 2015, showing the 4 extant study subareas   3
3. Northwest Bronx surface water and wetlands in October 1609   5
4. Harlem River and Spuyten Duyvil Creek saltmarshes in 1891   6
5. Details of the extant study area in 2014–15   11
6. Site of future Hillview Reservoir in 1891   12
7. Hillview Reservoir with conifer groves (1994)   13
8. Hillview Reservoir with no conifer groves (2014)   13
9. Jerome Meadows in 1891   14
10. Jerome Reservoir and Jerome Swamp in 1924   15
11. Kingsbridge Island, Kingsbridge Meadows, Van der Donck Meadows in 1891   16
12. Van Cortlandt Park and adjacent areas in 1924   17
13. Parade Ground, Dutch Gardens, and Van Cortlandt Lake in 1924   20
14. Harlem River Ship Canal, Spuyten Duyvil Creek, and Kingsbridge Meadows in 1895   20
15. Van der Donck Meadows and Kingsbridge Island in the 1700s   21
16. Van Cortlandt Park, Hillview Reservoir, Woodlawn Cemetery in 1940   23
17. Segments of Westchester County subsumed into Griscom's (1923) and Kuerzi's (1927) *Bronx Regions*   30
18. All nine Bronx NYS Breeding Bird Atlas blocks   43
19. The two study area NYS Breeding Bird Atlas blocks   44
20. Bronx-Westchester Christmas Bird Count participants 1924–2013   51

The 43 species accounts with scatterplots of Bronx-Westchester Christmas Bird Count annual data from 1924 to 2013

21. Canada Goose   91
22. Mute Swan   93

23. American Black Duck   98
24. Mallard   100
25. Northern Shoveler   103
26. Northern Pintail   104
27. Green-winged Teal   105
28. Canvasback   108
29. Greater Scaup   111
30. Hooded Merganser   119
31. Ruddy Duck   122
32. Ring-necked Pheasant   126
33. Double-crested Cormorant   136
34. Black-crowned Night-Heron   147
35. Sharp-shinned Hawk   157
36. Cooper's Hawk   159
37. Red-shouldered Hawk   161
38. Red-tailed Hawk   164
39. Wilson's Snipe   192
40. Ring-billed Gull   199
41. Mourning Dove   211
42. Red-bellied Woodpecker   231
43. Yellow-bellied Sapsucker   232
44. American Kestrel   238
45. Monk Parakeet   242
46. Blue Jay   268
47. American Crow   269
48. Fish Crow   271
49. Horned Lark   273
50. Tufted Titmouse   287
51. Carolina Wren   296
52. Eastern Bluebird   302
53. Northern Mockingbird   313
54. Cedar Waxwing   316
55. American Tree Sparrow   358
56. Field Sparrow   361
57. Vesper Sparrow   363
58. Savannah Sparrow   365
59. Northern Cardinal   381
60. Eastern Meadowlark   390
61. Rusty Blackbird   392
62. House Finch   400
63. Purple Finch   401

Tables

1. Sizes of important study area components   413
2. Approximate distances from Van Cortlandt Lake to locations mentioned in the text   414
3. Years by decades after 1941 when new species have been added to the study area cumulative list   415
4. Non–Long Island Sound New York State sites no farther than 3 smi (5 km) from Van Cortlandt Lake that have recorded species still unknown in the study area   416
5. Species found in only 1 of the 6 non–Van Cortlandt Park study subareas   416
6. Study area breeders (and their locations) that have nested only outside but still immediately adjacent to Van Cortlandt Park   417
7. Cumulative species lists for Van Cortlandt, Central, and Prospect Parks   417
8. Species unique to Van Cortlandt, Central, or Prospect Park   425
9a. Species recorded in both Central and Prospect Parks but not Van Cortlandt   425
9b. Species recorded in both Van Cortlandt and Prospect Parks but not Central   425
9c. Species recorded in both Van Cortlandt and Central Parks but not Prospect   425
10. Reasonably anticipated species never recorded in Central, Prospect, or Van Cortlandt Parks   426
11. Cumulative known and potential breeding species in Van Cortlandt, Central, and Prospect Parks   426
12. Study area cumulative breeding species that have never bred in either Central or Prospect Park   429
13. Synopsis of museum specimens from the study area and Riverdale in (largely) the late 1800s   430
14. HY Herring Gulls seen in Van Cortlandt Park in 1937 that were ascribable to their natal colonies by unique colorband combinations   433
15. The species of birds known to have bred in the study area after 1871   433
16. Unconfirmed or suspected past, and reasonably likely future, study area breeders with their probable sites   434
17. History of all species breeding in the study area after 1871   435
18. Study area breeders lost by decades after 1871, their probable year of last breeding, and likely reasons for their loss   440
19. New study area breeders gained by decades after 1871, probable year of first breeding, and likely provenance   441
20. Current, extirpated, or potential breeding species within the study area that regularly or occasionally nest on or nearly on the ground, or are oldfield succession breeders   442

21. Species believed to have bred in the study area more or less annually after 1871—primary breeders   443
22. Species believed to have bred in the study area more or less annually from the years indicated—secondary breeders   443
23. Known or potential study area breeding species that regularly or occasionally nest in or on artificial structures, especially nest boxes   444
24. Recent study area breeding species whose populations are increasing or decreasing   444
25. Breeding birds of Woodlawn Cemetery in 1960–61, plus species and numbers of likely pairs in late May 2014   445
26. Estimated numbers of species and pairs breeding in 2013–14 within Van Cortlandt Park and immediately adjacent urban areas   446
27. Estimated numbers of breeding pairs across 77 years of censuses of Van Cortlandt Swamp in the 1930s, 1960s, and 2013–14   448
28. Estimated numbers of breeding pairs (transformed to density per 100 acres [40 ha]) on censuses of Van Cortlandt Swamp in the 1930s, in the 1960s, and in 2013–14   449
29. Predicted ranges in the 1960s of breeding pairs of the 28 species that nested in Van Cortlandt Swamp in both the 1930s and 1960s if breeding densities of the 1930s censuses had remained stable after the swamp was reduced in size by 48%   450
30. Estimated numbers of occupied territories found on breeding bird censuses of a 25 ac (10 ha) mature forest plot in Van Cortlandt Park's Northwest Forest in 1991–92   451
31. Woodlawn Cemetery species found on a Bronx-Westchester Christmas Bird Count, 27 December 2015   452
32. Species believed to have occurred and probably overwintered in the study area annually after 1871—core winter residents   452
33. Estimated mean number of individuals per species found during Winter Bird Population Studies of Van Cortlandt Swamp in 1962–63, 1963–64, and 1965–66   453
34. Estimated mean number of individuals per waterbird species found during Winter Bird Population Studies of Jerome Reservoir in 1963–64 and 1965–66   454
35. Migrating shorebirds forced down to the recently drained Van Cortlandt Lake bed by severe weather conditions on 10 May 1951   455
36. Years with the highest study area daily spring counts for insectivorous/frugivorous Neotropical migrants before 1966 and after 1965   455
37. Earliest study area spring dates in chronological order for mostly Neotropical migrant landbirds before 1966 and after 1965   456
38. New York State and federal at-risk species known to have occurred in the study area   458

Preface

This book offers the first quantitative long-term historical analysis of the migratory, winter, and breeding avifaunas of any New York City natural area or park—Van Cortlandt Park and the Northwest Bronx—and spans the century and a half from 1872 to 2016. Only Manhattan's Central and Brooklyn's Prospect Parks have published even lightly annotated cumulative species lists, but they were last updated in 1967. Much has happened to the birdlife of New York City in the last 50 years.

Westchester County did not relinquish the Northwest Bronx to New York City until 1874. It was then heavily wooded with a low, dispersed human population except along the Broadway corridor. Consequently, its birdlife had remained rich and largely undisturbed. It came to the public's attention when in 1888 the City of New York took over the ancestral acreage of the Van Cortlandt family and created a large new park named for them. But apart from collecting and observations by two early ornithologists, Eugene Bicknell and Jonathan Dwight, its avifauna and that of adjacent Riverdale to its west had not been studied. As the city grew and its population expanded outward, attention gradually began to be focused on the birdlife of this gem of an area in the only one of the five New York City boroughs (counties) actually on the mainland of New York State. From 1872 to the present, coverage has been continuous and episodically intense, so it is an appropriate time for an analysis of its 150 years of ornithological information.

When we began gathering data for this book in the early 1960s, we had already known from our own fieldwork back to the early 1900s (Kieran) how complex the avifauna of the Northwest Bronx was, but not until later did we appreciate, for example, that the cumulative total of recorded Van Cortlandt Park bird species equaled those of Central Park and Prospect Park, respectively, despite the fact that their annual observer-hours had long exceeded those of Van Cortlandt by perhaps 50:1. Another eventual discovery was that Van Cortlandt's cumulative and current numbers of breeding species were nearly twice those of each of the other two parks and that of all the species that had ever bred in either Central or Prospect, only a single one had never bred with certainty in Van Cortlandt. This species richness we attribute to the physiographic diversity of Van Cortlandt and that it has always been largely a natural, not manufactured, park with strong ecological connectivity to adjacent Westchester County.

Our study area could never have been limited to just Van Cortlandt Park but had to embrace Woodlawn Cemetery, Jerome Reservoir, and Hillview Reservoir because they shared Van Cortlandt's avifauna daily, being contiguous with it (Woodlawn and Jerome) or only a few streets away (Hillview); the four had long been treated as a unit. We soon realized that the Jerome Swamp area immediately east of Jerome Reservoir also belonged, as did Jerome Meadows (the site of Jerome Reservoir) and the tidal Tibbett's Brook marshes from Van Cortlandt Park to the Harlem River/Spuyten Duyvil Creek known as Kingsbridge Meadows, even though these three were long gone. We briefly considered incorporating adjacent wooded Riverdale but chose instead to refer to it as needed to complement our study area analyses. And because they had not been updated since the 1960s, we ultimately elected to give the current and historical statuses of the 301 study area species across the Bronx, in Central and Prospect Parks, in New York City, and in the New York City area.

This book will enable students of urban birds to answer such questions as the following:

- Which bird species and in what numbers have been occurring in the Northwest Bronx since 1872?
- Why has this area long been so bird-rich?
- Which species and in what numbers currently breed in Van Cortlandt Park?
- Which breeding species have been lost and which gained in Van Cortlandt over the last 150 years, and why?
- What has been the 150-year history of all study area nonbreeding species?
- Which species have been restricted to Jerome Reservoir, Hillview Reservoir, Woodlawn Cemetery, Kingsbridge Meadows, Jerome Meadows, or Jerome Swamp?
- How has fresh- and saltwater wetland loss in the Northwest Bronx affected its birdlife?
- What has been the normal winter bird complement for the study area?
- How has this changed over the last 150 years, and why?
- Which migrants have declined, increased, or are arriving earlier than in years past?
- How do the breeding and nonbreeding avifaunas of mainland Van Cortlandt Park compare with those of island-based and densely urban Central and Prospect Parks?
- What urban park management conclusions can be drawn from such comparisons?
- What do they tell us about the resilience of urban forests, the value of New York City reservoirs to native waterbirds, and birds' need for open, undisturbed grasslands?

A wide array of readers will find this book timely and topical, among them New York City and other urban naturalists and observers; students of urban ecology, wildlife management, and conservation; landscape, restoration, and connectivity ecologists; undergraduate and graduate students in many disciplines; primary and secondary school students; urban environmental center users; and ordinary citizens concerned about the importance of urban natural and man-made areas to native birds. The baseline information it provides is unequaled for any other New York City natural area and will serve as a template for fresh studies there and elsewhere.

Acknowledgments

We owe singular thanks to many individuals and organizations for indispensable support at various stages in this project.

Richard Kane spent many days afield with all four authors and countless hours discussing what we knew and didn't know about birds in the Bronx. Among the earliest supporters of this book, he read various versions of the entire manuscript and never failed to offer insightful suggestions and tactful corrections. In particular, his comments on bird distribution and migration patterns in New Jersey illuminated and clarified many of our views.

Joseph L. Horowitz has been a proponent and constructive critic as far back as this book's inception and continuing through to closely reading the entire final draft and offering innumerable essential pointers. He has also been a frequent field companion of the authors and graciously provided timely copies of Tieck's incomparable Northwest Bronx histories (1968, 1971) during this book's writing.

John L. Young has long been an indefatigable Van Cortlandt Park observer and one of the best qualified to offer insights into changes in its avifauna over the last 60 years. He unstintingly made all his field notes available and unflinchingly responded to numerous queries about its recent birdlife. He too read the final manuscript cover to cover, as did his wife and Van Cortlandt Park champion, Catherine (Minty) O'Brien.

David Künstler has worked in and published about the birds of Van Cortlandt Park and Pelham Bay Park as a New York City Parks Department naturalist and resource manager for 30-plus years. He also answered countless questions about both parks with easy good humor and wry wit.

Ellen Pehek, from New York City Parks Department's Natural Resources Group, established a number of bird-related Van Cortlandt Park studies, the most important for our purposes being a series of quantified breeding bird censuses. She offered her findings for incorporation into this book, and they complemented and allowed comparison with similar Van Cortlandt censuses going back 80 years.

Susan Elbin provided background and other unpublished material from New York City Audubon's 2006 Van Cortlandt Park breeding bird census, Central Park breeding bird censuses, and New York harbor heron surveys. All of these helped place study area birds in a regional perspective.

Once the manuscript had been completed in its first final draft form, in addition to Richard Kane, Joseph Horowitz, John Young, and Minty O'Brien, it was read completely by Francine Buckley, Yolanda Garcia, Patricia Kane, Christopher Lyons, Peter Post, and Richard Veit. Collectively they offered invaluable corrections and suggestions that smoothed its message and obviated more than a few blunders.

We are indebted to the following for providing information and assistance of many kinds: American Museum of Natural History, *Bird Observer*, *Birds of North America* online, Bronx County Bird Club, Bronx County Historical Society, Brooklyn Bird Club, Florida Museum of Natural History, Friends of Van Cortlandt Park, Google Earth Pro, Cornell Lab of Ornithology and National Audubon Society's eBird program, Library of Congress, Linnaean Society of New York, Massachusetts Audubon Society, Museum of Comparative Zoology, Museum of the City of New York, National Audubon Society's Christmas Bird Count office, New Jersey Audubon Society's Scherman-Hoffman Sanctuary, New York Botanical Garden, New York City Audubon Society's harbor herons project, New York City Department of Parks and Recreation and its Natural Resources Group, New York State Breeding Bird Atlas program, New York State Department of Environmental Conservation, New York State Maritime College, New-York Historical Society, *North American Birds,* Patuxent Wildlife Research Center's Bird-banding Laboratory, Queens County Bird Club, Vassar College library, VertNet program, Wave Hill, and Western Foundation of Vertebrate Zoology.

Many other persons have been vital not only to this book's accuracy but to its very existence. It is a pleasure to thank all who

—furnished essential historical Bronx and study area background information, during often protracted discussions going back a full century: Ned Boyajian, John Bull, Irving Cantor, Geoffrey Carleton, Samuel Chubb, Allan Cruickshank, Eugene Eisenmann, William Ephraim, Richard Fischer, Ludlow Griscom, Richard Herbert, Joseph Hickey, George Hix, Thomas Imhof, George Komorowski, John Kuerzi, Richard Kuerzi, Nelson Nichols, Erik Petersen, Roger Peterson, Charles Rogers, Richard Ryan, Gus Schmidt, Charles Staloff, William Weber, Edward Whelen, William Wiegmann, Samuel Yeaton, and John Yrizarry;

—offered data or their diaries, journals, and field notebooks from which we extracted pertinent information: Debbie Becker, Irving Cantor, Frank Enders, Bill Ephraim, Stuart Friedman, Yolanda Garcia, Michael Gochfeld, Fred Heath, Joseph Horowitz, Tom Imhof, Carl Jaslowitz, Richard Kane, George Komorowski, Chris Lyons, Paul Mayer, Paul Meyer, Dan Rafferty, Marshall Russak, Gus Schmidt, Robert Scully, Paul Steineck, Si Stepinoff, Michael Teator, Guy Tudor, William Weber, Joel Weintraub, and Jeffrey Zupan;

—took the time to answer specific questions and requests about a variety of topics: John Askildsen, Scott Barnes, Matthieu Benoit, Barry Bermudez, Andrew Bernick, Lewis Bevier, Shane Blodgett, Ned Boyajian, Bill Boyle, Ned Brinkley, Malcolm Brown, John Bull, Barbara Butler, John Butler, Geoffrey Carleton, Scott Crocoll, Robert DeCandido, Stan DeOrsey, Robert Dieterich, Eugene Doggett, Kim Doggett, Wick Doggett, Alan Drogin, Susan Elbin, Walter Ellison, Tom Fiore, Arie Gilbert, Howard Ginsberg, Adele Gotlib, Jon Greenlaw, John Haas, Jennifer Hanson, Kevin Karlson, Bill Kornblum, Dave Krauss, Laurie Larson, Chris Leahy, Paul Lehman, Jean Loscalzo, Kevin McGowan, Hugh McGuinness, Ian McLaren, Shai Mitra, Steve Mlodinow, Ulrich Naf, Christopher Nagy, Todd Olson, John Ozard, Robert Paxton, Roger Pasquier, Ellen Pehek, Anders Peltomaa, Bryan Pfeiffer, Peter Post, Peter Pyle, Chris Rimmer, Laurel Rimmer, Betsy Rogers, Kenneth Rosenberg, Jack Rothman, Lee Schlesinger, Jessica Arcate Schuler, P.W. (Bill) Smith, Nadir Souirgi, Mark Szantyr, Scott Turner, John Waldman, Steve Walter, Seth Wollney, and Bruce Yolton;

—helped locate important references: Christian Artuso, Jim Berry, Louis Bevier, Ned Brinkley, Jeff Collins, René Corado, Joe DiCostanzo, Walter Ellison, Linnea Hall, Lois Horst, Jon Greenlaw, Elaine Kile, Ann Massa, Christopher Nagy, Chris Rimmer, Laurel Rimmer, Peter Oehlkers, Kim Peters, Wayne Petersen, P.W. (Bill) Smith, Richard Veit, Tom Webber, and Jim Wiley;

—devised the indexes and offered timely technical advice on numerous occasions: Jennifer W. Hanson;

—prepared the New York State Breeding Bird Atlas block figures for the Bronx: Ben Cacace, whose prowess with eBird files was magisterial;

—ensured that our assessments of the avifaunas of Central and Prospect Parks were current: Rick Cech, Peter Dorosh, Tom Fiore, Doug and Bob Gochfeld, Rob Jett, Dave Krauss, Peter Post, and John Yrizarry;

—answered questions and supplied critical Christmas Bird Count files, unpublished information on the two New York State Breeding Bird Atlases, and raw eBird data: Ian Davies and eBird staff, Geoff LeBaron, Kevin McGowan, and John Ozard;

—provided critical assistance with museum collections, pertinent specimens and bird-banding recoveries: Mike Anderson, David Boertmann, Peter Capainolo, Tony Fox, Kaj Kampp, Richard Kane, Bruce Peterjohn, Matthew Rogosky, Jeremiah Trimble, Tom Trombone, David Stroud, and Paul Sweet;

—most graciously gave permission to reproduce their images as figures: David Rumsey and Cartography Associates; Eric Sanderson, Christopher Spagnoli, and the Wildlife Conservation Society—who prepared a new Figure 3 expressly for our use; and Arthur Morris/BIRDS AS ART-Biog.com for his photograph of our cover Least Bittern.

—facilitated our image use: Google Maps, Google Earth Pro, the New-York Historical Society, the Museum of the City of New York, the New York City Department of Records, and the New York City Digital Map Atlas.

—furnished Bronx logistic support when needed: New York State Maritime College, Grace Tilger, and Lloyd Ultan;

—offered timely assistance on numerous occasions to ensure that the Macs used in the final preparation of the manuscript worked flawlessly and when problems arose immediately came to the rescue countless times: Joshua Clapper and Laurie Larson.

We particularly thank Walter Sedwitz's daughter Helene Salter and John Kieran's granddaughters Meg and Jane Kieran for acting on behalf of their deceased relatives.

Finally, we are in debt to Kitty Liu, Meagan Dermody, Susan Specter, Martyn Beeny, and the entire editorial, production, and marketing staff at Cornell University Press and to the anonymous reviewers of this book in manuscript form, all of whom were most generous with their time, patience, and constructive suggestions.

Notwithstanding the foregoing, if anyone has inadvertently been overlooked, we tender profound apology. All statements and opinions are solely those of the authors.

Urban Ornithology

Plate 1. Kingsbridge Meadows, 1906

Introduction

It was a dark and *very* stormy night.

On 18–19 November 1932 two large low-pressure systems—one well east of Cape Cod, the other southwest over the Appalachians—interacted with a high-pressure system over the Gulf of St. Lawrence to generate abnormally strong onshore winds and heavy coastal rains on the Atlantic Coast. In consequence, flocks of hundreds of thousands of small highly pelagic, plankton-eating seabirds named Dovekies (scientific names of most native birds appear in the Species Accounts, and of all other organisms in Appendix 2) were abruptly blown onshore, alongshore, or very near shore, from Nova Scotia to Cuba. Such an event had never been previously known and has never recurred.

The 19th was a Saturday and the 20th a Sunday, the only reason anyone was out and about. Had it been midweek, many of the storm's effects and most of the Dovekies would have arrived and departed, or died, unnoticed. But by Saturday and Sunday morning, they began popping up around New York City in the most amazing places, including the Hudson River, the Brooklyn Botanical Garden (Prospect Park), and Bronx Park. More to our point, they were even in the Northwest Bronx: two were on Jerome Reservoir on Sunday, unsurprising in that Jerome Reservoir is the largest freshwater body in the Bronx and so a major target for storm-blown waterbirds. It is inconceivable there were none on high-elevation Hillview Reservoir, the same size as Jerome, but nobody looked. It's also likely there was one on Van Cortlandt Lake, but nobody looked there either.

But a group of five Dovekies had made it to nearby Spuyten Duyvil, where too exhausted to fly or dive, they huddled low to the water and drifted with the currents. Befitting the depths of the Great Depression, they were noticed by some local boys—who quickly commandeered a rowboat, then chased, caught, cooked, and consumed all five.

The 1932 Dovekie wreck is among the most out-of-reality bird events in and near our study area, which is delimited and described below. Yet it is a fine example of birds' unpredictable propensity for appearing anywhere, anytime. The Northwest Bronx has seen its share of unexpected birds, but its avian importance lies in the regularity with which migrants of all kinds use it to feed, rest, and breed. It was there that the pioneering Bronx ornithologists Eugene P. Bicknell and Jonathan Dwight began to take scientific notice of them, documenting their occurrences and recording which bred where and for how long, which were migrants, and which were only winter visitors. They backed up their field notes with an array of museum specimens still extant. This when the area was still in Westchester County, less than 100 years after the American Revolution and only 20 after the death of John James Audubon in upper Manhattan. These two ornithologists began one of the longest continuous chronicles in American ornithology and the only one anywhere in, or even near, New York City.

THE PRECONTACT ENVIRONMENT

The northwestern corner of the modern borough (county) of the Bronx in New York City is bounded by the Hudson River on the west; the Harlem River Ship Canal, Harlem River, and Fordham Rd. on the south; the Bronx River on the east; and the city of Yonkers on the north (Figs. 1, 2). Note that all photos and maps

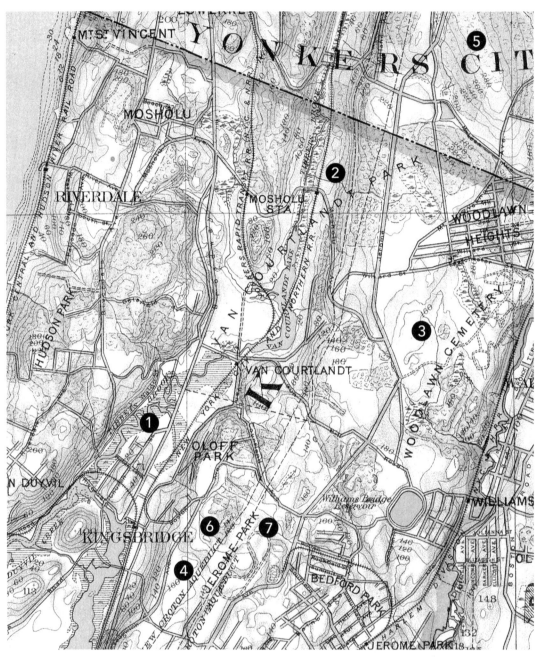

Figure 1. Northwest Bronx and adjacent Yonkers in 1891, showing the relative locations of all 7 study subareas. Only 4 appear on this map: Kingsbridge Meadows (1), Van Cortlandt Park (2), Woodlawn Cemetery (3), and Jerome Meadows (4), but the sites of future Hillview Reservoir (5), Jerome Reservoir (6), and Jerome Swamp (7) are indicated. From Bien and Vermeule 1891a. Courtesy of David Rumsey Map Collection.

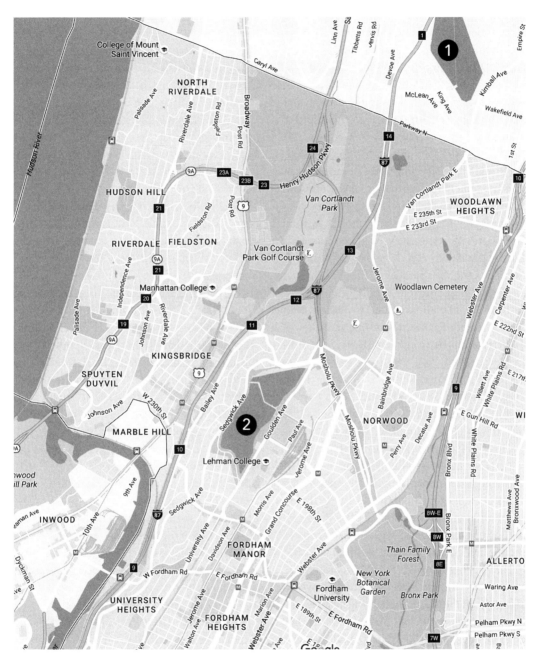

Figure 2. Northwest Bronx and adjacent Yonkers in 2015, indicating the 4 extant subareas: Van Cortlandt Park, Woodlawn Cemetery, Hillview Reservoir (1), and Jerome Reservoir (2). Image courtesy of Google Maps.

look northward unless specifically indicated otherwise.

Before the arrival of Europeans, the Northwest Bronx showed some of the greatest relief, most varied habitats, and elevational extremes within the borough, including its highest elevation: 282 ft (86 m), in Riverdale west of Fieldston Rd. and W. 250th St., in the woods past the end of Goodridge Ave. This is only 2 m higher than the highest natural point on the island of Manhattan in Fort Tryon Park. Geologically, the Northwest Bronx lies within the Manhattan Prong of the New England Geomorphic (or Physiographic) Province that extends southward from the Hudson Highlands through Westchester, the Bronx, Manhattan, and extreme western Queens, finally terminating on Staten Island. In the Northwest Bronx it is expressed as two long north–south ridges paralleling the Hudson River: Riverdale (or Spuyten Duyvil) Ridge just east of the river and Fordham Ridge to the east, with Van Cortlandt Park surrounding the valley of Tibbett's Brook between them (Figs. 1, 3). Within Van Cortlandt Park, maximum elevations are about 210 ft (64 m) (Northwest and Northeast Forests), 190 ft (58 m) (the Ridge, aka Croton Forest), 180 ft (55 m) (Shandler Area), and 140 ft (43 m)(Vault Hill). Woodlawn Cemetery reaches 210 ft (64 m) and Hillview Reservoir 310 ft (95 m). The basal rocks on the ridges are schists and gneisses, with occasional outcrops like the Yonkers granite on Vault Hill in Van Cortlandt, and marble underlies most of the NE–SW trending valleys. Other visible geological features include fault lines like the one outlining Mosholu Parkway between Van Cortlandt and Bronx Parks. Away from the ridges, the coastal lowland areas within the Manhattan Prong lie generally at an elevation of 20–25 ft (6–8 m), just as they do along the entirety of Tibbett's Valley in Van Cortlandt Park. Ecologically, all of the mainland south of the Hudson Highlands east of the Hudson River is considered the Manhattan Hills ecozone or province, while all of Long Island falls in the Coastal Lowlands ecozone.

In the Northwest Bronx, permanent, year-round freshwater streams of consequence were few (Fig. 3): the Bronx River on its eastern boundary and Tibbett's Brook flowing south from southern Yonkers through the valley described above and emptying via Kingsbridge Meadows into Spuyten Duyvil Creek. A handful of probably seasonal streams drained Riverdale Ridge west into the Hudson River and east into Tibbett's Brook and then south to Kingsbridge Meadows. Similarly, a few small streams drained Fordham Ridge west into Tibbett's Brook and east into the Bronx River and Jerome Meadows south-southeast into Mill Brook (now gone, occupied by Webster Ave; Fig. 3) and east into an unnamed stream (now gone) that ran the length of modern Mosholu Parkway, emptying to the Bronx River at Bronx Park. No large bodies of freshwater existed, but a few small ponds were scattered throughout Riverdale and the future Van Cortlandt Park (Fig. 3).

Freshwater marshes existed in patches along Tibbett's Brook and the Bronx River (Fig. 3), but the only major one was Van Cortlandt Swamp, which ran the length of Tibbett's Valley from the modern Yonkers line south into Kingsbridge Meadows. South of today's Van Cortlandt Park, Tibbett's Brook became brackish, then salty and tidal as it flowed through Kingsbridge Meadows on its way to the northwestern corner of Spuyten Duyvil Creek at modern W. 230th St. At the northeastern corner of Spuyten Duyvil Creek, Van der Donck Meadows with its brackish and freshwater marshes extended north to the east of modern Broadway, approximately following the track bed of the New York Central Railroad's Putnam Division, to about modern W. 237th St., where it merged with Kingsbridge Meadows, bracketing the dry land between them known as Kingsbridge Island (Figs. 1, 3, 11, 15).

The few saltmarshes were in isolated patches along the Harlem River, the most prominent being at Sherman Creek (modern Dyckman St. in Manhattan) and then along Spuyten Duyvil Creek from the Harlem to the Hudson Rivers. Kingsbridge Meadows was also largely in saltmarshes, but the Hudson River shore in Riverdale lacked any even before the railroad was built there in the late 1840s (Figs. 3, 4).

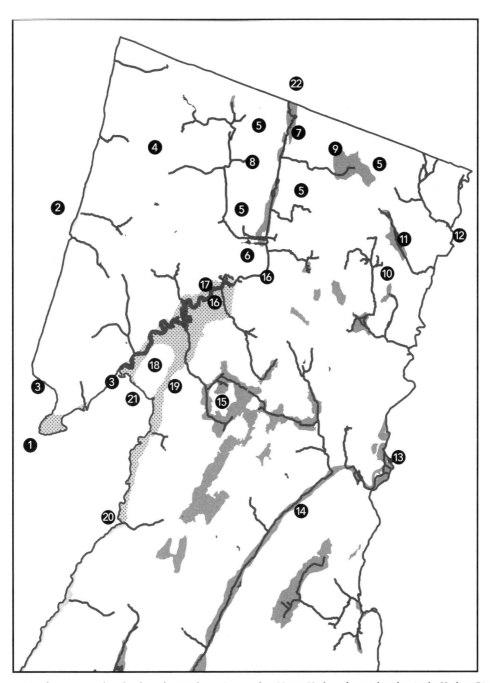

Figure 3. Surface water and wetlands in the Northwest Bronx when Henry Hudson dropped anchor in the Hudson River at Spuyten Duyvil in October 1609. Note the absence of any freshwater bodies larger than tiny ponds; Jerome Meadows draining west to Tibbett's Brook and east to the Bronx River; prominent north–south Mill Brook drainage; limited stream drainage and absence of marshes in Riverdale; and the few saltmarshes only on the Harlem River, Tibbett's Brook, and Spuyten Duyvil Creek. Location of 19th- and 20th-century features, clockwise from the lower left: Inwood Park (1), Hudson River (2), Spuyten Duyvil Creek (3), Riverdale (4), Van Cortlandt Park (5), Parade Ground (6), Lincoln Marsh section of Van Cortlandt Swamp (7), Van Cortlandt Swamp without Van Cortlandt Lake (8), Sycamore Swamp (9), Woodlawn Cemetery (10), Woodlawn Cemetery ponds (11), Bronx River (12), Williamsbridge marshes on Bronx River (13), Mill Brook (14), Jerome Meadows (15), Tibbett's Brook (16), Kingsbridge Meadows (17), Kingsbridge Island (18), Van der Donck Meadows (19), Harlem River (20), Marble Hill (21), Yonkers (22). Original image courtesy of Chris Spagnoli and Eric Sanderson, the Welikia Project, and the Wildlife Conservation Society, modified for use here.

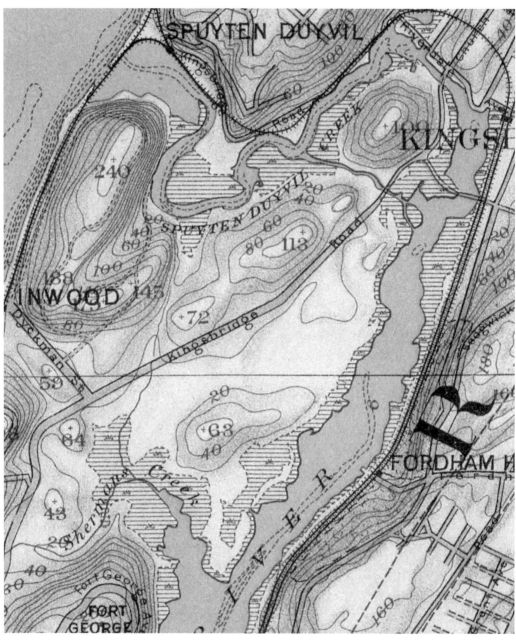

Figure 4. Spuyten Duyvil Creek in 1891 before the Harlem River Ship Canal was built, with Hudson Division railroad tracks following the creek north around Marble Hill, across modern W. 231st St. (not shown), then back south to Spuyten Duyvil. Kingsbridge Meadows drains into Spuyten Duyvil Creek (top right), with undiked saltmarshes in horizontal hatching along the Harlem River, Spuyten Duyvil Creek, and Kingsbridge Meadows. From Bien and Vermeule 1891a. Courtesy of David Rumsey Map Collection.

Most of the precontact Northwest Bronx was covered by old-growth Eastern Deciduous Forest, and its composition was eloquently described for Manhattan Island by Sanderson (2009). There is no reason to believe that the Northwest Bronx differed materially.

The king of kings was American Chestnut; it is estimated that more than half the wood of the forest was chestnut and that individual trees might have been as much as 4 feet wide and 120 feet tall. But the dukes, counts, and princes of the forest were many: Tuliptrees, oaks of several colors, hickories, maples, birches, Eastern White Pines, Eastern Hemlocks, Sweet Gums, and American Beeches.

The hilltops were the domain of the oaks—Chestnut, American White, Eastern Black, and Scarlet. Where the soil was sandy and fire frequent, Pitch Pines grew instead, small outposts from the vast Pine Barrens of southern New Jersey and Long Island.

The mid-slopes were dominated by chestnuts, attended by Red Maples, hickories, and birches, and retainers from up- and downslope. The bases between the hills were the Red Oaks' realm, with Tuliptrees, White Ashes, Tupelos, and American Hornbeams standing tall where the ground was not too wet. Majestic Eastern White Pines grew where the glaciers had left sandy lenses in flat places and on rocky slopes and in shadowy ravines. Nearby, hemlock's associate, the magnificent American Beech, sometimes 8 feet in diameter, sought out the forest coves where it could protect its fragile skin from fire.

Local Lenni-Lenape Indians did not live in a prelapsarian world. They actively managed their landscape for horticulture, fields, trails, game habitat, hunting areas, and villages. Sanderson's (2009) computer simulations indicated that 80–100% of Manhattan island had been purposely burned by the Lenape during the 1409–1609 period, and there are no reasons to believe the Northwest Bronx had not been also.

Non-saltmarsh grasslands were probably less common in the Northwest Bronx than on Manhattan, and most of them may have been maintained by prescribed burns. Into the 19th century, those of consequence would have been at Jerome Meadows, along the upland edges of Kingsbridge Meadows, and on the Van Cortlandt Parade Ground.

HISTORY AND INFRASTRUCTURE

When the first Bronx European explorer, Henry Hudson, dropped anchor in his eponymous river, he did so off Spuyten Duyvil in Riverdale. On 2–3 October 1609 the Bronx was occupied, sometimes only seasonally, by several different groups of Lenni-Lenape Indians in the Mohegan tribe of Algonquian speakers, present since at least 1000 CE. Some were remarkably placid, others more bellicose, but after an initial savage encounter with Hudson's ship and crew at Spuyten Duyvil, any intrinsic ingenuousness vanished. Numerous seasonal and permanent Indian settlements were clustered along Spuyten Duyvil Creek, where hunting, shellfishing, and finfishing were productive, and a permanent settlement existed on the Van Cortlandt Park Parade Ground, which may have been intentionally created ages earlier by burning, because it was a major Lenape corn-growing site.

Reginald Bolton (1922) discusses the Lenape village covering several acres on the eastern edge of the Parade Ground on the west bank of Mosholu (= Tibbett's) Brook. The tribe it belonged to was known as the Weckquaesgeeks (English spellings vary widely), and the village was probably named Mosholu, after a Delaware Indian word meaning shining stones (referring to its characteristic large, smooth rocks in the cascade in front of the village) or clear, not turbid (for its bright waters). A historic Indian trail (the Westchester Path) also crossed the Parade

Ground area, connecting the village site with areas elsewhere in the Bronx and Manhattan.

The government of Holland founded the colony of New Netherland in southern Manhattan in 1614, and in 1646 Adriaen van der Donck purchased a parcel of land soon called Colen Donck (Donck's Colony). This tract of land was the first New York Dutch settlement north of Manhattan and extended well into southern Westchester County. It was bounded on the west by the Hudson River, on the east by the Bronx River, and on the south by Spuyten Duyvil Creek. Van der Donck built a house below the future Van Cortlandt Mansion site and eventually erected a sawmill, laid out a farm, built a barn and stockade, and brought in colonists. In 1668, four years after the English took control of New Amsterdam and renamed it New York, that part of Donck's Colony which included the Parade Ground was sold to William Betts and George Tippit (= Tibbett), with the latter then living in Van der Donck's house.

Jacobus van Cortlandt next bought the tract in 1694–99. He was a prominent New Yorker who twice served terms as the city's mayor in the early 1700s. He established a large plantation on the Parade Ground, and in December 1699 he dammed Tibbett's Brook (forming Van Cortlandt Lake) to create a millpond and gristmill. The sawmill stood until about 1900, when it burned down, and the gristmill lasted until about 1917, when the Parks Department removed it. The small millpond below the Van Cortlandt Lake spillway is still there, little changed. The Van Cortlandt farmhouse seems to have been located near the existing grove of locusts and the small cemetery just north of the millpond in the southeast corner of the Parade Ground. It may have been destroyed and buried during railroad building and the Parade Ground reconstruction.

By 1732 Jacobus van Cortlandt had purchased much of the Bronx portion of the original Van der Donck's lands in the Bronx and Westchester. After his death in 1739, his son Frederick, who inherited the lands, built a large stone dwelling house on the plantation in 1748–49. This is the Van Cortlandt Mansion that still stands in the park east of Broadway at about W. 244th St. and is the oldest extant Bronx house. On Frederick's death in 1749, his brother Augustus inherited the property.

The Bronx and Manhattan Van Cortlandts were revolutionaries, and during the war, Augustus (who was then City Clerk of New York) was ordered to hide all city records from the British. They were stored in his backyard in Manhattan until 1776 but then moved to the Van Cortlandt family burial plot on Vault Hill when the Continental army housed troops on the family grounds. In October 1776, George Washington used the mansion as headquarters prior to the Battle of White Plains. The Continental and British armies were so close to each other that at one point in 1781, it is said that Washington kept large fires blazing on Vault Hill for several nights to deceive the enemy and allow colonial troops to escape. When New York City fell to the British late that year, General Howe then moved *his* headquarters to the Van Cortlandt Mansion, and its immediate area remained near or behind British lines until the end of the war.

The Van Cortlandt family continued to own the property throughout the nineteenth century. During that time they operated a working farm and mill. Robert Bolton (1848) describes it thus: "To the east of [Van Cortlandt's] house, the Mosholu [Tibbett's Brook], pent up by the mill dam, forms an extensive sheet of water, which is greatly enhanced by the vicinity of green meadows, orchards and neighboring hills. South of the pond is situated an old mill. Amid the grove of locusts on George's Point, a little north of the old mill, stood the original residence of the Van Cortlandts. On the Van Cortlandt estate is situated an Indian bridge and field; the former crossed Tippet's Brook, the latter forms a portion of the Cortlandt woods, an extensive range of woodland to the northeast of the mansion [and the Lake]."

By the mid-1700s, much of the southern area of Donck's land was owned by Jacobus van Cortlandt. His son Frederick built the Van Cortlandt Mansion in 1748, but he died the next year and the house went to his son James, then to James's brother Augustus. After he died in 1823, the property was to pass to his grandson Augustus White, provided that thereafter all who inherited the Van Cortlandt estate would take the Van Cortlandt name. In 1839 upon the death of Augustus van Cortlandt White, it passed to his brother Henry, who died only six months later. The estate then went to his nephew Augustus van Cortlandt Bibby. Under Bibby's ownership, the house was renovated, and much of the land in and around the area was intensively farmed. This was the Bibby of Bibby's Pond, the name by which Bicknell knew Van Cortlandt Lake. In June 1912, the very last parcels of the land that had been in the Van Cortlandt family for 212 years—719 city lots covering 60 acres south of the park and extending east to Jerome Reservoir—were finally sold by a Manhattan auctioneer.

THE LATE 19TH-CENTURY MILIEU

In the last 30 years of the 1800s, even though New York City was expanding rapidly, an enormous amount of natural area (much of it subject to various degrees of disturbance and degradation) still persisted, even in Manhattan. But in the four outer boroughs there were large swaths of such lands, most of them dotted with small farms. Consequently, there was little pressure to preserve and protect natural areas there. Only a few visionaries foresaw what would happen to the future New York City (Brooklyn and Queens were not even part of it until 1898, and until 1875 the West Bronx was still in Westchester County) and took the initiative.

In 1884 the New York State legislature passed the New Parks Act, which, inter alia, authorized three major parks in the Bronx: Van Cortlandt, Bronx, and Pelham Bay. These were the first new large-area parks within New York City since Central Park (1857) and Prospect Park (1867) but with an important difference. Central and Prospect were manufactured parks created by some of the very first American landscape architects from existing pieces of vestigial and often badly degraded natural areas, while the new Bronx parks set out to protect existing still-forested natural areas without the guidance of landscape architects, whose work in them was limited to a few very small public-use sections. In Van Cortlandt this probably happened only at the Dutch Gardens.

By the start of our study period in the early 1870s, moving any distance beyond one's immediate neighborhood was not easy. Local transport was still only by foot, horseback, or stagecoach and other horse-drawn vehicles. Most roads were unpaved and of dubious quality. For example, travel to downtown Manhattan was by railroad, occasional ferry-like vessels on the Harlem and Hudson Rivers, and private boats. Eugene Bicknell lived in Riverdale, but when commuting to Manhattan he walked from his house about a block west of Broadway at W. 253rd St. to the Putnam Division's Van Cortlandt Park train station, which gave him some field time every day.

Early Van Cortlandt Park (Fig. 1) was crossed by a few minor unpaved roads that were barely more than paths, and the Putnam Division ran on its current track bed with a spur for the Yonkers Rapid Transit line departing from Van Cortlandt Lake and then running northwest across the northeastern corner of the Parade Ground, the west side of Vault Hill, and through the Northwest Forest, with a station at Mosholu Ave. just east of Broadway.

The New York Central and Hudson River Railroad, constructed between 1847 and 1849 and squeezed in along the river's eastern shore, ran in the Bronx from Mt. St. Vincent station at the (now) Yonkers line, with stops at Riverdale (W. 254th St.) and Spuyten Duyvil. Heading into Manhattan from the Spuyten

Duyvil station, in a wide ⌒–shaped loop the tracks turned abruptly *northward* along the west bank of Spuyten Duyvil Creek, crossed Tibbett's Brook near its junction with Spuyten Duyvil Creek and then the mainland of today's Kingsbridge on W. 231st St., before finally turning back south again after joining the Putnam Division tracks at W. 230th St. (Figs. 1, 4, 11).

The late 1800s Northwest Bronx was strikingly different from today's, and Riverdale was a wilderness, with thick forested vegetation, scattered large houses/estates, no paved and few through roads. In Kingsbridge Heights, Kieran remembered well the street he grew up on, which into the 1890s was still unpaved; bereft of electricity, telephones and cars; lit by gas lamps; scattered with isolated frame houses, small farms, and numerous apple orchards; and populated by many horses, cows, and chickens. It was *rus in urbe*—the country in the city.

Few New Yorkers today realize that malaria was an everyday problem facing residents of the study area into the early 20th century. *Plasmodium vivax* is the protozoan that causes benign tertian malaria—oxymoronically *benign* because it kills only a few, *tertian* because fevers recur every third day. This was New York's endemic malaria and its major vector was Four-spotted Mosquito, a species regularly found in standing water around human habitation. Without doubt, *vivax* malaria was endemic in Van Cortlandt Park and elsewhere in the study area until its eradication in the early 1900s. Kieran was a victim as a child around the turn of the 20th century, when it was rampant at the Jerome Reservoir construction site, which was viewable from a window in his house. Fortunately, its symptoms are mild even though persisting throughout life, unlike those of frequently lethal but nonetheless one-off pantropical *P. falciparum* malaria.

Intra-Bronx horse drawn trolleys known as horsecars did not begin until >1820s, and they gradually replaced the horse-drawn oversize stagecoaches called omnibuses (the origin of today's word *buses*) that had provided intra- and interborough connections. Horsecar routes were scattered across the Bronx, but little information is available about them in the West Bronx beyond their use on Broadway and Riverdale Ave.

By the late 1800s they in turn were gradually replaced by electric streetcars (trolleys), especially after a massive outbreak of equine influenza in the early 1870s. By the early 1900s, streetcars ran to most outlying parts of the Bronx, including City Island and Clason Point, and on Broadway and Jerome/Central Aves. well into Yonkers. They were all eventually replaced by diesel-belching buses in 1948, with the last streetcars in the study area being those run by Yonkers Transit down Broadway to the subway at W. 242nd St.; they ceased operation in 1952. There were no subways in the Northwest Bronx until the Broadway Division of the IRT reached its terminus at W. 242nd St. and Broadway in 1907. To the east, the IRT's Jerome Ave. line did not reach Woodlawn Cemetery until 1918.

Electric automobiles were prominent but not numerous into the early 1900s, but it was not until Henry Ford's Model T in 1910 that gasoline-powered cars began their race to ubiquity. Roads to accommodate them in and around New York City took some time to catch up, and it was not until Robert Moses began to build commercial traffic-free parkways in the 1920s that traffic light– and stop sign–free highways began to appear.

THE STUDY AREA: CREATION AND LOSS

Even though residentially developed for small houses and quite recent high-rise apartments, the Northwest Bronx area is still one of the more natural parts of New York City. Large areas exceptionally favorable to birdlife included Riverdale, Van Cortlandt Park, Woodlawn Cemetery, Jerome Reservoir, Hillview Reservoir, the Hudson River, Spuyten Duyvil Creek, and the Harlem River.

Our study area, almost entirely within the Northwest Bronx, lies to the east of Broadway and comprises seven subareas. These are **Van Cortlandt Park** (Fig. 5); **Woodlawn Cemetery** (Fig. 5) immediately east of Van Cortlandt Park; **Jerome Reservoir** (Figs. 2, 10) to

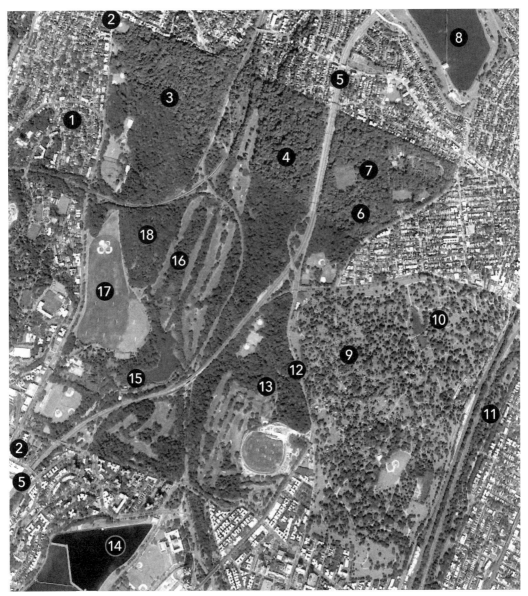

Figure 5. Details of the extant study area in 2014–15. Labeled features, clockwise from the upper left: Riverdale (1), Broadway (2), Northwest Forest (3), the Ridge (Croton Forest) (4), Major Deegan Expressway (5), Northeast Forest (6), Sycamore Swamp (7), Hillview Reservoir (8), Woodlawn Cemetery (9), Woodlawn Cemetery ponds (10), Bronx River (11), Jerome Avenue (12), Shandler Area (13), Jerome Reservoir (14), Van Cortlandt Lake (15), Van Cortlandt Swamp (16), Parade Ground (17), and Vault Hill (18). Note greatly reduced Van Cortlandt Swamp, the two-toned aspect of the Parade Ground discussed in the Avifaunal Overview, and the distribution of major forested areas (forest contrast change line is an artifact). Image courtesy of Google Earth Pro.

its south; **Hillview Reservoir** (Figs. 6, 7, 8) in immediately adjacent southwestern Yonkers; and three that were obliterated about 1930 by New York City's expansion: **Jerome Meadows** (Figs. 3, 9), wet, grassy, and brushy areas occupying the site of Jerome Reservoir and Jerome Swamp prior to the construction of their predecessor, Jerome Park Racetrack; **Jerome Swamp** (Figs. 3, 10), another wet area east of Jerome Reservoir and west of Jerome Ave. created during the reservoir's construction; and **Kingsbridge Meadows** (Plate 1, Figs. 1, 3, 11)—the expansive and variably fresh, brackish, and salt marshes enveloping Tibbett's Brook as it flowed first through the Dutch Gardens Marsh, then southwest for 1.2 smi (1.9 km) to its junction with Spuyten Duyvil Creek. Figures 1 and 2 depict and Table 1 gives vital statistics for each subarea. Table 2 offers distances from Van Cortlandt Lake to places frequently mentioned in the text.

Immediately west of the study area—thus east of the Hudson River and adjacent to Kingsbridge Meadows and Van Cortlandt Park—is the residential area of **Riverdale** (Figs. 1, 3, 5), extending from the Harlem River Ship Canal to the Yonkers city line. It is still one of the most rural parts of New York City, and its birdlife, geology, and vegetation are shared with our study area, so we make frequent reference to it.

At 1146 ac (463 ha), **Van Cortlandt Park** (Figs. 2, 3, 5, 12, 16) is the third largest in New York City and boasts several important ecosystems and natural areas, whose dimensions are also given in Table 1.

Tibbett's Brook (Plate 1, Figs. 1, 2, 3, 6, 11, 12, 16) is the dominant freshwater stream in the Northwest Bronx, and even

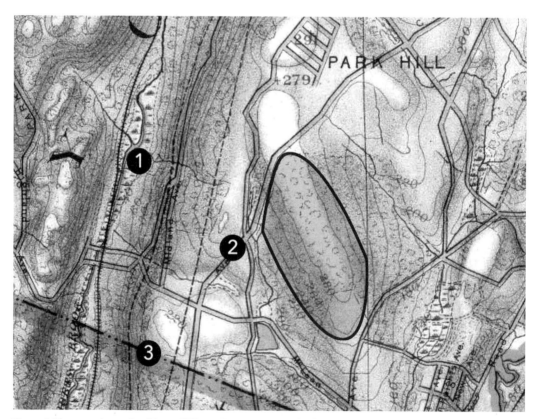

Figure 6. The approximate site of future Hillview Reservoir in 1891 (outlined). Note the widespread drainage into Tibbett's Brook (1) at the far left past Central Avenue (2), and the proximity of the New York City–Yonkers border (3). From Bien and Vermeule 1891b, courtesy of David Rumsey Map Collection.

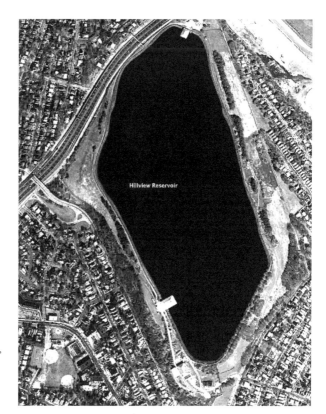

Figure 7. Hillview Reservoir in December 1994, showing the extensive conifer groves (dark treed areas) on all sides but especially its steep southwesterly slopes and the large grassy areas; cf. Fig. 8. Image courtesy of Google Earth Pro.

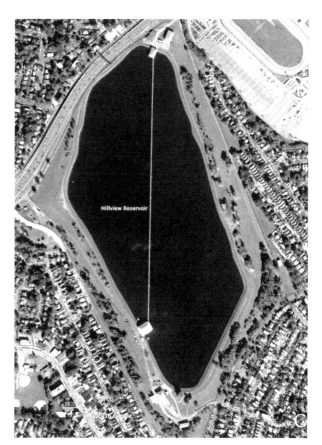

Figure 8. Hillview Reservoir in October 2014, showing loss of conifer groves, so that nearly all remaining trees are deciduous; cf. Fig. 7. Image courtesy of Google Earth Pro.

though it persists as only a shadow of its former glory, it has been and still is the essence of Van Cortlandt Swamp, Van Cortlandt Lake, and Kingsbridge Meadows and made each of them an avian treasure. Tibbett's Brook is said by some to arise in Yonkers at Valentine's Hill near the Yonkers Racetrack, by others from somewhere in Tibbett's Brook Park or slightly farther north, and one has it coming from an unnamed stream arising farther north near Bryn Mawr. Much of its watershed has been built up, and a good deal of the stream north of Tibbett's Brook Park is today in culverts, some apparently quite long which makes tracing its origins that much more difficult. It originally flowed through today's Van Cortlandt Park, creating the large cattail **Dutch Gardens Marsh** (Figs. 2, 11, 14), and at W. 240th St. Tibbett's Brook crossed Broadway (in culverts after the 1870s) and expanded out into the main Kingsbridge Meadows until merging with Spuyten Duyvil Creek. The entire marsh-swamp complex along Tibbett's Brook, which formerly ran from north of present-day Cross County Parkway south to Van Cortlandt Lake, comprised Van Cortlandt Swamp.

Van Cortlandt Swamp (Figs. 1, 2, 3, 6, 12, 16) and Van Cortlandt Lake were created simultaneously when Jacobus van Cortlandt dammed Tibbett's Brook for his gristmill in December 1699. Water backed up for more than 2 smi (3.2 km) into modern Westchester County, and as open water near Van Cortlandt Lake gradually sedimented, a large and productive cattail marsh developed at the northwest corner of the lake east of the railroad tracks and in the swamp's

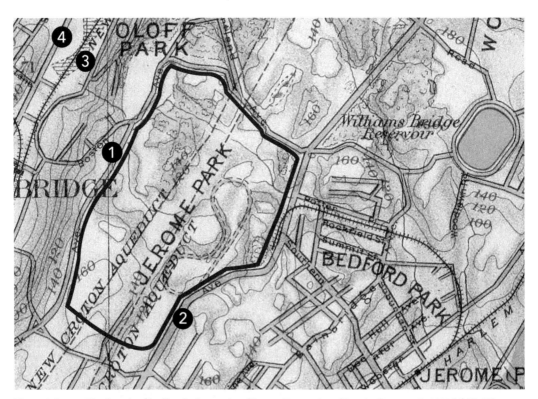

Figure 9. Jerome Meadows (outlined)—the future site of Jerome Reservoir and Jerome Swamp—in 1891 (cf. Fig 10). The figure 8–like oval along Jerome Ave. is Jerome Park Racetrack, removed for construction of the reservoir, whose eastern half was never built but which developed into Jerome Swamp. Bordering streets are Sedgwick (1) and Jerome (2) Aves., with Broadway (4) and persisting Van der Donck Meadows (3) in the upper left. From Bien and Vermeule (1891a). Courtesy of David Rumsey Map Collection.

Figure 10. Jerome Reservoir and Jerome Swamp (right) in 1924. The swamp's shape mirrored the reservoir's and was intended to be its right half, but when abandoned after preliminary excavation it developed into Jerome Swamp. It existed from about 1900 to about 1930, during which time its marshbird habitat was almost as good as that of nearby Van Cortlandt Swamp. Bordering streets are Sedgwick (1) and Jerome (2) Aves. Image from New York City Digital Map Atlas.

southern third west of the tracks. The swamp's northern two-thirds became a wooded swamp, and the entire area supported a complex suite of marsh birds, the largest anywhere in New York City. Even though we have no detailed descriptions, Van Cortlandt Swamp must have been a magnificent wetland during the years between 1700 and its first disruption for construction of the Putnam Division track bed in 1869. It was probably several hundred yards wide in places and sloped gradually up to the Ridge (Croton Forest) on the east and to Vault Hill/Northwest Forest on the west, with damp weedy fields on both sides of it.

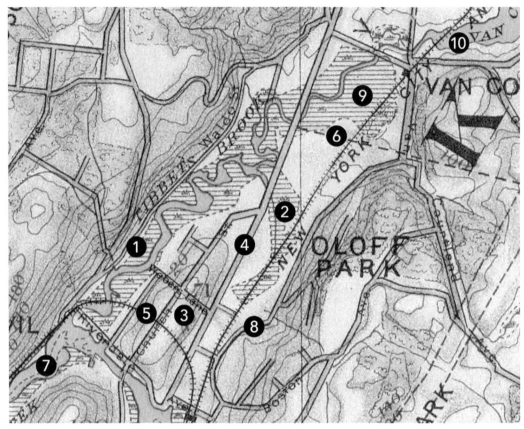

Figure 11. Kingsbridge Meadows (1) and Van der Donck Meadows (2) bracketing Kingsbridge Island (3) and Broadway (4) in 1891. Even though the southern half of Van der Donck Meadows began to be filled in the 1700s, its northern half was still largely natural in 1891, as was Kingsbridge Meadows, within which Tibbett's Brook flowed unimpeded from Van Cortlandt Lake to Spuyten Duyvil Creek. It was crossed only by the Hudson Division track bed (5) and what would become modern W. 230th St. (Riverdale Ave.) at the lower left. The dashed line east of Broadway and south of Van Cortlandt Park is the future W. 240th St. (6). Also shown are Spuyten Duyvil Creek (7), Bailey Avenue (8), Dutch Gardens Marsh (9), and Van Cortlandt Lake (10). From Bien and Vermeule 1891a. Courtesy of David Rumsey Map Collection.

This idyllic Van Cortlandt Swamp was not to last. Once the major part of its eastern side had been filled for the Putnam Division track bed, the next insult was building the Van Cortlandt golf course in 1895, which removed all of the remaining swamp east of the railroad track bed, leaving only 2 vestigial ponds. The eastern fields connecting the Ridge (Croton Forest) to the swamp became manicured fairways. In 1913, mosquito control landfilling removed another 10 ac (4 ha) of swamp east of the railroad. By the 1930s, Kieran was incensed at WPA (Works Progress Administration) vandals mindlessly cutting down dead trees in make-work projects, and Cruickshank lamented landfilling for even more make-work mosquito control.

Van Cortlandt Swamp's northernmost portion, south to today's Mosholu Parkway Extension, was known as **Lincoln Marsh** (Figs. 3, 12) and ran south from today's Tibbett's Brook (Westchester County) Park until construction of the Putnam Division track bed (1869–81) and the Henry Hudson Parkway (1934–37) eliminated most of it. In the process, Tibbett's Brook itself was channelized south of the county park

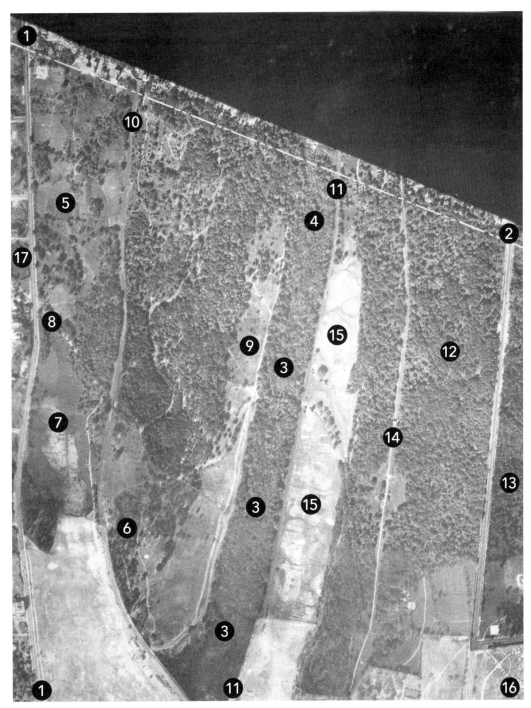

Figure 12. Van Cortlandt Park in 1924, from Broadway (1) on the west to Jerome Ave. (2) on the east, before the Henry Hudson Parkway made the first major inroads on Van Cortlandt Swamp (3) and its northernmost Lincoln Marsh section (4). Note the openness of the Northwest Forest (5) and Vault Hill (6); the Parade Ground (7) extending north to Mosholu Ave. (8); and the natural fields (9) sloping down to the west side of Van Cortlandt Swamp from the south end of Vault Hill to the Yonkers city line. Also shown are the Yonkers Rapid Transit (10) and Putnam Division (11) track beds, the Ridge (Croton Forest) (12), Northeast Forest (13), Old Croton Aqueduct (14), golf course segments (15), Woodlawn Cemetery (16), and Riverdale (17). Image from New York City Digital Map Atlas.

and funneled through four new small, bare ponds around and between the parkway's traffic lanes. Today these are heavily vegetated with little open water, but the middle two, surrounded by high-speed roads and nearly inaccessible, offer protected breeding sites, even if tiny, for study area birds. These ponds seem never to have had formal names on topographic maps, so the New York District of the US Army Corps of Engineers (USACE 2000: A-21), in a planning document addressing possible restoration schemes for various Bronx freshwater streams, has whimsically named them, north to south, Elm, Maple, Birch, and Sycamore Ponds—notwithstanding the absence of such trees near them.

Van Cortlandt Swamp's southern section today is the last vestige of the original Van Cortlandt Swamp and runs only from Mosholu Parkway Extension south to Van Cortlandt Lake. All traces of Tibbett's Brook south of the lake were buried underground in 1918, and Tibbett's Brook itself was diverted and dumped into the main Broadway combined storm sewer at W. 242nd St. From that year on, the remaining Tibbett's Brook marshes south of the Van Cortlandt Mansion (Dutch Gardens Marsh and Kingsbridge Meadows) received water only by runoff, rainfall, and tidal exchange via Kingsbridge Meadows. Eventually Kingsbridge Meadows itself was platted and filled, paved and built.

The most devastating ecological insult to an already amputated Van Cortlandt Swamp was construction of the Major Deegan Expressway (1949–55). The construction transformed quiet Jerome Ave. between Woodlawn Cemetery and the Yonkers city line (two lanes of traffic plus the Yonkers streetcar line that ended at the Woodlawn subway station) into a 10-lane limited access truck way and then cut southwest downslope across Van Cortlandt Park toward W. 240th St. alongside the Putnam Division track bed. In the process it destroyed large portions of the original Van Cortlandt Golf Course, so to make up for his action there, Robert Moses obliterated the half of Van Cortlandt Swamp immediately west of the Putnam track bed and the large natural field west of the remaining swamp sloping down from Vault Hill. These two prime year-round bird habitats became his replacement golf course sections.

Sedwitz, Norse, and Kieran had known Van Cortlandt Swamp intimately for decades and were so appalled and demoralized by the ecological destruction Moses had wrought that they never really got over it. Buckley arrived on the scene just as the last filling was ending, when the entire eastern half of the swamp was a sea of tree stumps and landfill. It has been his biggest regret that he had never known it in its glory, but perhaps it was just as well he hadn't.

From probably an initial size of 100 ac (40 ha) in the early 1800s (imagined from Figs. 3, 12), Van Cortlandt Swamp had been reduced to 19 ac (7 ha) by 1950, and at its southern end the cattail marsh was bisected by a new causeway allowing golfers to cross it dry-shod. From that time forward, the swamp was also fenced in from and by the surrounding golf course segments, but limited pedestrian access to the swamp was afforded at its southwest corner via a footbridge over the causeway. Eventually, that bridge was removed after falling into disrepair, and in the late 1990s a replacement was built. But by 2007 it too had been completely removed for no apparent reason, and since then there has been no easy access to the entire swamp north of the causeway, both a good and a bad situation. Adding insult to injury, in late February 2017 the causeway pedestrian bridge crossing Tibbett's Brook itself was abruptly closed indefinitely to pedestrian traffic (but not golfers), yet another slap in the face of John Kieran Trail users and those seeking Van Cortlandt Swamp birds.

Through all assaults, many breeding, wintering, and migrating marsh birds have persisted in Van Cortlandt Swamp but in diminishing numbers of species and individuals, reaching a nadir in the 1960s from which they have never recovered. Much of the drop in birdlife followed first the replacement of native cattails by exotic *Phragmites* and then an inexorable shift from an open marsh to a mostly closed wooded swamp as even *Phragmites* was replaced by Buttonbushes, Red Maples, and willows.

Van Cortlandt Lake (Figs. 1, 2, 5, 11, 13) begins at the south end of Van Cortlandt Swamp. It was attractive to all sorts of waterfowl from its creation, being the only large naturalistic freshwater body in the Northwest Bronx. Even today only Jerome Reservoir is larger. **The Triangle** or **Lake Marsh** (Fig. 13)—of cattails until they were overwhelmed by *Phragmites*—at its northwestern corner alongside the railroad was a guaranteed location for breeding and overwintering Virginia Rails until it was gouged out in 2001–3 when Van Cortlandt Lake was dredged following heavy sedimentation. The wooded area on the west side of Van Cortlandt Lake east of the railroad has always been a major spring migration site, known to many generations of observers as **the Island** (Fig. 13).

The **Dutch Gardens** (Fig. 13), with canals and a pond south, and below the grade, of the Van Cortlandt Mansion were built in 1902–3 and lasted until 1969 when Robert Moses extirpated their last vestiges during construction of an Olympic-sized swimming pool near the Van Cortlandt subway station. But with wonderful ecological irony, shortly thereafter Tibbett's Brook finally began to reassert itself after having been buried in the sewer line for 60+ years. First noticed as a seasonally slightly wet area just west of the former Putnam Division's Van Cortlandt station, it gradually expanded and was colonized by exotic *Phragmites*, which spread quickly, and wetland birds soon appeared. To their credit, Parks Department naturalists realized what was occurring and set about protecting and expanding this nascent marsh. Native cattails soon colonized it, and as water depth increased and marsh vegetation spread out, boardwalks and a viewing platform with a few interpretive signs were emplaced in 2005. The area is now called **Tibbett's Marsh** (Figs. 5, 13), a useful name to distinguish it from the much larger and older Van Cortlandt Swamp to its north. Depressingly, the viewing platform (was ?) burned to the ground in December 2015, and after repeated vandalism to area benches, the Parks Department may not replace any of it.

Without doubt the most profound landform changes in the Northwest Bronx and study area affected **Kingsbridge Meadows** (Plate 1, Figs. 1, 3, 11) and occurred from 1890 to the late 1920s. These involved cutting the Harlem River Ship Canal (1892–95) south of Marble Hill and the shifting of all ship traffic away from Spuyten Duyvil Creek north around Marble Hill to directly west *before* Marble Hill (Figs. 3, 4, 14). This led to the piecemeal filling of Spuyten Duyvil Creek from 1903 to 1917, in part with material from the construction of Grand Central Station directly delivered by rail. In turn, the old King's Bridge (Fig. 14) over Spuyten Duyvil Creek at modern W. 230th St. and Kingsbridge Ave.—the first and until 1900 the only direct connection from Marble Hill north into the Bronx—fell into disuse and was finally buried in situ in 1916, when at the old Indian ford (modern Broadway at W. 230th St.) the last small gap in Broadway was finally removed. The remainder of Spuyten Duyvil Creek was completely filled by 1917. Once Tibbett's Brook's direct tidal connection with Spuyten Duyvil Creek was severed, Kingsbridge Meadows was doomed, and filling began rapidly. The remaining marsh segments west of Broadway vanished in the late 1920s. The very last piece, the Dutch Gardens Marsh (Figs. 11, 14), was filled by a garbage dump from 1933 to 1937, then in 1937 by construction for the Van Cortlandt Stadium and sports facilities east of Broadway and north of W. 240th St., which opened in 1939.

Farther southeast of Broadway, the loss of **Van der Donck Meadows** (Figs. 3, 11, 15),

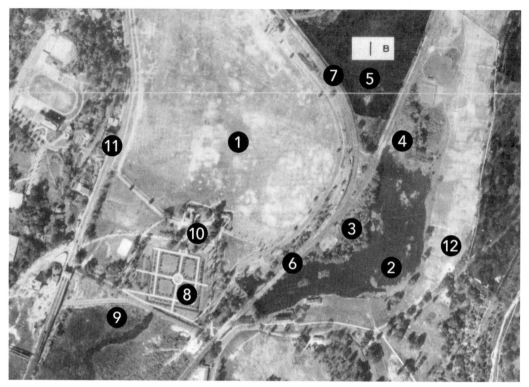

Figure 13. Southern Van Cortlandt Park in 1924 showing Parade Ground (1), Van Cortlandt Lake (2) with the Island (3) and the Triangle (4), the south end of Van Cortlandt Swamp (5), Putnam Division (6) and Yonkers Rapid Transit (7) track beds, Dutch Gardens (8), a corner of Dutch Gardens Marsh (9), Van Cortlandt Mansion (10), Broadway (11), and golf course (12). Image from New York City Digital Map Atlas.

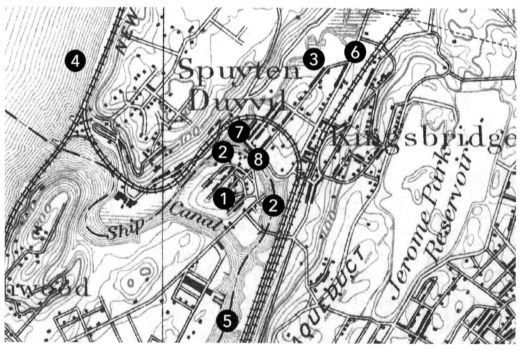

Figure 14. The new Harlem River Ship Canal in 1895 south of Marble Hill (1), with Spuyten Duyvil Creek (2) still unfilled north of it, and Kingsbridge Meadows (3) connecting to its northwest corner. Compare with Fig. 4. Also shown are the Hudson (4) and Harlem (5) Rivers, Broadway (6), W. 230th St. (7), and the old King's Bridge (8). Image from 1905 Harlem 15' topographic quadrangle.

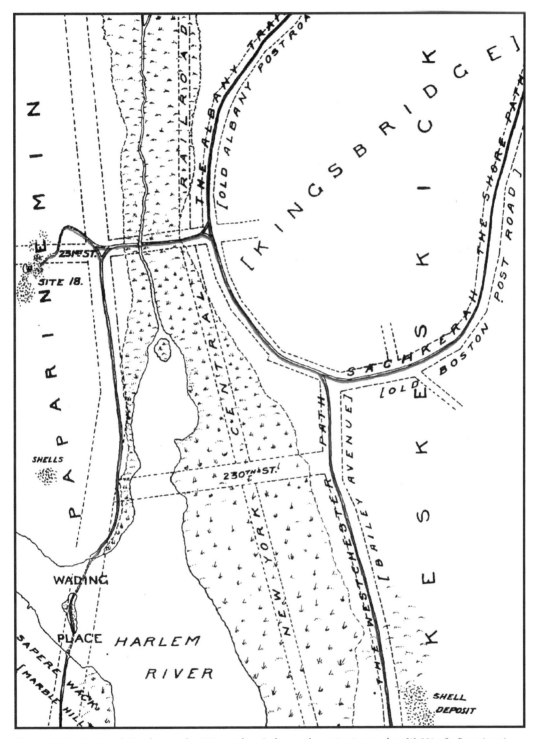

Figure 15. Van Der Donck Meadows in the 1700s, with an Indian trail crossing it at modern W. 231st St. *Paparinemin* was the Lenni–Lenape name for Kingsbridge Island. Image from Bolton 1922.

running from modern W. 230th St. to W. 238th St., where the Putnam Division track bed would eventually sit, and connecting with Tibbett's Brook and Kingsbridge Meadows around modern W. 237th St., began early, probably in the late 1700s, but then proceeded slowly with the last isolated marsh pieces near W. 238th St. not being filled until the late 1920s.

The geological/ecological origins of the 150 ac (60 ha) **Parade Ground** (Figs. 1, 2, 3, 5, 13) are unreported, but it seems to have been in use by Lenni-Lenape Indians since about 1000 CE. They may have kept it in grasses by prescribed burns and used it for agriculture and a village, but apparently it was already a naturally large flat area when they arrived on the scene. Dutch, English, and American colonists continued farming it. Well into the nineteenth century, the Van Cortlandt estate was a working farm with eight structures and supporting roads on the Parade Ground itself.

In 1889–90, the Parade Ground was formally reconstructed as a camp and drill ground for the New York National Guard by plowing and leveling the cornfields to the northwest of the Van Cortlandt Mansion and the orchard east of the carriage house, then filling the roadway that extended north from the carriage house to Vault Hill. Little additional grading was required beyond the filling of some marshes and ponds. During this process, the Indian village of Mosholu was first uncovered. The National Guard used the Parade Ground for camping, drilling, and horse races, as well as polo and cricket. During World War I it was an army training camp, but the site was abandoned by the military thereafter.

The Parade Ground then remained largely unmolested except for intermittent track and field events and was heavily used by breeding, migrating, and overwintering birds until its next insult. This was its decapitation by Henry Hudson Parkway construction in 1935–37, which isolated a small section along Broadway south of Mosholu Ave. This piece was left alone until 1955, when it was all converted to horse stables and paddocks for public riding.

The remaining Parade Ground, now reduced to about 60 ac (24 ha), stumbled along, gradually seeing increased human use but owing to its large percentage of native grasses and flowers also still attracting significant numbers of native birds seasonally for 100+ years: Killdeer, sandpipers, Horned Larks, Snow Buntings, Lapland Longspurs, sparrows, Meadowlarks and other ground-loving species. Then its ecological bottom was cut away when nearly all native grasses were removed and replaced by sterile hybrid turfgrasses in 2010–12. Only a corner in the southeast, ironically nearest the Van der Donck and Van Cortlandt homesteads, remained relatively undisturbed, and native birds still flock there year-round, but even that is slated for replacement by the turfgrass required in 4 (of 10) desperately needed cricket pitches.

Most of Van Cortlandt Park is still in native forest, but the ages of the various stands apparently have never been examined systematically by coring unless outside arborists or forest ecologists have done so without publishing their findings (Künstler, pers. comm.). Large trees grow in profusion in the **Northwest Forest, the Ridge (Croton Forest)**, and **Northeast Forest** (Figs. 1, 2, 3, 5, 16) in most places with a healthy and diverse herbaceous layer flora. The absence of grazing deer and very slight human use demonstrate well that urban forests can maintain their ecological health under the right conditions. Even fragile ground-nesting forest birds, always among the first to vanish, are still hanging on, but pressure from feral and domestic cats may be exacting a toll on their numbers. Within the Northeast Forest, an ephemeral vernal wooded swamp known as the **Sycamore Swamp** (Figs. 3, 5, 16) has proved to be a striking avian magnet for spring migrant and breeding insectivores as it has succeeded from an open cattail marsh to a vernally flooded wooded swamp.

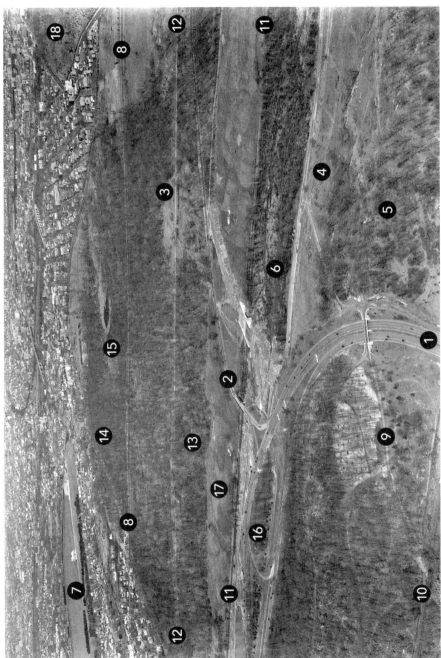

Figure 16. Aerial view of Van Cortlandt Park and environs looking ENE in March 1940. Note especially the new Henry Hudson Parkway (1); the Mosholu Parkway Extension (2) still under construction; the Ridge site of 1940s–50s hawkwatches (3); the open fields (4) sloping down from Vault Hill (5) into the north end of recently truncated Van Cortlandt Swamp (6); and the very dark conifer groves (7) surrounding Hillview Reservoir. Also shown are Jerome (8) and Mosholu (9) Aves.; Yonkers Rapid Transit (10) and Putnam Division (11) track beds; the Old Croton Aqueduct (12); Ridge (Croton Forest) (13); Northeast Forest (14); Sycamore Swamp (15); the former Lincoln Marsh section of Van Cortlandt Swamp now converted into highway-side ponds (16); golf course segments (17); and Woodlawn Cemetery (18). Image from New York City Dept. of Records.

The most pervasive ecological forest changes have involved the loss of native American Chestnuts to Chestnut Blight in the early 1900s and then loss of American Elms to Dutch Elm Fungi beginning in the 1920s and peaking in the 1960s. Corollary effects on study area birds are unknown. Exotic trees, shrubs, grasses, and flowers are unfortunately widespread in Van Cortlandt Park, and Parks Department ecologists are considering their extirpation and replacement by native species (Pehek, pers. comm.). However, see Avifaunal Overview, "Van Cortlandt's One Million Trees Involvement," for an example of unbridled extirpation of exotic plants that has had untoward effects on native birds.

It is believed by many that there are no trees in Van Cortlandt Park older than about 100 years. Yet photos taken around 1900 unambiguously show very large trees, and many of the park's native trees are reasonably long-lived, so it is unlikely they all died and have been replaced by new growth. Very many mature trees throughout each of the major forest areas today are large and surely older than 116 years. Kieran measured two fallen oaks in Van Cortlandt after the 1938 hurricane and found them to have reached 109 and 115 ft (33 and 35 m) in length. He estimated they were 200–300 years old. In Pelham Bay Park his tree-ring count of a freshly cut Chestnut Oak showed an age of 215 years, yet it was only 76 ft (23 m) tall. Tieck (1968) described numerous huge and still healthy trees in Riverdale in the mid-1960s that probably did predate the Revolution.

However, it does appear that most of the forests on northern Manhattan and in the west Bronx were laid bare for firewood during the Revolutionary War. Two of those so devastated were labeled on contemporaneous maps Cortlandt's Wood and Tippet's Wood (Sanderson 2009). These were almost certainly forests that included Van Cortlandt Park, so it is probable that few if any stands (perhaps even only the occasional tree) will be found older than about 235 years.

Within Van Cortlandt forests, it was relatively quiet for the 137 years after Van Cortlandt Lake's creation until the Old Croton Aqueduct's construction cut a large swath through the Ridge (Croton Forest) south toward future Jerome Reservoir (1837–42), but that forest has long since regrown. In the late 1880s the New Croton Aqueduct was also built on the Ridge but closer to Jerome Ave. than the Old Aqueduct, with a more modest loss of forest that has also subsequently regenerated. In 1895, the first Van Cortlandt golf course was built and then expanded in 1899, followed in 1904–15 by the Mosholu golf course. All told, golf course construction destroyed about 200 ac (80 ha) of Van Cortlandt Swamp, upland, and Ridge forest, and while it does provide some useful pond, wood, and edge avian habitat, their combined 200 acres are inaccessible to anyone not golfing. In 1936 a bridle path for horseback riding was cut through Vault Hill and the Northwest Forest with minor loss of woods.

Close examination and contrast of Figures 12 and 16 with Figure 5 indicate clearly how much forest growth and infilling has occurred in the last 100 years, particularly in the Northwest Forest, on Vault Hill, and on the Ridge. In this natural process, open-area bird communities and species have inevitably been replaced by forest counterparts, with an overall diminution in study area species richness.

Vault Hill (Figs. 1, 2, 3, 5, 12, 16) contained much open area until the 1970s (maintained by the most frequent burning anywhere in the park) when succession replaced its native grasses and wildflowers with a rapidly expanding young forest, including one of the finest stands of Sassafras in the park. Earlier, chunks of mature woodlands on northern Vault Hill and in the Northwest Forest had been lost when the Yonkers Rapid Transit Line was cut through in 1888. In 1895, Vault Hill was enclosed by a wire fence to restrain 25 wild American Bison—among the very last—that had been captured on the Great Plains by William Hornaday for a captive breeding program of the New York Zoological Society, proprietors of the Bronx Zoo. The zoo was running out of room, so the extra animals were moved to Van Cortlandt Park, where they did not fare well; later that year they were shipped to prairie land in Oklahoma. The subsequently successful work of the Bronx Zoo in managing another stock of bison in 1899 profited from the mistakes of the Van Cortlandt affair.

What might have been called a **Southern Forest** seems to have been largely treeless by the time the Van Cortlandt and Mosholu golf courses were constructed (Figs. 1, 2, 5). Its northeastern corner, today retaining a patch of reasonably mature woods, much second growth, and a few open areas, has been designated the Allen Shandler Recreation Area by the Parks Department, but it is severely disturbed and heavily used for mass recreation.

To its credit, the Parks Department has created a series of formal trails that made the major forest areas much more accessible to observers (but also to joggers and dog walkers): the John Kieran, John Muir, Cass Gallagher, and Old Croton Aqueduct Trails. But the jewel is the Putnam Trail that runs the length of the old Putnam Division track bed from the south end of Van Cortlandt Lake past Van Cortlandt Swamp to the Yonkers city line. However, see Avifaunal Overview, "Van Cortlandt Putnam Trail," for the brand-new threats it is now coming under from the Parks Department/Van Cortlandt Park Conservancy themselves.

Alone among study subareas, only **Woodlawn Cemetery** (Figs. 1, 2, 3, 5, 16) was never assaulted after its opening in 1865. But it did continue to reduce its naturally vegetated areas, especially wetlands, as burial sections expanded, and today only a single naturalistic work area (and so, maybe safe) remains, along with a small two-part pond. On the other hand, its trees have matured even if at the expense of little native herbaceous layer amid the manicured mausoleums.

Jerome Reservoir (Figs. 1, 2, 5, 9, 10) was built in the early 1900s and finally filled with water in 1905. Native grassland edges surrounding its east side along Goulden Ave. were slowly whittled away for landscaped parks and parking lots (1970s–2015) and are now all gone. **Hillview Reservoir** (Figs. 2, 6, 7, 8) was also built in the early 1900s and finally filled with water in 1915. Once constructed, both reservoirs were largely left alone. Extensive conifers were planted at Hillview in the 1920s–30s (Fig. 16), matured, and were then lost to insect damage and removed in the late 1990s. Contrast Figure 7 taken in December 1994, when nearly all trees were still conifers, with Figure 8 taken in October 2014, when the only trees left were deciduous and still in leaf.

A large part of **Jerome Meadows** (Figs. 3, 9) was erased when replaced by Jerome Race Track (1866) and its vestiges, then by Jerome Reservoir construction (about 1895). **Jerome Swamp** (Fig. 10) was formed in 1900 when the contractor building Jerome Reservoir finished its western (modern) half but only dug the hole for its planned mirror-image eastern half before abandoning it. This 20-ft deep, irregular declivity quickly filled with water and wetland vegetation, and Jerome Swamp was born. Then *its* eastern half up to Jerome Ave. became the IRT subway Jerome Yards in 1922 and the IND subway Concourse Yards in 1933. The western half lasted intact until DeWitt Clinton High School was built in 1928–29 on Mosholu Parkway and the Bronx Campus of Hunter College was built farther south in 1930–31. The area in between those two persisted unbuilt until modern Goulden Ave. (1939–40) and Harris Park (1941–42) jointly extirpated the last traces of Jerome Swamp.

In sum, the most severe study area landform changes occurred before 1955, so it has been 60 years without massive, irremediable disruption to area habitats, and those positive changes most important for birds happened more than 100 years ago: the creation of Jerome (1895–1905) and Hillview (1909–15) Reservoirs. Thus, apart from natural successional changes to Van Cortlandt Park and Woodlawn Cemetery, the only adverse changes of any consequence since the 1950s have been dredging and marsh removal in Van Cortlandt Lake (2003), eradication of native plants from most of the Parade Ground (2008–12), and loss of the extensive conifer groves at Hillview Reservoir in the late 1990s. Still, all the major changes occurring before 1955 did profoundly and permanently alter the study area's physiography, habitats, and birdlife.

ANCILLARY CHANGES

Coincident with the above insults to area habitats, additional New York City infrastructure growth, park management practices, and other changes also occurred as the city continued its ineluctable expansion and growth. Most directly affected or changed study area birdlife and habitats in various ways. A few did not affect the study area directly but did increase population and development in the Northwest Bronx, which then affected the study area indirectly. A handful actually benefited study area birdlife.

- Mosholu Parkway Greenbelt linking Bronx and Van Cortlandt Parks designed by Olmstead in the 1860s, built in 1888, and upgraded in 1935–37
- Van Cortlandt Park gazetted in 1888 and formally opened to public use in 1897
- Yonkers Rapid Transit service through Van Cortlandt Park ended in 1943, rails removed in 1944
- Putnam Division steam engines replaced by diesel locomotives in 1951
- Putnam Division passenger service ended in 1958
- Yonkers Hudson River sewage outfall at Mount St. Vincent closed in 1961 when adjacent treatment plant opened
- Pelham Bay Park garbage dump, the Bronx's last, opened in 1963 and closed in 1979
- Old Croton Aqueduct hiking trail (along the Ridge) established by Parks Department in 1974
- Rental rowboats removed from Van Cortlandt Lake because of heavy siltation in 1976
- Urban Park Ranger program established in Van Cortlandt Park in 1979
- Former Lincoln Marsh portion of Tibbett's Brook, alongside the Henry Hudson Parkway just south of the Yonkers line, lost its last open water in the 1970s
- Putnam Division freight service ended in 1982
- Putnam Division rails, ties, and telephone poles (including swallow wires) removed in the 1980s–90s
- Hillview Reservoir drained, resurfaced in 1993
- Public access to Hillview Reservoir interdicted following 9/11/2001
- 50,000 yd^3 (38,228 m^3) of sediment dredged from Van Cortlandt Lake in 2001–3 to restore its depth uniformly to 9 ft (2.7 m) (from 2 ft [0.6 m]) with spoil deposited on the Parade Ground; in the process the Triangle (= Lake Marsh) marsh vegetation was gouged out
- Native grass and wildflower surface of all but the southeast corner of the Parade Ground removed, replaced with a hybrid-turfgrass monoculture in 2008–12
- Jerome Reservoir drained, scraped clean, and reconfigured in 2008–14.

Avifaunal Overview

It is useful before delving into the status of all birds in the Northwest Bronx to take a broader view of what has been happening to its birdlife, to understand the concepts and terms used, data sources, and how its birds respond to environmental variables.

TERMINOLOGY

Writers of regional avifaunas have over the years developed a nomenclature of terms in widespread use. Many of these are now obsolete, having been emended or supplanted by new ones, yet they can still found in the published literature. Those in italics have been replaced in this book by more modern counterparts in parentheses, among them: *visitant* (visitor); *transient, transient visitant* (migrant); *casual, accidental* (vagrant); *race, form* (subspecies, taxon, taxa); *color-phase* (morph); *escapee* (escape); *immature* (hatching-year, or HY, second-year, or SY); *hybridize* (interbreed, hybridize, intergrade—as appropriate); *first-summer* (a confusing term that has been used for both HY and SY; it should denote only HY). A few unambiguous terms do continue in wide use—*breeding plumage, winter plumage, juvenile, first-winter*—as do some others not wholly unambiguous. Two more with a long history of ambiguity and misuse are also seen frequently in the literature (*subadult, nonbreeding plumage*), but we do not employ them. Newer descriptive terms include *trans-Gulf overshoot, spring range-prospector, blowback drift-migrant, floater,* and *nonbreeder*. These and other terms and concepts are defined and discussed in the Glossary (Appendix 3).

Our seasonal convention for dates and counts is *spring* (March–May), *summer* (June–July), *fall* (August–November), and *winter* (December–February). While arbitrary, it allows easy comparison across species, and the few instances of transgression were to admit dates barely into adjacent seasons.

SUBAREA COVERAGE

In the 1870s–90s, Van Cortlandt Swamp was known to the ornithologist and botanist Eugene P. Bicknell as Tibbett's Swamp and Van Cortlandt Lake as Bibby's Pond, and he worked both locations almost every Sunday for many decades. Living only a block west of Broadway at W. 253rd St. he was also constantly afield there as well and in those early days collected birds and plants extensively. The ornithologist Jonathan Dwight, a contemporary of Bicknell's, also collected birds extensively throughout the Bronx, including many in Van Cortlandt Park. A broad array of fieldwork by other early Bronx naturalists—and beginning in the 1910s, near-daily bird observations by some of the amateur observers in the New York City area—expanded their findings. Many contemporaneous maps notwithstanding, Van Cortlandt Swamp was indeed a swamp and marsh 140 years ago and had increased in size and complexity once the Putnam Division track bed construction in the 1840s had further constricted Tibbett's Brook flow, but Van Cortlandt

Swamp was born when Jacobus van Cortlandt created Van Cortlandt Lake to power his mill in December 1699.

In 1949–50, Robert Moses, the New York City and State planning dictator, gutted Van Cortlandt Swamp, New York City's sole remaining natural freshwater swamp-marsh, 40 ac (16 ha) in size. He did so without environmental thought to provide fresh land for a new golf course to replace the one he was about to squash for a new expressway. He overrode the weak opposition to this desecration of the third-largest park in New York City—one rich in Indian and colonial era deposits and sites and with the most breeding bird species and the largest mature native forests. The environmental details of this event have never been chronicled, nor has any assessment, historical or contemporary, of Van Cortlandt Park's avifauna ever been attempted. Surprisingly, Van Cortlandt is the only New York City park for which the latter has now been done.

Except for bursts of intense activity from the 1920s to the 1960s, study area coverage has been episodic, light, and very area-selective. Van Cortlandt Swamp and Lake have always been the center of attention, along with the Parade Ground. The Ridge (Croton Forest) was next in coverage, followed by the Sycamore Swamp and Northeast Forest, then Vault Hill, and finally the Northwest Forest, which received essentially no coverage at all until the 1980s. Woodlawn Cemetery has also long been in a coverage vacuum (and not only because of no access before 0830 or after 1630), for decades almost wholly unknown and then beginning in the 1950s being worked intensely on Christmas Bird Counts (still the case) but ignored the rest of the year. The late 1950s through the mid-1960s comprised the only exception to this dismal Woodlawn scenario. Similar restrictions—less critical earlier, draconian after 9/11—have affected Hillview.

Jerome and Hillview Reservoirs were examined for waterbirds on a regular basis beginning at Jerome in 1914 and at Hillview in 1929. Except for Christmas Bird Counts, Jerome's coverage dropped to near zero in the mid-1980s and remains there, abetted by no water from 2006 to 2014. Hillview was worked heavily from about 1930 to 1965 but thereafter only on Christmas Bird Counts, and since 9/11 only by NYCDEC-contracted wildlife technicians. Bicknell was the only observer who scoured Kingsbridge Meadows despite its great productivity and unique habitats, all of which were obliterated in the late 1920s. We also have little information about Jerome Meadows beyond Bicknell's notes, but Cruickshank and other Bronx County Bird Club members covered Jerome Swamp from about 1922 until about 1930, when it too was all but obliterated and unique, superb marshbird habitat was lost.

Most expected New York City area landbird species have occurred and still do occur in appropriate habitats throughout the study area. Table 3 records the 51 species added to the Van Cortlandt cumulative list since 1942, the publication date of Cruickshank's book. But at least 48 species unrecorded in the study area have occurred within 3 smi (5 km) of Van Cortlandt Lake (Table 4), indicating the potential for more new Van Cortlandt Park species. Many study area species have occurred just in Van Cortlandt Park itself (not shown) but only 6 have occurred exclusively in a single non-Van Cortlandt subarea, all of them water- or shorebirds on the reservoirs (Table 5). And just 9 species have bred in only a single non-Van Cortlandt subarea, all but 2 of them in grassland or saltmarsh sites long ago destroyed (Table 6).

HISTORICAL AND RECENT DATA SOURCES

Published Literature

Since Bicknell's time there has not been a single paper devoted solely to the status of all birds in the Northwest Bronx. Hundreds of individual records have appeared in *Forest and Stream*, *Bulletin of the Nuttall Ornithological Club*, *Auk*, *Bird Banding*, *Journal of Field Ornithology*, *Bird-Lore*, *Proceedings of the Linnaean Society of New York*, *Linnaean Newsletter*, *Audubon Magazine*, *Audubon Field Notes*, *American Birds*, *North American Birds*, and *Kingbird*. Many contained records and other information about small parts of the study area or of the Bronx that were often pertinent to the study area.

The last complete treatment of the birds of the Bronx was by Kuerzi (1927), and the last source of information about the birdlife of the West Bronx was Bicknell's field notes from 1872 to 1901 during his residence in Riverdale. After Bicknell's death these notes were finally excerpted and published by Griscom, coincidentally also in 1927. Almost 100 years have now passed, and the physical changes that have occurred in the area and to its birdlife have been profound. This book describes them and at the same time updates all species' statuses there and in Riverdale and throughout the Bronx.

In each species account we distill published historical information on birds in two parts where available and pertinent. The first part treats Riverdale and several versions of a "Greater Bronx Region" that are most applicable to the study area and to the Bronx itself. We are fortunate to have these historical sources, as they offer detailed and useful historical information, restricted as they were to the study area and its environs. They include publications by Bicknell for Riverdale and adjacent parts of the study area (1872–1905) extracted and edited by Griscom (1927) shortly after Bicknell's death; by Eaton in his New York County (of which the Bronx was part until 1914) in the status and distribution tables in Volume 1 of his monumental *Birds of New York* (1910); by Hix in 1905 describing an intensive but selective year (1903) actively afield around New York City; by Griscom in 1923 for the Bronx Region in his groundbreaking *Birds of the New York City Region*; and finally by Kuerzi in his intensive description of a larger Bronx Region (1920–26). Figure 17 contrasts the two Bronx Regions. There have been no comparable studies since then, a gap of 90 years. But fortunately, the start of that time span coincided with a burst of intensive fieldwork lasting into the 1940s by the Bronx County Bird Club, whose members included the Kuerzi brothers and Cruickshank, and into the 1950s by the younger Sialis Bird Club members, among them Norse, Irving Cantor, and William Weber.

The other part of the historical information we give for most species expands outward from Riverdale and the various Bronx Regions to cover the entire New York City area. And while the books in this section lack the local details offered by those above, they do enable placing the study area and then the Bronx within a modern regional perspective. Cruickshank's (1942) *Birds around New York City* was rich in data from the entire Bronx but especially the study area, since he lived very close to Jerome Reservoir and Van Cortlandt Park and examined them both frequently. His book was also the first to describe in detail the avifauna of the New York City area by means of complete seasonal timings, migration peaks, selected daily counts, and nest and egg dates from hundreds of widely scattered observers—all collated by him. The only subsequent New York City area book, Bull's (1964) *Birds of the New York Area* and supplement (1970), relied heavily on Cruickshank's data but also updated migration periods and provided maxima for most species season by season, which Cruickshank had not.

During the writing of this book it became clear that 23 taxa that have been found nearby or that might reasonably be expected have never been formally recorded in the Bronx—even though it is likely that most of them have already occurred. They are Great, Sooty, and Manx Shearwaters, Western Grebe, Anhinga, Magnificent Frigatebird, White Ibis, Black Brant, Purple Gallinule, Black-necked Stilt, Long-tailed Jaeger,

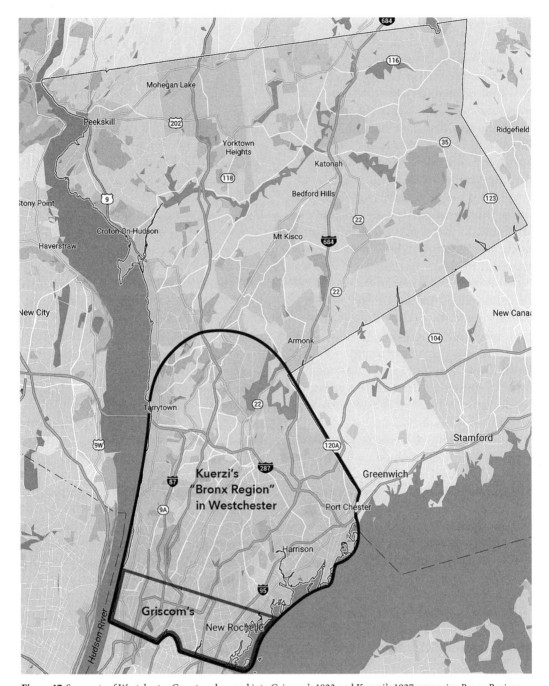

Figure 17. Segments of Westchester County subsumed into Griscom's 1923 and Kuerzi's 1927 expansive *Bronx Regions*. Image courtesy of Google Maps.

Franklin's, Thayer's, and Sabine's Gulls, Royal and Sandwich Terns, Black Guillemot, Varied Thrush, Bohemian Waxwing, Audubon's and Black-throated Gray Warblers, Western Tanager, and Bullock's Oriole.

The only two New York State–wide bird books since Eaton's (1910, 1914) two volumes were Bull's *Birds of New York State* (1974) and supplement (1976) and Levine's update (1998). Covering the entirety of New York State, they were unable to generate New York City area profiles for most species, let alone for the Bronx or for the study area. Rather, pertinent pieces of information have been extracted from these sources and inserted directly into the appropriate species accounts.

Until Cruickshank (1942), those writing about the status and distribution of New York birds almost never reported daily counts by species, rarely distinguished between frequency of occurrence and numerical abundance—in fact, usually conflated both concepts—and did not define words like *abundant, common, rare, regular*. Admittedly, defining such terms is a quagmire, and thoughtful students recognize that many of them are relative, not absolute, and whose use and definition vary greatly between species and authors.

Individual Observers' Notes

Regional books are not designed to provide local information, so extensive data mining from unpublished sources has been required; much of its results are published here for the first time. Assessments of various species' history in the study area, in the Bronx, in Central and Prospect Parks, and in the New York City area—placing study area statuses in a regional perspective—drew on the knowledge and intensive local fieldwork of the authors dating back to 1914 (Kieran), 1928 (Sedwitz), 1930 (Norse), and 1950 (Buckley). We also relied on the following observers' notes and personal discussions for detailed decadal study area and Bronx historical status information: the 1910s (Griscom, Kieran, Hix), the 1920s (Kieran, Cruickshank, Kuerzi), the 1930s (Kieran, Cruickshank, Norse), the 1940s (Kieran, Norse, Komorowski), and the 1950s (Kieran, Norse, Komorowski, Buckley, Kane, Horowitz). From the 1960s onward, our own experience, verbal accounts from others, various New York City area rare bird alerts, and long question-and-answer sessions with several hundred additional New York City area observers complemented published information in *Linnaean Newsletter, American/North American Birds*, and *Kingbird*.

Museum Specimens

VertNet (VertebrateNetwork) is a National Science Foundation–funded online collaborative project designed to help people discover, collect, and publish vertebrate biodiversity data from hundreds of biocollections. It enabled us to scan nearly all US and Canadian major bird collections for specimens taken by Bicknell, Dwight, and others in our study subareas, in Riverdale, and in the Bronx. Had we attempted this as recently as only 10–15 years ago, we would have uncovered very few pertinent specimens. Because digitizing of all bird specimens in cooperating museums is not yet complete, some pertinent specimens remain unlocated.

From the study area we know of 456 usable museum specimens, 99% of them taken before 1900, embracing 105 species—367 specimens of 77 species from Van Cortlandt Park, 53 of 22 species from Kingsbridge Meadows, 34 of 22 species from Woodlawn Cemetery, and singles of 2 species from Jerome Reservoir. We have located no specimens from Hillview Reservoir. Most were originally in the Dwight collection, which after his death in 1929 was absorbed into the main American Museum of Natural History collections, as was the small Woodlawn collection amassed by L. S. Foster. A few eventually made their way to the Museum of Comparative Zoology at Harvard.

From 1874 to 1900, the Mosholu resident E. P. Bicknell had been collecting heavily in adjacent Riverdale, amassing 311 specimens of 106 species that we have located. In addition, he collected the Kingsbridge Meadows specimens above at one of our study area subsites. (It would seem that Dwight was unaware of this area's richness, since we know of no specimens of his taken there.) Bicknell's collection was after his death in 1925 housed at Vassar College, but eventually the bulk of it went to the New York State Museum in Albany. We have tracked a handful of other Bicknell Riverdale skins to the American Museum of Natural History in

New York City, Harvard's Museum of Comparative Zoology, the US National Museum at the Smithsonian Institution, and the Academy of Natural Sciences in Philadelphia. Any others still undigitized will have escaped our purview.

Bicknell kept his own counsel, and while he and Dwight knew (of?) each other, each seemed unaware of the details of the other's collecting and specimens. (Foster seems to have been even more of a loner.) Neither Griscom when he was writing his *Birds of the New York City Region* nor Eaton when writing his *Birds of New York [State]* knew anything about most of Bicknell's specimens. Even when Griscom finally had access to Bicknell's Riverdale diaries in the mid-1920s, he was still unaware of many of Bicknell's specimens and clearly never saw them. The information on Bicknell's complete collection of birds taken in the greater Riverdale area and Dwight's in the study area is for the most part published here for the first time—nearly 150 years later.

Individual specimens of interest for one reason or another, as well as synopses of museum holdings, are noted in pertinent species accounts. Full details for all 767 specimens of 136 species are given in Appendix 6, while Table 13 synopsizes the distribution of known specimens from the 4 study subareas and Riverdale.

We take identifications of bird specimens in VertNet at face value, and except in a few cases, we have not attempted to confirm species identities. Consequently, there may be a few specimens that future work could reveal to be misidentified. In particular, we have in mind Bicknell's Thrush (q.v.), whose identification, even of specimens, remains a work in progress, so any references to Bicknell's Thrush specimens should not be accepted without reservation. Equally, there may be some Gray-cheeked Thrushes that are actually Bicknell's.

Perusal of the nearly 800 study area and Riverdale skins is informative and in a few cases surprising. For example, one cannot help but be struck by the large numbers of skins for several species that today are uncommon migrants but that back then had to have been far more numerous and locally breeding. Blue-winged and Worm-eating Warblers come quickly to mind, and there are others. It was also unexpected that there are some spring migration dates that even today might invite comment.

The other striking aspect of these collections is that they almost exclusively involved landbirds, in fact almost only passerines. We assume this represents their collectors' choices because while certain birds had become quite scarce in the market-gunning era, many others were present. Griscom, in commenting on Bicknell's Riverdale diaries, noted that he seemed never to have paid attention to birds on the Hudson River, and many waterbirds and raptors he ought to have encountered were simply unmentioned.

Bird Banding

The Bird Banding Laboratory provided information on all birds banded or recovered in the study area up to 31 December 2015, but the data are scanty. Much of the problem turns on the fact that recoveries are reported only to 10-minute latitude-longitude blocks, which in our case (405–0735) includes large pieces of Bergen County, New Jersey, and southern Westchester County. Even though the data sent by the Bird Banding Lab filtered out all New Jersey recoveries, intra-block locations for those in New York were routinely digitized only if recovered after 2005—and then only if sites had been volunteered. Bandings and recoveries before 2006 are in the process of being digitized. A total of only 221 useful records materialized, and 128 of those were feral Canada Geese banded, recovered, or both within our block or nearby, except for 2 wild breeders banded in the Canadian arctic but lacking any intra-block recovery location specifics. There was only a single study area goose recovery of one not banded locally: an arctic-breeding Canada banded near the southeastern corner of James Bay, Ontario, on 5 August 2003 and recovered in Van Cortlandt Park on 18 January 2007. By definition it would have thus belonged to *interior*; see the Canada Goose species account.

No neck-collared Canadas were reported in the study area before 2007. Of the 74 resighted in Van Cortlandt Park from 2007–15, 8 had been banded in spring or summer in the local 10-minute block (405-0735) but at unnamed locations, perhaps the Bronx Zoo, where 63 had definitely been banded in July. Thus all were ferals. Of the 3 banded outside the local 10-minute block, the first was a feral that had been banded in June in Lambertville, New Jersey (63 smi [101 km] away). The second had been banded on 7 July 2009 July in Jericho, Vermont (255 smi [410 km] away) and resighted on 30 December 2014; the third had been banded on 2 July 2013 in Varennes, Quebec (336 smi [541 km] away) and resighted on 18 January 2014. The second was most likely feral, and the third may also have been, but that is less certain.

Several of these neck-collared ferals were in the Parade Ground flock with the Barnacle Goose during the winter of 2012–13, doubtless the source of rumors that the geese had all been neck-collared in Greenland and so buttressing a Greenland origin for the recent Pink-footed, Cackling, Greater White-fronted, and Barnacle Geese there. While it is likely those 4 species were Greenland breeders, all Canadas had been collared in the Bronx Zoo in July.

Also lacking local details, a handful of common landbirds, ducks, and raptors have been banded or recovered within our block, but all were trivial. One truly tantalizing was the block-only recovery of an HY Common Tern on 28 July 1961 that had been banded as a chick at Moriches Inlet on 29 June 1961 by our old friend Leroy Wilcox. We'd love to know exactly where it was found within Block 405-0735.

A multiyear study of Herring Gull chicks banded with colony-specific color combinations in various New York–New England–Atlantic Canada colonies was designed to track their movements, arrival dates, and overwintering areas. HYs from 6 different colonies were recorded in Van Cortlandt Park in fall 1937 and are shown in Table 14.

Other Sources

Especially for recent study area data (since the early 1990s) but even back many decades, we were fortunate to be able to evaluate a 20,000 eBird record database through 30 November 2017. This proved essential in updating seasonal maxima and early/late dates for the commoner species often glossed over by active observers and for locating obscure but locally significant records that otherwise might never have seen the light of day.

Our efforts to place the status of study area species in a larger regional scale were hampered by the absence of published treatments of the status and distribution—current as well as historical—of birds in Westchester and Putnam Counties, but we were fortunate in finding Rockland (Deed and Wells 2010) and Dutchess (DeOrsey and Butler 2014) Counties' excellent recent assessments online. Censuses, atlases, and other aspects of assaying study area breeders and their trends are discussed in "Breeding Species," below.

The annual Christmas Bird Count database maintained by the National Audubon Society made available all Bronx-Westchester Christmas Bird Count data from the first in 1924 to the last we used, the 90th in 2013. Analysis of count-wide trends shown in 43 species accounts offered unusual insight into study area birds' winter numbers and changes, which are discussed in the relevant species accounts.

Owing to the ephemeral nature of many websites, web citations have been used only when there are no alternatives. The only electronic sources cited are those accessible to anyone, being neither password-protected nor

VAN CORTLANDT CONTRASTED WITH CENTRAL AND PROSPECT PARKS

Cumulative Species

One of the oldest New York City parks (1888) and the third-largest (1146 acres [464 ha]), Van Cortlandt has a topographic and vegetative diversity that remains unmatched anywhere within New York City. The park boasts information about its birds going back 143 years—almost to Audubon's day—a span unmatched in New York City and among the longest records of continual bird observations in any US park.

Unlike Central Park (843 ac [341 ha]) and Prospect Park (585 ac [236 ha]), Van Cortlandt Park is not a man-made park carved and molded out of existing disturbed areas by the early landscape architects Frederick Law Olmstead and Calvert Vaux. However, their company may have worked on the Dutch Gardens below the Van Cortlandt Mansion and Olmstead's son on the design for Woodlawn Cemetery, and they did prepare elaborate proposals for Jerome Reservoir that were never realized. Although assaulted and warped by roads dissecting it, two internal golf courses, and an enormous urban athletic complex on its southwest corner, Van Cortlandt Park nonetheless remains in a reasonably natural state, particularly its hundreds of acres of native forest, freshwater marsh, and wooded swamp. Despite insult from logging, agriculture, war, fires, homeless squatters, off-road vehicles, hunting and trapping, and Robert Moses, it endures. In many respects, it thrives.

To place Van Cortlandt Park's avifauna in a New York City park perspective, we contrast it with those of the most heavily worked and best known New York City parks, Central and Prospect. What constitute official Central Park or Prospect Park species lists is an open question. There have been no attempts to produce a formal list of their cumulative species since Carleton's (1958, 1970) Central and Prospect Park lists and Pasquier's (1974) short Central Park addendum—almost 50 exceedingly active years ago. Consequently, the authors of this book have combed the literature, talked to countless active observers, and consulted with those most knowledgeable about Central Park and Prospect Park observations, particularly since 1970 (see Acknowledgments). Our resulting species lists are offered without comment in Table 7. The Central and Prospect Park cumulative species and breeder lists we proffer do not claim to be official, since their main purpose was to enable first-cut contrasts with Van Cortlandt Park's avifauna. In the process it became clear that many significant records from both parks have fallen through the reporting cracks. We hope that knowledgeable observers will make concerted efforts to locate and archive as many as they can and eventually update and expand Carleton's 1958 and 1970 annotated lists.

We have not incorporated the following 17 candidate species into our Central Park cumulative list because we have been unable to locate any firm records for them, or they have been reported or rumored but we are unaware of any supporting descriptions, photographs, or observers: Greater White-fronted and Cackling Geese; Swallow-tailed Kite; Sandhill Crane; American Golden-Plover; Eastern and Western Willets; Sanderling; Short-billed Dowitcher; Red-necked Phalarope; Dovekie; Least, Black, and Forster's Terns; Say's Phoebe; Scissor-tailed Flycatcher; and Audubon's Warbler. Regrettably, no group/entity has assumed formal responsibility for tracking the Central Park avifauna accurately.

The Prospect Park avifauna, on the other hand, has since about 1990 been closely tracked by the Brooklyn Bird Club, although some information from 1967 to 1990 has yet to find its way into their files and may be lost. We have not incorporated the following 5 candidate species into our Prospect Park cumulative list because we have been unable to locate any firm records for them: Sandhill Crane, Boreal Chickadee, Audubon's Warbler, Yellow-headed and Brewer's Blackbirds.

Even though Van Cortlandt Park has been consistently productive as a migration monitoring point, it has always marshaled only a fraction of the number of observers in spring and fall that Central and Prospect Parks routinely do. Despite having been outnumbered by them 50:1 or more in annual observer-hours for many scores of years, Van Cortlandt has amassed almost as many cumulative species and almost twice as many cumulative breeding species, with a far more diverse winter population than either Central Park or Prospect Park. These comprise an apt index of its inherent year-round value for native birds and reflect its enormous diversity of terrestrial and aquatic habitats, most of them natural despite having been savaged for parkways and expressways since the mid-1930s.

Of the 337 species recorded among Van Cortlandt, Central, and Prospect Parks, 265 have been found in all 3 parks. Van Cortlandt's cumulative total is 301, Central's 303, Prospect's 298 (Table 7). Van Cortlandt has 12 unique species, Central 14, and Prospect 11 (Table 8). Another 11 have been recorded in both Central and Prospect Parks but not Van Cortlandt, 11 from both Van Cortlandt and Prospect Parks but not Central, and 12 from both Van Cortlandt and Central Parks but not Prospect (Table 9). In this respect, the 3 parks are quite similar, and only 10 or so reasonably anticipated species have never been found in at least one of the 3 parks (Table 10). New species continue to be added to all 3 parks on a steady, albeit episodic, basis.

Breeders

Cumulative breeding species after 1871 in Van Cortlandt Park (Table 15) total 123 (with Ruffed Grouse and Pileated Woodpecker breeding until earlier in the 19th century) plus 13 potential breeders (see Glossary in Appendix 3 for a definition); for Central Park it is 73 plus 7 potential; and for Prospect Park 71 plus 6 potential (Table 11), but these are best viewed as minima owing to intermittent historical breeding information for Central and Prospect Parks. Notwithstanding, there are no known reasons to question the three parks' relative rankings. Collectively, 124 species have bred across all 3 parks, and another 19 are potential past (>1871) or future breeders in one or more parks.

Indicative of Van Cortlandt's rich habitat, 39 study area breeding species (12% of the cumulative total) have never bred in either Central or Prospect Park (Table 12). Conversely, only a single species, Pied-billed Grebe, has bred in one of those parks but never in Van Cortlandt, although it was strongly suspected of having done so in Van Cortlandt Swamp from the 1930s to the 1950s. Tellingly, while about 70 species are breeding in Van Cortlandt Park as of 2016, corresponding numbers for Central and Prospect Parks are 20–30, and many species breed there for only a year or two before disappearing. Human disturbance, dogs off leash, and lack of herbaceous layer in both of those parks are high, expressed in their low numbers of potential new breeders.

Van Cortlandt Park has recorded about the same cumulative number of landbird species as Central Park and Prospect Park, but it has had and still has nearly double the total number of breeding species in either of those parks—or for that matter of *any* New York City park. Why? Because even though Van Cortlandt Park is not the largest New York City Park (Pelham Bay Park is), it retains the largest area of relatively natural terrestrial and aquatic habitats of any New York City park. This simultaneously increases the number of breeders while diluting the total number of nonbreeders because migrant landbirds are more easily lost in larger Van Cortlandt, which also has always had far fewer observers than Central Park or Prospect Park.

ECOLOGICAL CONNECTIVITY

Connectivity in birds is often less important than in mammals, given that most birds easily move about, at least during migration. For study area breeders, habitat is arguably more important, especially large forested areas with a good herbaceous layer, a large undisturbed swamp/marsh complex, the lake, the Tibbett's Brook ecosystem, and when empty of people, the Parade Ground. This is particularly true of migratory breeders that come and go twice yearly, assessing available habitat in the process.

Nonmigratory resident species are even more reliant on adequate connectivity, given that their range expansions occur most often through postbreeding dispersal, when nearby sources of suitable colonizers/recolonizers become paramount. Study area examples include Ring-necked Pheasant and Wild Turkey, Pileated Woodpecker, accipiters and buteos, Barred and Great Horned Owls. Eventual loss of connectivity with Sharp-tailed and Seaside Sparrows along the Harlem and Hudson Rivers probably was important in their disappearance as breeders at Piermont Marsh despite the persistence of suitable habitat.

Connectivity to southern Westchester County breeding sites has probably played a significant role in maintaining the study area's high species richness for the last 150 years despite some losses. Lack of connectivity, heavy human use, manicured sites, diminished understories, and many exotic (nonnative) plants have all played roles in the reduced breeding avifaunas of Central and Prospect Parks, where annual turnover of breeding species is far higher than in the study area.

BREEDING SPECIES

Overview

Owing to Van Cortlandt Park's great habitat diversity, of all New York City parks it has long had the richest breeding avifauna—cumulative or current. And of the 13 unconfirmed but expected historical or future breeders (Table 16), 3 were historical only and are unlikely to ever be confirmed. But 10 more, including another 3 that were also historical, can be anticipated at any time in the near future. Of these, Barn Owl and Purple Martin are contingent on emplacement of properly maintained nest boxes; and Yellow-crowned Night-Heron, Black Vulture, Barn Owl, and Common Raven are potential breeders at any time in/on structures at Jerome and Hillview Reservoirs and Woodlawn Cemetery.

Table 17 chronicles all study area breeders from 1872 to 2015, with presence/absence recorded qualitatively across 7 eras 14–30 years long that coincided roughly with published accounts or periods of census activity. One might predict that the greatest species richness would have been in the first eras, and it was, but only slightly. This high richness (87–97 firm, 91–101 firm + uncertain) was maintained across the first 5 eras and 93 years to 1965. As the table caption notes, this difference was genuine.

A clear drop in species richness to 74–79 (78–80) then occurred during the 3 most recent eras from 1966 to 2015. This was not a sudden break at 1965–66 but one that began slowly in the early 1940s and increased through the 1950s to early 1960s. The major cause was the filling of more than half of Van Cortlandt Swamp in tandem with destruction of the fields and edge between it and Vault Hill in 1949–50. In turn, that action accelerated the ongoing transformation of the remaining few acres (hectares) of cattail marsh into first a *Phragmites* marsh and then a Buttonbush–Red Maple–willow swamp. But numerous woodland breeders also disappeared at this time, and the combination was overwhelming.

Fortunately there has been enough study area field activity over the decades for us to chronicle breeding species turnover. Tables 18 and 19 note when old breeders disappeared and when new ones arrived and offer possible, likely, or certain reasons for each of these changes.

From 1872 to 2015, 50 breeding species vanished (Table 18); the loss of 22 from 1940 to 1967 comprised the largest cluster in the history of the study area. Reasons for the loss of the 50 break out as follows: 1 species was at its brief postintroduction apex, 2 were lost for unknown reasons, 3 were part of widespread northerly range contractions, 4 were lost because of adventitious breeding events, and 7 were part of area-wide population crashes or declines. Of the remaining 33, 4 were lost from disturbance, 8 from urbanization, and 21 to habitat loss including both obliteration and natural plant succession. Viewed another way, 25 of the 50 disappeared as direct results of human interference and urbanization. Even more may have been lost from indirect human actions. Though a great many of the species lost were ground- or near-ground nesters (Table 20), for some these losses may only have been coincidentally associated with destroyed habitats. Their open field or marsh habitats were typically the first destroyed during urbanization by an expanding New York City. That said, as a group they have been and remain at greater risk than any other group.

At the same time, 32 new breeders were arriving between 1872 and 2015 (Table 19); the gain of 17 between the early 1920s and 1959 was the largest in the history of the study area. Origins of the 32 break out as follows: 1 was relict or adventitious, and 3 were erratic New York City area breeding events; 2 may have been present earlier but unmentioned; 4 were adventitious breeders; 3 exhibited breeding behavior changes bringing them into the study area; 6 were recently introduced exotics (nonnatives); and 5 followed New York City area population increases and 8 widespread range expansions, all but one from the south.

In the almost 6 decades since 1959 there have been only 11 new breeders (2 of which were nonnative) against 15 lost breeders (3 of which were nonnative), not terribly far from a draw. The last new native breeding species involving more than a single pair arrived in the 1990s, and the last lost native species with more than a single pair disappeared perhaps 10 years later. Only 6 (2 nonnative) have been lost after the 1960s, so the decline in breeders has itself declined.

Perusal of Table 17 shows breeding activity for the entire 1872–2015 period for each of the 123 known study area breeders, indicating where uncertain or lacking information diminished the number of core species; how long each breeder persisted; and when it started, stopped, or restarted; and it enables the reader to contrast species composition and richness during each of the 7 time periods, including the most recent.

Despite some turnover in breeders, there has been surprising stability. There have been just 4 species that bred only once: Turkey Vulture, American Coot, Brown Creeper, and Nashville Warbler. There have been 5 additional species with anachronistic one- or two-year breedings: Black-crowned Night-Heron, Alder Flycatcher, Cliff Swallow, Horned Lark, and Purple Finch. The heron was suspected of nesting for decades, but only a single pair has ever been found (for 2 consecutive years), and the lark may have been only a postbreeding wanderer. Only 2 species were lost and then returned as breeders: Cooper's Hawk and Osprey, but as of early 2017, there are 4 more poised to do so: Acadian Flycatcher, Cliff Swallow, Eastern Bluebird, and Kentucky Warbler.

Thirteen study area breeding species remain unique in the Bronx post-1871, including Turkey Vulture, Sharp-shinned Hawk, Broad-winged Hawk, King Rail, Sora, Common Gallinule, American Coot, Eastern Whip-poor-will, Alder Flycatcher, Brown Creeper, Louisiana Waterthrush, Nashville Warbler, and European

Goldfinch. Within New York City, Turkey Vulture, Alder Flycatcher, Brown Creeper, Louisiana Waterthrush, and Nashville Warbler were unique breeders.

Just 9 species have nested only outside Van Cortlandt Park proper (Table 6). Of them, 6 were extirpated when Kingsbridge Meadows and Jerome Swamp were destroyed, 2 were singular breeders at Woodlawn Cemetery and Hillview Reservoir, respectively, and Common Nighthawk still breeds on extrapark rooftops and may also have been nesting within the park.

Table 21 lists 55 *primary breeders*—those that have nested more or less annually in the study area from 1872 to 2015—and Table 22 offers 18 *secondary breeders*—those we believe to have been nesting more or less annually since various dates back to the early 1900s. Together they comprise the 73 study area *core breeders.*

If post-1965 trends endure, the number of currently breeding species will level off at about 70. Several do seem ready to establish themselves as new breeders, but beyond Eastern Towhee and Brown Thrasher no current breeders appear in imminent danger of extirpation. Still, even acknowledging the enormous physical changes that have occurred in the Northwest Bronx over the last 143 years, it is sad that the breeding species in the study area have dropped from about 100 to about 70.

While firm estimates of the numbers of breeding pairs within the study area are few (see below), since 2000 the populations of most study area breeding species seem to have been stable. Against this, numbers of another 13 species have been increasing, while those of another 8 have been declining (Table 24).

An unexpected aspect of study area breeders is the 30 that make use of or are dependent on human structures of many sorts (Table 23). Absent or precarious populations of several—including Osprey, Barn Owl, American Kestrel, Purple Martin, Cliff Swallow, and Eastern Bluebird—would greatly benefit from carefully crafted habitats emplaced with foresight. Each of these has already proved responsive to artificial structures throughout the New York City area.

Some 25 of the 45 study area species that regularly or occasionally nest on or nearly on the ground have been extirpated, leading to suspicions of quadruped predation. Another 8 second-growth/oldfield successional breeders have also suffered disproportionate loss (Table 20). But most in both groups had been using swamp, marsh, or open field habitat that was destroyed, is greatly reduced, or is heavily disturbed. Some forest ground-nesters already extirpated or under threat (Veery; Ovenbird; Worm-eating, Black-and-white, and Kentucky Warblers; Eastern Towhee) may also have been adversely affected by quadrupeds like domestic and feral cats and dogs, Eastern Coyotes, Red and Gray Foxes, Raccoons, Virginia Opossums, American Minks, and Long- and Short-tailed Weasels—all of which occur and breed in Van Cortlandt Park at least. Even Fisher may now be resident: one was videoed at night in June 2014 in Morris Heights in the Bronx along the Harlem River, the first ever in New York City and <2 smi (3.2 km) south of Jerome Reservoir.

Woodlawn Cemetery has always held variable but small subsets of the landbirds breeding within Van Cortlandt Park. However, coverage there has been marginal, except during the late 1950s–early 1960s. Horowitz covered it then several times a week and prepared a tentative list of numbers of breeders. Table 25 records his findings, as well as those from an intense but brief survey in late May 2014 by Buckley. In both instances, presence should carry more weight than absence. The most notable breeding species ever found in Woodlawn Cemetery was the pair of Nashville Warblers in 1892. One can only wonder how long they may have been breeding there at a time when nearly no one visited it for birds but when much of it was still undeveloped and with excellent habitat for many species. In view of the recent breeding by

Pine Warbler in both Bronx Park and the Bronx Zoo, we expect that it will also breed in Woodlawn's extensive conifer plantings if it has not already done so. We also suspect that Common Nighthawk may breed atop buildings there and that Black Vulture, Common Raven, and Yellow-crowned Night-Heron could do so at any time. The major deterrent to fieldwork there has always been access in the very early morning before cemetery visitors arrive and when birds are most active and singing regularly. Regrettably, unlike those at Mt. Auburn Cemetery in Cambridge, Massachusetts, and Swan Point in Providence, Rhode Island, Woodlawn officials have never seen fit to offer early morning access. Moreover, there is no evidence from New York State Breeding Bird Atlas files that Woodlawn Cemetery was ever visited during either atlas period (see below).

Jerome Reservoir has been examined for breeders by various people over the years, always from outside its fences and often only before June. Rough-winged and Barn Swallows have been known there since the early 1900s; Bank could have been long breeding but overlooked among the Rough-wings; and in the last 10 years Cliff may have moved into buildings there, also overlooked. We believe that Spotted Sandpiper is still nesting there unmolested by pedestrians, and that American Kestrel, Barn Owl, Common Nighthawk, Chimney Swift, and Eastern Phoebe may also be using reservoir structures. Several species of urban landbirds (Blue Jay, Eastern Kingbird, Red-eyed Vireo, Tufted Titmouse, American Robin, Song Sparrow, House Finch) have long bred in woods, shrubs, and open areas immediately adjacent to the reservoir, but many of these sites no longer exist, the area having been recently "improved." There is also no evidence from New York State Breeding Bird Atlas files that Jerome Reservoir was ever visited during either atlas period.

Hillview Reservoir was worked regularly in the summer in the 1930s and the 1940s, when the only study area breeding of Eastern Meadowlark occurred; and it is also likely that Horned Lark nested in the 1950s and perhaps Savannah Sparrow in the 1930s and 1940s. But from the 1950s through the 1980s if anyone went to Hillview Reservoir looking for breeding birds, they never reported what they found. This is a shame because its extensive conifer groves may have supported unexpected breeders, especially after winter irruptions of Red-breasted Nuthatches, Golden-crowned Kinglets, Red Crossbills, and Pine Siskins. But the conifer groves are now gone and will not be replaced, so they and any unknown breeders have now faded into the fog of history. After September 11, 2001, the unbridled access enjoyed for 85 years was terminated, but even before that, drastic measures against all waterbirds had been in place since 1993 (see "Bird Control" below). It was only when bird control reports appeared online that we learned that Osprey had recently nested there and that Cliff Swallow had also recently established breeding colonies. Ospreys and swallows alike were swiftly extirpated, although the swallows are back.

Kingsbridge Meadows was regularly worked by Bicknell, but he seemed to have had little time for breeders, concentrating only on migrants and winter residents. The area must have been extraordinary judging by the species he collected there in the late 1880s, but slowly it began to be nibbled at from the sides, filled in, and built upon. In the late 1920s the last marsh pieces were eradicated. When the meadows existed, they supported breeding Saltmarsh (and we also suspect, Seaside) Sparrows, Clapper Rails, and Spotted Sandpipers. Sedge Wrens and Savannah Sparrows may also have nested there but during months of the year when Bicknell shifted his attention to plants. We know of no other information on birds using Kingsbridge Meadows beyond Bicknell's tantalizing glimpses.

As to breeders at Jerome Meadows, the precursor to Jerome Reservoir, we know even less, beyond that Bicknell found Alder Flycatchers there in the 1880s, one of only two breeding sites ever known for them within New York City. (The other was Van Cortlandt Swamp in 1977.) He also found Grasshopper and Vesper Sparrows and Bobolinks breeding in their drier portions in the late 1880s. After Jerome Meadows was obliterated by Jerome Reservoir, the remaining piece east of the reservoir, largely undisturbed, became Jerome Swamp, which was not examined for breeders by anyone until Cruickshank and other Bronx County Bird Club members did so from the early 1920s to its disappearance about 1930. The Virginia, Sora, and King Rails as well as Least Bitterns and Common Gallinules found there may have been breeding, but existing information is too vague to be certain.

Breeding Bird Surveys and Censuses

For landbirds (other than a few species on a few occasions and during New York State Breeding Bird Atlases and various breeding bird surveys), very few hard estimates of breeding pairs (total or density) are available anywhere in the study area. This is typical of nearly every similar location anywhere in the New York City area because obtaining these data is extremely time- and labor-intensive.

Considering just species richness, our overriding desire has always been to locate all species breeding in the study area, a goal quite different from that of breeding bird atlases, which is to document the broad-brush distribution of species across a wide geographic area. It is customary to attempt to find only 75% of the species (but not any numbers) breeding within a given block, and by definition, the missing 25% will be the rarest, least abundant, and least regularly breeding—the very ones that an analysis of a small geographic study area like ours does not want to overlook. Consequently, one should not necessarily look to large-area surveys for information on all species breeding in Van Cortlandt Park.

The study area has enjoyed only a few assays of its numbers of breeding *individuals* as well as species in Van Cortlandt Swamp (Feigin et al. 1937; Feigin et al. 1938; Enders et al. 1962; Zupan et al. 1963; Zupan et al. 1964; Zupan et al. 1965, Gera and Gera 1966, 1967); in a large wooded plot in the Northwest Forest (Jaslowitz and Künstler 1992; Jaslowitz 1993); in the Shandler Area and Van Cortlandt Swamp in 2013; and park-wide in 2014. None accessed the several sections of the 2 large golf courses. Heath and Zupan (1971–72), in a series of 10 articles in the *Linnaean Newsletter*, analyzed and discussed methodology and incidental observations from the first few 1960s Van Cortlandt Swamp censuses. A handful of other Van Cortlandt Park censuses attempted to detect breeding species, but most addressed portions of the park only episodically, especially between the 1980s and 2014 (Turner 1984; Delacretaz 2007; Künstler and Young 2010). Only the work done for the first New York State Breeding Bird Atlas in 1980–85 and the second in 2000–2005 purported to cover the entire study area, but in fact they did not. Regrettably, their data were merged with those for all species detected in the 2 large blocks that embraced the study area; teasing out study subarea data has proved to be exceedingly difficult (see discussion below). But whenever practical, data from individual atlas workers have been entered into pertinent species accounts.

Censusing Issues

Every breeding bird census attempt in the study area has been limited for one reason or another, a situation their organizers readily acknowledged. These included too few censusers who were knowledgeable observers proficient and comfortable with calls and songs (especially alternate songs) of local breeders; too great reliance on only observing (not identifying territorial songs of) potential breeders; too few hours censusing; too little time afield during the most critical daily hours; coverage of only parts of some study subareas; starting too late in the season or ending too early; and not censusing all areas simultaneously.

Demonstrable breeder detection problems exist for crepuscular/nocturnal species like American Woodcock, Common Nighthawk, all owls, and all rails; for most diurnal raptors except Red-tailed Hawk (Turkey Vulture, American Kestrel, Broad-winged, Red-shouldered, Cooper's, and Sharp-shinned Hawks); and for very early nesters like Great Horned Owl, American Woodcock, Horned Lark, and Killdeer.

There remains the classic sampling problem of locating the tail of the breeding species distribution—the least common species in the study area. This can be done only by rigorous, intensively repeated, near-daily censusing by experienced observers with superb audial identification skills—a standard only rarely approached in study area censuses. The truth is that no intensive, multi-observer, dawn–dusk–night census of the numbers and breeders of all species in the 4 extant study areas has ever been done.

Finally, because we are dealing with a moving target, there really is no unassailable single number of breeders of any species for an area as large as the study area, yet some total needs to be derived for each species. Consequently, it is always best to regard all estimated numbers of breeding pairs as *tentative* values to be confirmed or amended in subsequent years.

A major challenge for those seeking to quantify breeding species and numbers in any area is to eliminate any nonbreeders present, even when they are singing lustily. These can be numerous and fall into several different categories, with known examples from the study area: abnormally late migrants (many examples); nonadult prebreeders (Black-crowned Night-Heron); prospecting (future) breeders (Gadwall, Kentucky Warbler); out-of-range prebreeders (second-calendar-years that normally don't breed until their third calendar years—Lesser Scaup, Ruddy Duck); out-of-range nonbreeders (second-calendar-years that did not reach their normal breeding range (Yellow-bellied Flycatcher, Northern Waterthrush); floaters within usual breeding species (numerous nonbreeders that fail to obtain mates or territories). Another breeding census complication, albeit a welcome one, is the occasional adventitious breeding species that one must always be alert for (American Coot, Nashville Warbler).

The equally important converse of the foregoing is that any breeding species can be missed on censuses if ♂♂ happen not to be singing when a censuser passes by, if the census route misses their territories, or if the actual breeding population is very small. These problems are amplified when coverage is sporadic or starts too late in the true breeding season in a misguided attempt to avoid migrants. Further, it is not uncommon for genuine breeders to be erroneously dismissed as migrants. Consequently, failure to report known breeders can in most cases be safely ignored, and no valid conclusions about the completeness of breeding avifaunas can be drawn when there is such a high probability of false negatives.

Species that both breed in and migrate through the study area arrive in spring at different times, with local breeders arriving first, often well before expected migrants' arrival times. Days (or sometimes weeks) later, the usual migrants, heading to breeding sites often far to the north, will appear. In fall, local breeders also depart sooner than those from farther north. These patterns can also confound breeding bird surveys and also make it difficult to discern true migration dates from singing ♂♂ (normally the most effective census technique).

A breeding bird census gold standard for achieving robust numbers and finding the rare species at the tail of the distribution has never been applied to the entire study area simultaneously. Admittedly this would involve a massive onslaught by observers thoroughly familiar with the vocalizations of all remotely likely breeding species in the study area, because most breeders are initially detected vocally, not visually. Such a census would have to begin early enough in the season to detect Great Horned Owls and American Woodcock (February–March), and continue late enough to quantify American Goldfinches and Cedar Waxwings (August). It should cover every square yard (at least vocally) of every part of the study area repeatedly throughout the breeding season—at least twice every week from at least mid-April, optimally in groups of at least 2 qualified observers. Most important, observers should be on station from about an hour before first light (= 2 hours before local official sunrise) to 3–4 hours after first light (= 2–3 hours after local official sunrise). For marsh birds that are both nocturnal and crepuscular, starting 2–3 hours before first light (= 3–4 hours before local official sunrise) is rarely unproductive and is the only reliable way for locating most owls and nightjars. Last, to avoid migrant-–breeder confusion, breeding bird censusing should always be postponed on heavy migration days.

This complex, daunting protocol is the only sure way to detect all or nearly all species and numbers breeding in the study area in a given year. Every breeding bird census effort that has ever been made in the study area has fallen short of this Utopian standard—most woefully so—which is why more than a few species believed to still breed in the study area remain to be detected by recent searchers and why we have rather a soft comprehension of exactly how many of each species have been breeding throughout.

New York State Breeding Bird Atlases

The 2 New York State Breeding Bird Atlas efforts between 1980 and 1985 and 2000 and 2005 were massive and effective attempts to refine our knowledge of the exact distribution of all species breeding within New York State. They did this job well indeed, and the book (McGowan and Corwin 2008) presenting the results of the second New York State Breeding Bird Atlas in 2000–2005 also contrasted the results of both censuses 20 years apart. It is a monumental and thoroughly professional scientific analysis that will be a landmark mined for information for decades to come. But because the atlases were designed to a scale allowing their own realistic goals to be met, for purposes of uncovering all species breeding with the study area, they were of limited use to us. This is to be expected and should in no way be read as a criticism of the 2 atlas efforts and results.

Both atlases desired to uncover all species breeding within their mapping units, called blocks (unrelated to Bird Banding Lab reporting blocks)—each of which was a 5 km × 5 km square (3.1 smi × 3.1 smi = 6177 ac = or 9.7 smi^2)—by recording only presence and absence on visits to "each habitat" in the block, including once at night. Field effort for 8 hours in an entire block met the census minimum for adequate coverage, although coverage in most blocks greatly exceeded the minimum. A block was considered adequately censused once 76 species had been recorded, only half of which needed to be confirmed breeders. Census procedures candidly recognize that absence from a block's species list was far less biologically significant than presence. Further, they did not call for any attempts to be made to estimate the numbers of breeding pairs in a block.

The entire Bronx was encompassed by 9 atlas blocks (Fig. 18), 2 of which addressed the study area (Fig. 19). Block 5852B included all of Van Cortlandt Park save a tiny northeastern corner, all of Riverdale, southern Yonkers, a slice of Hillview Reservoir, half of Woodlawn Cemetery, Jerome Reservoir, a wedge of the Bronx east to the Bronx River Parkway,

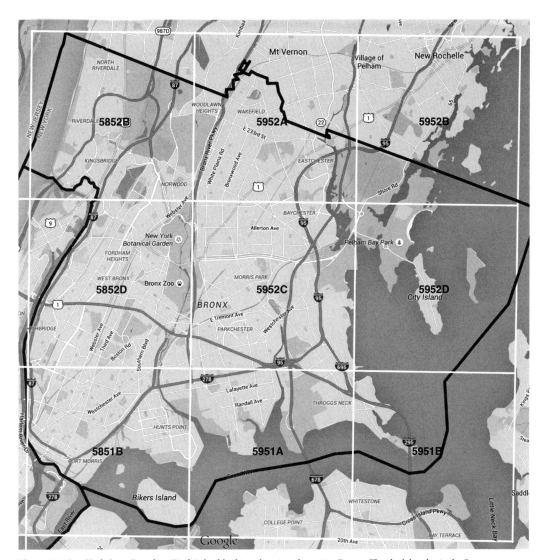

Figure 18. New York State Breeding Bird Atlas blocks embracing the entire Bronx. The dark border is the Bronx boundary. Image courtesy of Google Maps, as modified courtesy of B. C. Cacace.

and even the northern end of Inwood Park. Adjacent Block 5952A included the remainder of Van Cortlandt Park, Hillview Reservoir, and Woodlawn Cemetery, plus the Northeast Bronx east across the Hutchinson River and Parkway into Pelham Bay Park, and southern Westchester from Yonkers east to eastern Mt. Vernon. This is an incredibly varied bag covering a wide range of habitats over a large area.

In 5952A particularly, sorting out the few species that may have actually been found within the study area proved nearly impossible, even with access to the original field data sheets for both blocks, thanks to the good offices of the New York State Department of Environmental Conservation (NYSDEC).

Unfortunately, while field work in Atlas 1 in 1980–85 was strong in Van Cortlandt Park and

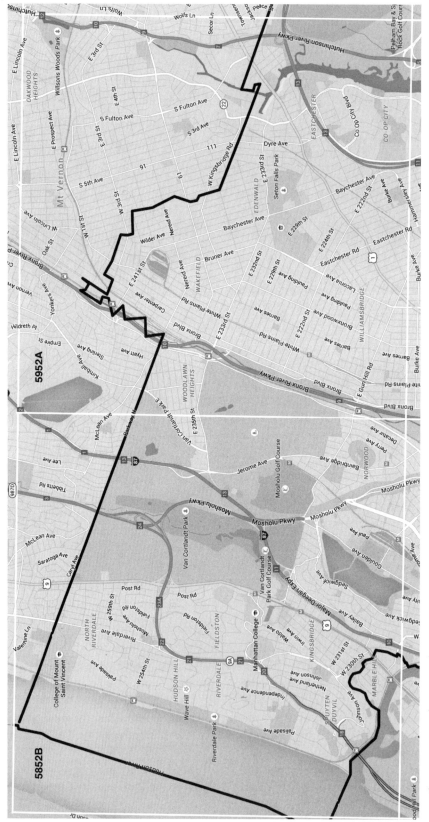

Figure 19. New York State Breeding Bird Atlas blocks embracing the study area. The dark border is the Bronx boundary. Image courtesy of Google Maps, as modified courtesy of B. C. Cacace.

Riverdale, no data were submitted for Woodlawn Cemetery, Hillview Reservoir, Tibbett's Brook Park, or the northern half of Inwood Park and nearly nothing from Jerome Reservoir. In Atlas II in 2000–2005 coverage was barely adequate in Van Cortlandt Park, token at Hillview Reservoir, and missing from Riverdale, Inwood Park, Jerome Reservoir, Woodlawn Cemetery, and Tibbett's Brook Park. Yet both atlases did meet the adequate coverage criterion for both Blocks 5852B and 5952A.

Breeding Bird Census Results

On the other hand, the Van Cortlandt Swamp and Northwest Forest study area breeding bird censuses published first in *Bird-Lore*, then *Audubon Field Notes*, and finally *Journal of Field Ornithology* were designed to quantify the *density* of breeders' territories but again, not to uncover all breeding species. To that end, they were by design *intensive*, not *extensive* like the atlases, and small plots of 20–50 ac (8-20 ha), not 9–square mile blocks, were preferred. It may have been unnoticed by readers of these annually published breeding bird censuses that no migrants were customarily listed. While the intent of this action was consonant with censusing goals—to treat only actual breeders within the study plot and to avoid nonbreeding migrants—it required census takers to eliminate even potential breeders from consideration once they had been deemed migrants. Then, because all birds in eliminated species were never specified, located, dated, or quantified, readers were precluded from drawing their own conclusions about overlooked unanticipated breeders.

The longest and most intensive study area Breeding Bird Censuses were done by teams of enthusiastic young observers in 1937 (the inaugural year for published Breeding Bird Censuses) and 1938 in Van Cortlandt Swamp, well before it was butchered in 1949–50. Some 25 years later, from 1961 to 1967, another generation repeated it, after the Swamp had been reduced to 48% of its former size in 1949–50. In 2013 and 2014 2 different teams using quite different methods censused it again, and their pooled results are given in Table 26, combined as if for a single census year.

Table 27 records the numbers of pairs of all breeders found during the 9 Van Cortlandt Swamp censuses. Note particularly those species that either disappeared completely or whose numbers dropped and those that were extirpated before 1966: Northern Bobwhite, Least Bittern, Virginia Rail, Sora, Common Gallinule, Eastern Bluebird, and Swamp Sparrow. Not a one has returned as a breeder. House Wren, Veery, and House Sparrow also nested in the 1930s but not in the 1960s. Yet some new species appeared in the 1960s that were absent in the 1930s: Black-billed Cuckoo, Willow Flycatcher, Least Flycatcher, Eastern Towhee, Northern Cardinal, Rose-breasted Grosbeak, and Orchard Oriole. Several of these doubtless took advantage of the Swamp's continuing transition from a cattail/*Phragmites* marsh to a largely wooded swamp. Species breeding richness declined from 30–31 in the 1930s to 20–30 in the 1960s, but the true decline occurred between the 1930s and early 1960s (27–32) and the late 1960s (20–24). While it is possible that something adverse may have happened to Van Cortlandt Swamp between 1963 and 1964, diminished censusing effort might also have contributed.

One unforeseen finding was that even though the Swamp's area had been cut in half, many species maintained or even increased their breeding populations. In other words, their breeding *densities* changed. Table 28 transforms all counts from the 1930s and 1960s to a common density of pairs per 100 ac (40 ha) to facilitate contrasts. This revealed that while some species shrank their breeding population to match area loss, some held their breeding density constant, and the majority actually increased their densities. Moreover, density per 100 acres summed across all species was greater each year in the post-landfilled Swamp than before landfilling. Alas, by 2013 the density per 100 acres had plunged to almost half that of the 1960s, with approximately the same species richness.

Table 29 offers another way to view the Van Cortlandt Swamp density question: 1960s counts were predicted as if reduced by 48% but maintaining 1930s densities. These values can then be contrasted with those actually observed in the 1960s, and it is immediately apparent that of the 28 species nesting in both census periods, 15 (54%) increased their breeding densities, 10 (36%) were stable, and 3 (11%) drastically reduced them—and 2 of those 3 no longer breed there.

By 2013–14, additional breeder changes had occurred, owing mostly to increased density of trees and reduced marshy areas. New breeding species since the 1960s were 8: Mourning Dove, Red-bellied Woodpecker, Blue Jay, Tree Swallow, Black-capped Chickadee, Carolina Wren, Blue-gray Gnatcatcher, and American Redstart, offset by the loss of 13 others: American Black Duck, Ring-necked Pheasant, Least Bittern, Virginia Rail, Sora, Yellow- and Black-billed Cuckoos, Least Flycatcher, Red-eyed Vireo, Marsh Wren, Brown Thrasher, Eastern Towhee, and Swamp Sparrow. Admittedly, some of these had begun to disappear during the early 1960s, and American Black Ducks have long been fitful breeders, but it is sad to confirm the permanent loss of so many archetypal freshwater marsh breeders. The censuses also noted Barn and Northern Rough-winged Swallows, but they may not have been breeding within Van Cortlandt Swamp and in any case have been Van Cortlandt Lake breeders and swamp foragers for 100+ years.

The most numerous study area breeding species continue to be Red-winged Blackbird, Warbling Vireo, Yellow Warbler, American Robin, and Gray Catbird. The vireo's increase has been striking and has occurred throughout Van Cortlandt Park and Woodlawn Cemetery. Yellow Warbler's numbers have also been growing, while those of robin and catbird have held their own and the blackbird's have dropped considerably.

Perhaps the most notable change has been the pervasive drop in densities between 1966 and 2014 (Table 28). Of the 20 species breeding in 1966, only 4 (20%) increased in density (Warbling Vireo especially so), 3 (15%) were stable, and 13 (75%) dropped in density, some strongly (Gray Catbird, Song Sparrow, Yellow Warbler, Red-winged Blackbird), and 3 have disappeared completely as breeders there—Red-eyed Vireo, Marsh Wren, and Brown Thrasher.

So, after 77 years and having been cut in half, Van Cortlandt Swamp alone continues to support about 30 breeding species—about as many as in all of Central or Prospect Parks—even though their mix and abundance have changed in tandem with that of the Swamp's vegetation and physiography. Whether these reflect normal changes as a large freshwater marsh succeeds to a treed shrub-swamp or augur ill for the future health of the wooded Swamp is unclear from only a single data point after 50 years.

Another quantitative Breeding Bird Census of a very small area was undertaken in one of the most heavily wooded and least disturbed woodlands, the Northwest Forest (Table 30). There, in 1991 and 1992, a 25 ac (10 ha) plot was found to support 24–26 species occupying 74–75 territories—again, almost as many species as breed annually in all of Central or Prospect Parks. There were few differences between the years, and only Baltimore Oriole did not nest in both years. There were no surprises, and the numbers of Wood Thrushes and Northern Cardinals were healthy and American Robins and Gray Catbirds were thriving. Alas, the single pair of Brown Thrashers and the paltry 5 pairs of Eastern Towhees failed to alert censusers to what would eventually happen to these 2 species across the study area.

Tempting as it was to consider extrapolating small-area quantitative results to all of Van Cortlandt Park and then to Woodlawn Cemetery, habitat types are by no means uniformly distributed across the study area; there have been too many known changes to vegetation and avifauna and too many gaping coverage holes for extrapolation to yield reliable, replicable, and useful results.

Finally, between 30 May and 12 August 2013 Pehek et al. of the New York City Parks Department Natural Resources Group undertook their quantitative breeding bird censuses of the Shandler Area and of Van Cortlandt Swamp. Then between 25 May and 5 July 2014, Young conducted his own Van Cortlandt Park quantitative census that covered all park areas except the golf courses and the southern and southeastern areas. Both data sets were graciously made available to us, and with those of other ad hoc sources, were all merged by the current authors to derive a subjective best estimate of the *minimum* number of pairs and species breeding park-wide between 2013 and 2014. Note that some of the cumulative 73 species may not breed annually and that others may also be breeding in the uncensused golf course areas. Numbers of breeding pairs should in most cases be considered only estimates for confirmation and correction in ensuing years.

A number of presence/absence Van Cortlandt Park breeding censuses addressing only small areas have also been undertaken, including those of Turner (1984), Delacretaz (2007), and Künstler and Young (2010). Their results have been folded into the appropriate species accounts. Most surprising of all, given the accessibility and road-and-path network of Woodlawn Cemetery, an all-species quantitative breeding census has never been done there but would be easy to establish and repeat within and between years.

Beyond Van Cortlandt Park, special techniques may be required to locate and quantify certain potential study area and adjacent breeders. Some that would pay dividends include kayak surveys of the Harlem and Bronx Rivers for Gadwall, Yellow-crowned Night-Heron, Osprey, Spotted Sandpiper, Barn Owl, Belted Kingfisher, Cliff Swallow, and Bank Swallow; Woodlawn Cemetery dawn, dusk, and nocturnal searches for Yellow-crowned Night-Heron, Eastern Screech-Owl, Great Horned Owl, and Common Nighthawk; city-street patrolling for Common Nighthawk, American Kestrel, and Monk Parakeet; and emplacement and monitoring of a hummingbird feeder array throughout Van Cortlandt Park and Woodlawn Cemetery for Ruby-throated Hummingbird.

New York City's Maritime Island Breeders

Until the mid-1970s, nothing was known about colonially breeding waterbirds (or for that matter, any birds) that may have been nesting on the various saltwater islands within and adjacent to New York City. The only exception to this statement was irregular censusing of the wading birds breeding on a few large islands in Jamaica Bay. Then from 1974 to 1978, annual helicopter survey–censuses were done of all waterbirds breeding coastally within New York City and on the periphery of Long Island (Buckley and Buckley 1980). Their discoveries of greatest relevance to this book were previously unknown heronries on Huckleberry Island off Pelham Bay Park and on North and South Brother Islands in the East River north of the Triboro Bridge, as well as breeding Herring and Great Black-backed Gulls at all of these sites. In 1983, New York City's first breeding Double-crested Cormorants were also found on South Brother Island (Buckley and Buckley 1984), the vanguard of an areal explosion continuing to this day. The origins of various herons, Glossy Ibis, and cormorants in the study area became more understandable.

Prompted by these findings, Manomet Bird Observatory and later New York City Audubon

Society began nearly annual on-ground censuses of all saltwater islands in and around New York City. These tracked wading bird, gull, tern, and cormorant populations after 1985. Two of their more interesting ancillary observations were breeding Gadwalls on numerous islands as close to the study area as North and South Brother, and that Spotted Sandpiper was established on almost all islands, a very great surprise about a bird many believed nearly extirpated as a New York City breeder. Pertinent data from all these surveys and censuses are given in the respective species accounts, where their implications for study area statuses are discussed.

WINTER SPECIES

Across Subareas

Van Cortlandt also supports a far more diverse winter population than Central or Prospect Park, another index of its inherent value for native birds year-round.

Few habitats are quite so birdless as eastern deciduous forests in winter, and Van Cortlandt Park and Woodlawn Cemetery are not exceptional, as 90 years of Christmas Bird Counts have demonstrated. Small mixed-species flocks of Black-capped Chickadees, Tufted Titmice, Hairy, Downy, and Red-bellied Woodpeckers, White-breasted Nuthatches, Brown Creepers, Northern Cardinals, and a few camp followers like Yellow-bellied Sapsuckers, Carolina Wrens, American Goldfinches, Slate-colored Juncos, White-throated Sparrows, Cedar Waxwings, American Robins, Golden-crowned Kinglets, and Hermit Thrushes roam throughout the woods. They occasionally encounter the handful of raptors seeking them (Cooper's, Sharp-shinned, and Red-shouldered Hawks) and cross paths with the less socially inclined Blue Jays, American and Fish Crows, Winter Wrens, and Song Sparrows, although on occasion almost any small songbird may temporarily join a mixed flock. Sparrows and juncos typically form their own large flocks that take up residence in weedy locations. Any blackbirds present also tend to flock by themselves, sometimes with European Starlings, and have episodically formed roosts in Van Cortlandt Swamp, sometimes in the thousands, their occurrence and persistence a function of winter severity and snow cover.

Waterfowl and species preferring Van Cortlandt Swamp, Van Cortlandt Lake, and the Woodlawn Cemetery ponds vary greatly from year to year, and species have periodically waxed and waned as dominant winter residents there. Numbers are largely dependent on the mildness/severity of the winter and how much open water is available in the Swamp or on the lake. When frozen out, waterfowl move to Jerome Reservoir (formerly also to Hillview Reservoir before it became aggressively waterbird-unfriendly in the 2010s) or other nearby locations.

Some archetypal Van Cortlandt Swamp winter residents in small numbers like Wood Duck, Ring-necked Pheasant, Black-crowned Night-Heron, Virginia Rail, American Kestrel, and Marsh Wren have now been seasonally (or completely) extirpated, and a few others like Wilson's Snipe, Rusty Blackbird, and American Tree Sparrow are currently hanging on by a feather. Notwithstanding, the Van Cortlandt Park winter bird species population still exceeds those of Central and Prospect Parks, but the lead is diminishing.

Most changes in Van Cortlandt's overwintering species have been unremarkable. Relatively recently arrived permanent residents like Red-bellied Woodpecker and House Finch continue to increase, and a few formerly uncommon winter visitors like Northern Shoveler, Hooded Merganser, Yellow-bellied Sapsucker, and Merlin are now expected in winter. A handful of others like Great Egret and Double-crested Cormorant are poised to be next. And since about 2010, up to several thousand arctic-breeding (not local feral) Canada Geese have found the Parade Ground to their liking, bringing with them Pink-footed, Cackling, Greater White-fronted, and Barnacle Geese from probable Greenland breeding grounds.

Woodlawn was nearly completely ignored in winter until the early 1950s, although there are no reasons to believe this was because it was birdless. Since that time, owing to its many conifers of various species, Woodlawn has become a mainstay for certain overwintering species scarce in adjacent Van Cortlandt Park. On many a Christmas Bird Count it provided the only Wood Duck, Black-crowned Night-Heron, Red-breasted Nuthatch, Ruby-crowned Kinglet, Brown Thrasher, Pine Siskin, or crossbill and now supports at least 1–2 overwintering Merlins, Yellow-bellied Sapsuckers, Hermit Thrushes, Eastern Towhees, Red Fox Sparrows, occasional Chipping Sparrows, and every now and then something unexpected like a Northern Harrier, Eastern Phoebe, Pine Warbler, Yellow-breasted Chat, Lincoln's Sparrow, or Rose-breasted Grosbeak. Birds on a very recent Bronx-Westchester Christmas Bird Count in Woodlawn Cemetery are listed in Table 31. In the major Common Redpoll incursion in March 1956, several feeders no longer active supported up to several hundred redpolls into April. Two species of accipiters and 2 of buteos regularly patrol its woods by day, and owls of several species including Long-eared and Saw-whet do so by night. As the cemetery continues to grow, undisturbed natural areas supporting sparrows of many species are shrinking as manicured and landscaped vegetation expands.

Beginning in 1913 at Jerome and in 1929 at Hillview, local observers began to appreciate the large numbers of waterfowl using both sites, especially in winter but also during spring and fall migration. Hillview became famous for large flocks of Common Mergansers (up to 300) and American Black Ducks, whose counts exceeding 2000 in the 1930s have never been repeated. In the early 1950s, Jerome became known as the easiest New York City location to see large Ruddy Duck flocks, previously doable only on eastern Long Island. Goldeneyes were also regular in winter on both reservoirs, but that ceased in the late 1940s. Sedwitz (1974, 1975, 1977, 1979) examined the populations and movements of (especially winter) gulls and waterfowl there.

Scaup of both species also overwintered, but Lessers outnumbered Greaters 10 or 20 to 1. It was always assumed they and other diving ducks were actively feeding when diving, but it was not until the 1970s that we learned on just what. Enders (1975) found that a thick muck/sediment layer (greater than 10 cm) on the bottom of Jerome Reservoir, accessible at the northern sloping seawall when it was drained from December 1965 to May 1967, supported a large population of chironomid midge larvae and many small (1–2 mm) freshwater clams. Mergansers are piscivores, but their reservoir diet was unknown until work at Hillview in the early 2010s revealed the presence of large numbers of Alewives (NYCDEP 2011). Both clams and fish presumably slipped through various New York City water supply filters when tiny and then grew to good size once in both reservoirs. Every time one of the reservoirs was drained it was also scraped clean of all sediment and food, so once refilled it always took a number of years before clam and fish recolonization and growth yielded enough food to support overwintering diving ducks (Sedwitz 1977).

Despite numerous invertebrates, there were never enough (or of the right sort) to support the large Canvasback flocks that overwintered regularly at Jerome Reservoir but were never seen diving. They would typically spend the day sleeping and preening but always left the reservoir at dusk to feed in saltwater on the East, Harlem, and Hudson Rivers and their embayments, returning only at dawn (Enders 1976). Greater (few) and Lesser Scaup (most) also fed and slept on Jerome Reservoir but periodically did leave to feed on the Hudson River in northern

Manhattan (Boyajian 1969a). Reservoir waterfowl have also long routinely moved to the East, Harlem, and Hudson Rivers, the Bronx Zoo, Pelham Bay Park, and Clason Point during the winter for feeding, some at night, some by day, and during icy bouts on the reservoir.

During the 1930s, '40s, and '50s, when numerous garbage dumps were operating in the East and Southeast Bronx, large numbers of gulls flew to Jerome and Hillview Reservoirs daily to bathe, preen, and drink in freshwater. With them regularly came small numbers of Glaucous and Iceland Gulls, and both sites were regarded as among the most reliable close to New York City in which to find these species. After the last dump closed in the late 1950s, gulls became scarcer on the two reservoirs, but when the Eastchester Bay garbage dump in Pelham Bay Park opened in 1963, the influx of gulls began all over again, this time bringing a new arrival into the study area: Lesser Black-backed Gull. And once again, after this dump closed in 1979, nearly all the unusual gulls disappeared.

During the intervals when Jerome Reservoir was drained (1965–67, 2008–14), thousands of Great Black-backed, Herring, Ring-billed, and Laughing Gulls routinely used the area for loafing, preening, and drinking at the freshwater pools that had formed on the bare concrete floor. Once water levels were restored, daily gull numbers dropped by an order of magnitude.

At Hillview Reservoir, as the many Scots, Austrian, and Japanese Black Pine groves planted in the 1920s–30s matured, they became reliable sites for overwintering species favoring conifers—Red-breasted Nuthatch, Golden-crowned Kinglet, Northern Goshawk, Northern Saw-whet Owl, and during irruption years, both crossbills and even Boreal Chickadee. Their loss in the late 1990s has left a huge habitat hole in the study area only partly filled by Woodlawn Cemetery. The open grassy areas around the reservoir have also long been reliable for winter Horned Larks, American Pipits, Snow Buntings, and Savannah Sparrows, study area flyovers nowhere else offered respite except on the Parade Ground.

Table 32 lists the 32 species we consider core winter residents in the study area—those that have been present every winter since 1872—and another 38 that have done so annually since at least 2000, many well before that date.

Winter Bird Population Studies

Attempts were made during the 1960s to quantify winter residents in Van Cortlandt Swamp (Goll et al. 1963, 1964; Enders 1966) and on Jerome Reservoir (Heath and Enders 1964; Zupan and Enders 1966), with results published in the annual Winter Bird Population Studies of *Audubon Field Notes*. Table 33 summarizes those for Van Cortlandt Swamp during winters from 1962–63 to 1965–66, and Table 34 those for Jerome Reservoir during 1963–64 and 1965–66.

Note that Winter Bird Population Studies are *not* analogous to Breeding Bird Censuses, which cumulatively tally breeding species and pairs to calculate breeding densities. Winter studies, rather, are sampling procedures to determine the mean number of individuals of each species seen per trip, rounded to the nearest digit. But as in Breeding Bird Censuses, the more samples taken over a longer period, the closer the sample mean will be to the true mean. Regrettably, ranges are never given, so we lack the highest and lowest counts or when they occurred. Also regrettable was the decision to exclude any birds merely overflying the plots, although when they were mentioned by the authors (sometimes with counts, often without) we have incorporated them in our tables. The results of both sets are somewhat unsatisfying and generally unremarkable, but they are all that exist.

Christmas Bird Counts

Christmas Bird Counts have greatly increased simultaneous coverage of places rarely visited in winter, especially deciduous woodlands in the study area. They also facilitate detection of landbird winter migration and movements that are often unnoticed, particularly weather-mediated local and long-distance hardship movements by land- and waterbirds.

The modern Bronx-Westchester Christmas Bird Count began in the Bronx in 1924, although it had some modest predecessors. On 24 December 1906, F. Huberta Fuste and Alice R. Northrop covered some portion of Van Cortlandt between 1030–1230 and 1430–1600, recording 13 species: Herring Gull 1, Downy Woodpecker 4, Blue Jay 8, American Crow 2, Song Sparrow 8, Brown Creeper 1, White-breasted Nuthatch 2, and Black-capped Chickadee 6. On 24 December 1920, Alvah Bessie, Eugene Eisenmann, and the Biological Field Club of DeWitt Clinton High School (at the time still in Manhattan) also worked some portion of Van Cortlandt and also found 13 species: Herring Gull 3, hawk sp. 4, American Kestrel 1, Hairy Woodpecker 2, Downy Woodpecker 1, American Crow 1, European Starling 2, White-throated Sparrow 10, American Tree Sparrow 25, Slate-colored Junco 2, sparrow sp. 2, Brown Creeper 1, White-breasted Nuthatch 2. It is clear that Van Cortlandt Swamp was not included in these early censuses and that observers' identification skills were as nascent as their optics were primitive.

For some time it was customary for there to be more than a single annual Christmas Bird Count in the same area, and this practice continued through 1922, when there were 3 different Bronx counts, with one of them covering Van Cortlandt Park, Bronx Park, the Bronx Zoo, Pelham Bay Park, and the Baychester Marshes; 70 Black-crowned Night-Herons in a Bronx Zoo roost was a highlight. But in 1924, Hickey and 8 others in the new Bronx County Bird Club pulled them all together and initiated a single Bronx Christmas Bird Count, although Westchester County coverage was in the future. From the beginning, Riverdale, Van Cortlandt Park, and Jerome Reservoir were included, alongside East Bronx locations from Clason Point to Pelham Bay Park; Woodlawn Cemetery and Hillview Reservoir were quickly added.

Over the first 90 years of its life, Bronx-Westchester Christmas Bird Count observers have increased in number. However, that increase has been strikingly linear at slightly less than a single observer per year (Fig. 20). Consequently,

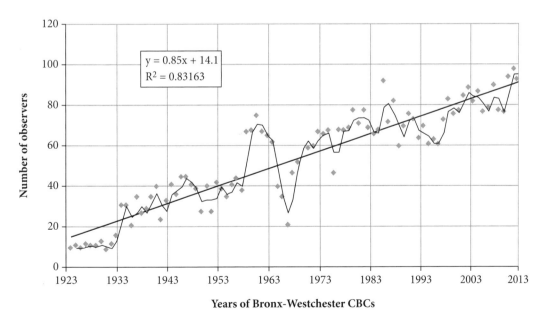

Figure 20. Increase in numbers of participants on Bronx-Westchester Christmas Bird Counts from 1924 to 2013. The slope of the line (0.85) indicates that the number increased each year by slightly less than 1, a slow, steady effect with some wobble as tracked by the 2-year moving-average trendline. Data courtesy of National Audubon Society Christmas Bird Count database.

it is unlikely that the Christmas Bird Count trends illustrated in this book were materially affected by the slow increase in numbers of observers, and even though the coverage area for the Bronx-Westchester Christmas Bird Count has varied dramatically since 1924—expanding, contracting, shifting—all study subareas have always remained within it.

Some winter daily maxima for study area landbirds derive from Christmas Bird Counts, when there was often simultaneous, nearly complete coverage of most of the study area. Nonetheless, many such maxima are missing because over those 90 years most such Christmas Bird Count numbers were pooled in a single West Bronx party total, with subarea counts then lost.

MIGRATION

Background

Migration is the major avian event in the study area in spring and fall, far more important than local, short movements in summer and winter. It is the main attraction for most study area observers and has been back to Bicknell's and Dwight's active days in the late 1800s.

It is often surprising to each succeeding generation of observers that on-ground detection of migrants captures only the tiniest fraction of the birds actually passing overhead, even by day. Daily maxima from (especially) coastal barrier beach sites or structure kills at the tallest and brightest-lit buildings hint at this and are often an order of magnitude greater than the highest observer-only daily maxima. But they in turn are eclipsed by the numbers of calling nocturnal migrants for readily identifiable species like the thrushes. On rare occasions, weather conditions coincide to produce on-ground numbers that beggar comprehension.

At Cape May, New Jersey, where very large counts are normal, a few stand out: up to a *million* Myrtle Warblers on 18 October 2005 and 10,000 Eastern Kingbirds on 30 August 1926. A striking spring example also comes from Cape May. On 5 April 2017, at least 1037 Blue-gray Gnatcatchers passed the famous Higbee Beach dike between about 0630 and 1015 (Johnson et al.). Winds the previous night had shifted from SW to W/NW, presumably drifting the gnatcatcher flock(s) offshore at some point south of Cape May, which then became the nearest land at first light. The previous spring daily maximum at Cape May had been 25–30.

Hurricanes also intercept and entrap within their eyes large numbers of migrating landbirds that are not released until dropped, exhausted, on the first landmass. At Jones Beach, Long Island, during Hurricane Gloria on 27 September 1985, 3000 Cape May Warblers astounded and delighted observers expecting Sooty Terns, no less so than the 15,000–20,000 Bobolinks deposited on Bermuda by Hurricane Emily on 25 September 1987 and the 10,000 by Wilma on 22 October 2005, as well as Emily's 75+ Connecticut Warblers.

Even these are mere samples of the total number of birds migrating north in spring and south in fall each year in North America. It was not until the advent of what has been termed radar ornithology (optimally using NEXRAD Doppler weather radar) that we began to understand that *billions* of birds are involved. In-field daily maxima at that point become trivial, yet in most locations they are all we have although they are replicable year after year. So on those rare occasions when even a few hundred warblers of a single species are seen in the daytime, it is incorrect for these counts to be dismissed as too high.

Spring Neotropical migrant songbirds frequently occur in big, concentrated, warm-weather–triggered events concentrated in May. Now often called *fallouts* (formerly *flights* or *waves*), they feature many singing ♂♂ that enable high daily counts. But in fall when migrants are spread from June to November and even December, trickle through even in the absence of northwest winds, and are dispersed in nonsinging mixed-species flocks, they always seem less numerous. However, very large autumn fallouts can occur during extreme weather conditions, especially from September to November following passage of major cold fronts with strong northwest winds.

Fall shorebirds are often recorded immediately after having been forced down to fields by thunderstorms or other severe weather events that pool standing water, or when lakes shrink in droughts or are drained (Van Cortlandt Park, Central Park, Prospect Park). But observers have to be there when this happens, because such fallouts can be fugitive. In 1951 there occurred the coincidence of Van Cortlandt Lake's having just been drained and extremely strong southwest winds with very heavy precipitation in advance of a cold front—all of which concentrated 15 species of migrating shorebirds in a major fallout on the afternoon of 10 May (Table 35), including several unique species or all-time study-area maxima. They were all but gone the next day. These species (except Stilt Sandpiper) are also regularly seen in spring outside the study area, if in limited numbers, as close as Piermont Pier and Pelham Bay Park, and certainly overfly the study area annually, so available habitat is the key to occurrence, as it always has been with migrating shorebirds. Another rainstorm fallout at nearby Piermont Pier on 28 May 2005 yielded a migrating Whimbrel and 300 Dunlin, the former the first Rockland County spring record and the latter supplanting the previous county maximum of 13. In the fall, a similar weather-induced shorebird fallout occurred at nearby Inwood Park on 12 August 2013, when 300–400 Semipalmated and 100–200 Least Sandpipers suddenly dropped onto a mudflat, tenfold more for both species than one would have expected there.

Because only suitable ephemeral habitat or severe weather will induce some migrants to land, many other species and groups that routinely overfly the study area even in large flocks are only rarely seen on the ground. Typical examples include Red-necked and Horned Grebes, Common Loons in spring, Snow Geese, Brant, all 3 scoters and many other waterfowl, southbound hawks in fall, most shorebirds (including Upland Sandpipers, which can occasionally be heard overhead at night in May and August giving their diagnostic bubbling calls), southbound Common Nighthawks, 6 species of swallows, and Chimney Swifts.

For species that migrate by day in very large flocks (cormorants, geese, Broad-winged Hawks, Common Nighthawks) it is hit-or-miss whether they will pass over the study area, or if they do whether anybody will be there to count them. Thus the highest New York City count for nighthawks comes from Parkchester, and the highest for Broad-winged Hawks from Pelham Bay Park—both only 3–5 smi (5–8 km) east of the study area. Some authors speak of hawk flightlines as if they were fixed in granite. In a geological sense they are, along the Kittatinny Mountains in New Jersey and Pennsylvania, but in most cases and even in the Kittatinnys, hawk migration occurs with favorable winds along a broad front with fluid concentration lines. Which exact route hawks may follow on any given day, week, or month varies with wind and other aspects of daily weather as well as their starting point each day. The major ridges in and near the study area—in Riverdale, the Northwest Forest, Vault Hill, and the Ridge (Croton Forest)—have all experienced major raptor flights and will continue to do so.

The topographical features called *leading lines* often play major roles in determining where migrants will appear and what routes they customarily follow. Hawks do follow mountain chains and ridges to take advantage of updrafts and other winds facilitating their migration. Large rivers also serve as leading lines, and the Hudson is one for diurnal landbird migration, as well as for many waterbirds by day and by night, both spring and fall. Vegetational greenbelts (usually trees but sometimes grasslands or streams) allow nocturnal migrants safe diurnal passage once they are safely on the ground. One such is the Bronx River, long used by birds moving into and out of the Bronx from Westchester. Unexpectedly, it has also facilitated the natural reinvasion of the Bronx after an absence of several hundred years by North American Beavers (now breeding along the Bronx River in the zoo), Eastern Coyotes (now resident throughout the Bronx), and River Otters.

The species and numbers of nocturnally and diurnally migratory landbirds seen routinely in Riverdale and in Inwood Park indicate movement in fall on a regular basis south along the east side of the Hudson River. These would seem to be in addition to and quite different in origin from those largely diurnal migrants detected by Boyajian (1968, 1969a, 1969b, 1970, 1972) in the late 1960s–early 1970s routinely crossing the river to the Palisades east to west in the fall or migrating along them in fall and spring.

Late summer, fall, and winter see the vanguard of any northern irruptive species. Red-breasted Nuthatches and Black-capped Chickadees can show up as early as July, and even crossbills occasionally appear in August, usually presaging a major irruption. Most irruptives arrive or peak within their normal migration periods, from October to December, and then settle in to overwinter unless they are just passing through, the way Pine Siskins and Evening Grosbeaks often do. Redpolls irrupt to their own drummer, frequently well into the winter and sometimes not even until March. In heavy irruptions, individuals of many species will remain far later than normal spring departure dates, often an indication of impending out-of-range breeding. Evening Grosbeak, Pine Siskin, and Red Crossbill are especially prone to this behavior, but rarely do any permanent range extensions ensue.

A last point concerns the ages of certain migrants. In spring, late arrivals, range extenders, colonists, and nonbreeders are usually breeding-inexperienced SYs. Likewise, fall out-of-range species, blowback drift-migrants, late lingerers, and abnormal overwinterers are normally hatching-years. Because avian mortality is greatest in young birds, in some species routinely reaching 90% or more, the preponderance of inexperienced, migration-untested hatching-years and second-years is predictable. But in both seasons it is just these individuals, in the wrong places or at the wrong times, that are ultimately responsible for major changes in species' ranges and migration patterns if they survive.

Spring Migration

Northbound raptor migration does occur, but it is normally so light as to pass unnoticed in the study area. Waterfowl migration can be impressive, especially on the reservoirs but also on Van Cortlandt Lake, although nearly all arctic goose flocks move northward well to the west of the Hudson River. Common Loons overfly the study area in May and might be missed except that they frequently attract attention by yodeling as they go—a most unexpected sound in a wooded New York City park. Except for a handful of Spotted, Solitary, Least, and Semipalmated Sandpipers; Greater Yellowlegs; and Wilson's Snipe, northbound shorebirds are normally scarce. Weather-induced fallouts of Horned and Red-necked Grebes in March–April have been occurring for decades, especially on the reservoirs. In March 2014, heavy rains and strong winds displaced northbound Tundra Swans east from their usual Pennsylvania routes, and one dropped into the tidal basin at Inwood Park, not far from the study area.

As everywhere, spring migration is the highlight of the avian year in the study area. More observers are afield then than at any other time, and excepting infrequent breeding bird censuses, locations are visited that otherwise see no observers except on Christmas Bird Counts. Still, most efforts have been concentrated in

Van Cortlandt Swamp, so observers miss most of the Neotropical migrant warblers, vireos, flycatchers, thrushes, orioles, and tanagers in the woods on the Ridge (Croton Forest) and in the Northeast Forest. For some remarkable warbler numbers during intense coverage there in the early 1960s, see Friedman and Steineck 1963.

By mid- to late April, there are usually some migrants around every day, but it is not until warm weather and rain combine to cause a fallout that species richness jumps up. Until the early 1960s, it was not unusual for active observers to record more than 100 species in a day in just Van Cortlandt Park, but we are not aware that this has happened since then.

On-ground spring warbler maxima from the study area in some cases have been quite high by New York City area expectations, approaching maxima from coastal mist-netting sites: Ovenbird 50, Black-and-white 200, Nashville 75, Hooded 10, American Redstart 100+, Northern Parula 400, Magnolia 100+, Chestnut-sided 40, Blackpoll 50, Bay-breasted 40, Yellow Palm 300+, Black-throated Green 50+, Canada 200, and Wilson's 15. But are these numbers truly abnormally high, or is it merely that these birds are infrequently met on the ground?

On 13 May 1972 between 0430 and 0730 Boyajian (1972) and colleagues encountered a large movement of warblers engaged in a morning flight on a cloudless day with light northwest winds at Alpine on the northern end of the Palisades, almost opposite Riverdale. Here in May, there was often a morning flight between first light and sunrise before nocturnal migrants had begun their descent to ground for the day. Most days they were high enough that only 5–10% were identifiable to species, but this day they were near the ground and many landed in treetops for 10–30 seconds before resuming northward flight. Two numbers are shown for each of the most numerous species during the 3-hour observation period, those actually counted and in parentheses the estimated total of those passing that were referable to particular species: Black-and-white 325 (650), Hooded 16 (25), American Redstart 520 (1000) Cape May 78 (100), Northern Parula 285 (500), Bay-breasted 52 (75), Blackpoll 62 (100), Black-throated Blue 185 (250), Myrtle 610 (1300), Canada 25 (50), and Wilson's 25 (55). Another 2000 were unidentifiable to species. The author commented wisely that these counts represented not any abnormal concentration of birds but rather abnormal observation conditions.

Most spring landbirds are nocturnal migrants. They travel in species-specific flocks or as species-specific clusters within large multi-species flocks. Different species keeping track of each other is the prime reason for differences in the nocturnal calling so commonly heard at night but rarely once birds are back on the ground. Upon landing, the clusters tend to break up, which is why both small conspecific flocks of migrants as well as individuals scattered across small distances can often be found early in the morning after their arrival.

As a group, spring migrants tend to arrive on relatively firm schedules, and New York City area books over the years (Chapman 1906; Eaton 1910, 1914; Griscom 1923; Cruickshank 1942) were able to array arrival dates for common migrant species clusters. But since the 1950s it has become increasingly apparent that arrivals of out-of-range species and of normal species at anomalously early times occur frequently but for different and sometimes overlapping reasons. The most frequent is simply highly favorable weather conditions over a broad area to the southwest: above-normal temperatures, sustained strong southwesterly airflows below about 2000 ft (610 m) altitude, and often precipitation just before first light, concentrating migrants overhead and then forcing them abruptly to the ground. These conditions usually result in moderate early dates for expected species. Next most frequent is the occurrence of prospecting or colonizing individuals (often SYs making their first northbound return trips) of more southern species that have been expanding their breeding ranges northward. In the New York City area this has involved Red-bellied Woodpecker; Blue-gray Gnatcatcher; Prothonotary, Kentucky, Hooded, and Yellow-throated Warblers; Summer Tanager; Blue Grosbeak; Orchard Oriole; and perhaps Swainson's Warbler.

Least frequent but most dramatic are what have been termed trans-Gulf overshoots, which involve Neotropical migrants that cross the Gulf of Mexico on their way back to North American breeding grounds. Under certain weather conditions first recognized by Bagg (1956, 1957, 1958), perfectly positioned, strong low-pressure systems with intense counterclockwise winds may entrain migrants about to make landfall along the north shore of the Gulf and then impel them far to the northeast. These conditions can deposit many greatly out-of-range, exceedingly early, and often weakened migrants at abnormal locations, especially along the immediate coast and on islands. They have brought early April Painted Buntings to Nova Scotia and Yellow-throated Warblers to Cape Cod and bedecked Long Island barrier beaches with Blue Grosbeaks and Indigo Buntings. In the study area, such trans-Gulf overshoots (many of which are also SYs) have included extremely early occurrences for Eastern Kingbird, Yellow-throated Vireo, Blue-gray Gnatcatcher, Gray-cheeked Thrush, Northern Waterthrush, Black-and-white Warbler, Northern Parula, Yellow and Prairie Warblers, Summer Tanager, Blue Grosbeak, Indigo Bunting, and Bobolink.

Since the publication of Terborgh's (1989) seminal book *Where Have All the Birds Gone?*, observers in many locations have also commented on and documented much lower North American numbers of Neotropical migrant songbirds in the past 50 years than before. And even within the 51 species from the study area, the effect is clearly demonstrable: only 5 species reached their maxima after 1965, vs. 46 before 1966 (Table 36). Most scientists blame habitat degradation or outright loss of Middle and South American overwintering habitat.

A related change, that of increasingly earlier first arrival dates in spring tracking global climate warming, is also demonstrable in study area data for 83 species. If the earliest Neotropical migrant songbirds were still arriving on roughly the same dates they had been before 1966, there would be almost no earliest arrival dates after 1965. Yet as Table 37 shows, the numbers are almost equal: 37 species whose earliest arrival date is after 1965 compared with 46 whose earliest arrival date is before 1966, confirming earlier and earlier first arrivals.

The distressing take-home message is that while Neotropical overwintering and temperate-nesting songbirds are responding to global warming by advancing their arrival times, their populations are simultaneously drastically diminishing.

Summer Migration

Apart from late northbound warbler, thrush, and flycatcher migration up to mid-June and early southbound migration by the same groups from late June through July, little true summer migration is palpable in the study area beyond the occasional southbound adult Least Sandpiper. Most summer movements are local or regional post-breeding dispersion of adults and hatching-years involving almost any New York City area breeder or summer resident.

Fall Migration

Southbound migration is the most protracted of all in the study area, with different groups of species arriving serially from mid-June to mid-December, occasionally later.

For many Neotropical migrants, study area daily fall maxima are much lower than spring maxima, yet because the species passing through in fall also contain young of the year, their total numbers should be much higher—ignoring those few species with different spring and fall migration routes. This happens because in spring migrants are arriving near their destinations, are running out of fat, and need to feed, whereas in fall they are just starting out, their fat reserves are high, so they pass over and beyond the New York City area before making their first landfalls. This is true more of Neotropical migrants early in the fall than for sparrows and others later in the fall already closer to their final destinations. However, this spring versus fall scenario is sometimes violated spectacularly, such as by the Cape May Warblers at Jones Beach during Hurricane Gloria (see above).

A common fall migration strategy, especially in shorebirds, has adults departing well before hatching-years and traveling separately from them. In other groups like geese, entire families travel in the same flock. And like other species with specialized habitat needs, many shorebirds, diving ducks, loons, grebes, and Bonaparte's Gulls common or even abundant on Long Island routinely overfly the study area without stopping unless forced down by stormy weather.

Diurnal raptor migration involving up to 15 species occurs every fall along the study area ridges, but its detection and the numbers involved vary greatly year to year with the occurrence of appropriate weather conditions, typically strong northwest winds after cold fronts. When these are lacking or weak, the hawks trickle through, barely noticed. But when conditions are perfect, amazing numbers may occur even in the Bronx, like the 15,000+ Broad-winged Hawks at Pelham Bay Park on 11 September 1990. And though they are not raptors, the 5000+ Common Nighthawks over Parkchester on 9 September 1945 responded to the same conditions. With its eastern Canada breeding population continuing to increase, more Bronx and even study area Golden Eagles are being seen, and if regular hawkwatches were reinstated in Van Cortlandt Park, they might increase by an order of magnitude. Minor hawkwatches on the Ridge (Croton Forest) from the mid-1940s to early 1950s (Fig. 16) uncovered the intensity of study area migration, but they were never expanded. Those at Pelham Bay Park in the late 1990s were more systematic and hint at what a new study area effort might yield.

The first southbound landbirds begin to trickle through with cold fronts in July or even late June, but few observers are afield in city parks then and so most miss study area early migrants like Rough-winged and Bank Swallows; Louisiana Waterthrush; Worm-eating, Blue-winged, Golden-winged, Yellow, Hooded, and Mourning Warblers; and Orchard Oriole. The peak of fall landbird migration by numbers of species and individuals comes later, in August–September for insect and fruit eaters and October–November for seed eaters.

Another migration phenomenon is responsible for the appearance of out-of-range species and extremely late fall dates for local species in the study area. Until the early 1990s, when the first Cave Swallows, Violet-green Swallows, and Ash-throated Flycatchers showed up in November at Cape May, New Jersey, it had not been appreciated that the November–January Yellow-breasted Chats and Western Kingbirds that had puzzled observers in the 1940s and '50s were all manifestations of a single phenomenon. Despite speculation about origins back to at least the early 1950s (Alperin 1951; Bergstrom 1951), it was not until Baird et al. (1959) that the first pieces were put together, although their paper's importance has too often been overlooked by subsequent authors. It set the stage for subsequent work so that we now understand how the system works.

Sustained southwesterly airflows from September to December in advance of strong southeasterly moving cold fronts often bring massive numbers of **blowback drift-migrants** (not reverse migrants, which are functionally different even though they also go in the wrong direction; see Newton 2008; Howell et al. 2014 for distinctions) to the northeast and the study area and are responsible for recent November Franklin's Gull and Cave Swallow incursions as well as for increasing numbers of other long-distance fall wanderers northeast as far as Newfoundland, including White-winged Dove, Rufous Hummingbird, Pacific-slope Flycatcher, Say's Phoebe, Ash-throated Flycatcher, Western Kingbird, Bell's Vireo, Mountain Bluebird, Townsend's Solitaire, Le Conte's Sparrow, Western Tanager, Black-headed Grosbeak, and Painted Bunting, among many species.

Originating in the far southwest or southern Great Plains and moving north with sustained warm-air southwesterly flows in advance of cold fronts, these birds can arrive anywhere from the Great Lakes east, and with the onset of northwest winds after cold front passage many of them then move east/southeast to the coast, where they are concentrated and more easily detected. Thus in fall, they appear on the coasts (especially barrier beaches) under two different but related weather conditions: in warm air with southwest winds *before* frontal passage and then in cold air with northwest winds *after* frontal passage. So with more numerous species like Western Kingbird, it is not uncommon to find some before and then more after cold front passage, when birds from the two groups then gather in common flocks. Now that observers understand this system, they are increasingly going out in October–November to look for them and, coastwise at least, are finding them every year. It is now understood that extremely late dates for certain birds in eastern North America are flags that western North American taxa may be involved, especially in nighthawks; *Empidonax*, *Myiarchus*, and *Contopus* flycatchers; hummingbirds; orioles; tanagers; grosbeaks; Cliff Swallows; Orange-crowned, Nashville, and Wilson's Warblers.

In the same manner, many fall migrants of normally occurring eastern species have long been appearing (especially coastally) 1, 2, or even 3 months later than the last normal inland migrants. Formerly these had all been considered just (north)eastern breeders that occasionally happened to linger. Now we know that they too are blowback drift-migrants but *eastern* breeders that were intercepted in their southbound migration somewhere to the south or southwest and then blown back northward—exactly as the more exotic western vagrants are, *and often with them*. In most cases, hatching-years make up the bulk of such late migrants. The September to December 2017 period exhibited what was probably the most widespread and persisting species diversity and distribution of blowback drift-migrants in memory, ranging from at least the Carolinas to Newfoundland.

Even though it was an extreme storm event, a massive fallout detected most heavily in Nova Scotia in early October 1998 (McLaren et al. 2000) offers the strongest direct evidence of the blowback drift-migrant origin of late-lingering eastern species in the Northeast and the study area. Owing to this storm's origins east of Florida, none of the 55 or so species involved were from the Great Plains, the West, or the Southwest—unlike the typical October–November blowback events that produce Cave Swallows and Ash-throated Flycatchers annually in the Northeast. It is also worth noting that blowback drift-migrant events are normally *not* associated with storms like that described by McLaren et al. but rather with very strong pre-cold front southwest winds followed by strong post-cold front northwest winds. Unusually, the Nova Scotia birds arrived there with *northeast* winds after having been spun counterclockwise around a low not far southeast of Nova Scotia.

In the study area, many routine diurnal landbird migrants—Chimney Swift, Northern Flicker, Eastern Kingbird, Eastern Phoebe, American and Fish Crows, Blue Jay, all swallows, Black-capped Chickadee, Eastern Bluebird, American Robin, Cedar Waxwing, American Pipit, Lapland Longspur, Snow Bunting, Bobolink, blackbirds and winter finches of several species—follow the ridges southbound, as do hawks, and like hawks they routinely cross the Hudson River to the Palisades. Crossing sites vary, but east-to-west movement is frequent and obvious, especially in October–November after passage of cold fronts.

Winter Migration

Most winter migration occurs early in the season, often continuing what had begun in the late fall. Southbound arrivals in January and February are frequently weather-induced hardship movers, sometimes locally over short distances following Van Cortlandt Lake or Swamp freezing, or from great distances inland when severe freeze-ups or deep snow cover forces all birds out. However, a few genuine overwintering species do not arrive until winter is well along, among them Long-eared, Short-eared, Snowy, and Saw-whet Owls, and Rough-legged Hawks. Wild arctic Canada Geese increasingly don't arrive in the study area in numbers until early December.

Passerine migration usually ends in early December, but a few migrants (Horned Lark, American Pipit, Lapland Longspur, Snow Bunting, Tree and Red Fox Sparrow) may still be arriving later. Gulls of many species increase throughout the winter, and while some may be true migrants, many (Ring-billed, Bonaparte's, Black-headed) are hardship movers frozen out inland. Winter waterfowl fall in the same category, with Hudson River and Jerome and Hillview Reservoir diving ducks often not arriving for the winter until December.

Most landbird irruptions begun in fall continue into winter, many like Common Redpoll not peaking until mid- or even late winter. And when raptors like Rough-legged Hawk, owls, and Northern Shrike are irrupting, they also often don't reach maximum numbers until midwinter. By winter's end, especially in warm Februaries, they may meet the first northbound spring migrants like Killdeer and American Woodcock on their way north.

RESOURCE CONCERNS

Even though the area including and surrounding Van Cortlandt Park has been subject to regular and episodic disturbance dating back to Indian clearing and burning, the arrival of Europeans accelerated disturbance and assumed new dimensions. Once the Van Cortlandts began to operate an active plantation, the formerly wooded area was forced to endure expanded and accelerated plowing, fertilizing, clear-cutting, stream diversion and damming, and wetland filling. The American Revolution inflicted its own injuries, and logging continued relentlessly into the early 1800s but eventually ceased for lack of suitable trees. Even so, we still do not know the ages or locations of the oldest living tree stands.

Soils in the entire area were routinely turned, compacted, fertilized, and polluted; the area's most prominent soil type, Greenbelt, typically exhibits buried horizons of anthropogenic materials to a depth of a meter. In Van Cortlandt Park, soil compaction manifested 300 years of active land-use change, especially filling of wetlands. The changes included extensive surface leveling, initially for grain fields in the 1700s; then golf course construction from the 1890s to the 1950s; the Dutch Gardens in the early 1900s; and finally, athletic fields, cross-country trails, parking areas, and playgrounds down to the present day. The park has also suffered other uniquely urban insults already described.

Despite having been heavily used and abused, old forests and wetlands persist, new ones appear, plant succession continues, and a resident and migratory avifauna thrives and even grows. Species have come and gone for various reasons, but Van Cortlandt Park still boasts the largest breeding avifauna of any New York City park—if not quite the one in 1644, a strong one nonetheless. However, it has had and still has tractable resource concerns whose successful addressing can only enhance the park's avifauna. The most important follow.

Keystone Species

Eastern Coyote (or Coywolf) has the potential to become a study area keystone predator. If its density increases enough to reduce numbers of actual and potential mammalian predators of ground-nesting birds (domestic/feral cats, Raccoon, Striped Skunk, Opossum) and herbivores like White-tailed Deer and Eastern Cottontail that diminish or eliminate native herbaceous layer vegetation, ground-nesting birds now reduced in numbers or extirpated from the study area (Eastern Towhee; Veery; Brown Thrasher; Black-and-white, Worm-eating, and Blue-winged Warblers) may return to previous levels of abundance.

American Chestnut was assuredly a keystone forest tree species whose loss dramatically affected Eastern Deciduous Forests. Yet in the study area only Acadian Flycatcher disappeared when Chestnuts did (this may have been only coincidence), but in any event the birds are now recolonizing the New York City area. There are no study area data on any effects the loss of Chestnuts may have had on numbers of other breeding birds. Broad- and Narrow-leaved Cattails are keystone species for native freshwater marsh-nesting birds (American and Least Bitterns, Sora, Virginia and King Rails, Common Gallinule, Marsh Wren, Swamp Sparrow), which will not return to Van Cortlandt Swamp until a large area of cattails has been restored. *Spartina alterniflora* was a keystone species for native saltmarsh nesting birds (Clapper Rail, Saltmarsh and Seaside Sparrows) when the Kingsbridge Meadows still existed.

Loss of keystone habitat types (*Typha* freshwater marshes; native grasslands, Kingsbridge Meadow saltmarshes) by human manipulation or eradication and widespread biocide contamination (Osprey, Bald Eagle, Peregrine Falcon) have played significant biological roles in the decline of avian breeding species richness in the study area.

Threatened and Endangered Birds

Of only 3 *federally* Threatened and Endangered Species that have occurred in the study area, Bald Eagle and Peregrine Falcon have been delisted and Wood Stork was a one-time vagrant. Of 33 *New York State* at-risk species that have been found in the study area, all have occurred in migration but several are now all but unknown there, 14 were historical breeders only (2 more may also have bred), and only 2 are current breeders: Cooper's Hawk and Common Nighthawk (Table 38).

Exotic (Nonnative) Birds

Nine exotic taxa are on the study area cumulative list: Mute Swan, Ring-necked Pheasant, Rock Pigeon, Eurasian Collared-Dove, Monk Parakeet, European Starling, House Finch, European Goldfinch, and House Sparrow; all but the dove have bred. Unestablished exotic species we know to have been seen in or near the study area include Muscovy Duck, Mandarin Duck, Helmeted Guineafowl, Chukar Partridge, Andean Gull (see Appendix 5), African Collared-Dove (= Ringed Turtle-Dove, Barbary Dove), Rose-ringed Parakeet (= Ring-necked Parakeet), Nanday Parakeet (= Black-hooded Parakeet, Nanday Conure), Budgerigar, Red-vented Bulbul, Red Bishop sp., Scaly-breasted Munia (= Nutmeg Mannikin), and Tricolored Munia (= Tricolored Mannikin). Given the rate at which exotics are appearing throughout the New York City area, this list is only preliminary. Fortunately, few if any

Bird Control

In the 1970s, Van Cortlandt Park officials attempted to eradicate newly established Monk Parakeet breeders near the Van Cortlandt Mansion. The birds did depart that site, but they or their descendants remain in the area but outside the park. In the 1990s similar efforts were made periodically to control the numbers of feral Canada Geese that frequented the two Van Cortlandt Park golf courses and the Parade Ground. These have not been strikingly effective.

More successful have been Hillview Reservoir's attempts to eliminate (not merely control) waterbirds using the reservoir, which is part of the New York City water supply. The effort was initiated after a US Environmental Protection Agency order that would have required installation of a concrete cover over the entire reservoir to eliminate fecal contamination by birds, mostly gulls. Yonkers residents objected strenuously, and a bird eradication program was the compromise. Begun about 1993, it entailed installing wire grids across the entire reservoir and on the concrete divider running its length and favored by roosting gulls. This program was augmented by teams that harassed, shot, or captured any waterbirds that succeeded in landing on the water (largely gulls and Ruddy Ducks). It also included removing the first Osprey nest in the area in 100+ years and nests of all swallows breeding on reservoir buildings, among them Barn and the first Cliff Swallows breeding in the study area in 55 years (NYC-DEP 2011, 2012). All were obviously deemed dangerous waterbirds. The program is ongoing.

Nest Boxes and Platforms

Breeding populations of several hole- or platform-nesting species are limited throughout the study area by scarcity of opportunities (*Wood Duck*, Northern Flicker, *American Kestrel*, *Eastern Phoebe*, Tree Swallow, House Wren), and a few that ought to be nesting are currently completely precluded from doing so (*Osprey, Barn Owl, Purple Martin, Eastern Bluebird*). These, as well as other obligate hole-nesters like Eastern Screech-Owl, Hairy Woodpecker, Great Crested Flycatcher, and White-breasted Nuthatch, would benefit greatly from a program of species-specific artificial structure emplacements. Wood Duck, Osprey, and Eastern Phoebe often prefer platforms in or over water as in Van Cortlandt Swamp and Van Cortlandt Lake. Italicized species should receive highest priority.

A few nest boxes were installed in some Van Cortlandt Park locations not long ago, but most were poorly constructed or badly sited and failed to attract residents. Many nest boxes available commercially are correctly crafted, but many are not, and some even kill their inhabitants, so professional advice is clearly called for. Protection from mammalian predators, especially Raccoons, is also critical to success, as are annual cleaning and maintenance of structural integrity. A well thought-out plan for artificial nesting structures, prepared in consultation with appropriate experts, should also target the 6 golf course sections throughout Van Cortlandt Park, as well as Woodlawn Cemetery, all of which offer unparalleled sites for American Kestrel, Eastern Phoebe, Purple Martin, Tree Swallow, and Eastern Bluebird structures. A splendid side benefit is that nest box programs enhance other uncommon to rare native wildlife, especially Southern Flying Squirrel, and they offer prime opportunities for involving carefully supervised volunteer workers.

Mammalian Disrupters

While hard data are lacking on common exotic (nonnative) quadrupeds like feral/domestic dogs and Brown Rats causing trouble for native study area birds, constant vigilance is required, given their proven adverse history elsewhere. Domestic/feral cats, however, are certainly lethal study area predators on birds. Their numbers are increasing, abetted by feeding sites on public lands that no agency will interfere with for fear of public backlash. These animals are known throughout the world to cause widespread damage to wild birds and native small mammals (Marra and Santella 2016), and there can be no doubt that they are playing biologically significant adverse roles in the study area's continuing loss of ground- and near-ground nesting species.

There is no specific study area evidence that *native* quadruped predators like Red and Gray Fox, recently arrived Eastern Coyote, American Mink, possibly Fisher, several weasels, Striped Skunk, Raccoon, and Virginia Opossum are having widespread predatory effects on birds, although they may have been adversely affecting some ground-nesters. However, in the city, captured nuisance mammals like Raccoons are often liberated in the nearest wooded area. The practice has occurred in the study area, and such artificially augmented populations can cause abnormal predation on breeding birds—not to mention spreading diseases like rabies. Exotic (nonnative) pets are also released or escape into urban parks, but most survive only for short periods thanks to Great Horned Owls and coyotes. Also only recently arrived in Van Cortlandt Park, White-tailed Deer are unexpectedly scarce (coyote predation?) and do not seem to have demonstrated the herbaceous layer elimination that is epidemic in locations with high deer densities.

Van Cortlandt Surface Water Quality

In the 1960s–70s, raw sewage leaking into Tibbett's Brook in southern Yonkers led to eutrophication of the entire system south to and including Van Cortlandt Lake. In 1962, 1963, and 1964, major fish die-offs occurred in Van Cortlandt Lake, where sediment runoff from the Major Deegan Expressway containing pollutants like toxaphenes, oil, and gas was an added insult. By 1976 sedimentation had become so intense that an earlier mean lake depth of 9–12 ft (2.7–3.7 m) had been reduced to only 2–4 ft (0.61–1.2 m), leading in 1976 to elimination of the hallmark rowboat rental active since the early 1900s. These insults only accelerated the replacement of native cattails by exotic (nonnative) *Phragmites* and loss of the marsh birds for which Van Cortlandt Swamp had been famous for more than a century. In 2001–3 the lake was finally dredged again to a mean depth of 9 ft (2.7 m). Since then, the recent spread of exotic (nonnative) Water Chestnut has significantly curtailed water flow, and the resulting sedimentation rate has increased so rapidly that water has now returned to pre-dredging depths and is worsening. It is unknown whether water quality has improved, but since December 2015 the nonprofit Friends of Van Cortlandt Park has spearheaded ongoing monitoring projects to qualify and quantify the health of Tibbetts Brook/Van Cortlandt Lake. This is designed to yield a new comprehensive water quality baseline.

Van Cortlandt Vandalism and Fires

During the 1970s–90s nadir of civic comity in New York City, Van Cortlandt Park was not immune. Clusters of structures built by the homeless littered the Putnam track bed north of Van Cortlandt Lake, and while their inhabitants were generally nonintrusive, several appeared to have been deinstitutionalized mental patients, one of whom attacked a local observer with a bottle, injuring him so severely that he subsequently died from the wounds. Other homeless encampments were scattered along the edges of the Northwest, Croton, and Northeast Forests. In this same period a few dozen cars, usually stolen, were driven into the same nominally vehicle-free areas (especially in the Northeast Forest) and then abandoned, sometimes after having been set ablaze. Other fires were periodically set in woods and brush throughout the park, including Van Cortlandt Swamp, some scarring large areas before being extinguished. Small groups, some of them Boy Scouts, camped in the forests and even in Van Cortlandt Swamp, chopping down trees for fires in the process. Kids setting off firecrackers in mid-breeding season, dog walkers' off-leash pets harassing any reachable wildlife, peripheral trash dumping, cars and all-terrain vehicles being driven off road, and casual structural vandalism were so much harassment lagniappe. All these incidents went largely overlooked officially and therefore unpoliced, but protests by observers and other concerned park visitors finally led to establishment in 1979 of the splendid New York City Park Ranger Program and a greatly increased NYPD presence, both welcomed.

Van Cortlandt Illicit Trapping, Shooting, and Grazing

Hunting of waterfowl and shorebirds occurred into the late 1800s on Van Cortlandt Lake, at the time called Tremper's Lake or Bibby's Pond, until Van Cortlandt Park was finally gazetted. But old habits died hard, and as late as the early 1960s, a breeding ♂ Wood Duck was shot one summer. Harassment finally drove off Van Cortlandt Lake's sole breeding pair of Mute Swans in 2009 (they returned in 2016, failed, and tried again in 2017) that had been rendering successful feral Canada Goose nesting a sometime thing. Illegal shooting, usually of birds and small mammals by boys with slingshots and BB or pellet guns, occurred in Van Cortlandt Park since the early 1900s at least and may still be a problem. Illegal trapping of mammals in Van Cortlandt Swamp for food or fur (Eastern Cottontails and especially Muskrats, even the occasional American Mink) was also evident. From the 1920s to at least the 1950s, Kieran, Cruickshank, and other observers would find and destroy traps regularly. Their mammalian prey may have been a source of food during the Depression, but traps in the past also captured rodents, songbirds, and numerous marsh birds, including one of 2 King Rails attempting to overwinter in 1955–56.

Some incursions have even been commercial. In Riverdale in the mid-1930s, across Broadway from the Northwest Forest, two farmers raised small herds of cows and goats in what was known as Swamp Hollow, bounded by Mosholu Ave. and W. 260th St., Fieldston Rd. and Broadway. It was crossed by a small creek and tributaries, some of which persisted into the 1950s, that eventually drained across the Parade Ground and into Tibbett's Brook. Brazenly, the farmers had been pasturing their cows and goats in adjacent Van Cortlandt Park for decades until it anonymously came to the attention of Van Cortlandt officials. The farmers' livestock were abruptly removed from the park, and they quickly went out of business.

Van Cortlandt Plant Ecology

Unquestionably, the most important biological research to date in Van Cortlandt Park and the study area has been summarized in Henning's (2015) PhD dissertation. It was a massive long-term assay of the vascular flora of Van Cortlandt Park and an intensive ecological dissection of its plant communities and the factors affecting them. Like any fine study, it asked as many questions as it answered, and we can only hope it will launch a surge of ecological research in the park.

Despite earlier efforts to establish a comprehensive Van Cortlandt Park flora, Henning was the first to do so. In his 6-year study, he surveyed park-wide vascular plants weekly from April through September 2008–13 using east-west and north-south transects and random quadrats throughout what he termed the five major Van Cortlandt Park sections (Parade

Ground/Vault Hill, Northwest Forest, the Ridge (Croton Forest), Northeast Forest, and Shandler Area), as well as all park ornamental beds. Even though historic **Van Cortlandt Swamp** was not named anywhere in his entire dissertation despite his frequent discussion of Kieran's (1959) lifelong plant work there, he did sample it but using the Planner-Newspeak neologism of Tibbetts Wetland. Notwithstanding this disappointing oversight, for woody plants Van Cortlandt Swamp proved to be the most biodiverse region within the park, although quadrat data on herbaceous plants throughout the park have yet to be analyzed.

Henning found 1102 plant species, 3 times more than previously thought and proof that the park had always been seriously undersampled. Of the 1102, 71% were herbaceous, and, sad but not unexpected, more than 50% were nonnatives (exotics). Nonetheless, he also found 31 New York State at-risk species, indicating that despite frequent burning, soil compaction and disruption, vandalism, and invasions by exotics, Van Cortlandt Park remains a refugium for endangered plants. He found the 3 most abundant trees to be Black Cherry, Norway Maple, and Red Oak, with only the maple nonnative, widespread, and increasing, and he reaffirmed that hard data on the ages of extant park forest stands remain unknown. Floristically, the northern end of the park (Northwest Forest, the Ridge [Croton Forest], and Northeast Forest) was richer and more diverse than the southern end, and unexpectedly, herbaceous species in general exhibited more stability than woody species. The increase in known park plants from the herbaceous layer confirms the hypothesis that this is where most richness is concentrated in eastern woodlands. At the same time their higher extirpation/extinction rates and the large number of state-listed herbaceous taxa mandate that park personnel pay far more attention to them than they have to date—of particular concern to us because they are especially important to forest birds.

Henning also highlighted a major Van Cortlandt Park unintended-consequences problem—the widespread use of Monsanto's herbicide Roundup (glyphosate/glyphosphate) in the Parks Department's aggressive "renovation" process of extirpating invasive plants, since it seems their fixation on preserving Van Cortlandt's woody flora comes at the expense of its herbaceous layer. Of the native herbaceous plant species discovered in Henning's survey, 29 were New York State at-risk or unprotected taxa. Further, a rank of S1 in New York State indicates a critically imperiled taxon found in low numbers in 5 or fewer sites statewide—the most vulnerable ranking a plant can have and one that 11 herbaceous species in Van Cortlandt now hold. Of the remaining ranked taxa, some were represented by only single plants and so were not only rare statewide but excessively scarce within the park, yet Roundup blindly targets them all. Their collectively rare status demands their recognition for forceful protection by the Parks Department in all planning activities and any actions taken to eradicate invasives.

Van Cortlandt Forest Openings

A minor bright spot in this dim picture has been the opening of the forest canopy in several locations throughout Van Cortlandt Park by Hurricanes Irene (2011) and Sandy (2012) as well as 2 small tornadoes[!] during which many large trees fell. This has allowed weedy and shrubby patches of second-growth vegetation to crop up, in turn quickly occupied by two species almost gone as breeders following loss of habitat, Chestnut-sided Warbler and Indigo Bunting. No attempts must be allowed to eliminate these ecologically essential forest gaps.

Van Cortlandt's One Million Trees Involvement

Announced by New York City Mayor Bloomberg in 2007, the aim of this largely public relations program was to plant 1 million trees throughout New York City, even in areas where trees already naturally grew in profusion. Van Cortlandt Park was too tempting a target, so in 2010–14 an unknown number were planted in several park locations. The largest was in the Northwest Forest just east of a baseball field at Broadway and W. 260th St. This site held an existing forest and herbaceous layer composed of Black Locust trees, weedy fields, vine-covered shrubs, and a thick mixed groundcover including the Jewelweed favored by migrating Ruby-throated Hummingbirds (up to 40 were there across August–September at its peak around 2000). The site was of sufficiently good habitat quality to have yielded many migrants over the last 15 years, among them Acadian Flycatcher and Golden-winged and Mourning Warblers. It was assuredly not wasteland.

Nevertheless, 10 ac (4 ha) were cleared of existing vegetation and then replanted with tiny trees (of native wild stock, one hopes), including Sugar Maple, Red Oak, and Tuliptree. They were densely packed together, many hundreds if not thousands of trees in a limited area bereft of herbaceous layer plants. When queried about this density, Park employees suggested that if 1 tree out of 4 survived, the effort would be considered a success. An existing natural area well used by native birds year-round is now gone, and until the new trees mature it will be nearly birdless for a long time.

Van Cortlandt Parade Ground

After the Parade Ground was reseeded with a hybrid turfgrass monoculture in 2009 (except in the southeastern corner), the numbers and variety of native birds using it year-round plummeted. Native grasses and small wildflowers were replaced by sterile golf-course grasses that birds have been avoiding. Counts of migrating Killdeer, Snow Bunting, Lapland Longspur, Vesper Sparrow, Horned Lark, Eastern Meadowlark, and American Pipits fell to near zero in those areas, and maxima dropped to one-10th or less of those from the 1950s through the 1980s. Killdeer almost vanished as a breeding species. But at the same time, grass-eating Canada Geese exploded from previously small feral groups (under a dozen) to the flocks of 2000+ wild arctic breeders that began appearing around 2010. Once again, an area that has been affording unique habitat for vanished grassland birds for more than a century was nearly destroyed because plans were made and actions taken without any input from persons with explicit Van Cortlandt ecological and ornithological knowledge.

Van Cortlandt Swamp

The extraordinary and unique New York City resource that is Van Cortlandt Swamp with its formerly large cattail marsh has been slowly changing since its creation and growth between 1700 and 1840. But the replacement of native cattails by exotic (nonnative) *Phragmites* and growth of wet-swamp woody shrubs and trees accelerated when the Swamp was reduced by half in 1949–50 and again when sedimentation and pollution favored exotic (nonnative), weedy species 20 years later. When Van Cortlandt Lake was dredged in 2001–3, the last remaining marsh supporting overwintering Virginia Rails at the head of the lake was peremptorily excavated, and no attempt has been made to restore it. And even though the last patches of cattails in the main Swamp had been engulfed by *Phragmites* several decades ago, in the 2014 Van Cortlandt Park Master Plan (below) there is only passing reference to its removal and not a word about replacing it with native cattails in the "upper basin of Van Cortlandt Lake," another Planner-Newspeak neologism for **Van Cortlandt Swamp**. Cattail marsh restoration there is the sine qua non for resurrecting the suite of missing freshwater marsh birds that made Van Cortlandt Swamp famous throughout the Northeast for more than 150 years. Removal of invasives including *Phragmites* at recently emergent Tibbett's Marsh has allowed natural recolonization by volunteering cattails and is to be applauded. But there is still a long way to go, and there seems to be no evidence that park master planners were aware of cattails' importance to the entire Van Cortlandt Park ecosystem.

Van Cortlandt Putnam Trail

A current effort by the New York City Parks Department/Van Cortlandt Park Conservancy to degrade the habitat they are charged with preserving and protecting is their proposal to insinuate a wide paved bicycle trail through one of its oft-touted natural features, the Putnam Trail—the old Putnam Division track bed that runs the length of Van Cortlandt Park alongside the lake and the Swamp. Having been left alone for well over two decades since the last tracks and ties were removed, it functions as an avian spinal cord linking the entire Tibbett's Brook watershed ecosystem and birdlife by providing superb natural migration and breeding habitat for almost 2 smi (3 km). It is also the breeding site for 1–2 pairs of the park's handful of Blue-gray Gnatcatchers, and of the last (and next?) Eastern Bluebird nest. Cherishing this tranquil ambiance, thousands of passive users have been quietly enjoying its soil roadbed and expanding vegetation and wildlife for decades. Botanically, in the 30 years following its closure the Putnam rail bed has filled in autonomously with an impressive array of spring flowering plants (Henning 2015), confirming the richness of its native plant seedbank—one that would certainly not survive paving.

Notwithstanding, the Parks Department/Van Cortlandt Park Conservancy continue to press their proposal, even as its terms constantly shift. As of early 2017, these have included

- paving the Putnam Trail with nonporous asphalt
- mutating it into a predominately biking boulevard
- widening it from 8 ft (2.4 m) to 10 ft (3.1 m) or 11 ft (3.4 m)
- removing anywhere from 7 to 300+ trees along its length
- and in the process wasting several million dollars of funds badly needed for park-wide resource management while marginalizing those who profit most from its natural, health, and aesthetic benefits

Opposition to this initiative has coalesced around an organization called Save the Putnam Trail. It is an amalgam of concerned users from observers to cyclists to walkers who have banded together to provide defensible, ecologically sound counterproposals on their website: https://www.savetheputnamtrailnow.wordpresscom/. Their main suggestions have been to resurface the Putnam Trail with a permeable ADA-compliant surface instead of impervious asphalt and to retain its existing width with only an absolute minimum of closely monitored vegetation trimming. Their version's success will hinge in part on convincing the New York State Department of Environmental Conservation (NYSDEC) that it should not issue a required wetlands permit, given how heavily the park/conservancy's proposal will adversely affect immediately adjacent wetlands. Notwithstanding this opposition, the department and conservancy refuse to abandon their proposal. *Quis custodiet ipsos custodes?*

Van Cortlandt Master Plan

Released in July 2014, the New York City Department of Parks' newest master plan for Van Cortlandt Park does not even pay lip service to the breeding, overwintering, and migratory birds that have made the park famous for more than 150 years. Among a host of deficiencies, it completely ignores the existence and ecological importance of the seasonally flooded Sycamore Swamp in the Northeast Forest. But the most egregious measure of how out of touch it is with the park's history, avifauna, and naturalist users is that once again the historic name ***Van Cortlandt Swamp*** does not appear anywhere—not in text, not on the myriad of multicolored maps and charts. It simply does not exist. The plan is accessible online at http://issuu.com/nycparksplanning/docs/vcp_master_plan_july17__2014.

Daylighting Tibbett's Brook

Unofficial proposals have been circulating since about 2010 to redress the 1910s burial of Tibbett's Brook southwest of Van Cortlandt Lake in a tunnel that connects to the Broadway Combined Sewer. A new proposal would result in Tibbett's Brook's streamflow instead remaining above ground downstream from Van Cortlandt Lake, flowing south and then emptying directly into the Harlem River as it did formerly, only now not far east of the Broadway subway bridge instead of to its west near Spuyten Duyvil. It would entail creating a new streambed along the old Putnam track bed south to the Harlem River, as part of a southward extension of the Putnam Trail. The new stream would more or less follow the route of Van der Donck Meadows that ran north from the Harlem River at W. 230th St. to W. 238th St. just east of Broadway, where it merged with Tibbett's Brook west of Broadway, creating Kingsbridge Island (see Figs. 1, 3). Assuming it would be landscaped with native plants in sufficient density, never paved over with asphalt, kept free of trash and debris, and patrolled heavily, such a restored area could become an important avian movement corridor. Detailed plans have not been finalized, but information on various aspects can be found online at http://www.thenatureofcities.com/2015/03/17/daylighting and on pp. 67–69 and 81 of the 2014 Van Cortlandt Park Master Plan (above), which has curiously avoided the widely used technical term *daylighting*—uncovering and naturalizing previously buried watercourses and streams.

THE FUTURE

What Isn't Known

Assembling and analyzing all the information about birds in the study area for almost a century and a half inevitably highlighted gaps in our knowledge. Many are small, but a few are profound. They are a mix of elementary questions about distribution, frequency of occurrence, and numerical abundance. But there are also those relating to habitat restoration and management that profoundly affect Van Cortlandt Park's birdlife. Some involve the entire study area, some involve just Van Cortlandt Park, some are almost microsite-specific, but all are in various ways important.

- Do any of the following, and how many pairs of each, continue to breed in the study area even if not annually: Sharp-shinned, Red-shouldered, and Broad-winged Hawks; Virginia Rail; Sora; Least Flycatcher; White-eyed and Yellow-throated Vireos; Worm-eating, Blue-winged, Black-and-white, Kentucky, and Hooded Warblers; Yellow-breasted Chat; Field and Swamp Sparrows?
- How many pairs of Killdeer, American Woodcock, Spotted Sandpiper, Ruby-throated Hummingbird, Eastern Phoebe, and Brown Thrasher breed throughout the study area and exactly where?
- How many of these species breed in or on buildings in or close to each of the 4 extant study subareas: Black Vulture; Barn Owl; Common Nighthawk;

Chimney Swift; American Kestrel; Monk Parakeet; Eastern Phoebe; Common Raven; Bank, Cliff, and Northern Rough-winged Swallows?
- What percentage of nominal Gray-cheeked Thrushes using the study area in spring and in fall are actually Bicknell's Thrushes? What are each species' migration dates?
- What percentage of Alder/Willow Flycatchers using the study area in spring and in fall are actually Alders? What are Alder's and Willow's respective migration dates?
- What is the current study area status of spring Western Palm Warblers?
- Where did all the Bronx Rose-ringed and Nanday Parakeets go, when, and why?
- Which additional exotic (nonnative) bird species have occurred in the study area?
- What are the dimensions of current spring and fall hawk migrations (species, numbers, dates) in Van Cortlandt?
- Where do southbound study area and Riverdale hawks eventually cross to New Jersey?
- How many pairs of Wild Turkey breed in Van Cortlandt Park?
- Is the southeastern Parade Ground now the only section routinely used by ground-loving birds other than geese? Why?
- How many pairs of Eastern Screech and Great Horned Owls breed in Van Cortlandt?
- What breeding, migratory, and overwintering species, and in what numbers, have moved into the *Phragmites*/cattail patches at Tibbett's Marsh, at Lincoln Marsh, and immediately west of the Sycamore Swamp?
- What are the current winter bird populations for all species in the study area?
- What species of feral mammals and in what numbers live in or use the study area, and where?
- What is the study area Eastern Coyote population? Are they having any demonstrably deleterious effects on any study area birds?
- What are the sources of possible predation on Van Cortlandt ground-nesters?
- What is the Van Cortlandt deer population? Are they having any deleterious effects on herbaceous layer vegetation?
- What are the ages of each of the many forest stands in Van Cortlandt? How old, and where, are the oldest individuals of each of the important tree species?
- What areas of Van Cortlandt have the healthiest, and which the most degenerate, herbaceous layer vegetation?
- How long do individual waterbirds remain on Jerome and Hillview Reservoirs, and which species continue to feed there after bottom cleanings?
- How long does it take for various waterfowl to repopulate Jerome and Hillview after draining?
- What species and in what numbers now breed at Hillview?
- What species and in what numbers now breed at Jerome?
- Do any of these species ever breed in Woodlawn Cemetery, and if so where, and how frequently: Black Vulture, Eastern Screech and Great Horned Owls, Common Raven, Pine Warbler, Red-breasted Nuthatch, Golden-crowned Kinglet, Pine Siskin, Red and White-winged Crossbills?

- What is the complete breeding bird population of Woodlawn?
- What is the complete winter bird population of Woodlawn?
- How many pairs of Wild Turkey breed in Woodlawn?
- What nonbreeding species of owls, in what numbers and locations, and when, occur annually in Woodlawn?

What Needs to Be Done

Equally important, and where the research rubber meets the wildlife management road, are the recommendations for actions that in part stem from research done in preparation for this book and in part from certain management actions already taken, especially within Van Cortlandt Park, that have had both deleterious and enhancing effects on the area's birdlife (see "Resource Concerns," above).

As in many public areas, there has long been a Van Cortlandt Park disconnect between actions planned and undertaken by maintenance staff on the one hand and research needed or already under way in the park on the other. An equally vital contrast exists between the explicit mandate for public recreation and the often only implicit mandate for enhancing and conserving native wildlife that might never become known to most park visitors. All these issues need to be identified and folded into the planning and budgeting for all publicly owned areas, be they parks or reservoirs. Woodlawn Cemetery is a private entity open to the public, but the same concerns apply to management of its resources in ways that not only will not damage wildlife but will also enhance it whenever compatible with the cemetery's intrinsic function.

Among the most important systemic changes to benefit Van Cortlandt Park's birdlife would be establishment of an entity composed of knowledgeable and experienced local observers, botanists, and ecologists whose task it would be to make recommendations, to evaluate landscape-altering actions and their timing, to obviate irreversibly damaging habitat alterations, and in general serve as a watchdog on Parks Department/Van Cortlandt Park Conservancy actions, proposed and ongoing.

For credibility, this entity must be wholly independent of the Parks Department and the Van Cortlandt Park Conservancy as well as any other groups formally related to Van Cortlandt Park management. It should issue an annual report on how Van Cortlandt is being managed, and it should assertively promote its own existence and findings. It should be undergirded by a coordinated cluster of volunteers (retirees are an untapped resource) who would inspect all areas of the park regularly and report their findings on a dedicated website to keep all interested parties aware of park management actions or inactions in timely fashion.

Another essential action to be taken is a widely advertised, open-to-the-public schema for preparation of a proactive Van Cortlandt Park Avian Conservation Plan, to supplement and correct errors and deficiencies in the 2014 Van Cortlandt Park Master Plan. It should be rigorously integrated with all park planning and day-to-day maintenance activities and should explicitly itemize actions and timings prohibited, allowed, actively supported, and implemented to conserve and enhance habitat for breeding, migrating, and overwintering birds throughout all sections of the park, including especially all golf courses and the Parade Ground.

Systemic actions to fill gaps in our knowledge should include the following:

- When New York State Breeding Bird Atlas III (2020–25) begins, ensure that there is intensive coverage of *all* of Van Cortlandt Park, including all golf course sections, plus Woodlawn Cemetery, and all of Jerome and Hillview Reservoirs *inside* the fences,

including buildings and other structures, especially at night.
- Regularly verify estimated current populations for all breeding species in the entire study area, subarea by subarea.
- Ensure that breeding bird census activity begins in mid-April and runs through late July or early August; record for later evaluation species, numbers, and locations of all migrants that could even remotely be area breeders.
- Identify all study area habitats that are critical to various bird species but that may also be under current or future threat from various sources.
- Establish, maintain, and monitor hummingbird feeders throughout Van Cortlandt Park to assay Ruby-throated's true breeding population.
- Quantify Parade Ground bird use year-round.
- Search for shorebirds on the Parade Ground after heavy rains, carefully localizing them.
- Establish a permanent pseudo–Breeding Bird Survey route in Woodlawn Cemetery; repeat it for 5 years every 20 years, in sync with New York State Breeding Bird Atlases.
- Reestablish, maintain, and regularly census winter bird feeders in Woodlawn Cemetery.
- Establish regular first-light Woodlawn Cemetery field work year-round, augmented by nocturnal owl censuses.
- Work with Van Cortlandt Park management to provide very early morning golf course access during breeding and winter bird censuses and Christmas Bird Counts.
- Work with Woodlawn Cemetery management to provide very early access during spring and fall migration, breeding and winter bird censuses, and Christmas Bird Counts.

The following management actions would immediately benefit and protect avian habitats in the study area:

- In Van Cortlandt Park, establish rigorous, directed law enforcement patrols to control, then end, numerous resource and wildlife-damaging violations known to occur, including taking endangered/threatened native plants, capturing and killing turtles and other herps, trapping mammals, shooting with slingshots or guns of any kind, harassing birds with off-leash dogs, riding noisy illegal ATVs on pedestrian trails, camping, setting fires, cutting trees, and abandoning fishhooks.
- Monitor and maintain the recently installed sewer and drainage systems (Parade Ground, Van Cortlandt Lake) so as not to compromise or pollute recently upgraded park waters.
- Prohibit the use of herbicides/pesticides on the Parade Ground and rodenticides everywhere, especially those known to cause secondary poisoning in raptors.
- Apply no chemical treatments to kill vegetation or flowering plants of any kind without consultation with local botanists and ornithologists.
- Preserve any remaining flowering "weeds" occurring on the Parade Ground. These are actually wildflowers whose seed heads provide food and cover for small open-country birds, protect the soil itself, and are self-maintaining areas of beauty.
- For grassland birds, let all vegetation grow on the large, fenced-off southeastern segment of the Parade

- Ground, mowing it only once every year or two to arrest succession and maintain grasses.
- Remove no native aquatic vegetation beyond algae for stream and lake cleaning.
- Allow selected exotic (nonnative) plants to remain if their avian value is established and they are not aggressively invasive (e.g., Jewelweed for hummingbirds).
- Remove all *Phragmites* in Van Cortlandt Swamp and replace it with cattails.
- Remove *Phragmites* west of the Sycamore Swamp and replace it with cattails.
- Whenever Van Cortlandt Lake needs to be drained, with adequate advance timing consult with local observers to optimize drawdown to provide migrating shorebird habitat.
- At Van Cortlandt Swamp, maintain closed access but create a path along the *inside* edge of the golf course allowing observers close approach to the Swamp.
- Restore wires as perches for swallows and other species along the Van Cortlandt Lake/Swamp junction by the former railroad bridge.
- Construct or uncover, then maintain and protect, large soil piles or embankments near water in Van Cortlandt Park for use by nesting Bank Swallows and Belted Kingfishers.
- Develop and implement a park-wide plan for installation and maintenance of professional-grade, predator-proofed nest boxes and platforms for Wood Duck, American Kestrel, Eastern Phoebe, Osprey, Barn Owl, Purple Martin, and Eastern Bluebird as discussed in "Nest Boxes and Platforms" above.
- Install and maintain Purple Martin houses at Van Cortlandt Lake.
- Implement and rigorously enforce a Van Cortlandt park-wide plan for elimination of all exotic (nonnative) quadrupeds.
- Immediately shut down any domestic-cat feeding stations and rigorously prevent their reestablishment.
- On Hillview Reservoir slopes and inside the fences at Jerome Reservoir, let grass and forbs grow for grassland birds, mowing only once every year or two to arrest succession.
- Reestablish Hillview conifer plantations as windbreaks for year-round songbird cover.
- Assist Woodlawn Cemetery management in maintaining and restoring native herbaceous layer vegetation in less visible locations.

Introduction to the Species Accounts

NEW YORK CITY AREA AVIAN STATUS CHANGES AFTER 1964

In the 50+ years since Bull's (1964) book—the last treatment of the New York City area avifauna—there have been considerable changes in its birdlife. Some have been positive, some negative, some sudden, some gradual, some predictable but many unexpected by active observers. Prior to our examination of the birds of a corner of New York City, it is useful to review the more notable area-wide changes, many of which are manifest in study area species.

Changes affecting many species simultaneously include the following:

- Great increase in knowledge of the year-round status of pelagic birds
- Understanding of shorebird age-class plumages and differential fall migration timing
- Understanding of gull and tern age-class plumages and molts
- Refining *Empidonax* flycatcher statuses with major advances in their in-field identification
- Increasing appreciation of in-field songbird ageing and sexing techniques
- Evaluation of the extent of the spring trans-Gulf overshoot phenomenon
- Discovery of major spring migration events on barrier beaches
- Discovery of the fall blowback drift-migrant phenomenon
- Severe declines in spring numbers of Neotropical migrant flycatchers, vireos, thrushes, warblers, tanagers
- Severe declines in breeding grassland birds: Upland Sandpipers; Horned Larks; Field, Vesper, Grasshopper, and Savannah Sparrows; Eastern Meadowlarks; Bobolinks

Species-specific changes include the following:

- Colonization by breeding Double-crested Cormorants, Cattle Egrets, Glossy Ibises, Black Vultures, Mississippi Kites, Gull-billed and Forster's Terns, Chuck-will's-widows, Red-bellied Woodpeckers, Monk Parakeets, Tufted Titmice, Red-breasted Nuthatches, Blue-gray Gnatcatchers, Golden-crowned Kinglets, Yellow-throated Warblers, Summer Tanagers, Blue Grosbeaks, Boat-tailed Grackles
- Incipient colonization by White-faced Ibis, Eurasian Collared-Doves
- Recolonization after (near-) extirpation by Great Blue Herons, Wild Turkeys, Bald Eagles, Ospreys, Cooper's Hawks, American Oystercatchers, Eastern Willets, Laughing Gulls, Pileated Woodpeckers, Peregrine Falcons, Acadian Flycatchers, Cliff Swallows, Eastern Bluebirds, Orchard Orioles
- Incipient recolonization by Prothonotary and Kentucky Warblers
- Increasing migrants from Sandhill Cranes now breeding in the Northeast, Trumpeter Swans introduced in upstate New York and Canada

- Great increase in House Finches, then population collapse from epidemic conjunctivitis, followed by slow recovery
- Strong urban spread and breeding by Red-tailed Hawks, Great Horned Owls, Common Nighthawks, American Kestrels, Willow Flycatchers, Common Ravens, Northern Mockingbirds, Northern Cardinals
- First overwintering Pink-footed, Greater White-fronted, Ross's, Barnacle, and Cackling Geese; Tufted Ducks
- Increase in overwintering Snow Geese, arctic Canada Geese, Northern Shovelers, Hooded Mergansers, Common Eiders, Harlequin Ducks, Double-crested and Great Cormorants, Ring-billed Gulls, Yellow-bellied Sapsuckers, Merlins, Fish Crows
- Increase, then decline, in overwintering Canvasbacks, Greater Scaup
- Increase in hummingbird overwintering attempts at feeders
- Decreases in overwintering American Black Ducks, Lesser Scaup, White-winged Scoters, Red-breasted Mergansers, Horned Grebes, Bonaparte's Gulls, Short-eared Owls, American Kestrels, Field and American Tree Sparrows
- Disappearance of overwintering Pine Grosbeaks, Evening Grosbeaks
- (Near-) extirpation of breeding American Bitterns, Northern Harriers, Northern Bobwhites, Ring-necked Pheasants, Ruffed Grouse, King Rails, Black Rails, Short-eared Owls, Sedge Wrens, Henslow's Sparrows
- Near-extirpation of migrant and overwintering Loggerhead Shrikes
- Great decrease in breeding Black-crowned Night-Herons, Least Bitterns, Sharp-shinned and Red-shouldered Hawks, Piping Plovers, Spotted Sandpipers, Roseate Terns, Marsh Wrens, Brown Thrashers, Chestnut-sided Warblers, Eastern Towhees, Purple Finches
- Great decrease in migrant and overwintering Rusty Blackbirds, in migrant Black Terns, Common Nighthawks, Bank Swallows, Golden-winged and Connecticut Warblers
- Great increase in breeding Herring and Great Black-backed Gulls, then decline
- Precipitous decline in American Crows from West Nile Virus, then recovery

The species below are being seen or detected more frequently now than in 1964 (omitting most pelagics, plus the recent and incipient colonizers and recolonizers above that are also occurring with greater frequency as migrants). This increased detection obviously tracks bird range and population changes, but it is abetted by new field guides (many addressing only particular groups of birds) and journals with cutting-edge identification techniques; excellent new binoculars, telescopes, and digital cameras with telephoto lenses; GPS, cell phones, iPods, and similar electronic devices; e-mail, texting, Listservers; and more mobile and widely traveled, experienced observers.

Fulvous Whistling-Duck	Brown Booby	Purple Gallinule
Barrow's Goldeneye	American White Pelican	Black-necked Stilt
Pacific Loon	Brown Pelican	American Avocet
Eared Grebe	White Ibis	Western Willet
Western Grebe	Swallow-tailed Kite	Buff-breasted Sandpiper
Manx Shearwater	Golden Eagle	Long-billed Dowitcher

Wilson's Phalarope	Royal Tern	Swainson's Warbler
Black Guillemot	Sandwich Tern	Audubon's Warbler
Black-headed Gull	White-winged Dove	Clay-colored Sparrow
Little Gull	Rufous Hummingbird	Lark Sparrow
Franklin's Gull	Alder Flycatcher	Le Conte's Sparrow
Thayer's Gull	Say's Phoebe	Oregon Junco
Lesser Black-backed Gull	Ash-throated Flycatcher	Painted Bunting
Bridled Tern	Cave Swallow	Dickcissel
Arctic Tern	Varied Thrush	Yellow-headed Blackbird

HISTORICAL DATA

Like most ornithological histories, that of our study area is a palimpsest. It is our intention to highlight the layers so that the chronological whole becomes clearer to readers. We hope that future authors, whether operating in the study area, the Bronx, or the New York City region, will thereby find their tasks easier.

We presuppose that all species occurring or breeding regularly in Riverdale in Bicknell's time also did so then in the study area, habitat considered. Bicknell's extensive Riverdale specimen collection complemented Dwight's from Van Cortlandt Park and together provided fascinating insight into the landbirds breeding and occurring in the Northwest Bronx in the late 1800s. Moreover, Bicknell uniquely collected extensively in Kingsbridge Meadows, also one of our own study subareas.

Griscom (1923) and Kuerzi (1927) both used the term *Bronx Region* in their publications, but they meant different areas and included considerably more than just Bronx County. Griscom's Bronx Region was "the whole Borough of the Bronx north to a line connecting Yonkers and New Rochelle." But he did not specify where in Yonkers or New Rochelle, so we don't know precisely how far north of the New York City border his line was located. Kuerzi's Bronx Region was almost four times larger than Griscom's but just as imprecise: "the Borough of the Bronx [extending] north to the upper reaches of the Bronx, Saw Mill, and Grassy Sprain Rivers; or approximately north to a parabola connecting Tarrytown, Kensico, and Rye." His readers would never have had any way of knowing which records discussed were in Bronx County or in Westchester County except in those few instances when exact locations were given. Figure 17 offers our understanding of those lines' approximate positions.

There was a great disparity in status, distribution, and abundance information between these two publications that were only four years apart. This happened because Griscom was largely reporting only information new since Eaton (1910, 1914); just a handful of carefully selected persons provided data to him; he rejected many data sources, especially those of the new Bronx County Bird Club's youthful members; Kuerzi's Bronx region embraced much more of Westchester County than did Griscom's; and the greatly increased fieldwork by Bronx County Bird Club members from 1923 to 1926 vigorously seeking to fill in blanks and address errors they found in Griscom's 1923 book found its way into Kuerzi's 1927 paper. It is highly likely that many records subsumed in either of these 2 Bronx Regions were actually from our study area. Alas, at a remove of almost 100 years, there has been no way to uncover and disentangle them—except occasionally from *Bird-Lore* and *Audubon Magazine* seasonal accounts, Linnaean Society meeting notes, and the current authors' discussions over the decades with older observers.

In the historical accounts, *unmentioned* does not mean *unrecorded* unless so stated; many times unmentioned birds were common and regular in several seasons. Where Bicknell's entry says a species was unmentioned, that implies it was actually so regular or common that Griscom chose to exclude mention of it in his 1927 synopsis of Bicknell's records.

Other historical accounts often omitted common or expected species or birds in winter, or discussed only breeders. Also, *summer resident* was used for decades to mean breeders, even though it encompasses nonbreeding residents in many species. We have usually not changed it because it complements *winter resident* while offsetting *migrant*. Earlier authors evidently thought it unimportant to give daily maxima for any species until Cruickshank (1942); commendably, Bull (1964) was the first to do so season by season for most species. With eBird, future writers at least ought to have an easier time quantifying seasonal occurrence patterns and abundances for the often underreported common birds in their subject areas.

NEWER DATA

When we evaluated original field notes and notebooks, it was quickly apparent that few observers meticulously recorded the numbers of every species seen in every location on a given day. Often, on a trip to multiple sites in the Bronx only a single list of species and counts would be given for the entire day. There were many exceptions, of course, but all too frequently daily maxima by locations were wanting.

Every record presented to readers is a distributional data point. As such, each should contain complete metadata: species, location, date, number, observer—and if known and pertinent, elevation, subspecies, sex, and age. Sight reports lacking metadata are compromised to the same degree as a specimen without a date or a photograph missing a location.

Many records in Bull (1964, 1970, 1974, 1976) and nearly all in Levine (1998)—but not earlier authors—omitted locations, observers, and often both, making it difficult or impossible for anyone to allocate them geographically, contact the original observers for additional details, or evaluate records based on observer experience and ability. These omissions have also occurred too often in *Kingbird* and in *Bird-Lore, Audubon Field Notes, American Birds*, and *North American Birds* seasonal reporting columns covering the New York City area. Thus the 53 records in the species accounts labeled "unknown observer" are shown with observers' names omitted just as they originally appeared. We must assume that each writer had vetted every record before promulgation.

Even though few if any qualitative seasonal statuses would be materially changed by additional data, we are confident the following species have occurred more frequently than their limited study area data indicate: Cackling Goose; Eurasian Wigeon; Redhead; Long-tailed Duck; Red-breasted Merganser; Red-throated Loon; Horned and Red-necked Grebes; Snowy Egret; Little Blue Heron; Yellow-crowned Night-Heron; Glossy Ibis; Black Vulture; Barn, Barred, Long-eared, and Northern Saw-whet Owls (all 4 may occur annually); Eastern Whip-poor-will; Red-headed Woodpecker; Acadian and Alder Flycatchers (both must occur annually); Northern Shrike; Purple Martin; Sedge Wren; Bicknell's Thrush; Lapland Longspur; Prothonotary, Kentucky, Cerulean, and Yellow-throated Warblers; Clay-colored, Grasshopper, Henslow's, and Nelson's Sparrows; Summer Tanager; Blue Grosbeak; Dickcissel; Pine Grosbeak; Red and White-winged Crossbills. There being no formal repository for such information, it has long fallen through the reportability cracks in *Kingbird* and *North American Birds* or has been subsumed (and thereby lost forever) in summary statements like the "24 Boreal Chickadees in Westchester and Bronx counties during the fall 1975 incursion"—the largest on record and an exact quotation from *Kingbird*, but no more details were forthcoming. This lack of information is even a problem for Central and Prospect Parks with their multitudes of observers, and most other New York City parks suffer an even greater loss of data on a regular basis. Area bird clubs would perform a singular service by archiving individual (especially historical) records for the area of their purview and never assuming it is being done elsewhere. It probably isn't.

CHRISTMAS BIRD COUNTS

Some winter daily maxima for study area landbirds derive from Bronx-Westchester Christmas Bird Counts, when there was often simultaneous, nearly complete coverage of most of the study area. Nonetheless, many such maxima are lacking because over 90 years most such count numbers were pooled into single West Bronx party totals, and thus most subarea counts have been lost.

Long-term count trends for 43 study area species (Figs. 21–63) are shown in appropriate species accounts. Their scatterplots give the total count-area numbers (the Y-axis) found each year (the X-axis); atop these data points is superimposed a 2-year moving-mean (= rolling-average) trend line for each species, commonly used with time-series data to smooth out short-term fluctuations while highlighting longer-term trends. In all species selected for analysis, the count-wide trends tracked equivalent study area trends remarkably well.

BOUNDS

Our interchangeable terms *New York City area* and *New York City region* exclude New Jersey, various parts of which Griscom (1923), Cruickshank (1942), and Bull (1964) incorporated in their accounts. Our New York City area includes only New York City, Long Island, Westchester, and Rockland Counties unless specified otherwise, even though for discussion of breeding ranges and migration routes, it extends to about 75 smi (120 km) northwest to northeast of Van Cortlandt Lake and occasionally into Connecticut, New Jersey, or Pennsylvania.

In our species accounts, Cruickshank (1942) served as our distribution and abundance base owing to its breadth and depth of information about birds in the Bronx. Data from subsequent publications (Bull 1964, 1970, 1974, 1976; Levine 1998) are given only when they changed or updated Cruickshank's information.

The nominal cutoff date for species account information was 31 December 2016, but this was waived when circumstances warranted.

Taxonomy and nomenclature follow the 7th edition (1998) of the AOU's *Checklist of North American Birds* and its supplements through No. 56 in 2015, except that as do many non-AOU classifications, we treat each of the following as separate biological species: Eurasian and Green-winged Teals; Eastern and Western Willets; Myrtle and Audubon's Warblers; Savannah and Ipswich Sparrows; Red, Slate-colored, and Sooty Fox Sparrows; and Slate-colored, Oregon, and Pink-sided Juncos. In scientific name terminology, parentheses indicate disagreement among taxonomists about specific vs. subspecific status. Thus *Anas (crecca) carolinensis* indicates that some regard Green-winged (*carolinensis*) as a subspecies of Eurasian Teal (*crecca*), while others treat it as a full species.

Except as noted, all species are considered to have been continuously present (some only seasonally) in Van Cortlandt Park from 1872 to 2016; every species occurs annually and is common in migration, winter, or summer; and except as noted and excluding breeders, all species known from Van Cortlandt Park have also occurred in Central and Prospect Parks, whose seasonal daily maxima (none later than June 1967) are from Carleton (1958, 1970). No newer park-wide statuses for the birds of those parks exist.

The few available study area nest and egg dates are omitted because they are not recent and so should not differ materially from those already published for the entire New York City region. A few unique situations are discussed in appropriate species accounts.

SPECIES ACCOUNT FORMAT

NYC area, current. Describes in capsule form each species' current status in the New York City area as this book uses that term.

Bronx region, historical. Describes in capsule form each species' historical status in the various Bronx regions, back to Bicknell if data are available. Note that use of the phrase *as of* signals data cutoff dates for various sources, not their years of publication.

NYC area, historical. Describes in capsule form each species' historical status in the New York City area as this book uses that term.

Study area, historical and current. The core of every species account. Details each species' historical and then current status in the study area and its subareas. It may also include information about a species' current status in Riverdale, elsewhere in the Bronx, Central or Prospect Park, all New York City parks, New York City, Westchester County, Long Island, or the entire New York City area. Information on nearest breeders may include locations in Pennsylvania, New Jersey, or Connecticut.

Comments. Addresses such topics as closely related species that have occurred near but not in the study area; taxonomy; identification; banding and specimens; etc.

OTHER CONVENTIONS

Readers are referred to Appendix 3 for definition of terms, abbreviations, and initialisms we have used. The following usages also apply to the species accounts:

- Study area species names are underlined if breeding has ever been proved in the study area, and dash-underlined if breeding is only possible, probable, or likely—in the past or the near future.
- Van Cortlandt Park, Woodlawn Cemetery, Jerome Reservoir, and Hillview Reservoir are often reduced to Van Cortlandt, Woodlawn, Jerome, and Hillview, respectively.
- Van Cortlandt Park is always implied whenever a study subarea is omitted from a record, just as 1 is implied whenever a count is omitted.
- Inarguably redundant, the word *birds* has been excluded from all counts.
- A *resident* is present year-round and breeds; formerly called a *permanent* or *year-round resident*. The term may be modified by *winter* or *summer* to denote seasonal residency, and *summer* is occasionally further modified by *nonbreeding*.
- Extreme migration dates for common species present year-round are often meaningless and/or difficult to obtain, just as they are for late fall/early spring migrants that also overwinter locally and for late spring/early fall migrants that also breed locally.
- *Significant* when referring to population trends normally means *statistically* (or occasionally *biologically*) significant.
- *Maximum, maxima* denote highest single-day counts.
- *Incursion, irruption,* and *invasion* are not considered synonyms; see Appendix 3.
- Names of all cited observers are given in Appendix 4.

- All records for Allen are for *Deborah* Allen, for Friedman are for *Stuart* Friedman, for Horowitz are for *Joseph* Horowitz, for Kuerzi are for *John* Kuerzi, for Mayer are for *Paul* Mayer, and for Young are for *John L.* Young—except when these surnames are preceded by an initial indicating someone else.
- Out-of-state counties show both county and state abbreviations, but for well-known New Jersey locations like Cape May, the Great Swamp, the Hackensack Meadowlands, the Palisades, Sandy Hook, and Troy Meadows, New Jersey is always implied.
- For all nonbreeders recorded in the study area, locations are given for likely or known *nearest breeders* anywhere on Long Island or within 100 smi (160 km) of Van Cortlandt Lake in New Jersey, New York, or Connecticut. All counties are in New York State unless stated otherwise.
- Nearest breeders should always be taken to mean *nearest most recently known breeders from a variety of sources and time periods*. Some may be 50 years old, some only 1 year old; many derive from New York State Breeding Bird Atlases I (NYSBBA I) and II (NYSBBA II) in 1980–85 and 2000–2005, respectively.
- Even though *Bronx Park* is the formal name for the lands comprising both the New York Botanical Garden and the Wildlife Conservation Society (formerly the New York Zoological Society), this book preserves the tradition that refers to the former as *Bronx Park* and the latter as the *Bronx Zoo*.
- What we and others term Inwood Park is formally Inwood *Hill* Park, just as Jerome Reservoir is formally Jerome *Park* Reservoir, and in Yonkers Central Avenue is formally Central *Park* Avenue.
- The East Bronx landform has long been known as Throgg's Neck, but the bridge crossing it was named the Throgs [sic] Neck Bridge when built in the early 1960s.

SPECIES ACCOUNTS

Ducks, Geese, Swans: Anatidae

Fulvous Whistling-Duck *Dendrocygna bicolor*

NYC area, current
An erratic and perhaps increasing vagrant from populations in Florida (where introduced) and Louisiana to Texas that for uncertain reasons began to explode to the northeast as far as New Brunswick, Canada, in flocks to 30 in fall and winter 1961–62 and 1962–63. This continued every year until 1966, when it stopped as abruptly as it had begun. It resumed intermittently in the late 1960s until tapering off in the late 1980s–early 1990s, only to pick up once more in the early 2000s. There are now 15–20 records, the most recent in the early 2010s.

NYC area, historical
Bull's *NYC Area* as of 1968: flock of 6–8 in Dec 1962, then 6 records in flocks of 3–8 in 1965, single in 1966

Bull's *New York State* as of 1975: flocks of 11 in 1972, 6 in 1975

Levine's *New York State* as of 1996: flock of 18 in 1977

Study area, historical and current
The sole record is of a single in Van Cortlandt Swamp on 31 Oct–1 Nov 1965 (Norse, Stepinoff), part of an influx to the New York City area when between Apr 1965 and Dec 1966 there were 6 records, some involving small flocks. The second in the Bronx, in Pelham Bay Park from 12–16 Nov 1990 (Rodewald), was apparently killed by a Red-tailed Hawk (DeCandido 1991b). Unrecorded in Central or Prospect Parks.

Pink-footed Goose *Anser brachyrhynchus*

NYC and study areas, current
A recent European winter vagrant to the Northeast, mid-Atlantic states, and the New York City area. With greatly increased numbers of Greenland breeders, some Pink-foots have been following recently colonized Canada Geese back to their North American overwintering grounds rather than to northwestern Europe—exactly as Barnacle and Greater White-fronted Geese have been doing.

First recorded in North America when 1 was collected in MA in 1924, singles then overwintered in DE in 1953–54 and (the first in New York State) at Babylon, Long Island, in 1971–72 (unknown observers, *Kingbird*), followed by 1 in the late 1970s at Timber Point Golf Course, Heckscher State Park (fide Cooper), and at Middle Island from 16–31 Jan 1991 (Clinton, Ruscica et al.). All were deprecated as escapes even though each was embedded in presumed arctic Canada Goose flocks.

It was not until the first in NF in 1980 and QC in 1988 and 1989 that some voices began to suggest that these may have been genuine vagrants. The following (modern) regional firsts began to appear not long thereafter: PA 1997, CT 1998, MA 1999, VT 1999, NS 2005, NY 2007, RI 2007, ME 2009, NB 2010, NH

2011, NJ 2011, and MD 2011—the farthest south to date. Nearly all were singles but 2 appeared in NJ in 2016, 3 in ME in 2009, and 5 in NF in spring 1995 during a large fallout of Icelandic breeding species. Singles remain the rule, but increasingly (as in the winter of 2016–17 around New York City) multiple singles are scattered among various Canada Goose flocks, especially on Long Island. Unknown from Central or Prospect Parks but 1 moved between Kissena and Flushing Meadow Parks in Queens from 27 Dec 2008–12 Jan 2009 (E. Miller et al.).

There is but a single study area record, the first in the Bronx: Van Cortlandt Parade Ground 22–29 Dec 2016 (Fiore, Dolan et al.; photos). It is unrecorded in Westchester and Rockland Counties, but singles were in Orange County in Mar 2013 and fall 2016. Without doubt, increased scrutiny of arctic Canada Goose flocks will detect additional Pink-footeds elsewhere in New York State.

Greater White-fronted Goose   *Anser albifrons*

NYC area, current
A winter visitor/resident in very small numbers in the Northeast and the New York City area from the Greenland breeding population (*flavirostris*) that has been increasing markedly since the 1980s but that is known in the Bronx only from Van Cortlandt Park. Recorded in Prospect but not Central Park.

NYC area, historical
Cruickshank's *NYC Region* as of 1941: western vagrant, shot 5 times on Long Island 1846–89

Bull's *NYC Area* as of 1968: 2 recent Long Island records: Apr 1944, Dec 1948 in flocks of arctic Canadas; others in Feb, Mar–Jun deprecated as escapes

Bull's *New York State* as of 1975: wild status of most now accepted

Study area, historical and current
A recent winter resident first seen on the Parade Ground and then Van Cortlandt Lake on 28 Oct 2010 (McGee) when it had just begun overwintering with a large arctic Canada Goose flock. It was last seen on 26 Feb 2011 (Bochnik). The same bird overwintered in Van Cortlandt a year later from 11 Nov 2011 (Bochnik)–5 May 2012 (Van Doren et al.) but spent its time elsewhere in the winter of 2012–13 and then returned to Van Cortlandt from 1 Dec 2013–mid-Feb 2014 (DiCostanzo et al.) and perhaps also on 19–20 Feb 2017 (King et al.).

Comments
All study area individuals have been *flavirostris*, the Greenland breeding taxon comprising the vast majority of northeastern North America White-fronts.

Snow Goose   *Chen caerulescens*

NYC area, current
An increasing fall migrant and winter resident inland and recently in large numbers coastally. Spring flocks tend to move well west of the Hudson River.

Bronx region, historical
Bicknell's *Riverdale* 1872–1901: unrecorded

Eaton's *Bronx + Manhattan* 1910–14: rare migrant

Griscom's *Bronx Region* as of 1922: unrecorded, but once in Croton Point area

Kuerzi's *Bronx Region* as of 1926: no Bronx records

NYC area, historical
Cruickshank's *NYC Region* as of 1941: on Long Island uncommon to rare in spring/fall (maximum 700) migration, extremely rare in winter; 24 Sep–27 Apr; away from coast often unnoticed in spring/fall migration. 3 Bronx records, once each in Van Cortlandt and on the Hudson

Bull's *NYC Area* as of 1968: increased in numbers during spring/fall migration, but also increased numbers of observers; 21 Sep–24 May; Long Island maxima fall 5500+, spring 700; no overwintering flocks, per site maximum 4

Bull's *New York State* as of 1975: status unchanged

Study area, historical and current
A regular migrant as hundreds pass overhead north- and southbound, even though most do so unnoticed; nearly all are Baffin Island-breeding white-morph *atlantica* (Greater Snow Goose). Occasionally a few or a small flock will alight and 1–2 sometimes remain for short periods in all 4 extant subareas.

In fall, first migrants arrive in early Oct, peak in late Oct–early Nov, and depart by mid-Nov, with extreme dates of 10 on 8 Oct 1999 (Garcia) and 2 on 30 Nov 2011 (Kravatz) and flocks to 40 on several occasions, with a maximum of 120 on 4 Oct 2000 (Young). One on Jerome 27 Oct–10 Nov 1974 (Sedwitz) was the only blue-morph *caerulescens*. Until the late 1970s all Blue Geese were by definition *caerulescens* (Lesser Snow Goose) and most still are, but blue alleles now occur in *atlantica* (Greater Snow Goose) and also in Ross's Goose, though quite rare in both.

We know of 13 winter records: 1 may have overwintered on Jerome in 1964–65 (Enders et al.); Jerome 16 Jan 1966 (Zupan, Enders); Van Cortlandt 27 Dec 1970 (Buckley, Sedwitz); 29 Dec 2006 (Young); 2 from 25 Nov–12 Dec 2011 (Scully, Kravatz et al.); 8 Dec 2012 (Baksh); 2 from 20–28 Dec 2014 (McGee et al.); 12 Dec 2015 (McGee); 20 over Woodlawn on 27 Dec 2015 (Gotlib et al.); 16–17 Jan 2016 (Fung et al.); 22 Dec 2016–20 Jan 2017 (Fiore et al.); 13 over Jerome on 12 Dec 2017 (Ward); 120 over Hillview on 14 Dec 2017 (Camillieri).

There are only 3 spring records: Van Cortlandt Lake on 24 Apr 1957 (Mayer), Woodlawn on 12–13 Mar 1966 (Horowitz et al.), and, quite late, on the Parade Ground on 30 May 1994 (Lyons). The 200 over the Bronx Zoo on 23 Mar 2016 (Olson) and the 250 over Riverdale on 27 Mar 1982 (Sedwitz) were during the typical peak northbound flight period, when most migrate well to the west of the Hudson River.

Brant   *Branta bernicla*

NYC area, current
A common but local winter visitor/resident and migrant, restricted to bay and sound saltmarshes as close as Pelham Bay Park. Thousands move down the Hudson River and Long Island Sound from late Oct–late Nov, and in spring migration overfly the study area by the hundreds on nights in May when they vocalize noisily, but few ever alight in any season.

Bronx region, historical
Bicknell's *Riverdale* 1872–1901: unrecorded

Eaton's *Bronx + Manhattan* 1910–14: common migrant

Griscom's *Bronx Region* as of 1922: 2 singles Croton Point area; no other records

Kuerzi's *Bronx Region* as of 1926: rare spring/fall migrant, winter visitor/resident (14 Nov–17 Apr)

NYC area, historical
Cruickshank's *NYC Region* as of 1941: uncommon migrant, rare winter resident coastally; irregularly strays to western Long Island Sound in Bronx, seen twice on Hudson in Riverdale

Bull's *NYC Area* as of 1968: greatly increased in numbers; now regular spring migrant in large flocks in Hudson Valley; Long Island maxima winter 10,000–20,000, spring 10,000, summer 40

Bull's *New York State* as of 1975: Long Island maxima fall 10,000, summer 400

Levine's *New York State* as of 1996: Long Island winter maximum 48,000

Study area, historical and current
A now nearly annual spring migrant first recorded on the Van Cortlandt Park Parade Ground 19 Apr 1932 (L. N. Nichols) but not again until several hundred were calling low overhead on the evening of 25 May 1960 (Buckley)—and the next day all had left Jamaica Bay. If this time gap is genuine, it might reflect the plunge in East Coast Brant numbers following the fungal blight that attacked Eelgrass, Brants' formerly preferred marine plant food, in 1931. They were occasional but scarce on the Hudson River in Riverdale, with 21 there on 4 Jan 1932 (Cruickshank), a high count.

In fall, first migrants arrive in early Oct, peak in late Oct–early Nov, and depart by late Nov, with extreme dates of 2 Oct 2001 and 13 Nov 2010 (Baksh) and a maximum of 110 on 13 Oct 2009 (Young). They are most often seen over Van Cortlandt but occasionally also Hillview.

There are few winter records of flyovers: 28 Dec 2003 (Lyons et al.), 125 on 8 Jan 2011 (Baksh), 9 on 19 Feb 2011 (Baksh), 12 on 26 Feb 2011 (Klein), and 15 on 11 Jan 2014 (Souirgi).

In spring, first migrants arrive in Mar, peak in mid–late May, and depart by late May, with extreme dates of 4 Mar 1999 on Van Cortlandt Lake (Chavez) and 31 May 2003 (Young). Since the early 1960s, hundreds routinely pass overhead day and night in Apr–May, especially toward the end of May, when Jamaica Bay winter residents are departing. The maximum is 1375 on 21 May 1984 over Van Cortlandt/Riverdale (Sedwitz).

A few dozen summer every year on Long Island Sound from Pelham Bay Park to the Whitestone Bridge, but the only West Bronx flock was 3 in Riverdale 30 Jun 2005 (unknown observer, *Kingbird*).

Brant rarely alight on the Parade Ground, and if so only singly and usually departing quickly. Exceptional single southbound juveniles feeding on the grass for extended periods include 16–26 Nov 2003 (Garcia, Lyons, Künstler) and 13 Oct–1 Nov 2007 (Young), and northbound on Van Cortlandt Lake from 4–13 Mar 1999 (Chavez, Künstler, Young).

Barnacle Goose   *Branta leucopsis*

NYC area, current
A European winter vagrant to the Northeast and the New York City area that has been increasing markedly since 2000. Several are now recorded around the New York City area every winter, nearly always singles. It is known from Central and Prospect Parks, but there are only 3 Bronx records.

NYC area, historical
Cruickshank's *NYC Region* as of 1941: 6 Long Island singles in Mar, Oct, Nov 1876–1926, of which 2 deprecated as escapes

Bull's *NYC Area* as of 1968: wild status of many now accepted

Bull's *New York State* as of 1975: status unchanged; Mar 1975 on Long Island

Study area, current
A single study area record: Van Cortlandt Parade Ground 23 Nov 2012–1 Feb 2013 (Mako et al.; photos), after which it moved to Larchmont Reservoir, Westchester County—doubtless the same bird that was first at Inwood Park 11–13 Nov 2012 (Barrett, B. Purcell et al.). Probably the same individual was refound in Mar 2013 on reservoirs in Eastchester/Larchmont. However, this was obviously *not* the one that had been banded on Islay in Scotland's Inner Hebrides on 13 Nov 2002 and that showed up at the Orchard Beach parking lot on 26–27 Nov 2010 (Michael, Rothman). Alas, it too then moved—north, to spend the 2010–11 winter along the CT coast. Another was at Randall's Island (Triboro Bridge) on 25 Jan 2014 (Auerbach) and subsequently found its way onto Central Park Reservoir. A single with Canada Geese at the Orchard Beach parking lot on 5 Jan 1969 (Stepinoff) was deprecated in Bull (1974) "on good authority" as an escape, but because details supporting that decision have never surfaced, this first Bronx record stands. Yet another was at Piermont Pier on 20 Dec 2009 (Ciganek, Knoecklein). With greatly increased numbers of Greenland breeders, it is believed that Barnacles have been following colonizing Canada Geese there back to their North American overwintering grounds rather than to northwestern Europe, where most Barnacles overwinter.

Cackling Goose *Branta hutchinsii*

NYC area, current
Before it was split from Canada Goose in 2004, it was generally regarded as a vagrant that was identifiable only from specimens. Yet by the late 1990s critical observers were regularly finding them in the single digits on eastern Long Island in large flocks of arctic-breeding Canada Geese. After 2004, when searching for them was sanctioned by their split from Canada Goose, they were detected throughout the New York City area, usually singles but occasionally family groups of 3–6, and they often overwintered. This is their current status, which may also have been true, at least in part, before 2004. Recorded in Prospect Park but not Central, although that will change.

NYC area, historical
Cruickshank's *NYC Region* as of 1941: annual reports of small Canada Geese, but absent specimens all deprecated as unidentifiable

Bull's *NYC Area* as of 1968: no specimens; all observations deprecated as unidentifiable

Bull's *New York State* as of 1975: single recent specimen and photograph; all observations deprecated as unidentifiable

Study area, historical and current
Currently an infrequent but increasing Van Cortlandt Parade Ground winter visitor and resident only recently detected. Because it had little or no profile with New York City area observers until its split from Canada Goose, it may have been occurring in the study area for decades in very small numbers—overlooked, ignored, or deprecated as unidentifiable by regional compilers.

All study area records date from the Bronx's first in Van Cortlandt Park on 10 Dec 1996 (Young), followed by 2–6 Jan 2006 (Young), 14 Jan 2007 (Bochnik), 12 Mar 2011

(Mitra et al.), 25 Nov 2012 (Young)–11 Feb 2013 (Russ), with 2 there from 5–11 Jan 2013 (Baksh et al.), 6–16 Nov 2013 (Allen, Souirgi et al.), and in 2014, 2–3 from mid-Nov (m. ob.) to 25 Dec (Baksh), peaking at 6 on 29–30 Nov (Dancis et al.)—the Bronx and study area maximum—after which it has been seen there each winter.

Otherwise, in the Bronx it has been recorded at the Bronx Zoo from 13 Nov–16 Dec 2016 (Haluska, Olson) and on 28 Nov 2017 (Olson), and at Pelham Bay Park on 17 Feb 2005 (DiCostanzo), possibly the same individual that overwintered from 20 Nov 2005–9 Mar 2006 (Ott et al.); a noisily calling adult with 2 HYs that landed on the Orchard Beach parking lot on 30 Oct 2010 (Buckley); and 2 more there on 24 Oct 2015 (Benoit et al.).

All but 1 of the study area Cackling Goose records postdate the split because only since then have most observers been routinely scrutinizing flocks of wild Canadas for Cacklers. Still, there has been a genuine increase in *hutchinsii* in the Northeast, especially in extreme western New York State, where flocks of 20–40 are now being seen, with a photographed flock of 33 as close as Bridgewater, NJ, on 2 Jan 2012. The very large flocks of arctic Canadas on the Parade Ground each winter have appeared only since about 2010, and they will continue to repay close examination as long as they continue.

The recent increase in New York City area records is real and only coincidentally follows the split. Careful observers over the years saw almost none (ignoring the intense discouragement of their identification) until the mid-1990s despite painstaking dissection of countless arctic Canada Goose flocks.

Comments
Several times in late fall and winter since 2000, observers have reported larger Cackling Geese on the Van Cortlandt Parade Ground that seemed not to be *hutchinsii*, the expected eastern subspecies. Some may have been photographed. But despite statements or belief that these individuals were identifiable as a more western subspecies, *taverneri*, this conclusion is not tenable. The last formal treatment of Cackling Goose subspecies dates to the 5th edition of the *AOU Checklist* (1957). Not only was this long before the Canada–Cackling Goose split, but it also could not have anticipated the enormous advances in our knowledge of the arctic tundra distribution of the various Cackling subspecies. It is not inaccurate to assert that understanding of Cackling Goose subspecies—how many; their names; their identification; their breeding distribution, migration routes, and overwintering areas—is in disarray. Meantime, observers can familiarize themselves with the identification material on Cackling Goose subspecies (plus *parvipes* Canada Goose) admirably presented in Mlodinow et al. (2008)—but only with the strong caveat that what was known when that paper was written may be substantially altered in the near future. See also Comments in the Canada Goose account, next.

Canada Goose   *Branta canadensis*

NYC area, current
Two discrete and presumably non-interbreeding populations occur: wild, wary migratory breeders from the arctic that until recently were found largely on the north and south forks of eastern Long Island from Nov–Mar; and a feral, "nonmigratory" population derived from midcontinent breeders and resident throughout

the entire New York City area, breeding in New York City parks, on golf courses, and at industrial parks. These were established in the early 1930s by escapes from waterfowl collections and from introduction by various governmental waterfowl biologists that unfortunately continued in some areas into the 1960s.

Bronx region, historical

Bicknell's *Riverdale* 1872–1901: overflying every spring/fall, flocks alighting only twice

Eaton's *Bronx + Manhattan* 1910–14: common migrant

Griscom's *Bronx Region* as of 1922: rare migrant occasionally overhead, seldom alighting; no longer common in migration along Hudson

Kuerzi's *Bronx Region* as of 1926: uncommon migrant in spring (27 Feb–12 May)/fall (5 Oct–28 Dec), infrequently alighting

NYC area, historical

Cruickshank's *NYC Region* as of 1941: arctic breeders common coastal migrants, rare in winter; inland uncommon flyovers, rarely alighting in Mar/Nov; feral breeders ubiquitous

Bull's *NYC Area* as of 1968: Long Island wild maxima fall 4000, winter 5000, spring 8000, 12,000

Bull's *New York State* as of 1975: status unchanged

Study area, historical and current

A spring/fall migrant and winter resident, and individuals from wild arctic and feral populations occur. Only the latter are present year-round, augmented by other New York City area ferals in winter. In the past, small flocks of wild Canadas rarely stopped on the Parade Ground during migration, perhaps owing to a heavy human presence. And they never alighted or overwintered in 4-digit flocks until 2010, when they swarmed over and devoured the newly installed Parade Ground turfgrass monoculture that had replaced the rougher sod's scattered native grasses and forbs. Since then, 4-digit wild flocks have become regular, with maxima in fall of 2400 on 25 Nov 2013 (Baksh) and in winter of 3650 on 20 Jan 2013 (Bochnik). These arctic breeders (most belonging to *interior*: Fox et al. 2012) usually arrive in mid–late Oct, peak in Dec–Jan, but depart with the first significant snow cover. In mild years they may remain until early Mar when they abruptly decamp for the arctic.

On 15 Jan 2009 the famous jet that was forced to make an emergency landing in the ice-choked Hudson River off lower Manhattan had to do so because its engines had been damaged shortly after takeoff when it encountered a flock of Canada Geese. In response, the Port Authority of NY and NJ, which operates JF Kennedy, Newark, and LaGuardia airports, began an intensive goose eradication program that was fully operational by the fall of 2010. LaGuardia is only 7.7 smi (12.4 km) south of Van Cortlandt, and it is tempting to conjecture that the unprecedented arrival in Nov 2010 of the first-ever 4-digit flocks of wild Canada Geese on the Parade Ground was related to LaGuardia's bird control activities. Supporting this idea was the presence with them of multiple Cackling and single Pink-footed, Greenland Greater White-fronted, and Barnacle Geese. Pointedly, these species had been occurring with wild Canadas in preceding winters at Flushing Meadows, Kissena, and other parks close to LaGuardia.

As of 31 Dec 2015, the sole study area banding recovery of a wild arctic goose was on 18 Jan 2007. It had been banded as an HY on 5 Aug 2003 on the east side of James Bay, Ontario, Canada and thus also belonged to *interior*. Some in the recent large Parade Ground flocks were reputedly wearing neck collars affixed on Greenland, but information from Greenland banders does not support this belief (A.D. Fox, pers. comm.). See the discussion of neck-collared geese's origins in the Avifaunal Overview "Bird Banding" section.

Griscom, writing about his Bronx Region in 1923, called Canada Goose a rare migrant seldom alighting, with fall dates of 9 Oct 1915 (Hix, L.N. Nichols)–22 Dec 1909 (Griscom, La Dow) and only 2 in spring, on 13 Mar 1915 and 15 Mar 1926 (Nichols, Nichols), all singles. Only 4 years later, Kuerzi (1927), writing about *his* Bronx Region, termed Canada Goose an uncommon migrant, infrequently alighting with fall dates of 5 Oct 1925–2 Dec 1919 (all Coles in southern Westchester County) and spring dates of 27 Feb 1922–12 May 1916 (again Coles), all singles given that Kuerzi pointedly mentioned 20 sitting on Hudson River ice in Riverdale on 2 Jan 1893 (Bicknell). All would have been wild arctic breeders.

Cruickshank (1942) did not discuss study area or Bronx Canada Goose numbers, so we have no later published information on their frequency of occurrence, but by the mid-1930s they were being seen in spring/fall migration in small flocks—again, rarely alighting, so the 2 on the Parade Ground on 1 Apr 1932 (L. N. Nichols) were deemed reportable. Likewise, in fall 1950 small flocks overflying the study area—32 on 12 Oct, 275 on 15 Oct, and 85 on 26 Oct (Komorowski)—warranted mention, although it is uncertain whether by then they were (all) wild arctic breeders (see below). This pattern continued through the next 50 years, with slight increases in numbers overflying, but alighting was unusual. Until Jun 1967, the spring maxima in Central and Prospect Parks were, respectively, 200 on 25 Apr 1953 (Skelton) and 181 on 13 Apr 1944 (Nathan, Soll)—well before New York area feral populations of consequence existed.

Feral Canadas in the Northeast (the Atlantic Flyway Resident Population of waterfowl biologists) are commonly believed to derive from several sources to various degrees: releases/escapes from waterfowl collections and hunters' live decoys, misguided state and federal stocking efforts, and occasional wild migrants lured into oversummering by feral residents. US Department of Interior Breeding Bird Surveys (DOIBBS) data for this species in New York State showed a striking increase in breeders beginning about 1985 and continuing unabated (Sauer et al. 2007; McGowan and Corwin 2008). NYS–DEC waterfowl biologists estimated the New York State resident population at 5000 pairs in 1978, 20,000 by 1990, and 90,000 by 2005 (McGowan and Corwin 2008). It is derived from a mélange of mostly midcontinent *maxima* (the so-called Giant Canada Goose), some northeastern breeders (*canadensis*), and perhaps others. Layered atop these mixed populations are the extensive intraregional movements by all ferals, facilitated by their collective molt-migrations north into the low arctic, where they mix with local breeders while they all undergo group molting.

The spring Canada Goose maximum is 379 on 12 Mar 2011 (Baksh), but this count may involve only ferals. Most ferals depart in late May–early Jun, long after wild arctic breeders have left in Feb–Mar, and return in Sep–early Oct well before wild birds in late Oct–Nov. Thus, the 700 in Van Cortlandt on 14 Sep 2013 (Souirgi) would have been ferals returning from their molt-migration.

The first study area breeding feral geese appeared in Van Cortlandt about 1932 (Kieran, Sedwitz, Norse), and by 1935 1 or 2 pairs were nesting (Weber et al.), continuing to 2016 (Buckley, Young et al.). During the 1930s Canada Geese were not resident in the study area even though breeding; 0–5 was normal, maxima rarely exceeded 10, and 25 generated comment. In many years none were found on the Bronx-Westchester Christmas Bird Count, but in the early 1960s ferals began to occur in increasing numbers, augmented 40 years later by arctic breeders (Fig. 21).

Figure 21

By the early 1940s numbers were increasing slightly, and 1942 was the last year Canada Goose was missed on the entire count. In the 1950–60s, fall/winter Van Cortlandt maxima began to increase, rarely reaching 200, with 20–40 normal (Buckley, Kane, Enders). From the 1970s onward, feral numbers began to increase rapidly to a fall maximum of 800 in Nov 1972 (Young) and by 2013 to a late summer maximum of 250 when local young of the year were present. Feral flocks have long moved daily between Van Cortlandt, Woodlawn (where 2–3 pairs were also nesting: Horowitz), and Jerome, occasionally reaching 200+ in winter. In winter they usually flew into Van Cortlandt Swamp for roosting at night, a behavior that continues to this day, but now ferals are sometimes accompanied by wild arctic breeders.

Comments
A few times in late fall and winter since 2000, observers have reported Canada Geese on the Van Cortlandt Parade Ground they suspected or believed belonged to a smaller western subspecies, *parvipes*. Some may have been photographed, but this subspecific identification is not tenable. As with Cackling Goose (see above), understanding of Canada Goose subspecies—how many; their names; their identification; their breeding distribution, migration routes, and overwintering areas—is in disarray. In particular, some experts doubt the identification of all eastern specimens of *parvipes*, including the 1 from the North Jersey shore that had been claimed as the only New York City area record of this taxon. Given this confusion, that body size in Canada (and Cackling?) Goose can be drastically affected by diet, age, and sex, and that Canada and Cackling may hybridize in the wild in the arctic, subspecific identification of vagrant odd-sized Canada Geese is a bad idea. Photographic documentation of such birds for eventual analysis by goose experts is the best we can hope for. See also Comments in the Cackling Goose account, above.

As counts of wild Canada Geese with Greenland connections increased in the study area, with them have come such great rarities away from Long Island as Greater White-fronted, Cackling, and Barnacle Geese. This being the case, we anticipate the future Parade Ground occurrence of **Ross's Goose**, *C. rossii*, which has been appearing annually for some time in small but growing numbers in the New York City area, including the first in the Bronx at the Orchard Beach parking lot 14–21 Jan 2017 (Aracil et al.; photos), possibly the one seen at adjacent Tibbett's Brook Park on 5 Feb 2017 (Ellen; photos). It has not yet been recorded in Central or Prospect Parks.

Mute Swan   *Cygnus olor*

NYC area, current
An all-too-well established exotic (nonnative) in the New York City area, especially on Long Island, that grew from releases at what is now Connetquot State Park in Oakdale, Long Island, and in Southampton, around 1910. Others may have escaped from waterfowl fanciers at various times to augment this burgeoning population.

Bronx region, historical
Bicknell's *Riverdale* 1872–1901: unrecorded

Eaton's *Bronx + Manhattan* 1910–14: unrecorded

Griscom's *Bronx Region* as of 1922: unrecorded

Kuerzi's *Bronx Region* as of 1926: unrecorded

NYC area, historical
Cruickshank's *NYC Region* as of 1941: several hundred Long Island pairs concentrated in Great South Bay, Hamptons, whence they wander through New York City area

Bull's *NYC Area* as of 1968: established throughout but most numerous on Long Island; maximum 500 Mecox Bay Dec 1959

Bull's *New York State* as of 1975: status unchanged

Levine's *New York State* as of 1996: on Hudson 216 released at Rhinebeck in 1910, followed on Long Island 328 at Southampton, Oakdale in 1912

Study area, historical and current
A recent resident breeder extirpated in 2009 but that returned in 2017. Occasional others may wander in and out of the study area at any time of the year.

Until the late 1950s Mute Swan was unrecorded in the study area but for a single record: Hillview on 14 Dec 1935 (Norse, Cantor). The next was the first in Van Cortlandt, from 5 Mar–16 Apr 1958 (Mayer, Rafferty et al.), and the third not until 6 May 1965 (Norse). Only by late 1965 did singles and occasionally pairs begin to appear regularly in Van Cortlandt and at Jerome. Between Oct and Mar Van Cortlandt Lake residents move to Jerome, Spuyten Duyvil, or the Hudson River during freeze-ups.

The first and for many years the only Bronx breeding occurred at Baxter Creek in 1954 (Pappalardi, Russak), following first reports of adults there on 5 May 1951. In the study area, 1–2 were more or less resident on Van Cortlandt Lake (moving to Jerome occasionally even in warm weather) from Apr 1980, but not until 1982 did a pair finally attempt to nest although unsuccessfully. Then in 1983 they brought off 4 young and since then nested regularly if intermittently (Sedwitz 1985), until the recent winter population explosion of arctic Canada Geese that drove them away after nesting failure in 2008 and 2009. They next attempted breeding in 2016 but failed, and in 2017, when they succeeded (Young). The maximum study area count is 8 at Hillview on 10 Jan 2017 (Camillieri). They have been intentionally harassed out of Woodlawn and perhaps at both reservoirs. Nearest recent breeders have been at Pelham Bay Park, at numerous locations along the East River—and possibly also on the Hutchinson, Bronx, and Harlem Rivers—and on North Brother Island.

On the Bronx-Westchester Christmas Bird Count (see Fig. 22) they were all but unrecorded until the early 1950s–60s, when their numbers increased briefly, dropped back again in the early '70s, peaked at 275 in the early '90s, then dropped again to 100–150, where they remain.

DOIBBS data for this species in New York State showed a slow but steady increase beginning about 1985 but leveling off in the early 2000s (Sauer et al. 2007; McGowan and Corwin 2008). Alas, NYSBBA II in 2000–2005 found

Figure 22

this species to have increased its number of occupied blocks by 87% since 1980–85 (McGowan and Corwin 2008), and it continues to spread within the New York City area.

Comments
A few native **Tundra Swans**, *C. columbianus*, overfly the New York City area annually but are infrequent in the Bronx, with no study area records. However, 1 was as close as Spuyten Duyvil on 14 Mar 2014 (Gaillard), and in an unprecedented late Feb 2018 fallout of multiple hundreds in an arc from Jamaica Bay north and west to Westchester Co., only a flock of 6 going west over the Zoo on 25 Feb (Olson) made it to the Bronx. Historically, the only others have been 5 at Baxter Creek on 20 Nov 1933 (Kuerzi, Kuerzi), 1 at Pelham Bay Park on 24 Nov 2002 (Collier), 6 at Pelham Bay Park 7 Mar 2004 (unknown observer, *Kingbird*), and 6 at Pelham Bay Park 4 Nov 2005 (Bourque). In recent falls and winters it has been regular as close as Croton Point. With New York State now having an established population of introduced **Trumpeter Swans**, *C. buccinator*, that are showing up in the New York City area (already found in Central but not Prospect Park), this is another reason why it can never be assumed that that every Bronx swan is a Mute.

Wood Duck   *Aix sponsa*

NYC area, current
A widespread and locally very common spring/fall migrant and a very local but widespread breeder and scarce winter visitor/resident. Continues to breed at multiple locations in the Bronx, Brooklyn, Queens, and on Staten Island.

Bronx region, historical
Bicknell's *Riverdale* 1872–1901: common spring/fall migrant, breeder; single Jan record

Eaton's *Bronx + Manhattan* 1910–14: uncommon migrant, rare breeder

Griscom's *Bronx Region* as of 1922: Van Cortlandt Swamp best location in New York City area; still breeds regularly; maximum 40

Kuerzi's *Bronx Region* as of 1926: fairly common spring/fall migrant (13 Feb–23 Dec), decreasing as a breeder

NYC area, historical
Cruickshank's *NYC Region* as of 1941: fairly common spring/fall (maximum 200) migrant, rare but regular winter visitor, maximum 18; few still nest within New York City

Bull's *NYC Area* as of 1968: status unchanged

Bull's *New York State* as of 1975: status unchanged

Study area, historical and current
A longtime breeder, common migrant, and frequent winter visitor/resident in Van Cortlandt and Woodlawn, occasionally appearing on the reservoirs after severe freezes.

In spring, first migrants arrive in mid-Mar, peak in late Mar–early Apr, and depart by early May, with probable early arrivals in Woodlawn on 12 Feb 1977 (Teator) and in Van Cortlandt on 13 Feb 1925 (Cruickshank), with a maximum of 16 on 19 Mar 1932 (Cruickshank).

Wood Ducks have bred in Van Cortlandt Swamp since the first study area records in the early 1870s and do so routinely in other scattered and overlooked sites throughout Van Cortlandt and Woodlawn. They are sometimes seen perched high up in trees in the middle of the woods, as on the Ridge. If they routinely nest in such locations, their breeding population would be easily underestimated. They bred in the Sycamore Swamp in 1964 (Russak), 5 were seen there in Apr–May 2006 (Klein) and May 2014 (Buckley, Young). Their center of abundance has always been Van Cortlandt Swamp, where 5 pairs bred in 1936–38 (Sialis Bird Club) but only 2–4 pairs since the 1960s.

Within New York City, Wood Ducks remain very local and both NYSBBA I in 1980–85 and II in 2000–5 found them only on Staten Island, in the Bronx, and in Queens with only minor shifts in location (McGowan and Corwin 2008). DOIBBS statewide data showed no significant changes between 1966 and 2005 (Sauer et al. 2005).

In fall, first migrants arrive in late Aug, peak in Oct, and depart by early Nov, with extreme dates of 17 on 27 Aug 2013 (Young) and 14 Nov 1996 (Young), and maxima of 48 on 3 Oct 1981 (Sedwitz) and 65 on 21 Oct 1951 (Linnaean Society field trip).

The success of nearly annual attempts at overwintering by 1–10 individuals (maximum 14 on 9 Dec 1928: Cruickshank) varies with winter severity and heavy freezes, after which the ducks sometimes move to Hillview and Jerome. Singles/pairs also attempt to overwinter irregularly at Woodlawn but are quickly frozen out. In the record-breaking cold winter of 1933–34 a group of 18 did overwinter at the Bronx Zoo Wildfowl Pond (Cruickshank et al.).

Gadwall   *Anas strepera*

NYC area, current
Formerly a scarce migrant in very small numbers on Long Island until its unexpected increase followed by first Long Island and third East Coast breeding at Jones Beach in 1947 (Sedwitz et al. 1948). Since that beachhead, it has increased greatly in numbers and as a breeder throughout the New York City area, although it favors tidewater sites.

Bronx region, historical
Bicknell's *Riverdale* 1872–1901: unrecorded

Eaton's *Bronx + Manhattan* 1910–14: rare migrant

Griscom's *Bronx Region* as of 1922: unrecorded

Kuerzi's *Bronx Region* as of 1926: unrecorded

NYC area, historical
Cruickshank's *NYC Region* as of 1941: uncommon but annual Long Island fall migrant, winter resident (25 Aug–24 May, maximum 50); elsewhere very rare migrant, almost never in winter

Bull's *NYC Area* as of 1968: 25 Aug–24 May; uncommon to rare except on Long Island, where winter maxima now 77, 100; first bred 1947 Jones Beach (maximum 20 pairs), 1961 Jamaica Bay Wildlife Refuge (maximum 3 pairs); Long Island winter maxima 130–150; now breeding 7 sites with 200+ pairs

Bull's *New York State* as of 1975: 28 pairs Jamaica Bay Wildlife Refuge, 100 pairs Gardiner's Island

Study area, historical and current
An increasing spring/fall migrant, winter resident, and summer resident that must be considered a potential breeder given that it nests nearby in New York and NJ. Yet it is puzzling that no study area count has ever exceeded 9. Unrecorded in Central Park until 1954 and as of Jun 1967 never in Prospect Park; they are now numerous in both places and multiple pairs are remaining into Jun (without breeding) in Central.

Until the 1950s, they were seen exceedingly infrequently in the study area. The first were 2 on 21 Apr 1922 (R. Friedman); then there were none until singles at Hillview on 6 Mar 1938 (Norse) and 23 Oct 1938 (B. Gell-Mann); in Van Cortlandt on 13 Mar and 28 Sep 1951 (Kane) and 21 Jun 1952 (Buckley). They finally became annual in very small numbers in the mid-1950s, beginning with 2 or more moving between Van Cortlandt and Jerome, occasionally Spuyten Duyvil, 11 Sep–31 Dec 1955 (Kieran, Buckley et al.); then on Van Cortlandt Lake 20 Sep 1963 (Buckley). By 1967 they were being seen annually in fall and early winter in Van Cortlandt and Jerome, but usually only 2–5 at a time, peaking at 7 on Jerome from 27–31 Oct 1971 (Sedwitz).

The fall flight is light, beginning in mid-Sep and peaking in Oct–Nov, with the earliest on 18 Sep 1998 (Young) and study area maxima of only 9 in Van Cortlandt on 29 Nov 2014 (Baksh) and 7 at Jerome 27–31 Oct 1971 (Sedwitz). After mid-Nov they prefer Jerome to Van Cortlandt but do not overwinter, the latest date for the former site being 6 on 20 Dec 1967 (Sedwitz), and for the latter 2 on 17 Feb 2016 (McGee). During the early 1990s their already small numbers dropped even lower, and it is now unusual to see more than 1–2 at Jerome or Van Cortlandt anytime between Sep–Apr. Recently, there has been a light spring movement in Apr, the latest date being 2 (the maximum) in Van Cortlandt on 28 Apr 2012 (Baksh), but see below.

Well illustrated by NYSBBA I in 1980–85 and II in 2000–2005, Gadwall's New York State breeding clusters are on Long and Staten Islands in the southeast, along the St. Lawrence River in the northwest, and inland in large isolated upstate marshes like Montezuma, Iroquois, and Oak Orchard NWRs (McGowan and Corwin 2008). In addition to this distribution, a few pairs may also breed somewhere in the Northeast Bronx. NYSBBA I in 1980–85 and II in 2000–2005 found 1–2 pairs breeding in the Southeast Bronx on North and South Brother Islands but not yet in/near Pelham Bay Park. Singles in Van Cortlandt on 5 Jun 1980 (Ephraim), 15 Jun 2000 (Jaslowitz), 9 Jul 2006 (Delacretaz 2007), 12 May 2012 (Baksh), 24 May 2015 (Bochnik) and 12 Aug 2015 (Hudda)—in the Northwest Forest!—may have been local area breeders or prospectors. In the 1990s hundreds of pairs were breeding in the Hackensack Meadowlands, and many still do (Walsh et al. 1999), with smaller numbers at several Staten Island locations for years. They also breed (irregularly) not far to the northwest of the study area at the Piermont Marshes, so breeding expansion into the Northwest and Northeast Bronx is anticipated. Multiple pairs in Central Park in recent summers are most likely breeders from nearby East River islands.

Eurasian Wigeon   *Anas penelope*

NYC area, current
Greenland and Iceland breeders are seen annually on Long Island in spring/fall and sometimes overwinter. Recorded with some regularity in Prospect Park but in Central Park only between 1951–56. Far less common in the

Bronx, where it has been seen at Pelham Bay Park, the Bronx Zoo, and Van Cortlandt. Until fairly recently ♀♀ and eclipse ♂♂ were considered inseparable from American Wigeons, and many have been overlooked.

Bronx region, historical

Bicknell's *Riverdale* 1872–1901: unrecorded

Eaton's *Bronx + Manhattan* 1910–14: unrecorded

Griscom's *Bronx Region* as of 1922: pair Pelham Bay Park 9 Feb 1922 only record

Kuerzi's *Bronx Region* as of 1926: status unchanged

NYC area, historical

Cruickshank's *NYC Region* as of 1941: Long Island spring/fall migrant, winter resident (12 Sep–29 Apr; maximum 14 ♂♂); elsewhere very rare Mar/Oct migrant

Bull's *NYC Area* as of 1968: 4 Sep–14 May; less common than earlier, no maxima >2 >1952

Bull's *New York State* as of 1975: late spring date 1 Jun

Study area, historical and current

An infrequent fall/spring migrant, usually with American Wigeon.

Only 13 records exist, scattered between Oct and Apr: 19 Nov 1927, 28 Mar 1943, 17 Nov 1943, 21 Mar 1952, 6 Nov 1952, 9 Oct 1954 (all Kieran); 21–24 Jan 1944 (Eisenmann, Komorowski); 25 Mar 1944 (Norse); 2 on 23 Oct 1955 (Birnbaum); 13 Apr 1956 (Mayer); 2 on 29 Jan 1983 (Ephraim); 11 Nov 1990 (Young); 4 Oct 1992 (Young). All were ♂♂ but some were accompanied by likely ♀♀, formerly regarded as unidentifiable but not at all difficult if seen reasonably well by experienced observers. The Central Park maximum was 3, also ♂♂.

American Wigeon   *Anas americana*

NYC area, current

An often abundant spring/fall migrant, and a regular winter visitor and resident, most numerous coastally.

Bronx region, historical

Bicknell's *Riverdale* 1872–1901: unmentioned

Eaton's *Bronx + Manhattan* 1910–14: irregular migrant

Griscom's *Bronx Region* as of 1922: wild migrants reported on Bronx Zoo Wildfowl Pond; no other records

Kuerzi's *Bronx Region* as of 1926: uncommon migrant in spring (28 Feb–21 Apr)/fall (30 Jul–30 Nov)

NYC area, historical

Cruickshank's *NYC Region* as of 1941: Long Island spring/fall migrant, winter resident (30 Jul–9 Jul; maximum 100+); away from Long Island spring/fall migrants recorded more frequently than 1920s

Bull's *NYC Area* as of 1968: increased greatly since mid-1950s, with recent Long Island maxima fall 8000, winter 2500; bred Jamaica Bay Wildlife Refuge, Flushing Meadows 1961

Bull's *New York State* as of 1975: bred Piermont Marshes 1974

Study area, historical and current

A spring/fall migrant and winter resident but in much smaller numbers than are seen coastally.

In fall, first migrants arrive in late Sep, peak in mid-Nov, and depart by late Nov, with extreme dates of 7 Sep 1964 on Jerome (Sedwitz) and 21 Nov 2008 (Young), and a Van Cortlandt maximum 150 on 7 Nov 1957 (Mayer) and 28 Nov 1963 (Norse). The Woodlawn maximum is 18 on 6 Nov 1960 (Horowitz) and Jerome's is 85 on 25 Nov 1962 (Sedwitz).

Overwinters in Van Cortlandt Swamp, typically moving to Jerome and Hillview with freeze-ups; the winter maximum is 16 on Jerome on 11 Dec 1966 (Sedwitz).

In spring, first migrants arrive in mid-Mar, peak in mid-Apr, and depart by early May, with extreme dates of 1 Mar 2000 (Young) and 25 May 1980 (Ephraim) and a maximum of 20 on Jerome on 17 Mar 1964 and 27 on Van Cortlandt Lake on 3 Mar 1958 (Buckley).

Bred as close as the Piermont Marshes in 1974 and on several occasions at Jamaica Bay Wildlife refuge. NYSBBA II in 2000–2005 found this species to have decreased its number of occupied blocks by 48% since NYSBBA I in 1980–85, and it has disappeared from the handful of its intermittently occupied New York City area breeding sites (McGowan and Corwin 2008).

American Black Duck   *Anas rubripes*

NYC area, current
In the New York City area a widespread, abundant coastal (especially) and inland spring/fall migrant, breeder, and overwinterer. Numbers have declined in the last 50 years (all study area maxima were before the 1960s), but this may reflect the loss of saltmarsh habitat in the New York City area that accelerated after World War II.

Bronx region, historical
Bicknell's *Riverdale* 1872–1901: common spring/fall migrant, overwintered 1877–78

Eaton's *Bronx + Manhattan* 1910–14: common winter visitor

Griscom's *Bronx Region* as of 1922: feral descendants of wild ancestors resident throughout; those truly wild do not breed along Hudson as formerly [but cf. comments below]

Kuerzi's *Bronx Region* as of 1926: common, occasionally abundant migrant (16 Aug–15 May), fairly common in winter, few scattered breeders

NYC area, historical
Cruickshank's *NYC Region* as of 1941: residents breed widely, augmented in fall, winter (maximum 1000+), spring by migrants from north of New York City area

Bull's *NYC Area* as of 1968: Long Island winter maximum 5000, plus 25,000 on all 1953–54 New York City area Christmas Bird Counts including NJ, CT

Bull's *New York State* as of 1975: status unchanged

Levine's *New York State* as of 1996: Long Island counts 12,000–23,000 from 1985–96

Study area, historical and current
A common spring/fall migrant, breeder, and winter resident.

For reasons never articulated, Griscom (1923) held the singular view—never supported by any other authors and quickly forgotten—that the only wild native Black Ducks in the New York City area were those breeding on eastern Long Island and that all others were their feral descendants or from other unspecified releases.

Yet only 4 years later this notion was being politely ignored (Kuerzi 1927) and mercifully soon died. Strangely, it was unrelated to the long-contentious recognition of two subspecies of Black Duck—New York City area breeding *tristis* and northern, wilder *rubripes*, the so-called Red-legged Black Ducks that were only winter visitors/residents and identifiable by their bright(er) red legs. Once it had been demonstrated that leg color in Black Ducks varied only with age and sex, *rubripes* was merged with *tristis*, and observers (including this book's authors) happily gave up scrutinizing Black Duck flocks (see Farrand 1990). But where Griscom got his idea that most New York City area breeding Black Ducks were feral remains a mystery.

In the study area, Black Duck is a breeder, spring/fall migrant, and winter resident—when greatest numbers occur. They have always been most numerous on the 2 reservoirs and regular only in small numbers in Van Cortlandt, where they are always outnumbered by Mallards. For reasons uncertain, most subarea numbers peaked before World War II, perhaps related to later loss of preferred saltmarsh breeding habitat. First breeders were found in the study area by about 1915.

In fall, first migrants arrive in late Aug, peak in Oct–Nov, and depart by late Nov unless overwintering, with an early fall date of 15 Aug 2004 (Young) and maxima of 80 on 22 Nov 1953 (Phelan, Buckley), 201 in Woodlawn on 27 Nov 1959 (Horowitz), and 135 on Jerome on 28 Oct 1955 (Buckley).

Winter residents are highly freeze-dependent and move readily between Van Cortlandt, Jerome, Hillview, Spuyten Duyvil, etc. During the 1930s, winter counts of 2000+ were routine at Hillview (Norse et al.), and Jerome's high was 800+ on 30 Jan 1937 (Imhof). Maxima in the 1950s were down to 200 at Jerome on 4 Feb 1956 (Kane) and 200–400 there during the 1960s–70s, peaking at 750 on 17 Jan 1963 (Sedwitz) and 1000+ across Jerome, Hillview, and the Hudson River on 23 Dec 1956 (Sedwitz et al.). Since the 1980s, numbers at those locations have dropped, with maxima now only in the 30–50 range; it may no longer occur at Hillview. Numbers recorded on the Bronx-Westchester Christmas Bird Count (Fig. 23) show this trend well. Van Cortlandt numbers have never been high, with a maximum of 60 on 30 Jan 1956 (Scully).

In spring, winter residents depart and new migrants arrive in mid-Mar, local numbers stabilizing by early May, with a Van Cortlandt maximum of 20 on 21 Mar 1954 (Phelan, Buckley), 150 at Jerome on 20 Mar 1955 (Kane), and 300 on 7 Mar 1936 at Hillview (Weber et al.).

Breeding begins by mid-Apr in Van Cortlandt and Woodlawn and regularly if not annually at Jerome, but no longer at Hillview now that its bird control measures are in force. First

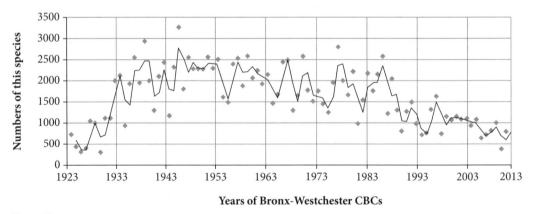

Figure 23

breeding in Van Cortlandt Swamp may not have been until 1933 (Norse) but 1–2 pairs have bred continually (if not always successfully) since then, with about 10–15 pairs throughout the entire study area. Summer numbers fluctuate, with recent maxima of 8 in Van Cortlandt on 31 Jul 2013 (McGee) and 4 at Jerome on several occasions.

Statewide and in the New York City area, NYSBBA I in 1980–85 and II in 2000–2005 indicated clear declines in the number of occupied blocks, especially within New York City and in the Bronx (McGowan and Corwin 2008), and DOIBBS data for this species in New York State affirmed a precipitous decline from 1966–2005 (Sauer et al. 2007). At the same time that Black Duck was decreasing, Mallard was increasing, and though many have seen this as cause and effect, the data are lacking. Yes, there has been increased interbreeding between native Black and feral Mallards, but no, there isn't any evidence to indicate this led to anything deleterious to Blacks, including postulated genetic swamping. A simpler and more plausible explanation is that Mallard's greater tolerance of humans and Black's distrust of them (especially in hunting season) has led to the former's increase and the latter's decrease—simultaneously but independently.

Mallard   *Anas platyrhynchos*

NYC area, current
Throughout, 2 discrete populations occur: resident breeders of dubious authenticity, and genuine prairie pothole breeders that arrive in fall, overwinter, and depart in spring.

Bronx region, historical
Bicknell's *Riverdale* 1872–1901: only once, 20 Nov 1876

Eaton's *Bronx + Manhattan* 1910–14: uncommon migrant; no mention of feral residents; no breeders closer than central New York State

Griscom's *Bronx Region* as of 1922: tame residents originating in Bronx Zoo occur throughout year in Van Cortlandt; truly wild migrants inseparable from them

Kuerzi's *Bronx Region* as of 1926: uncommon but regular migrant in spring (7 Feb–15 May)/fall (1 Sep–22 Jan) on Hudson, otherwise ferals tracing to Bronx Zoo cannot be distinguished from wild ones

NYC area, historical
Cruickshank's *NYC Region* as of 1941: feral residents breed widely, augmented fall, winter, spring by midwestern migrants (maximum 100+)

Bull's *NYC Area* as of 1968: Long Island winter maximum 450, plus 2500 on all 1949–50 New York City area Christmas Bird Counts (including NJ and CT), increasing to 4000 by 1953–54—totals difficult to apportion among increased observers, genuine population increases, and wild:feral ratio shifts

Bull's *New York State* as of 1975: conventional wisdom that feral Mallards replacing or genetically swamping native Black Ducks untrue in coastal marsh stronghold

Levine's *New York State* as of 1996: status unchanged

Study area, historical and current
A common spring/fall migrant, breeder, and winter resident.

Feral Mallards are resident and commonly believed to derive from several sources dating back to the early 1900s: releases/escapes from waterfowl collections and hunters' live decoys, misguided state and federal restocking efforts, and perhaps occasional wild migrants lured into oversummering by feral residents. Layered atop these are extensive intraregional movements by all ferals.

Ferals were unknown to Eaton (1910–14) but were already widespread in the New York City area by Griscom's (1923) time, with Van Cortlandt explicitly mentioned as hosting many whose origins he ascribed to the Bronx Zoo. He noted that feral birds were common throughout the year, so the first study area breeding can be inferred to have occurred by about 1915. The less confiding and bread-avoiding individuals that appear after Sep, overwinter, and then depart by Apr are by and large genuinely wild Mallards from the center of the continent and thus do not breed locally.

Even though not breeding in Kuerzi's (1927) Bronx region, by the mid-1930s Van Cortlandt Swamp breeding censuses 3 nesting pairs were located, continuing through the mid-1960s (Heath et al.) and later (Young et al.). In the study area as of 2016, feral Mallards were nesting in all 4 extant subareas, with a current population of 9–14 pairs: 3–5 in Van Cortlandt, 2–3 in Woodlawn, 3–4 at Jerome, and 1–2 at Hillview.

Few changes were noted within New York City between NYSBBA I in 1980–85 and II in 2000–2005, and Mallards were widely distributed except in the most urban parts (McGowan and Corwin 2008). DOIBBS data for this species in New York State showed a steady increase between 1966 and 2005 (Sauer et al. 2005).

Despite the fact that feral Mallards have been interbreeding with native Black Ducks in the New York City area since the 1920s and 1930s (Sedwitz, Norse, Kieran), neither Griscom (1923) nor Cruickshank (1942) nor Bull (1964, 1975) mentioned mixed pairs or hybrids; Bull (1976) was the first to do so.

On the Bronx-Westchester Christmas Bird Count Mallard numbers built up gradually to a peak in the 1980s–90s, then declined

Figure 24

somewhat, as shown in Figure 24. At the same time, Black Ducks reached their peak quickly in the 1930s, maintained numbers until the 1990s, when they began a slow decline (which may not have ended) to about 20% of their peak. While it is suspected that the rapid Mallard peaking was due to growth of its feral population, reasons for its recent decline are unclear, as they are for Black's more recent decline. Assertions that Mallard × Black interbreeding and resulting swamping by Black Duck genes have been responsible for the decline are unsupported. Mallard's much greater tolerance of humans may have abetted their increase.

Sorting wild from feral Mallards can be difficult, but the main fall/winter influx of the former usually follows early cold fronts in late Sep, peaking in Nov. Subsequent pulses throughout the winter are related to inland freeze-ups. Winter residents quietly begin to disappear in Mar, when northbound returnees also pass through. By late Apr usually only ferals are left. Sample mixed maxima by season and location include *spring* Van Cortlandt 90 on 20 Mar 2005 (Klein), Jerome 20 on 15 Apr 1971 (Sedwitz); *summer* 8–10 study area pairs breeding (this book) + 33 in Van Cortlandt on 22 Aug 2004 (Klein), 21 on Jerome on 12 Aug 1977 (Sedwitz); *fall* 73 in Woodlawn on 27 Nov 1959 (Horowitz), 48 in Van Cortlandt on 8 Sep and 23 Nov 2005 (Klein), 8 on Jerome 27 Oct 1979 (Sedwitz); *winter* 161 in Van Cortlandt on 27 Dec 2009 (Lyons et al.), 1020 at Jerome on 5 Feb 1971 (Sedwitz).

NYSBBA results indicated that Mallards have been increasing as breeders statewide, occupying 27% more blocks by NYSBBA II in 2000–2005 than during NYSBBA I in 1980–85 (McGowan and Corwin 2008). DOIBBS data for this species in New York State indicated that Mallard was twice as numerous by 2005 as it had been in the early 1980s, and 5 times more numerous than in the mid-1960s (Sauer et al. 2005). Most of the increases related to (originally) feral birds, given that the New York State Dept. of Environmental Conservation released 30,000+ pen-reared Mallards between 1934 and 1954, augmented by untallied private efforts. It was these that are believed responsible for the resident breeders in and around New York City.

Are *feral* study area Mallards now stable, increasing, or decreasing? It is difficult to say with certainty, but they seem to be increasing slightly in summer. And are *native* study area Mallards now stable, increasing, or decreasing? Also difficult to say with certainty, but highest fall, winter, and spring counts may have passed their peak.

Blue-winged Teal   *Anas discors*

NYC area, current
A spring/fall migrant, very local breeder in small numbers, but almost absent in winter.

Bronx region, historical
Bicknell's *Riverdale* 1872–1901: 6 shot Van Cortlandt Lake, favored hunting site, 15 Sep 1880

Eaton's *Bronx + Manhattan* 1910–14: fairly common migrant

Griscom's *Bronx Region* as of 1922: formerly common on Hudson, but no recent records; otherwise very rare

Kuerzi's *Bronx Region* as of 1926: rare migrant in spring (18 Mar–26 Apr)/fall (29 Aug–14 Nov)

NYC area, historical
Cruickshank's *NYC Region* as of 1941: common spring/fall (maximum 200) migrant throughout, but absent in winter; rare, erratic Long Island breeder

Bull's *NYC Area* as of 1968: fall maximum 250; exceedingly infrequent in winter, maximum 30; recent breeding at Orient, Easthampton, Baldwin, Staten Island,

Jamaica Bay Wildlife Refuge, where 6 pairs

Bull's *New York State* as of 1975: a few more scattered Long Island breeding sites

Study area, historical and current
A migrant in small numbers in Van Cortlandt, through the 1930s–40s at Hillview, and very rarely at Jerome or Woodlawn. It is the only study area dabbling duck that does not normally occur in winter.

In spring, first migrants arrive in late Mar, peak in mid-Apr, and depart by early May, with extreme dates of 9 Mar 1944 (Komorowski) and 25 May 1938 (Feigin et al.), and a maximum of 7 on 25–26 Apr 1953 (Buckley, Scully). Until Jun 1967, only singles had been seen in Central and Prospect Parks.

Blue-winged Teal is a potential Van Cortlandt Swamp breeder any year water levels are attractive, although there is only a single Jun–Jul record: a pair on 4 Jun 1981 (Sedwitz). The nearest breeders are in the Hackensack Meadowlands and Jamaica Bay Wildlife Refuge, but in the Bronx a pair nested at Baxter Creek in 1957 (Pappalardi).

The New York City area is at the margin of its New York State breeding range, and the about 30 occupied blocks on Long and Staten Islands during NYSBBA I in 1980–85 plummeted to 3 by NYSBBA II in 2000–2005 (McGowan and Corwin 2008). Some should return once currently saltwater Tobay Pond at Jones Beach and the West Pond at Jamaica Bay Wildlife Refuge are fully restored to their previous freshwater conditions.

In fall, first migrants arrive in late Aug, peak in late Sep–early Oct, and depart by early Nov, with extreme dates of 27 Aug 2005 (Morales) and 31 Oct 1940 (Norse), and a maximum of 7 on 14 Sep 1935 (Imhof).

After Nov they are exceedingly infrequent, with only 4 records of singles, none overwintering: 7 Dec 1933 (Cruickshank); 25 Nov–17 Dec 1938 (Norse, Cantor); 20 Jan 1952 (Norse, Cantor); crippled ♀ on 23 Jan 1956 (Buckley).

Northern Shoveler *Anas clypeata*

NYC area, current
An exceedingly local fall migrant and winter visitor/resident that until the 1960s was known reliably from a few Long Island sites and never in large numbers. Since then it has increased greatly, especially in the Bronx, where it is now erratic in small flocks in winter.

Bronx region, historical
Bicknell's *Riverdale* 1872–1901: unrecorded

Eaton's *Bronx + Manhattan* 1910–14: rare migrant

Griscom's *Bronx Region* as of 1922: exceedingly rare, 1 recent record Baychester marshes

Kuerzi's *Bronx Region* as of 1926: status unchanged

NYC area, historical
Cruickshank's *NYC Region* as of 1941: very local Long Island spring/fall migrant, winter resident (26 Aug–18 May; maximum 43); elsewhere, scarce spring/fall migrant, barely known in winter

Bull's *NYC Area* as of 1968: recent Long Island maxima fall 300, winter 100, spring 70; occasional breeding Jones Beach, Jamaica Bay Wildlife Refuge

Bull's *New York State* as of 1975: fall maximum 350

Study area, historical and current
A scarce but annual fall migrant, winter visitor/resident, and spring migrant on Van Cortlandt Lake and Jerome (the Bronx distribution center), formerly on Hillview.

For many decades unrecorded in the study area. Kieran did not see it in Van Cortlandt between 1914 and 1956, and there are only 2 study area records in that period: Van Cortlandt 18 Mar–15 Apr 1939 (Norse, Cantor) and Hillview 9–16 Dec 1939 (Norse, Cantor). Shovelers did not become regular until 12 appeared in Van Cortlandt on 21 Oct 1956 (Herbert et al.), and by 1959 some began moving regularly into Van Cortlandt from the nucleus of a small overwintering flock at the Bronx Zoo's Wildfowl Pond—5 overwintered in Van Cortlandt 1959–60 until the lake froze in Mar. After the 1950s they were most numerous on Jerome. Numbers peaked in both areas in the 1960s and '70s, then dropped to almost zero until a hint of resurgence a few years before 2010. They are unrecorded from May–Aug. Unrecorded in Central and Prospect Parks until Nov 1955.

Fall migrants arrive in Oct and peak in Nov unless overwintering, with extreme dates in Van Cortlandt of 14 Sep 2014 (Gomes) and 29 Nov 1012 (Finger), and maxima of 58 in Van Cortlandt on 10 Nov 2017 (Cole) and 177 at Jerome on 25 Nov 2017 (Cole et al.) It seems the 55 at Jerome on 22 Oct 2016, reaching 95 on 8 Nov (Karlson), were presaging a new uptick.

In winter they were scarce in Van Cortlandt through the 1960s and '70s, a time when they were regular on Jerome. Since then, occasional birds have dropped into Van Cortlandt Lake from Jerome, and in recent years a few have stayed until frozen out in late Jan or early Feb. Overwintering has never occurred in Van Cortlandt even under mild conditions, and the 2 in Van Cortlandt Swamp on 28 Feb 2006 (Künstler) probably came from Jerome. The recent Van Cortlandt winter maximum was 24 on 15 Dec 2015 (Young).

Most study area records have come from Jerome, where during the period 1962–76 they were unrecorded until the winter of 1964–65 when there were 4 records and a maximum of 11. Numbers slowly increased over the next few years, always peaking in Dec–Jan: 11 on 11 Jan 1966; 65 on 10 Jan 1970, and a maximum of 178 on 30 Jan 1971. They then dropped back to 76 on 6 Jan 1972, finally tapering off to 6 on 29 Jan (all Sedwitz).

On the Bronx-Westchester Christmas Bird Count (Fig. 25) they were unrecorded until the mid-1960s, increased quickly to 223 in 1982, after which they gradually dropped back to lower numbers.

There is only a hint of spring migration in Mar on Van Cortlandt Lake, which is also when Bronx overwinterers melt away, with the earliest on 18 Mar 2000 (DiCostanzo), the latest on 25 May 1980 (Ephraim), and a maximum of 9 at Jerome on 30 Mar 2018 (O'Reilly).

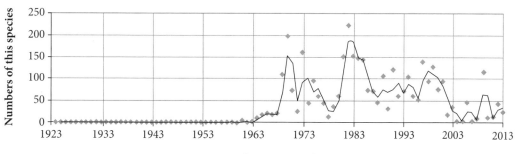

Years of Bronx-Westchester CBCs

Figure 25

Shoveler is a very sparse breeder in New York State, found in only 19 blocks in 1980–85 and 23 blocks in 2000–2005 (McGowan and Corwin 2008). Nearest (irregular) breeders may be at the Hackensack Meadowlands and Jamaica Bay Wildlife Refuge.

Northern Pintail *Anas acuta*

NYC area, current
A common spring/fall migrant and a locally abundant winter visitor/resident on Long Island; always scarce on the mainland.

Bronx region, historical
Bicknell's *Riverdale* 1872–1901: unrecorded

Eaton's *Bronx + Manhattan* 1910–14: common migrant

Griscom's *Bronx Region* as of 1922: very rare migrant, 1 recent (1918) record

Kuerzi's *Bronx Region* as of 1926: uncommon migrant in spring (22 Feb–25 Apr)/fall (24 Oct–30 Dec)

NYC area, historical
Cruickshank's *NYC Region* as of 1941: widespread migrant in spring/fall (maximum 200), mostly coastal winter resident (maximum 600); much commoner than 1920s

Bull's *NYC Area* as of 1968: bred Jamaica Bay Wildlife Refuge 1962; maxima fall 200, winter 600

Bull's *New York State* as of 1975: fall maximum 600

Study area, historical and current
A not quite annual spring and fall migrant and winter resident, although always in smaller numbers and less predictably than any other regular dabbling duck. Appears in all 4 extant subareas, usually only singles or twos.

In fall, first migrants arrive in mid-Sep, peak in late Oct, and depart by early Nov, with extreme dates of 29 Aug 1937 (Norse) and 28 Nov 1958 (Horowitz) and a maximum of only 4 on 24 Oct 1925 (Cruickshank).

From 1935–40 as many as 5 overwintered annually at Hillview (Norse, Imhof), but only singles occasionally wandered into Van Cortlandt Swamp and Jerome, never remaining very long—the same story as in 2016. One did overwinter on Jerome during the winter of 1963–64 (Enders et al.). The all-time study area maximum was only 8 there on 11 Dec 1966 (Sedwitz). It has always been the least common dabbling duck in the study area and is not recorded annually. It has been equally scarce on the Bronx-Westchester Christmas Bird Count (Fig. 26).

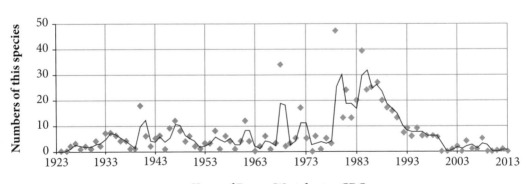

Figure 26

In spring, when usually unrecorded, first migrants arrive in early Mar, peak in late Mar, and depart by early Apr, with extremes of 4 (the maximum) at Jerome on 24 Feb 1955 (Kane, C. Young) and Van Cortlandt Lake on 26 Apr 1952 (Kane, Kieran).

Green-winged Teal   *Anas (crecca) carolinensis*

NYC area, current
A widespread freshwater spring/fall migrant and winter resident locally in small numbers.

Bronx region, historical
Bicknell's *Riverdale* 1872–1901: 14 Apr 1886 only

Eaton's *Bronx + Manhattan* 1910–14: uncommon migrant

Griscom's *Bronx Region* as of 1922: almost unknown—several records from Bronx Zoo Wildfowl Pond

Kuerzi's *Bronx Region* as of 1926: uncommon migrant in spring (20 Mar–29 Apr)/fall (15 Oct–21 Dec)

NYC area, historical
Cruickshank's *NYC Region* as of 1941: common migrant in spring (maximum 500+)/fall, rare, local on Long Island in winter; far less numerous inland

Bull's *NYC Area* as of 1968: in migration now rarely reported in flocks >100; Long Island maxima fall 210, winter 450, spring 450; recent breeding Easthampton, Jamaica Bay Wildlife Refuge

Bull's *New York State* as of 1975: status unchanged

Study area, historical and current
A spring/fall migrant and winter resident in small numbers in all 4 extant subareas.

In fall, the first migrants arrive in early Sep, peak in mid-Nov, and depart after major freeze-ups in early Dec, with extreme dates of 25 (the maximum) on 5 Sep 1950 (Komorowski) and 19 Nov 2011 (Baksh).

Small flocks of fewer than a dozen often attempt to overwinter in Van Cortlandt Swamp, but after freeze-ups they move out, with a maximum of 28 on Van Cortlandt + Hillview on 25 Dec 1953 (Buckley). They are scarce on Jerome, where from 1962–76 only 3 single ♂♂ were seen, usually for only a day or two, in Oct, Jan, and Mar (Sedwitz). On the Bronx-Westchester Christmas Bird Count (Fig. 27) they

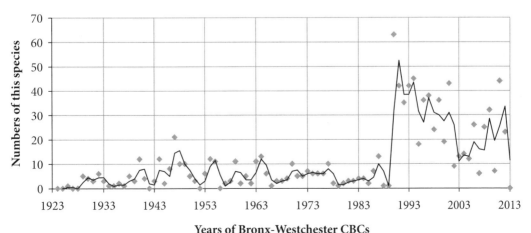

Figure 27

were exceedingly scarce until the early 1990s, when they increased only to drop back to near zero 25 years later.

In spring, the first migrants arrive in early Mar, peak in mid-Mar, and depart by late Apr, with extreme dates of 28 Feb 2013 (Young) and 2 on 14 May 1953 (Buckley et al.), and a maximum of 5 on 31 Mar 2015 (Hannay). A flock of 16 in Van Cortlandt Swamp on 17 Feb 1955 (Kane) may also have been migrants.

While there are no Jun–Aug records, it is a remotely possible breeder in Van Cortlandt Swamp. Nearest breeders are in the Hackensack Meadowlands, on Staten Island, and at Jamaica Bay Wildlife Refuge, single digits in each location. NYSBBA II in 2000–2005 found this species to have decreased its number of occupied blocks by 45% since NYSBBA I in 1980–85, especially in the New York City area (McGowan and Corwin 2008).

Eurasian Teal   *Anas (crecca) crecca*

NYC area, current
Greenland and Iceland breeders are seen annually on Long Island but rarely more than 1–2 at a time, usually in Nov and then in Mar–May, and occasionally overwintering. Until very recently ♀♀ were considered inseparable from Green-winged Teals, and many have been overlooked. Recorded in Central Park but not Prospect and almost unknown away from Long Island.

Bronx region, historical
Griscom's *Bronx Region* as of 1922: unrecorded

Kuerzi's *Bronx Region* as of 1926: unrecorded

NYC area, historical
Cruickshank's *NYC Region* as of 1941: rare but local Long Island winter visitor (13 Oct–13 May); maxima 4–6; infrequent inland with several Bronx records [locations not given]

Bull's *NYC Area* as of 1968: decreased in frequency, abundance >early 1940s; daily maximum 2; late Nov–13 May; single Bronx record

Bull's *New York State* as of 1975: status unchanged

Study area, historical and current
An exceedingly infrequent Bronx and study area fall/winter visitor/resident.

Only 3 Van Cortlandt Lake records, all ♂♂. The first was on 11 Nov 1937 (Kieran), and the next 2 not until 63 years later, from mid-Dec 2000 (Pirko)–mid-Apr 2001 (Young, Buckley et al.), and then presumably the same ♂ from 27 Nov 2001 (unknown observer, *Kingbird*)–28 Feb 2002 (unknown observer, *Kingbird*).

Otherwise in the Bronx there were several records from the famous Waterfowl Pond in the Bronx Zoo in the 1930s (A.D. Cruickshank, pers. comm.), but only mentioned in passing as in Bronx County in Cruickshank (1942) with the dates and observers unstated. There were also 2 ♂♂ at Pelham Bay Park from 8–11 Jan 1956 (Penberthy, Buckley, Mudge). A ♂ at Tuckahoe/ Yonkers reservoirs from 10 Feb–19 Mar 2005 (Block et al.) was probably the 2000–2002 Van Cortlandt bird.

Canvasback   *Aythya valisineria*

NYC area, current
A regular but episodically numerous/scarce winter resident on fresh- and saltwater in favored locations.

Bronx region, historical
Bicknell's *Riverdale* 1872–1901: unrecorded

Eaton's *Bronx + Manhattan* 1910–14: rare migrant

Griscom's *Bronx Region* as of 1922: almost regular occurrence on Jerome Reservoir phenomenon of local ornithology

Kuerzi's *Bronx Region* as of 1926: regular spring/fall migrant, winter resident (23 Oct–11 Apr) Long Island Sound, East River. Occasional on Hudson, no longer regular on Jerome

NYC area, historical

Cruickshank's *NYC Region* as of 1941: a rare spring/fall migrant; 2 Aug–1 May; very local winter resident; most regular on East River in Bronx, but 100 anywhere memorable

Bull's *NYC Area* as of 1968: increased greatly since 1920s, now widespread, abundant winter resident, maximum 5300; late spring date 22 May; occasional in summer

Bull's *New York State* as of 1975: status unchanged

Levine's *New York State* as of 1996: winter maximum 6175

Study area, historical and current

A migrant and winter resident/visitor, especially on Jerome and Hillview.

Exceedingly infrequent on Van Cortlandt Lake, with singles on 7 Nov 1954 (Kane, Buckley); 21–29 Apr 1956 (Buckley, Kane); 28 Mar 1959 (Buckley); 24 Feb 1961 (Heath); and 3 on 13 Jan 2008 (Bochnik). Found only once in Woodlawn, on 18 Oct 1978 (Teator).

After a few on Jerome in the early 1900s, Canvasbacks were no longer of regular occurrence by 1926 (Kuerzi), and Cruickshank (1942) does not even mention them there. Yet by the 1950s they had returned and were seen frequently there and at Hillview for the next 30–40 years. The earliest fall arrival was on 23 Sep 1973 (Sedwitz), with 150 as early as 6 Nov 1965 (Kane) and the latest on 20 May 1935 (Cantor). Large numbers usually arrived in early Dec, peaked shortly thereafter (maximum 4235 on 11 Dec 1969: Sedwitz), then dwindled throughout the winter, eventually moving completely to local saltwater once the reservoir froze. Peak spring count was 1255 on 7 Mar 1980 (Sedwitz). Highest counts generally were in the late 1960s and during the severe winter of 1977–78, and by the early 1980s peaks were down to the 1300s (Sedwitz). Numbers slowly dropped throughout the late 1980s–90s, bottoming out around the turn of the century. The 23 there on 17 Feb 1995 (Garcia) was one of the last flocks of any size. Canvasbacks never feed on Jerome but only roost there by day, departing at dusk or later and returning about a half-hour after sunrise (Enders 1976). These particular birds fed variously on the Hudson River, the East River, and around the Triboro Bridge.

During the brutally cold winter of 1976–77 with the heaviest ice in the Hudson River in decades, the oil-laden barge *Ethel H.* struck ground in the Hudson south of Bear Mountain on 4 Feb, spilling 420,000 gallons of heavy fuel oil. Canvasbacks winter in good numbers on the river in Rockland County, and by 27 Feb 100+ oiled individuals were preening vigorously on the slope at the north end of Jerome (Sedwitz).

The decline shown on Jerome was mirrored across the entire Bronx-Westchester Christmas Bird Count (Fig. 28). Until the 1950s Canvasbacks were almost unrecorded, became increasingly numerous from the 1960s–80s—peaking at 8745 in 1985—then by the late 1990s plunged to near zero, their status as of 2016. No accepted explanation for this decline has been forthcoming, although it may be no more than a routine shift in preferred winter concentration areas.

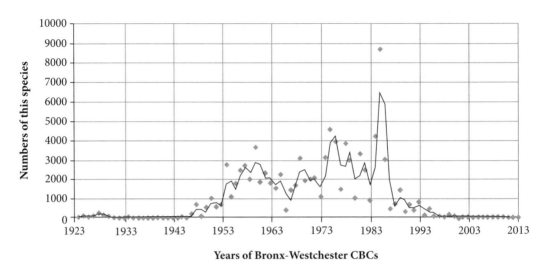

Figure 28

Redhead   *Aythya americana*

NYC area, current
The most local and least numerous bay diving duck, most often found overwintering at favored Long Island sites and exceedingly rare elsewhere; infrequently noted in migration.

Bronx region, historical
Bicknell's *Riverdale* 1872–1901: unrecorded

Eaton's *Bronx + Manhattan* 1910–14: uncommon migrant

Griscom's *Bronx Region* as of 1922: very rare; 2 Jerome records 1914, 1915

Kuerzi's *Bronx Region* as of 1926: rare migrant, winter visitor (1 Nov–4 Apr)

NYC area, historical
Cruickshank's *NYC Region* as of 1941: exceedingly local rare Long Island fall migrant, winter visitor (18 Sep–22 May; maximum 200+); less frequently inland on migration

Bull's *NYC Area* as of 1968: 6 Sep–5 Jun; restricted to eastern Long Island, where less common than early 1900s when fall maximum 2000, now 160; introduced 1960 Jamaica Bay Wildlife Refuge by New York State DEC, maximum of 10 pairs bred intermittently until 1983

Bull's *New York State* as of 1975: status unchanged

Study area, historical and current
The least common regular diving duck, usually in tiny groups with Lesser Scaup on Jerome (and formerly on Hillview). When present, they move routinely between Van Cortlandt Lake, Jerome, and Spuyten Duyvil.

About 15 records: Jerome 21 Mar–4 Apr 1914 (Kieran et al.); 10 Jan 1915 (Pangburn); Hillview 4 Apr 1937 (Norse); Jerome 21 Mar 1951 (Kane); 2 on Van Cortlandt Lake and 2 more on Jerome 3 Dec 1955–30 Jan 1956 (Sedwitz et al.); Jerome 25–27 Dec 1964 (Sedwitz,

Buckley); 5 Feb 1966 (Zupan, Enders); plus a pair on 4 Dec 1969, singles on 5 Mar 1972 and 15 Mar 1976 (all Sedwitz); Van Cortlandt 1 Jan 1997 (Lyons, Garcia); Hillview on 21 Jan 2015 (Camillieri). During an exceptionally cold winter, 1–2 were present on Jerome from 2 Jan–9 Mar 1977, peaking at an unprecedented 34 on 9 Feb (Sedwitz)—the highest Bronx count known.

Ring-necked Duck   *Aythya collaris*

NYC area, current
A relatively common but extremely local spring/fall freshwater migrant and rare winter visitor/resident. Much commoner in northern NJ at all seasons than in nearby New York.

Bronx region, historical
Bicknell's *Riverdale* 1872–1901: unrecorded

Eaton's *Bronx + Manhattan* 1910–14: rare migrant

Griscom's *Bronx Region* as of 1922: unrecorded

Kuerzi's *Bronx Region* as of 1926: 2 Jan Hunt's Point records only ones in Bronx

NYC area, historical
Cruickshank's *NYC Region* as of 1941: fairly common migrant in spring (maximum 300)/fall (24 Sep–15 May) especially inland; now much commoner than Canvasback; ♂ at Tuckahoe, Westchester County, in Jun 1941

Bull's *NYC Area* as of 1968: 20 Sep–7 Jun; greatly increased, no longer much commoner than Canvasback; winter maximum 340

Bull's *New York State* as of 1975: status unchanged

Study area, historical and current
A spring/fall migrant on Van Cortlandt Lake, Jerome, and Hillview in very small numbers, occasionally attempting to overwinter but usually forced by freezing freshwater to Spuyten Duyvil or the East and Hudson Rivers.

In fall, first migrants arrive in early Nov, peak in mid-Nov, and depart by late Nov, with extreme dates of 18 Oct 1936 (Weber et al.) and at Hillview on 22 Nov 1936 (Norse), singles all. The Jerome fall maximum was 2 on 12 Nov 1961 (Horowitz). Unrecorded in Prospect Park until 1951.

In winter on Van Cortlandt Lake normally only singles or pairs, with a maximum of 8 on 1 Feb 2013 (Young). Much more regular on Jerome; for example, between 1962 and 1976, 1–4 in scaup flocks from Dec–Feb, some successfully overwintering (Sedwitz, Enders).

In spring, first migrants trickle in during early Mar, peak in late Mar, and depart by early Apr, with extreme dates of 2 on 3 Mar 1961 (Horowitz) and 18 May 2014 (Gaillard, Alvarez), and maxima of 5 in Van Cortlandt from 8–22 Mar 1952 (Kane, Buckley, Kieran) and 12 on Jerome on 10 Mar 1968 (Sedwitz).

Tufted Duck   *Aythya fuligula*

NYC area, current
This Eurasian native is actively colonizing North America from Iceland, and 1 indicator of its success was the 74 on Quidi Vidi Lake in St John's, Newfoundland, on 6 Feb 2015. In the New York City area they are now seen not

quite annually in winter on the Hudson and East Rivers and Central Park Reservoir, as well as other scattered locations as far north as Rye and east to Southampton, Long Island. Singles or pairs are usually in scaup flocks.

NYC area, historical
Bull's *NYC Area* as of 1968: 2 recent records deprecated as escapes

Bull's *New York State* as of 1975: increasing records but all deprecated as escapes

Levine's *New York State* as of 1996: many more records; still viewed as escapes

Study area, historical and current
A fairly recent winter/spring vagrant to the Bronx and study area.

There are 4 study area occurrences. The first of this species in New York State or on a Bronx-Westchester Christmas Bird Count was also the first ♀ detected in North America. An HY, it was found in Spuyten Duyvil Creek on 26 Dec 1955 (Buckley, Sedwitz, Kane, Scully, Lee). Because it was a duck and because a ♀ had been reputedly lost from a Boonton, NJ, collection a few months prior, some assumed the Spuyten Duyvil had to have been that ♀. Fortunately, the NY individual was seen very well by many observers who noted its lack of bands and clipped nails; moreover, the NJ ♀ was an adult, not an HY, and had a long crest. Thus that rumor died. It was eventually seen by dozens of observers and photographed in color by several, including movie footage taken by E. T. Gilliard. It moved with overwintering scaup between Spuyten Duyvil, Jerome, and Van Cortlandt Lake, where it was also detected as follows: Jerome 29 Dec 1955 (C. Young, Russak), Van Cortlandt Lake 22 Jan (Buckley) and 4 and 24 Feb 1956 (Kane), Jerome 13 Apr 1956 (Buckley)—the last date it was seen.

Second was an adult ♂ at Jerome on 15 Mar 1962 (Sedwitz), New York State's second. Given waterfowl's relatively long lives, it may have been the same ♂ that overwintered on, and moved between, the Hudson River north and south of the George Washington Bridge (also on the NJ side at Edgewater), and the East River near Randall's Island and Hell Gate Bridge, from Jan–Apr 1966, and then through the winters of 1966–67, 1967–68, 1968–69 (m. ob.). Third was also an adult ♂ at Jerome on 19 Feb 1972 (Friton). The fourth—adult ♂ and ♀ at Jerome on 7 Feb 1977 (Sedwitz)—may have been the pair that spent the winter of 1974–75 on the East River by the Triboro Bridge.

In this same period a ♂ was at Connetquot River State Park on Long Island in Feb 1970 and 1971, at Bay Head, NJ, during the winter of 1971–72, a ♂ and ♀ frequented the wildfowl pond in the Bronx Zoo between 24 Feb and 15 Apr 1972 (Maguire, Peszel et al.), and a ♀ at Spuyten Duyvil 5 Feb 1984 (Sedwitz) may have been the same one overwintering in Central Park. There were several of both sexes and occasional ♂-♀ pairs overwintering on Central Park Reservoir from 1979–91, affirming that Tufted Ducks began their winter invasion into the northeastern United States in the mid-1950s. Unrecorded in Prospect Park.

Greater Scaup *Aythya marila*

NYC area, current
A common spring/fall migrant and overwinterer in large but shrinking flocks on saltwater bays and rivers, including Long Island Sound and the Hudson River.

Bronx region, historical
Bicknell's *Riverdale* 1872–1901: unmentioned

Eaton's *Bronx + Manhattan* 1910–14: common migrant

Griscom's *Bronx Region* as of 1922: rare on Hudson, regular but uncommon on Long Island Sound; Jerome in spring 1914

Kuerzi's *Bronx Region* as of 1926: common spring/fall migrant and winter visitor/resident (17 Oct–23 May)

NYC area, historical
Cruickshank's *NYC Region* as of 1941: very common winter resident on Long Island Sound, East River in East Bronx, regular spring/fall migrant on inland lakes and reservoirs

Bull's *NYC Area* as of 1968: maxima fall 75,000, winter 85,000, 250,000

Bull's *New York State* as of 1975: 60,000 Pelham Bay Park 1953

Study area, historical and current
A winter resident/visitor and spring migrant in very small numbers on Jerome and Hillview but also in Van Cortlandt. In all 4 extant subareas it is much less frequent and numerous than Lesser. Until the 1940s, many individuals and flocks of scaup were left unidentified, but because recent data showed that Greater is by far the rarer of the two in the study area, the bulk of unidentified scaup after the early 1900s on both reservoirs would have been Lessers.

There are only 3 fall study area records: on Van Cortlandt Lake on 5 Nov 1930 (Sedwitz) and found dead in Woodlawn on 14 Oct 1976 (Teator), and at Hillview on 24 Nov 2017 (Camillieri)—which suggests that southbound they bypass freshwater unless forced down.

In winter there are only these Van Cortlandt Lake records: 4 from 23–28 Feb 1955 (Buckley, Sedwitz), 2–4 from 3–16 Feb 1956 (Scully), on 23 Jan 1960 (Heath), and 2 on 26 Feb 1973 (DiCostanzo). During the 1962–63 and 1964–65 Winter Bird Censuses, singles were seen on Van Cortlandt Lake several times (Enders et al.) A few Greaters follow Lessers onto Jerome to roost most winters, and up to 8 were there from 15 Mar–4 Apr 1914 (Kieran, Griscom et al.). On Jerome between 1962 and 1977, Sedwitz recorded them 4 times in Dec, twice each in Jan and Feb, 12 times in Mar, and 4 times in Apr. The latter 2 months are the peak of Lesser's spring migration, and Greaters were accompanying them. Most were singles and pairs, with maxima of 10 on 10 Mar 1965 and 12 on 7 Mar 1953 (Sedwitz). Otherwise the maximum on Jerome was a flock of 250 on 30 Jan 1937 (Imhof, Norse), pushed in by harsh weather. Their numbers on the Bronx-Westchester Christmas Bird Count, especially in Pelham Bay Park's Eastchester Bay, peaked between the 1940s and 1970s but then dropped to far lower levels, probably a result of garbage dump pollution of the Bay (Fig. 29).

The handful of other Van Cortlandt Lake records are during spring migration: 19 Mar 1936 (Weber, D. Lehrman), 4 Mar 1946 (Ephraim), 10 on 3 Mar 1956 (Scully), 3 May 1956 (Scully), 3–5 Mar 1958 (Buckley, Mayer), 4 on 28 Mar 1959 (Buckley), 7 on 3 Mar 1961 (Horowitz),

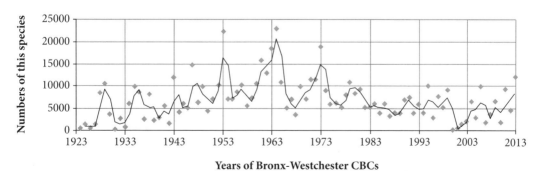

Figure 29

and 2 on 26 Feb 1973 (DiCostanzo). No study area Jun–Oct records.

It is now clear in the study and New York City areas that on open ocean and the larger bays and sounds, Greater is the default scaup most of the year. On sheltered bays, tidal creeks, lagoons, and broad rivers, both scaups occur with local differences that need to be learned. Greaters tend to be more widely distributed, while Lessers tend to have favorite locations. In most places where both occur, they rarely form mixed flocks. On freshwater year-round, Lesser is the default scaup, especially in spring/fall migration, although Greaters may appear when forced down by bad weather. By the same token overwintering Lessers are often forced to saltwater by heavy freezes, when they typically favor selected, sheltered sites. In and around the Bronx, Greaters are the Long Island Sound/East River default scaup, but after freezes Lessers may appear in quiet tidal creeks. Flocks of several hundred Greaters formerly overwintered in quiet waters around the Triboro Bridge and above and below the George Washington Bridge on both sides of the Hudson River until they were replaced by Lessers (Post, pers. comm.). But in Van Cortlandt, and on Hillview and Jerome, Lessers are the default scaup, Greaters occurring only rarely (Boyajian 1969c; Sedwitz 1975, 1977).

Lesser Scaup   *Aythya affinis*

NYC area, current
A common spring/fall migrant and local overwinterer in flocks on freshwater and selected saltwater ponds and rivers, including Long Island Sound and the Hudson River. Far more often found on reservoirs than Greater.

Bronx region, historical
Bicknell's *Riverdale* 1872–1901: unmentioned

Eaton's *Bronx + Manhattan* 1910–14: fairly common migrant

Griscom's *Bronx Region* as of 1922: status unclear; Jerome spring 1914

Kuerzi's *Bronx Region* as of 1926: uncommon migrant in spring (3 Mar–18 Apr)/fall (18 Oct–20 Nov), once 14 Jul 1925

NYC area, historical
Cruickshank's *NYC Region* as of 1941: fairly common freshwater spring/fall migrant, very uncommon in winter, field separation from Greater Scaup still problematic

Bull's *NYC Area* as of 1968: status unchanged

Bull's *New York State* as of 1975: status unchanged

Study area, historical and current
A winter resident/visitor and spring migrant in very small numbers on Jerome and Hillview, but also in Van Cortlandt. In all 4 extant subareas it is far more frequent and numerous than Greater. Until the 1940s, many individual and flocks of scaup were left unidentified, but because more recent data have showed that Greater is by far the rarer of the two in the study area, the bulk of unidentified scaup after the early 1900s on both reservoirs would have been Lessers.

The only large flocks have been on Jerome, and though data are sparse, on Hillview in the 1930s and perhaps later. Since the 1950s, in the study area this species has been most reliable on Jerome in early spring.

In fall a few usually appear on Van Cortlandt Lake in Oct–Nov (Kieran), with the earliest on 28 Sep 1951 (Kane) and the maximum of 15 on 10 Nov 1954 (Sedwitz, Buckley). They normally depart in late Nov, but in mild winters persist until the lake freezes, after which they move to Jerome.

Between 1962 and 1977, Sedwitz found Lessers on Jerome almost every fall in Oct–Nov, with the earliest on 9 Oct 1963 and 10 Oct 1964, and a fall maximum of 60 on 22 Oct 1962. They normally increased slowly through the winter, usually in flocks <50, peaking at 190 on 11 Feb 1977. Quite a few do follow other ducks from external sites onto Jerome to roost; historically, the first were 4 present from 24 Mar–4 Apr 1914 with 23 Greaters (Griscom, Kieran et al.).

On Van Cortlandt Lake in winter there are only 5 records: 5 Jan 1952 (Kane); 3 from 13 Feb–3 Mar 1955 (Buckley et al.); singles on 23 Jan 1956 (moribund, Scully) and from 16 Feb–3 Mar 1956 (Scully), plus 3–12 from 3 Feb–3 Mar 1958 (Buckley et al.).

In spring, first reservoir migrants arrive in mid-Mar, peak in late Mar–early Apr, and depart by late Apr, with extreme dates of 3 Mar 1951 on Jerome (Kane) and 21 May 1938 on Hillview (Norse). On Jerome between 1962 and 1976, each year numbers ticked upward in Mar, generally peaking at 100+ later that month: 300 on 11 Mar 1964 and 320 on 17 Mar 1965, reaching 395 on 27 Mar (Sedwitz). The highest spring count ever on Jerome was 515 on 5 Apr 1964 (Sedwitz), which dropped to 350 on 15 Apr, 95 on 17 Apr, and none on 7 May (Sedwitz).

On Van Cortlandt Lake, the few migrants follow much the same spring timetable but in far smaller numbers, usually <6 and with maxima of 30 on 6 Apr 1960 (Buckley) and 75 on 19 Mar 1958 (Mayer). Extreme dates there are 3 on 3 Mar 1961 (Horowitz) and 1 from 12 Apr–23 May 1959 (Zupan). In 1938, a ♂ oversummered at Hillview (Norse, Cantor).

Similar numbers (perhaps the same birds) were counted on the Manhattan side of the Hudson River just above and below the George Washington Bridge during the early and mid-1960s (Post et al.). At the same time on the NJ side at Edgewater, Boyajian (1969c) found Lessers regular or resident in fall, winter, and early spring in flocks up 500. See the preceding species for a discussion of both scaups' habitat preferences.

Surf Scoter *Melanitta perspicillata*

NYC area, current
A spring/fall migrant and winter visitor/resident on saltwater around Long Island that migrates coastwise and down the Hudson River, where it is rarely seen in winter. The least frequent of the scoters inland during migration, but second most common in winter around Long Island. Recorded infrequently in Central and Prospect Parks.

Bronx region, historical
Bicknell's *Riverdale* 1872–1901: unrecorded

Eaton's *Bronx + Manhattan* 1910–14: common winter visitor

Griscom's *Bronx Region* as of 1922: occurs on Hudson at Tappan Zee but otherwise unrecorded

Kuerzi's *Bronx Region* as of 1926: irregular winter visitor, sometimes in small flocks (1 Nov–2 May)

NYC area, historical
Cruickshank's *NYC Region* as of 1941: second most numerous scoter in New York City area; common eastern Long Island winter resident (maximum several thousand), rare on extreme west end of Long Island Sound, Hudson; all but unknown inland

Bull's *NYC Area* as of 1968: 1930 Long Island winter maxima 25,000, 120,000; occasional in summer

Bull's *New York State* as of 1975: status unchanged

Study area, historical and current
Near the study area, an erratic fall (e.g., 200 on 15 Oct 2000: unknown observer, *Kingbird*) and spring migrant on the Hudson River and a very infrequent, perhaps only former, winter resident/visitor (Kieran). Has occurred in the study area only when forced down by severe weather, which has happened twice on Hillview: in Nov 1936, but date misplaced (Corrigan fide Norse), and on 6 Nov 1951 (Kane), and once on Jerome: from 24–29 Oct 2017 (Karlson, Cole; photos). An amazing total of 1200 at City Island on 26 Dec 2016 (Walter et al.) is by far the highest 1-day Bronx total ever, enhanced by the 130 Black Scoters with them on 3 Dec (Benoit, Aracil).

Comments
King Eider, *Somateria spectabilis*, has occurred twice on the Hudson River in Riverdale, from 23 Dec 1928–late Jan 1929 and from 22 Dec 1929–29 Jan 1930 (Cruickshank et al.), and also at Piermont Pier from 23 Nov–6 Dec 2008 (Ciganek et al.). There are several 1930s Bronx records from Long Island Sound, and recently as follows: Ferry Point Park 2 Dec 1973–28 Feb 1974 (Friton), and at Pelham Bay Park on 5 Dec 2006 (Aracil). It has never been recorded in Central or Prospect Park but is known to appear on inland lakes after bad weather during spring migration. **Common Eider**, *S. mollissima*, is unrecorded in Prospect Park but was found once in Central Park on 22 Nov 1986 (Krauss et al.) and 4 times in the Pelham Bay Park area: ♀ on 15 Nov 1952 (Kane, Whelen et al.), the same ♀ sitting on the shore on 15 Nov 1953 (Buckley, Phelan, Lee), 3 on 29 Sep 1999 (Aracil), and adult ♂,♀ on the tantalizing date of 23 May 2012 (Künstler). A ♀ was in the Hudson River off Riverdale on 19 Nov 1935 (Wiegmann), and another was at Piermont Pier on 6 Mar 2015 (Edelbaum, Weiss). At Sands Point, directly across Long Island Sound from Pelham Bay Park, 2 were there on the unexpected date of 2 Jun 1994, and 2–3 (the same?) from 23 Sep–11 Oct 1995 (Quinn). **Harlequin Duck**, *Histrionicus histrionicus*, is regular in winter on rocky Long Island shorelines, especially at jetties and groins, but is otherwise unknown. It has never been seen on the lower Hudson River, but in the Bronx it has occurred 3 times at Throgg's Neck: 12 Mar 1993 (Walter), 21 Jan 1996 (Walter), and 2 on 15 Dec 2013 (Buckley), and may be overlooked on Long Island Sound. Unrecorded in Central and Prospect Parks.

White-winged Scoter *Melanitta fusca*

NYC area, current
A spring/fall migrant and winter visitor/resident on saltwater around Long Island that migrates coastwise and down the Hudson River, where it is rarely found in winter. The second least frequent of the scoters inland during migration but the most common in winter around Long Island. Recorded infrequently in Central and Prospect Parks.

Bronx region, historical
Bicknell's *Riverdale* 1872–1901: unrecorded

Eaton's *Bronx + Manhattan* 1910–14: common winter visitor

Griscom's *Bronx Region* as of 1922: regular in Lower New York Bay, rare (?) on Long Island Sound; common Hudson migrant at Tappan Zee

Kuerzi's *Bronx Region* as of 1926: common, occasionally abundant winter visitor/resident (24 Sep–15 May) on Long Island Sound, rare migrant on Hudson

NYC area, historical
Cruickshank's *NYC Region* as of 1941: most abundant New York City area scoter on ocean, Long Island Sound (maximum 100,000), uncommon but regular on East,

Hudson Rivers, rarely on lakes, reservoirs east of Hudson [locations not given]

Bull's *NYC Area* as of 1968: Long Island maxima 25,000–180,000; oversummers annually

Bull's *New York State* as of 1975: status unchanged

Study area, historical and current
Near the study area a fall and spring migrant, and an erratic, perhaps only former winter resident/visitor on the Hudson River (e.g., 40 at Dyckman St. on 14 Oct 1973: Sedwitz) or Long Island Sound, but occurring in the study area only when forced down by severe weather.

In the study area there are only 8 records: Jerome 21 Mar–14 Apr 1914 (Kieran et al.); 5 on Jerome on 27 Oct 1930 (Sedwitz); 3 on Hillview 31 Oct 1936 (Imhof); Jerome on 16 Mar 1947 (Norse, Cantor); 2 in Van Cortlandt on 21 Mar 1951 (Kane); 8 overhead in Van Cortlandt on 18 Oct 2001 (Young); Hillview from 10–15 Feb 2017 and on 19 Oct 2017 (Camillieri et al.).

Black Scoter   *Melanitta americana*

NYC area, current
A spring/fall migrant and winter visitor/resident on saltwater around Long Island that migrates coastwise and down the Hudson River, where it is rarely found in winter. The most frequent of the scoters inland during migration but the least common in winter around Long Island. Unrecorded in Central Park.

Bronx region, historical
Bicknell's *Riverdale* 1872–1901: unrecorded

Eaton's *Bronx + Manhattan* 1910–14: fairly common winter visitor

Griscom's *Bronx Region* as of 1922: fairly common migrant on Hudson at Tappan Zee, may occur on Long Island Sound

Kuerzi's *Bronx Region* as of 1926: irregular winter visitor/resident (2 Sep–18 May), occasionally in flocks on Long Island Sound, rare migrant on Hudson

NYC area, historical
Cruickshank's *NYC Region* as of 1941: least numerous scoter in New York City area, preferring open ocean (maximum thousands); uncommon on western Long Island Sound, irregular on Hudson, very rare inland

Bull's *NYC Area* as of 1968: Long Island winter maxima 10,000, 18,000; occasional summer

Bull's *New York State* as of 1975: status unchanged

Study area, historical and current
Near the study area a fall and spring migrant in considerable numbers along the Hudson River (e.g., 35 at Dyckman St. on 14 Oct 1973: Sedwitz, 100 in Riverdale 11 Oct 1964: Cantor) and Long Island Sound that alights on freshwater only when forced down by severe weather. Within the study area there is only a single record: at Jerome on 24 Nov 1939 (Norse, Cantor).

Long-tailed Duck   *Clangula hyemalis*

NYC area, current
A regular winter visitor/resident on Long Island Sound, Great South Bay, the ocean, with a few on the Hudson River. Most migrate coastwise, although some use the Hudson River in both spring and fall. Very scarce away from saltwater.

In the Bronx, a migrant and rare winter visitor on the Hudson River and at Pelham Bay Park, where small numbers overwinter.

Bronx region, historical

Bicknell's *Riverdale* 1872–1901: thrice on Hudson

Eaton's *Bronx + Manhattan* 1910–14: common winter visitor

Griscom's *Bronx Region* as of 1922: regular in Lower New York Bay, once in Upper; rare on Hudson at Croton Point, no Bronx records

Kuerzi's *Bronx Region* as of 1926: uncommon but regular winter visitor/resident (12 Oct–15 May)

NYC area, historical

Cruickshank's *NYC Region* as of 1941: irregular winter visitor (maximum 1000+) on eastern Long Island, Long Island Sound, Hudson, very rarely on freshwater reservoirs

Bull's *NYC Area* as of 1968: Long Island maxima winter 2500, spring 5000; occasional summer

Bull's *New York State* as of 1975: status unchanged

Study area, historical and current

A scarce reservoir visitor that has never been found in Van Cortlandt.

There are only 7 records, all of migrants or those frozen out farther north: at Hillview 23 on Apr 1939 (Norse, Komorowski) and from 8–19 Jan 2017 (Camillieri), with the rest on Jerome: 1 Jan 1949 (Norse); 2 on 4 Jan 1953 (Kane, Buckley); 12 Nov 1955 (Scully, Buckley); 3 Nov 1969 after a storm (Sedwitz); and 28 Dec 1975 (Sedwitz). Rumors of another there in late Dec 1995 were unconfirmable. Known from Central and Prospect Parks, but until Jun 1967 only once in each.

Bicknell recorded it in the Hudson River off Riverdale on 30 Nov 1877, 12 Oct 1879, and a flock on 26 Dec 1879. From the 1930s through the 1950s, it was seen there annually but irregularly between Nov and Mar, usually singles or flocks <10 that rarely lingered (Cruickshank, Kieran, Kane, Buckley). Since the 1950s few have been reported on the nearby Hudson River except at Piermont Pier, where it is infrequent but perhaps increasing, occasionally in small flocks. Yet 46 on the Hudson River at Inwood Park on 30 Oct 2014 (Farnsworth, Barrett) indicate that flocks still do occur nearby.

Bufflehead   *Bucephala albeola*

NYC area, current

A common spring/fall migrant and overwinterer on saltwater, including Long Island Sound and the Hudson River.

Bronx region, historical

Bicknell's *Riverdale* 1872–1901: unmentioned

Eaton's *Bronx + Manhattan* 1910–14: fairly common winter visitor

Griscom's *Bronx Region* as of 1922: rare on Long Island Sound, decreasing

Kuerzi's *Bronx Region* as of 1926: regular winter visitor/resident, increasing recently

NYC area, historical

Cruickshank's *NYC Region* as of 1941: winter resident (maximum 100+), commonest on eastern Long Island but also on Long Island Sound and larger inland lakes, reservoirs

Bull's *NYC Area* as of 1968: 16 Sep–late May, occasional summer; winter maximum 500

Bull's *New York State* as of 1975: Long Island maxima fall 1000, spring 400; winter 600 in Bronx

Study area, historical and current
A nearly annual fall migrant and early winter resident/visitor in very small numbers, especially on Jerome and Hillview but also on Van Cortlandt Lake.

In fall, when most often seen (especially on Van Cortlandt Lake), records range between Hillview on 26 Oct 2016 (Camillieri) and 2 on Van Cortlandt Lake on 8 Dec 1953 (Kane, Buckley), with a maximum of 14 on Jerome on 25 Nov 2017 (Cole et al.).

In winter most have been found on Jerome and Hillview, very rarely Van Cortlandt Lake, and usually left quickly; the maximum was 4 on Jerome on 27 Feb 1955 (Sedwitz). Successful overwintering within the study area occurred only once: on Jerome from 27 Nov 1930–14 Apr 1931 (Sedwitz), although singles seen irregularly on Van Cortlandt Lake during the winter of 1964–65 (Enders et al.) and at Hillview during Jan–Feb 2017 and Jan 2018 (Camillieri et al.) may also have done so.

Spring migrants are sparse, with 9 records: at Hillview 2 on 29 Mar 1964 (Horowitz), on 26 Apr 2016 and from 18–28 Mar 2017 (Camillieri et al.); at Jerome on 7 Mar 1994 (Lyons, Garcia) and 17 Mar 2016 (Karlson), with up to 11 from 4 Mar–18 Apr 2018 (Cole et al.); and on Van Cortlandt Lake on 26 Mar 1993 (Young), 1 Apr 2014 (Young), and 3–4 May 2014 (Furgoch, McGee).

Common Goldeneye *Bucephala clangula*

NYC area, current
A common spring/fall migrant and overwinterer on saltwater, including Long Island Sound and the Hudson River.

Bronx region, historical
Bicknell's *Riverdale* 1872–1901: unmentioned

Eaton's *Bronx + Manhattan* 1910–14: fairly common winter visitor

Griscom's *Bronx Region* as of 1922: now rare on Hudson, Long Island Sound; occasional on Jerome

Kuerzi's *Bronx Region* as of 1926: common winter visitor/resident (1 Nov–19 May) on Long Island Sound and East River, fairly common on larger reservoirs and rivers

NYC area, historical
Cruickshank's *NYC Region* as of 1941: very common fall migrant, winter resident (1 Sep–early Jun; maximum 8000), most numerous [eastern] Long Island Sound but also on larger inland lakes, reservoirs

Bull's *NYC Area* as of 1968: winter maximum 10,000

Bull's *New York State* as of 1975: status unchanged

Study area, historical and current
Now only an exceedingly infrequent fall and spring migrant and winter resident/visitor in very small numbers on Jerome and Hillview, rarely on Van Cortlandt Lake; study area overwintering is a thing of the past.

In fall, only 2 records known: 11 Nov 1938 on Hillview (Norse, Cantor) and 3 Nov 1959 on Jerome (Heath).

In the 1930s–40s goldeneyes overwintered from Dec–Mar at Hillview in small flocks (maximum 48 in 1936–37: Norse), but not since, although 1 from 9 Jan–20 Feb 2018 (Camillieri et al.) almost did. Otherwise only singles have been intermittently found there; the last was on 21 Jan 1952 (Buckley). At Jerome, where 1 was seen on 31 Dec 1914 (Griscom) and 10 on 2 Feb 1915 (Chubb), none were found again until singles on 30 Jan 1937 (Imhof), 23 Dec 1961 (Enders, Heath), and 24 Feb 1968 (Sedwitz). Another seen irregularly on Jerome during the winter of 1963–64 may also have overwintered (Enders et al.). Fewest of all have occurred on Van Cortlandt Lake: on 7 Feb–1 Mar 1956 (Buckley et al.), 24 Dec 1961 (Enders), 22 Feb 2008 (Finger), and from 10 Dec 2014–2 Jan 2015 (Young, Le Strange). None have certainly overwintered within the study area since the 1940s (Norse, Sedwitz), and the last large Hudson River count was 150 moving south on 8 Dec 1954 (Buckley).

In spring, the few singles are also grouped tightly: 15 Apr 1914 on Jerome (Kieran), 11 Apr 1939 Hillview (Norse), found dead at Van Cortlandt Lake on 6 Apr 1960 (Buckley), 5 Apr 1970 on Van Cortlandt Lake (Kane), 21 Mar 1976 on Jerome (Sedwitz), and 5 Apr 1976 on Van Cortlandt Lake (Kane).

Comments
Arctic **Barrow's Goldeneye**, *B. islandica*, was always an exceedingly rare winter visitor to eastern Long Island, even until the time of Bull (1964). But in recent years it has been occurring more frequently, as far west (or south) as the Bronx at Ferry Point Park/Throgg's Neck in the winters of 1990–91, 1991–92, and 1992–3 (the same ♀), with another ♀ at City Island on 26 Dec 2011 (all Walter et al.) and a ♂ at Pelham Bay Park from 7–13 Mar 2016 (Jamison, Rothman; photos). It was also seen once in Orange County on inland Lake Tiorati from 25–29 Apr 1954—an SY ♂ heading back north after having overwintered somewhere farther south, itself an amazing occurrence in those days (Steffens, Orth et al.).

Hooded Merganser *Lophodytes cucullatus*

NYC area, current
A locally common spring/fall migrant and erratic overwinterer on fresh water until frozen out to sheltered saltwater. An extremely scarce but perhaps overlooked breeder.

Bronx region, historical
Bicknell's *Riverdale* 1872–1901: once only: 21 Nov 1880

Eaton's *Bronx + Manhattan* 1910–14: uncommon migrant

Griscom's *Bronx Region* as of 1922: very rare on Jerome, with 2 records in 1914, 1915

Kuerzi's *Bronx Region* as of 1926: rare but regular migrant, winter resident (24 Oct–6 Apr)

NYC area, historical
Cruickshank's *NYC Region* as of 1941: uncommon migrant in spring/fall (maximum 48) on freshwater, often attempting to overwinter until frozen out to saltwater

Bull's *NYC Area* as of 1968: coastal winter maxima 100–135; bred Westchester County 1966

Bull's *New York State* as of 1975: status unchanged

Study area, historical and current
A spring, fall, and early winter migrant, and winter resident.

In fall, first migrants arrive in late Sep, peak in early Nov at 10–15 and depart after the first heavy freeze by late Nov, with extreme dates of 9 Sep 1953 (Buckley) and 22 (the maximum) on 25 Nov 2017 (Cole et al.)

During the 1930s and 1940s Hoodeds overwintered at Hillview in small flocks to 25 (Norse) and in groups to 5–6 in Van Cortlandt (Kieran) but never thereafter in numbers on Hillview. Since the 1980s, overwintering in Van Cortlandt has increased, and a few move to Jerome after freeze-ups but never stay because of continuing prey unavailability. Recent Van Cortlandt maxima have been increasing, with a current peak of 40 on 18 Jan 2008 (Klein), changes mirrored on the Bronx-Westchester Christmas Bird Count where since the early 1990s they have been steadily increasing (Fig. 30).

In spring, first migrants arrive in mid-Mar, peak in late Mar, and depart by mid-Apr, with extreme dates of 3 Mar 2012 (Rofe) and 23 Apr 1961 (Tudor) and a maximum of 10 on 22 Mar 2014 (DiCostanzo).

Hooded might breed if predator-proofed, properly crafted nest boxes were installed in Van Cortlandt Swamp, even though there are no May–Aug study area records. The nearest breeders are at the Great Swamp or in extreme southwestern CT, but ♀♀ and young were seen at Cross River Reservoir, Westchester County, in Jun 1966 (Grierson) and May 2012 (Bochnik), and just-fledged HYs in Bronx Park from 17–19 Jun 1955 (C. Young et al.) and at Hillview on 11 Jul 2017 (Camillieri) also indicate breeding not far away.

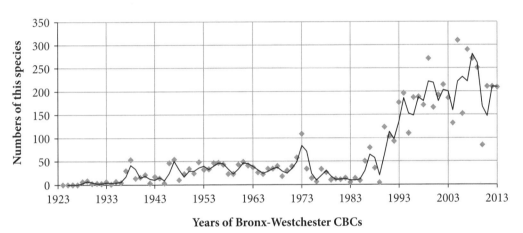

Figure 30

Common Merganser *Mergus merganser*

NYC area, current
A regular and locally numerous winter resident on freshwater in the Bronx and Westchester County but in far smaller numbers and highly locally on Long Island unless forced out from inland lakes by severe freeze-ups. Most migrate directly overland, when they are sometimes forced down onto freshwater by bad weather.

Bronx region, historical
Bicknell's *Riverdale* 1872–1901: unmentioned

Eaton's *Bronx + Manhattan* 1910–14: uncommon winter visitor

Griscom's *Bronx Region* as of 1922: very common on Hudson, especially north of [future George Washington Bridge], where sometimes abundant during severe cold waves if river frozen solid farther north

Kuerzi's *Bronx Region* as of 1926: regular, common freshwater winter visitor (1 Nov– 13 May)

NYC area, historical
Cruickshank's *NYC Region* as of 1941: common winter visitor (10 Sep–mid-May; maximum 1500 on Hudson at Croton Point) on large inland lakes until frozen out, scarcest on Long Island; commonest on Hudson north of Yonkers but regular in small numbers on inland lakes, reservoirs in spring/fall migration, winter

Bull's *NYC Area* as of 1968: winter maximum 2300 on Hudson in Rockland County; breeding unknown

Bull's *New York State* as of 1975: bred Orange County in 1973

Study area, historical and current
A spring/fall migrant and winter resident when numbers were much higher in the 1930s and 1940s than since, with largest counts on Hillview and Jerome. It has always been infrequent in Van Cortlandt.

In fall, first migrants arrive in late Oct and peak in Nov, after which they depart if not overwintering, with the earliest in Van Cortlandt on 13 Sep 1946 (Komorowski) and a maximum of 7 on Jerome on 9 Oct 1953 (Buckley).

In winter, all but restricted to Jerome <2008 when it was drained, in flocks <25 (Winter Bird Population Study means of 20 in 1963–64 and 9 in 1965–66), but much larger flocks overwintered during the 1930s there and at Hillview, where the maximum was 300 on 22 Feb 1938 (Norse). This contrasts sharply with winter maxima of 48 at Jerome on 5 Feb 1955 (Kane) and 52 at Hillview on 21 Feb 2018 (Camillieri)—the highest counts there after the 1930s, although 180 on Grassy Sprain Reservoir on 26 Dec 2016 (Christmas Bird Count) was nostalgic. In mild winters some tried to stay on Van Cortlandt Lake all winter, to a maximum of 30 on 10 Jan 1953 (Buckley), but none have done so since the 1970s.

Numbers on Jerome dropped to zero after the draining and refilling in 1966–67 because the small fishes they were catching had been extirpated. Before, an average of about 8 overwintered annually from Dec–Apr, diving frequently and successfully. Numbers never recovered from the 1966–67 draining, as only the occasional single migrant was ever seen.

In spring, first migrants arrive in early Mar, peak in late Mar, and depart by mid-Apr, with extreme dates in Van Cortlandt of 28 Feb 2009 (Young) and 3 from 15 Apr–14 May 1953 (Buckley et al.), with maxima of 18 on 14 Mar 2005 (Young) and 20 on 21 Mar 1952 (Kieran). The only spring migration reservoir counts are from Hillview: 65 on 13 Mar 1937 (Imhof) and 200 on 14 Mar 1936 (Weber et al.), although some of those may have overwintered.

Nearest breeders are now along the Croton River in Westchester County as well as in an expanding population along the Delaware River and adjacent inland areas in north-central NJ and in central Orange and Rockland Counties.

Red-breasted Merganser *Mergus serrator*

NYC area, current
A common saltwater spring/fall migrant and winter resident around Long Island and New York City, including Long Island Sound and the Hudson River. Except when forced down by extreme weather, it scrupulously avoids freshwater, even in migration. It is also a very rare but longtime breeder in Long Island saltmarshes.

Bronx region, historical
Bicknell's *Riverdale* 1872–1901: unmentioned

Eaton's *Bronx + Manhattan* 1910–14: common migrant

Griscom's *Bronx Region* as of 1922: rare winter visitor to Long Island Sound; common on Hudson 50 years ago but no longer.

Kuerzi's *Bronx Region* as of 1926: fairly common migrant, winter visitor (15 Oct–7 Jun), sometimes numerous in saltwater

NYC area, historical
Cruickshank's *NYC Region* as of 1941: abundant spring/fall migrant, winter resident (maximum 400) on saltwater into East, Hudson Rivers; extremely rare on freshwater in migration

Bull's *NYC Area* as of 1968: 1937–41 Nov maxima 10,000–40,000 from eastern Long Island not approached since

Bull's *New York State* as of 1975: status unchanged

Study area, historical and current
A scarce migrant and nearly unrecorded winter visitor to Jerome and Hillview, with a handful of records from Woodlawn and Van Cortlandt. Notwithstanding, it is a common migrant and winter resident as close as Pelham Bay Park and in smaller numbers on the Hudson River.

There are only 13 study area records, all in winter/spring: at Hillview on 22 Feb 1938 (Norse), 1–4 Apr 1939 (Norse, Cantor), 18 Jan (Camillieri) and 15 Feb 2017 (Adams); at Jerome 20 ♂♂ and ♀♀ on 1 Apr 1953 (Buckley), on 13 Dec 1969 (Kane, Buckley), 27 Feb 1977 (Sedwitz), 5 from 17 Mar–4 Apr 1994 (Lyons, Garcia); and on 14 Mar 2018 (Karlson); low over Woodlawn on 6 May 1961 (Horowitz); and on Van Cortlandt Lake on 20 Dec 1993 (Garcia), 13–16 Apr 2007 (Klein), and from 31 Jan–6 Feb 2013 (Drake, Young, McGee). Daily in-season reservoir coverage should yield additional records. More frequently reported in Central and Prospect Parks, but only 2 winter/spring records at the nearby Bronx Zoo: 17 Jan 1980 (J. Gillen) and 1–9 Mar 2013 (Prelich, Olson).

Ruddy Duck *Oxyura jamaicensis*

NYC area, current
A common but local fall migrant and winter visitor/resident on fresh- and saltwater.

Bronx region, historical
Bicknell's *Riverdale* 1872–1901: unmentioned

Eaton's *Bronx + Manhattan* 1910–14: fairly common migrant

Griscom's *Bronx Region* as of 1922: common only in Croton Point area; single Bronx records in 1915, 1921

Kuerzi's *Bronx Region* as of 1926: irregular and uncommon migrant, rare in midwinter (24 Oct–15 May)

NYC area, historical
Cruickshank's *NYC Region* as of 1941: common eastern Long Island fall migrant (maximum 400), rare in winter; elsewhere uncommon in migration on freshwater, occasionally attempting to overwinter

Bull's *NYC Area* as of 1968: greatly increased >1945, Long Island maxima fall 1200, winter 2000; bred Jamaica Bay Wildlife Refuge 1955–63 (maximum 40 pairs)

Bull's *New York State* as of 1975: status unchanged

Study area, historical and current
A migrant and winter resident in all 4 extant subareas but only singles in Woodlawn and usually

<10 in Van Cortlandt. Most numerous and regular at Jerome and formerly at Hillview, where the handful still using it were eliminated for "pollution control" by the New York City Department of Environmental Protection from 2010 on.

the early 1950s and its subsequently fluctuating numbers (Fig. 31).

In fall, first Jerome migrants arrive in mid–late Sep, peak in Nov, and if not overwintering, depart by late Nov, with 3 records in Aug, the

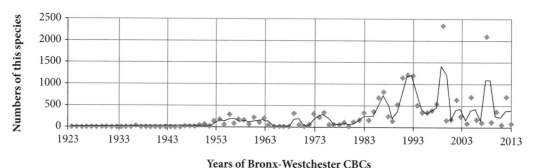

Years of Bronx-Westchester CBCs

Figure 31

Large numbers have favored Jerome since the early 1950s except when it was drained from Dec 1966–May 1967 and from 2008–14, but it did not take them long to return to Jerome after refilling: 140 were there on 17 Mar 2016 (Karlson). Believed by many to only rest at Jerome, Ruddies nonetheless actively dive there, and finally Enders (1975) solved the mystery of what they may have been eating: small freshwater clams, midges, and a variety of other invertebrates in a rich bottom organic layer. Ruddies on Jerome also feed elsewhere but arrive and depart only by dark, as no one has seen them coming or going there or anywhere else in the study area. It is assumed they feed on the Hudson and East Rivers and on Long Island Sound, but this is conjecture. Oddly, before the 1950s no one mentioned Ruddies occurring regularly on Jerome, and the 2 that overwintered there in 1929–30 (Bronx County Bird Club) were worthy of note. Even Cruickshank (1942) called it essentially a bird of only eastern Long Island, but by 1950 that had clearly changed. The first large count was 150+ on 4 Jan 1953 (Kane, Buckley)—the start of recent high numbers. Bronx-Westchester Christmas Bird Count totals confirmed its absence until

earliest on 28 Aug 1963 (Sedwitz), and maxima of 400+ in early Nov 1957, 520 on 15 Nov 1963 (Enders), and 620 on 17 Nov 1963 (Sedwitz)—the most ever on Jerome. More typical counts have been in the low 100s, although some years only a handful ever appeared. In the 1930s–40s, it occurred at Hillview on the same timetable, but counts rarely exceeded 50 (Norse). On Van Cortlandt Lake first migrants arrive in early Nov, peak in Nov, and depart by late Nov, with early dates of 3 Oct 1936 (Norse) and 2 on 9 Oct 1955 (Sedwitz, C. Young), and a maximum of 7 on several occasions. Highest counts from nearby are even larger: 2000 on the Hudson River at Nyack on 12 Dec 1965 (Hopper, Polhemus) and 2500 on Central Park Reservoir on 16 Dec 1997 (m. ob.).

In winter the bulk of study area Ruddies are on Jerome, although Nov high counts drop off in Dec to 100–300, but by Jan, depending on icing, none may appear until the first trickle of spring migrants in Mar (Sedwitz). On Hillview the maximum was 24 on 13 Jan 1951 (Komorowski). On Van Cortlandt Lake 6–8 may try to overwinter, but they are usually frozen out during Jan and move elsewhere.

In spring, the very light northbound flight begins in early Mar, peaks in late Mar, and is over by early Apr, with extremes in Van Cortlandt of 5 on 3 Mar 2012 (Rofe) and on 1 Jun 1959 (Zupan), and a maximum of 14 on 14 Mar 1993 (Young). Often only 1–2 are on Jerome from late Mar into May.

On Van Cortlandt Lake, Ruddies sometimes linger late (1 May 1955: Kane et al., 2 May 1998: Young) and in 1956 two nonbreeders on Van Cortlandt Lake from 29 Mar–26 Jul 1956 (Kieran et al.) constitute the only Van Cortlandt oversummering attempt. On Jerome, Sedwitz (1977, 1979) found Ruddies once in May, 3 remained there until 15 May 1951 (Russak), and in 1981 and 1982 a ♂ and ♀ also oversummered. Nearest regular breeders (since the late 1950s) have been in the Hackensack Meadowlands and at Jamaica Bay Wildlife Refuge.

New-World Quail: Odontophoridae

Northern Bobwhite *Colinus virginianus*

NYC area, current
A native resident game bird that is close to the northern edge of its range and that has declined severely with urbanization since the 1930s–40s. It has also been subject to periodic supplements released by state and sometimes private groups, although most of these have been less hardy than native quail and perished quickly from disease, weather, local quadrupeds, and other predators. Some reported in the early 2000s on Staten Island were released.

Bronx region, historical
Bicknell's *Riverdale* 1872–1901: common resident

Eaton's *Bronx + Manhattan* 1910–14: uncommon resident

Griscom's *Bronx Region* as of 1922: formerly common resident, remnant survives in Van Cortlandt Swamp

Griscom's *Riverdale* as of 1926: now extirpated in Riverdale

Kuerzi's *Bronx Region* as of 1926: fairly common resident in Pelham section, 1–2 pairs still in Van Cortlandt

NYC area, historical
Cruickshank's *NYC Region* as of 1941: disappearing resident throughout, still breeds in Bronx; 12–48 on recent Bronx-Westchester Christmas Bird Counts

Bull's *NYC Area* as of 1968: uncommon to rare resident everywhere, greatly decreased recently

Bull's *New York State* as of 1975: status unchanged

Study area, historical and current
A former resident throughout, long extirpated. It was last recorded in Van Cortlandt (which had been supporting 1–2 dozen) on 20 May 1942 (S. Horowitz), was never recorded or was overlooked at Hillview, was last recorded at Jerome on 16 Jan 1935 (Weber, Stephenson), at nearby Inwood Park on 24 Jan 1936 (Weber, Banner), where 18 were feeding on the snow, and in Riverdale on 24 Apr 1942 (Kieran).

It has long been assumed or believed that native Bobwhite numbers in and around New York City had declined by the mid–late 1800s and that southern and southwestern stock had been released in an effort to bolster them. A Bronx resident writing as "JWD" (actual name unknown)

contributed his beliefs and recollections to a *Forest and Stream* feature (vol. 60, no. 1:7) entitled "Game in Greater New York" in 1903:

> I should estimate the number of quail in the entire borough of the Bronx at between 150 and 200. This is a very conservative estimate. There were to my personal knowledge at least three bevies at Eastchester, another between Williamsbridge and Westchester, and a fifth at the extreme southern end of Throgg's Neck. I do not know how many birds [were] located in Pelham [Bay] Park this year; but there are always quail there in considerable numbers, and the same statement is true of Van Cortlandt Park and the adjacent country. [But t]hese are not the original quail of [Bronx] county. While there have always been a few of the birds here, they are from year to year becoming more plenty. Furthermore, intelligent old sportsmen of these parts assure me present birds are not so large as the former occupants. The present quail unquestionably entered the Bronx from a New Rochelle preserve by way of Pelham [Bay] Park. They have spread very generally throughout the [county] but never until this year have they come so far south as Throgg's Neck.

By the early 1950s quail were also rumored to be surviving in the Northeast Bronx, an area then supporting large truck farms and splendid quail habitat. Nonetheless, there is only a single Bronx mainland record since the 1940s: 2 that were calling on 19 Sep 1954 in a 10-acre dense sunflower field at Baxter Creek, in the East Bronx near the Whitestone Bridge (Phelan, Bauer, Bauer, Buckley). This site was absorbed into St Raymond's Cemetery in the late 1960s and the habitat obliterated. The area was remarkably wild in the mid-1950s, and local residents were unaware of anyone with captive quail; the area easily could have supported the last Bronx mainland population of quail.

Bobwhites had been rumored to be surviving on Riker's Island until the late 1940s, and so on 7 Dec 1969 Sedwitz and Buckley received permission to drive there. With perplexed but nonetheless intrigued Corrections Department escorts, to our great surprise we quickly located 20–30 Bobwhite, were assured they had always been there, and were told they were wild. Riker's Island had lots of open habitat and food then (we also saw a Red-tailed Hawk, 2 Rough-legged Hawks, 5 American Kestrels, and 40+ pheasants that day), but on Google Earth it now looks as if it would barely support starlings, so the quail and pheasants are long gone. Even so, NYSBBA I in 1980–85 showed quail breeding in a square very close to but not embracing Riker's Island. This suggests they survived somewhere in northern Queens until 1980–85, although details are lacking, and recently released birds may have been involved.

Bobwhites in Pelham Bay Park in Sep 1986 and Nov 2011 and in Bronx Park in Aug 2009 and the zoo in Jun 2013 were not wild, but could they have survived from the NYC Parks Department introductions at Pelham in the late 1990s? Surely. Were the much earlier Baxter Creek and Rikers Island quail also recent releases? Unlikely; current information indicates they were probably the last of the earlier population.

During NYSBBA I in 1980–85, presumed relict populations of native Bobwhite were still in New York City in 2 Queens blocks and in 5 in northern Westchester County. But by NYSBBA II in 2000–2005 none were in New York City (those at Jamaica Bay Wildlife Refuge were first stocked in the 1950s), Westchester, Putnam, or Rockland Counties, and almost none in Orange County or (only) extreme northern Nassau County. Fortunately, Suffolk County's population seemed to be holding its own, even if somewhat diminished. DOIBBS data for this species in New York State showed a precipitous decline beginning as early as 1966 and continuing unabated (Sauer et al. 2005; McGowan and Corwin 2008). This drop was mirrored on statewide Christmas Bird Counts, and Bobwhite has not been found even in Westchester County on a Bronx-Westchester count since 1941, excepting 3 in 1964 (new releases?).

Pheasants, Grouse: Phasianidae

Ring-necked Pheasant *Phasianus colchicus*

NYC area, current
Formerly a widespread resident breeding throughout, even in New York City parks and along barrier beaches. Populations peaked in the 1960s and then began a decline continuing to this day, generally ascribed to habitat loss from succession and development, feral cats and Eastern Coyotes, and increasing numbers of urban-breeding Red-tailed Hawks and Great Horned Owls. It is now absent from locations where it was common in the 1940s–60s. Only likely recent releases have ever been found in Central or Prospect Parks.

Bronx region, historical
Bicknell's *Riverdale* 1872–1901: unrecorded

Eaton's *Bronx + Manhattan* 1910–14: unrecorded

Griscom's *Bronx Region* as of 1922: first noted in 1916; uncommon but increasing

Kuerzi's *Bronx Region* as of 1926: increased markedly since introduction several years ago; most common in eastern section, well established in interior

NYC area, historical
Cruickshank's *NYC Region* as of 1941: common resident throughout, well established in Bronx [locations not given]

Bull's *NYC Area* as of 1968: status unchanged

Bull's *New York State* as of 1975: status unchanged

Study area, historical and current
A formerly well-established study area exotic (nonnative) breeder but probably extirpated about 2007.

First introduced to North America in New York between 1730 and 1733 but failed to become established. Other early introductions occurred in New York in 1877 and 1886–91, NJ in 1887, MA in 1897–98. Releases of various subspecies and intergrades continued into the 1970s in established and new locations. Distribution was relatively stable since 1941, although abundance has declined throughout much of its North American range (Guidice and Ratti 2001).

An exotic (nonnative) species of mixed origins and taxa introduced in the New York City area on multiple occasions but established by about 1890. "JWD" (see the Bobwhite species account) wrote in 1903 that "pheasants, like quail, have found their way out of a New Rochelle preserve and are spreading over a considerable area. They are mostly seen near New Rochelle, Mt. Vernon, and Bronxville, but several have been potted [sic] near Pelham [Bay] Park and Eastchester," earlier than the 1916 first report in Griscom's Bronx Region (which did include lower Westchester County from Yonkers east to New Rochelle). They were on the first Bronx-Westchester Christmas Bird Count in 1924, with numbers gradually increasing (perhaps amplified by periodic restocking or escapes) to a peak in 1960, then declining—a pattern duplicated in the study area.

There, they first bred in Van Cortlandt, Woodlawn, and maybe also Jerome beginning about 1932 (Norse, Kieran), peaking 30 years later with an estimated 100 in Van Cortlandt in the early 1960s (Meyer), and a daily maximum of 43 across Van Cortlandt and Woodlawn on 26 Dec 1960 (Tudor et al.). None have been there on a Bronx-Westchester Christmas Bird Count since a single from 2001–6, and the very last eBird record was 1 on 29 Apr 2007 (Klein). Area-wide data from the Bronx-Westchester Christmas Bird Count (Fig. 32) confirmed that their numbers peaked in the late 1950s–early 1960s, only to begin a steady decline thereafter to (near) extirpation by 2010 or so.

Counties, they were down to <20 blocks, and in eastern Suffolk County were doing better but greatly reduced (McGowan and Corwin 2008). DOIBBS data for this species supported a consistent statewide decrease between 1966 and 2005 (Sauer et al. 2005). Nearest breeders

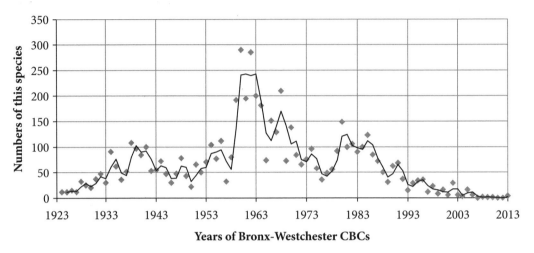

Figure 32

Even though NYSBBA I in 1980–85 and II in 2000–2005 recorded breeders in Blocks 5852B and 5952A, which include the study area (McGowan and Corwin 2008), none have been seen in the breeding season in the study area since about 2008 (Young), so it now appears that pheasants have been extirpated. They were widely distributed within New York City during NYSBBA I, but by NYSBBA II were in many fewer blocks. In Nassau and western Suffolk

to the study area are unknown but may be in Pelham Bay Park, where they are also almost gone (Künstler), or possibly on nearby Hart's Island or somewhere in the Southeast Bronx—like the single at Pugsley Creek Park on 18 Jun 2015 (Victor). Reasons for the decline are uncertain but may include some combination of habitat changes and increasing numbers of resident Great Horned Owls, Red-tailed Hawks, Raccoons, and Eastern Coyotes.

Ruffed Grouse   *Bonasa umbellus*

NYC area, current
Now a declining resident almost gone from eastern Long Island and Westchester County, the nearest healthy population being on the Orange-Sullivan County border.

Bronx region, historical
Bicknell's *Riverdale* 1872–1901: unrecorded

Eaton's *Bronx + Manhattan* 1910–14: formerly resident

Griscom's *Bronx Region* as of 1922: long extirpated

Kuerzi's *Bronx Region* as of 1926: 1–2 pairs may survive on Sprain, Elmsford Ridges

NYC area, historical
Cruickshank's *NYC Region* as of 1941: occasional at Grassy Sprain; long extirpated from Bronx

Bull's *NYC Area* as of 1968: status unchanged

Bull's *New York State* as of 1975: status unchanged

Study area, historical and current
A long-extirpated resident still breeding in northeastern Westchester County, where it is rapidly declining.

Eaton (1910) explicitly stated that it was probably extirpated from Richmond, Kings, and New York (which then included the Bronx) Counties while pointedly omitting Queens, but Bicknell (in Griscom 1927) does not mention it occurring in his greater Riverdale area from 1875–1902—unless he deemed its occurrence trivial, which seems unlikely.

"JWD" (see the Bobwhite species account) expressed his feelings about Bronx grouse in 1903 thus:

> [Since 1892] I have never seen a ruffed grouse anywhere within [the Bronx] and I do not believe there have been any here in many years. The last bird seen nearby, so far as I can tell, was killed above New Rochelle—out of the Bronx—[in 1887]. If by any possibility there should be a bird or two here, it would likely be in the upper part of Van Cortlandt Park, or between that point and the Mt. Vernon line. Old residents tell me they have not killed or seen grouse east of the Bronx River [since about 1868]. But at least a few of the birds remained in or near Van Cortlandt Park much later than this.

Nearest breeders to Van Cortlandt were only 5 smi (8 km) away at Grassy Sprain Reservoir from the 1950s to the 1980s. Even closer, in 1971 2 pairs bred on the Palisades between the George Washington Bridge and Alpine (Boyajian 1971), still there through the mid-1990s (Walsh et al. 1999), so occurrence in Van Cortlandt becomes less surprising. On the Bronx-Westchester Christmas Bird Count after 1985 in southern Westchester County they were recorded only once, in 1998, but until then had been fairly regular.

In the study area, the first from the Bronx in many, many scores of years was on the Ridge in Van Cortlandt on 28 Dec 1955 (S. Friedman, B. Friedman, Steineck) and was carefully described by all observers. Even more unexpected given the location was 1 at Baxter Creek on 21 Aug 1966 (Kane), but equally out of place were the 3 seen in the Hackensack Meadowlands on 3 Aug 1974 (Sturm) and another at Pelham Bay Park in (no date) Aug 1985 (Morrison, Sisinni). These dates all indicate postfledging dispersal.

This species is not so rock-sedentary as generally believed, being well known to wander postbreeding especially in years when its population is very high—during so-called crazy seasons. During NYSBBA I in 1980–85 they were widespread, if thin, in Westchester County, but by NYSBBA II in 2000–2005 they were found in only 7 blocks in the extreme northeast (McGowan and Corwin 2008). DOIBBS data for this species in New York State affirmed a consistent decrease between 1966 and 2005 (Sauer et al. 2005).

Wild Turkey   *Meleagris gallopavo*

NYC area, current
A recent recolonizer following reintroductions in upstate New York. Amazingly, now recorded in all 5 New York City boroughs and in Central and Prospect Parks. Small established breeding populations exist in several Bronx locations and perhaps also in Queens. Well established in Westchester County and eastern Suffolk County but very scarce farther west.

NYC area, historical
Cruickshank's *NYC Region* as of 1941: extirpated since colonial times

Bull's *NYC Area* as of 1968: 42 released in Putnam County 1959; 500 released Gardiner's Island 1954; no others known

Bull's *New York State* as of 1975: restocked northern Westchester, Putnam Counties 1959

Levine's *New York State* as of 1996: restocked birds spreading south to southern Westchester County, Bronx

Historically, turkeys were resident throughout the immediate New York City area through the 1600s but disappeared no later than the early 1700s. They were gradually pushed farther inland to wilder areas, and although in remote parts of Orange and Rockland Counties until 1840, they vanished from New York State not long after 1850. They persisted in the wildest parts of central PA into the mid-20th century, by which time they began to reinvade New York State. These formed the stock from which all successful northeastern introductions sprang (Eaton 1964).

The first main New York State introduction using wild birds ran from 1952–57 in the Southern Tier and the Catskills, with turkeys establishing themselves as close to New York City as the Orange-Sullivan County border. Attempts in the late 1950s to reintroduce farm-raised birds in Putnam and northern Westchester Counties failed, as had all similar previous attempts (Eaton 1988). However, wild-caught birds released in 1974 in Dutchess County predictably did survive and are believed to be the source of those that quickly spread throughout Westchester County and eventually into New York City (Künstler 2000). During NYSBBA I in 1980–85, turkeys were no closer than central Westchester County, but by NYSBBA II in 2000–2005 they were in 5 Bronx blocks and 1 on Staten Island. The invasion was well under way. DOIBBS data for New York State identified a consistent increase beginning about 1985 (Sauer et al. 2005; McGowan and Corwin 2008).

Study area, historical and current
An introduced, now breeding native that recently recolonized the study area from the north.

The first within New York City since colonial times seems to have been a ♀ in Bronx Park in summer 1986 that stayed until 1989, and another was in Pelham Bay Park from 1988. The Bronx Park bird was eventually joined by a second, and they seemed to move between there and the Bronx Zoo, where 2 nests with eggs were finally found in 1999, although no ♂♂ have been mentioned. The Pelham Bay Park population was growing rapidly and had reached 35 by 1999 (Künstler 2000).

In the study area, the first was a ♀ in Van Cortlandt Swamp on 30 Nov 1993 (Young), followed quickly by another in Riverdale on 26 Dec 1993 (Buckley, Roche—who were told by residents that a ♂ and several ♀♀ had been there since May.) The first Van Cortlandt ♂ startled the driver when it flew over his car on the Major Deegan Expressway in Van Cortlandt on 3 Apr 1994 (Schutz). The Van Cortlandt ♀ remained until Mar 1997 (Lyons, Garcia), and then others began to be reported in the Shandler Area, starting with 2 on 18 Oct 1998 and reaching 6 on 21 Jan 2001 (unknown observer, *Kingbird*). They have been breeding in the study area since the late 1990s, although there are no estimates of total population size, as many disperse and emigrate after fledging. At least 1–2 displaying ♂♂ and an unknown number of breeding ♀♀ are resident in Van Cortlandt and Woodlawn. The first proven breeding was a ♀ with 6 young on the Ridge in Jul 2004, so there were clearly more and earlier than we know about. The only actual nest found so far, with a ♀ incubating 12 eggs on 10 Aug 2005 (Künstler et al.), was along Mosholu Parkway Extension.

Not content with conquering the Bronx, the first for Manhattan Island since the 1600s made it to Inwood Park on 21 Apr 1994 (unknown observer, NYRBA), and there were 2 ♀♀ and 1 ♂ there on 18 Dec 1999 (unknown observer, *Kingbird*). Subsequently, attacks by dogs eliminated any breeding population that may have established itself there (J. DiCostanzo, pers. comm.). But Zelda, the ♀ that arrived at Battery Park on the south tip of Manhattan Island in 2003, survived 11 years until struck by a car not far from the South Street Seaport in 2014. If you can make it in New York . . .

Loons: Gaviidae

Red-throated Loon *Gavia stellata*

NYC area, current
A common spring/fall coastwise migrant that also occurs on Long Island Sound and the Hudson River. Overwinters commonly on all regional saltwater bodies.

Bronx region, historical
Bicknell's *Riverdale* 1872–1901: unrecorded

Eaton's *Bronx + Manhattan* 1910–14: rare winter visitor

Griscom's *Bronx Region* as of 1922: 2 old Hudson records, otherwise unknown in interior

Kuerzi's *Bronx Region* as of 1926: uncommon but regular fall Long Island Sound migrant (11 Oct–22 Feb); unrecorded elsewhere

NYC area, historical
Cruickshank's *NYC Region* as of 1941: common migrant in spring (maximum 9)/fall (maximum hundreds), winter resident around Long Island but rare elsewhere (maximum only 6), several Hudson records; oversummered Hillview Jul–Nov 1936

Bull's *NYC Area* as of 1968: maxima spring 75, fall 100, winter 480

Bull's *New York State* as of 1975: status unchanged

Levine's *New York State* as of 1996: Long Island winter maximum 686

Study area, historical and current
An irregular spring migrant that has oversummered once and is occasional in winter.

Migrates up and down the Hudson River (e.g., singles on 7 and 22 Oct 1982: Sedwitz) and heavily on Long Island Sound (dozens in early Nov between 2000 and 2014 at Ft. Schuyler: Buckley), yet never recorded as a fall migrant in the study area.

To date, 11 records: at Hillview from late Jul–3 Nov 1936 (Norse, Cantor et al.); then on 25 Feb 1951 (Komorowski—oiled), 20 Jan 2017 (Adams, Batren), and 2 Feb 2018 (Camillieri), but only once on Jerome: on 9 Apr 2018 (Horan). On Van Cortlandt Lake it occurred on 27 Apr 1941 (Norse), 15 Mar 1942 (Norse, Cantor), 4 Apr 1945 (Ryan), 3–17 Mar 1946 (Komorowski, Ephraim—oiled), 4–16 Mar 1951 (Kane, Buckley et al.), and 17–19 Apr 1996 (Young, Garcia, Lyons). Another unexpected nonbreeder was at Spuyten Duyvil from 11–14 Jun 2009 (Alomeri, Allen).

Common Loon *Gavia immer*

NYC area, current
A regular and locally numerous winter resident on saltwater around Long Island and in much smaller numbers in the Bronx and Westchester County. Migrates coastwise in fall and spring but also overland in spring when they are sometimes forced down onto freshwater by bad weather. It is a New York State Species of Special Concern.

Bronx region, historical
Bicknell's *Riverdale* 1872–1901: unmentioned

Eaton's *Bronx + Manhattan* 1910–14: occasional winter visitor, fairly common migrant

Griscom's *Bronx Region* as of 1922: rare migrant on Hudson, larger reservoirs, Long Island Sound

Kuerzi's *Bronx Region* as of 1926: regular, fairly common spring/fall migrant, winter visitor (16 Aug–2 Jun)

NYC area, historical
Cruickshank's *NYC Region* as of 1941: common migrant in spring/fall (maximum 100+), winter resident, rare on inland reservoirs

Bull's *NYC Area* as of 1968: seen with increasing frequency, larger numbers on inland waters; maxima fall 470, winter 200, spring 300

Bull's *New York State* as of 1975: spring maximum 600

Study area, historical and current
An annual but uncommon spring migrant and sparse winter resident/visitor, apparently unknown on the water at Jerome Reservoir. Migrants routinely pass overhead in Apr–May, frequently yodeling as they go—an eerie but wondrous experience.

In spring, first migrants arrive in late Apr, peak in early May, and depart by mid-May, with extreme dates of 12 Apr 1953 (Norse, Cantor) and 2 adults flying over Van Cortlandt on 7 Jun 2014 (Souirgi et al.), and a maximum of 30 on 7 May 1955 (Norse, Cantor). Until Jun 1967, the spring maximum in Prospect Park was 100 overflying on 7 May 1950 (Jacobson, Sedwitz).

The sole study area fall migrant was at Hillview on 31 Oct 2016 (Camillieri), and there are only 4 winter records: Van Cortlandt Lake 17 Dec 1938 (Norse, Cantor, M. Gell-Mann) and Hillview on 22 Feb 1939 (Norse), 27 Jan and 15 Feb 2017 (Adams).

DOIBBS data for this species in New York State depicted a consistent increase between 1966 and 2005 (McGowan and Corwin 2008).

Grebes: Podicipedidae

Pied-billed Grebe *Podilymbus podiceps*

NYC area, current
Always a very local breeder in small numbers but a widespread and numerous spring/fall migrant, and erratic winter visitor/resident in small numbers. It has been declining as a breeder since the 1940s, yet it bred in Alley Pond Park

in 1951 and Prospect Park in 1983. It is a New York State Threatened Species.

Bronx region, historical

Bicknell's *Riverdale* 1872–1901: unmentioned

Eaton's *Bronx + Manhattan* 1910–14: common migrant, occasional resident

Griscom's *Bronx Region* as of 1922: rare migrant, only 2 recent spring records.

Kuerzi's *Bronx Region* as of 1926: formerly bred [locations not given], now fairly common migrant, chiefly fall; rare in winter on Long Island

NYC area, historical

Cruickshank's *NYC Region* as of 1941: fairly common migrant in spring/fall (maximum 17), does not breed within New York City

Bull's *NYC Area* as of 1968: Long Island fall maximum 125; 250 on 1953–54 Long Island Christmas Bird Counts

Bull's *New York State* as of 1975: status unchanged

Study area, historical and current

A spring/fall migrant, irregular in winter. Recorded every month of the year and may have bred from the 1930s to the 1950s.

In fall, first migrants arrive in late Aug, peak in late Sep–Oct, and depart by early Nov if not attempting to overwinter, with extreme dates of 3 Aug 1952 (Kane, Buckley) and 25 Nov 2012 (Baksh) and a maximum of 14 on 17 Nov 1943 and 12 on 19 Oct 1951 (Kieran), but only 2–3 since the late 1950s.

One successfully overwintered on Van Cortlandt Lake in 1959–60 (Meyer), and the early winter maximum there was 8 on several occasions. Van Cortlandt residents once frozen out sometimes moved to Jerome and Hillview but have never successfully overwintered there except on Jerome in 1965–66 (Enders et al.). The odd bird may show up on Van Cortlandt Lake almost anytime in the winter if there is open water—like the 2 in Van Cortlandt Swamp on 23 Jan 1997 (Künstler).

In spring, first migrants arrive in mid-Mar, peak in early–mid-Apr, and depart by late Apr, with extreme dates of 12 Mar 1951 (Kane) and 12 May 1973 (DiCostanzo) and a maximum of 4 on 10 Apr 1954 (Scully) and 14 Mar 2006 (Young). It has become far less regular in spring since the 1980s.

Until the late 1950s ♂♂ were often heard giving their unmistakable breeding song in Mar–Apr and were long suspected of being Van Cortlandt breeders, but this was never proved. Suspicions were fueled by summer reports (among them occasional fresh but flying juveniles) on 28 Jul 1934 (Weber, Banner), 18 Jul 1937 (Norse), and 14 Jun 2004 (Klein), but these were more likely failed breeders or postbreeding dispersers from Grassy Sprain Reservoir. Still, the pair that bred only once at Oakland Lake in Alley Pond Park had young by 21 Apr 1951 (Yeaton), which gives one pause about disregarded Van Cortlandt Swamp breeding. If they had been breeding there then, the Dyker Beach golf course pond and Jamaica Bay Wildlife Refuge would have been the only other contemporary New York City sites, plus Grassy Sprain Reservoir and the Hackensack Meadowlands.

During NYSBBA I in 1980–85 New York City area blocks with breeders were found in Westchester County (3), Rockland County (3), Staten Island (1), Brooklyn (2), Nassau County (1), and Suffolk County (5). By NYSBBA II in 2000–2005, corresponding counts were Westchester County (1), Rockland County none, Staten Island (1), Brooklyn (1), Nassau County none, and Suffolk County (5 new sites). They continue to disappear, and the nearest current breeders are in the Hackensack Meadowlands.

Horned Grebe   *Podiceps auritus*

NYC area, current
A relatively common spring/fall migrant that is most frequent on saltwater, where it may be declining as a winter resident. Recorded in Central and Prospect Parks.

Bronx region, historical
Bicknell's *Riverdale* 1872–1901: unrecorded

Eaton's *Bronx + Manhattan* 1910–14: fairly common winter visitor

Griscom's *Bronx Region* as of 1922: rare visitor on Long Island Sound, Jerome, not recorded on Hudson since 1880; only 3 recent records 1914–19

Kuerzi's *Bronx Region* as of 1926: common migrant, fairly common in winter (6 Oct–23 May) on Long Island Sound; regular on reservoirs and rivers

NYC area, historical
Cruickshank's *NYC Region* as of 1941: Long Island common spring/fall migrant, winter resident (maximum 100+); annual but rare migrant on inland lakes, reservoirs

Bull's *NYC Area* as of 1968: maxima fall 250, winter 1200, spring 48

Bull's *New York State* as of 1975: status unchanged

Levine's *New York State* as of 1996: Long Island winter maximum 2000+

Study area, historical and current
An infrequent migrant on Jerome and Hillview and Van Cortlandt Lake but a spring/fall migrant in small numbers on the Hudson River.

Even though appearing in winter after having been frozen out farther north, they have made only 2 possible overwintering attempts, in 1924–25 and 1963–64, perhaps because of greatly reduced food following periodic attempts to eliminate fish, etc. on both reservoirs. There are 22 study area records, most during spring migration.

About 14 have been seen on Jerome: on 19 Mar 1914 (Hix); on 4–5 Jan 1925 (Cruickshank); 2 on 5 Mar 1925 (Ruff); on 22 Nov 1930 (Sedwitz); 2 from 22–31 Oct 1934 (Cantor); on 30 Nov 1957 (Mayer); on 22 Jan 1963 (Heath, Enders); on 5 Feb 1971 (Sedwitz); on 14 Mar 1971 (Sedwitz)—dead and desiccated, possibly the Feb bird that had been unable to find adequate food; on 4 Jan 1984 (Meyer); different singles from 7–18 Mar and 2–7 Apr 1994 (Lyons, Garcia); and on 30 Jan 2017 (Karlson). One seen irregularly on Jerome during the winter of 1963–64 may have successfully overwintered (Enders et al.).

Only 6 have dropped onto Van Cortlandt Lake: on 5 Jan 1952 (Kieran); on 30 Mar 1957 (Mayer); extremely early on 21 Sep 1980 (Ephraim); 13–20 Apr 1989 (Künstler et al.); 8–18 Mar 2000 (Garcia, DiCostanzo); and on 12 Apr 2014 (Young, Souirgi et al.).

It has been seen twice on Hillview: 30 in breeding plumage on 22 Mar 1925 (Kuerzi) and a single on 19 Jan 2018 (Camillieri).

Comments
A very rare but annual winter visitor to Long Island, western **Eared Grebe**, *P. nigricollis*, has occurred twice in the Bronx at City Island on 27 Dec 2007 (Walter) and at Pelham Bay Park from 27 Jan–1 Feb 2018 (Benoit et al.; photos). On the Hudson River found no closer than Dutchess County and unrecorded in Central or Prospect Parks.

Red-necked Grebe   *Podiceps grisegena*

NYC area, current
A very local winter visitor/resident in small numbers at Long Island saltwater sites, occasionally appearing in 3-figure numbers in Mar after stormy weather. Very rare inland, where usually occurring only when forced to ground in migration.

Bronx region, historical
Bicknell's *Riverdale* 1872–1901: unrecorded

Eaton's *Bronx + Manhattan* 1910–14: occasional winter visitor

Griscom's *Bronx Region* as of 1922: very rare, 3 East Bronx records; flights occasional in late winter/spring near New York City

Kuerzi's *Bronx Region* as of 1926: irregular and uncommon migrant, winter visitor (15 Oct–28 Apr), occasionally fairly numerous in late winter/early spring; chiefly on Long Island Sound, rare on reservoirs, rivers

NYC area, historical
Cruickshank's *NYC Region* as of 1941: fairly common spring/fall migrant (10 Sep–late May, once Jun), with occasional weather-induced groundings of dozens, hundreds in late winter/early spring; variably numerous Long Island saltwater winter resident (maximum 12+); rare fall migrant on Hudson, more so on reservoirs [locations not given]

Bull's *NYC Area* as of 1968: maxima winter 35, spring 115

Bull's *New York State* as of 1975: status unchanged

Study area, historical and current
An infrequent spring migrant and occasional winter visitor on Jerome and Hillview, after having been frozen out of the Great Lakes.

There are only 11 occurrences, all but 3 singles, although daily in-season Jerome and Hillview coverage should yield more, especially during its Mar–Apr spring migration peak. They are: at Jerome on 8 Jan 1963 (Sedwitz), 3 from 15–21 Mar 1982 (Sedwitz), 23 Mar 1994 (Young), and 2 from 4 Mar–4 Apr 2004 (Lyons); at Hillview on 31 Mar–25 Apr 1937 (Weber, Norse et al.), 4 in late Mar 1994 (unknown observer, *Kingbird*)—the study area maximum—and 11 Jan 2017 (Magill) and 1 Mar 2018 (Camillieri). Singles at Hillview on 18 Oct (Camillieri) and at Jerome on 22 Oct 2017 (Karlson) are the sole fall records. Only 1 has been seen on Van Cortlandt Lake, from 13–19 Apr 2012 (Young et al.), but the uncorroborated rumor persists that it had actually been found in St. James Park, on Jerome Ave. south of Kingsbridge Rd., and then released on Van Cortlandt Lake on 13 Apr after rehabilitation.

Comments
Extremely infrequently seen in the New York City area, usually on Long Island waters and never in the Bronx or Westchester County, vagrant **Western Grebe**, *Aechmophorus occidentalis*, has occurred at Piermont Pier from 10–13 Nov 2009 (Weiss et al.) and from 7–10 Jan 2016 (Haas et al.). Unrecorded in Central or Prospect Park.

Storks: Ciconiidae

Wood Stork   *Mycteria americana*

NYC area, current
A vagrant to the New York City area from the Southeast, usually postbreeding, with <10 records dating back to 1890. Never recorded in Central or Prospect Park.

NYC area, historical
Griscom's *NYC Region* as of 1922: Long Island Jun 1890

Cruickshank's *NYC Region* as of 1941: status unchanged

Bull's *NYC Area* as of 1968: 4 more Apr–Aug Long Island records, including flocks of 10–13, 11, 15

Bull's *New York State* as of 1975: Staten Island 11–12 Oct 1973

Levine's *New York State* as of 1996: status unchanged

Study area, historical and current
A study area vagrant with only a single record: 2 HYs on 2 Oct 2003 (Young) that were studied through a telescope for 10 minutes as they circled in thermals heading south over Van Cortlandt, where they descended and slowed prior to landing in the Swamp, only to abort at the last moment, regaining altitude and continuing on toward Manhattan. Seen so close as to fill the telescope's field of view, when their naked heads and pale bills were obvious.

This is the first Bronx record and the 5th and most recent for New York City, the other 4 being 11 on 10 Jun 1961 at Jamaica Bay Wildlife Refuge (Johnson); plus singles on 16 May 1965 at Jamaica Bay Wildlife Refuge (F. Davis); Fresh Kills, Staten Island on 11–12 Oct 1973 (Deppe); Breezy Point 7 Aug 1994 (Hake). Also southbound, 1 circled the Mt. Peter, Orange County hawkwatch on 27 Oct 1991 (A. Martin).

Northward Wood Stork vagrancy is usually associated with failed breeding in the Southeast, especially during prolonged droughts or when winter breeding season freshwater is too high. Twice as many have occurred in upstate New York as coastally. However, there are a dozen or more NJ records of singles or small flocks, the most recent at Sandy Hook in Aug 2017.

Cormorants: Phalacrocoracidae

Double-crested Cormorant   *Phalacrocorax auritus*

NYC area, current
An abundant and increasing spring/fall migrant and relatively recent colonial breeder. Migrates largely coastwise in large goose-like skeins on Long Island Sound and along Long Island's south shore. Formerly all but unrecorded

in winter but now routinely seen in flocks. First bred within New York City in the late 1970s, now breeds in all 5 boroughs and continues to increase. An abundant migrant on Long Island Sound and along the Hudson River, a recent (1983) and increasing Bronx breeder and an increasing winter resident but only as singles or small flocks.

Bronx region, historical

Bicknell's *Riverdale* 1872–1901: unmentioned

Eaton's *Bronx + Manhattan* 1910–14: fairly common migrant

Griscom's *Bronx Region* as of 1922: very rare on Hudson in Croton Point area, elsewhere inland

Kuerzi's *Bronx Region* as of 1926: rare migrant in spring (24 Apr–23 Jun)/fall (31 Aug–7 Nov), first in 1923

NYC area, historical

Cruickshank's *NYC Region* as of 1941: common coastal spring/fall migrant; inland usually recorded only overhead in Apr–May (maximum several hundred)

Bull's *NYC Area* as of 1968: eastern Long Island maxima fall 15,000, spring 5000, summer 300; almost unknown in winter

Bull's *New York State* as of 1975: status unchanged

Levine's *New York State* as of 1996: 3500+ pairs breeding New York City area 1995; 400+ on 1995 area Christmas Bird Counts

Study area, historical and current

A regular spring/fall migrant often seen overflying Van Cortlandt and Woodlawn, yet they rarely land on Jerome.

One seen by Griscom in Riverdale on 31 Aug 1923 he believed to be the first Bronx record, and another on Van Cortlandt Lake on 26 Oct 1935 (Norse) may have been the first study area record. One on 22 Sep 1946 (Komorowski) was the next, then 4 on 12 Apr 1953 (Buckley et al.), 2 on 6 May 1961 (Heath, Enders), and 7 on 6 May 1962 (Horowitz et al.). By the mid-1960s they were occurring regularly in Apr–May in very small numbers, but there were none on either reservoir. Through about the late 1970s singles or small flocks were seen flying over Van Cortlandt occasionally in Apr or Sep, very rarely in Jun–Aug; there were no records after Sep.

By the early 1990s the first singles or twos began appearing on Van Cortlandt Lake between Apr and Sep, a few occasionally lingering to late Nov, especially by the early 2000s. By about 2005 a few could be found on Van Cortlandt Lake or Hillview almost any time from Mar–Nov.

In spring, first migrants arrive in late Mar, peak in Apr–May, and depart by late May, with an overhead maximum of 125 on 26 Apr 1998 (Young) and of 10 on Van Cortlandt Lake on 6 May 2012 (Bochnik).

In summer a few SYs and other non-breeders come in and out of Van Cortlandt Lake, with a maximum of 4 on 13 Aug 2006 (Klein).

During NYSBBA I in 1980–85 breeding cormorants were reported in blocks in Rockland County (1), the Bronx (2), Nassau County (2), and Suffolk County (6), but following initial range expansion and exponential growth, new colony establishment began that continues to this day. During NYSBBA II in 2000–2005 newly occupied blocks were all along the Hudson River from Piermont Pier to the VT border and inland as follows: Putnam County (8), Rockland County (6), Westchester County (17), Bronx (2), Manhattan (1), Staten Island (3), Brooklyn (1), Queens none, Nassau County (5), and Suffolk County (11) (McGowan and Corwin 2008). They continue to expand their occupied blocks as of 2016. Nearest breeders are on Huckleberry Island near Pelham Bay Park and on South Brother Island.

In fall, first migrants arrive in late Aug, peak in late Sep–early Oct, and depart by early Nov, with extreme dates of 28 Aug 1972 (Sedwitz) and 1 Dec 2005 (Young), with an overhead maximum of 115 on 16 Oct 2007 (Young) and of 9 on Van Cortlandt Lake on 24 Sep 2009 (Young).

It was not until the winter of 2005–6 that the first study area overwintering attempt occurred, when 2 remained on Van Cortlandt Lake from 1 Dec 2005 until 4 Feb 2006, when it finally froze over (Young). A few have been seen since during periods of open water, but it has not occurred in the study area on a Christmas Bird Count. Its striking increase on that count since 1981 is evident in Figure 33.

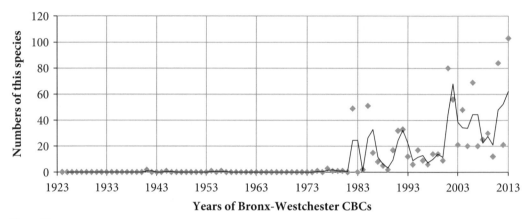

Figure 33

Comments
Recorded as a spring range-prospector in Central and Prospect Parks and elsewhere in the New York City area, southern **Anhinga**, *Anhinga anhinga*, has been slowly expanding its breeding range northward, now as close as MD. Its future study area occurrence is likely.

Great Cormorant *Phalacrocorax carbo*

NYC area, current
An increasing saltwater winter resident favoring Long Island and Long Island Sound rocks and jetties. An uncommon spring/fall migrant on Long Island that has also been recently found to be regular then on the Hudson River. A winter resident as close as Pelham Bay Park and increasing along the Hudson River south to Piermont Pier. Recorded in Central and Prospect Parks.

Bronx region, historical
Bicknell's *Riverdale* 1872–1901: unrecorded

Eaton's *Bronx + Manhattan* 1910–14: rare migrant

Griscom's *Bronx Region* as of 1922: unrecorded

Kuerzi's *Bronx Region* as of 1926: unrecorded

NYC area, historical
Cruickshank's *NYC Region* as of 1941: status uncertain owing to misidentifications, generally restricted to eastern Long Island

Bull's *NYC Area* as of 1968: great increase in recent years, especially on western Long Island Sound (maximum 40); 17 Pelham Bay Park Jan 1964; 14 Sep–3 May

Bull's *New York State* as of 1975: status unchanged

Levine's *New York State* as of 1996: increasing in winter on lower Hudson south to Piermont Pier

Study area, historical and current
A regular spring/fall migrant on the Hudson River and Long Island Sound but only a single study area record: 2 in Van Cortlandt on 13 Nov 2010 (Baksh).

To be sought from Nov–Mar and during migration overhead in the increasing flocks of Double-cresteds, as well as among those roosting frequently on Van Cortlandt Lake. The above 2 Greats and a few Riverdale Christmas Bird Count singles beginning in 1992 are part of the population first found overwintering along the Hudson in Westchester and Dutchess Counties in 1991–92. These may have been there undetected for some time, given that a few had been identified at Piermont Pier back to 1934. Regular in winter at Pelham Bay Park since the 1940s, reaching 35–40 there in Feb 1975 (DiCostanzo); a boat covering all Long Island Sound islands, islets, and rocks from Ft. Schuyler to New Rochelle located 109 Greats with 46 unexpected Double-cresteds on 26 Dec 1982 (Buckley).

Pelicans: Pelecanidae

American White Pelican *Pelecanus erythrorhynchos*

NYC area, current
An irregular but increasing and now nearly annual visitor from the High Plains or Southeast likely to occur at almost any time of the year, usually along the coast. Almost unknown in or near the Bronx. Unrecorded in Prospect Park.

NYC area, historical
Cruickshank's *NYC Region* as of 1941: 2 1800s specimens, 3 recent Oct, Feb records

Bull's *NYC Area* as of 1968: 8 recent records in Sep–Oct

Bull's *New York State* as of 1975: status unchanged

Study area, current
One study area record: on Van Cortlandt Lake for several hours on 13 Dec 2006 (Künstler, Pendergrass; photos). Most New York City area records are flyovers near the immediate coast, sometimes in small flocks, but in the Bronx it has occurred only at the Throgs Neck Bridge on 3 Jul 1989 (Caspers, Caspers), at the Whitestone Bridge on 16 Dec 2012 (Schulman), and 5 at Pelham Bay Park on 8 Dec 2004 (Aracil) and 1 on 26 Dec 2017 (Keogh). One was in Yonkers on 28 Jul 2009 (unknown observer, *Kingbird*). More recently 1 at Piermont Pier on 11 Jun 2013 (Edelbaum et al.) and the next day was seen flying down the Hudson River at Dobbs Ferry (Benson).

Comments
Increasing in numbers on the Atlantic Coast and now breeding occasionally as close as Barnegat Inlet, NJ, **Brown Pelican**, *P. occidentalis*, has occurred once in the Bronx—Fort Schuyler 15 Nov 2011 (Buckley), a post-Irene HY working its way down Long Island Sound from New

Haven and which was seen again at Sands Point, Long Island on 12 Dec 2011 (Quinn). However, this species has also been seen several times along the Hudson River in Manhattan and Westchester County: 2 at Croton Point on 30 Jul 1992 (Chapman); a vague report of 1 on the Yonkers and Hastings waterfronts sometime in the early 1990s—local newspapers may have more information; 2 on the Hudson River in Manhattan on 23 May 1993 (J. Caspers, Machover); another there on 7 Sep 1993 (Giunta); and 1 reported anonymously from the Hudson River at Scarborough on 18 Aug 1996. There have been rumors of others from the Hudson in Riverdale but lacking details. Unrecorded in Central or Prospect Park.

Herons: Ardeidae

American Bittern *Botaurus lentiginosus*

NYC area, current
A spring/fall migrant, formerly a thinly distributed but highly local breeder now almost extirpated, and a regular winter visitor and resident, largely in coastal saltmarshes. It is a New York State Species of Special Concern.

Bronx region, historical
Bicknell's *Riverdale* as of 1872–1901: bred Van Cortlandt Swamp to 1879, spring/fall migrant at Kingsbridge Meadows

Eaton's *Bronx + Manhattan* as of 1910–14: uncommon breeder

Griscom's *Bronx Region* as of 1922: rare migrant, probably extirpated breeder; bred Bronx River marshes near Gun Hill Rd. 1917

Kuerzi's *Bronx Region* as of 1926: extirpated breeder, fairly common migrant in spring (5 Apr–19 May)/fall (27 Jul–26 Dec)

NYC area, historical
Cruickshank's *NYC Region* as of 1941: widespread common spring/fall migrant (maximum 10), winter visitor

Bull's *NYC Area* as of 1968: status unchanged

Bull's *New York State* as of 1975: status unchanged

Study area, historical and current
A longtime spring/fall migrant and occasional winter visitor/winter resident in Van Cortlandt Swamp, although with so few observers these days that it may no longer be noted annually. Some extreme migration dates are equivocal owing to possible breeding and frequent overwintering attempts.

In spring, first migrants arrive in mid-Mar, peak in Apr, and depart by mid-May, with extremes of 14 Mar 1953 (Buckley) and 18 May 1951 (Kane), and many maximum of 2.

It bred in Van Cortlandt Swamp through 1879 (Bicknell) and in 1917 in a marsh on the Bronx River adjacent to Woodlawn, long before the Bronx River Parkway was built (Griscom 1923) and was a regular migrant. It was strongly suspected of breeding in Van Cortlandt Swamp in the 1920s–30s and 1950–55 but this was frustratingly never proved (Kieran, Cruickshank, Norse, Buckley). Through the 1960s ♂♂ routinely boomed in Mar–May, but we are nonetheless aware of no Jul records in that period, and although 1 was seen on 1 Jun 1983 (Naf) it would have been only a nonbreeder or very late migrant. However, peak egg dates for this species in the New York City area are 1–14 May, so they could have been breeding in Van Cortlandt Swamp into the 1960s. In 1950, Komorowski firmly believed them to have been breeding there based on pervasive booming, and they may have been. Two on 17 May 1940 (Norse) were also suspicious. The last booming in Van Cortlandt Swamp was on 27 Mar 1996 (Garcia), but it may still be heard occasionally.

By the time of NYSBBA I in 1980–85 the bittern decline in the New York City area was well under way, although they were recorded in 8 blocks on Long Island, 2 on Staten Island, 4 in Rockland, but none in Westchester and Putnam Counties. By NYSBBA II in 2000–2005, in the same areas only 2 possible blocks were found along Great South Bay on Long Island and another at the Piermont Marshes. There were no others (McGowan and Corwin 2008). DOIBBS efforts for this species in New York State missed most occupied blocks, while showing a (nonsignificant) decrease between 1966 and 2005 (Sauer et al. 2007). The nearest current breeders are in the Hackensack Meadowlands.

In fall, first migrants arrive in late Aug–early Sep, peak in Oct, and depart by early Nov, with extremes of 3 Aug 1953 (Buckley) and 25 Nov 1952 (Kane), and a maximum of 3 on several occasions.

Singles and twos routinely lingered into Dec and occasionally Jan through the 1960s, but rarely since then. The last successful overwintering was in 1962–63 (Enders et al.), as lingerers typically move into saltmarshes once frozen out of freshwater.

Specimens
Bicknell collected 1 at Kingsbridge Meadows on 21 Oct 1876 (NYSM).

Least Bittern   *Ixobrychus exilis*

NYC area, current
Now an almost unknown breeder except at Jamaica Bay and a few sites on eastern Long Island, and a rarely seen spring/fall migrant. Recorded a handful of times in Central and Prospect Parks in spring. It is a New York State Threatened Species.

Bronx region, historical
Bicknell's *Riverdale* 1872–1901: unmentioned

Eaton's *Bronx + Manhattan* 1910–14: local breeder

Griscom's *Bronx Region* as of 1922: bred Van Cortlandt Swamp 1918, in swamp north of Van Cortlandt 1921; formerly bred on destroyed Dyckman St. marshes

Kuerzi's *Bronx Region* as of 1926: rare migrant (30 May–26 Sep); breeds Tarrytown and Van Cortlandt

NYC area, historical
Cruickshank's *NYC Region* as of 1941: widespread but exceedingly local breeder, until recently in small colonies in Bronx, Brooklyn, Queens, and Nassau Co., where few pairs persist; uncommon (?)migrant in spring/fall (23 Apr–2 Oct, plus Dec specimen)

Bull's *NYC Area* as of 1968: 16 Apr–22 Nov; 3 more winter records; now breeding Jamaica Bay Wildlife Refuge

Bull's *New York State* as of 1975: status unchanged

Study area, historical and current
Now an extirpated breeder in Van Cortlandt Swamp and an occasional spring/fall migrant.

It was first reported breeding in Van Cortlandt Swamp in 1918 (Chubb, C. Lewis) but may have been previously overlooked. As many as 5 pairs nested in 1938 (Norse et al.), down to 1–2 pairs through 1961, the last year they bred (Zupan, Heath). It was also seen in Jerome Swamp in May 1927, but breeding was unmentioned (Cruickshank). A pair did breed in 1955 at Turtle Cove in Pelham Bay Park but never returned (Young 1958).

In spring, they first arrived in Van Cortlandt Swamp in mid–late Apr, with the earliest on 16 Apr 1955 (Norse, Buckley), and after 1962 there are only 4 records, all of presumed migrants: 6 Sep 1976 (Ephraim), 8 Oct 1999 (Garcia), 4–6 Sep 2000 (Jaslowitz), and 22 May 2010 (Baksh).

The only other post-1962 West Bronx records are from the Bronx Zoo: 9 Sep 1999 and 18 May 2003 (Olson). It may occur annually in migration in Van Cortlandt Swamp, but these days so few observers are there at first light in late Apr and May that bitterns could be easily overlooked, especially when silent. Loss of its favored cattail nesting and feeding habitat when replaced by exotic (nonnative) and aggressive *Phragmites* was responsible for its Van Cortlandt Swamp departure.

In fall, breeders normally departed by late Sep but there are several Nov records, the latest on 22 Nov 1953, when a ♂ was flying around the Swamp (Phelan, Buckley).

Least Bitterns have always been scarce and exceedingly local around New York City, and NYSBBA I in 1980–85 found occupied blocks in Putnam County (3), Westchester County (4), Brooklyn (1), Queens (1), Nassau County (2), and Suffolk County (7). By NYSBBA II in 2000–2005 corresponding data were Putnam County (1), Westchester County (2), Brooklyn (1), Queens (none), Nassau County (none), and Suffolk County (3) (McGowan and Corwin 2008). Time will tell if this trend persists. Nearest breeders as of 2016 would seem to be at the Piermont Marshes and the Hackensack Meadowlands, although occasional one-off breeding events will occur, like the 2 possibly breeding in a small fresh marsh in Pelham Bay Park in Jun–Jul 1989 (Arrowsmith et al.).

Great Blue Heron *Ardea herodias*

NYC area, current
A widespread spring/fall migrant and winter visitor/resident, especially in tidal areas. It may now be starting to extend its breeding range back to the coast from nearby inland wetlands, but it has not recolonized Long Island.

Bronx region, historical
Bicknell's *Riverdale* 1872–1901: spring/fall migrant

Eaton's *Bronx + Manhattan* 1910–14: common migrant

Griscom's *Bronx Region* as of 1922: rare spring/fall migrant

Kuerzi's *Bronx Region* as of 1926: regular, fairly common migrant in spring/fall (maximum 24+), few remain into Feb

NYC area, historical
Cruickshank's *NYC Region* as of 1941: common migrant in spring/fall (maximum 24), rare winter

Bull's *NYC Area* as of 1968: new colony at Sandy Hook contained 100 pairs 1957, with others inland in Sussex County NJ, and Orange and Putnam Counties.; maxima fall 200, winter 100

Bull's *New York State* as of 1975: status unchanged

Levine's *New York State* as of 1996: winter Christmas Bird Count maxima 116 (Central Suffolk), 72 (Bronx-Westchester)

Study area, historical and current
An annual but possibly decreasing spring/fall migrant, irregular winter visitor in all 4 extant subareas, and a potential Van Cortlandt Swamp breeder. May appear almost any time of the year, and recorded every month.

In spring, first migrants arrive in early Mar, peak late Mar–early Apr, and depart by early May, with extreme dates of 6 Mar 2005 (Klein) and 31 May 2014 (Winston) and a maximum of 14 on 2 Apr 2011 (Young). Until Jun 1967, the spring maximum in Prospect Park was 7 on 19 Apr 1954 (Usin).

NYSBBA I in 1980–85 found no breeders within New York City or on Long Island, and in Westchester County no closer than the northernmost third. By NYSBBA II in 2000–2005, they had filled in Westchester County blocks south to the White Plains area but no farther. DOIBBS data for this species depicted a slightly rising statewide trend line between 1966 and 2005 (McGowan and Corwin 2008).

The last coastal breeding pairs anywhere in the immediate New York City area had been on Sandy Hook until 1976 (Buckley and Buckley 1980), so the single pairs on Goose Island, Pelham Bay Park, in 2011 and 2012 (Künstler) and at Clove Lakes Park on Staten Island since 2013 (Barron) were the first in 34 years. A just-fledged juvenile retaining down on its head at Huckleberry Island on 9 Jun 1975 (Buckley and Buckley 1980) was at the time considered not locally reared, but that dismissal may have been premature. One perched on North Brother Island on 20 Jul 2016 (Willow) was also suspect. A pair nesting at Tarrytown Reservoir Westchester County in 2006 (Mark) was part of this coastward movement, but they have not yet returned as Long Island breeders, where last known from Gardiner's Island in 1900.

In fall, first migrants arrive in mid-Aug, peak in Oct, and depart by mid-Nov, with extreme dates of 25 Jul 2004 (Klein) and 30 Nov 2012 (Baksh), and several maxima of 3.

Every year 1–3 attempt to overwinter, and some often succeed but only by regularly moving to and from extra-study area sites for feeding during freeze-ups.

Great Egret *Ardea alba*

NYC area, current
A widespread spring/fall migrant, colonial breeder, and increasing winter resident. It is the most wide-ranging and most frequently encountered white heron, commonest in Long Island saltmarshes.

Bronx region, historical
Bicknell's *Riverdale* 1872–1901: unrecorded

Eaton's *Bronx + Manhattan* 1910–14: rare summer visitor

Griscom's *Bronx Region* as of 1922: rare visitor; Van Cortlandt 1916, 1917

Kuerzi's *Bronx Region* as of 1926: also Baychester Marshes Aug–Sep 1925

NYC area, historical
Cruickshank's *NYC Region* as of 1941: locally common to uncommon Long Island summer visitor (31 Mar–8 Nov; maximum 50+); very local inland

Bull's *NYC Area* as of 1968: great increase in recent years; now breeding Sandy Hook, nearby CT, 3 sites on western Long Island; new Long Island roost maxima 165, 280; regularly lingering to Dec, occasionally overwintering

Bull's *New York State* as of 1975: increasing as breeder, regularly overwintering

Levine's *New York State* as of 1996: winter maximum 15 on Queens Christmas Bird Count

Study area, historical and current
A visitor that is almost resident in spring, summer, and fall, and recently even in winter in Van Cortlandt, but also at Woodlawn and at both reservoirs. Most are almost certainly from the established colonies at South Brother Island and Pelham Bay Park, where they are now all but resident.

Yet until the early 1980s it was only an infrequently recorded visitor in the study area, with 9 records and a maximum of 3 in the Dutch Gardens from 16 Jul–9 Oct 1916 and from 19 Jul–5 Aug 1917 (Chubb et al.); then none until 20 Sep 1952 (Buckley); 25–28 May 1953 (Buckley et al.); 20 Aug 1953 (Phelan, Penberthy, Buckley); 14 Sep 1953 (Buckley); 8–22 Sep 1957 (Rafferty, Horowitz); 23 Sep 1962 (Sedwitz); 8 Jul 1984 (Sedwitz). By the 1990s they no longer elicited comment but were enjoyed whenever seen.

Great Egrets are now seen almost year-round, though still among the missing in Feb. First arrivals now appear in Apr, the earliest on 17 Mar 2001 (Collins), with a study area maximum of 4 on 7 Apr 1995 (Lyons, Garcia). Throughout the summer and fall 1–3 wander in and out of the study area with no recent detectable postbreeding pulses. Potential breeding is only an outside possibility. They thin out by Nov–Dec, and while none has yet been found in the study area on a Christmas Bird Count, the latest Van Cortlandt date currently is 3 Jan 2013 (Schwartz-Weinstein et al.), and 1 was in Bronx Park on 23 Dec 2001 (Krauss, Olson).

Snowy Egret *Egretta thula*

NYC area, current
A colonial breeder concentrated around Long Island and New York Harbor saltwater sites. In spring and after breeding, single individuals often appear inland, though not as frequently as Little Blue Herons and Great Egrets.

Bronx region, historical
Bicknell's *Riverdale* 1872–1901: unrecorded

Eaton's *Bronx + Manhattan* 1910–14: unrecorded

Griscom's *Bronx Region* as of 1922: unrecorded

Kuerzi's *Bronx Region* as of 1926: unrecorded

NYC area, historical
Cruickshank's *NYC Region* as of 1941: uncommon but regular Long Island nonbreeding visitor (29 Apr–early Dec, maximum 12), but increasing records away from coast

Bull's *NYC Area* as of 1968: great increase in recent years; now breeding Sandy Hook, nearby CT, and at 8 western Long Island sites; Long Island nonbreeding maximum 288; regularly lingering to Dec

Bull's *New York State* as of 1975: 15 Mar–23 Dec, plus Jan; winter maximum 9

Levine's *New York State* as of 1996: winter maximum 25 on Brooklyn Christmas Bird Count

Study area, historical and current
Despite a large breeding population as close as South Brother Island and Jamaica Bay (with a few pairs now in the Pelham Bay Park area), in Van Cortlandt this species is only slightly less

frequent than Little Blue Heron (which prefers freshwater) but unlike that species has been unrecorded since 2001.

About 16 records: 31 Jul 1922, 29 Aug 1925, 4 Sep 1927, 18 Jul 1933 (Kieran); then none until 20 Aug 1953 (Phelan, Penberthy, Buckley); 20 Jul 1962 (Rafferty); 17 Sep 1964 (Steineck); 4 Aug 1975 (Young); 14 May 1981 (Sedwitz); 4 Aug 1991 (Collerton); Van Cortlandt Lake 22 Aug 1992 (Jaslowitz); 19 Apr 1994 (Lyons); 27 Aug 1994 (Lyons); 4 Apr 1995 (Young); 2 May 1996 (Young); 4 Nov 2001 (Young).

It is increasingly being seen in small groups as close as Spuyten Duyvil (e.g., 3 on 15 Jun 1975: Sedwitz), after the first there on 29 May 1966 (Norse), and 1–2 are now being seen with some regularity in the Bronx Zoo in summer (Olson). One at Pelham Bay Park from 23 Dec 2001–26 Jan 2002 (Steineck et al.) is the latest in the Bronx so far, although like Great Egrets they can be expected to linger later and later, eventually overwintering.

Nearest breeders are at Huckleberry Island, Pelham Bay Park, and South Brother Island. See Buckley and Buckley (1980) and McGowan and Corwin (2008) for locations of New York City area heronry sites and breeding numbers.

Little Blue Heron   *Egretta caerulea*

NYC area, current
A thinly distributed colonial breeder that is near the northern edge of its breeding range. Nonbreeders and postbreeding dispersers occur widely in spring, summer, and fall, and are quite prone to occur inland and on other freshwater sites.

Bronx region, historical
Bicknell's *Riverdale* 1872–1901: unrecorded

Eaton's *Bronx + Manhattan* 1910–14: unrecorded

Griscom's *Bronx Region* as of 1922: only single Staten Island record

Kuerzi's *Bronx Region* as of 1926: 2 records: Aug 1924, Aug–Sep 1925

NYC area, historical
Cruickshank's *NYC Region* as of 1941: locally common Long Island nonbreeding visitor (19 Apr–6 Nov; maximum 100+) plus large inland freshwater marshes

Bull's *NYC Area* as of 1968: early spring date 16 Mar; increasing, especially inland; now breeding at 3 western Long Island sites

Bull's *New York State* as of 1975: late fall date 27 Dec; maxima summer 120, fall 54, winter 8

Study area, historical and current
An irregular Van Cortlandt Swamp spring migrant and postbreeding disperser recently nesting in 2 locations in the East Bronx.

About 20 records, all of white HYs except as noted: 27 Aug 1922, 1 Aug 1924, 3 Sep 1925, 2 on 7 Sep 1927, 12 Aug 1929, 3 on 21 Aug 1930, adult 23 May 1936, adult 28 May 1952 (all Kieran); 29 May 1943 (S. Horowitz); 10 Aug 1965 (Gera, Gera); adult 27 Jul 1977 (Steineck); 2 Aug 1980 (Young); adult 15 Apr 1994 (Lyons, Garcia); HY and adult 1–18 Sep 1994 (Lyons, Garcia, Young); adult 2 Jun 1996 (Lyons, Garcia); 15 Sep 2007 (Young); adult 29 Apr 2009 (Young); adult 22 May 2010 (Baksh). An HY was also in Riverdale 16 Aug 1925 (Kuerzi, Kuerzi).

Nearest recent but exceedingly irregular breeders have been single pairs on Goose Island, Pelham Bay Park (1999–2008) and on South Brother Island from 1997 on (NYCA surveys); some from either site could have reached the study area.

Tricolored Heron   *Egretta tricolor*

NYC area, current
The least numerous breeding heron and all but restricted to the South Shore of Long Island. Wanders occasionally postbreeding, and recorded in Central and Prospect Parks.

Bronx region, historical
Bicknell's *Riverdale* 1872–1901: unrecorded

Eaton's *Bronx + Manhattan* 1910–14: unrecorded

Griscom's *Bronx Region* as of 1922: unrecorded

Kuerzi's *Bronx Region* as of 1926: unrecorded

NYC area, historical
Cruickshank's *NYC Region* as of 1941: 10 recent Long Island/Staten Island records but never in Bronx

Bull's *NYC Area* as of 1968: 9 Mar–1 Dec; increased in recent years; first Westchester County record; maximum 8; bred Jamaica Bay Wildlife Refuge 1955

Bull's *New York State* as of 1975: maximum 11; 3 new Long Island breeding sites; late fall date 17 Dec

Study area, historical and current
There is but a single record, an HY in Van Cortlandt Swamp on 17 Aug 1973 (Ephraim) that may be the one at Croton Point from 20–25 Aug 1973 (Howe).

This is the scarcest heron away from Long Island and is otherwise known in the Bronx only from Pelham Bay Park: summer 1984 (Cassidy); 21 Jul 1986, 31 Aug 1988 (Künstler); most unexpectedly on 26 Dec 1988 (Rucht, J. Caspers); and from 2–16 Jun 2014 (Rothman, Benoit et al.). There are several records north of the Bronx, including Westchester County's second at Rye on 4 Sep 1958 (Kane), 28 May 1967 in Larchmont (Bahrt), Piermont Pier and other Rockland County sites in May–Aug 1969–76; 2 at Cornwall in Jul–Aug 1972 (Treacy); Croton Point 17 Nov 1975, 2–8 Aug 1976 (Howe); Piermont Pier 30 Aug 1998 (unknown observer, Lower Hudson Valley Bird Line); Tarrytown Lake 29 Aug 2010 (unknown observer, NYRBA), among others.

The nearest breeders are at Jamaica Bay where 1–2 pairs are barely hanging on at what may be the northernmost point in their breeding range. It has neither bred nor ever been seen in the heronries at South Brother Island, Pelham Bay Park, or Huckleberry Island.

Cattle Egret   *Bubulcus ibis*

NYC area, current
A relatively recent arrival in the New York City area (the first was in 1954), which quickly peaked as a migrant and breeder and has now become exceedingly scarce. It has occurred in Central and Prospect Parks.

NYC area, historical
Bull's *NYC Area* as of 1968: 4 Apr–9 Dec; spring maximum 40; spring 1962 invasion

Bull's *New York State* as of 1975: 26 Mar–31 Dec; maximum 60 Jamaica Bay Wildlife Refuge 26 Sep 1973; first breeding sites Gardiner's Island 1970, Jones Beach 1973

Levine's *New York State* as of 1996: coastal colonies peaked at 350 pairs in 1985, then down to 42 pairs by 1995; nonbreeders increasingly scarcer

Study area, historical and current
An exceedingly erratic spring/summer visitor. In the study area, Cattle Egrets first occurred when they were expanding into and within the NYC area, where they are now almost gone as breeders and migrants; study area records have correspondingly shrunk.

There are but 7 Van Cortlandt records: Van Cortlandt Swamp 2 May 1962 (Rafferty, Maguire); Van Cortlandt Swamp 23 May 1962 (Steineck); Van Cortlandt Swamp 11 May 1964 (C. Young et al.); Van Cortlandt Lake 15 Aug 1989 (Jaslowitz); Van Cortlandt Swamp 12 Apr 1997 (fide Pirko); Parade Ground 20 Oct 2002 (Young) and 19–22 Oct 2016 (Young et al.). Another was flying north on the Hudson River in Riverdale on 5 May 1982 (Sedwitz).

Elsewhere in the Bronx, there are few nonbreeding records: at Pelham Bay Park on 23 May 1962 (Stepinoff) and exceedingly late there on 26 Dec 1976 (Steineck et al.). One at Hunt's Point from 7–27 Oct 2004 (unknown observer, *Kingbird*), 2 on 13 Jul 2005 (Dadone), and during Aug 2007 (unknown observer, *Kingbird*) were probably from the nearby heronry on South Brother Island, where the last single nest was in 2007 (NYCA surveys).

Following first breeding at Gardiner's Island in 1970, Cattle Egrets nested widely but thinly in New York City area heronries from Jamaica Bay east to Jones Beach but also on South Brother Island off Hunts Point in the Southeast Bronx from 1978 when first found, and 30 pairs nested there (Buckley and Buckley 1980) until the last pair in 2007 (NYCA surveys). This colony is the likely source of the few study area records. A single pair also nested at Huckleberry Island off Pelham Bay Park in 1988 (Künstler 1989) but not again. They then began to disappear as New York City harbor breeders, declining from a 1985 peak of 266 pairs on Prall's and Shooter's Islands (Staten Island) alone, plus uncounted numbers on South Brother Island and in Jamaica Bay that same year (NYCA surveys). By 2011, there were none breeding anywhere in the NYCA surveys, the nearest colony then being on Pea Patch Island in the upper end of Delaware Bay with an estimated 700 adults (= pairs) in 2011 (Bennett) and almost 900 adults (= pairs) in 2012 (Bennett). As of 2016, in the New York City area they have become exceedingly scarce at any time of the year and have retreated nearly completely as a breeder from NJ northeastward.

Green Heron *Butorides virescens*

NYC area, current
A widespread but very local noncolonial breeder and spring/fall migrant in small numbers.

Bronx region, historical
Bicknell's *Riverdale* 1872–1901: spring/fall migrant, breeder

Eaton's *Bronx + Manhattan* 1910–14: fairly common breeder

Griscom's *Bronx Region* as of 1922: local breeder, regular migrant

Kuerzi's *Bronx Region* as of 1926: common spring/fall migrant (13 Apr–12 Oct), breeder

NYC area, historical
Cruickshank's *NYC Region* as of 1941: widespread common spring/fall migrant, breeder (1 Apr–12 Dec, maximum 37)

Bull's *NYC Area* as of 1968: 1 Apr–26 Nov, plus increasing Dec–Jan records; summer maximum 143

Bull's *New York State* as of 1975: status unchanged

Study area, historical and current
A breeder and spring/fall migrant in Van Cortlandt, Woodlawn, and Jerome.

In spring, first migrants arrive in mid-Apr, peak in early May, and depart by late May, with extremes of 9 Apr 1928 (Kessler) and in Woodlawn on 22 May 1961 (Horowitz), and maxima of 5 on 5 May 1953 (Buckley) and 8 May 2009 (Young). Until Jun 1967, the spring maximum in Central Park was 25 on 20 May 1966 (Plunkett) and in Prospect Park 14 on 5 May 1950 (Whelen).

It has long bred in Van Cortlandt Swamp (5–6 pairs) and Woodlawn (single pair), perhaps also at Jerome, with a total population of 7–8 pairs, incubating as early as 3 May 1930 (Sedwitz). Before its 1949–50 devastation, Van Cortlandt Swamp supported 3–4 pairs, but afterward only 1–2 pairs. From 2008–13, 2 pairs actually nested high and dry up on Vault Hill, flying down to the Swamp to feed (Young). It is unknown whether there were other nests in Van Cortlandt Swamp itself at the same time, or if not, what pressures there drove them to nest in dry woods. Other wetland sites within the study area (notably Woodlawn, the Sycamore Swamp, and the former Lincoln Marsh) may also harbor additional pairs when water levels are high. Also breeds erratically in Central and Prospect Parks.

Between NYSBBA I in 1980–85 and NYSBBA II in 2000–2005 many blocks around New York City became unoccupied, although new ones were also found. None were located in the East/Southeast Bronx, surely a sampling error. Sparse DOIBBS data showed a level statewide trend line between 1966 and 2005 (Sauer et al. 2005; McGowan and Corwin 2008).

In fall, first migrants arrive in late Aug, peak in early Sep, and depart by early Oct, with extreme dates of 5 on 26 Aug 2004 (Klein) and 28 Oct 1959 (Heath), 24 Oct–1 Nov 1955 (Birnbaum, Buckley), and 11 Nov 2012 (Stuart et al.), with a maximum of 6 on 7 Sep 2002 (Young). Bronx lingerers have been seen in the Bronx Zoo as late as 27 Nov 2011 (Fazzino), and at Pelham Bay Park as follows: 2 on 22 Dec 1974, 2 on 28 Dec 1986, and 1 on 22 Dec 1991 (all Steineck et al.).

Specimens
Bicknell collected 1 at Kingsbridge Meadows on 29 May 1877 (NYSM).

Black-crowned Night-Heron *Nycticorax nycticorax*

NYC area, current
A spring/fall migrant, formerly a widespread colonial breeder whose numbers have dropped greatly from peaks in the 1930s, and a regular though diminished winter visitor and resident, especially in tidewater areas

Bronx region, historical
Bicknell's *Riverdale* 1872–1901: common spring, less common fall migrant, occasional summer; pair flushed from wooded swamp on 13 Apr 1876 possibly breeding

Eaton's *Bronx + Manhattan* 1910–14: common migrant

Griscom's *Bronx Region* as of 1922: common resident, not breeding

Kuerzi's *Bronx Region* as of 1926: common resident; large numbers overwintered in Bronx Zoo until 1925–26, when new 150-bird roost established in Pelham Bay Park. No large breeding colony believed to exist locally

NYC area, historical
Cruickshank's *NYC Region* as of 1941: common spring/fall migrant, widespread breeder (3000+ pairs in 1935), uncommon, local in winter; 10 pairs nested near Tuckahoe

Bull's *NYC Area* as of 1968: fewer large colonies than formerly but Sandy Hook colony increased to 700+ pairs; winter maximum 150

Bull's *New York State* as of 1975: breeding decline continues

Study area, historical and current

Formerly a frequent visitor year-round in Van Cortlandt and Woodlawn and the commonest heron breeding in the New York City area. But since the 1960s its numbers as both breeders and visitors have dropped precipitously, in part from biocide contamination from which it has not recovered.

In the decades leading to the 1960s, it was a year-round visitor/resident in Van Cortlandt Swamp, 5–10 per day not being unusual with a maximum of 54 from 21–25 Apr 1952 (Norse, Kieran). During the 1950s–60s, 4–6 roosted in Woodlawn conifers all year, with a maximum of 20 there on 1 Nov 1955 (Buckley). In spring 1951, the 8–10 adults consistently in Van Cortlandt Swamp led Komorowski to believe they were breeding, but no nests or fledglings were ever found. Indeed, since the early 1900s they had been so regular and numerous in Van Cortlandt Swamp that many took it for granted they were breeding. And while the closest breeders at the time were those in Pelham Bay Park, Van Cortlandt breeding was never able to be proved—until Naf finally located a pair on a nest with 2 young in the Lincoln Marsh area south of McLean Ave, Yonkers in May 1983 and 1984. One or more are seen every spring and summer in Van Cortlandt Swamp, but no other breeding is known. Numbers of pairs and occupied colony sites in the New York City area have been declining steadily since the 1940s after peaking at 3400 pairs in 20 colonies in 1935, and even though they were recorded in 5 blocks in the Bronx during NYSBBA II in 2000–2005, we believe most did not involve breeding birds (McGowan and Corwin 2008). The nearest current breeders would be at Pelham Bay Park or North Brother Island.

In spring, the few first migrants arrive in mid-Mar, peak in mid-Apr, and depart by late Apr, with an extreme date of 7 on 24 Mar 1994 (Young) and 54 (the maximum) on 25 Apr 1952 (Norse, Kieran). Through Jun 1967, the spring maximum in Prospect Park was 24 on 5 May 1950 (Whelen).

Up to 6 non–adults can appear almost any time in the summer, but they rarely remain long. No southbound arrivals have been noted, but any breeders not attempting to overwinter usually depart by mid-Nov. Apart from the counts above, study area winter maxima rarely exceeded 4–6, and they have not been there on a Christmas Bird Count since the 1970s. We are a very long way from the winter roosts of 150 at Pelham Bay Park in 1925–26, the 70 at the Bronx Zoo in 1922–23, and even the 60+ at Pelham Bay Park on 31 Oct 1953. Large Black-crowned Night-Heron roosts are history in the study area and throughout the Bronx, and this decline has been all too clear on the Bronx-Westchester Count (Fig. 34). Until Jun 1967, the fall maximum in Central Park was 12 on 25 Sep 1952 (Post).

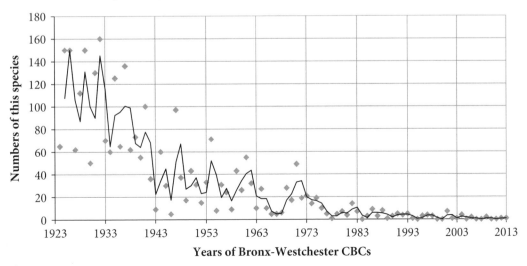

Figure 34

Yellow-crowned Night-Heron *Nyctanassa violacea*

NYC area, current
A cryptic breeder in very small, mostly Long Island colonies (rarely exceeding 20 pairs), by themselves or occasionally at edges of large mixed species heronries. Normally present only from Apr–Sep, it is the rarest winter heron after Least Bittern and Tricolored.

Bronx region, historical
Bicknell's *Riverdale* 1872–1901: unrecorded

Eaton's *Bronx + Manhattan* 1910–14: unrecorded

Griscom's *Bronx Region* as of 1922: unrecorded

Kuerzi's *Bronx Region* as of 1926: unrecorded

NYC area, historical
Cruickshank's *NYC Region* as of 1941: uncommon but annual Long Island summer visitor (19 Mar–16 Dec), increasing inland; 2 breeding sites

Bull's *NYC Area* as of 1968: 19 Mar–1 Nov, 5+ winter; greatly increased breeder, known from >12 sites, including 2 pairs Pelham Bay Park

Bull's *New York State* as of 1975: first Westchester County breeding

Study area, historical and current
An irregular spring, summer, or fall visitor to Van Cortlandt that rarely lingers. Unrecorded in the Bronx and study area until 1930.

About 18 mostly spring records of HYs or SYs (extreme dates of 27 Apr and 20 Sep) as follows: 25 May 1930 (Kuerzi)—apparently the first in the Bronx; assuming breeding plumage May–Jun 1931 (Herbert, Kassoy, Kuerzi); 1–3 May 1932 (Kieran); 28 May 1933 (Kieran); 7 Oct 1935 (Weber et al.); 23 May 1936 (Kieran); 27 Apr 1938 (Kieran); 21 May 1938 (Norse, Cantor); 6 May 1939 (Bull et al.); 2 Jun 1943 (Kieran); 26 May 1946 (Kieran); 8 May 1948 (Kieran); 2–7 May 1951 (Norse); 10 May 1953 (Kane, Kieran, Buckley); adult 6 Aug 1978 (Sedwitz); adult 29 Apr 1993 (Young); 13–20 Sep 1994 (Young); adult roosting along Van Cortlandt Lake on 20 May 2004 (Künstler et al.); HY on the Parade Ground (!) on 15 Oct 2005 (Lyons, Garcia, Abramson)—the latest and last occurrence.

Although its nearly obligate crustacean feeding might seem to limit its Bronx breeding distribution to the shore of Long Island Sound, it also feeds on fishes, insects, herps, even earthworms, and nests at the Piermont Marshes, and so it might breed in the study area. An early adult and an SY at Pelham Bay Park on 19 Apr 1953 and for a month thereafter (Buckley et al.) may have been prospectors or even breeders, given that the first nest in the Bronx was there on 22 Apr 1962 (W. Hastings). Subsequently they have been found breeding in the Bronx at Clason Point and on North and South Brother Islands, and there were 3 nests in nearby Mt. Vernon in Jun–Jul 2008 (unknown observer, *Kingbird*). They are amazingly easy to overlook when nesting in residential areas as they do regularly, and their preference for conifers and solitary nesting renders Woodlawn worth careful investigation.

Ibises: Threskiornithidae

Glossy Ibis *Plegadis falcinellus*

NYC area, current
Until 1944 a nearly unrecorded vagrant from Florida with 6 records, after which singles were recorded almost annually until a major invasion in 1959. It quickly became more numerous and widespread, finally nesting at Jamaica Bay Wildlife Refuge in 1961. As of 2016 it can appear anywhere in the area at any time and is increasingly reported in early winter. Most breeders are in heron colonies on Long Island and in the New York Harbor area.

NYC area, historical
Cruickshank's *NYC Region* as of 1941: Long Island specimens Sep 1847, Oct 1848; Van Cortlandt May 1935; Jones Beach Aug 1937

Bull's *NYC Area* as of 1968: expansion from southeastern US >1943 almost annual on Long Island with 1955 Cape May breeding; explosion into New York City area in flocks 1959, 1962, with first breeding in Jamaica Bay Wildlife Refuge 1961 and expanding outward thereafter; 14 Apr–31 Dec; many inland records >1964

Bull's *New York State* as of 1975: expansion continues, about 300 pairs breeding at 6+ Long Island sites; nonbreeding maximum 225 by 1969; reported every month but Feb

Levine's *New York State* as of 1996: spring maximum 250

Study area, historical and current
Until 1950 only 2 Van Cortlandt Swamp records: 13–15 May 1935 (Wiegmann et al.) and 4 May 1949 (Ephraim). Since the 1950s it went from an erratic straggler to an annual spring/summer visitor, coinciding with its establishment and increase as a New York City area breeder, especially on South Brother Island.

There are at least 24 additional records, all singles except as noted: 19 Aug 1953 (Ephraim); 29 Apr–1 May 1959 (Heath, Honig et al.); 3 on 3 May 1967 (Norse); 7 May 1971 (Ephraim); 29 Apr–6 May 1973 (Sedwitz, DiCostanzo); 29 Apr 1978 (Ephraim); 16 Aug 1980 (Ephraim); 27 May 1981 (Sedwitz); 12 Sep 1986 (Young); 17 May 1991 (Young); 18 Aug 1993 (Lyons, Garcia); 5–7 May 1994 (Young, Garcia); 2–3 on 6 May 1995 (Garcia); 25 Aug 1995 (Garcia); 12 May 1996 (Garcia); 29 May 1996 (Young); 3 on 6 Jul 1996 (Lyons, Garcia); 3 on the Van Cortlandt golf course on 19 Jun (McKeller), then 4 at Tibbett's Marsh on 30 Jun 1998 (Kilanowski); 17 May 1999 (Garcia, Lyons, Young); 7–22 Aug 1999 (Young); 24 Apr 2001 (Young); and at Hillview 3 on 10 May 2016 and 6 on 11 Apr 2017 (Camillieri). Before the 1959 invasion, 3 in Bronx Park on 11 May 1957 (Gilbert) were most unexpected, and it remains barely known there.

In the New York City area it was first found breeding in 1961 at Jamaica Bay Wildlife Refuge and increased fast, peaking in the early 1980s at about 900 pairs. Since then breeding pairs have fallen to half that number and may be dropping. Breeders nearest to the study area have been on Huckleberry Island off Pelham Bay Park (in summer 1989, the first in Westchester County: Künstler 2007), on Goose Island in

Pelham Bay Park in 2004 (Künstler 2007), and on South Brother Island since the first pair in 1992 (NYCA surveys).

Comments
Vagrant southeastern **White Ibis**, *Eudocimus albus*, is extremely infrequent in the New York City area, although it has occurred in Prospect Park and at Jamaica Bay Wildlife Refuge. It is unrecorded in the Bronx, but a juvenile was in the Piermont Pier area on 10 Aug 2008 (Weiss et al.). It is increasing greatly as a northward-moving breeder and disperser, one recent measure being an impressive flock of 1700+ at Chincoteague VA, in Jul 2015 (H. T. Armistead, pers. comm.).

New-World Vultures: Cathartidae

Black Vulture *Coragyps atratus*

NYC area, current
An increasing resident and breeder, scarcest on Long Island but now seen regularly in Central and Prospect Parks. First breeding was on Staten Island in 2009.

NYC area, historical
Cruickshank's *NYC Region* as of 1941: vagrant to New York City area, 14 mostly summer Long Island records 1877–1940

Bull's *NYC Area* as of 1968: very scarce but increasing; singles every month but Feb, Nov

Bull's *New York State* as of 1975: status unchanged

Study area, historical and current
A recent southern arrival, rapidly increasing in the Bronx and the study area and an impending breeder.

Black Vulture is possible any time of the year even though we know of only 9 Van Cortlandt records, all very recent: 2 on 11 Feb 2008, 29 Apr 2010, 29 Apr 2011, 4 on 4 Nov 2011, 4 on 22 Dec 2012 (Duran-Ruiz), 2 on 22 Sep 2013, 2 Oct 2014, 5 on 17 Mar 2015 (all Young), plus an impressive 11 on 22 Mar 2015 (Young et al.)—a New York City maximum certain to be eclipsed. At Hillview where it is a potential breeder, twos have been found several times between 2016–18 (Camillieri et al.), and it is also being seen with growing frequency as close as Inwood Park while breeding on the Palisades opposite Riverdale.

It has been steadily pushing northward in the last 50 years, first breeding in NJ in 1981 and then in New York State in 1997 in Ulster County. It is generally only a wanderer east of the Hudson River, but since the early 2000s the number of records has increased greatly, particularly during Apr. Since the early 2000s it has been seen annually even in small flocks as close as northwestern Yonkers, where it may also be breeding. In 2010 a nest was found on a TV tower in White Plains (Vellozzi). Apart from the Palisades, that is the closest to the study area.

While breeding in Queens, Brooklyn, and Manhattan remains undetected, it is expected momentarily, given the nest inside Fort Wadsworth at the Staten Island base of the Verrazano Bridge in Jul 2009. A recent series of Apr–May records from Bronx Park and the Bronx Zoo are intriguing enough to hint at a possible breeding site somewhere along the Bronx River. Mausoleums in Woodlawn are a possibility warranting investigation, as are abandoned buildings on North Brother Island and elsewhere in the Southeast Bronx, derelict vessels along the Bronx and Harlem Rivers, or even in active Bronx Zoo buildings themselves. In NJ, unanticipated breeding sites include old barns, sheds, and even duck blinds (R. Kane, pers. comm.), so out-of-the-box thinking is clearly in order. During NYSBBA II in 2000–2005 it was increasing greatly in the Hudson Valley, with 93 occupied blocks between Dutchess, Putnam, Westchester, Rockland, and Ulster Counties, although 82 were only possible (McGowan and Corwin 2008). This is an amazing distribution considering that it was *unrecorded* during NYSBBA I in 1980–85.

Turkey Vulture   *Cathartes aura*

NYC area, current

Formerly only a southern vagrant but now a widespread resident breeder, many of whose numbers depart in winter, others forming winter roosts. Until very recently bred only from Westchester County northward.

Bronx region, historical

Bicknell's *Riverdale* 1872–1901: only once, 19 Jun 1895

Eaton's *Bronx + Manhattan* 1910–14: rare summer visitor

Griscom's *Bronx Region* as of 1922: very rare visitor, several times over Bronx Zoo

Kuerzi's *Bronx Region* as of 1926: as Griscom, with 2 additional Jun, Jul records

NYC area, historical

Cruickshank's *NYC Region* as of 1941: uncommon visitor south of extreme northern Westchester County, slowly increasing in spring, summer, fall

Bull's *NYC Area* as of 1968: increased greatly >1950, except Long Island where still rare

Bull's *New York State* as of 1975: still increasing

Levine's *New York State* as of 1996: fall maximum 100 Orange County; winter maximum 115 Rockland County.

Study area, historical and current

An increasing wanderer from Westchester County that may occur at any time in all 4 extant subareas, but least frequently in winter. To our knowledge, none have ever been seen feeding on the ground in the study area. Also a regular spring/fall migrant in ever-growing numbers.

First recorded 18 May 1923 (Chubb), then on 9 Mar 1927 (Kieran), 3 May 1930 (Sedwitz), 23 Apr 1935 (Norse), several on 15–17 Apr 1938 (R. Allen et al.), 4 Oct 1946 (Komorowski), 2 on 10 Apr 1950 (Kane), 25 Oct 1955 (Birnbaum), 21 Jun 1958 (Mayer), 22 Sep 1963 (Post), and 11 Apr 1966 (Kane). By the early 1970s singles and twos were being seen from Apr–Nov, and 2 on 27 Jan 1971 (Lehman) were among the first in winter.

In spring, first migrants arrive in late Mar, peak in mid-Apr, and dissipate by late May, with a maximum of 11 on 10 Apr 1993 (Young).

In summer 3–5 can occur at almost any time, and the nearest regular breeders are now on the Palisades. NYSBBA II in 2000–2005 found the nearest cluster of occupied blocks in central southern Westchester County, essentially unchanged from NYSBBA I in 1980–85 (McGowan and Corwin 2008). DOIBBS data showed a consistently rising statewide trend line between 1966 and 2005, with an upward inflection point about 1980 (Sauer et al. 2005).

In view of this study area summer status, completely unpredicted was the pair that nested on the ground in a rock crevice in Van Cortlandt in May 2009 with a single egg found on 31 May but which may have been laid in Apr, judging by the adults' activities (DeCandido and Allen 2014). Unfortunately, the nesting failed from human disturbance or quadruped predation, the eggshell pieces being found in mid-Jun with no adults in attendance. Subsequent nesting in that location may have been precluded when the site was occupied by a homeless person—an unanticipated downside to an exceptionally long incubation period. This is the first nesting of a Turkey Vulture within the city of New York; time will tell if it recurs.

During NYSBBA I in 1980–85 the closest occupied block was in central Westchester County, and by NYSBBA II in 2000–2005 it was only slightly closer in Westchester County (McGowan and Corwin 2008). NYSBBA II reported that Turkey Vulture had bred on Staten Island, but investigation revealed that a volant juvenile seen on western Staten Island in Aug 2000 (the source of that record) was believed by its finder *not* to have been raised locally because neither it nor adults had been previously seen there that spring or early summer (A. Purcell, Wollney).

In fall, first migrants arrive in Sep, peak in mid-Oct, and depart by Nov, with a maximum of 87 on 9 Oct 2011 (Young). Extreme dates are increasingly irrelevant.

They became expected in winter by the late 1980s when 1–3 was the rule, but they did not begin to linger in the study area until the mid-2000s. The current winter maximum seems to be 5 at Hillview on 9 Dec 2016 (Adams). The nearest winter roosts are on the Palisades and in northwestern Yonkers and White Plains.

Ospreys: Pandionidae

Osprey   *Pandion haliaetus*

NYC area, current
A spring/fall migrant and burgeoning breeder whose numbers are recovering from near extirpation in the 1960s following runaway pesticide use. Earlier restricted to eastern Long Island, they are now reoccupying their 1800s breeding range, and are again nesting in all 5 New York City boroughs on poles erected for them but also taking advantage of blossoming cell phone towers everywhere. It remains a New York State Species of Special Concern.

Bronx region, historical
Bicknell's *Riverdale* 1872–1901: spring/fall migrant

Eaton's *Bronx + Manhattan* 1910–14: fairly common migrant

Griscom's *Bronx Region* as of 1922: scarce Hudson Valley migrant

Kuerzi's *Bronx Region* as of 1926: very common migrant in spring (19 Mar–12 Jun)/fall (5 Aug–14 Nov), occasional in summer

NYC area, historical
Cruickshank's *NYC Region* as of 1941: regular spring/fall migrant (14 Feb–7 Dec; maximum 54), most numerous on Long Island, in Sep; absent in winter

Bull's *NYC Area* as of 1968: 10 Mar–13 Jan; precipitous decline in main breeding population on eastern Long Island, where maxima fall 135, spring 50

Bull's *New York State* as of 1975: slow breeding recovery under way

Levine's *New York State* as of 1996: status unchanged

Study area, historical and current
A spring/fall migrant, often stopping to feed for a few days or even several weeks at Van Cortlandt Lake or Woodlawn, occasionally at Jerome and Hillview.

In spring, first migrants arrive in late Mar, peak in Apr, and depart by early May, with extreme dates of 14 Mar 2006 (Young) and 2 Jun 1953 (Buckley) and a maximum of 4 on 16 Apr 2006 (Young).

Ospreys bred for many years in a large American Chestnut tree on a hill northeast of Van Cortlandt Lake until 1878 (Grinnell, Herrick) and along the Bronx River in Bronx Park for an unstated period during the mid–late 1800s (Griscom 1927). This species has recovered from its nearly fatal nadir in the 1960s following widespread DDT poisoning, and about 2005 it began nesting on David's Island in Long Island Sound north of Pelham Bay Park, and starting in 2013 on a cell tower along the Hutchinson River within Pelham Bay Park (Künstler)—for the moment the only breeding pair within the Bronx. One or more pairs bred around Little Neck Bay in 1998 (Walter), and within the study area, a pair was reported breeding at Hillview in 2010 by reservoir personnel who removed its nest to obviate "drinking water contamination" (NYCDEP 2011). It would not be surprising were a pair to take up residence in Van Cortlandt if a raccoon-proofed tower were provided. Singles at Van Cortlandt Lake on 25 Jun 2011 (Baksh) and on 22 Jun 2013 (Morgan) may have been displaced Hillview or Pelham breeders or prospectors. Otherwise, the nearest inland breeders are at reservoirs or on cell phone towers in southern Westchester County or in Pelham Bay Park.

During NYSBBA I in 1980–85 the closest occupied block was in extreme northeastern Nassau County, and there were none confirmed in Westchester County. By NYSBBA II in 2000–2005 confirmed breeding was occurring on Staten Island, in Pelham Bay Park, near Rye, and on the Hudson River in Rockland County. Others were appearing at suitable inland sites in Westchester County, Rockland County, and various Hudson Valley locations.

In fall, first migrants arrive in Sep with an extreme date of 2 Aug 2009 (Young) and a peak in late Sep–early Oct. During the few postwar years hawkwatches were operating, typically 4–10 were seen daily; in 1946, 96 were counted from 12 Sep–9 Oct, with a maximum of 28 on 22 Sep (Komorowski). Most are gone by mid-Oct, the latest being singles on Van Cortlandt Lake on 4 Nov 2007 (Young), 7 Nov 1950 (Norse), 26 Nov 2011 (Finger), and at Hillview on 14 Dec 2017 (Camillieri).

Kites, Eagles, Hawks: Accipitridae

Swallow-tailed Kite *Elanoides forficatus*

NYC area, current
A southern spring range-prospector that until 2000 or so had been recorded in the New York City area only once every 20–30 years, if that often. It is slowly increasing in frequency throughout the Northeast, where it is now annual in spring, and multiples are also seen annually as close as Sandy Hook. Nearest breeders are only

in southern North Carolina, but with global warming they are expected to move northward.

NYC area, historical
Cruickshank's *NYC Region* as of 1941: vagrant, 6 records; once in Bronx

Bull's *NYC Area* as of 1968: additional record Jamaica Bay Wildlife Refuge 1956

Bull's *New York State* as of 1975: another Orange County 1974

Study area, historical and current
There are only two study area records. The first was on 22 Apr 1928 low over Jerome during a cloudburst (Kassoy)—which was unlikely to have been the same bird seen on 30 Apr a short distance south over the NYU Bronx campus (Hickey, Cruickshank). The second was on 25 Apr 2013 flying west midday over the Van Cortlandt Parade Ground (Young). What may have been the same individual was seen that evening over Hartsdale, Westchester County. Another was in downtown Yonkers flying south along the Hudson on 11 May 2015 (Block). Elsewhere in the Bronx, one was at Pelham Bay Park on 26 May 1982 (Root), and it has been recorded in Prospect but not yet Central Park.

Comments
Another southern raptor advancing northward is **Mississippi Kite**, *Ictinia mississippiensis*. It is now known to be breeding (or to have bred) in NH, MA, RI, CT, NY, and NJ and is seen almost annually in the New York City area. In 2012 a pair nested as close as Sterling Forest, Orange County, and others are currently breeding in Ocean County, NJ. In Prospect Park on 9 May 1950 a bird believed to have been an adult Mississippi Kite was watched by group of veteran observers (Sedwitz, Carleton, Jacobson, Alperin, Whelen et al.). But it was never seen again and was unrecorded at the time anywhere in the Northeast, so they were reluctant to press the issue even though convinced of its identity. It was 29 years until the next New York occurrence, involving 2 that appeared on Staten Island during a major Periodic Cicada emergence and remained there from 28 May–8 Jun 1979 (m. ob.). There are now multiple Central and Prospect Park occurrences, as well as 1 at nearby Inwood Park on 25 May 2008 (Allaire). It is unrecorded but anticipated in the study area, with 2 in the Bronx: at the Pelham Bay Park hawkwatch on 27 Aug 1989 (DeCandido) and over the Bronx Zoo on 21 Sep 2016 (Olson).

Bald Eagle   *Haliaeetus leucocephalus*

NYC area, current
Formerly a spring/fall migrant, and a regular winter visitor and resident, particularly along the Hudson River. Last bred on Long Island in 1930, and by the 1950–60s had been all but extirpated in the Northeast by DDT contamination. Recovery programs have been unbelievably successful to the point where by 2012 it was breeding in every NJ county, after having been reduced to a single pair in the 1970s. It finally bred on Gardiner's Island, Long Island about 2006, then along the Carmans River in 2013, and on Shelter Island in 2014. In 2015 a fourth pair was on the Great River near Connetquot State Park, with a fifth on the William Floyd Estate in Mastic. Prospectors are now also being seen at other Long Island sites.

Bronx region, historical
Bicknell's *Riverdale* 1872–1901: common in winter on the Hudson, seen occasionally in summer until 1888

Eaton's *Bronx + Manhattan* 1910–14: uncommon migrant

Griscom's *Bronx Region* as of 1922: only in winter as wanderers from Hudson

Kuerzi's *Bronx Region* as of 1926: occasional, usually in winter, but may occur anytime as it nests nearby (21 Aug–27 May)

NYC area, historical
Cruickshank's *NYC Region* as of 1941: uncommon spring/fall migrant, winter resident (maximum 14); a possible anytime, most regular on Hudson north of Inwood in winter (maximum 6); no longer breeds in New York City area

Bull's *NYC Area* as of 1968: recent maximum 18 Croton Point 1951; last Long Island breeding Gardiner's Island 1930, northern NJ 1952 (Morris County) and 1957 (Monmouth County)

Bull's *New York State* as of 1975: breeding decline accelerates; winter residents now only historical

Levine's *New York State* as of 1996: population crash starting to be reversed by systematic hacking, reaching 37 young in 1996 from none fledged in 1979; first Hudson winter aggregations reappearing

Study area, historical and current
Not certainly known to have ever bred in the Bronx but probably did until the 1700s and then not until 2017. Formerly a year-round visitor from the Hudson River until its population crash in the 1950s.

Until the mid-1950s, eagles were expected in winter along the Hudson River, and singles wandered into Van Cortlandt anytime from Aug–May, but especially from Oct–Mar (Cruickshank, Kieran, Norse). Migrants were formerly regular at Sep–Oct Van Cortlandt hawkwatches: e.g., in 1946, 10 were seen on 7 days from 12 Sep–1 Oct. Daily maxima were 5 on 24 Sep 1950 and 6 on 17 Sep 1950; 2s and 3s were often seen (all Komorowski, Russak).

Except for 2 not quite adults in Woodlawn in Jan–Feb 1977 (Capodilupo, Teator), none were seen after 1960 until population recovery took effect by the 1980s, when, for example, Young's first in many decades was on Hudson River ice on 3 Mar 1998. Local reports then began to be augmented by 4 prefledglings hacked at nearby Inwood Park in 2002 and again in 2003. Migrants were also occasionally seen, like the 2 on 15 Sep 2001 (unknown observer, *Kingbird*) and the single on 14 Sep 2013 (Souirgi). By 2010, eagles had become more or less resident in the Riverdale area, and once again singles and pairs are wandering into the study area at any time, augmented by occasional but increasing fall migrants. New York State DEC data for this species in New York State showed a flat line until about 1985, when a steep increase began, continuing to this day (McGowan and Corwin 2008).

In Mar 2017, the first eagle nesting attempt within New York City since the 1700s occurred when a pair usurped an empty Osprey nest in Pelham Bay Park (Rothman). One eagle seemed to be incubating, but the pair was displaced when the Ospreys returned. By 28 Dec 2017 what seems to be the same pair had returned and was spending time at Turtle Cove with a 3-month lead on the Ospreys.

Northern Harrier   *Circus cyaneus*

NYC area, current
A widespread fall/spring coastwise migrant and winter resident in open grassy areas, especially saltmarshes. Formerly a widespread breeder, now restricted to North and South Shore saltmarshes and scattered inland in open areas on Long Island and Staten Island. May be overlooked nesting on abandoned garbage dumps. It is a New York State Threatened Species.

Bronx region, historical
Bicknell's *Riverdale* 1872–1901: unmentioned

Eaton's *Bronx + Manhattan* 1910–14: common migrant, rare breeder

Griscom's *Bronx Region* as of 1922: uncommon all times of year near New York City; bred Van Cortlandt Swamp until 1896, now uncommon migrant

Kuerzi's *Bronx Region* as of 1926: fairly common migrant in spring (late Mar–mid-May)/fall (mid-Aug–Jan)

NYC area, historical
Cruickshank's *NYC Region* as of 1941: fairly common Long Island spring/fall migrant, regular winter resident (maximum 6); uncommon migrant away from coastal marshes, scarce in winter

Bull's *NYC Area* as of 1968: decreased as a breeder, now relatively rare and local; Long Island fall maximum 46, winter 40 on 1961–62 South Shore Christmas Bird Counts

Bull's *New York State* as of 1975: Long Island fall maximum 150+

Study area, historical and current
A spring/fall migrant, nearly unknown winter visitor, and extirpated breeder. Kieran recorded 1–2 every year between 1914 and 1956. Known from all 4 extant subareas.

In fall, first migrants arrive in late Aug, peak in late Sep, and depart by late Oct, with extreme dates of 29 Aug 2010 (Young), 3 on 18 Nov 1934 (Weber, Jove), and 19 Nov 1995 (Garcia), and a maximum of 11 on 18 Sep 1951 (Komorowski). Until Jun 1967, the fall maxima in Central and Prospect Parks were, respectively, 7 on 13 Sep 1952 (Aronoff) and 4 on 23 Sep 1951 (Whelen et al.).

Only 3 winter singles: 1 eating Brown Rats in Woodlawn on 17 Dec 1960 (Horowitz, Heath) and seen again in Van Cortlandt Swamp on 26 Dec (Tudor, Zupan); 26 Dec 2004 (Lyons et al.); and 12 Feb 2011 (Baksh).

In spring, first migrants arrive in late Mar, peak in early Apr, and depart by mid-Apr, with extreme dates of 16 Mar 1994 (Garcia, Lyons) and at Hillview 28 Apr 2013 (S. Martin), all singles.

Harriers bred in Van Cortlandt Swamp until 1896 (Dwight), and there are no indications they ever resumed breeding there. Latham (1957) noted that on eastern Long Island, nests were always in cattails, which is what former Van Cortlandt Swamp breeders were using.

Upstate, harrier blocks increased between NYSBBA I in 1980–85 and II in 2000–2005, but near New York City they continued to disappear. On Long Island alone they vanished from about 45 blocks and within New York City were only around Jamaica Bay and on Staten Island. East of the Hudson River none were detected in Westchester, Putnam, or Dutchess Counties, and only 3 occupied blocks were in Orange and Ulster Counties (McGowan and Corwin 2008). The nearest breeders as of 2016 are in the Hackensack Meadowlands.

Sharp-shinned Hawk *Accipiter striatus*

NYC area, current
A common fall migrant, a rare spring migrant, an exceedingly rare, local, and declining breeder only beyond New York City and Long Island, and a widely but thinly distributed winter visitor/resident. It is a New York State Species of Special Concern.

Bronx region, historical
Bicknell's *Riverdale* 1872–1901: uncommon fall migrant, occasional in winter

Eaton's *Bronx + Manhattan* 1910–14: fairly common migrant, rare breeder

Griscom's *Bronx Region* as of 1922: common migrant, last bred 1916, no winter records

Kuerzi's *Bronx Region* as of 1926: fairly common migrant in spring (10 Mar–30 May)/fall (3 Aug–28 Dec); rare in winter. 1–2 pairs still breed [locations not given]

NYC area, historical

Cruickshank's *NYC Region* as of 1941: common fall (maximum 11/hour)/uncommon spring migrant but rare in winter; nested within New York City in recent years [locations not given]

Bull's *NYC Area* as of 1968: fall Long Island maximum 50, inland 110 in Van Cortlandt; 1920s horrific shooting numbers from Fisher's Island

In fall, first migrants normally arrive in early Sep, but there are numerous Aug records back to the 1930s and continuing to 2016. These are well before normal migration dates, and we believe they represent nearby breeders. Sharp-shins peak in mid-Oct, and depart by late Oct, and in 1946, 424 were tallied between 4 Sep and 7 Oct with a single-day maximum of 110 on 18 Sep 1951. Daily totals of 30s–60s were frequent at hawkwatches (all Komorowski, Russak)

In winter, a few occur in all 4 extant subareas, but usually only singles are seen with multiple maxima of 3 throughout the study area. Sharp-shin has never been numerous on the Bronx-Westchester Christmas Bird Count, and that was the case until about 1975. Then their numbers took off and reached high levels for a top predator (Fig. 35). This was puzzling until Duncan (1996) argued convincingly that the great winter increase throughout the North-

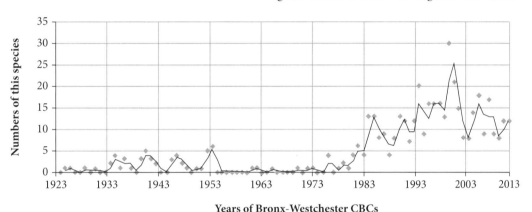

Figure 35

Bull's *New York State* as of 1975: Long Island maximum 175 Fire Island Lighthouse 1 Oct 1974, with 385 same day at Hook Mountain, just above Tappan Zee Bridge

Levine's *New York State* as of 1996: winter maximum 17

Study area, historical and current

A spring/fall migrant, overwinterer, and extirpated breeder.

east was correlated nicely with the equally great increase in numbers of winter feeders—which Sharp-shins found to be irresistible larders. Subsequently their winter numbers, at least in the New York City area, leveled off.

In spring, first migrants arrive in late Mar, peak in mid-Apr, and depart by late Apr, but extreme dates are unclear as May individuals may be breeders. The maximum is 2 on 12 Mar 2011 (Baksh).

The regular occurrence of this species in May–Aug from the 1930s–50s indicated its breeding, although the last pair in Van Cortlandt was in 1952 (Komorowski, Buckley et al.). Longtime Grassy Sprain breeders now gone may have helped maintain the 1–2 pairs in the study area. They become so secretive when nesting that it is extremely easy to overlook them unless very close to a nest. An assumed Sharp-shinned Hawk nest was reported without any details from Mill Rock, in the East River adjacent to Randall's Island and the Triboro Bridge, on 22 May 2013 (NYCA surveys). If correct, this would be the first breeding within New York City in 60 years and in a most improbable location.

NYSBBA II in 2000–2005 found this species to have increased its number of recorded upstate blocks by 68% since NYSBBA I in 1980–85, but in the New York City area there were none, including on Long Island. Only 5 occupied blocks (mostly) in central and northern Westchester County were located, so closest breeders to the study area were on the Palisades near Piermont Pier (McGowan and Corwin 2008). DOIBBS data showed a 6.9% statewide increase between 1966 and 2005, leveling off after 1979 (Sauer et al. 2005).

Specimens
Dwight collected a possible local breeder in Van Cortlandt on 3 May 1887 (AMNH) and Bicknell 2 in Riverdale in Sep, Oct 1876–79 (USNM).

Cooper's Hawk   *Accipiter cooperii*

NYC area, current
A spring/fall migrant, overwinterer, and scarce but increasing breeder. Almost extirpated in the 1960s, numbers began to recover by the 1980s, and as of 2016 they may even be more numerous locally than before their crash. They are again breeding in probably all New York City boroughs but Manhattan and are regularly seen at area hawkwatches in double digits. It is a New York State Species of Special Concern.

Bronx region, historical
Bicknell's *Riverdale* 1872–1901: bred pre-Van Cortlandt 1883; single Riverdale winter record

Eaton's *Bronx + Manhattan* 1910–14: fairly common migrant, rare breeder

Griscom's *Bronx Region* as of 1922: no longer breeding; uncommon migrant, rare in winter

Kuerzi's *Bronx Region* as of 1926: fairly common migrant in spring (12 Mar–30 May)/fall (13 Aug–2 Jan); rare in winter [breeding unreported]

NYC area, historical
Cruickshank's *NYC Region* as of 1941: common spring/fall migrant, rare but regular in winter; few pairs still nest within New York City, including Staten Island [no other location details]

Bull's *NYC Area* as of 1968: compared with Sharp-shinned, more widespread in summer, commoner in winter, much less numerous in migration

Bull's *New York State* as of 1975: status unchanged

Levine's *New York State* as of 1996: fall maximum 37 at Butler hawkwatch, Westchester County

Study area, historical and current
A spring/fall migrant, overwinterer, and probable breeder.

In fall, first migrants arrive in early Sep, peak in mid-Oct, and depart by late Oct, but those on extreme dates are uncertainly separable from breeders and winter residents. A maximum of 3–4 was the norm in Sep–Oct 1946–53 at hawkwatches (Komorowski, Russak); the recent maximum is 6 on 15 Oct 2007 (Young).

In winter found in all 4 extant subareas, with a current maximum of <10 throughout the study area. They have increased dramatically in winter since the mid-1980s, and study area daily Christmas Bird Count maxima of 10 are now occurring, up from the 1970s nadir when nearly none were seen (Fig. 36).

overwinterers beginning in the 1980s, we assume that 1–2 pairs are breeding again in the study area—a ♂ was seen 3 times in Van Cortlandt in the same area through 13 Jun 2001 (Jaslowitz) and another was at Hillview on 12 Jul 2017 (Camillieri); an HY on 29 Jul and 7 Aug 2013 (Young) is also evidence of resumed study area breeding. It has also recently bred in Bronx Park from 2001–2003 (DeCandido and Allen 2005b), does so probably annually in Pelham Bay Park, has done so on Staten Island since at least 1999, and was suspected in 2 Brooklyn locations in 2007 (unknown observer, *North American Birds*). Like all accipiters, an

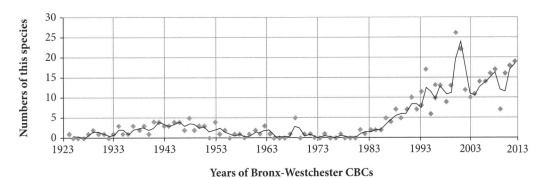

Figure 36

In spring first migrants, usually singles, arrive in late Mar, peak in mid-Apr, and depart by late Apr, but extreme dates are uncertain. The maximum was 3 on 22 May 1954 (Scully), some or all of which may have been local breeders.

A few pairs bred continuously in Van Cortlandt from the mid-1800s–1958 (Bicknell, Kieran, Norse, Buckley, Kane), even though missed by Kuerzi, Cruickshank, and Bull, after which the species' population in the Northeast crashed from biocides. With recovery and greatly increased numbers of migrants and

adult Cooper's becomes exceedingly quiet and stealthy when nesting and is easily overlooked.

During NYSBBA I in 1980–85, no occupied Blocks were found on Long Island, in New York City, or Rockland County, and only 1 in northeastern Westchester County. But by NYSBBA II in 2000–2005 this had leapt to <50 occupied Blocks in Nassau and Suffolk Counties, plus 4 on Staten Island, 2 in the Bronx, and 15 in Westchester County (McGowan and Corwin 2008). Cooper's Hawk breeding populations in and near New York City have now returned to 1940s levels.

Northern Goshawk *Accipiter gentilis*

NYC area, current
An annual but rare fall migrant and winter visitor/resident in very small numbers that is subject to infrequent irruptions. In the last 30 years it has been regular in winter on Long Island barrier beaches, but it is a great rarity elsewhere. It is a New York State Species of Special Concern.

Bronx region, historical
Bicknell's *Riverdale* 1872–1901: once, on 12 Jan 1890

Eaton's *Bronx + Manhattan* 1910–1914: rare winter visitor

Griscom's *Bronx Region* as of 1922: only single record, Bronx Zoo

Kuerzi's *Bronx Region* as of 1926: no new Bronx records

NYC area, historical
Cruickshank's *NYC Region* as of 1941: erratic, rare winter visitor (3 Oct–28 Apr), last irruption 1926–27 when recorded in all 5 New York City boroughs

Bull's *NYC Area* as of 1968: maxima >1 only at NJ hawkwatches

Bull's *New York State* as of 1975: status unchanged

Study area, historical and current
An erratic but slowly increasing mostly fall migrant and winter visitor.

In fall, first migrants arrive in early Oct, peak in early Nov, and depart by late Nov. There is a hint of a spring movement in Mar, coinciding with the usual departure dates of the 1–2 that irregularly (?) overwinter in the study area. The range of dates spans 25 Sep 1999 (Jaslowitz, Künstler)–26 Apr 1936 (D. Lehrman, Stephenson), with the bulk between Dec–Mar and all singles except 2 hawk-ridge migrants on 5 Nov 1944 (Komorowski).

Formerly fairly frequent in the now-eradicated conifer groves at Hillview and as of 2016 probably regular only in Woodlawn. If annual fall hawkwatches were reinstated in Van Cortlandt, this species should prove to be annual in small numbers. With the recent recovery of regional Cooper's Hawk breeding populations, competition between the two can be expected to increase. Statewide, NYSBBA II in 2000–2005 found this species to have decreased its number of occupied blocks by 20% since NYSBBA I in 1980–85, and the number of reported blocks within Westchester (1), Putnam (3), Orange (3), and Rockland (1) Counties was also down (McGowan and Corwin 2008). The nearest current breeders may be in Morris County, NJ.

Red-shouldered Hawk *Buteo lineatus*

NYC area, current
A regular fall migrant in small numbers, usually inland but occasionally along the beaches. Rare in spring, and a decreasing inland breeder possibly now extirpated from Long Island and New York City. It is a New York State Species of Special Concern.

Bronx region, historical
Bicknell's *Riverdale* 1872–1901: common breeder, regular spring/fall migrant

Hix's *New York City* as of 1904: Bronx Park resident

Eaton's *Bronx + Manhattan* 1910–14: fairly common resident

Griscom's *Bronx Region* as of 1922: common winter resident, few pairs still nesting [locations not given]

Griscom's *Riverdale* as of 1926: now locally extirpated

Kuerzi's *Bronx Region* as of 1926: common spring/fall migrant, winter resident; few pairs breed [locations not given]

NYC area, historical

Cruickshank's *NYC Region* as of 1941: fairly common resident (maximum 6), in recent years breeding in Van Cortlandt, Inwood Parks However, 122 were seen as close as Tuckahoe on 29 Oct 1944 (Komorowski).

During most winters 3–4 Red-shoulders were seen in the study area through the 1950s, but since then they have nearly disappeared at that season, singles now being found on Christmas Bird Counts only once every few years. They have always been scarce on Bronx-Westchester Christmas Bird Counts, and after the early 1960s were even less frequently encountered (Fig. 37).

Few spring migrants have been detected, but some should move through in Mar–Apr, like recent singles on 27 Apr 1998 (Young) and 13 Apr 2013 (Baksh).

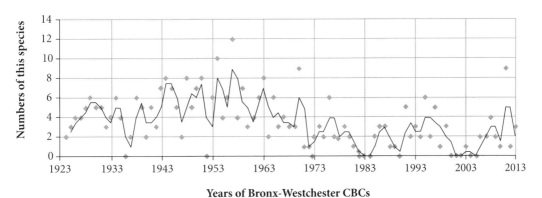

Figure 37

Bull's *NYC Area* as of 1968: most numerous in fall along inside coast while relatively uncommon on Kittatinny hawkwatches; maximum 122

Bull's *New York State* as of 1975: greatly diminished breeder in recent years

Study area, historical and current

A spring/fall migrant, erratic winter resident, and former breeder in Van Cortlandt and perhaps Woodlawn.

In fall, first migrants arrive in mid-Sep, peak in mid-Oct, and depart by early Nov, with extreme dates of 17 Aug 2006 (Young) and 5 (the maximum) on 3 Nov 2012 (Young).

Eaton (1910, 1914) listed it as a breeder in his Bronx + Manhattan county with no additional details and Hix (1906) called it a common resident in Bronx Park. A pair or two bred regularly on the Ridge in Van Cortlandt (as well as at nearby Grassy Sprain Reservoir) through the 1930s (Cruickshank 1942) and the 1950s (Norse, Kieran, Komorowski, Buckley) until 1963–64 (Meyer). Since then there have been no summer reports but the study area's coverage at that season diminished precipitously until the 1980s.

NYSBBA I in 1980–85 found occupied blocks in Suffolk County (7); none in Nassau Co.; 1 each on Staten Island, in Brooklyn, and in the Bronx; Westchester County (8); Rockland County (10). By NYSBBA II in 2000–2005 these had all fallen to zero except 4 in Westchester County and 1 in Rockland County. Despite this drastic decline, a pair bred in Bronx Park and/or the Bronx Zoo in 1974 and 1975 (Maguire et al.) and 1980 (Roderick). Even more surprising, 2 pairs bred in Prospect Park in 1981 (Yrizarry). It was recorded as a breeder in the block south of the study area on NYSBBA I, 1980–85, but not in NYSBBA II 2000–2005. The nearest reliable breeders may be on the Palisades.

Specimens
Dwight collected 1 in Van Cortlandt on 24 Nov 1887 and Foster another in Woodlawn on 1 Sep 1890 (both AMNH).

Broad-winged Hawk *Buteo platypterus*

NYC area, current
A fall migrant (often in the thousands) and decreasing breeder that occurs almost unseen as a spring migrant in small numbers.

Bronx region, historical
Bicknell's *Riverdale* 1872–1901: huge flight 16 Sep 1878

Eaton's *Bronx + Manhattan* 1910–14: uncommon resident

Griscom's *Bronx Region* as of 1922: rare spring/fall migrant

Kuerzi's *Bronx Region* as of 1926: uncommon migrant in spring (21 Apr–23 May)/fall (22 Aug–14 Oct) when sometimes abundant; may still breed Sprain Ridge

NYC area, historical
Cruickshank's *NYC Region* as of 1941: common spring and abundant fall (maximum 1000+) migrant; breeds no closer than central Westchester County; 7 recent winter records

Bull's *NYC Area* as of 1968: status unchanged

Bull's *New York State* as of 1975: maxima fall 2500, 2700, spring 30 in Prospect Park; breeds on Palisades

Study area, historical and current
A fall migrant in episodically large flocks, former (only?) breeder, and barely noticed spring migrant.

In fall, first migrants arrive in late Aug (earliest 11 Aug 1950: Sedwitz), peak in late Sep, and depart by mid-Oct. In many years 3-digit flocks appear following cold fronts in mid-Sep, with daily maxima of 1000+ on 19 Sep 1925 (Cruickshank), 1500+ on 18 Sep 1956 (Kane, Post, Scully), 2500+ on 20 Sep 1962 (Meyer), 2520 on 21 Sep 1948 (Komorowski), 2700 on 22 Sep 1945 (Komorowski), and 5000+ between Bronx and Van Cortlandt Parks on 14 Sep 1975 (Maguire et al.), although routine hawkwatches have not been maintained since 1953 (Komorowski, Russak). Nearby, there were 1500+ over Bronx Park on 19 Sep 1970 (Maguire et al.), and 1200 on 27 Sep 1975 (Maguire). An impressive count of 15,549 at Pelham Bay Park on 11 Sep 1990 (DeCandido et al.) is the highest ever recorded in the Bronx and New York City.

Broad-wings are infrequent after Oct, with only 5 records: 2 on 6 Nov 1935 (Cantor); 3 in a kettle on 18 Nov 1936 (Karsch); 5 in a kettle on 20 Nov 1955 (Birnbaum); 12 Feb 1921 (Watson); plus an HY that survived from 7 Dec 1933–7 Feb 1934 during one of the coldest winters on record in the New York City area by feeding on Brown Rats at the garbage dump on the site of the Van Cortlandt stadium (Cruickshank, Norse et al.).

In spring, the 1–2 migrants detected per day arrive in mid-Apr and peak in late Apr, with extremes of 9 Apr 2003 (Lyons) and 29 May 1996 (Young), but cf. the following.

Since the late 1930s 1–2 pairs may have bred in the Northwest Forest, on the Ridge, and perhaps also in Woodlawn (Kieran). We are not aware of any nests having been found, as they become very secretive when breeding and their Wood Pewee-like calls are often overlooked by those unfamiliar with them. One carrying a long stick through the canopy in the Northwest Forest on 23 Apr 1953 (Buckley), a pair from 8 May–17 Jun 1955 (Kane et al.), and 1 low in the forest in Riverdale being pursued by grackles on 8 May 2016 (Willow) were potential breeders; a recent single in the Northwest Forest on 19 May 2012 (Baksh) may also have been. SY Broad-wings are regular if not well known late May–Jun migrants, so any then, especially small flocks high overhead like the 6 over White Plains on 2 Jun 1964 (Robben), are just that, whereas solitary birds flying low through forested areas intimate breeding. A pair or two may nest at Grassy Sprain Reservoir, but otherwise the closest breeders may be along the Palisades, where there were 9 pairs in the late 1960s (Boyajian 1971). Broad-wing was recorded in Block 5852D, south of a study area block, during NYSBBA I in 1980–85 but not during NYSBBA II in 2000–2005.

NYSBBA I also recorded many occupied blocks in Nassau, Suffolk, and Westchester Counties, but by NYSBBA II they had been reduced to 7 (1 confirmed) in Suffolk County, 2 in Nassau County, 1 (confirmed) on Staten Island, and 12 (5 confirmed) in Westchester County (McGowan and Corwin 2008). Sparse DOIBBS data showed a nonsignificant statewide increase from 1966–2005 (Sauer et al. 2007).

Comments
Now almost annual in the New York City area and in multiples every fall at Cape May, vagrant western **Swainson's Hawk**, *B. swainsoni*, has occurred with southbound migrating Broad-wings once in the Bronx—at the Pelham Bay Park hawkwatch on 24 Sep 1988 (DeCandido et al.)—but with increasing frequency in central Westchester County as well as in Rockland and Orange Counties. Unrecorded in Central or Prospect Parks.

Red-tailed Hawk   *Buteo jamaicensis*

NYC area, current
Resident and seeable almost any day of the year throughout, and especially obvious in winter. They have adapted to urbanity, are now breeding in most large New York City parks or on adjacent buildings, and are familiar sights on overhead light fixtures along highways. Despite local increases, their main fall migration route remains well to the west in NJ.

Bronx region, historical
Bicknell's *Riverdale* 1872–1901: arrived as early as 22 Aug 1880

Eaton's *Bronx + Manhattan* 1910–14: fairly common resident

Griscom's *Bronx Region* as of 1922: common winter resident; not known to nest

Kuerzi's *Bronx Region* as of 1926: common in winter, occasional summer [locations not given]

NYC area, historical
Cruickshank's *NYC Region* as of 1941: greatly declined since 1920s, now fairly common migrant in spring, especially fall (maximum hundreds); local but not uncommon winter visitor; breeds no closer than northern Westchester County

Bull's *NYC Area* as of 1968: decreased considerably as breeder since 1930s; maxima 24 in winter Pelham Bay Park, 30 spring Prospect Park

Bull's *New York State* as of 1975: status unchanged

Study area, historical and current
A resident breeder, spring/fall migrant, and winter resident.

Long before they became Freeway Buzzards and copulated shamelessly in Central Park for TV cameras, this species had a status and distribution that would baffle observers today. From Griscom (1923): not definitely known to breed south of Putnam County; Cruickshank (1942): very few pairs nest in the highlands of Rockland County and extreme northern Westchester County; Bull (1964): has decreased considerably [since the 1930s and now breeds] chiefly in the highlands of northwestern NJ.

Yet by the time of NYSBBA I in 1980–85, it was nesting across Westchester County, in the Bronx, on Staten Island, and along Nassau County's north shore (McGowan and Corwin 2008), and the first Central Park nest was in 1992. They now breed in Van Cortlandt (2–3 pairs), Woodlawn (2–3 pairs), and Riverdale, as well as in Bronx, Inwood, and Pelham Bay Parks, and becoming typically reclusive when nesting, allow us only to estimate a total of 4–6 pairs for the study area. It is unclear when they first began breeding there, although summer adults in 1952–54 (Buckley et al.) indicate established breeding.

NYSBBA II in 2000–2005 recorded major increases over NYSBBA I in 1980–85 in the number of occupied Blocks in New York City, Long Island, and Westchester County, with multiple breeding now occurring in all 5 New York City boroughs (McGowan and Corwin 2008). Statewide, DOIBBS data showed a non-significant statewide increase from 1966–2005 (Sauer et al. 2005).

First fall migrants usually appear in late Aug, peak in Oct, and depart by mid-Nov, with extremes of 17 Aug 1936 at Hillview (Norse) and 9 on 21 Nov 2010 (Young). Days with 20+ are frequent, with a maximum of 37 on 16 Oct 2010 (Young). This pattern contrasts sharply with that in 1946, when Van Cortlandt hawk-watchers saw only 2 over 24 days in Sep–Oct (Komorowski), indicating the species' great increase since then.

They are the most easily found *Buteo* in winter, when 5–10/day are routine, with a maximum of 13 on 22 Dec 2013 (Lyons et al.). The Bronx-Westchester Christmas Bird Count increase since the 1930s (Fig. 38) does not of

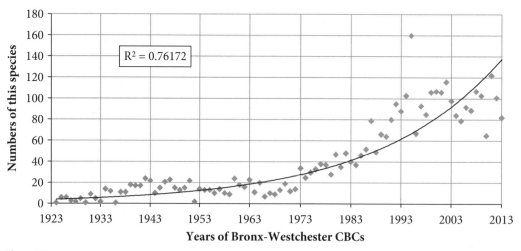

Figure 38

course represent local breeders, but wherever they are coming from, that population has been increasing exponentially. This has been mirrored in the study area but 1 order of magnitude lower.

In spring, any northbound movement is so slight in Mar–Apr that in most years none are noted; if any are, only singles and twos overfly the study area and would be difficult to distinguish from local breeders.

Rough-legged Hawk *Buteo lagopus*

NYC area, current
An irregular, occasionally irruptive winter visitor/resident, favoring coastal saltmarshes and large inland open areas.

Bronx region, historical
Bicknell's *Riverdale* 1872–1901: unrecorded

Eaton's *Bronx + Manhattan* 1910–14: rare winter visitor

Griscom's *Bronx Region* as of 1922: rare winter visitor

Kuerzi's *Bronx Region* as of 1926: uncommon winter visitor/resident (11 Oct–4 Apr)

NYC area, historical
Cruickshank's *NYC Region* as of 1941: fairly common fall/uncommon spring migrant, exceedingly local winter resident (6 Oct–12 May; maximum 12); very uncommon away from coastal saltmarshes

Bull's *NYC Area* as of 1968: status unchanged

Bull's *New York State* as of 1975: late spring date 21 May; Long Island winter maxima 15, 25

Study area, historical and current
An irregular Van Cortlandt and Hillview winter visitor and (mostly) fall migrant, formerly regular but now barely recorded. These arctic breeders, like Snowy Owls, are subject to periodic irruptions, when they can be widespread and favor large coastal saltmarshes and active or abandoned garbage dumps. It can be difficult to distinguish migrants from wandering winter residents, and only singles have been detected.

Study area migrants typically are among the latest raptors at hawkwatches, sometimes not appearing until late Nov or Dec but often by late Oct. If regular hawkwatches were reinstated in Van Cortlandt, this species would prove to be nearly annual in low numbers. Most do not stop, but singles have overwintered in Van Cortlandt Swamp in 1951–52 (Kane et al.), 1962–63 and 1963–64 (Enders, Heath et al.). They may have done so in other years, judging by the number of scattered midwinter records there, in Woodlawn, and at Hillview, but more pervasive data are not at hand. Especially in irruption years, a slight return flight is sometimes detected in Mar. Study area earliest fall dates are 26 Sep 1996 (Young) and 12 Oct in 1956 (Steineck, Yeganian) and in 1959 (Heath); in spring 18 Apr 1965 (Kane) and 1 May 1993 (Young). The most recent occurrence was on 19 Jan 2008 (Garcia).

A few Rough-legs occur almost annually in marshes and dumps on the Bronx shore of Long Island Sound and the East River, increasing to high single digits in major irruption years. But with the extensive habitat loss there since the 1960s, they are now much harder to find. From the 1920s to 1960s they were also seen irregularly along the Hudson River in Riverdale (Cruickshank, Kieran, Buckley).

Golden Eagle *Aquila chrysaetos*

NYC area, current
A fall migrant in small but increasing numbers at inland hawkwatches, and an increasing but extremely scarce fall migrant and very occasional winter visitor/resident on Long Island. It is a New York State Endangered Species.

Bronx region, historical
Bicknell's *Riverdale* 1872–1901: unrecorded

Eaton's *Bronx + Manhattan* 1910–14: unrecorded

Griscom's *Bronx Region* as of 1922: unrecorded

Kuerzi's *Bronx Region* as of 1926: unrecorded

NYC area, historical
Cruickshank's *NYC Region* as of 1941: rare but annual fall migrant but mostly on NJ hawk ridges

Bull's *NYC Area* as of 1968: 18 Sep–2 Mar; very rare, only singles away from NJ hawkwatches

Bull's *New York State* as of 1975: status unchanged

Study area, historical and current
An extremely scarce fall migrant in late Oct–Nov, with only 4 records. From the Ridge during hawkwatches, single adults were seen on 22 Sep and 1 Oct 1946 (Komorowski), an immature was over Vault Hill on 10 Oct 1956 (Buckley), and another immature over the Parade Ground on 30 Nov 2013 (Souirgi et al.). If regular hawkwatches were reinstated in Van Cortlandt, this species would prove to be routine in small numbers, as they may be at Pelham Bay and Central Parks (see below). Extremely unusual Bronx spring Goldens were studied in Bronx Park on 27 Apr 1955 (C. Young et al.) and at Pelham Bay Park on 26 Mar 2007 (Rothman).

Only a handful of New York State blocks support breeding Goldens (8 in 1980–85, 9 in 2000–2005), so most of those seen in New York State and in the study area come from eastern Canada (McGowan and Corwin 2008).

Golden Eagle's great increase in the Northeast has also been detected nearby, as follows: crossing the Hudson River from Yonkers to Alpine on 8 Nov 1969 (Boyajian); in Bronx Park 13 Nov 1980 (Hait); at Pelham Bay Park on 21 Nov 1971 (Kane), 30 Oct 1988 (DeCandido et al.), 2 on 22 Oct 1989 (DeCandido 1991b), 2 juveniles on 29 Oct 1990 (DeCandido 1990); 7 in Central Park from 8 Oct–24 Nov 1995, 2 each on 30 Oct, 24 Nov 1995 (Freedman et al.).

Rails, Coots, Gallinules: Rallidae

Yellow Rail *Coturnicops noveboracensis*

NYC area, historical and today
An exceedingly infrequently detected but doubtless annual spring and especially fall migrant, occasional in winter in coastal saltmarshes. It has never been detected in Central or Prospect Park. It may appear almost

anywhere, and there is no known location where one might listen for them in spring migration, as in Cape May.

Bronx region, historical

Bicknell's *Riverdale* 1872–1901: 2 fall specimens

Eaton's *Bronx + Manhattan* 1910–1914: rare

Griscom's *Bronx Region* as of 1922: unrecorded

Kuerzi's *Bronx Region* as of 1926: as per Bicknell

NYC area, historical

Cruickshank's *NYC Region* as of 1941: extremely rare migrant but overlooked; 30 Aug–28 Nov, 29 Mar–29 Apr, plus 2 in winter; 34 Long Island records plus Bicknell's 2

Bull's *NYC Area* as of 1968: late fall date of 4 Dec

Bull's *New York State* as of 1975: status unchanged

Study area, historical and current

This is another species that passes overhead every fall (and maybe spring) yet is almost never detected on the ground. There are 3 study area records: the first 2 were found dead—on the edge of the Parade Ground at Broadway and W. 246th St. on 2 Oct 1881 (Bicknell) and along the Van Cortlandt Swamp Causeway fence on 19 Oct 1951 (Kieran). The third was quite alive, in the cattail-*Phragmites* marsh at the head of Van Cortlandt Lake on the morning of 9 May 1991 (Jaslowitz). It was repeatedly giving the characteristic 2-3, 2-3 *tic* couplets with the unique timbre and staccato cadence unmatched by any North American frog, toad, or insect (for those so concerned) and made only by singing Yellow Rails during their breeding season. As they often do, it immediately responded to mimicking *tics* the hearer made with stones as it moved about the marsh. This appears to be the first confirmed instance of a singing Yellow Rail anywhere in southern New York State.

Elsewhere in the Bronx, 3 others in fall were captured alive and taken to the Bronx Zoo: 2 from Pelham Bay Park, on 31 Aug 1955 (fide Kane) and 27 Aug 1959 (Sanford), and another from an apartment house on the Grand Concourse at E. 187th St. on 14 Nov 1966 (Yrizarry). Also nearby, 1 was collected in the Dyckman St. marshes on 29 Sep 1880 (Bicknell).

Spring Yellow Rails have been heard singing in northern NJ in the Great Swamp NWR on 13 May 1995 and regularly, if not annually, in southern NJ in Cumberland County's Delaware bayshore marshes in Apr–May from 2006–2014 (*North American Birds*; Boyle 2011). In passing, it is useful to note that while Yellow Rails do vocalize on northbound migration (presumably with increasing frequency as they approach their breeding grounds and as spring progresses), they seem never to do so on fall migration or where they are overwintering.

Comments

From extensive discussions with scores of observers, we are personally unaware of any Yellow Rails (except that above) from the immediate New York City area that were heard giving their unique vocalization, despite a statement in Bull (1964) about possible Yellow Rails that were only heard. Until very recently, observers reporting Yellow Rails rarely described exactly what they heard, and apparently few had any idea what genuine Yellow Rails actually sounded like.

This confusion is unsurprising given the field guide information available at the time. What misled listeners was that the first (1934) and second (1939) editions of Peterson's *Field Guide to the Birds* described Virginia Rail's advertising song (a very fast *kik-kik-kik-kik-m'greeeerrr*) as being given by Yellow Rails, as did Pough and Eckelberry's *Audubon Water Bird Guide* (1951). The latter guide and Peterson's third edition (1947) blunted this error somewhat by noting that the 2+3 *tic*-couplets were *also* given by Yellows. Robbins et al.'s Golden Field Guide (1966) was the first to unequivocally describe Yellow Rail's sole, unique vocalization correctly. Virginia Rail's advertising song is heard many times more frequently than true Yellow Rail vocalizations (and there are surely many more Virginia Rails around), so authors of regional avifaunas began to doubt heard-only Yellow Rail records—unfortunately then concluding that Yellow Rails were not identifiable audibly. The waters were also considerably muddied by the long-standing debate in MA over whether the Virginia's advertising song (given by the unidentified rail locally called The Kicker) was given by Black Rails, *Laterallus jamaicensis*; Yellow Rails; or, as some authors even contended, *both* species. Even the Black Rail's striking *kee-kee-kerrr* vocalization was not well known; Peterson (1947) did not mention it and Pough and Eckelberry (1951) just noted it in passing after first describing other calls that are actually given only infrequently. Again, it was not until Robbins et al. (1966), who knew Black Rails well from the Delmarva Peninsula, that word did get out about what advertising ♂ Blacks actually sounded like.

Until the late 1960s Virginia Rail's song could be heard in spring in Van Cortlandt Swamp and especially easily at NJ's Troy Meadows, where unhelpfully, Yellow Rails have also been *seen* in spring on several occasions. But no one had ever watched a Yellow Rail giving its call locally, nor had anyone even seen a Virginia Rail giving *its* vocalization until very recently. This was a great pity, for Yellow Rail vocalizations are unique if heard clearly. Modern observers now have access to excellent recordings of both Black and Yellow Rails on their smart phones, so the days (and nights) of confusion should be history.

Specimens
Bicknell collected 1 at the study area edge, on Broadway on 2 Oct 1881 (NYSM).

Clapper Rail   *Rallus crepitans*

NYC area, current
This obligate saltwater breeder is common in Long Island *Spartina alterniflora* marshes from Jamaica Bay east to Shinnecock Inlet, locally on the North Shore, and in those on Long Island Sound in the Bronx and Westchester County. All but a handful usually depart for the winter. Recorded in both Central and Prospect Parks.

Bronx region, historical
Bicknell's *Riverdale* 1872–1901: Kingsbridge Meadows breeder; specimen 4 May 1878

Eaton's *Bronx + Manhattan* as of 1907: rare summer resident, does not breed

Griscom's *Bronx Region* as of 1922: unrecorded in Bronx/Westchester County saltmarshes

Kuerzi's *Bronx Region* as of 1926: no breeding sites known, transients rarely recorded

NYC area, historical
Cruickshank's *NYC Region* as of 1941: most numerous on Long Island South Shore saltmarshes, where very rare in winter; scarce in Bronx/Westchester County saltmarshes but bred Baxter Creek 1934, Rye 1941; bred Spuyten Duyvil to 1924

Bull's *NYC Area* as of 1968: bred Piermont Marshes 1959; much more abundant Long Island breeder than previously believed; more frequent in winter

Bull's *New York State* as of 1975: status unchanged

Levine's *New York State* as of 1996: 1–2 on 11 of 30 1960–89 Christmas Bird Counts in lower Hudson Valley, Long Island

Study area, historical and current
An extirpated breeder and occasional winter resident at former Kingsbridge Meadows.

Bicknell found Clappers breeding in Harlem River saltmarshes at Sherman Creek until 1883, but Kuerzi (1927) and Cruickshank (1942) reported 1–2 pairs hanging on in the marshes at the mouth of Tibbett's Brook (Kingsbridge Meadows) until 1924, not long after which Tibbett's Brook was buried and Kingsbridge Meadows ceased to exist. Clappers bred to the north of the study area at Piermont in 1959, and they persist as close as the Hutchinson River marshes in Pelham Bay Park.

Specimens
Bicknell collected 1 on 4 May 1878 (MCZ) on the wharf at Riverdale, which site may actually have been what was formally known as the Riverdale Pier, along the Hudson River not far above Spuyten Duyvil.

King Rail   *Rallus elegans*

NYC area, current
This freshwater marsh breeder is an extremely scarce spring/fall migrant and a very rare winter resident in coastal saltmarshes. It is a New York State Threatened Species and is unrecorded in Central and Prospect Parks.

Bronx region, historical
Bicknell's *Riverdale* 1872–1901: unrecorded

Eaton's *Bronx + Manhattan* 1910–14: rare breeder

Griscom's *Bronx Region* as of 1922: unrecorded

Kuerzi's *Bronx Region* as of 1926: rare migrant Apr–Sep, no evidence of breeding but may have bred formerly [locations not given]; twice in winter: Hunt's Point 1 Feb 1926 (Kuerzi), Jerome Swamp 26–28 Dec 1926 (Cruickshank, Ruff, Kuerzi)

NYC area, historical
Cruickshank's *NYC Region* as of 1941: rare breeder, extremely rare spring/fall migrant, winter visitor to saltmarshes; bred Bayside, Queens 1924, Orient 1925, Van Cortlandt Swamp 1927

Bull's *NYC Area* as of 1968: bred Montauk 1941, Orient 1952, 1954, Lawrence 1954; may be more frequent in winter saltmarshes than known

Bull's *New York State* as of 1975: status unchanged

Study area, historical and current
This southern species was intermittently resident in Van Cortlandt Swamp for 40 years but was extirpated in the early 1960s.

Unknown to Bicknell (in Griscom 1927) or Griscom (1923), King Rail was first recorded on 19–25 Apr 1925 at Jerome Swamp (Kuerzi, Cruickshank, et al.), then in Van Cortlandt 23–26 Apr 1926 (Cruickshank, Kuerzi), the start of repeated, year-round Van Cortlandt records lasting until the 1960s despite the Swamp's near destruction in 1949–50 (Kieran, Norse, Buckley, Kane). A nest on 26 May 1927 (Cruickshank) fledged 10 young, a family group of 8 was seen 12 Sep 1936 (Imhof), as were 2 on numerous other occasions; displaying ♂♂ were seen and heard calling almost every Apr–Jun from 1925–1956.

Depending on winter severity and freezing, 1–2 also regularly overwintered in Van

Cortlandt Swamp, and 1 of 2 doing so in Jan 1956 was found moribund on 13 Jan 1956 (Buckley, Scully); the second survived the winter. Two were in Van Cortlandt on 24 Dec 1933 (Cruickshank) and on 22 Dec 1957 (Buckley et al.). The last records there are 2 from 12 Sep–4 Oct 1959 (Tudor, Zupan, Young); 14 May 1960 (Young); 30 Apr 1965 (Norse); 18 Nov 1997 (Fiore).

Van Cortlandt Swamp was for many decades the only established 20th-century King Rail breeding site within New York City and southern New York State, although 1 may have bred (paired with a Clapper?) in a brackish marsh at Pelham Bay Park in 1955 (Young 1958). We believe King Rail's demise as a Van Cortlandt breeder was due to its preferred native cattail habitat's having been supplanted by exotic (nonnative) *Phragmites* and the gradual succession of Van Cortlandt Swamp's southern marsh to a Red Maple–willow–Buttonbush swamp, the same stressors that led to extirpation of Least Bittern, Virginia Rail, Sora, and Marsh Wren—and also American Bittern and Black-crowned Night-Heron.

In southern New York State on both NYSBBA I in 1980–85 and II in 2000–2005, King Rail was recorded breeding only in Rye, Westchester County (McGowan and Corwin 2008). That site having been abandoned, the nearest current breeders would be in NJ's Great Swamp and Troy Meadows. From the 1960s–80s it may also have been breeding at the East and West Ponds at Jamaica Bay Wildlife Refuge, where singles were seen or heard calling intermittently.

Virginia Rail   *Rallus limicola*

NYC area, current
A vanishing breeder now found near New York City with certainty only at Pelham Bay Park, Jamaica Bay Wildlife Refuge, along Jones Beach, and at a few sites in eastern Long Island and Westchester County. It is a regular but infrequently detected spring/fall migrant—sometimes in densely urban parks—and a scarce and local winter resident along the coast.

Bronx region, historical
Bicknell's *Riverdale* 1872–1901: common spring/fall migrant, breeds Kingsbridge Meadows, occasional elsewhere

Eaton's *Bronx + Manhattan* 1910–14: uncommon breeder

Griscom's *Bronx Region* as of 1922: still breeds Van Cortlandt Swamp; formerly in destroyed Dyckman St. marshes

Griscom's *Riverdale* as of 1926: no longer breeds

Kuerzi's *Bronx Region* as of 1926: formerly common breeder (5 Apr–19 Dec) now restricted to a few remaining favorable localities [unnamed]

NYC area, historical
Cruickshank's *NYC Region* as of 1941: common but local breeder, uncommon spring/fall migrant, very local winter resident

Bull's *NYC Area* as of 1968: maxima fall 7, winter 4

Bull's *New York State* as of 1975: status unchanged

Study area, historical and current
A spring/fall migrant and former resident breeder and winter resident in Van Cortlandt

Swamp, Kingsbridge Meadows, and Jerome Swamp.

For many generations of observers, Virginia Rail was *the* emblematic Van Cortlandt Swamp bird. From the late 1800s–1962, the Swamp was famous as the most accessible and reliable place within New York City to see this engaging creature. Many observers saw their very first rail of any kind there. Nowadays, only a handful are heard and occasionally seen on migration, although it is doubtless still an annual migrant. It bred at Jerome Swamp in 1925 (Cruickshank et al.) and also at Kingsbridge Meadows in the late 1800s.

In spring, judging by vocalization increases, first migrants normally arrived in mid-Apr, peaked in early May, and departed by mid-May. But when there was a large overwintering and healthy breeding population, it was rarely possible to separate migrants from overwinterers or breeders. There is no reason to believe they don't still occur annually in Van Cortlandt Swamp in spring/fall, but the last one there in spring was on 10 Apr 2006 (Young) and in fall on 3 Oct 2005 (Young). Another was calling in the new Tibbett's Marsh on 17 May 2014 (Young), the first in that location.

Until 1961, Virginia Rails were resident breeders in the southern, cattail marsh portion of Van Cortlandt Swamp and in the Triangle at the head of Van Cortlandt Lake; some also bred (1955–56 at least) in the *Phragmites* marsh west of the Sycamore Swamp (Buckley et al.). In 1937, 9 pairs bred in the main Van Cortlandt Swamp (Norse, Weber), but after its 1949–51 decimation, only 4 pairs were breeding by 1961 (Heath, Zupan)—the last year of certain Van Cortlandt Swamp nesting. Virginia Rail's demise as a breeder there was due to its preferred native cattail habitat's having been supplanted by exotic (nonnative) *Phragmites* coupled with the gradual succession of the Swamp's southern marshy portion to a Red Maple–willow–Buttonbush swamp.

Virginia Rail is another marsh bird consistently overlooked and underestimated during NYSBBA I in 1980–85 and II in 2000–2005 and by DOIBBS workers. For instance, NYSBBA II found them in only 3 blocks within New York City (McGowan and Corwin 2008), surely far off target. Nearest breeders are at Pelham Bay Park.

In fall, first migrants normally arrived in mid-Aug, peaked in late Sep, and if not overwintering, departed by mid-Nov. Anywhere from 1–6 could be heard and seen on many days, with maxima of 9 on 9 Nov 1954 (Buckley) and 16 on 12 Nov 1955 (Scully, Buckley).

In times past, Virginia Rails overwintered every year, even with heavy freezes when they could be enjoyed treading carefully on ice as they oozed through the cattails. Sample winter maxima include 6 on 11 Jan 1954, and 10 were believed to have overwintered in 1955–56 and 1957–58 (Scully, Buckley). It is likely that a count of 7 on 23 Mar 1956 (Scully) comprised winter residents, but some may have been early migrants. The last on the Christmas Bird Count were 2 on 23 Dec 2001 (Lyons et al.) but 1 vocalizing on 19 Nov 2017 (Souirgi) was a potential winter resident.

It is uncertain whether local breeders were resident or if some or all decamped for the winter only to be replaced by more northerly breeders. In any case, before their sad departure, Van Cortlandt Swamp was their most reliable winter site in New York City, but this is now history. Their final insult was the abrupt dredging in 2002 of their last redoubt, the cattail-*Phragmites* Triangle at the head of Van Cortlandt Lake between the Putnam Division track bed and the golf course.

Specimens

Bicknell collected 1 at Kingsbridge Meadows on 18 Sep 1898 (AMNH).

Sora *Porzana carolina*

NYC area, current
An uncommon spring and a common (formerly abundant) fall migrant in freshwater marshes and near the edges of saltwater marshes. A very rare and declining breeder as cattail marshes become *Phragmites* forests. Exceedingly scarce in winter.

Bronx region, historical
Bicknell's *Riverdale* 1872–1901: common fall migrant at Kingsbridge Meadows

Eaton's *Bronx + Manhattan* 1910–14: common migrant, rare breeder

Griscom's *Bronx Region* as of 1922: extirpated breeder, one 1917 record

Kuerzi's *Bronx Region* as of 1926: uncommon spring, common fall migrant (17 Apr–26 Dec); 1–2 breeding pairs [locations not given]

NYC area, historical
Cruickshank's *NYC Region* as of 1941: uncommon spring/common fall migrant (3 Apr–late Nov), only 5 winter records; breeds in Bronx [locations not given], several Van Cortlandt Swamp winter records

Bull's *NYC Area* as of 1968: recent fall maximum 18; about 24 winter records

Bull's *New York State* as of 1975: unrecorded as Long Island breeder since 1930s; recent fall maximum 27

Study area, historical and current
A spring/fall migrant, former breeder, and occasional winter resident in the southern cattail marsh portion of Van Cortlandt Swamp.

It may also have bred in Jerome Swamp and at Kingsbridge Meadows, where Bicknell found it regular, especially in fall migration; he collected 2 there in Sep–Oct 1876 (NYSM). It is far less frequently encountered now than it was until the early 1960s. The last Van Cortlandt Swamp record was 2 Oct 2006 (Young), but there is no reason to believe that it does not occur annually in migration.

Sora is another marsh bird consistently overlooked and underestimated during NYSBBA I in 1980–85 and II in 2000–2005 and by DOIBBS workers. Even though it is nowhere near as widespread and numerous as Virginia Rail, NYSBBA II located none within New York City (McGowan and Corwin 2008), despite its possibly breeding at Jamaica Bay Wildlife Refuge.

In fall, first migrants arrived in mid-Aug, peaked in late Sep–early Oct, and departed by early Nov, with extreme dates of 4 Aug 1991 (Collerton) and 2 on 12 Nov 1956 (Scully), and a maximum of 20 on 2 Oct 1955 (Sedwitz, Buckley).

There are now too many winter records to list, and until the early 1960s Sora may have been more regular in winter in Van Cortlandt Swamp than at any other New York City area site. At least 1–2 successfully overwintered every 5–6 years between 1926 and 1927 (Cruickshank) and 1961 and 1962 (Kane, Buckley et al.) but not since then, with a maximum of 3 on 13 Jan 1952 (Buckley). The first ever on the Bronx-Westchester Christmas Bird Count was in Jerome Swamp on 26 Dec 1926 (Cruickshank, Ruff).

In spring, first migrants appeared in mid-Apr, peaked in late Apr–early May, and departed by mid-May, with extremes of 3 Apr 1938 (Norse) and 9 May 1932 (Cruickshank), and a daily maximum of 4—the 2 pairs nesting in 1937 (Norse, Weber et al.). Two on 22 Mar 1952 (Buckley) were probably those overwintering that season (Komorowski).

It may have bred at Jerome Swamp in 1925 (Cruickshank et al.). In Van Cortlandt Swamp breeding was always intermittent—2 pairs in 1937 (Norse, Weber), 1 pair in 1951 (Komorowski), 1 pair in 1963 (Heath, Zupan), the last we know of—and it is unclear to what environmental variables they responded but with hints they needed high-water years. Sora's demise as a breeder and overwinterer there was also due to its preferred native cattail habitat's having been supplanted by exotic (nonnative) *Phragmites* and the gradual succession of the Swamp's southern cattail marsh to a more heavily wooded Red Maple–willow–Buttonbush swamp.

NYSBBA II in 2000–2005 found this species to have increased its number of occupied blocks by 15% since 1980–85, except that most close to the New York City area have disappeared (McGowan and Corwin 2008). Nearest regular breeders are in NJ's Great Swamp and Troy Meadows, in Rockland County, and perhaps at Jamaica Bay Wildlife Refuge.

Specimens
Bicknell collected 2 at Kingsbridge Meadows in Sep, Oct 1876 (NYSM).

Common Gallinule   *Gallinula chloropus*

NYC area, current
Gallinules had always been very rare migrants and marginal breeders in very small numbers at a handful of ephemeral Long Island locations until the East and West Ponds were created at Jamaica Bay Wildlife Refuge in 1953. In the Hackensack Meadowlands an unexpected and enormous population (200+ pairs) was discovered in 1962 but may have been there for 10 years. The only New York State population of similar size has been at Montezuma NWR since the early 1900s.

Bronx region, historical
Bicknell's *Riverdale* 1872–1901: unrecorded

Eaton's *Bronx + Manhattan* 1910–114: local breeder [locations not given]

Griscom's *Bronx Region* as of 1922: single Oct West Farms migrant; no suitable [breeding] habitat now

Kuerzi's *Bronx Region* as of 1926: uncommon, decreasing migrant (16 Apr–24 Oct); several pairs still breed [locations not given]

NYC area, historical
Cruickshank's *NYC Region* as of 1941: former common breeder in large marshes in, near New York City, now all but extirpated; otherwise uncommon spring/fall migrant (29 Mar–early Dec; maximum 24+), exceedingly rare in midwinter. Still breeds Van Cortlandt Swamp, perhaps source of breeding pair at Grassy Sprain Reservoir 1939

Bull's *NYC Area* as of 1968: increasingly reported in winter on Long Island (7 in 1954–55); newly established cluster Jamaica Bay Wildlife Refuge 1959–62 (sole known on Long Island) ranged from 4–9 pairs

Bull's *New York State* as of 1975: status unchanged

Study area, historical and current
A former Van Cortlandt Swamp breeder, now only a very infrequent spring/fall migrant.

Until first seen building a nest in Van Cortlandt Swamp on 26 May 1927 (Cruickshank), it had been a very rare migrant there with few records: 23 Sep 1923 (Hix), 6 May 1925 (Kassoy), and 16 Apr 1926 (Cruickshank), plus another at Jerome Swamp on 17 Apr 1925 (Cruickshank) where it may also have been breeding undetected. Through 1942, its last year (Norse), 2–6 pairs bred in the Swamp (3 pairs in 1937, 6 pairs in 1938: Norse, Weber), arriving in Apr (earliest 12 Apr 1930: Cruickshank) and departing in Oct (latest 25 Oct 1936: Norse). Not far away,

Kieran found a freshly dead migrant on a Riverdale sidewalk on 29 Apr 1933. The reasons for their abandonment of this breeding site are unknown, as it occurred 8 years *before* half of Van Cortlandt Swamp was obliterated. In 1942, the only other reliable breeding site within New York City was the pond/marsh inside Dyker Beach Golf Course in Brooklyn, where they persisted, almost unknown, until the 1960s.

NYSBBA II in 2000–2005 determined this species to have decreased its number of occupied blocks by one-third since 1980–85, and almost all of the few sites around New York City were no longer occupied (McGowan and Corwin 2008). Nearest current breeders may include an occasional pair at Jamaica Bay (8–10 pairs in 1962 but gone as an annual breeder by 1986) and in the Hackensack Meadowlands, where there were an impressive 200+ pairs in 1962 but little information on them subsequently.

Since 1942, single gallinules (2 on one occasion) have appeared in the study area as migrants on Van Cortlandt Lake or in Van Cortlandt Swamp. In spring on 2 Apr 1955 (Kane); 19–23 Apr 1964 (Horowitz et al.); 19 Apr 1965 (Kane); 25 May 1981 (Sedwitz); in a vernal pool in the Northwest Forest for several days in late Apr 2002 (fide Young); and in fall on 13 Oct 1951 (Norse); 17 Sep 1964 (Steineck); 7 Sep 1969 (Kane); 2 on 1 Oct 1969 (Sedwitz); Sep in late 1970s but date misplaced (Young); 20–30 Sep 1994 (Young, Lyons, Garcia), 12–15 Sep 1999 (Garcia, Young). One on 3 Jul 1953 (Phelan, Buckley) was a nonbreeding or postbreeding wanderer. Unrecorded Nov–Mar, even between 1927 and 1942, when they were breeding.

Comments
Southeastern **Purple Gallinule**, *Porphyrio martinicus*, is an almost annual vagrant to the New York City area and has occurred in Central and Prospect Parks. It has never been found in the Bronx or the study area but should be looked for: one was very close at Tibbett's Brook Park in Yonkers from 4–7 Sep 2013 (Whitt et al.).

American Coot   *Fulica americana*

NYC area, current
In the New York City area a locally abundant spring/fall (especially) migrant that also may overwinter even more locally in large numbers. It also breeds, but usually just single pairs at irregular sites.

Bronx region, historical
Bicknell's *Riverdale* 1872–1901: regular fall migrant until 1880, then only occasional; twice in spring

Eaton's *Bronx + Manhattan* 1910–14: fairly common migrant

Griscom's *Bronx Region* as of 1922: very rare migrant; favorable habitat destroyed or ruined

Kuerzi's *Bronx Region* as of 1926: rare migrant in spring (2–28 Apr)/fall (7 Oct–12 Nov); no Bronx winter records

NYC area, historical
Cruickshank's *NYC Region* as of 1941: rare spring/uncommon fall (maximum 300) migrant, very rare breeder (only for 2 years at 2 sites); very rare in winter

Bull's *NYC Area* as of 1968: recent fall maxima 2300; in 1950s single pairs bred Hewlett, Alley Pond Park, Mecox Bay; Jamaica Bay Wildlife Refuge 25+ pairs 1959 reached 50 pairs 1961 but only 15 pairs 1965

Bull's *New York State* as of 1975: status unchanged

Study area, historical and current
A spring/fall migrant in all 4 extant subareas that occasionally overwinters.

In fall, first migrants arrive in late Oct, peak in early Nov, and depart by mid-Nov unless overwintering, with extremes of 15 Oct 1951 (Kane) and 30 Nov 2013 (Baksh), and a maximum of 6 at Hillview on 30 Nov 1950 (Russak et al.).

Coots attempt to overwinter in Van Cortlandt most years (never more than 2) and are occasionally successful, but if frozen out they often move to Jerome or Hillview and then return to Van Cortlandt after thawing. During an especially harsh winter, a flock of 25, the most ever in the study area, appeared on Jerome on 26 Dec 1976 (Sedwitz) and 1 overwintered in 1963–64 and 2 in 1965–66 (Enders et al.).

In spring, the few migrants first appear in Mar, peak in early Apr, and depart in early May, with extremes of 22–24 Feb 1952 (Kane, Buckley) and 14 May 1953 (Buckley), and a maximum of 2 from 9–15 Apr 1954 (Scully, Kane) and on 10 Mar 2013 (McGee).

Breeding has occurred once in Van Cortlandt Swamp (adults with a brood on 9 Aug 1936: Petersen) but is possible again should coots find water levels and vegetation that are suitable. In 1936, there was no other New York City or Long Island breeding site, but they were probably breeding in the Hackensack Meadowlands. However, in 1951, a pair had young on Oakland Lake in Alley Pond Park on 10 Jun (Yeaton). The nearest current breeders are in the Hackensack Meadowlands, where as many as 800 birds were in just the Kearny Marsh portion in 1980 (R. Kane, pers. comm.) and also at Jamaica Bay Wildlife Refuge (50+ pairs in 1961).

NYSBBA II in 2000–2005 found this species to have increased its number of occupied blocks by 10% since NYSBBA I in 1980–85 at the same time that it had all but disappeared from its few New York City area locations (McGowan and Corwin 2008).

Cranes: Gruidae

Sandhill Crane *Grus canadensis*

NYC area, current
A rapidly increasing and now annual spring/fall migrant and winter visitor/resident most often seen in migration overhead, formerly only singles but now also in flocks. Sandhills may have occurred in colonial times but were otherwise unrecorded before 1970 and remain unrecorded in Central or Prospect Park.

NYC area, historical
Bull's *NYC Area* as of 1968: unrecorded

Bull's *New York State* as of 1975: single record, Hook Mountain, Rockland County 1970; not seen on Long Island until Montauk in 1975

Levine's *New York State* as of 1996: no additional downstate records

NYC and study areas, historical and current
The first that remained around the New York City area for a protracted period was on eastern Long Island throughout early 1995, and others soon followed. It is now being seen regularly, typically on migration in Oct–Nov and Mar–May but increasingly in winter and

occasionally in summer. In spectacularly rapid fashion it has also (re-?) colonized the Northeast since about 2000 and is now breeding in ME, NH, VT, MA, RI (suspected), CT (suspected), NY, NJ, and PA, albeit only a few pairs in each state. As northeastern populations have increased, so too has the size of regional flocks encountered in migration, and double-digit groups are now often tracked as they cross multiple states southward from northern New England to VA and beyond. In the winter of 2016–17 at least one flock of 7 was overwintering as close as central NJ. Moreover, since the late 1990s a flock (possibly peaking at 16 in 2009) originally derived from a wild Sandhill paired with an escaped Common Crane (*G. grus*) has attracted additional wild Sandhills, and that flock and possibly a few offshoot subgroups are now resident and breeding along Delaware Bay in Cumberland County, NJ (W. Boyle, pers. comm.)

There is but a single study area occurrence, a flock of 3 passing Hillview Reservoir on their way into Van Cortlandt Park on 9 Jan 2017 (Camillieri), and not far away, 7 went south over Ft. Tryon Park—perhaps after overflying Riverdale unseen—on 3 Dec 2016 (Sadock), possibly the first seen in Manhattan in centuries. In the Bronx there have been only 2 prior records: 2 at Pelham Bay Park on 24 Nov 2003 (Brash) and a single over the Bronx Zoo on 9 May 2016 (Olson). There will be more.

Plovers: Charadriidae

Black-bellied Plover *Pluvialis squatarola*

NYC area, current
A common spring/fall migrant, restricted to coastal areas except when habitat appears in open, grassy areas inland and after heavy rains. Recorded in Central and Prospect Parks.

Bronx region, historical
Bicknell's *Riverdale* 1872–1901: unrecorded

Eaton's *Bronx + Manhattan* 1910–14: fairly common migrant

Griscom's *Bronx Region* as of 1922: barely detected in Croton Point area; no other records

Kuerzi's *Bronx Region* as of 1926: rare migrant in spring (12 May–6 Jun), uncommon but regular in fall (9 Aug–15 Nov)

NYC area, historical
Cruickshank's *NYC Region* as of 1941: abundant Long Island migrant in spring (maximum several thousand)/fall, rare winter resident; irregular on Hudson, exceedingly rare at freshwater sites

Bull's *NYC Area* as of 1968: Long Island maxima spring 6000, fall 600; regular in winter coastally, maximum 150 on 1953–54 Brooklyn-Queens-Nassau Christmas Bird Counts

Bull's *New York State* as of 1975: status unchanged

Levine's *New York State* as of 1996: Long Island fall maximum 1203; winter maximum 547 on 1991 Southern Nassau Christmas Bird Count

Study area, historical and current
A common migrant and uncommon winter resident along saltwater on Long Island, in the East Bronx, and along the Hudson River that

overflies the study area regularly but that has been recorded only thrice: on the Van Cortlandt Parade Ground on 10 Aug 1947 (Norse) and 19 on 7 Oct 1950 (Sedwitz) plus 2 breeding-plumaged adults on drained Van Cortlandt Lake 10 May 1951 (Buckley). During and immediately after heavy rains anytime from late Jul–early Nov, many shorebirds that normally overfly inland sites are regularly forced to the ground, and this species is classic in that regard.

American Golden-Plover *Pluvialis dominica*

NYC area, current
An infrequent spring and annual fall migrant, largely restricted to immediate coastal areas except when adventitious habitat appears in open, grassy areas inland. Recorded once in Prospect Park, never in Central Park.

Bronx region, historical
Bicknell's *Riverdale* 1872–1901: unrecorded

Eaton's *Bronx + Manhattan* 1910–14: fairly common fall migrant

Griscom's *Bronx Region* as of 1922: unrecorded

Kuerzi's *Bronx Region* as of 1926: rare fall migrant 31 Aug–25 Oct [no locations given]

NYC area, historical
Cruickshank's *NYC Region* as of 1941: regular fall migrant (maximum 76) on Long Island, Long Island Sound shores in Bronx and Westchester County; almost unrecorded inland or in spring

Bull's *NYC Area* as of 1968: 31 Mar–9 Jun, 9 Jul–30 Dec; Long Island maxima fall 200, spring 9, when more regular >1947

Bull's *New York State* as of 1975: Long Island fall maximum 300

Study area, historical and current
An uncommon fall migrant that overflies the Bronx annually, evidenced by the number of records from East Bronx coastal areas from the 1920s–70s, although it is only rarely detected on the ground elsewhere in the county.

There are 5 study area records, all but 1 singles, on the Van Cortlandt Parade Ground, where rough ground with native grasses was available from 2005–9 before being improved by an exotic (nonnative) turfgrass monoculture in early 2010: 30 Oct 2005 (Young, Lyons, Jett; photos), 2 Oct 2006, 2 from 18–25 Oct 2007 (photos), 22 Sep 2008, 25 Sep 2009 (all Young). Daily fall coverage of the Parade Ground under the right grass conditions (should they ever recur), especially after torrential rains on Aug–Oct weekdays, could show this and other shorebirds to occur far more frequently than the few records imply.

Semipalmated Plover *Charadrius semipalmatus*

NYC area, current
A common spring/fall migrant along saltwater on Long Island, Long Island Sound, and less frequently but regularly, along the Hudson River.

Bronx region, historical
Bicknell's *Riverdale* 1872–1901: unrecorded

Eaton's *Bronx + Manhattan* 1910–14: common migrant

Griscom's *Bronx Region* as of 1922: Hudson Valley regular migration route; formerly common migrant at destroyed West Farms, otherwise unrecorded

Kuerzi's *Bronx Region* as of 1926: regular but uncommon migrant in spring (3 May–14 Jun)/fall (17 Jul–17 Oct) [locations not given]

NYC area, historical
Cruickshank's *NYC Region* as of 1941: abundant spring (maximum 1000+)/fall (maximum 1000+) migrant (23 Mar–9 Dec) on Long Island/Long Island Sound shores; annually in small numbers along inland pond and lake edges

Bull's *NYC Area* as of 1968: 23 Mar–10 Dec; Long Island maxima 6000 spring, 4000 fall; occasionally overwintering

Bull's *New York State* as of 1975: early spring date 18 Mar

Study area, historical and current
This mudflat shorebird in common on Long Island Sound and regular along the Hudson River. It overflies the study area regularly but has only rarely been detected.

There are 7 records, all but 1 in spring: 26 Apr 1936 (Kramer); 5 on 10 May 1951 (Buckley); 11 May 1952 (Norse); Jerome 26 May 1994 (Lyons); 5 Aug 1999 (Young); 13 May 2001 (Young); 10 May 2003 (Young). It is probably overlooked on the Parade Ground after heavy rains in Aug–Sep and has occurred in Central but not Prospect Park. Nearby, 4 dropped into Inwood Park on 24 Aug 2014 (Souirgi) and 3 on 20 Sep 2014 (Peltomaa).

Killdeer *Charadrius vociferus*

NYC area, current
A widespread, common spring/fall migrant, breeder, and erratic winter resident.

Bronx region, historical
Bicknell's *Riverdale* 1872–1901: only once, 3 at Kingsbridge Meadows 27 Mar 1888

Eaton's *Bronx + Manhattan* 1910–14: fairly common migrant, rare breeder

Griscom's *Bronx Region* as of 1922: breeds only Clason Pt.; otherwise scarce migrant recorded year-long

Kuerzi's *Bronx Region* as of 1926: common migrant, uncommon local breeder, especially at Clason and Hunt's Points in Bronx; rare midwinter

NYC area, historical
Cruickshank's *NYC Region* as of 1941: fairly common spring/fall (maximum 20) migrant, increasing but local breeder throughout, sparingly in Bronx [locations not given]; all but unknown in winter away from Long Island

Bull's *NYC Area* as of 1968: fall maximum 150; >early 1950s overwintering in larger numbers throughout: 700 on all [undefined] Jan 1954 Christmas Bird Counts

Bull's *New York State* as of 1975: status unchanged

Study area, historical and current
A spring/fall migrant, breeder, and occasional winter resident in all 4 extant subareas.

In spring, first migrants arrive in mid-Mar, peak in Apr, and depart by late May, with extreme dates of 25 Feb 2012 (Young) and 16 May 1998 (Young) and a maximum of 150 on 3 Apr 1997 (Young).

It may have been a very scarce study area breeder since Bicknell's time, but early information is sparse. It did not become an obvious breeder until about 1932, and despite habitat loss and disturbance 4–6 pairs still nest across all 4 extant subareas, often in unexpected/unlikely locations like the Van Cortlandt stables near Broadway and Mosholu Ave., at both reservoirs, and in Woodlawn. Study area egg/chick dates range from 7 May 1950 (4 just-hatched chicks from a mid-Apr laying on the Vault Hill grassland sloping into Van Cortlandt Swamp: Buckley) to 23 Aug 2009 (adults with 3–4 young on the Parade Ground: Garcia, Lyons). It is unknown whether any breed on study area rooftops as they do elsewhere.

NYSBBA I in 1980–85 and II in 2000–2005 (McGowan and Corwin 2008) recorded breeders throughout New York City and adjacent counties, as well as in study area block 5852B (they were missed in 5952A during NYSBBA II even though present). DOIBBS data showed a statewide decline beginning about 1975 and persisting (Sauer et al. 2005).

In fall, first migrants arrive in Jul, peak at various times thereafter depending on the year, and depart by early Nov, with extreme dates of 21 Jul 1953 (Buckley) and 30 Nov 1935 (Norse) and a maximum of 21 on 18 Oct 1996 (Young).

Single or 3–5 Killdeer may appear in winter at any time, but usually do not linger, especially if there is significant snow cover. The maximum of 8 on 19 Feb 2011 (Baksh) were locally overwintering or may have been early migrants.

Comments
Midwestern **American Avocet**, *Recurvirostra americana*, is now occurring annually in the New York City area, but there is only a single Bronx record, at the Orchard Beach construction site on 7–8 Oct 1939 (Sialis Bird Club). However, it has been seen recently at Piermont Pier on 7 Oct 2001 (D. Wells et al.), 15 Jun 2012 (Edelbaum et al.), 15 Jul 2014 (Edelbaum), and 13 Jun 2015 (Edelbaum). The 2014 bird was tracked as it moved south down the Hudson River from Croton Point to Piermont Pier, even stopping at Inwood Park on 15 Jul (Knox, DiCostanzo). Unrecorded in Central Park, southern **Black-necked Stilt**, *Himantopus mexicanus*, is an irregular, not-annual wanderer to the New York City area, usually along the South Shore of Long Island. Nonetheless, it has occurred once in Prospect Park on 2 May 2014 (Randall), and after Hurricane Irene another was found on 28 Aug 2011 in Sleepy Hollow, Westchester County (Edelbaum, Van Doren).

Sandpipers, Phalaropes: Scolopacidae

Spotted Sandpiper   *Actitis macularius*

NYC area, current
In the New York City area, a widespread freshwater spring/fall migrant in small numbers and a formerly widespread breeder—even in New York City, where numbers have diminished except on Staten Island.

Bronx region, historical
Bicknell's *Riverdale* 1872–1901: common breeder, spring/fall migrant, especially Kingsbridge Meadows

Eaton's *Bronx + Manhattan* 1910–14: common breeder

Griscom's *Bronx Region* as of 1922: common migrant, has not bred since 1917

Griscom's *Riverdale* as of 1926: now almost gone as Riverdale breeder

Kuerzi's *Bronx Region* as of 1926: common spring/fall migrant (18 Apr–23 Oct) and breeder [locations not given]

NYC area, historical
Cruickshank's *NYC Region* as of 1941: common migrant in spring/fall (maximum 30), and widespread breeder but almost gone from New York City [locations not given]

Bull's *NYC Area* as of 1968: 3 Apr–21 Dec

Bull's *New York State* as of 1975: status unchanged

Study area, historical and current
Historically, a spring/fall migrant and breeder in all 7 subareas. In Bicknell's time bred at Kingsbridge Meadows, Jerome Meadows, Van Cortlandt, and Riverdale. Later, Cantor found 4 nests just at Jerome on 25 May 1934. As of 2016, 1–2 pairs still breed irregularly among Van Cortlandt, Woodlawn, Jerome, and Hillview, even if not annually.

In spring, first migrants arrive in late Apr, peak in early May, and depart by late May, with extreme dates of 16 Apr 1944 (Komorowski) and 26 May 2008 (Young)—a local breeder?—with a maximum of 7 on 10 May 1951 (Buckley).

The most recent territorial pairs were in Van Cortlandt Swamp on 6 Jun 1980 (Ephraim), and on 1 Jun 1981, 12 Jun 1982, 17 Jun 1983, and 17 Jun 1984 (all Sedwitz). Other Van Cortlandt records indicative of breeders include 4 Jul 1962 (Zupan), 22 Jun 1980 (Ephraim), and 27 Jun 2015 (Willow). At Hillview 1–2 pairs in 2016 and 2017 were likely also breeding. Singles on 21 Jul 1953 (Buckley), 16 Jul 2006 (Delacretaz 2007), plus 2 on 21 Jul 1966 (Kane) and another in Woodlawn on 25 Jul 2014 (McGee, Sargent) could have been any of local breeders, regional postbreeding dispersers, or even early migrants.

This species is among the most overlooked breeders in the New York City area, preferring as they now do isolated islets, islands, and shorelines on the East River and Long Island Sound. For example, since 1980 (and probably long before) they have been breeding in or very close to the Bronx on Huckleberry Island, North and South Brother Islands, Goose Island in Pelham Bay Park, on the new Ferry Point golf course in Jun 2016 (Sime), as well as on many islands and islets in the East River (Künstler, NYCA). They may also breed undetected along inaccessible or rarely visited Bronx stretches of the Hudson and Harlem Rivers, and even on the Hutchinson and Bronx Rivers—a worthwhile search effort for spring/summer kayakers. Single pairs were present in Riverdale in summer from 1980–1884 (Sedwitz) and along the Bronx River in the zoo from May–Jul 2014 (Olson).

During NYSBBA I in 1980–85, breeders were widely distributed around New York City. Occupied blocks were in Westchester County (about 25), Bronx (7), Brooklyn (5), Staten Island (7), Queens (8), Nassau County (about 20), and Suffolk County (about 65). By NYSBBA II in 2000–2005, numbers had dropped considerably. Corresponding occupied blocks were in Westchester County (about 16), Bronx (1), Brooklyn (2), Staten Island (7), Queens (1), Nassau County (13), and Suffolk County (about 45) (McGowan and Corwin 2008). Sparse DOIBBS data showed a consistent statewide decline beginning in 1965, although many may have been missed for habitat sampling reasons (Sauer et al. 2005). The largest New York City area population may now be on Staten Island's Fresh Kills landfill (Veit, Wollney et al.).

In fall, first migrants arrive in early Aug, peak in mid-Aug, and depart by early Sep, with extreme dates of 2 on 21 Jul 1966 (Kane) and 31 Oct 1965 (Cantor) and a maximum of 5 on 26 Aug 2004 (Klein).

Spotted Sandpiper is so far unrecorded in winter in the Bronx, but winter records have been slowly increasing in the NYC area.

Specimens
Bicknell collected 2 at Kingsbridge Meadows in Apr, May 1876–77 (NYSM).

Solitary Sandpiper   *Tringa solitaria*

NYC area, current
A regular spring/fall migrant coastally and inland, usually only singles or small groups.

Bronx region, historical
Bicknell's *Riverdale* 1872–1901: scarce spring/fall migrant

Eaton's *Bronx + Manhattan* 1910–14: migrant

Griscom's *Bronx Region* as of 1922: uncommon spring/fall migrant

Kuerzi's *Bronx Region* as of 1926: fairly common migrant in spring (24 Apr–6 Jun)/fall (6 Jul–11 Nov)

NYC area, historical
Cruickshank's *NYC Region* as of 1941: fairly common spring/fall migrant, more regular inland than coastally

Bull's *NYC Area* as of 1968: 7 Apr–18 Nov; Long Island maxima spring 12, fall 9

Bull's *New York State* as of 1975: status unchanged

Study area, historical and current
A spring/fall migrant in Van Cortlandt and Woodlawn, and occasionally at Jerome and Hillview.

In spring, first migrants arrive in mid-Apr, peak in early May, and depart by mid-May, with extreme dates of 5 Apr 1929 (Black) and 27 May 1931 (Norse, Cantor) and a maximum of 21 on drained Van Cortlandt Lake on 10 May 1951 (Buckley).

In fall, first migrants arrive in early Aug, peak in late Aug–early Sep, and depart by early Oct, with extreme dates of 31 Jul 1966 (Kane) and 31 Oct 1965 (Norse) plus one injured on 18 Nov 1934 (Weber, Jove), and a maximum of 6 on 4 Sep 2013 (Young).

Specimens
Dwight collected 1 in Van Cortlandt on 30 Apr 1889 (AMNH).

Greater Yellowlegs   *Tringa melanoleuca*

NYC area, current
A common spring/fall migrant that irregularly overwinters in small flocks in Long Island saltmarshes.

Bronx region, historical
Bicknell's *Riverdale* 1872–1901: unmentioned

Eaton's *Bronx + Manhattan* 1910–14: common migrant

Griscom's *Bronx Region* as of 1922: formerly common migrant, now rare but annual

Kuerzi's *Bronx Region* as of 1926: regular, common migrant in spring (6 Apr–8 Jun)/fall (4 Jul–22 Nov)

NYC area, historical
Cruickshank's *NYC Region* as of 1941: common migrant in spring (maximum 100+

coastally)/fall, very rare winter resident, especially inland

Bull's *NYC Area* as of 1968: Long Island maxima 400 spring, 1000 fall; 10 in winter at Pelham Bay Park

Bull's *New York State* as of 1975: status unchanged

Levine's *New York State* as of 1996: Long Island winter maximum 45 on 1984 Southern Nassau Christmas Bird Count

Study area, historical and current
A spring/fall migrant in small numbers in Van Cortlandt, occasional in winter.

In spring, first migrants arrive in Apr, peak in May, and depart by late May, with extreme dates of 27 Mar 2017 at Hillview (Batren) and 3 Jun 1962 (Heath) and a maximum of 21 on drained Van Cortlandt Lake on 10 May 1951 (Buckley).

In fall, first migrants arrive in Aug, peak in early Sep, and depart by late Oct, with extreme dates of 22 Aug 1958 (Buckley) and at Hillview on 31 Oct 1936 (Hillview: Norse), singles all.

In winter, singles have appeared in Van Cortlandt, Woodlawn, or Jerome every 5–10 years, twice even when Van Cortlandt Lake was fully frozen, but they rarely persist into Jan before moving to saltwater. The only overwintering was in Van Cortlandt in 1961–62, where one calling on Van Cortlandt Lake ice before daylight on the Bronx-Westchester Christmas Bird Count caused the gathered observers to first suspect a Pine Grosbeak.

Western Willet   *Tringa (semipalmata) inornata*

NYC area, current
This is *the* late summer, fall, and recently winter willet, most regular on Long Island saltmarshes and mudflats but also occurring on Long Island Sound shores. It is a locally common fall migrant, with a handful recently overwintering, as well as occurring in spring. Unrecorded in Central or Prospect Park.

Bronx region, historical
Bicknell's *Riverdale* 1872–1901: 1 shot Kingsbridge Meadows

Eaton's *Bronx + Manhattan* 1910–14: rare migrant

Griscom's *Bronx Region* as of 1922: unrecorded

Kuerzi's *Bronx Region* as of 1926: rare fall migrant (4 Aug–12 Nov)

NYC area, historical
Cruickshank's *NYC Region* as of 1941: uncommon but regular Long Island fall migrant (maximum 46); late fall date 29 Oct; very rare in Bronx on Long Island Sound

Bull's *NYC Area* as of 1968: late fall date 27 Dec; Long Island fall maximum 80

Bull's *New York State* as of 1975: status unchanged

Study area, historical and current
Only a single study area record. Bicknell collected one at Kingsbridge Meadows on 7 Sep 1880, which was then a popular shooting site. This specimen was lost track of some time ago, but we suspect it is the otherwise unidentified HY mount in the New York State Museum, 7271, whose accession number and date are in accord with the NYSM's 300+ Bicknell late 1800s Riverdale specimens.

Willets *sensu lato* have occurred in both spring (Eastern only) and fall (Western only) in Pelham Bay Park, Baxter Creek, and other East Bronx tidal sites, and (Western) on freshwater in Bronx Park on 3 Oct 1953 (Maguire, Kane, Buckley).

One at Piermont Pier on 7 Aug 1981 (Deed) and 2 on 4 Sep 2005 (D. Wells et al.) were most likely Western by date, and one there on 14 Jul 2013 (Ciganek), was probably also a Western, but Easterns are nesting as close as Rye on Long Island Sound—only 12.4 smi (20 km) to the southeast.

Specimens
Bicknell collected 1 at Kingsbridge Meadows on 7 Sep 1880 (NYSM).

Comments
Eastern Willet, *T. (semipalmata) semipalmata*, is quite different in phenology from Western. In the Northeast, Eastern is a coastal saltmarsh breeder that generally disappears by midsummer, while Western is an inland breeding fall migrant appearing in mid–late summer and remaining late, increasingly into and through the winter. Eastern typically occurs in the Bronx on Long Island Sound in spring in small numbers (where it is a potential breeder), although not yet in the study area. It has been recorded in Prospect but not Central Park.

Eastern formerly bred along the entire Atlantic coast from Nova Scotia to Florida, but starting in the early 1800s market gunning systematically extirpated it from every state except NJ, where it persisted on the vast Delaware Bay marshes (Stone 1937). It last nested in MA in 1887, CT in 1873, but when in RI and New York is uncertain. A remnant breeding population persisted in Nova Scotia, abetting its eventual East Coast recolonization.

Willets along the Hudson in Yonkers on 6 Jun 1955 (Ephraim), at Piermont Pier (4 on 8 May 1973, 1 on 18 May 1976: both Amos), and in Dutchess Co. at Poughkeepsie on 18 May 1957 (Pink et al.) and Cruger Island on 13 May 1979 (Waterman et al.) were not photographed, described in detail, or identified further. They may all have been all Easterns but this possibility needs to be evaluated given that in the remainder of inland upstate New York, *irrespective of season*, there have never been *any* proven Easterns (J. Haas, K. McGowan, pers. comm.). That said, just when East Coast breeders were all but extirpated, A. K. Fisher did collect an adult ♂ willet at Croton Point on 22 May 1885 (MCZ 301869) that measurements and plumage confirm is *semipalmata* (Trimble, Buckley). This seems to be the farthest inland Eastern Willet occurrence in New York State.

Lesser Yellowlegs   *Tringa flavipes*

NYC area, current
A common spring and abundant fall migrant in the New York City area but far less frequently encountered away from saltwater than Greater; very scarce in winter.

Bronx region, historical
Bicknell's *Riverdale* 1872–1901: unrecorded

Eaton's *Bronx + Manhattan* 1910–14: common fall migrant

Griscom's *Bronx Region* as of 1922: formerly common fall migrant at destroyed West Farms; no other records

Kuerzi's *Bronx Region* as of 1926: rare migrant in spring (22 Apr–7 Jun)/common/abundant fall (27 Jun–22 Oct)

NYC area, historical
Cruickshank's *NYC Region* as of 1941: rare spring/abundant fall (maximum several hundred) Long Island migrant, much scarcer inland on freshwater

Bull's *NYC Area* as of 1968: Long Island maxima 6 spring, 5000 fall, 4 winter; possible decrease in numbers >late 1930s

Bull's *New York State* as of 1975: status unchanged

Levine's *New York State* as of 1996: Long Island winter maximum 12 on 1988 Southern Nassau Christmas Bird Count

Study area, historical and current
A much less frequent spring/fall migrant in Van Cortlandt than Greater, usually only singles. Also recorded at Jerome, Hillview, Woodlawn.

In spring, all records are in May, with extreme dates of 1 May 1934 (Cantor) and 27 May 1934 (Weber, Jove), and a maximum of 16 on drained Van Cortlandt Lake on 10 May 1951 (Buckley).

In fall, first migrants arrive in Aug, peak in Sep, and depart by mid-Oct, with extreme dates of 28 Jul 1938 at Hillview (Norse, Cantor) and 3 Nov 1935 (Weber, Norse), mostly singles but there are several records of 2, the fall maximum.

Upland Sandpiper   *Bartramia longicauda*

NYC area, current
Formerly an uncommon spring/fall migrant (least frequent within New York City proper) and a widespread but thinly scattered breeder in grasslands now almost all gone. Persisting on a few closed-access airports and landfills, it is a New York State Threatened Species.

Bronx region, historical
Bicknell's *Riverdale* 1872–1901: unrecorded

Eaton's *Bronx + Manhattan* 1910–14: rare migrant, breeder

Griscom's *Bronx Region* as of 1922: single late Jul Manhattan night record; no others

Kuerzi's *Bronx Region* as of 1926: occasionally heard flying over in late Jul/Aug, also 8 Aug 1924 Riverdale (Griscom), 9 Sep Baychester Marshes, and 3 Oct New Rochelle

NYC area, historical
Cruickshank's *NYC Region* as of 1941: rare Long Island migrant in spring/fairly common fall (maximum 12+), elsewhere extremely rare migrant

Bull's *NYC Area* as of 1968: 18 Mar–28 Oct; Long Island maxima spring 7, fall 45; almost gone as Long Island breeder

Bull's *New York State* as of 1975: 2 pairs bred Kennedy Airport 1969, to 12 by 1973

Study area, historical and current
A spring/fall migrant overflying the study area, sometimes giving its unmistakable calls in the middle of the night. Multiple Central Park and Prospect Park occurrences. This sandpiper may occasionally drop onto the Parade Ground very early in the morning, especially in Apr–May and Jul–Aug, should anyone be looking.

Ten study area and nearby records: overhead at night in immediately adjacent Riverdale on 8 Aug 1924 (Griscom); on the Parade Ground on 11–12 May 1951 (Kieran, Kane et al.), 19 Apr 1952 (Norse, Solomon), 18 Aug 1984 (Sedwitz), and 7 May 1991 (Saphir); at night over Jerome on 15 Apr 1962 (Sedwitz), and 1 found dead at Hillview on 16 Aug 2017 (Camillieri), probably killed by anti-bird wires. Heard at night were singles over adjacent Riverdale on 6 Sep 1957 and 24 Aug 1964 (Buckley), and 3 over nearby Inwood Park on 14 Aug 1965 (Norse).

DOIBBS data tracked a consistent statewide decline beginning in 1965 (McGowan and Corwin 2008). The nearest current breeders may still be at JFK Airport if they have not finally been harassed away but otherwise no closer than Orange

County. The 3 on the closed landfill at Pelham Bay Park in May–Jul 1987–2001 seem to have been only migrants (fide Künstler); they are annual in low numbers in fall in the East Bronx

Comments
Curlews of any species are normally rare away from Long Island, but a few **Whimbrel**, *Numenius phaeopus*, do move down the Hudson River and Long Island Sound and so occur regularly in very small numbers at Pelham Bay Park and other East Bronx locations. One was in Bronx Park on the very late date of 13 Nov 1955 (Ozard), and it has occurred twice in Central Park, including a flock of 7 on 14 Aug 1961 (Post). Unrecorded in Prospect Park. Most tantalizing was this unamplified statement by "JWD" in the Jan 1903 issue of *Forest and Stream* (see the Bobwhite species account): "I killed a long-billed curlew on the [East Bronx salt] meadows four or five years ago [= 1898–99], [my] only instance of meeting with this bird." Was he referring to just a Whimbrel, or a genuine **Long-billed Curlew**, *N. americanus*, at a time when they were still occurring in the Northeast? We'll never know.

Stilt Sandpiper *Calidris himantopus*

NYC area, current
On Long Island a very rare but slowly increasing spring but an uncommon and local fall migrant, favoring shallow fresh- and saltwater pools. Rare inland, which it overflies, landing only with bad weather or on adventitious optimal habitat with lake and reservoir low waters. Unrecorded in Central or Prospect Park.

Bronx region, historical
Bicknell's *Riverdale* 1872–1901: unrecorded

Eaton's *Bronx + Manhattan* 1910–14: unrecorded

Griscom's *Bronx Region* as of 1922: single fall 1909 West Farms migrant

Kuerzi's *Bronx Region* as of 1926: irregular and uncommon fall migrant [locations not given] from 14 Jul–5 Oct

NYC area, historical
Cruickshank's *NYC Region* as of 1941: excessively rare Long Island migrant in spring/fairly common fall (maximum 75), very rare on Long Island Sound shores in fall; unrecorded inland

Bull's *NYC Area* as of 1968: 13 Apr–10 Jun, 1 Jul–17 Nov; fall Long Island maxima 200–300; inland records increasing on low-water reservoirs

Bull's *New York State* as of 1975: status unchanged

Study area, historical and current
In the Bronx a rare but maybe annual fall migrant coastally when suitable habitat is available, such as happened at Baxter Creek from the late 1940s to the 1970s and also episodically at Pelham Bay Park.

In the study area, a single record when Van Cortlandt Lake was drawn down: 10 May 1951, in high breeding plumage with 16 Lesser Yellowlegs (Buckley) but gone the next day. This is the only Bronx spring record.

Comments
One of the least frequently encountered near-annual shorebirds around New York City is **Ruff**, *C. pugnax*. Only rarely recorded away from Long Island, it has occurred once in the Bronx: 2 studied closely in a weedy pool on the Orchard Beach parking lot, on 22 Aug 1952

(C. Young, Cap). It was also found inland on Lake Tappan, Rockland County, on 8 Aug 1980 (Schwartz). Unrecorded in Central or Prospect Park.

Sanderling   *Calidris alba*

NYC area, current
An abundant spring/fall migrant and winter resident on Long Island's South Shore, less so on Long Island Sound shores west to the Bronx. It is notoriously reluctant to come to ground inland or on freshwater, even though passing overland by the thousands every spring/fall. Recorded in Prospect Park but not Central and reasonably regular, if rare, in migration at Piermont Pier.

Bronx region, historical
Bicknell's *Riverdale* 1872–1901: unrecorded

Eaton's *Bronx + Manhattan* 1910–14: unrecorded

Griscom's *Bronx Region* as of 1922: single Pelham Bay Park 15 May record

Kuerzi's *Bronx Region* as of 1926: uncommon fall migrant (8 Jul–18 Nov)

NYC area, historical
Cruickshank's *NYC Region* as of 1941: abundant spring (maximum several hundred)/fall (maximum 1000) Long Island migrant, but very local in winter; rare spring/uncommon fall migrant on Long Island Sound shores, unrecorded at inland freshwater sites

Bull's *NYC Area* as of 1968: Long Island maxima spring 1500, fall 2000, winter 500

Bull's *New York State* as of 1975: status unchanged

Levine's *New York State* as of 1996: Long Island maxima fall 4000, winter 540

Study area, historical and current
Only a single record: 2 breeding plumaged adults on drained Van Cortlandt Lake on 10 May 1951 that were gone the next day (Buckley). Nearby, southbound adults appeared briefly on a drained lake in the Bronx Zoo on 31 Jul 2012 and 20 Aug 2015 (Olson).

Dunlin   *Calidris alpina*

NYC area, current
A spring/fall migrant and locally common winter resident in the New York City area but quite restricted to saltwater areas with extensive tidal flats. In the Bronx Dunlins are all but unrecorded away from Long Island Sound, where they occur in spring/fall migration but in very small groups; extremely rare and local in winter.

Bronx region, historical
Bicknell's *Riverdale* 1872–1901: unrecorded

Eaton's *Bronx + Manhattan* 1910–14: unrecorded

Griscom's *Bronx Region* as of 1922: single fall West Farms record; no others

Kuerzi's *Bronx Region* as of 1926: rare migrant in spring (15–24 May)/uncommon fall (20 Sep–3 Nov)

NYC area, historical
Cruickshank's *NYC Region* as of 1941: fairly common migrant in spring (maximum 100)/

common in fall (maximum hundreds) on Long Island/Long Island Sound, rare winter resident; nearly unrecorded at inland freshwater sites

Bull's *NYC Area* as of 1968: Long Island maxima 900 spring, 1800 fall, 600–900 winter

Bull's *New York State* as of 1975: Long Island winter maximum 5000

Levine's *New York State* as of 1996: Long Island maxima 15,000 fall, 4212 winter

Study area, historical and current
As so many shorebirds do, they overfly the study area but elect not to alight. One record on Van Cortlandt Lake with many other shorebirds after it had just been drained: 2 in breeding plumage on 10 May 1951 (Buckley), gone the next day. Unrecorded in Central or Prospect Park but regular on the Hudson River at Piermont Pier.

Comments
Highly maritime, rock-loving **Purple Sandpiper**, *C. maritima*, is widespread on Long Island and the sound but has also been found once in Prospect Park (12 on 11 Dec 1938: Breslau), on the Hudson River at Piermont Pier on 1 Nov 1977 (Schwartz), and is regular in winter at Pelham Bay Park. After winter storms it is possible on Jerome or Hillview.

Least Sandpiper   *Calidris minutilla*

NYC area, current
An abundant, common spring/fall migrant grasspiper on Long Island but also inland.

Bronx region, historical
Bicknell's *Riverdale* 1872–1901: spring/fall migrant Kingsbridge Meadows

Eaton's *Bronx + Manhattan* 1910–14: common migrant

Griscom's *Bronx Region* as of 1922: formerly common spring/fall migrant at West Farms; now rare

Kuerzi's *Bronx Region* as of 1926: regular, fairly common migrant in spring (27 Apr–16 Jun)/fall (4 Jul–22 Oct)

NYC area, historical
Cruickshank's *NYC Region* as of 1941: abundant migrant in spring (maximum 1000)/fall on Long Island/Long Island Sound but uncommon, irregular at inland freshwater sites

Bull's *NYC Area* as of 1968: 10 Mar–18 Nov plus 3 in Dec–Jan; Long Island maxima 5000 spring, 1200–1600 fall

Bull's *New York State* as of 1975: status unchanged

Study area, historical and current
A regular spring but irregular fall migrant in small flocks in Van Cortlandt.

In spring, first migrants arrive in early May, peak in the first half of May, with extreme dates of 1 May 2004 (Young) and 27 May 1934 (Weber, Jove) and a maximum of 34 on drained Van Cortlandt Lake on 10 May 1951 (Buckley). There are many fewer fall than spring records, anomalous for an abundant shorebird. First migrants arrive in late Jul, peak in Aug–Sep, and depart by late Sep, with extremes of 5 at Hillview on 11 Jul 2017 (Camillieri) and 2 on 17 Oct 1983 (Sedwitz) and a maximum of 7 on 19 Aug 2002 (Young). Many fall migrants are probably overlooked. Regular examination of the Parade Ground after heavy rains from Jul–Sep and in Van Cortlandt Swamp and Lake margins during periods of drought would produce additional records—as witnessed by the 100–200 at Inwood Park on 12 Aug 2012 (Souirgi).

Specimens
Bicknell collected 1 at Kingsbridge Meadows on 13 May 1880 (NYSM).

Buff-breasted Sandpiper   *Calidris subruficollis*

NYC area, current
A fall-only Long Island coastal grasspiper, usually in single digits, that is only irregularly seen inland in Aug–Sep on turf-farms or on low-water lakes and reservoirs.

NYC area, historical
Cruickshank's *NYC Region* as of 1941: Long Island rare fall migrant 7 Aug–15 Oct; single inland record

Bull's *NYC Area* as of 1968: early fall date 3 Aug, maximum 8

Bull's *New York State* as of 1975: fall maximum 55

Study area, historical and current
Recorded in the Bronx only from the Baxter Creek area in the mid-1950s (Buckley 1958), and a single and slightly late study area record: a tame juvenile on 4 Oct 1964 in grass alongside Jerome Reservoir (Sedwitz). Judging from records in 1954 (3 singles between 25 Sep and 14 Oct: Bauer, Bauer, Buckley et al.) and on 21 Aug 1955 (Schmidt) at Baxter Creek when coverage was intense, this species passes overhead southbound and would not be unexpected any year on the Parade Ground—although it would have to be at dawn on a rainy, windy Wednesday to avoid people. Unrecorded in Central or Prospect Park.

Pectoral Sandpiper   *Calidris melanotos*

NYC area, current
An uncommon spring and common fall grasspiper, more typically found on Long Island but occurring inland when suitable conditions appear or during and after hurricanes and other strong storms.

Bronx region, historical
Bicknell's *Riverdale* 1872–1901: unrecorded

Eaton's *Bronx + Manhattan* 1910–14: common migrant

Griscom's *Bronx Region* as of 1922: formerly exceedingly rare spring/fall migrant only at West Farms

Kuerzi's *Bronx Region* as of 1926: rare migrant in spring (7–29 May), fairly common fall (14 Jul–7 Nov)

NYC area, historical
Cruickshank's *NYC Region* as of 1941: uncommon to rare migrant on Long Island in spring (maximum 20+)/common fall (maximum 200), but only irregular on inland lakes, ponds [locations not given]

Bull's *NYC Area* as of 1968: 18 Mar–18 Jun, 6 Jul–11 Dec; Long Island maxima 40 spring, 1200 fall

Bull's *New York State* as of 1975: early spring date 2 Mar, Long Island spring maximum 72

Study area, historical and current
An irregular migrant, more frequent in fall than spring. Doubtless overlooked, especially

on the Parade Ground (which is rarely checked carefully after heavy rains) and along Van Cortlandt Lake and Swamp margins.

In spring only 3 records: 9 on drained Van Cortlandt Lake on 10 May 1951 (Buckley), 2 on 8 May 1955 (Kane et al.), and a single on 17 May 1956 (Kane).

In fall, first migrants arrive in Sep, peak in early Oct, and depart by late Oct, with extreme dates of 23 Aug 2003 (Young) and 3 in Nov: 13 Nov 1927 (Cruickshank), and on the Parade Ground 15 Nov 2001 (Garcia, Lyons) and 18 Nov 1955 (Post, Messing). All have been singles save 2 on 9 Oct 1954 (Scully, Buckley).

Semipalmated Sandpiper   *Calidris pusilla*

NYC area, current
A common spring/fall migrant along fresh- and saltwater on Long Island, Long Island Sound and along the Hudson River.

Bronx region, historical
Bicknell's *Riverdale* 1872–1901: unrecorded

Eaton's *Bronx + Manhattan* 1910–14: abundant migrant

Griscom's *Bronx Region* as of 1922: formerly regular spring/fall migrant at West Farms; no other records

Kuerzi's *Bronx Region* as of 1926: regular, fairly common migrant in spring (2 May–20 Jun)/abundant fall (8 Jul–2 Nov)

NYC area, historical
Cruickshank's *NYC Region* as of 1941: abundant Long Island/Long Island Sound migrant in spring (maximum several thousand)/fall, less common but regular at inland freshwater sites

Bull's *NYC Area* as of 1968: 3 Apr–7 Dec; Long Island maxima 25,000 spring, 6000 fall; increasing in winter (maximum 17)

Bull's *New York State* as of 1975: all winter records (old and new) deprecated as misidentified Western Sandpipers

Study area, historical and current
A regular spring and (so far) infrequently recorded yet probably annual fall migrant only in Van Cortlandt but that ought to occur in the other 3 extant subareas.

In spring, singles and small flocks have been recorded between 23 Apr 1952 (Buckley) and 23 May 2003 (Young), with maxima of 10 on 21 May 1956 (Steineck, Yeganian) and 12 on 10 May 1951 (Buckley).

Only 4 fall records: in Van Cortlandt on 9 Jul 1933 (Norse), 23 Aug 2003 (Young), and 8 on 28 Aug 2010 (Baksh), plus 1 at Hillview on 26 Jul 2016 (Camillieri). Regular examination of the Parade Ground after heavy rains from Jul–Oct and in Van Cortlandt Swamp or along the lake during periods of drought would produce additional records—as witnessed by the 300–400 at nearby Inwood Park on 12 Aug 2012 (Souirgi).

Comments
Two annually occurring fall shorebirds in the New York City area, in the Bronx, and occasionally as close as Spuyten Duyvil, have never been recorded in the study area: **Western**, *C. mauri*, and **White-rumped**, *C. fuscicollis*, **Sandpipers**. Both have occurred in Central Park but not Prospect. White-rumped also appeared at Grassy Sprain Reservoir on 12 Sep 1960 during Hurricane Donna (Steineck). Their study area absence largely owes to lack of suitable habitat, but if Van Cortlandt Lake drawdowns were to occur in Aug–Sep, both (and other species unrecorded in the study area) would eventually appear. Undisturbed rain pools on the Parade

Ground in the same period might also attract them. A third, **Baird's Sandpiper**, *C. bairdii*, is another grasspiper normally restricted to Long Island, but it has occurred in the Bronx at Pelham Bay Park (17 Aug 1996, 15 Aug 2002: Jaslowitz; 15–16 Aug 2017: Aracil et al.; photos), and at Baxter Creek on 23 Aug, 6 Sep 1947, and 21, 29 Aug 1948 (Norse et al.), 19 Sep–4 Oct 1954 (maximum 4: Norse, Buckley et al.), and 13 Aug–15 Sep 1955 (maximum 5: C. Young, Kane et al.). It has not been found in Central or Prospect Parks, but is to be looked for in the study area under the same conditions as the preceding 2 species.

Short-billed Dowitcher *Limnodromus griseus*

NYC area, current
A common spring/fall migrant along saltwater on Long Island, Long Island Sound, and the Hudson River. Tends to avoid freshwater; recorded in Prospect Park but not Central.

Bronx region, historical
Bicknell's *Riverdale* 1872–1901: unrecorded

Eaton's *Bronx + Manhattan* 1910–14: unrecorded

Griscom's *Bronx Region* as of 1922: unrecorded

Kuerzi's *Bronx Region* as of 1926: fairly common fall migrant 2 Jul–30 Sep on Long Island Sound

NYC area, historical
Cruickshank's *NYC Region* as of 1941: abundant migrant in spring (maximum 1000)/fall on Long Island, fairly common on Long Island Sound in Bronx; unrecorded at inland freshwater sites

Bull's *NYC Area* as of 1968: 24 Mar–19 Nov plus several in Dec–Jan; Long Island maxima 4000–8000 spring, 3000 fall

Bull's *New York State* as of 1975: early spring date 20 Mar

Levine's *New York State* as of 1996: Long Island fall maximum 3800

Study area, historical and current
Only 3 records, in part following severe weather: 5 May 1944 (Small et al.), 3 on drained Van Cortlandt Lake on 10 May 1951 (Buckley), and 2 at Jerome on 15 Sep 1964 (Ryan).

Comments
Long-billed Dowitcher, *L. scolopaceus*, prefers freshwater edges to salt, and although reported regularly at Baxter Creek in the 1950s–60s, it has not been found in the Bronx since then. But as it is a regular fall migrant in the New York City area with several Piermont Pier records and overflies the Bronx annually, it is a potential study area species; unrecorded in Central and Prospect Parks.

Normally only on the South Shore of Long Island, godwits are rarely seen inland even though they may overfly the whole area regularly in fall. **Marbled**, *Limosa fedoa*, has occurred only once in the Bronx, at Ferry Point Park on 13–14 Aug 1955, as Hurricane Connie was passing to the west (Buckley et al.). One at Sands Point remained directly opposite Pelham Bay Park from 10 Aug–4 Sep 1978 (Quinn). **Hudsonian**, *L. haemastica*, seems more willing to land at suitable inland sites, but except for a vague rumor of one at Clason Point in the 1970s there are no Bronx records. It has occurred at Haverstraw, Rockland County, on 18 Sep 1960 after Hurricane Donna, and at Piermont Pier on 22 Oct 2005

(Ciganek et al.), 25 Aug 2010 (Ciganek), and 20 Jun 2014 (Nutter, Edelbaum). No godwits have been recorded in Central or Prospect Park.

Wilson's Snipe   *Gallinago delicata*

NYC area, current
A fairly common, widespread spring/fall migrant, and a regular but highly local winter resident in small numbers, largely coastally.

Bronx region, historical
Bicknell's *Riverdale* 1872–1901: common spring, uncommon fall migrant

Eaton's *Bronx + Manhattan* 1910–14: fairly common migrant

Griscom's *Bronx Region* as of 1922: very rare migrant, formerly common; single Riverdale winter record Feb 1880

Kuerzi's *Bronx Region* as of 1926: regular and fairly common migrant in spring (8 Mar–22 May)/fall (7 Sep–23 Dec), occasional in winter; also 27 Jul 1923 (Kuerzi) (location not given)

NYC area, historical
Cruickshank's *NYC Region* as of 1941: common spring (maximum 100+)/fall migrant, erratic in winter along coast

Bull's *NYC Area* as of 1968: maxima fall 23, winter 30

Bull's *New York State* as of 1975: status unchanged

Levine's *New York State* as of 1996: winter maximum 35

Study area, historical and current
An annual but uncommon spring/fall migrant and regular but decreasing winter resident in Van Cortlandt and Woodlawn. Occurs today in numbers greatly reduced from the 1930s–60s, and much less frequently in winter.

In spring, first migrants arrive in early Mar, peak in early Apr, and depart by late Apr, with extreme dates of 25 Feb 2012 (Young) and 30 May 1981 (Sedwitz), and maxima of 18 on 3 Apr 1956 (Kane) and 20 on 10 Apr 1936 (Norse).

Unrecorded in Jun–Jul, although winnowing (♂ courtship display flights) was occasionally heard in Van Cortlandt Swamp in Apr–May from the 1930s–60s. Nearest current breeders are in NJ's Great Swamp/Troy Meadows area and in Orange County. DOIBBS data have been essentially flat statewide from 1966–2005, although sampling may not have been optimal (Sauer et al. 2005; McGowan and Corwin 2008).

In fall, first migrants arrive in late Aug, peak in Oct, and depart by early Nov, with extreme dates of 23 Aug 1994 (Lyons) and 14 Nov 2013 (Young), and a maximum of 5 on 18 Oct 1954 (Scully). One in Bronx Park on 4 Jul 1957 (Maguire, C. Young) would have been a postbreeding disperser from not too far away.

Until the 1970s, up to 10 attempted to overwinter annually in Van Cortlandt Swamp and succeeded in many years, but they were often frozen out to saltwater habitat. Since then only 1–2 have been found erratically. Reasons for the 1930s and 1950s–80s peaks on the Bronx-Westchester Christmas Bird Counts (Fig. 39) are unknown but they do match study area trends.

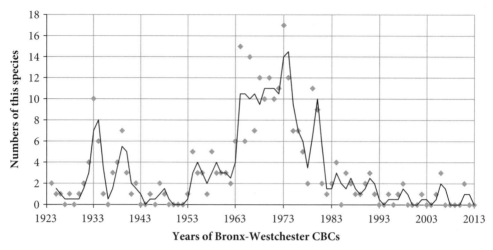

Figure 39

American Woodcock *Scolopax minor*

NYC area, current
A spring/fall migrant, an erratic but widespread and increasing breeder even within New York City, and a regular winter visitor and resident in very small numbers.

Bronx region, historical
Bicknell's *Riverdale* 1872–1901: common breeder, last 1886

Eaton's *Bronx + Manhattan* 1910–14: fairly common migrant, uncommon breeder

Griscom's *Bronx Region* as of 1922: formerly common breeder, still at Baychester marshes; rare migrant elsewhere

Kuerzi's *Bronx Region* as of 1926: still breeds locally in limited numbers; fairly common spring/fall migrant, rare in winter

NYC area, historical
Cruickshank's *NYC Region* as of 1941: spring/fall (maximum 12–36) migrant and erratic winter resident; regular in big New York City parks, where a few pairs still breed [locations not given]

Bull's *NYC Area* as of 1968: maxima spring 9, winter 9

Bull's *New York State* as of 1975: status unchanged

Study area, historical and current
A regular migrant, scarce in winter; bred until 1886 (Bicknell) and thereafter (e.g., an incubating ♀ found dead in Van Cortlandt Swamp 21 Apr 1908: Abbott). Judging from displaying ♂♂ on study area leks, it still breeds, although few observers listen for them in Mar–May on warm, quiet nights.

In spring, first migrants arrive in late Feb, peak in late Mar–early Apr, and depart by early May, with extreme dates of displaying ♂♂ on 3 Mar 1955 (Sedwitz, Buckley et al.) and 2 May 1971 (Kane), and a maximum of 12 on 13 Mar 2007 (Young). Until Jun 1967, the spring maximum in Central Park was 3 on 7 Apr 1958 (Post).

Despite being unrecorded on NYSBBA I in 1980–85 and II in 2000–2005, in Van Cortlandt displaying ♂♂ can be heard beginning in early Mar along the edges of the Parade Ground and parts of the Swamp, earlier in warm winters/early springs, but are rarely if ever sought elsewhere within the study area at this time. Wholly unanticipated was the Woodcock displaying in Van Cortlandt Swamp on 23 Dec 1995 (Garcia)—the earliest we know of. It is unknown what fraction of displaying ♂♂ are not (impending) breeders. Study area Woodcock are only occasionally located in summer (e.g., 19 Jun 1937, 25 Aug 1936: both Norse) even though doubtless always present. As many as 12–20 pairs may be breeding across the study area, with 8–16 in Van Cortlandt alone. Breeding also occurs in Pelham Bay Park (Künstler) and doubtless elsewhere in the Bronx in suitable habitat.

NYSBBA I in 1980–85 and II in 2000–2005 (McGowan and Corwin 2008) found this species widely but thinly distributed around New York City, breeding in less urban parts of every New York City borough except Manhattan, but the detection process has been far from optimal.

In fall, first migrants arrive in late Sep, peak in Oct–Nov, depart by late Nov, with extreme dates of 6 Sep in 1991 and 2006 (Young) and 4 (the maximum) on 8 Nov 1924 (Cruickshank).

Woodcock attempt to overwinter in Van Cortlandt most years and succeed often, with a daily maximum of 2 many times. They have a curious winter habit, unappreciated by many observers, of seeking out soft soil under conifers where they probe for worms. Consequently, most go undetected in such an unexpected habitat.

Specimens
One was found dead at Jerome on 11 Nov 1980 (Meyer: AMNH).

Red-necked Phalarope *Phalaropus lobatus*

NYC area, current
An offshore spring/fall migrant occasionally blown ashore by storms. Otherwise, a rare but annual spring/fall migrant, usually singles on Long Island, occasionally inland. Recorded in Prospect Park and not Central, but cf. "Comments" below.

Bronx region, historical
Bicknell's *Riverdale* 1872–1901: unrecorded

Hix's *NYC* in 1903: unrecorded

Eaton's *Bronx + Manhattan* as of 1907: unrecorded

Griscom's *Bronx Region* as of 1922: once, Aug 1911 at West Farms

Kuerzi's *Bronx Region* as of 1926: rare migrant 11 Aug–26 Oct

NYC area, historical
Cruickshank's *NYC Region* as of 1941: rare Long Island on-land spring/fall migrant; 2 Apr–12 Nov; very rare migrant on Hudson, along Bronx, Westchester County shores

Bull's *NYC Area* as of 1968: Long Island 29 Mar–23 Nov; maxima spring 900+, fall 165

Bull's *New York State* as of 1975: status unchanged

Study area, historical and current
A surprising visitor that has been seen only once in the study area: a juvenile on 22 Aug 1943, spinning and feeding on the water's surface in the south end of Van Cortlandt Swamp (Kieran). Kuerzi (1927) indicates multiple records in his expanded Bronx Region from 26 Aug–26 Oct, all but one without locations. The

only other firm Bronx records are single juveniles at West Farms (11 Aug 1923: Kuerzi) and Baxter Creek (several in Aug–Sep in the 1950s–60s: Norse, Russak, Buckley); a phalarope unable to be unidentified to species in Bronx Park on 18 May 1963, a day when both Red and Red-necked were blown ashore from Long Island to Cape May (Horowitz 1963); several Hurricane Irene-borne adult flocks on the Hudson in Riverdale and at Fort Schuyler on 28 Aug 2011 (Buckley); and 2 juveniles on the Orchard Beach Parking Lot on 23 Aug 2017 (Fogarty et al.). Other singles from Westchester, Rockland, and Dutchess Counties point to this species moving overland and down the Hudson River in fall in small numbers, usually unseen.

Comments
Red Phalarope, *P. fulicarius*, the other pelagic phalarope, has been found 4 times in the Bronx: at Baxter Creek from 4–7 Nov 1954 (Buckley 1958); twice at Spuyten Duyvil: on 30 Oct 2012 during Hurricane Sandy (DiCostanzo, Shaw) and on 1 Oct 2015 (Farnesworth); and at Pelham Bay Park on 17 Apr 2018 (Antonio; photos). It has occurred twice in Oct in Westchester County, never in Rockland County, and twice (May, Oct) in Dutchess County. In Central Park, singles have been seen as follows: on 5–6 Oct 1950 (Messing; originally thought to have been a Red-necked but subsequently reidentified as a Red); from 22–26 Sep 1983 (Clayton et al.; photos); and a fading breeding-plumaged ♀ on 14 Aug 1996 (Collin). Red Phalarope has not yet been found in Prospect Park. Non-pelagic **Wilson's Phalarope**, *P. tricolor*, is unrecorded in the study area or in Central and Prospect Parks but has occurred multiples times in the Bronx as close as Bronx Park and Baxter Creek, as well as in Westchester County. It is probably an annual Bronx fall migrant in single digits even if infrequently detected.

Auks: Alcidae

Dovekie *Alle alle*

NYC area, current
A normally highly pelagic, erratic winter visitor that is seen along the oceanfront in small numbers that vary greatly year to year. Very rarely, they have been blown ashore in Nov in extreme storms, sometimes far inland in great numbers. Recorded in Prospect Park (and nearby Green-Wood Cemetery) but not Central Park.

NYC area, historical
Cruickshank's *NYC Region* as of 1941: extremely rare Long Island winter visitor, unrecorded elsewhere until huge Nov 1932 incursion, when thousands found throughout entire region, even at many inland locations

Bull's *NYC Area* as of 1968: most frequent mid-Nov–early Mar but recorded year-round

Bull's *New York State* as of 1975: status unchanged

Study area, historical and current
During an extraordinary and unique wreck on 18–19 Nov 1932 (Murphy and Vogt 1933; Nichols 1935), Dovekies were scattered well inland along the northeastern Atlantic coast including throughout the Bronx. In the study area,

there was only one occurrence: 2 on Jerome on 20 Nov (Ephraim), but another was almost within Woodlawn in a backyard on Carpenter Ave. near E. 223rd St. (Lunt 1933). And not far away were the following: several in Bronx Park on 20 Nov (Lunt); 2 on Kensico Reservoir on 20 Nov (Logan); 2 in Scarsdale on 19–20 Nov (unknown finders; given to AMNH); 7 from the Bronx—locations not given but some may have been in the study area—on 19–21 Nov (Lunt; to AMNH); singles on 20 Nov in Ardsley (Lichtenauer) and in Harrison (Atherton). It is certain many more vanished after dropping into untenanted locations, like the 5 at Spuyten Duyvil on 20 Nov that were captured, cooked, and consumed by local boys (G. Hastings)—a poignant reminder that this was the depths of the Great Depression.

Lunt (1933) mentioned that there were only 2 Bronx records prior to 1932, without giving any details. If so, they were unknown to Griscom (1923), Kuerzi (1927), and Cruickshank (1942) and are now lost. There are 2 subsequent Bronx records: at Clason Pt. on 11 Nov 1974 (Russak) and on Lake Agassiz in the Bronx Zoo on 11 Nov 1983 (Krauss). Nearby, in 1951 singles were badly oiled in New Rochelle on 20 Dec and barely alive in Bronxville on 21 Dec (Darrow). In a minor incursion in Feb 1960 when 5 were found at New Rochelle on 21 Feb (A. Olsen et al.), there were probably some in Bronx waters that escaped detection. Dovekies may occur on western Long Island Sound more frequently than is currently believed, but if they remain well offshore, they will continue to be invisible.

Gulls, Terns, Skimmers: Laridae

Bonaparte's Gull *Chroicocephalus philadelphia*

NYC area, current
A spring/fall migrant on saltwater, including the Hudson River and Long Island Sound, but only occasionally on fresh. Overwinters on saltwater, largely in New York Harbor and environs, and on Long Island's South Shore, but regular in Central and Prospect Parks.

Bronx region, historical
Bicknell's *Riverdale* 1872–1901: unmentioned

Eaton's *Bronx + Manhattan* 1910–14: fairly common migrant, occasional winter visitor

Griscom's *Bronx Region* as of 1922: now very rare on lower Hudson; only 2 recent Long Island Sound records

Kuerzi's *Bronx Region* as of 1926: fairly common migrant [locations not given] in spring (5 Mar–13 Jun/fall (7 Aug–27 Dec), rare in winter

NYC area, historical
Cruickshank's *NYC Region* as of 1941: regular spring/fall migrant, uncommon to rare in winter on saltwater, but rare in fall, almost unrecorded in spring on reservoirs near coast

Bull's *NYC Area* as of 1968: most numerous on tidewater in/near New York City, but rare on inland lakes; maxima 2500 fall, 10,000 winter, 2500 spring

Bull's *New York State* as of 1975: status unchanged

Study area, historical and current

A frequent but local Bronx winter visitor/resident and spring/fall migrant, normally restricted to marine habitats. It does migrate down the Hudson River and Long Island Sound in fall in flocks, and formerly was frequent in small flocks on the Hudson all winter long. On occasion, large numbers from Upper New York Bay have reached the George Washington Bridge, the maximum being 5000+ there on 13 Dec 1959 (Post), when 1000+ also made it to the Mt. St. Vincent sewer (Buckley). While this pattern largely ended in the 1960s—the last flocks of any consequence were 150 on 8 Dec 1968 (Kane) and 200+ on 13 Dec 1969 (Kane, Buckley)—a half dozen may appear at Spuyten Duyvil on occasion. Like many marine birds that overfly inland areas and freshwater bodies without landing, they are also occasionally forced down onto large reservoirs by bad weather (Post 1964).

There are only 2 study area records, both from Jerome: 25–26 Dec 1954 (Buckley et al.) and 28 Nov 1963 (Norse). They are far more regular in Prospect Park and especially in Central Park, where the maximum until Jun 1967 was 27 on 27 Nov 1959 (Messing). Their infrequent occurrence in the study area mirrors both its lower intensity of coverage and the diminished Hudson River overwintering population that is often attributed to these gulls' intimidation by the Verrazano and George Washington Bridges.

Comments

Largely pelagic **Black-legged Kittiwake**, *Rissa tridactyla*, can appear almost anywhere after strong storms. An adult was found in Prospect Park on 13 Mar 1918 (Vietor), and HYs on the Hudson River at Cornwall Bay, Rockland County on 5–6 Dec 1981 (fide Deed), off W. 105th St. in Manhattan on 23 Jan 1963 (Sedwitz) off lower Manhattan on 4 May 1932 (an SY; Urner), and at Pelham Bay Park on 26 Dec 1927 (Kuerzi). Cruickshank (1942) referred to 3 more reports by experienced Bronx County Bird Club observers from the shores of Westchester and the Bronx between 1928 and 1941, but we have been unable to locate any information about them. Kuerzi's was thus the only Bronx occurrence until an HY at Orchard Beach on 5 Nov 2017 (Miller, Ritter, Horan; photos).

Black-headed Gull *Chroicocephalus ridibundus*

NYC area, current

An annual but very rare and local winter visitor/resident to tidewater in the New York City area since the late 1940s, this originally Eurasian gull was initially restricted to marine habitat, particularly Upper New York Bay, where it was often seen from the Staten Island Ferry in Bonaparte's Gull flocks. It was first noted in Prospect Park in 1957 and Central Park in 1958.

NYC area, historical

Cruickshank's *NYC Region* as of 1941: first Long Island records 1937, 1940

Bull's *NYC Area* as of 1968: rare but annual winter visitor to New York Harbor (18 Aug–14 May), maximum 9

Bull's *New York State* as of 1975: status unchanged

Study area and the Bronx

Both adults and HYs have been seen in the Bronx, where the first was in a flock of Bonaparte's Gulls at the former Mt. St. Vincent sewer outlet on the Hudson on 22 Jan 1955 (Kane). The first in the study area was the next in the

Bronx, an HY at Jerome on 25 Dec 1965 (Buckley), followed by adults there from 23 Dec 1973–2 Mar 1974 (Buckley, Sedwitz et al.) and on 29 Dec 1974 (Sedwitz). The fourth, in breeding plumage, was on the Parade Ground with Ringbills on 20–22 Mar 2003 (Saphir et al.). There have been 3 more Riverdale records, all adults: at Spuyten Duyvil on 10 Mar 1994 (unknown observer, *Kingbird*) and on 23 Dec 2000 (Veit, Buckley), and south of the sewage treatment plant at Mt. St. Vincent on 30 Oct 2010 (Buckley). Additional nearby Hudson River birds were at the Kennedy Marina in North Yonkers on 16 Mar 1991 (unknown observer, *Kingbird*) and at Piermont Pier, in breeding plumage on 31 Jul 2007 (Dow).

The only others in the Bronx were single adults at Clason Point on 27 Dec 1985 (Cantor, Russak) and Orchard Beach from 2–13 Nov 2017 (Allen et al.), plus 2 in breeding plumage there on 17 Apr 2007 with a breeding-plumaged **Little Gull**, *Hydrocoloeus minutus*, the only Bronx record for that species, all within a flock of about 500 Bonaparte's. This was the day after an extremely strong Apr northeaster that had sprinkled Sooty Terns, Red Phalaropes, Little Gulls, and other pelagic migrants throughout coastal New England and New York. Note the 7+ Little Gulls in Southport, CT, on 8 Apr 2015 (Mueller) with a similar cluster of migrating Bonaparte's Gulls.

Because of proximity to their normal New York Harbor overwintering areas, Black-headed Gulls are more frequently seen in Central (especially) and Prospect Parks and along the Hudson River as far as the George Washington Bridge. Parade Ground examination of spring Ring-billed Gull flocks during and after rainstorms and regular in-season reservoir coverage might show this species to occur more frequently than is known. See Appendix 5 for discussion of a study area Andean Gull, *C. serranus*.

Laughing Gull   *Leucophaeus atricilla*

NYC area, current
A widespread, common spring and (especially) fall migrant, and recently recolonizing colonial breeder at Jamaica Bay; extremely scarce past early winter.

Bronx region, historical
Bicknell's *Riverdale* 1872–1901: unrecorded

Eaton's *Bronx + Manhattan* 1910–14: uncommon breeder

Griscom's *Bronx Region* as of 1922: great flight fall 1921, 1922, with many on Hudson to Rockland County for first time

Kuerzi's *Bronx Region* as of 1926: unrecorded <1916; fairly common spring, abundant fall migrant (24 Apr–5 Dec), normally gone by late Nov

NYC area, historical
Cruickshank's *NYC Region* as of 1941: fairly common spring/abundant fall (maximum 1000+) migrant on Long Island Sound, Hudson, inland reservoirs; very rare in winter

Bull's *NYC Area* as of 1968: common to locally abundant; maxima spring 800, fall 2000 at Pelham Bay Park 1938, winter 75

Bull's *New York State* as of 1975: 700 at Piermont Pier Aug 1973

Study area, historical and current
A spring, summer, and fall visitor to all 4 extant subareas but especially Jerome (and Hillview before bird control measures were initiated in 1993), the Hudson River, and Spuyten Duyvil. The first was 9 at Jerome on 21 Jul 1925

(Kuerzi, Kuerzi). It was then seen throughout the area regularly in small numbers until the late 1950s, e.g., 20 on the Parade Ground on 29 May 1937 (Imhof); 16 Jun 1936 (Weber). Between 1960 and 1990 for reasons unknown its status changed from the most common warm-weather gull (e.g., 80 on the Parade Ground on 16 Apr 1958: Young) to the least common on the Van Cortlandt Parade Ground, while maintaining its abundance on Jerome, especially in fall. It is unrecorded in Jan–Feb.

In spring, first migrants arrive in late Mar–early Apr and may be seen anytime until late Sep on Jerome. Their numbers there peak in Sep, and they depart by mid-Nov, with extreme dates of 8 Mar 1970 (Sedwitz) and 4 on 12 Dec 1925 (Cruickshank), and a spring maximum of 500 on 5 May 1967 (Sedwitz). Van Cortlandt maxima are now normally <6, but occasionally after storms and rain in Aug–Sep, 50+ may appear on the Parade Ground. In recent periods when Jerome was drained (9 Dec 1966–28 May 1967, and 2008–14) fall counts of 350+ in Sep were not uncommon (Sedwitz, Buckley), with 101 as late as 18 Oct 1980 and 233 on 15 Nov 1963 (Sedwitz).

Laughing Gull remains a winter rarity in the study area, with none more recent than 3 at Jerome on 1 Dec and 20 on 7 Dec 1963 (Sedwitz). They are notorious for departing the day before Christmas Bird Counts no matter the date, but they frequently lingered later in the 1950s, when as many as 5 could be found on Bronx-Westchester Christmas Bird Counts at the Mt. St. Vincent sewer outlet (Kane, Buckley et al.).

Comments
Midwestern North America's **Franklin's Gull**, *L. pipixcan*, is a nearly annual vagrant in the New York City area but has never been recorded in the Bronx or Central and Prospect Parks. Rockland County's first was at Piermont Pier for several weeks from 7 Oct 1999 (Weiss et al.). During a major blowback drift-incursion on 13–14 Nov 2015, they began appearing in flocks from NH to VA, with 350+ at Cape May on a single day. Around New York City they were seen along the south shore from Brooklyn east to Shinnecock Bay, as well as on the lower Hudson River, the East River, and Staten Island. None were seen (was anyone looking?) on Long Island Sound west of CT, on the Hudson River above midtown, or anywhere in the Bronx. Central and Prospect Parks didn't yield any either, but not for want of searching. Except for a few on Staten Island and in northern NJ, all around New York City were gone the next day. The only other massive incursion, also in Nov, occurred under much the same storm/wind conditions in 1998 (Brinkley 1999) and yielded Westchester County's sole record, at Croton Point on 25 Nov 1998 (Bickford).

Ring-billed Gull   *Larus delawarensis*

NYC area, current
An increasing, abundant spring/fall migrant and winter visitor/resident. Its breeding range is moving eastward, and it could breed at any time on inland reservoirs (as in MA) or in coastal Herring Gull colonies.

Bronx region, historical
Bicknell's *Riverdale* 1872–1901: unrecorded

Eaton's *Bronx + Manhattan* 1910–14: fairly common winter visitor

Griscom's *Bronx Region* as of 1922: unrecorded on Hudson lately; one recent Feb Long Island Sound record

Kuerzi's *Bronx Region* as of 1926: fairly common migrant (18 Jul–2 Jun), occasionally overwintering

NYC area, historical

Cruickshank's *NYC Region* as of 1941: regular spring/fall (maximum 1000) migrant, winter visitor on Long Island Sound, Hudson, rare on reservoirs in winter

Bull's *NYC Area* as of 1968: great increase >1949, maxima 3000 fall, 2300–2800 winter

Bull's *New York State* as of 1975: great increase in recent years

Levine's *New York State* as of 1996: Long Island winter maximum 7190 on 1985 Northern Nassau Christmas Bird Count

Study area, historical and current

Now the commonest spring/fall migrant/winter gull on the Van Cortlandt Parade Ground and Jerome, but only since the 1950s, before which it was scarce enough to warrant comment. Following its explosive population growth in the eastern Great Lakes, it has become the commonest freshwater gull in the New York City area in winter, and even at some saltwater sites.

In fall, the first juveniles and postbreeding adults appear in mid–late Jul. Numbers build up until peaking in Sep–Nov, when maximum counts occur on Jerome. After major rainstorms, flocks appear on the Parade Ground and forage actively for earthworms. Maxima there have been in the 250+ range since the 1960s but much higher on Jerome: e.g., 1013 on 21 Sep 1980 (Sedwitz), 1100 on 6 Nov 1984 (Sedwitz).

In winter most depart by mid-Nov, and when freshwaters finally freeze, the remainder go, not returning until spring migration. Winter maxima are 500 in Van Cortlandt on 21 Dec 2011 (Berthoud) and 250 on Jerome on numerous occasions.

In spring, first migrants arrive in Mar, peak in Apr, and depart by May, with maxima of 300+ on Van Cortlandt Lake on 16 May 1964 (Russak) and 300 on Jerome on 5 May 1967 (Sedwitz).

In Van Cortlandt in summer a half-dozen SYs may be present between 1 Jun and 15 Jul, with a maximum of 80 on Jerome on 13 Aug 1964 (Sedwitz). Nearest breeders seem to be in Delaware County (McGowan and Corwin 2008)—the first intimation that they could be working their way toward the New York coastline as breeders. They are now almost to the coast in NH and MA on inland reservoirs.

Ring-billed Gull may have been unrecorded in the study area until one at Jerome on 22 Jan 1925 (Hickey). A few did occur on the reservoirs in spring/fall migration but not elsewhere at the time, and it did not become regular in Van Cortlandt until the 1950s. In winter through the 1930s they were all but unrecorded; 3 at Hillview on 14 Jan 1936 (Weber et al.) were notable at the time, and the entire 1936 Bronx-Westchester Christmas Bird Count recorded it only at Piermont Pier. It was genuinely scarce on the count until the mid-1950s, after which numbers quickly took off until peaking and then plateauing 30 years later (Fig. 40).

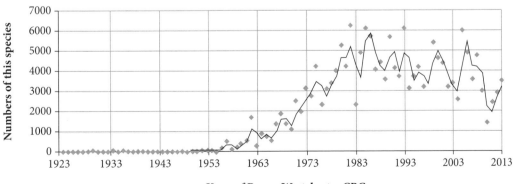

Figure 40

Herring Gull   *Larus argentatus*

NYC area, current
A colonial breeder on Long Island and around New York Harbor whose numbers peaked in the late 1970s and have been declining since. When many garbage dumps were open around New York City from the 1920s to the 1980s, they crested in winter at 200,000, but now are down to 10% of that number. Most numerous on Long Island but widespread throughout.

Bronx region, historical
Bicknell's *Riverdale* 1872–1901: unmentioned

Hix's *New York City* as of 1904: abundant winter resident

Eaton's *Bronx + Manhattan* 1910–14: abundant

Griscom's *Bronx Region* as of 1922: year-round on Long Island Sound and Hudson

Kuerzi's *Bronx Region* as of 1926: abundant winter resident, few oversummering on Long Island Sound; fall migrants arrive Aug–Sep

NYC area, historical
Cruickshank's *NYC Region* as of 1941: abundant tidewater winter resident (to 30,000), common spring/fall migrant but also regular on reservoirs

Bull's *NYC Area* as of 1968: breeding increasing, now west to Jamaica Bay

Bull's *New York State* as of 1975: status unchanged

Levine's *New York State* as of 1996: Long Island winter maximum 135,600 on 1971 Brooklyn Christmas Bird Count

Study area, historical and current
A common and widespread spring/fall migrant and resident, least numerous in summer. Usually seen overflying the area but often in large numbers on both reservoirs.

Even though it had long been occurring, we find no explicit study area records until 200 on the Parade Ground on 17 Nov 1934 (Weber, Jove) and 5 there on 29 Sep 1935 (Weber, Jove). But from Eaton's time forward (Bicknell chose not to mention Herring Gull), they were considered common to abundant migrants and especially winter visitors/residents prone to overfly the study area at any time except in summer (Kieran, Norse et al.).

Postbreeding dispersers tend to arrive in early Oct, peak in Nov, and if not overwintering depart by late Nov. In spring, numbers begin to thin in mid-Mar, and most breeders have left by early May, with varying numbers of prebreeders oversummering but moving about a great deal.

Van Cortlandt seasonal maxima are *spring* 14 on 26 Mar 2011 (Baksh); *summer* 200; *fall* 200 on 2 Nov 1954 (Scully); *winter* 500+ on the Van Cortlandt Parade Ground numerous times, although since about 2000, only a handful among the Ring-bills has been normal. Jerome seasonal maxima are considerably greater: *spring* 2100, *summer* 200, *fall* 4800; *winter* 7000 on 19 Jan 1977 during a particularly hard freeze (all Sedwitz). Hillview seasonal maxima peaked at 5000+ many times in winter during the 1930s (Norse, Weber, Imhof et al.)

Once breeders began to reach western Long Island (to Jamaica Bay Wildlife Refuge by 1953), study area summer records increased. Breeders then began to colonize New York Harbor (Hoffman/Swinburn Islands in 1964) and the East River/western Long Island Sound (South Brother Island by 1978 or earlier, Huckleberry Island off Pelham Bay Park by 1976), and eventually backfilled down the East River to even the tiniest islets and rocks by the late 1990s/early 2000s. As of 2016, the nearest breeders are on Huckleberry Island off Pelham Bay Park, but in 2014, 663 Herring Gull nests were on Riker's Island rooftops and in 2016 atop the Javits Center in midtown Manhattan near the Hudson River (NYCA). Buckley and Buckley (1980) presented extensive discussion of New York City area gull colony sites and breeding numbers.

Iceland Gull   *Larus glaucoides*

NYC area, current
A quite uncommon but annual winter visitor/resident especially during severe winters, whose numbers were closely tied to the many garbage dumps that are now all closed. Glaucous, Iceland, and Lesser Black–backed Gulls favor freshwater for drinking and bathing at reservoirs near their feeding sites, and Iceland is usually found more frequently than Glaucous.

Bronx region, historical
Bicknell's *Riverdale* 1872–1901: unrecorded

Eaton's *Bronx + Manhattan* 1910–14: unrecorded

Griscom's *Bronx Region* as of 1922: numerous lower Hudson Valley observations 1918–23

Kuerzi's *Bronx Region* as of 1926: more frequent winter visitor than Glaucous Gull (22 Nov–15 May; 31 Jul)

NYC area, historical
Cruickshank's *NYC Region* as of 1941: uncommon to rare winter visitor, rare on interior lakes, reservoirs but regular at coastal garbage dumps

Bull's *NYC Area* as of 1968: Long Island maxima 7–10, 12 Central Park Reservoir; 8 Clason Point dump in Bronx

Bull's *New York State* as of 1975: status unchanged

Study area, historical and current
In the study area an infrequent winter visitor to Hillview, Jerome, and Van Cortlandt (formerly), notably from the 1930s–70s when garbage dumps were active in the East Bronx and in the Hackensack Meadowlands. They were also regular in winter on the Hudson River, frequent at the Mt. St. Vincent sewer outlet, and occasional at Spuyten Duyvil.

Highest counts always occurred during severe winters; in the record-cold winter of 1933–34, 12 different individuals were at the Westchester Creek dump not far upriver from the Whitestone Bridge. Only 3 study area records have not involved singles: at Hillview, 2 from 4–11 Jan 1936 (Norse, Weber) and 3 from 13–21 Feb 1937 (Norse, Weber et al.), and at Jerome, 3 from late Dec 1973–late Mar 1974 (Sedwitz et al.). Of all those seen, only 2 were not HYs/SYs, both at Jerome: an adult on 5 Jan 1966 (Sedwitz) and a third-winter on 21 Jan 1966 (Sedwitz). Until Jun 1967, the winter maximum in Central Park was 7 on 7 Feb 1957 (Messing).

The earliest was at Jerome on 28 Nov 1963 (Sedwitz) and the latest there on 14 May 1967 (Sedwitz), with most in Dec–Feb but several in Apr and even 2 in May. On the Van Cortlandt Parade Ground singles have been found only thrice: 16 Dec 1939 (Kieran), 28 Mar 1963 (Heath), and 30 Apr 1966 (Norse). None have been seen in the study area since 1971.

Comments
Until the 1960s observers assumed that most Icelands were nominate *glaucoides* and that *kumlieni* was exceedingly rare, which we now know to be exactly opposite the true condition. All before 1960 recorded as nominate *glaucoides* should now be considered *kumlieni*, and records for both have been merged in this account. Although nominate *glaucoides* may occur, its reliable field separation from *kumlieni* in all

plumages is still being worked out. See Malling Olsen and Larsson 2004 and Howell and Dunn 2007 for current information and comprehensive color photographs.

Experienced New York City area observers have only recently become comfortable separating vagrant (?) **Thayer's Gull**, *L. thayeri*, from its close relatives Iceland and Herring. While it is an unlikely winter visitor on the reservoirs now that all area garbage dumps are closed, the Hudson River or Long Island Sound is where the first in the Bronx will be detected. And indeed, it was on Inwood Park's Dyckman St. pier in the Hudson River that the first in the New York City area away from Long Island/New York harbor recently showed up, an HY photographed on 14 Dec 2014 (Messer). Unrecorded in Central or Prospect Park.

Lesser Black-backed Gull   *Larus fuscus*

NYC area, current
An increasing but uncommon spring/fall migrant and winter visitor/resident, whose numbers were formerly tied to the many garbage dumps that are now closed. This European species began invading North America in the 1930s, and the first for North America was at Beach Haven, NJ, on 9 Sep 1934 (Urner), followed closely by one at Baxter Creek on 9 Dec 1934 (Kuerzi, Kuerzi). It began to be seen regularly in very small numbers in the late 1950s–early 1960s, when New York City area garbage dumps were at their apex (or nadir). Overwintering numbers to the south have increased greatly, and in spring when birds follow the Susquehanna River northward through PA, flocks have now reached almost 1000, heading north to unknown but assumed North American breeding sites in central/eastern Canada.

NYC area, historical
Cruickshank's *NYC Region* as of 1941: 6 New York City area records 1934–40, 4 on Long Island

Bull's *NYC Area* as of 1968: very rare but annual winter visitor, all singles but 3–4 during the 1956–57 winter (30 Aug–14 Apr)

Bull's *New York State* as of 1975: Long Island winter maximum 2

Levine's *New York State* as of 1996: Long Island winter maximum 6

Study area, historical and current
An infrequently recorded winter visitor when area garbage dumps were operating but not since.

All study area records date from the period when the Pelham Bay Park and Hackensack Meadowlands garbage dumps were open and gulls routinely bathed, drank, and roosted at Jerome and Hillview. The first was an HY on Hillview on 27–28 Mar 1964 (Carleton, Horowitz, Buckley)—the first of that age-class ever detected in the New York City area—followed by an adult at Jerome 27 Dec 1970 (Buckley, Sedwitz et al.). During the winters of 1973–74 (23 Dec–7 Jan) and 1974–75 (28 Nov–16 Mar) there were 6 identifiably different individuals (SYs, TYs, adults) from the Pelham Bay Park dump that regularly came into Jerome and Hillview even though only singles were ever seen (Buckley, Sedwitz et al.). On a few occasions several singles of different ages were seen on the same day at Jerome and Hillview (Sedwitz, Buckley).

The last time this species was seen in the study area was in Mar 1975, even though the Pelham Bay Park dump did not close until 1979. Surprising at the time was a breeding-plumaged adult at Jerome on 21 Jul 1970 (Sedwitz), but 47 years later we recognize this as an early (failed?) southbound breeder, that date being only slightly earlier than adult arrival dates along the south shore of Long Island in recent years.

Nowadays, this species should be looked for on the Van Cortlandt Parade Ground after heavy rains from Aug–Apr, where it remains unrecorded. The only other nearby records were different individuals at Spuyten Duyvil on 5 Mar and on 9 Dec 2013 (DiCostanzo). As expected, all adults have been *graellsii*, the Iceland, Greenland, and undiscovered Canada breeding subspecies. Despite area dump closures, it is seen regularly in migration and in winter as close as Piermont Pier.

Glaucous Gull   *Larus hyperboreus*

NYC area, current
A quite uncommon but annual winter visitor/resident especially during severe winters, whose numbers were closely tied to the many garbage dumps that are now almost all closed. Glaucous, Iceland, and Lesser Black–backed Gulls favor freshwater for drinking and bathing at reservoirs near their feeding sites, and Glaucous is usually found less frequently and in smaller numbers than Iceland.

Bronx region, historical
Bicknell's *Riverdale* 1872–1901: unrecorded

Eaton's *Bronx + Manhattan* 1910–14: rare winter visitor

Griscom's *Bronx Region* as of 1922: unrecorded

Kuerzi's *Bronx Region* as of 1926: irregular and uncommon in winter (13 Oct–25 Apr)

NYC area, historical
Cruickshank's *NYC Region* as of 1941: uncommon to rare winter visitor on Long Island, extremely rare on interior lakes, reservoirs but regular at coastal garbage dumps

Bull's *NYC Area* as of 1968: coastal winter maxima 5–6

Bull's *New York State* as of 1975: status unchanged

Study area, historical and current
In the study area an infrequent winter visitor to Hillview and very rarely to Van Cortlandt and Jerome (formerly), notably from the 1930s–70s, when garbage dumps were active in the East Bronx and in the Hackensack Meadowlands.

Highest counts always occurred during severe winters. All records but 2 have involved single HYs or SYs: 4 at Hillview on 13 Feb 1937 (Weber, Norse et al.), and an adult there on 14 Jan 1951 (Komorowski). Glaucous Gulls in the New York City area usually arrive later than Icelands, although in the study area the earliest was at Jerome on 5 Nov 1953 (Buckley) and the latest at Hillview on 21 Apr 1934 (Rich, Rich), with the bulk in Jan–Feb and only a handful in Mar–Apr. Away from Hillview, the only other study area records were at Jerome on 30 Jan 1937 (Imhof, Norse) and on the Van Cortlandt Parade Ground on 19 Apr 1965 (Kane).

Glaucous has also been regular in very small numbers on the Hudson River at Riverdale as late as 14 May 1982 (Sedwitz), but there have been no study area records since 1965. Kuerzi's 5 at the Baychester dump on 17 Feb 1935 remains among the highest counts in the New York City area.

Great Black-backed Gull *Larus marinus*

NYC area, current
A localized colonial breeder with Herring Gulls, mostly on Long Island, and a widespread, common fall and spring migrant and winter resident. Breeding numbers have diminished since their peak in the late 1980s, but nonbreeders may be increasing.

Bronx region, historical
Bicknell's *Riverdale* 1872–1901: unmentioned

Eaton's *Bronx + Manhattan* 1910–14: uncommon winter visitor

Griscom's *Bronx Region* as of 1922: scarce in New York Harbor, lower Hudson; very rare winter visitor, 6 Long Island Sound records, 1 from Jerome

Kuerzi's *Bronx Region* as of 1926: regular and fairly common winter visitor (15 Sep–17 Apr) on Long Island Sound (maximum 200), uncommon on Hudson, rare on reservoirs

NYC area, historical
Cruickshank's *NYC Region* as of 1941: common winter resident coastally, numerous Jerome records

Bull's *NYC Area* as of 1968: breeding west to Jamaica Bay; maxima fall 900, winter 3000

Bull's *New York State* as of 1975: breeding in New York Harbor; summer maximum 500 at Piermont Pier

Levine's *New York State* as of 1996: Long Island winter maximum 15,865 on 1976 Brooklyn Christmas Bird Count

Study area, historical and current
A widespread, common fall/spring migrant and winter resident, largely on both reservoirs.

First recorded at Jerome on 16 Nov 1920 (Kieran), followed by another there 4–5 Jan 1925 (Cruickshank). They did not become regular in the West Bronx until the mid-1930s (Kieran, Cruickshank, Norse), when seen most frequently on Jerome. In the study area it is now an annual visitor, especially on Jerome and Hillview, as well as the Van Cortlandt Parade Ground after rains. Numbers peaked in the 1930s–70s when multiple East Bronx and Hackensack Meadowlands garbage dumps were operating.

Most frequent from Nov–Mar, with Jerome maxima of *spring* 13, *summer* 7, *fall* 57, *winter* 250, and Van Cortlandt maxima (where they are far less frequent) of *spring* 3, *summer* 5, *fall* 5, *winter* 7. No Hillview seasonal counts are available.

It may be that considerable numbers migrate south (and north?) along the Hudson River, exemplified by the 1500 moving south at Dobbs Ferry, Westchester County, sometime in fall 1979 (unknown observer, *North American Birds*).

Without doubt the nearest breeders were the pair that raised 4 young in 2001 (and earlier?) at the Yonkers sewage treatment plant adjacent to Mt. St. Vincent in the extreme Northwest Bronx, a most unusual location (unknown observer, NYRBA). Otherwise the closest are a few pairs at Goose Island in Pelham Bay Park (first in 1996: Künstler 2007), Huckleberry Island (first in 1977: Buckley and Buckley 1980), and on North/South Brother Islands (first in 1977: Buckley and Buckley 1980). In 2014 8 Great Black-backed Gull nests were on Riker's Island rooftops. Also breeding at Rye, Westchester County, in 1984, although this was not the first breeding site away from the oceanfront, as some thought.

Least Tern   *Sternula antillarum*

NYC area, current
A Long Island colonial breeder restricted to fresh dredge spoil sites or undisturbed beaches, and a widespread but uncommon spring/fall migrant that regularly moves up the Hudson River to Piermont Pier, especially postbreeding. It is a New York State Threatened Species. Recorded repeatedly in Prospect Park but never in Central.

Bronx region, historical
Bicknell's *Riverdale* 1872–1901: unrecorded

Eaton's *Bronx + Manhattan* 1910–14: uncommon migrant

Griscom's *Bronx Region* as of 1922: unrecorded

Kuerzi's *Bronx Region* as of 1926: unrecorded

NYC area, historical
Cruickshank's *NYC Region* as of 1941: local but increasing Long Island breeder; 20 Apr–5 Nov; uncommon spring/fairly common fall (maximum 200) migrant; only 6 Long Island Sound records west of Eaton's Neck [locations not given]; no freshwater records

Bull's *NYC Area* as of 1968: Long Island maximum 500, 5 days after tropical storm; 6 at Piermont Pier after Hurricane Connie Aug 1955

Bull's *New York State* as of 1975: 70 at Piermont Pier Jul 1974

Study area, historical and current
Most infrequent away from tidewater, this species has been recorded only twice in the study area: an adult on Van Cortlandt Lake 13–17 Aug 1956 (Young) and 2 adults at Jerome on 6 Jul 1960 (Wallman).

Even though not breeding in the Bronx as of 2016, it almost certainly bred at Baxter Creek in 1955 and 1956 (C. Young, Russak), maybe in other years, but nowhere else, largely for lack of suitable habitat. However, this species is a well-known rooftop breeder in the southern United States and might resort to such sites along the Bronx's Long Island Sound and even East River edges. Nearest breeders are at Sands Point, Long Island—only 2.9 smi(4.6 km) across Long Island Sound from Pelham Bay Park, where they forage regularly—and in the Hackensack Meadowlands.

Comments
Now breeding on, but remarkably restricted to, the South Shore of Long Island, **Gull-billed Tern**, *Gelochelidon nilotica*, has occurred once in the Bronx, at Pelham Bay Park from 6–19 Jul 1995 (Künstler). As its numbers increase, it will be prone to additional postbreeding wandering but is unrecorded in Central or Prospect Park.

Caspian Tern   *Hydroprogne caspia*

NYC area, current
A regular but uncommon spring/fall migrant most often found on saltwater, typically on Long Island. Recorded in both Central and Prospect Parks.

Bronx region, historical
Bicknell's *Riverdale* 1872–1901: unrecorded

Eaton's *Bronx + Manhattan* 1910–14: rare migrant

Griscom's *Bronx Region* as of 1922: unrecorded

Kuerzi's *Bronx Region* as of 1926: unrecorded

NYC area, historical
Cruickshank's *NYC Region* as of 1941: rare migrant in spring (maximum 5)/uncommon in fall (maximum 14); 3 Mar–24 Oct; extremely rare on Long Island Sound; once Piermont Pier, once Jerome

Bull's *NYC Area* as of 1968: Long Island maxima 19 spring, 400+ in Hurricane Donna; late fall dates 23 Nov, 28 Dec

Bull's *New York State* as of 1975: status unchanged

Study area, historical and current
Even though an irregular but possibly annual fall migrant at Pelham Bay Park, there are only 2 study area records: 30 Sep 1934 on Jerome (Cruickshank) and 2 on 8 Sep 1956 over Van Cortlandt during a large Broad-winged Hawk movement (Kane). However, on 28 Jul 2009 1 was moving down the Hudson River in Yonkers (unknown observer, *Kingbird*), as were 3 at Mt. St. Vincent on 29 Aug 1956 and another on 14 Sep 1956 (Buckley); 1 extremely late was just below the George Washington Bridge in Edgewater, NJ, on 6 Dec 1967 (Boyajian). Southbound Hudson River movement in fall in Riverdale is annual, and it is regular at Piermont Pier.

Common Tern   *Sterna hirundo*

NYC area, current
A coastal spring/fall migrant and colonial breeder in marine portions of the New York City area, including sites in Jamaica Bay, the East River, and Long Island Sound. It is also annual in small numbers on the Hudson River in Riverdale, mostly postbreeding and during and after tropical storms and is not infrequent in Central and Prospect Parks.

Bronx region, historical
Bicknell's *Riverdale* 1872–1901: unmentioned

Eaton's *Bronx + Manhattan* 1910–14: common breeder, migrant

Griscom's *Bronx Region* as of 1922: regular in fall on lower Hudson below West 125th St.

Kuerzi's *Bronx Region* as of 1926: rare migrant in spring (7 May–10 Jun), fairly common fall (9 Jul–30 Oct) [locations not given]

NYC area, historical
Cruickshank's *NYC Region* as of 1941: increasing Long Island breeder, uncommon spring/common fall migrant on saltwater, coastal fall maximum 5000+; inland maximum 70 on Hudson at Riverdale

Bull's *NYC Area* as of 1968: 15 Apr–5 Dec, plus late Dec, Mar; breeding west to Jones Beach; western Long Island Sound maximum 150 Rye

Bull's *New York State* as of 1975: status unchanged

Study area, historical and current
There seems to be but a single study area record: Van Cortlandt Lake on 31 Aug 1939 (Norse, Cantor). And even though Cruickshank (1942) spoke of quite a few records from lakes and reservoirs in Bronx and Westchester Counties, they seem never to have been published anywhere. One would expect it to occur far more frequently than the sole record, and perhaps it does—in midsummer when few observers are afield in the study area. In 1961, a chick banded at Moriches Inlet on 29 Jun was recovered as a fledged juvenile on 28 Jul in the 10-minute

Bird Banding Lab Block that embraces the study area, but exactly where in it remains unknown. The nearest current breeders are on islands in the East River, or as of 2016 at City Island (fide Rothman). Recorded regularly from both Central and Prospect Parks.

Comments

Two more species that regularly occur nearby are the study area's most surprising unrecorded species. **Forster's Tern**, *S. forsteri*, is now breeding in several locations on Long Island including on the North Shore, is regular from Apr–Oct as close as Pelham Bay Park, where 200 unrelated to a hurricane on 27 Sep 2015 (Rothman et al.) is the Bronx maximum. It is probably annual on the Hudson River in summer/fall, when it also occurs in the 100s on the Hackensack River (R. Kane, pers. comm.). It is regular in Prospect Park but unrecorded in Central Park. One studied on nearby Grassy Sprain Reservoir on 20 Aug 1953 (Penberthy, Phelan, Buckley) was the first ever seen on inland fresh water in Westchester County. Inland breeding **Black Tern**, *Chlidonias niger*, is also probably annual in summer/fall (especially during and after tropical storms) on the Hudson River and on Long Island Sound, and has also been found as close as Spuyten Duyvil on several occasions in both spring/fall. It has been seen several times in Prospect Park but never in Central.

Black Skimmer   *Rynchops niger*

NYC area, current
A locally abundant saltwater colonial breeder along the south shore of Long Island that ventures inland on occasion, with multiple records in both Central and Prospect Parks, typically toward dusk in Jun–Sep and possibly increasing. Regular in small numbers on western Long Island Sound from May–Sep. It is a New York State Species of Special Concern.

Bronx region, historical
Bicknell's *Riverdale* 1872–1901: unrecorded

Eaton's *Bronx + Manhattan* 1910–14: unrecorded

Griscom's *Bronx Region* as of 1922: unrecorded

Kuerzi's *Bronx Region* as of 1926: once each New Rochelle Jul, Hunt's Point Sep

NYC area, historical
Cruickshank's *NYC Region* as of 1941: very rare Long Island breeder, spring migrant, uncommon late summer visitor (maximum 200+ after tropical storms); 30 Apr–24 Nov; very rare on western Long Island Sound, on Hudson at Riverdale

Bull's *NYC Area* as of 1968: 26 Apr–Jan (after hurricanes); now widespread, numerous, especially after tropical storms when on Hudson north to Piermont Pier; breeds west to Jamaica Bay; fall maximum 2000+

Bull's *New York State* as of 1975: early spring date 14 Apr

Study area, historical and current
A typical Long Island saltwater species, it has occurred only once in the study area: 4 adults skimming on Van Cortlandt Lake on 25 Jun 1963 (Rigerman). This unique occurrence mirrors the park's relatively low intensity of coverage in midsummer when skimmers seem most prone to wander, like the adults flying over the nearby Cross County Parkway in Yonkers on 6 Jun 2012 (unknown observer, *Kingbird*). Two more on the Hudson River off Spuyten Duyvil on 26 Dec 1964 (Barber) were

most unexpected as to location and date, yet only a few miles farther up the Hudson 1 was at Piermont Pier on 23 Nov 1992 (Harten), an equally abnormal time. Others there on 18 May 2000 (B. Hall, Klose) and 25 May 2001 (Lutter, Previdi) were less surprising. The only other nearby record was on the Hudson River at Riverdale on 3 Sep 1934 (Cruickshank). The sight of recent feeding skimmers at dusk and at night in Central and Prospect Park lakes is not surprising, given that skimmers' catlike vertical pupils are adapted for low-light feeding. We would not be surprised if they also occasionally fed at night on Van Cortlandt Lake, but who would be there to see them?

Pigeons, Doves: Columbidae

Rock Pigeon   *Columba livia*

NYC and study areas, historical and current
A ubiquitous urban and suburban resident conspicuously absent from wilder areas like undeveloped barrier beaches. No reliable population trends are apparent as no population estimates have been undertaken, but their numbers are not in jeopardy.

An exotic (nonnative) species established in the Northeast in the 1600s for food by the earliest colonists, its populations have been augmented countless undocumented times by escapes and releases, and continuing unabated. Unmentioned by DeKay (1844), Giraud (1844), Bicknell (1872–1901), Eaton (1910, 1914), and Griscom (1923)—even in his "Introduced Species" section—their presence was finally acknowledged by Cruickshank (1942) and Bull (1964, 1974), although with little information beyond occurring almost everywhere and breeding year-round with a clutch of 2.

In the study area the few seasonal maxima include *spring* 50; *summer* 25; *fall* 250; and *winter* 150.

As of 2016, there are 200–300 in 5–6 flocks immediately surrounding Van Cortlandt, plus half that number near Jerome, Woodlawn, and Hillview. Probably <10 pairs may be nesting on buildings within in each of these 4 subareas.

Despite occurring in all blocks in and around New York City, NYSBBA and DOIBBS data suggested a slow statewide increase until about 1985 followed by a slow decline that may now be leveling off near 1966 numbers (Sauer et al. 2005; McGowan and Corwin 2008).

Eurasian Collared-Dove   *Streptopelia decaocto*

NYC and study areas, historical and current
After Collared-Doves had colonized Florida unassisted from a feral population in the Bahamas in the 1970s, they quickly began to spread to the far Northwest—exactly as they had done in Europe in the middle of the 20th century, when they expanded out from the Balkans to the British Isles in 20 years. Today in North America they have reached Alaska, are widespread on the American and Canadian Great

Plains, and only since the 1990s have the first finally been recorded in New England and the Maritimes, reaching Nova Scotia in 2008.

Since about 2000, recorded in NJ and in New York where a small breeding population seems established upstate. A pair may have bred once in CT, where there are <12 records; there are multiple records for MA, a few from ME, and apparently only 1 each in RI, VT, and NH. There are no proven established breeding populations in the Northeast yet, but these should follow. Unrecorded in Prospect Park, but in 2000, there were 2 reports from New York City: Staten Island on 7 Feb (Stetson; photos) and Central Park on 19 May (m. ob.).

In the study area, the first for the Bronx was examined closely and carefully distinguished from feral African Collared-Doves, *S. roseogrisea* (= Ring-necked Dove, Ringed Turtle-Dove, Barbary Dove), that also have been seen occasionally in Van Cortlandt. It was sitting quietly on the causeway in Van Cortlandt Swamp on 20 May 2014 (Buckley) and allowed close approach but eventually flushed and flew strongly away to the northeast and was not seen again.

In 2014 other Eurasian Collared-Doves were reported on 22 Jun (Ringer) from the Hudson River Greenway (between W. 22nd and 29th Sts.) and then in the same general area by many observers through Dec, and at Inwood Park on 16 Jul (Clugston, Souirgi) and again on 3 Sep (Auerbach). Another was reported in Hastings on 9 May 2015 (Bannerman), so they may be finally shifting away from their North American northwest/southeast axis and attempting to colonize the New York City area and northeast.

Passenger Pigeon *Ectopistes migratorius*

NYC and study areas, historical
This extraordinary creature became extinct in the wild just before 1910 but was once not only common but sometimes abundant in the New York City area. It reputedly did not breed any closer than Orange County (in Ulster County as late as 1871) but did occur at New York City in the fall as single migrants and small flocks until the early 1880s.

Close to the study area, Bicknell found it in his greater Riverdale area only once in spring (30 Mar 1875) but considered it regular in the fall through the 1870s, and there is no reason to doubt that it was also occurring then in the study area. He saw them in Riverdale as follows:

1 on 20 Oct **1871**

1 on 25 Oct **1873**

13 on 11 Oct **1874**

1 on 11 Sep **1875**

15 on 2 Sep, **1** on 3 Sep, and **1** on 17 Oct **1876**

2 on 30 Aug and **30** on 6 Oct **1877**

4 on 6 Sep, **8** on 13 Sep, **small flock** on 14 Sep, **9** and then **20** on 15 Sep, **30** on 16 Sep

8–10 on 18 Sep, flocks of **4** and **10** on 20 Sep, **15+** on 22 Sep, **1** on 23 Sep, **1** on 24 Sep, and **4–5** on 15 Oct **1878**—seemed to be more abundant this fall than for some years previously

2 on 21 Sep **1879**

3 on 5 Sep, **1** on 11 Sep, **25–30** on 19 Sep, 3 Oct, and **2** on 9 Oct **1880**

3 on 8 Sep, and **2** on 30 Sep **1881**

And then there were none.

In the New York City area the last Passenger Pigeons ever shot were in Sep–Oct 1889 in Queens, in Westchester County, and in

Rockland County, and the last ever seen were A. H. Helme's flock of 6 at Miller Place, Suffolk County, in the fall of 1890.

Specimens
Bicknell collected 1 in Riverdale on 23 Sep 1878 (MCZ).

White-winged Dove *Zenaida asiatica*

NYC area, current
A recent vagrant from Florida and the Southwest but now almost annual, with numerous records >1996. Most appear on Long Island, usually in 2 pulses: May–Jul, perhaps post-breeding dispersals from the newly established FL population or the Southwest, and Aug–Dec, more likely (lingering) blowback drift-migrants. While they have not been recorded every month, they soon will be, just as in NJ. Recorded in Prospect but not Central Park.

NYC area, historical
Cruickshank's *NYC Region* as of 1941: vagrant Nov 1929 [blowback drift-migrant]

Bull's *NYC Area* as of 1968: status unchanged

Bull's *New York State* as of 1975: second record Dec 1973 [blowback drift-migrant]

Levine's *New York State* as of 1996: 3 more records [but others overlooked]; great increase in records throughout Northeast

Study area, historical and current
A single record, the first for the Bronx but not unanticipated, near the tennis courts by Van Cortlandt Lake on 14 Sep 2000 (Young). This individual, alone and silent at 0630, was examined for 15 minutes at close range. The second in the Bronx was at Pelham Bay Park on 23 Nov 2015 (Aracil; photos). It has occurred once in Prospect Park on 11 Sep 2016 (Howard) and at Piermont Pier on 11 May 2000 (Herskovics).

Mourning Dove *Zenaida macroura*

NYC area, current
A widespread spring/fall migrant, breeder, and winter resident that gathers in large winter flocks in open areas. Around 2000 its numbers peaked and leveled off.

Bronx region, historical
Bicknell's *Riverdale* 1872–1901: common breeder throughout; unrecorded in winter

Eaton's *Bronx + Manhattan* 1910–14: common breeder

Griscom's *Bronx Region* as of 1922: scarce in Greater New York City; uncommon migrant

Griscom's *Riverdale* as of 1926: extirpated breeder

Kuerzi's *Bronx Region* as of 1926: fairly common migrant, uncommon breeder, occasional in winter (6 Mar–22 Nov; 12 Feb and 27 Feb)

NYC area, historical
Cruickshank's *NYC Region* as of 1941: regular migrant in spring (maximum several dozen)/fall (maximum 200), widespread breeder including wilder sections of Bronx; only very locally common (maximum 100+) winter

Bull's *NYC Area* as of 1968: fall maximum 600; increasingly numerous in winter

Bull's *New York State* as of 1975: winter maximum 400

Study area, historical and current

A widespread spring/fall migrant, breeder, and winter resident that has increased in late fall/winter since about 2000.

In spring, first migrants arrive in early Mar, peak in Apr, and depart by mid-May, with a maximum of 35 on 2 Apr 1956 (Scully). Extreme dates are meaningless.

It breeds in all 4 study areas, with an estimated population of 40 pairs, <30 of them in Van Cortlandt. Both NYSBBA I in 1980–85 and II in 2000–2005 recorded them in all blocks in and near New York City (McGowan and Corwin 2008), and DOIBBS data showed a 2% annual statewide increase from 1966–2005 (Sauer et al. 2005).

In fall, first migrants arrive in Jul, peak in Sep–Oct, and depart by Nov, with a maximum of 110 on 6 Nov 2013 (Young). Again, extreme dates are meaningless.

In winter they are locally numerous especially at feeders, granaries, and other sources of supplemental seeds. On the Bronx-Westchester Christmas Bird Count they were extremely rare until the mid-1950s when numbers shot up, peaked in the 1970s at 1600, and then erratically dropped to the upper hundreds where they remain (Fig. 41). The study area maximum was 141 across Van Cortlandt and Woodlawn on 26 Dec 2004 (Lyons et al.).

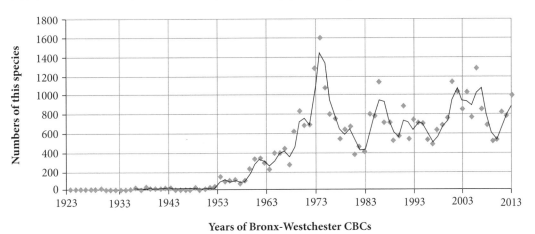

Figure 41

Cuckoos: Cuculidae

Yellow-billed Cuckoo *Coccyzus americanus*

NYC area, current
An annual but uncommon spring/fall Neotropical migrant, and a widespread breeder that is disappearing within New York City.

Bronx region, historical
Bicknell's *Riverdale* 1872–1901: regular spring/fall migrant, breeder

Hix's *New York City* as of 1904: common breeder in the parks

Eaton's *Bronx + Manhattan* 1910–19: fairly common breeder

Griscom's *Bronx Region* as of 1922: common breeder

Kuerzi's *Bronx Region* as of 1926: common breeder (3 May–18 Oct)

NYC area, historical
Cruickshank's *NYC Region* as of 1941: common breeder, regular spring/fall migrant; 29 Apr–2 Nov; unusual to rare migrant in parks of large urban developments [locations not given]

Bull's *NYC Area* as of 1968: 16 Apr–29 Nov; mind-bending 1954 Long Island fall maximum 1000+ associated with 2 hurricanes: Edna in Sep, known transport of thousands of landbirds north to New England, Hazel in Oct

Bull's *New York State* as of 1975: early spring date 11 Apr [trans-Gulf overshoot]

Study area, historical and current
In small numbers, a spring/fall migrant and longtime breeder.

In spring, first migrants arrive in early May, peak in mid–late May, and depart by early Jun, with extremes of 6 May 2001 (Young) and 22 May 1992 (Young), and a maximum of only 2 on numerous occasions.

Much the commoner of the 2 cuckoos in migration, in the late 1800s, it was a common breeder in Riverdale (Bicknell) and in Van Cortlandt (Dwight). It breeds in Van Cortlandt, sometimes also Woodlawn (Horowitz), but <5 pairs in all.

NYSBBA II in 2000–2005 found breeders in only 5 New York City blocks on Staten Island, in Brooklyn and Queens, and study Block 5852B. This is a significant drop from about 20 Blocks during NYSBBA I in 1980–85, but cuckoos are easily overlooked when silent (McGowan and Corwin 2008). Sparse DOIBBS statewide data showed no significant trend from 1966–2005 (Sauer et al. 2005).

In fall, first migrants arrive in early Aug, peak in late Aug–early Sep, and usually depart by late Sep, with extreme dates of 26 Jul 2017 at Hillview (Camillieri) and 14 Oct 1959 (Mayer), with a maximum of 2 on several occasions. However, there is often a second pulse in late Oct that can contain both late migrants (local breeders can be heard vocalizing in mid-Oct in southern New England) and lingering blowback drift-migrants. There are 2 such late study area records: 22 Oct 2007 (Young) and 23 Oct 1937 (Kessler, R. Kuerzi).

Extraordinarily large numbers of cuckoos are sometimes transported within the eyes of hurricanes, like the 1000+ between Orient (Latham), Fisher's Island (H. L. Ferguson), and Sandy Hook (Stout) following Hurricane Edna on 11 Sep 1954. It was estimated that about one-third were Yellow-billed and two-thirds Black-billed (Bull 1964). That same storm also carried many other southeastern landbird migrants in addition to cuckoos to Cape Cod and nearby areas (Griscom and Snyder 1955). After the 21 Sep 1938 Great Atlantic Hurricane, Latham also reported large numbers of Yellow-billed Cuckoos at Orient. In such years some often linger well into Nov.

Specimens
Dwight collected 4 in Van Cortlandt in Aug, Sep 1888–90 (AMNH) and D. W. Smith another on 5 Jun 1933 (MCZ).

Black-billed Cuckoo *Coccyzus erythropthalmus*

NYC area, current
A spring/fall Neotropical migrant and a thinly but highly local breeder now almost extirpated from New York City except perhaps in the study area and on Staten Island.

Bronx region, historical
Bicknell's *Riverdale* 1872–1901: common breeder

Hix's *New York City* as of 1904: uncommon breeder in the parks

Eaton's *Bronx + Manhattan* 1910–14: common breeder

Griscom's *Bronx Region* as of 1922: uncommon breeder

Griscom's *Riverdale* as of 1926: now extirpated breeder

Kuerzi's *Bronx Region* as of 1926: less common than Yellow-billed as breeder (2 May–19 Oct)

NYC area, historical
Cruickshank's *NYC Region* as of 1941: fairly common breeder, regular spring/fall migrant; 28 Apr–13 Nov; unusual to rare migrant in parks of large urban developments [locations not given]

Bull's *NYC Area* as of 1968: early spring date 15 Apr; fall maximum 9 dead Fire Island Lighthouse 1883

Bull's *New York State* as of 1975: status unchanged

Study area, historical and current
A scarce but annual spring/fall migrant and breeder. Generally many fewer are found than Yellow-billeds.

In spring, first migrants arrive in early May, peak in mid–late May, and depart by early Jun, with extremes of 4 May 2000 (Garcia) and 31 May 1958 (Young) and a maximum of 7 on 17 May 1953 (Linnaean Society field trip). One in Bronx Park on 15 Apr 1961 (Maguire, Hackett) was a classic trans-Gulf overshoot. Until Jun 1967, the spring maximum in Prospect Park was 3 on 21 May 1944 (Soll).

In the late 1800s it was a regular breeder in Riverdale (Bicknell) and in Van Cortlandt (Dwight). There is no reason to believe it does not still breed in Van Cortlandt and also Woodlawn (Horowitz), but <5 pairs in total.

NYSBBA II in 2000–2005 reported breeders in only 9 New York City Blocks (Bronx, Brooklyn, Queens 1 each, Staten Island (6), down from about 16 blocks during NYSBBA I in 1980–85, but silent cuckoos are easily overlooked (McGowan and Corwin 2008). Sparse DOIBBS statewide data showed no significant trend from 1966–2005 (Sauer et al. 2007).

In fall, first migrants arrive in early Aug, peak in late Aug, and depart by late Sep, with extreme dates of 8 Aug 2004 (Young) and 22 Oct 2011 (Baksh), and a maximum of 2 on several occasions. See Comments in the Yellow-billed Cuckoo account above about hurricane transport of hundreds of Black-billeds to the Long Island area.

Specimens
Dwight collected 2 in Van Cortlandt in May 1886, 1890 (AMNH).

Barn Owls: Tytonidae

Barn Owl   *Tyto alba*

NYC area, current
A widespread, rarely encountered resident that breeds in abandoned buildings and structures but that readily takes to predator-proofed, properly crafted nest boxes. Individuals can show up in nonbreeding locations at any time but typically in fall and winter, when they also move into dense conifer groves to roost by day.

Well known to incur heavy mortality during severe winters.

Bronx region, historical

Bicknell's *Riverdale* 1872–1901: unrecorded

Eaton's *Bronx + Manhattan* 1910–14: rare, breeds

Griscom's *Bronx Region* as of 1922: unrecorded

Kuerzi's *Bronx Region* as of 1926: formerly bred; numerous recent records indicate it still does [locations not given]

NYC area, historical

Cruickshank's *NYC Region* as of 1941: uncommon resident throughout, possible anywhere, anytime but usually rare/unrecorded

Bull's *NYC Area* as of 1968: winter maximum 4 roosting Pelham Bay Park

Bull's *New York State* as of 1975: status unchanged

Study area, historical and current

Almost undetected yet may be a resident breeder in or near the study area. Even though Barn Owls may have been breeding in scattered nearby churches, rooftop water towers, and abandoned buildings in the Bronx for 100+ years, there are only 3 Van Cortlandt records of nonbreeders: 24 Apr 1937 (Imhof); 20 Sep 1946 (Norse); 30 Oct 1954 (Buckley). In 1958, one was calling loudly at 0200 in nearby Riverdale on 6 May, at 0015 on 12 Jun, and at 0130 on 24 Jun (Buckley) and was presumably a local breeder, probably in some church steeple. There are no significant dense conifer patches for winter roosts in the study area except in Woodlawn, where Barn Owl remains unrecorded but surely overlooked.

It was unmentioned by Bicknell and by Kuerzi but has been breeding in many out-of-the-way Bronx locations since the early 1900s. Griscom (1923) put it well: "If the student could search church steeples, belfries, old farm buildings, and barns as zealously as conifer groves and hollows in trees, he would undoubtedly see many more Barn Owls"— to which may confidently be added cemetery vaults, water towers, and abandoned commercial structures. What may have been the first known nesting pairs in the Bronx were in 1933 at Pelham Bay Park (Kassoy), where it still breeds (Künstler), and at Hunt's Point in 1938 (Cruickshank, Kassoy, Hickey). Otherwise, the only other breeding pairs recorded were in abandoned buildings on North Brother Island, although they are probably on (all?) other East River/Long Island Sound islands with structures. A pair bred in Yankee Stadium in 1984 (DiCostanzo). At least 11 overwintered (and bred?) in abandoned buildings on David's Island, Westchester County in Long Island Sound off Pelham Bay Park in the winter of 1975–76 (O'Dell), and at Jamaica Bay Wildlife Refuge almost every nest box has been continually occupied since the early 1980s with 400+ young banded (D. Riepe, pers. comm.). The take-home message is clear— build predator-proofed, properly crafted and emplaced nest boxes, and they will come.

On 19 Aug 2003, in an attempt to establish the species in Van Cortlandt (breeding had never been recorded previously), to great media attention the New York City Parks Department hacked 6 young Barn Owls (origins unknown) there. The only information as to their fate is that 2 were subsequently found dead near the East River (Künstler), and the failed effort was quietly buried.

NYSBBA I in 1980–85 and II in 2000–2005 considered this species to be widely but thinly distributed around New York City, but Barn Owls are consistently overlooked in all general censuses. NYSBBA II found it to have decreased its number of occupied blocks by 78% since 1980–85, especially upstate (McGowan and Corwin 2008). It is far more common, if largely overlooked, all around New York City, as evidenced by the speed with which it has occupied nest boxes in secure locations like Jamaica Bay. The nearest current breeders may be in structures visible from the study area if not actually within it.

Typical Owls: Strigidae

Eastern Screech-Owl *Megascops asio*

NYC area, current
A widespread but declining resident that is easily overlooked and whose status has only rarely been systematically assayed by rigorous nocturnal sampling. Nothing is known about any aspects of its local migration, or even whether it occurs. Information on its historical and current status in the 5 New York City boroughs was offered in DeCandido (2005).

Bronx region, historical
Bicknell's *Riverdale* 1872–1901: common resident Riverdale, Van Cortlandt

Eaton's *Bronx + Manhattan* 1910–14: fairly common resident

Griscom's *Bronx Region* as of 1922: common resident throughout

Kuerzi's *Bronx Region* as of 1926: common resident

NYC area, historical
Cruickshank's *NYC Region* as of 1941: fairly common resident breeding Central Park, along Riverside Drive; hints of immigrants during severe winters

Bull's *NYC Area* as of 1968: 18 on same-day Queens, Southern Nassau Christmas Bird Counts

Bull's *New York State* as of 1975: status unchanged

Study area, historical and current
Resident and common back to the late 1800s (Bicknell) in Riverdale and Van Cortlandt. As of 2016 it is a widespread but scarce resident in Van Cortlandt, where the first systematic nocturnal playback/calling censusing was done once a month from Apr–Oct, 2008–2010 (Nagy 2012). In the Northwest Forest, Putnam Trail, Ridge, and Northeast Forest, he estimated a breeding population of 7–12 pairs and a density of about 2.5 pairs/km^2. We are certain there are more pairs in Woodlawn, the Shandler Area, Van Cortlandt Swamp, on/near the golf courses, and in other sections he did not sample.

Screech-Owl densities were estimated by various methods over all seasons at 1.2 pairs/km^2 in suburban CT and 0.1–0.6 in rural CT and rural midwestern United States (Gehlbach 1995). The total Van Cortlandt wooded area is about 800 ac (324 ha), plus another 400 ac (162 ha) in Woodlawn, totaling c. 1200 acres (≈ 5.0 km^2). Using Gehlbach's extreme densities of 0.1 and 1.2/km^2 yields a range 5–12 pairs of Van Cortlandt Screech-Owls. But extrapolating from densities in nearby Riverdale (8–12/night along only a single mile of Palisade Ave. in the early 1950s: Buckley et al.), 5–12 pairs for all of Van Cortlandt should be too low by a factor of 2, so we proposed a tentative Van Cortlandt + Woodlawn breeding population of 10–24 pairs—gratifyingly close to the 7–12 pairs recently estimated for a large part of Van Cortlandt by Nagy.

Nagy and Rockwell's (2013) long-term study of breeding Screech-Owls in various New York City parks and southern Westchester County found that extirpation probability was low in parks larger than 1 km² but conversely, that those larger than 3 km² could also be less suitable when they supported resident Great Horned or Barred Owls (known Screech-Owl predators). In Van Cortlandt (4.7 km²) Great Horneds are resident, but Screeches nonetheless seem to be holding their own. Nagy and Rockwell also noted that the only 3 New York City parks where they located Screech-Owl populations (Inwood, Riverdale, and Van Cortlandt) were within 4 km of each other, and they suggested that the 3 areas functioned as a metapopulation, with the somewhat-less-urban habitat matrix in Riverdale being permeable to Screech-Owls. We are sure these owls are also resident in Woodlawn, Bronx Park, and probably the Bronx Zoo, and we suggest that these breeders also belong to that same metapopulation. Treed Mosholu Parkway and the Bronx River Parkway would be effective greenbelts facilitating subpopulation connectivity.

There is no study area information on postbreeding dispersal and numbers or even whether local breeders are augmented by winter arrivals. Both NYSBBA I in 1980–85 and II in 2000–2005 recorded breeders in Block 5852B, which includes the study area (McGowan and Corwin 2008), although following intensive residential development there, nightly maxima in one small section of Riverdale have dropped from the low teens in the early 1950s to the low single digits in the early 2000s (Buckley et al.).

There is no evidence that any effort has been made in Van Cortlandt or Woodlawn to emplace properly crafted and predator-proofed nest boxes. Nonetheless, we are convinced that occupancy would be rapid were this to occur, and we urge its implementation. The only natural nest site located recently had adults bringing food to young on 28 Apr 2006 (Künstler, Frielich) in the Northwest Forest.

Both NYSBBA I in 1980–85 and II in 2000–2005 recorded potential breeders in every New York City county except Brooklyn, with 7–9 blocks on Staten Island alone (McGowan and Corwin 2008). DOIBBS data showed no statewide trends from 1966–2005 (Sauer et al. 2005, although DeCandido (2005) believed the New York City population had declined >1950.

Specimens
Bicknell collected 1 in Riverdale on 24 Jan 1900 (NYSM).

Great Horned Owl *Bubo virginianus*

NYC area, current
An overlooked but recently increasing resident now breeding in all 5 New York City boroughs and throughout Long Island. In much the manner of Red-tailed Hawks, Horned Owls have become urban denizens and are thriving on Rock Doves, Gray Squirrels, Eastern Cottontails, Striped Skunks, Virginia Opossums, Brown Rats, and other urban mammals.

Bronx region, historical
Bicknell's *Riverdale* 1872–1901: single 31 Dec 1891 record

Eaton's *Bronx + Manhattan* 1910–14: rare resident, does not breed

Griscom's *Bronx Region* as of 1922: cannot tolerate civilization; several shot in winter Bronx Zoo

Kuerzi's *Bronx Region* as of 1926: formerly bred, now only winter visitor/resident (20 Nov–23 Apr)

NYC area, historical
Cruickshank's *NYC Region* as of 1941: very rare resident in northern Westchester Co.; 1–2 pairs on Palisades only breeders in metropolitan area, where otherwise erratic winter visitor

Bull's *NYC Area* as of 1968: very recently bred Bronx, Pelham Bay Parks, confirming distributional sea change

Bull's *New York State* as of 1975: status unchanged

Study area, historical and current
Bicknell only ever saw one in Riverdale; Griscom, then Kuerzi, then Cruickshank none in the Northwest Bronx, and we have been unable to locate a single study area occurrence before 26 Dec 1952 (Buckley), followed by a few more in winter into the early 1960s. Not long thereafter (about 1970) they became resident throughout Van Cortlandt, although nocturnal calling counts have never been attempted (a maximum of only 2, seen). In Woodlawn, which has always been closed at night, the 3 local breeders there on 29 Apr 2012 (McGee) were seen, not heard. Since the 1970s 1–2 have been found in Van Cortlandt on most Bronx-Westchester Christmas Bird Counts.

A total of 5–6 pairs may breed throughout Van Cortlandt and Woodlawn, but unless targeted efforts are made to locate their nests in mid–late winter, they are usually overlooked. Horned Owls are notoriously early nesters, Cruickshank (1942) giving New York City area egg dates from 22 Jan–20 Apr, and even though a pair incubating in Van Cortlandt on 27 Dec 1998 was quite unexpected on a Christmas Bird Count (Lyons 2000), others in Van Cortlandt in 2 different locations and years on 13 Jan were not that far behind (Künstler 1994). They are also assumed to be breeding in areas with big trees in Riverdale. Both NYSBBA I in 1980–85 and II in 2000–2005 recorded breeders in Block 5852B, which includes the study area (McGowan and Corwin 2008).

Snowy Owl *Bubo scandiacus*

NYC area, current
A normally scarce winter visitor to coastal areas on Long Island and Long Island Sound, unrecorded most years. Irruptions related to arctic lemming population cycles sometimes bring 100 or more to Long Island as in the winter of 2013–14, but even then they are only rarely found away from immediate coastal areas. Unrecorded in Prospect Park but seen in Central Park in mid-Dec 1890 (Foster). Details of singles reported in Central on 6 Mar 1992 and during the winter of 2013–14 have not surfaced. Nonetheless, singles have been seen several times atop Riverside Church and elsewhere in Manhattan and on the East River during recent irruptions.

Bronx region, historical
Bicknell's *Riverdale* 1872–1901: unrecorded

Eaton's *Bronx + Manhattan* 1910–14: occasional winter visitor

Griscom's *Bronx Region* as of 1922: unrecorded

Kuerzi's *Bronx Region* as of 1926: rare, irregular winter visitor (4 Nov–5 Apr) [locations not given]

NYC area, historical
Cruickshank's *NYC Region* as of 1941: on Long Island fairly common in winter during heavy irruption years, otherwise unrecorded; most Bronx records from Long Island Sound saltmarshes [details and locations not given]

Bull's *NYC Area* as of 1968: 12 Oct–4 May; during largest irruptions hundreds on Long Island, scores killed

Bull's *New York State* as of 1975: status unchanged

Study area, historical and current
There are 3 study area records: on the Van Cortlandt golf course on 29 Nov 1919 (Kieran), on the Parade Ground on 19 Apr 1946 (Gershon), and at Hillview on 19 Dec 1954 (Sedwitz, Phelan, Buckley)—seen at nearby Grassy Sprain Reservoir 1 week later on the Bronx-Westchester Christmas Bird Count. In the Bronx this species has been seen almost exclusively around Long Island Sound, especially at Pelham Bay Park and in the Baxter Creek/Clason Point area, typically during large irruption years. In 2013 one was at Soundview on 18–22 Dec (fide Walter), an irruption year when singles were also recorded in southern Westchester County in Larchmont on 23 Nov (MacKinnon), Irvington on 4 Dec, Tappan Zee on 12–18 Dec, and Piermont Pier on 22–23 Feb (all unknown observers, *Kingbird*). Two of the most recent in the Bronx were also at Clason Point, on 24 Jan 2014 (Aracil) and 27 Dec 2015 (Keogh et al.), with a third at Orchard Beach on 3 Feb 2018 (Cole et al.).

Barred Owl   *Strix varia*

NYC area, current
An increasingly scarce winter visitor/resident given to minor irruptions when it often appears in even the most urban of New York City parks. If it does breed on Long Island only 1–2 pairs are involved, and it has been decreasing as a breeder in Westchester County since the 1960s. Up to 8 Staten Island breeding pairs disappeared during the 1940s.

Bronx region, historical
Bicknell's *Riverdale* 1872–1901: resident breeder

Eaton's *Bronx + Manhattan* 1910–14: uncommon resident; does not breed

Griscom's *Bronx Region* as of 1922: perhaps 1–2 pairs still resident; usually found only in winter

Griscom's *Riverdale* as of 1926: long-extirpated breeder

Kuerzi's *Bronx Region* as of 1926: still fairly common resident [locations not given]

NYC area, historical
Cruickshank's *NYC Region* as of 1941: highly local resident all but absent from Long Island; decreased greatly since mid-1920s, now only few pairs in southern Westchester County; until 1926 bred Van Cortlandt; frequent winter visitor in New York City parks, maximum 5 in Bronx Park

Bull's *NYC Area* as of 1968: status unchanged

Bull's *New York State* as of 1975: status unchanged

Study area, historical and current
A historical breeder that is now only a very infrequently encountered winter visitor or occasional resident despite its being nearly annual in winter in Bronx and Pelham Bay Parks.

Even though it was a resident breeder in Van Cortlandt through about 1926, we know of only 8 fall/winter records there despite its possibly occurring regularly: 26 Dec 1914 (La Dow); 31 Oct 1925 (Cruickshank); 26 Jan 1930 (Sedwitz); 25 Oct 1936 (Norse); 9 Dec 1951 (Kane); 4 Nov 1959 (Rafferty); 11 Mar 1996 (Zabouli fide Lyons); 23–24 Dec 2012 (Garcia et al.). We expect it also to be regular in Woodlawn, and possibly the same individual overwintered in northwestern Riverdale where seen in Jan 2014, 2015, and 2016 (Smart et al.). Despite its former status as a resident and breeder, no systematic nocturnal searching for this often noisy nightbird has been done in Van Cortlandt, Woodlawn, or Riverdale, although C. Nagy (pers. comm.) did not hear any in Van Cortlandt during his 2008–10 extensive nocturnal Screech-Owl surveys.

In 1944 a pair was assumed to have bred at New Rochelle (Bull), and through the late 1950s a small breeding population persisted as close as Grassy Sprain Reservoir (Norse, Buckley) but are probably gone. Both NYSBBA I in 1980–85 and II in 2000–2005 found none within New York City, and only a pair or 2 on Long Island where it has always been excessively rare. In Westchester County 23 occupied blocks fell to 14, but some of this loss may have been sampling error. The nearest current breeders may be on the Palisades.

Long-eared Owl   *Asio otus*

NYC area, current
An annual but rare and local winter visitor in conifer roosts at favored locations, less regular on Long Island than on the mainland. Also a very scarce and intermittent breeder on Long Island, and an occasionally detected migrant, especially in fall in barrier beach conifers and in New York City parks.

Bronx region, historical
Bicknell's *Riverdale* 1872–1901: unrecorded

Eaton's *Bronx + Manhattan* 1910–14: rare resident

Griscom's *Bronx Region* as of 1922: rare winter visitor, single recent record

Kuerzi's *Bronx Region* as of 1926: uncommon but regular winter resident (11 Oct–24 Apr) [locations not given]

NYC area, historical
Cruickshank's *NYC Region* as of 1941: irregular/uncommon winter visitor (maximum 16), extremely rare breeder throughout; most nonbreeding records from conifers at cemeteries, New York City parks, nurseries, reservoirs

Bull's *NYC Area* as of 1968: winter maximum 37 Pelham Bay Park 2 Jan 1961; bred Staten Island 1947 (first within New York City)

Bull's *New York State* as of 1975: status unchanged

Study area, historical and current
An infrequently reported winter visitor/resident, owing to the absence of conifer groves in Van Cortlandt.

There are only 11 study area records, 4 of them from Van Cortlandt: 19–23 Dec 1925 (Cruickshank, Kuerzi); 30 Oct 1954 (Buckley); during the mid-1980s when for several years there was a winter roost of 9 in Van Cortlandt Swamp (Naf) whose actual location was sensibly a well-kept secret; Vault Hill 28 Dec 1997 (Garcia et al.). Recorded thrice at Hillview: 26 Jan 1936 and 8 Dec 1946 (Norse, Cantor), and 21 Dec 1947 (Norse, Tiffany), and 4 times at Woodlawn: 2 Jan 1893 (Foster: AMNH), 12 Apr

1959 (Horowitz), 6 Nov 1960 (Horowitz et al.), and 4 on 8 Feb 1977 (Teator). Increased Woodlawn coverage would show Long-eared Owl to be a regular if not annual winter visitor/resident.

NYSBBA II in 2000–2005 found it to have decreased its number of occupied blocks by 41% since 1980–85, and it continues to be almost undetected as a New York City area breeder (McGowan and Corwin 2008). We suspect it is overlooked by those unfamiliar with its breeding vocalizations, so the closest breeding site is unknown. For several decades, the Scots Pine grove on the Pelham Bay Park golf course hosted the largest group of New York City area overwintering Long-eareds. The maxima were 36 on 18 Jan 1968 (Shore), 37 on 2 Jan 1961 (unknown observer, Bull 1964), and an amazing 54 in just 2 trees during Jan 1965 or 1966 (Kane), although the exact date has been misplaced.

Specimens
Foster collected 1 in Woodlawn on 2 Jan 1893 (AMNH).

Short-eared Owl *Asio flammeus*

NYC area, current
A declining winter visitor/resident on saltmarshes, large open areas, and closed garbage dumps that is occasionally recorded in spring or fall migration. Until the late 1950s it was regular and numerous in the coastal Southeast Bronx where suitable habit is now nearly gone. They are, however, seen regularly in fall migration along the Hudson River at Croton Point and at Piermont Pier. It is a New York State Endangered Species. Unrecorded in Central or Prospect Park.

Bronx region, historical
Bicknell's *Riverdale* 1872–1901: unrecorded

Eaton's *Bronx + Manhattan* 1910–14: fairly common migrant

Griscom's *Bronx Region* as of 1922: rare in winter, only 2 recent Nov records

Kuerzi's *Bronx Region* as of 1926: fairly common migrant, winter resident (1 Aug–26 Apr) [locations not given]

NYC area, historical
Cruickshank's *NYC Region* as of 1941: fairly common fall migrant, winter resident, local breeder on Long Island saltmarshes, dumps; very rare breeder on Long Island Sound Bronx marshes [locations not given], where otherwise only winter visitor/resident

Bull's *NYC Area* as of 1968: greatly decreased as Long Island breeder; winter maxima 16 in/near Pelham Bay Park 1934–35, 40 Jamaica Bay Wildlife Refuge 1959

Bull's *New York State* as of 1975: status unchanged

Study area, historical and current
Only an infrequent late fall/early winter visitor in the 1930s–40s, probably from the large regularly overwintering population in the Baychester–Pelham Bay and Clason Point marshes of the East Bronx.

There are 5 records: Van Cortlandt on 19 Oct 1934 (Cantor); Hillview on 21 Dec 1935 (Norse, Cantor); at Jerome in late Oct 1944 and late Oct 1945 (Ryan); 2 in Van Cortlandt on 20 Nov 1948 (Norse, Cantor).

NYSBBA II in 2000–2005 found this species to have decreased its number of occupied blocks by 33% since 1980–85, but the decline was especially acute in saltmarshes on the south shore of Long Island: down from 6 sites to only one (McGowan and Corwin 2008).

Northern Saw-whet Owl *Aegolius acadicus*

NYC area, current
A late fall migrant and winter visitor/resident with a widespread but exceedingly thin distribution, mostly in conifers and especially on barrier beaches.

Bronx region, historical
Bicknell's *Riverdale* 1872–1901: unrecorded

Eaton's *Bronx + Manhattan* 1910–14: uncommon winter visitor

Griscom's *Bronx Region* as of 1922: recent Oct record; undoubtedly overlooked

Kuerzi's *Bronx Region* as of 1926: regular but uncommon winter visitor/resident (11 Oct–early Apr)

NYC area, historical
Cruickshank's *NYC Region* as of 1941: irregular winter visitor/resident (10 Oct–28 Apr), plus 1–2 pairs breeding; largely overlooked, subject to periodic irruptions when they may occur almost anywhere

Bull's *NYC Area* as of 1968: 22 Sep–13 May, plus occasional summer records including Long Island breeding in 1966, 1968; Long Island winter maximum 8 on Jones Beach strip 1959–60, fall maximum 26 seen 29 Oct 1967 Cedar Beach to Oak Beach

Bull's *New York State* as of 1975: 45 banded Fire Island Lighthouse Oct–Dec 1974

Study area, historical and current
An exceedingly infrequently reported winter visitor/resident owing to the absence of conifer groves in Van Cortlandt, where there are only 3 records: Van Cortlandt Swamp on 21 Dec 1963 (Horowitz, Tudor), in the Northwest Forest on 4 Feb 1994 (Künstler), and on the Ridge in the first week of Jan 2013 (Murray).

Found thrice at Hillview in conifer groves that are now gone: 3 on 14 Jan 1936 (Weber et al.), and singles on 1 Dec 1937 (Carleton, Komorowski) and 1 Nov 1955 (Phelan, Buckley). It is a potential breeder in widespread Woodlawn conifers, especially if predator-proofed, properly crafted nest boxes were emplaced, even though there are no records there. One was calling/singing in Riverdale along the Hudson River nightly from 16–28 Mar 1956 (Rosin)—a local breeder, or a migrant getting an early start? One in Bronx Park on 1 May 1965 (Peszel) also raised questions about local breeding. The nearest regular breeders are in northern NJ (possibly on the Palisades) but in New York normally no closer than Sullivan and Ulster Counties (McGowan and Corwin 2008), although adventitious (only ?) breeding has occurred at 4 different Long Island locations over the years. Saw-whets are annual in winter at Pelham Bay Park, where the maximum was 7 on 23 Dec 1961 (Post), the highest Bronx single-day count.

Comments
Occasionally irruptive, northern **Boreal Owl**, *A. funereus*, is nearly unknown around New York City, but one made it to Central Park from 19 Dec 2004–13 Jan 2005 (Demes, Post et al.). Another was meticulously described by veteran observers J. Mayer and G. Rose during an hour's scrutiny from 15 ft (4.6 m) at Croton Point on 12 Feb 1951, but it was deprecated after the observers had wisely elected to suppress the details to preclude disturbance.

Nightjars: Caprimulgidae

Common Nighthawk   *Chordeiles minor*

NYC area, current
This ground-nester has been declining drastically since the 1940s in undisturbed habitat on Long Island but has paradoxically been increasing steadily in the most urban of New York City sections in all 5 counties, where they nest on rooftops. Nocturnal spring but diurnal fall Neotropical migrants, they are usually seen as singles roosting in trees by day in spring, but in fall hundreds and occasionally thousands move southward in large, loose flocks appearing toward dusk. It is a New York State Species of Special Concern.

Bronx region, historical
Bicknell's *Riverdale* 1872–1901: uncommon spring, common to abundant fall migrant; does not breed

Hix's *New York City* as of 1904: common breeder often seen flying over housetops at dusk

Eaton's *Bronx + Manhattan* 1910–14: fairly common breeder

Griscom's *Bronx Region* as of 1922: breeds New York City rooftops; nested 1916 near Van Cortlandt; rare spring, common fall migrant

Kuerzi's *Bronx Region* as of 1926: common breeder, abundant fall migrant (1 May–16 Oct)

NYC area, historical
Cruickshank's *NYC Region* as of 1941: fairly common migrant in spring/common to abundant fall (maximum 1000+), from 17 Apr–11 Nov; no longer breeds New York City or Westchester County.

Bull's *NYC Area* as of 1968: 12 Apr spring date; now rarely nests in New York City [locations not given] but still breeds occasionally on southern Westchester County rooftops

Bull's *New York State* as of 1975: fall maximum 1500

Study area, historical and current
A regular but rarely seen spring and a common fall migrant, and a scarce but annual breeder. Most study area spring migrants roosting in trees by day are never found, the antithesis of Central and Prospect Parks, where they are quickly located by the scores of observers combing the parks daily in May.

In spring, first migrants appear in early May, peak and depart in mid-May, with extreme dates of 5 May 2000 (Young) and 4 (the maximum) on 26 May 2000 (Young)—maybe including local breeders.

Nighthawks apparently did not breed in Riverdale in the late 1800s (Bicknell) but were reported to have nested in 1916 near Van Cortlandt (Griscom 1923) and characterized as one of the few species to have become adapted to metropolitan and suburban life in comparatively recent years (Kuerzi 1927). Yet even though Cruickshank (1942) noted that they no longer nested within New York City, they have long been breeding atop apartment houses immediately south and east of Van Cortlandt: around Jerome (since the 1960s: Buckley, Sedwitz), adjacent to Woodlawn (in the 1950s–60s: Kieran, Horowitz; and on 4 Jun 2002: Jaslowitz), near Hillview (in 1984, a nest was on a rooftop in adjacent Wakefield: Brinker), and in Riverdale from the 1960s–2002 (Sedwitz, Buckley, Jaslowitz). Those singing in

Van Cortlandt on 21 Jun 1955 (Steineck, Yeganian) and Woodlawn on 8 Jun 1961 (Horowitz) were also local breeders. We estimate that as of 2016, 6–10 pairs breed on rooftops adjacent to (and maybe within) the study area.

NYSBBA I in 1980–85 recorded 12 blocks with breeders within all New York City boroughs save Queens, plus 5 in Westchester County, 1 in Nassau County, and 20 in Suffolk Co. On NYSBBA II in 2000–2005 these had fallen to only 7 in 4 New York City boroughs (not Manhattan), plus none in Westchester County, 1 in Nassau County, and only 7 in Suffolk County (McGowan and Corwin 2008). However, much of this loss may only be sampling error.

In fall, first migrants arrive in early Aug, peak in late Aug–early Sep, and depart by late Sep, with extreme early dates of 17 Aug in 2004 and 2011 (Young) and late dates of 23 on 6 Oct 2015 (Aracil), 25 on 11 Oct 1935 (Cantor, Weber, L. Lehrman), and 18 Oct 2014 (Young).

A blowback drift-migrant hawking insects in adjacent Riverdale on 19 Nov 1971 (Lehman) is among the latest for the New York City area and is assumed to have been a Common.

In the past, southbound nighthawks routinely numbered in the 100s: e.g., 2–16 Sep 1953 (hundreds: Kane, Buckley), 200 on 4 Sep 1959 (Rafferty). Recent counts have been more modest and less frequent: e.g., 64 on 2 Sep 2013 but 254 on 5 Sep 2003 (Young). Occasionally, weather and birds converged and truly large numbers transited New York City, like the 5000+ counted over Parkchester in the east-central Bronx on 9 Sep 1945 (Russak). However, with declines in nighthawks throughout the Northeast since the 1960s, it is unlikely such numbers will ever again be witnessed within New York City.

Specimens
Bicknell collected 1 in Riverdale on 20 Oct 1876 (NYSM).

Eastern Whip-poor-will *Antrostomus vociferus*

NYC area, current
An annual but infrequently recorded spring/fall Neotropical migrant that has been declining as a breeder for 100+ years and is now largely restricted to eastern Suffolk County. It is a New York State Species of Special Concern.

Bronx region, historical
Bicknell's *Riverdale* 1872–1901: bred Riverdale 1870s, regular spring migrant until 1885, only 3 fall records, as late as 18 Oct 1876

Eaton's *Bronx + Manhattan* 1910–14: local breeder

Griscom's *Bronx Region* as of 1922: rare spring migrant

Kuerzi's *Bronx Region* as of 1926: formerly common breeder, now rare [locations not given], scarce migrant (16 Apr–18 Oct)

NYC area, historical
Cruickshank's *NYC Region* as of 1941: resident, regular but uncommon to rare spring/fall migrant, breeding no closer than central Westchester County; 15 Apr–18 Oct

Bull's *NYC Area* as of 1968: 2 Apr–6 Nov

Bull's *New York State* as of 1975: 3 pairs bred Staten Island 1969

Study area, historical and current
Almost certainly an annual spring/fall migrant. It was a regular breeder in the Riverdale area—and by extension Van Cortlandt—into the late 1800s, where it may have continued breeding after disappearing from Riverdale, but no contemporaneous data are known.

Within the study area it is otherwise only a rarely detected spring migrant, with 13 spring and

2 early fall records, the last in 1959. All but 1 were silent: 13 May 1928 (Kieran), 14 May 1930 (Kieran), 4 May 1935 (Weber), 26 Apr 1936 (Schmidt), 17 May 1947 (Kieran), 12 May 1951 (Kieran), 10 May 1952 (Kane), 4 May 1953 (Kane), 14 May 1953 (Penberthy), 22 May 1953 (Buckley), 29 Apr 1956 (Scully), 3 May 1956 (Phelan et al.), singing in Woodlawn on 28 Apr 1959 (Horowitz), and in Riverdale on 17 May 1956 (Kieran). In fall, seen only twice: 21 Aug 1930 (Kieran) and 1 Aug 1953 (Phelan), and in nearby Riverdale on 28 Aug 1925 (Griscom). Bicknell collected 1 in Riverdale on the late date of 18 Oct 1876 (fide Cruickshank), and one was found moribund in the Bronx Zoo on 25 Aug 1941 (AMNH).

Eaton (1910, 1914) listed it as a breeder in his Bronx + Manhattan county with no additional details. Nearest breeders were no closer than central Westchester County in the early 1950s but now may have retreated to the Rockland–Orange County line. Present in 6 blocks on Staten Island during NYSBBA I in 1980–85 but in only a single possible block during NYSBBA II in 2000–2005, when the 3–4 Westchester County blocks had dropped to none. Statewide, sites have diminished by >50%, but incomplete sampling may have played a major role (McGowan and Corwin 2008). Nonetheless, it is depressing to realize that as recently as 1976, 35 pairs were nesting on Sandy Hook—within sight of lower Manhattan.

This species has vanished as a breeder from many former urban/suburban/exurban sites in the Northeast, usually attributed to free-ranging or feral cats and dogs, plus human-garbage commensals like Raccoons, Striped Skunks, Virginia Opossums, possibly Red and Gray Foxes, and lately Eastern Coyotes—all of which prey heavily on ground-nesters. Also implicated is greatly increased use of biocides and industrial pollutants that may inordinately affect saturniid moths, a principal Whip-poor-will food (Robbins et al. 1986).

Specimens
Bicknell collected 1 in Riverdale on 10 May 1877 (NYSM).

Comments
Southeastern **Chuck-will's-widow**, *A. carolinensis*, has been expanding its range in the New York City area since the 1970s, although breeding is restricted to Staten Island and south coastal Long Island. It has been found several times in spring in Central, Prospect, and other New York City Parks but never in the Bronx. One was at Croton Point on 24 Apr 1999 (unknown observer, Lower Hudson Valley Bird Line).

Swifts: Apodidae

Chimney Swift *Chaetura pelagica*

NYC area, current
A common spring/fall Neotropical migrant and ubiquitous area breeder that can be locally abundant just before fall migration at staging sites in favored chimneys. Breeds widely in all 5 New York City boroughs.

Bronx region, historical
Bicknell's *Riverdale* 1872–1901: abundant breeder

Hix's *New York City* as of 1904: abundant breeder

Eaton's *Bronx + Manhattan* 1910–14: abundant breeder

Griscom's *Bronx Region* as of 1922: common breeder and migrant

Kuerzi's *Bronx Region* as of 1926: common breeder, spring/fall migrant

NYC area, historical

Cruickshank's *NYC Region* as of 1941: common migrant in spring (maximum 100+)/fall (maximum several thousand), from 3 Apr–28 Oct; breeds throughout

Bull's *NYC Area* as of 1968: late fall date 4 Nov

Bull's *New York State* as of 1975: status unchanged

Study area, historical and current

A common spring/fall Neotropical migrant and area breeder, occasionally abundant at staging sites in favored chimneys.

In spring, first migrants arrive in late Apr, peak in early May, and depart by late May, with extreme dates of 14 Apr 2013 (Young) and 9 on 22 May 1954 (Scully), and normal maxima of 40 in Van Cortlandt on 8 May 1955 (Kane) and 60 at Jerome on 5 May 2016 (Karlson) and at Hillview on 2 May 2016 (Sargent). However, on 13 May 1995 Lyons and Garcia watched hundreds pile into a Jerome Reservoir building at dusk—a most unexpected study area spring aggregation. Until Jun 1967, the spring maximum in Central Park was 15 on 10 May 1935 (Helmuth).

Although not known to use natural cavities or trees any longer, they do breed within buildings in all 4 extant subareas and in surrounding areas, ranging widely throughout when feeding. The estimated summer breeding population of 70–90 pairs embraces at least 30–40 pairs at Jerome, probably 30–40 pairs around Hillview, and 10 pairs in the chimney of the former boathouse/now golf course building on Van Cortlandt Lake (Sedwitz 1982).

Doubtless breeding throughout the most urban portions of all 5 New York City boroughs, even though not found (probably overlooked) in all blocks on NYSBBA I in 1980–85 or II in 2000–2005 (McGowan and Corwin 2008). Similar losses in Westchester, Nassau, and Suffolk Counties are probably also artifactual, despite DOIBBS data showing a 1.6% annual statewide decrease from 1966–2005 (Sauer et al. 2007).

In fall, first migrants appear in early Aug, peak in late Aug, and depart by mid-Sep, with normal extremes of 12 Aug 2012 (Aracil) and 2 on 25 Sep 2010 (Baksh), and a maximum of 600 on 4 Sep 2012 (Young). Like Yellow-billed Cuckoo, this species often shows an Oct–Nov secondary peak in the New York City area comprising blowback drift-migrants; 4 on 18 Oct 1936 (Weber et al.), 6 on 22 Oct 2011 (Baksh) and 1 on 24 Oct 2009 (Young) fall into this category. The sole major premigratory Aug chimney roost site at the study area was at St Ann's church and school near the intersection of Gun Hill Rd. and Bainbridge Ave.—230 ft (70 m) southwest of the southwestern corner of Woodlawn—where on a single occasion in early Oct 2007, Künstler counted 300 going to roost. At similar sites elsewhere in the New York City area thousands can also sometimes be seen in Aug. On 30 Aug 1947, 1500 diurnal migrants passed Fort Tryon Park in groups of 60 or so (Gilbert). Until Jun 1967, the fall maximum in Prospect Park was 35 on 29 Aug 1944 (Soll).

Specimens

Bicknell collected 1 in Riverdale in 30 Apr 1879 (NYSM).

Hummingbirds: Trochilidae

Ruby-throated Hummingbird   *Archilochus colubris*

NYC area, current
A regular but uncommon spring/fall trans-Gulf Neotropical migrant that is increasingly being found late in the fall and even into winter.

Bronx region, historical
Bicknell's *Riverdale* 1872–1901: common spring/fall migrant, breeder

Hix's *New York City* as of 1904: breeds Bronx, Van Cortlandt Parks

Eaton's *Bronx + Manhattan* 1910–14: common breeder

Griscom's *Bronx Region* as of 1922: bred formerly throughout, now no closer than Hastings; uncommon May, common Aug migrant

Kuerzi's *Bronx Region* as of 1926: common spring/fall migrant, fairly common breeder [locations not given]; 30 Apr–7 Oct

NYC area, historical
Cruickshank's *NYC Region* as of 1941: widespread breeder scarce or gone within New York City but has bred Bronx Park; fairly common migrant in spring/common fall (maximum 18), from 25 Apr–23 Oct

Bull's *NYC Area* as of 1968: 14 Apr–13 Dec

Bull's *New York State* as of 1975: status unchanged

Study area, historical and current
A spring/fall migrant breeding in at least Van Cortlandt. May and Jul migrants cannot be safely separated from local breeders.

In spring, first migrants arrive in late Apr, peak in early May, and depart by late May, with extreme dates of 29 Apr 1997 (Young) and Jerome 26 May 1944 (Russak), and maxima of only 2 on numerous occasions. Until Jun 1967, the corresponding maximum in Prospect Park was 4 on 16 May 1944 (Soll).

In the late 1800s, it was a common breeder in Riverdale (Bicknell) and in Van Cortlandt (Dwight), and it still nests in Van Cortlandt and Woodlawn but probably only 2–3 pairs. The last nest was found in 1955 (Buckley), and neither NYSBBA I in 1980–85 nor II in 2000–2005 recorded breeding anywhere in the Bronx, but they are remarkably easy to overlook. In Central Park the first nesting pair was in late May 2014 (DeCandido and Allen 2015), and in Prospect Park a ♀ was sitting on a completed nest in the first week of May 2015, a time when any Van Cortlandt hummingbird would be dismissed as a migrant. But given that they are customarily double-brooded, recent records of a pair on 19 Jun 1982 (Sedwitz), ♀♀ 15 Jul 2008 (Künstler, Young), 10 and 15 Jul 2013 (Pehek et al.), 21 Jun 2014 (Hannay), and 18 Jun 2016 (Ward) were all likely local breeders. The simplest way to quantify the numbers of breeders would be emplacement and regular monitoring of feeders from Apr–Jul at multiple Van Cortlandt and Woodlawn sites. Elsewhere, the nearest regular breeders are in Bronx Park and Pelham Bay Park, and they should also be in the Bronx Zoo and Riverdale.

NYSBBA I in 1980–85 found breeders in only 3 New York City blocks, down to only 1 on Staten Island in NYSBBA II in 2000–2005 (McGowan and Corwin 2008). However, it is almost certain these are gross underestimates, as hummingbirds are among the easiest breeders to overlook. DOIBBS data showed a slight but consistent statewide increase from 1966–2005 (Sauer et al. 2005).

In fall, first migrants arrive in early Aug, peak in late Aug–early Sep, and depart by mid–late Sep, with extreme dates of 1 Aug 1953 (Phelan et al.) and 1 Oct 1961 (Tudor). A dozen per day are not unusual (Young), with a maximum of 34 on 18 Sep 1956 (Post). One of this species' favorite study area fall foods is Jewelweed, but in a burst of misdirected enthusiasm, New York City Parks Department employees are striving to obliterate it from Van Cortlandt as an undesirable exotic (nonnative). Predictably, regular observations of fall hummingbirds and Mourning and Connecticut Warblers have also declined.

Specimens
Dwight collected 3 in Van Cortlandt in May, Aug, Sep 1886–90 (AMNH) and Bicknell 3 in Riverdale in Jul, Aug, Sep 1876–79 (NYSM).

Comments
The absence of late-flowering study area ornamentals has precluded any local blowback drift-migrant Nov records of late fall western vagrants like **Rufous Hummingbird**, *Selasphorus rufus*. This species has been found as close as Wave Hill in Riverdale from 16–24 Nov 1993 (Loeb et al.); at Lenoir Preserve in northwestern Yonkers from 17 Nov 2001–6 Jan 2002, 28 Oct–29 Nov 2002, 23 Nov 2006–6 Jan 2007, 6 Nov–18 Dec 2011 (Bochnik et al.); and in Bronx Park on 19 Sep 2007 (Bustamante) and 10–16 Nov 2007 (Rakowski, Fraza et al.). Much less frequent **Calliope Hummingbird**, *S. calliope*, has also been detected: 2 ♂♂ were in Ft. Tryon Park mid-Nov–27 Dec 2001 (m. ob.) and another in Larchmont 4–7 Nov 2004 (unknown observer, *Kingbird*). But not all late hummingbirds are western vagrants: a Ruby-throat was at Lenoir Preserve from 7–23 Dec 2001 (H. Martin et al.) alongside a Rufous, and other very late Ruby-throats have been at Inwood and Ft. Tryon parks. Calliope is unrecorded in Central or Prospect Park, as is Rufous in Prospect.

The first New York State Rufous was not detected until Sep 1980 in the Adirondacks, but the species is now annual. Have they been overlooked or is this a new phenomenon? Rufous and many of the other often lingering blowback drift-migrant hummingbird species so far recorded in the East have probably been occurring for some time but overlooked because (1) few observers paid attention to hummingbirds in eastern North America, where only one species bred; (2) until Williamson's (2001) and Howell's (2002) fine hummingbird guides, information permitting confident identification of the HYs that constitute the majority of vagrant hummers was all but nonexistent; (3) the proliferation of hummingbird feeders maintained after normal fall Ruby-throat departure dates is a very recent phenomenon; (4) observers are now seeking out late-flowering plants supporting hummingbirds in Nov and Dec; and (5) information about them is now spread rapidly and frequently via the Internet.

Kingfishers: Alcedinidae

Belted Kingfisher *Megaceryle alcyon*

NYC area, current
A spring/fall migrant, highly local New York City breeder almost extirpated except on Staten Island, and regular winter visitor and resident in small numbers.

Bronx region, historical
Bicknell's *Riverdale* 1872–1901: common breeder sometimes overwintering Spuyten Duyvil

Hix's *New York City* as of 1904: common breeder Bronx Park

Eaton's *Bronx + Manhattan* 1910–14: common breeder

Griscom's *Bronx Region* as of 1922: rare summer, winter resident, common migrant

Kuerzi's *Bronx Region* as of 1926: common spring/fall migrant, fairly common breeder, occasional in winter

NYC area, historical
Cruickshank's *NYC Region* as of 1941: widespread area breeder but decreasing in New York City; a common spring/fall (maximum 15) migrant, but scarce in winter

Bull's *NYC Area* as of 1968: maxima spring 6, fall 37, winter 6

Bull's *New York State* as of 1975: status unchanged

Study area, historical and current
A conspicuous spring/fall migrant throughout, and a persisting breeder and winter resident/visitor in Van Cortlandt Swamp and maybe Woodlawn.

In spring, first migrants arrive in mid-Mar, peak in Apr, and depart by mid-May, with probable extreme dates of 24 Feb 1952 (Kane) and 6 Jun 1995 (Young), and a maximum of 4 on 6 Apr 1952 (Kane).

Probably 1–2 pairs are breeding in Van Cortlandt Swamp, along Tibbett's Brook north of the Mosholu Parkway Extension, and perhaps in Woodlawn; also along the adjacent Bronx River, the Hudson River in Riverdale, and the Harlem River. The last study area nest was on 15 Apr 1960 in a stream bank on the golf course just south of the Mosholu Parkway Extension (Buckley), but pairs were present all summer and probably breeding in 1980 (Ephraim), 1981, 1984 (Sedwitz), and again in 2006 (Delacretaz 2007). Singles in Van Cortlandt on 30 Jul 1999 (Young) and 31 Jul 2013 (McGee) may also have been local breeders. The very few scattered suitable sites are rarely examined and could also support breeding Bank Swallows (q.v.). Both species are adventitious breeders that quickly take advantage of locations few observers would imagine they'd ever use; see Kiviat et al. (1985) for a fascinating discussion about such upstate Hudson River sites.

NYSBBA I in 1980–85 found breeders in about 20 blocks in all 5 New York City boroughs, the most in Staten Island, but by NYSBBA II in 2000–2005 this had fallen to 9, all on Staten Island (McGowan and Corwin 2008). It is unclear what fraction of the missing blocks is due to overlooked breeders or those misidentified as migrants. Habitat-limited DOIBBS data did show a consistent statewide decrease from 1966–2005 in New York State (Sauer et al. 2007).

In fall, first migrants arrive in early Aug, peak in late Aug, and depart by late Oct if they are not going to overwinter, with extreme dates of 8 Aug 2004 (Young) and 29 Nov 2012 (Finger), and a maximum of 4 on 14 Sep 1935 (Imhof).

Usually 1–3 individuals attempt to overwinter, often successfully, but if frozen out they move to Jerome or Hillview and return to the Swamp as soon as open water reappears.

Woodpeckers: Picidae

Red-headed Woodpecker *Melanerpes erythrocephalus*

NYC area, current
This midcontinent species was formerly a local breeder and migrant. Then about 1900, its numbers fell precipitously throughout the Northeast, sometimes attributed to nest holes lost to the exploding starling population and being struck roadside by newly arrived automobiles. Numbers remained low, and it was an exceedingly local breeder in scattered clusters until the 1920s–30 when they dropped again, this time to almost none breeding, with only a few migrating in fall, typically near the coast. It is a New York State Species of Special Concern.

Bronx region, historical
Bicknell's *Riverdale* 1872–1901: unmentioned

Eaton's *Bronx + Manhattan* 1910–14: local breeder

Griscom's *Bronx Region* as of 1922: rare migrant, none recently; bred Riverdale 1917, 1921

Kuerzi's *Bronx Region* as of 1926: uncommon migrant in spring (mid-Mar–late May)/fall (late Aug–late Nov), frequently overwintering, breeding locally [locations not given]

NYC area, historical
Cruickshank's *NYC Region* as of 1941: erratic throughout as scarce, ephemeral breeder, winter visitor/resident, erratic May/Sep migrant; nested Bronx 1930s [locations not given]

Bull's *NYC Area* as of 1968: no longer breeding east of Hudson River

Bull's *New York State* as of 1975: status unchanged

Levine's *New York State* as of 1996: maximum 9 on 1983 Queens Christmas Bird Count

Study area, historical and current
Formerly a local breeder, occasional spring/fall migrant, and winter resident, now only an infrequent spring/fall migrant.

One of the few remaining early 1900s New York City breeding sites was Van Cortlandt, where 3–4 pairs nested until the late 1930s (Norse, Kieran, Cruickshank), when they were usually present only from Jun–Sep, although seen as early as 6 May 1925 (Cruickshank), and 2 were seen on 18 Feb 1923 (Eisenmann). A pair or 2 also bred along Mosholu Parkway near Jerome in 1938 (Eisenmann), 3 pairs at Pelham Bay Park in 1934–35, and another pair in Scarsdale in 1934 (Malloy et al.). By the late 1930s all had disappeared, and it was 79 years until the next New York City breeding attempt, at Pelham Bay Park in Apr–May 2017 (Rothman).

Subsequent to their disappearance throughout the New York City area, they have been found in the study area only 7 times: 11 Apr 1943 (Levine); HY 25 Sep 1949 (Norse); HY 24 Sep 1951 (Kieran); 4 May 1962 (Steineck, Friedman); HY 5 Oct 1975 (Jaslowski); adult 14 Sep 2013 (Souirgi et al.); HY Woodlawn 22 Dec 2013 (Gotlib, McAlexander). Singles were in Riverdale on 22 May 1940 (Kieran) and 17–24 Apr 2016 (Trombone et al.) plus a nonbreeder or very late migrant in the Bronx Zoo on 19 Jun 2013 (Olson).

NYSBBA II in 2000–2005 found this species to have decreased its number of occupied Blocks by 76% since 1980–85 (McGowan and Corwin 2008), and it is an extraordinary and unpredictable breeder anywhere in the New York City area. The nearest regular breeders (few) are in Orange and Ulster Counties.

Red-bellied Woodpecker   *Melanerpes carolinus*

NYC area, current
Now the commonest and most widespread woodpecker after Downy, resident and breeding widely and regularly seen as spring/fall migrants, even on Long Island barrier beaches and in all 5 New York City boroughs.

Bronx region, historical
Bicknell's *Riverdale* 1872–1901: unrecorded

Eaton's *Bronx + Manhattan* 1910–14: rare in winter

Griscom's *Bronx Region* as of 1922: unrecorded

Kuerzi's *Bronx Region* as of 1926: unrecorded

NYC area, historical
Cruickshank's *NYC Region* as of 1941: vagrant to New York City region; single Prospect, Central, Van Cortlandt Park records

Bull's *NYC Area* as of 1968: >1954 rare to uncommon coastal visitor; does not breed; maximum 2

Bull's *New York State* as of 1975: 16 in spring 1969 Long Island invasion; bred Tarrytown 1975

Levine's *New York State* as of 1996: winter maximum 95 on 1993 Northern Nassau Christmas Bird Count

Study area, historical and current
A recent colonizer that is now a resident breeder, spring/fall migrant, and winter resident.

This expanding study area species was first recorded on 13 May 1928 (Herbert, Kassoy, Kuerzi) but not again until 23 Apr 1964 at the Sycamore Swamp (M. Gochfeld, Friedman). It was not seen regularly anywhere in the Bronx until the 1980s. It is also an irregular migrant in Apr–May (maximum of 10 on 21 May 2011: Bochnik) and Oct–Nov (maximum of 6 on 11 Nov 2012: Stuart).

Red-bellieds are most easily found in winter, when they are more often heard than seen. Predictably, the largest number was on a Christmas Bird Count: 43 across Van Cortlandt and Woodlawn on 22 Dec 2013 (Lyons et al.).

While it is uncertain exactly when they first bred in the study area, it was during the late 1980s. It is now a widespread and noisy breeder, with >25 pairs breeding throughout Van Cortlandt and another 4–5 in Woodlawn. On the minus side, their breeding success in the study area may have come at the expense of the Hairy Woodpecker nests they usurp (Young), counterbalanced by their easy eviction of poaching starlings.

NYSBBA and DOIBBS data for this species in New York State supported a moderate increase throughout its range, including New York State, beginning in the mid-1980s and ongoing (Sauer et al. 2005; McGowan and Corwin 2008).

Singles that overwintered at Pelham Bay Park from 21 Dec 1963–14 Feb 1964 (Heck et al.) and winter 1967–68 were the first resident in the Bronx for any period and the 1967 bird the first on a Bronx-Westchester Christmas Bird Count. Count numbers did not begin to accelerate until the mid-1980s, then more or less leveled off in the early 2000s (Fig. 42). In 2001 for the first time more Red-bellieds were seen than Downies.

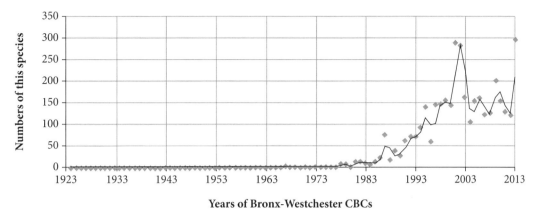

Figure 42

Red-bellied's colonization of the New York City area (there is no evidence that it had ever been extirpated) began without warning with 6 records on western Long Island in mid-May 1961, and then 7 that fall, including 2 in Westchester County. In spring 1962, the invasion expanded, with more than 30 reported throughout the New York City area. After that numbers leveled off through the 1960s as their new range edge consolidated. The second invasion wave occurred in Apr–May 1969, when 30+ were recorded on Long Island from Prospect Park to Westhampton, culminating in the long-awaited first breeding, at Old Field, Suffolk County, in May–Jun. From that point onward the population grew rapidly, in some areas now even outnumbering Downies.

Yellow-bellied Sapsucker *Sphyrapicus varius*

NYC area, current
A relatively common spring/fall migrant, and scarce but increasing winter resident.

Bronx region, historical
Bicknell's *Riverdale* 1872–1901: common spring/fall migrant, regular in winter

Eaton's *Bronx + Manhattan* 1910–1914: common migrant, occasional winter visitor

Griscom's *Bronx Region* as of 1922: uncommon spring, common fall migrant, rare in winter

Kuerzi's *Bronx Region* as of 1926: fairly common migrant in spring (1 Apr–20 May)/ sometimes abundant in fall (14 Sep–26 Nov); very rare in winter

NYC area, historical
Cruickshank's *NYC Region* as of 1941: uncommon but regular spring/common fall (maximum 27) migrant; 7 Sep–late May, plus 2 in Jul; rare, unpredictable in winter

Bull's *NYC Area* as of 1968: spring maximum 20

Bull's *New York State* as of 1975: 23 Aug–3 Jun; unique fall maximum 200–400 Riis Park to Fire Island 5 Oct 1974; in winter more common inland than coastally

Study area, historical and current

A relatively common spring/fall migrant and scarce but increasing winter resident.

In spring, first migrants arrive in early Apr, peak in late Apr, and depart by mid-May, with extreme dates of 8 Mar 2000 (Young) and 25 May 1954 (Scully), and a maximum of 3 on 20 Mar 2006 (Young). Until Jun 1967, the spring maximum in Central Park was 20 on 13 Apr 1927 (Griscom) and in Prospect Park 13 on 29 Apr 1965 (Raymond).

Breeding sapsuckers do not become regular until eastern Dutchess County and in the Catskills (McGowan and Corwin 2008) but recently have been as close as central Westchester County. DOIBBS data showed a slight but significant statewide decline from 1966–79, followed by a significant increase to 2005 (Sauer et al. 2005).

In fall, first migrants arrive in late Sep, peak in mid-Oct, and depart by early Nov, with extreme dates of 2 on 1 Sep 1951 (Buckley) and 10 Nov in 1946 (Norse, Cantor), and a maximum of 5 on 26 Sep 2004 (Young). Until Jun 1967, the fall maximum in Prospect Park was 8 on 3 Oct 1944 (Soll).

Up to 5–6 may attempt to overwinter in mild seasons, and numbers have been increasing since the early 1990s, peaking on 22 Dec 2013, when 16 were found between Van Cortlandt and Woodlawn (Lyons, Gotlib et al.). This trend is mirrored in Bronx-Westchester Christmas Bird Count totals for 90 years (Fig. 43). It is rarely appreciated that in the late 1800s Bicknell considered it regular in winter in Riverdale.

Specimens

Dwight collected 1 in Van Cortlandt on 14 Apr 1888 and Foster 2 in Woodlawn in Sep, Oct 1887–89 (all AMNH). Bicknell collected 2 in Riverdale in Oct 1875–76 (NYSM).

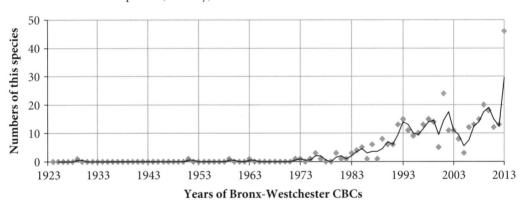

Figure 43

Downy Woodpecker   *Picoides pubescens*

NYC area, current
A widespread resident, spring/fall migrant, and winter visitor/resident throughout, even in the most urban parts of all 5 New York City boroughs.

Bronx region, historical
Bicknell's *Riverdale* 1872–1901: common migrant, resident

Eaton's *Bronx + Manhattan* 1910–14: common breeder

Griscom's *Bronx Region* as of 1922: common resident

Kuerzi's *Bronx Region* as of 1926: common resident

NYC area, historical
Cruickshank's *NYC Region* as of 1941: common resident, spring/fall migrant, still breeding Central Park; winter maximum 15

Bull's *NYC Area* as of 1968: spring maximum 28

Bull's *New York State* as of 1975: status unchanged

Levine's *New York State* as of 1996: winter maximum 286 on Bronx-Westchester Christmas Bird Count

Study area, historical and current
A widespread resident, spring/fall migrant, and winter visitor/resident.

There is a very light spring movement in Mar–May, with a maximum 5 on numerous occasions.

Downies breed in all 4 extant subareas, with an estimated population of 15–20 pairs, of which 10+ are in Van Cortlandt. NYSBBA and DOIBBS data showed its widespread statewide occurrence but no consistent trend from 1966–2005 (McGowan and Corwin 2008; Sauer et al. 2005).

There is a very light fall movement in Oct–Nov, with a maximum of 5 on 8 Nov 2005 (Klein). Winter numbers are higher when residents are augmented by northerly immigrants, the maximum being 30 on 22 Dec 2013 between Van Cortlandt and Woodlawn (Lyons, Gotlib et al.).

Specimens
Dwight collected 3 in Van Cortlandt in Feb, Apr 1884–86, Foster another in Woodlawn on 4 Jan 1890 (all AMNH), and Bicknell 3 in Riverdale in Jan, Oct 1876–79 (NYSM, MCZ).

Hairy Woodpecker *Picoides villosus*

NYC area, current
A resident but uncommon breeder in all 5 New York City boroughs in areas with mature deciduous forests, and a regular spring/fall migrant in small numbers; always outnumbered by Downy and Red-bellied Woodpeckers.

Bronx region, historical
Bicknell's *Riverdale* 1872–1901: uncommon migrant, breeder, winter resident

Eaton's *Bronx + Manhattan* 1910–14: uncommon breeder

Griscom's *Bronx Region* as of 1922: common resident

Griscom's *Riverdale* as of 1926: no longer surviving in Riverdale

Kuerzi's *Bronx Region* as of 1926: fairly common resident [locations not given]

NYC area, historical
Cruickshank's *NYC Region* as of 1941: fairly common resident, spring/fall migrant

Bull's *NYC Area* as of 1968: fall maximum 35 inland, 5 on beaches

Bull's *New York State* as of 1975: status unchanged

Levine's *New York State* as of 1996: winter maximum 142 on 1962 Bronx-Westchester Christmas Bird Count

Study area, historical and current
A resident but uncommon breeder in areas with mature deciduous forests and a regular spring/fall migrant in small numbers.

It has proved impossible to establish firm spring migration dates for this species, but there is a light spring movement in Mar with a maximum of 4 on 15 Mar 2014 (Alvarez).

A thinly scattered resident breeder, with 6–8 pairs in Van Cortlandt and another in Woodlawn; 3 on 5 Aug 20 2013 (Young) were probably a family group. They may have suffered from competition with the burgeoning Red-bellied Woodpecker population and have lost active nest holes to them in Van Cortlandt (Young).

Both NYSBBA I in 1980–85 and II in 2000–2005 verified a broad New York City area breeding distribution, where they were understandably absent only from mature-tree–free urban portions of all New York City boroughs and southern Nassau County (McGowan and Corwin 2008). DOIBBS data confirmed their widespread occurrence but showed no consistent statewide trend from 1966–2005 (Sauer et al. 2005).

There is a very light movement in fall in Oct–Nov, with infrequent irruption years; 1–2 is the daily norm, with a fall maximum of 4 on 22 Sep 2005 (Young).

The winter maximum is 10 across Van Cortlandt and Woodlawn on 26 Dec 1960 (Tudor et al.).

Specimens
Bicknell collected 1 in Riverdale on 22 Oct 1875 (NYSM).

Black-backed Woodpecker *Picoides arcticus*

NYC area, current
An exceedingly infrequent irruptive winter visitor. Until 1923, it was known only from 2 specimens taken during the winter of 1886–87 on Long Island.

Bronx region, historical
Griscom's *Bronx Region* as of 1922: unrecorded

Kuerzi's *Bronx Region* as of 1926: very rare winter visitor: 3 Bronx Park 1923–27, plus Van Cortlandt 1925–26, Mt. Kisco 1925–26, New Rochelle 1925–26, 1926–27; 2 other Westchester County records <1915

NYC area, historical
Cruickshank's *NYC Region* as of 1941: as in Kuerzi, plus singles from Easthampton 13 Oct 1936, Montauk 26 Dec 1937

Bull's *NYC Area* as of 1968: 28 Sep 1958 Central Park

Bull's *New York State* as of 1975: status unchanged

Study area, historical and current
One study area record: 10 Oct 1925 in Van Cortlandt: ♂ with 2 other woodpeckers also believed to have been this species (Cruickshank, Moss). This was during a historic incursion when 2–3 reappeared each winter in the New York Botanical Garden's Hemlock Grove in Bronx Park between 1923 and 1927. In the same period others were seen or overwintered in southern Westchester County and Bergen County, NJ. This bizarre pattern has repeated only once, when 3 reappeared between 1957 and 1961 in pine groves at Oradell Reservoir, again in Bergen County. Unrecorded in Prospect Park, but once in Central Park (28 Sep 1958: Bloom). New York City area records after that are few: Fort Tryon Park 14 Nov 1963 (Post, Raices), Scarsdale 16 Nov 1963 (Hirschberg), Great Neck 17 Oct 1965 (W. Davis), Bronx Park 21–22 Oct 1965 (Maguire, van Wert, Rafferty), Upton, LI on 16 Nov 1965 (Stepinoff), and Levittown, LI on 18 Oct 1970 (Leary). Since then, the only one anywhere in the New York City area was at a Mt. Kisco feeder for 15 minutes on 10 Jan 1995 (Kriegeskotte).

Northern Flicker — *Colaptes auratus*

NYC area, current
A widespread and common spring/fall migrant, especially along the coast, and a widespread breeder and local winter resident in all 5 New York City boroughs.

Bronx region, historical
Bicknell's *Riverdale* 1872–1901: spring/fall migrant, resident

Hix's *New York City* as of 1904: common breeder

Eaton's *Bronx + Manhattan* 1910–14: common breeder, uncommon resident

Griscom's *Bronx Region* as of 1922: common breeder, rare in winter

Griscom's *Riverdale* as of 1926: now only summer

Kuerzi's *Bronx Region* as of 1926: common breeder, frequently overwintering

NYC area, historical
Cruickshank's *NYC Region* as of 1941: common breeder, migrant in spring/fall (maximum 50+), most regular in winter near coast

Bull's *NYC Area* as of 1968: Long Island maxima 150 spring, 2000 fall, and 40 winter

Bull's *New York State* as of 1975: status unchanged

Levine's *New York State* as of 1996: winter maximum 323 on 1984 Montauk Christmas Bird Count

Study area, historical and current
A resident breeder in all 4 extant subareas and a conspicuous spring/fall migrant, overwintering in varying but low numbers depending on the year.

In spring, first migrants arrive in mid-Mar, peak in early Apr, and depart by mid-May, with extreme dates of 9 Mar 2008 (Finger) and 21 May 1953 (Buckley), and a maximum of 100 on 18 Apr 1965 (Horowitz). Until Jun 1967, the spring maximum in Prospect Park was 65 on 12 Apr 1953 (Usin).

About 20 pairs breed across all 4 extant subareas, with <10 in Van Cortlandt, and we are aware of no evidence that study area breeders are declining. Both NYSBBA I in 1980–85 and II in 2000–2005 recorded breeders in Blocks 5852B and 5952A, which include the study area (McGowan and Corwin 2008).

Both NYSBBA I in 1980–85 and II in 2000–2005 noted a broad New York City area breeding distribution, where they were understandably absent only from the most urbanized portions of Manhattan, Brooklyn, and Queens (McGowan and Corwin 2008). DOIBBS data indicated a 4.1% annual statewide decline from 1966 to the late 1970s, less so from 1980–2005, perhaps tracking forest maturation (Sauer et al. 2005).

In fall, first migrants arrive in Sep, peak in early Oct, and depart by early Nov, with probable extreme dates of 29 Aug 2013 (Aracil) and 1 Dec 2005 (Young) and a maximum of 28 on 14 Sep 1935 (Imhof). Until Jun 1967, the fall maximum in Central Park was 200 on 3 Oct 1959 (Bloom).

In winter, occurs in varying numbers with daily maxima of 20+ on Christmas Bird Counts, some surviving the winter. On 22 Dec 2013, 6 were found in Woodlawn alone (Gotlib, McAlexander).

Specimens
Foster collected 1 in Woodlawn on 5 Apr 1897 (AMNH) and Bicknell 2 in Riverdale in Oct 1875–78 (NYSM).

Pileated Woodpecker *Dryocopus pileatus*

NYC area, current
Absent from Long Island and New York City as a breeder since the mid-1800s, but records of wanderers/potential colonizers have been increasing since 1940 and especially since the 1960s. The increase has been most noticeable in the Bronx, where nearby Westchester County breeders have been slowly working their way south, paralleling behavior in nearby northern NJ, where Pileated is now often a suburban backyard bird. Evidently resident on Staten Island as of 2016, although it is unclear whether any are yet breeding. Recorded and probably breeding in the areas of the future Prospect and Central Parks but no later than the early 1800s.

Bronx region, historical
Bicknell's *Riverdale* 1872–1901: unrecorded

Eaton's *Bronx + Manhattan* 1910–1914: unrecorded

Griscom's *Bronx Region* as of 1922: unrecorded

Kuerzi's *Bronx Region* as of 1926: unrecorded

NYC area, historical
Cruickshank's *NYC Region* as of 1941: extirpated from Long Island, Bronx, southern Westchester County; occasionally seen as close as Grassy Sprain Reservoir

Bull's *NYC Area* as of 1968: greatly increased in recent years; seen in Bronx Park 1939, then none until Ft. Tryon Park 1959, Inwood Park 1961, Van Cortlandt 1962, all in Apr

Bull's *New York State* as of 1975: bred Tarrytown 1958 and near White Plains 1963

Levine's *New York State* as of 1996: overwintered Forest Park, Queens 1981; seen in several northern Nassau County sites spring/summer 1982 but no breeding evidence

Study area, historical and current
A resident species that bred into the early 1800s, and is now in the process of slowly recolonizing the study area, where it is not yet being seen annually.

First recorded nearby at Grassy Sprain on 26 Dec 1938 (Thomas, Van Deusen) but not established there as a breeder in 1941 (Cruickshank). However, it was resident there by the late 1940s (Norse) through the mid-1950s (Buckley et al.), and 3 pairs were breeding in southwestern Westchester County by 1964 (unknown observer, *Kingbird*). Reported there during NYSBBA I but missed on NYSBBA II. First found on the Bronx-Westchester Christmas Bird Count in 1940–41 at Grassy Sprain Reservoir, then not again until 1961, after which it was seen slightly more frequently although not (almost) annually until 1980, with a maximum of 7 in 2012. Otherwise, the nearest current breeders are on the Palisades.

There have been 16 study area records, all since 1962, scattered throughout Van Cortlandt: on the Ridge 9 Apr 1962 (Maguire et al.); in the Shandler Area in Apr 1984 (fide Turner); on the Ridge from Mar–Jun in 1985 (Naf); on the Ridge on 24 Apr 1990 (Naf, Abramson); Van Cortlandt Swamp on 18 Sep 1994 (Garcia); in the Northwest Forest on 31 Mar 1995 (Lyons, Jaslowitz) and 13 May 1998 (Künstler); in the Northwest Forest on 13 May 1998 (Künstler); on the Ridge 9 Apr 1999 (Jaslowitz); in the Sycamore Swamp in Apr 2011 (Parks Dept. Van Cortlandt website); in the Northwest Forest on 4 Sep 2013 (Duran-Ruiz); in Yonkers 1 short street from the Ridge main trail on 13 Aug 2013 and 2 Mar 2014 (Block); in Van Cortlandt Swamp on 3 Apr 2014 (Rodriguez); near Van Cortlandt Lake 12 Apr 2014 (Souirgi et al.); in the Northwest Forest on 20 May 2015, calling loudly (Young)—most likely the same one seen in adjacent Riverdale on 18 Apr (Young,

O'Brien). Singles are being seen increasingly by Van Cortlandt park rangers on routine patrol, so it would be unsurprising if Pileated Woodpeckers were soon resident and breeding there—which would be the first within New York City limits in 200 years unless Staten Island or Pelham Bay Park wins the race.

In the Bronx and adjacent Manhattan, modern records outside the study area include Bronx Park 21 Apr 1940 (Komorowski); Fort Tryon Park 12 Apr 1959 (Gilbert); Inwood Park 23 Apr 1962 (Norse, Kallmann); Bronx Park 24 Apr 1964 (Rafferty, Hackett); Riverdale 17 May 1971 (McGuinness); Bronx Park 4 May 1978 (D. Hall); Pelham Bay Park Apr 1986–Apr 1989 (Künstler, M. Brown); Riverdale 28 Apr 1999 (Ephraim); Bronx Park 5 May–15 Sep 2012 (Becker et al.); Inwood Park 19–21 Apr 2013 (DiCostanzo); Bronx Park 27 Jun 2015 (Allen, DeCandido et al.); Pelham Bay Park 18 Oct 2015 (Aracil, Benoit); Riverdale 16 Apr 2016 (Malina), Inwood Park 25 Mai 2017 (Waldron, photos); Fort Tryon 9 Apr 2017 (Skrentny); and 2 in Bronx Park on 21 Oct 2017 (van Zyl). Recent New York City area spring records probably involve potential colonizers.

If colonization of Riverdale and Van Cortlandt has not yet happened, it seems imminent. In southern Westchester County they bred in Tarrytown in 1958 and in Silver Lake in 1963, and by about 2000 had moved into New Rochelle and are now breeding immediately adjacent to (possibly even within) Pelham Bay Park north and northwest of the golf courses.

NYSBBA I in 1980–85 recorded breeding in 2 Westchester County blocks adjacent to Pelham Bay Park, but NYSBBA II in 2000–2005 did not. Otherwise, closest breeders were at Grassy Sprain Reservoir during both atlases (McGowan and Corwin 2008). DOIBBS data showed a statewide increase of 3.6% per year from 1966–2005 (Sauer et al. 2005).

Falcons: Falconidae

American Kestrel   *Falco sparverius*

NYC area, current
A spring/fall migrant, local breeder unexpectedly numerous in urban areas, and regular winter visitor and resident, although overall in greatly diminished numbers since the 1960s.

Bronx region, historical
Bicknell's *Riverdale* 1872–1901: spring/fall migrant, local breeder, winter resident

Eaton's *Bronx + Manhattan* 1910–14: uncommon breeder, occasional resident

Griscom's *Bronx Region* as of 1922: common resident

Kuerzi's *Bronx Region* as of 1926: common resident, spring/fall migrants in Apr and Oct

NYC area, historical
Cruickshank's *NYC Region* as of 1941: breeder throughout New York City area in small numbers, even on Manhattan buildings; regular fall coastal (maximum 100)/uncommon spring migrant, winter resident

Bull's *NYC Area* as of 1968: Long Island maxima spring 25, fall 500–550

Bull's *New York State* as of 1975: fall maximum 1000 Fire Island Lighthouse

Levine's *New York State* as of 1996: fall maximum 1386 Fire Island Lighthouse

Study area, historical and current
Formerly a widespread and often encountered resident, and 10+ pairs bred in all 4 extant subareas through the 1950s (Kieran, Norse, Buckley, Horowitz). Study area numbers in the breeding season have dropped steadily from the 1960s–2016, mirroring declines throughout the Northeast. Holding on as a breeder in the study area and neighborhood buildings but now much scarcer in migration and now resident in winter only at Hillview.

In spring, there is an extremely light movement in Mar–May with extremes of 26 Mar 2010 (Young) and 30 Apr 2011 (Baksh) and recently, only singles.

From NYSBBA I in 1980–85 to NYSBBA II in 2000–2005, kestrels in and around New York City almost disappeared. In Westchester County they dropped from about 37 to 5 occupied blocks, within New York City from about 29 to 16, in Nassau County from 29 to 4, and in Suffolk County from about 120 to 40 (McGowan and Corwin 2008). Urban and suburban kestrels are routinely overlooked, but DOIBBS data showed slight but significant declines throughout southern New England and New York State from 1966–2005 (Sauer et al. 2007).

Even though found in the 2 blocks south of the study area during NYSBBA II, they were not detected within it. Nonetheless, it is likely that 1–2 pairs each breed at Woodlawn, Jerome, Hillview, and in Van Cortlandt, but even more do so on buildings just outside the study area. Likely breeders were in Van Cortlandt on 6 Jun 1980, 15 Jun 1981, and 22 May 1984, and a pair with fledged young was there on 12 Jun 1982 (Sedwitz); another pair was on the Fordham University Campus on 19 Jun 2012 (Factor). In 2010 an estimated 60–100 pairs were breeding in strictly urban environments throughout New York City, with 25 pairs estimated in Manhattan alone (http://www.battaly.com/nehw/AmericanKestrel/news/). Breeders are significantly underestimated citywide, and there have been no specific attempts to assess the Bronx breeding population; it is a regular breeder at Pelham Bay Park (Künstler). Strategic placement of predator-proofed, properly crafted nest boxes in each of the 4 extant subareas would help mitigate local declines in breeders.

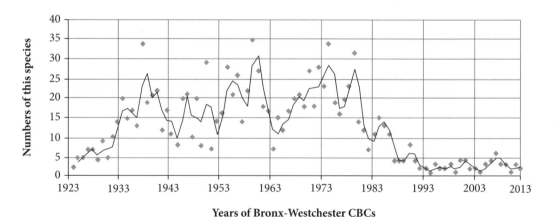

Figure 44

In fall, first migrants arrive in late Aug, peak in Oct, and depart by mid-Nov, with extreme dates of 24 Aug 1946 (Komorowski) and 13 Nov 2010 (Baksh), and maxima of 32 on 18 Sep 1951 and 43 on 22 Sep 1953 with 64 counted from 24 Aug–4 Oct 1946 (all Komorowski).

A few migrants (4–5+) formerly overwintered annually, but since the late 1960s only 1–2 have been scarce and erratic study area winter visitors, with a maximum of 3 resident at Hillview. This trend is mirrored clearly in the Bronx-Westchester Christmas Bird Count data (see Fig. 44). Winter kestrels may appear almost anywhere and often near habitation, but they rarely remain more than a few days.

Merlin   *Falco columbarius*

NYC area, current
An uncommon but annual spring and especially fall migrant, and increasing winter resident in small numbers, especially near the coast.

Bronx region, historical
Bicknell's *Riverdale* 1872–1901: single spring, 4 fall records

Eaton's *Bronx + Manhattan* 1910–14: common migrant

Griscom's *Bronx Region* as of 1922: only 2 recent May, Sep records

Kuerzi's *Bronx Region* as of 1926: uncommon migrant in spring (18 Apr–17 May)/fall (26 Aug–21 Oct)

NYC area, historical
Cruickshank's *NYC Region* as of 1941: uncommon migrant in spring/more common coastally fall (maximum 100) on Long Island; extremely rare in winter

Bull's *NYC Area* as of 1968: late Mar–28 May, 5 Aug–22 Nov; fall maximum 110; increasing in winter near coast

Bull's *New York State* as of 1975: status unchanged

Levine's *New York State* as of 1996: fall maximum 292; regular on coast in winter

Study area, historical and current
An uncommon but annual spring and especially fall migrant, and nearly annual winter resident in very small numbers.

In spring, first migrants arrive in late Mar, peak in mid–late Apr, and depart by early May, with extreme dates of 2 Mar 2007 and 2010 (Young) and 27 May 1934 (Weber, Jove), singles only.

In fall, first migrants arrive in mid-Aug, peak in Sep, and depart by early Oct, with extreme dates of 11 Aug 2002 (Young) and 9 Nov 1954 (Buckley), and a maximum of 5 on 19 Sep 2004 (Young). In 1946 during Komorowski's Van Cortlandt hawkwatches, only 6 were seen between 12 Sep and 1 Oct, all singles save 2 on the last date, typical of this coastwise fall migrant.

Increasingly seen in winter >1955, especially in Woodlawn, where overwintering is now regular, if not annual, and at Hillview. The maximum is 3 at Hillview on 25 Feb 2017 (Adams).

Within New York State, Merlin was unrecorded as a breeder during NYSBBA I in 1980–85, but during NYSBBA II in 2000–2005 was found in an unexpected 131 blocks, the closest in Greene County. With breeding also now occurring on Nantucket, Martha's Vineyard, Cape Cod, and Ulster County, when will they reach Bronx Park?

Peregrine Falcon   *Falco peregrinus*

NYC area, current
Formerly a native breeder on the Palisades and certain Manhattan skyscrapers through the early 1960s (*anatum*), as well as a fall migrant and scattered winter resident (*anatum* and *tundrius*). After genuine *anatum* breeders were extirpated from the entire Northeast by pesticide contamination in the 1960s, a synthetic population created from a genetic farrago was slowly reintroduced on buildings and bridges, where they have now established a thriving and growing local population. The arctic breeding *tundrius* forming the bulk of previous coastal migrants were not affected and occur annually on the South Shore of Long Island, mostly in fall. Peregrine is a New York State Endangered Species.

Bronx region, historical
Bicknell's *Riverdale* 1872–1901: unmentioned

Eaton's *Bronx + Manhattan* 1910–14: rare migrant, uncommon breeder

Griscom's *Bronx Region* as of 1922: 2 pairs on Palisades; not uncommon visitor at any time

Kuerzi's *Bronx Region* as of 1926: fairly common visitor anytime

NYC area, historical
Cruickshank's *NYC Region* as of 1941: limited spring/fall (maximum 12 in 1963) migrant, rare in winter; handful of nearby breeders can appear anywhere, anytime

Bull's *NYC Area* as of 1968: status unchanged

Bull's *New York State* as of 1975: native *anatum* extirpated as breeders 1960s by pesticides; Long Island fall maximum 23 [arctic *tundrius*] 1967

Levine's *New York State* as of 1996: Long Island daily fall maximum 49 [mostly arctic *tundrius*] 1989, seasonal maximum 249 in 1990; genetically heterogeneous captive-reared adults hacked at many coastal sites

Study area, historical and current
Prior to the 1960s Peregrines occurred at almost any time of the year in all 4 extant subareas, usually singles, occasionally 2. Being a coastwise migrant, they were only infrequently seen in fall at hawkwatches, with maxima of only 2 on 22 and 26 Sep 1946 (Komorowski) and at Woodlawn on 15 Sep 1964 (Horowitz). Following the recent hackings, singles are being seen year-round once again throughout the study area, but they have never overwintered exclusively in the study area. In the 1950s in Van Cortlandt Swamp one was seen carrying a snake (Kieran) and another stooping on a Virginia Rail (Buckley). A pair in Van Cortlandt on 12 Apr 1953 (Buckley) may have been migrants or among the few remaining New York City area breeders. Two juveniles, some of the very last raised in the New York City area, were in Van Cortlandt Swamp on 19 Jul 1964 (Zupan, Heath), and another spent the summer of 1965 there (Gera, Gera). On the other hand, one on 3 Sep 1978 (Ephraim) was more likely a migrant.

After that, they were absent as residents from the New York City area until the first captives were hacked in 1980 in midtown Manhattan, followed by pairs on the Throgs Neck and Verrazano Bridges in 1983, the Tappan Zee Bridge in 1988, and then almost in the Bronx in 2000 when a pair appeared on the subway trestle at W. 225th St. and Broadway and bred the following year. These birds began to appear in Van Cortlandt almost immediately and still nest on the W. 225th St. bridge. Another 5–6 pairs were hacked on the Palisades between the George Washington Bridge and Alpine and are now established there, with the first wild nest in 2003. Other nearby breeders, hacked or wild descendents thereof, today include pairs on the

George Washington, Tappan Zee, Throgs Neck, and Whitestone bridges.

No *tundrius* have ever been firmly identified in the study area or anywhere in the Bronx, even though they migrate in numbers down the Long Island outer coast in fall. Some, even all, of those recorded when the hawkwatch was active at Pelham Bay Park (28 in 1988, 20 in 1989, 27 in 1990: DeCandido 1990, 1991b) may have been *tundrius* but were not so identified.

Comments
High-arctic breeding **Gyrfalcon**, *F. rusticolus*, very rarely reaches south in winter to the NYC area. However, during the late fall and winter of 1989–90 in a major northeastern Gyrfalcon incursion, individuals appeared and took up residence for varying periods at Jones Beach, Jamaica Bay, with 2 at Sandy Hook. Possibly 1 of them was also the only one ever seen in the Bronx, at Pelham Bay Park from 26 Dec–2 Jan (J. Caspers). Almost in the Bronx was an HY black morph sitting on the bottom of the main cable of the Triboro Bridge and allowing approach to within 20 ft (6 m) at 0715 on Sunday, 9 Jan 1955 (Sedwitz). Just 7 smi (11 km) NNW of the study area, a gray morph adult was found on the NJ Palisades at Alpine on 21 Jan 2017 (Girone et al.; photos) and was subsequently enjoyed by many. It appeared most days in midafternoon, eventually departing east/southeast across the Hudson and northwestern Yonkers, then southeast toward Sprain Ridge before disappearing from view. Where it roosted at night and hunted in the morning were never determined, and it was last seen at Alpine on 2 Feb.

Parrots: Psittacidae

<u>Monk Parakeet</u>   *Myiopsitta monachus*

NYC and study areas, historical
This escaped (about 1968) exotic (nonnative) and communal nester is now locally well established around New York City, especially in the East Bronx where the first nest was at Pelham Bay Park in Dec 1970. Since the early 1970s, some have been seen more or less continuously resident and breeding around City Island, Pelham Bay Park, and Throgg's Neck. Colonizers, dispersers, and nonbreeders may appear almost anywhere and begin nest construction immediately, often in midwinter.

The first study area occurrence (date unknown) was probably shortly before the first proven breeding when a nest involving an unknown number of pairs was built at the Van Cortlandt Mansion in 1974–75 and then removed by park officials (fide Jaslowitz). These may have involved some of the 3+ Monks and 6–8 **Rose-ringed Parakeets**, *Psittacula krameri*, that were roosting, feeding, and (the Monks) nest building at Fieldston Rd. and W. 250th St. in Riverdale in the winter of 1973–74 (Kane 1974). They were probably some of the 46+ Rose-ringeds first seen on the 1974 Bronx-Westchester Christmas Bird Count at a roost on the grounds of the Kingsbridge Veterans' Administration Hospital (now the James J. Peters VA Medical Center) on Sedgwick Ave. between Fordham and Kingsbridge Rds. (Keil 1974). They may have been the same ones frequenting the Fordham University campus through the mid-1980s and possibly nesting there (fide Kane), but they disappeared (voluntarily?) shortly thereafter.

On 29 Dec 1996, 4 Monk Parakeets were in Riverdale along the Hudson River (Veit, Buckley), and there have been a number of vague Riverdale reports since then. On 31 Jan 2002 several were near the Van Cortlandt Mansion (unknown observer, *Kingbird*). Then between 2005 and 2009 Monks were seen on and off there (e.g., 9 Jul 2005: Garcia; 23 Apr 2007: Young; 23 Sep 2009: Young; 6 on 22 May 2010), but if nesting occurred, it went undetected. However, as of 2012 an unknown number had built a large communal nest in a light stanchion at Gaelic Park opposite the southwestern corner of Van Cortlandt at Broadway and W. 240th St., which was eventually usurped by Red-tailed Hawks (Young). It is not known whether the parakeets moved elsewhere, possibly west up the hill along Waldo Ave., where they have also been seen recently. In Mar 2016, 2 nests with 7 adults were located in Yonkers 0.85 smi (1.4 km) NNW of Hillview (Block). The closest other breeders may be in or near the Bronx Zoo.

Small groups (often reported without counts) have been subsequently seen occasionally throughout the study area, and although there have been no other nesting attempts, it is a potential visitor and breeder in/near all 4 extant subareas. On 22 Dec 2013, several were in Woodlawn (Gotlib, McAlexander), indicating another possible nearby breeding site. Christmas Bird Count data in Figure 45 display well their long-term Bronx status.

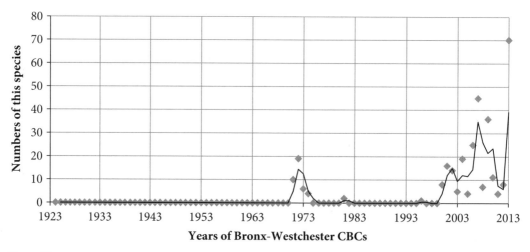

Figure 45

Tyrant Flycatchers: Tyrannidae

Olive-sided Flycatcher *Contopus cooperi*

NYC area, current
An uncommon but annual late spring and early fall Neotropical migrant, especially in New York City parks, but remarkably scarce on Long Island barrier beaches—a pattern matching Louisiana Waterthrush and Golden-winged Warbler, among others.

Bronx region, historical

Bicknell's *Riverdale* 1872–1901: 6 fall records

Eaton's *Bronx + Manhattan* 1910–1914: rare migrant

Griscom's *Bronx Region* as of 1922: 3 recent records but little early fall fieldwork

Kuerzi's *Bronx Region* as of 1926: uncommon migrant in spring (12 May–11 Jun)/fall (6 Aug–24 Sep)

NYC area, historical

Cruickshank's *NYC Region* as of 1941: uncommon migrant in spring (10 May–15 Jun; maximum 6)/fall (2 Aug–16 Oct) in Hudson Valley [locations not given] but elsewhere much scarcer

Bull's *NYC Area* as of 1968: early spring date 4 May, early fall date 27 Jul; fall maximum 5

Bull's *New York State* as of 1975: fall maximum 6, early fall date 22 Jul

Study area, historical and current

A scarce but annual spring/fall migrant in Van Cortlandt and Woodlawn, often missed when few observers are afield during its fall Aug peak.

In spring, first migrants arrive in mid-May, peak in and depart by late May, with extreme dates of 10 May 1979 (Young) and 26 May 1997 (Lyons, Garcia), mostly singles with a maximum of 4 on 16 May 1953 (Phelan). Until Jun 1967, the spring maximum in Central Park was 2 on 28 May 1967 (Plunkett).

DOIBBS data showed a severe statewide decline from 1966–2005 (Sauer et al. 2005; McGowan and Corwin 2008). In 1954 it bred as close as Newburgh, Orange County, but that was a singularity. The nearest current breeders are in Ulster County, where numbers have dropped, and it has never bred in NJ.

In fall, first migrants arrive in early–mid Aug, peak in late Aug, and depart by mid-Sep, with extreme dates of 14 Aug 2013 (Young) and 28 Sep 1951 (Kane) and normally only singles or 2s with a maximum of 4 on 1 Sep 1935 (Norse). Until Jun 1967, the fall maximum in Prospect Park was 5 on 19 Aug 1944 (Soll).

Specimens

Bicknell collected 1 in Riverdale on 11 Sep 1876 (NYSM).

Eastern Wood-Pewee *Contopus virens*

NYC area, current

A not uncommon spring/fall Neotropical migrant and local breeder in all 5 New York City boroughs.

Bronx region, historical

Bicknell's *Riverdale* 1872–1901: regular spring/fall migrant (3 May–6 Oct), common breeder

Eaton's *Bronx + Manhattan* 1910–14: common breeder

Griscom's *Bronx Region* as of 1922: uncommon breeder, common migrant

Kuerzi's *Bronx Region* as of 1926: common breeder, spring/fall migrant (3 May–28 Oct)

NYC area, historical

Cruickshank's *NYC Region* as of 1941: common spring (maximum 28)/fall migrant, breeder (30 Apr–28 Oct), slowly decreasing as New York City area breeder, although few pairs still nest within city [locations not given]

Bull's *NYC Area* as of 1968: late fall date 30 Oct; maxima spring 12, fall 15

Bull's *New York State* as of 1975: late fall date 15 Nov

Study area, historical and current
A not uncommon spring/fall Neotropical migrant and breeder.

In spring, first migrants arrive in early May, peak in mid-May, and depart by late May, with extreme dates of 27 Apr 1953 (Buckley) and 3 Jun 1960 (Horowitz, Heath) and a maximum of 5 on 28 May 1997 (Young). Until Jun 1967, the spring maximum in Prospect Park was 12 on 3 Jun 1945 (Soll).

This species has been breeding in Van Cortlandt, Woodlawn, and Riverdale since the late 1800s. As of 2016, there are 8–10 pairs throughout Van Cortlandt and intermittently another 1–2 in Woodlawn.

Both NYSBBA I in 1980–85 and II in 2000–2005 identified a broad New York City area breeding distribution, where they were understandably absent only from the most urbanized portions of all New York City boroughs (McGowan and Corwin 2008). DOIBBS data showed a 2.1% statewide decline from 1966–2005 (Sauer et al. 2005).

In fall, first migrants arrive in early Aug, peak in late Aug–early Sep, and depart by late Sep, with extreme dates of 2 Aug 2004 (Young) and 4 Oct in 1925 (Cruickshank) and 2015 (Gomes), and a maximum of 7 on 13 Sep 2013 (Young). Bicknell collected 1 in Riverdale on 6 Oct 1881. Until Jun 1967, the fall maximum in Central Park was 35 on 15 Sep 1956 (Post).

Specimens
Foster collected 1 in Woodlawn on 13 Jun 1885 (AMNH) and Bicknell 5 in Riverdale in May, Aug, Sep, Oct 1876–81 (NYSM).

Yellow-bellied Flycatcher *Empidonax flaviventris*

NYC area, current
An annual but quite uncommon spring/fall Neotropical migrant that is not seen as frequently or in the same numbers it was before the 1960s.

Bronx region, historical
Bicknell's *Riverdale* 1872–1901: regular spring/fall migrant

Eaton's *Bronx + Manhattan* 1910–14: uncommon migrant

Griscom's *Bronx Region* as of 1922: rare spring/fall migrant

Kuerzi's *Bronx Region* as of 1926: uncommon migrant in spring (13 May–8 Jun)/uncommon to fairly common in fall (11 Aug–6 Oct)

NYC area, historical
Cruickshank's *NYC Region* as of 1941: uncommon migrant in spring (7 May–19 Jun)/fall (4 Aug–6 Oct)

Bull's *NYC Area* as of 1968: early fall date 29 Jul; maxima 8 both spring/fall

Bull's *New York State* as of 1975: early fall date 22 Jul; maxima 9 spring, 13 fall

Study area, historical and current
An annual but quite uncommon spring/fall Neotropical migrant.

In spring, first migrants arrive in early May, peak in mid-May, and depart by late May–early Jun, with extreme dates of 5 May 2000 (Young) and 17 Jun 1927 (Cruickshank), and a maximum of 4 on 14 May 1953 (Penberthy, Phelan, Buckley). One studied closely in the Van

Cortlandt *Phragmites* marsh west of the Sycamore Swamp from 18 Jun–1 Jul 2010 (Young) was an SY prebreeder. Until Jun 1967, the spring maximum in Central and Prospect Park was 8 on 30 May 1953 (Aronoff).

NYSBBA I in 1980–85 and II in 2000–2005 showed that the nearest breeders are in the Catskills (McGowan and Corwin 2008). DOIBBS data evidenced a strong but nonsignificant statewide increase from 1966–2005 (Sauer et al. 2005).

In fall, first migrants arrive in early Aug, peak in late Aug–early Sep, and depart by late Sep, with extreme dates of 3 (the maximum) on 1 Aug 1953 (Buckley et al.) and 23 Sep 1954 (Buckley).

Specimens
Bicknell collected 4 in Riverdale in May, Aug, Sep, Oct 1879–81 (NYSM), and while 8 Dwight 1885–96 specimens were labeled as taken near New York City, none had more precise locations (AMNH).

Acadian Flycatcher *Empidonax virescens*

NYC area, current
A recolonizing species that disappeared as a breeder throughout New York State and the Northeast when native American Chestnut trees were extirpated by Chestnut Blight fungus in the late 1800s. Following an early 1970s range expansion, it has become a rare spring and overlooked fall migrant, but remains the scarcest local *Empidonax*. It has been breeding sporadically on Long Island and in Westchester, Rockland, and Dutchess Counties, where it is slowly increasing.

Bronx region, historical
Bicknell's *Riverdale* from 1872–1901: common breeder in Riverdale 1872–1902 but only twice in Aug, indicating very early departure

Eaton's *Bronx + Manhattan* 1910–14: fairly common breeder; map on p. 21, vol. 1 showed breeding in 1906–7 across study area

Griscom's *Bronx Region* as of 1922: extirpated from Riverdale and his entire Bronx Region, with only 3 singing 1915–1920 [locations not given]

Kuerzi's *Bronx Region* as of 1926: formerly common breeder with several pairs still at Grassy Sprain Reservoir though 1925 and another in Scarsdale

NYC area, historical
Cruickshank's *NYC Region* as of 1941: very rare migrant 5 May–30 Sep; few pairs still breeding in southern Westchester County, among very last in New York State

Bull's *NYC Area* as of 1968: not breeding in [the present book's] New York City region, but creeping slowly north in NJ

Bull's *New York State* as of 1975: Long Island invasion spring 1970 with 11 netted Fire Island Lighthouse; bred Dutchess County 1973–75

Levine's *New York State* as of 1996: very slowly recolonizing Suffolk, Nassau, Westchester, Dutchess, Rockland, adjacent counties, where now breeding in Eastern Hemlock groves

NYC and study areas, historical and current
A former breeder on the cusp of recolonizing the area, although a very scarce spring and an all but undetected fall migrant.

Through the early 1900s it was a regular breeder in Riverdale, Van Cortlandt, and elsewhere in the New York City area, especially in the Hudson Valley—even in Central Park until 1892. Then it began to disappear from the Northeast (gone from Riverdale in the 1890s, where 6 pairs had nested: Bicknell), an event believed by many related to disappearance of native American Chestnut trees. By 1915 Acadians had all but vanished near New York City, and Griscom (1923) reported the last 3 from his Bronx Region (locations unspecified) on 9 Jun 1915 (E. G. Nichols), 3 Jun 1917 (L. N. Nichols), and 3 Jun 1920 (Griscom). A small breeding population persisted at Grassy Sprain from 1926–41, and although there have been no certain breeders there subsequently, they could have been the source of singing Van Cortlandt singles in the Tibbett's Brook valley just north of Van Cortlandt on 17 May 1931 (Cruickshank), and in Van Cortlandt on 3 Jun 1938 (Feigin et al.), 10–19 May 1952 (Norse, Buckley), 4 Jun 1952 (Kane), 22 May 1953 (Kieran, Buckley), and 23 May 1954 (Phelan, Buckley).

Then there were none in or near the Northwest Bronx for 9 years until extremely early singles in Inwood Park on 8 May 1963 (Norse), in Van Cortlandt on 2 May 1964 (Norse), and in nearby Bronx Park on 6 May 1964 (Stepinoff et al.), with more normal later migrants in Inwood Park on 24 May 1967 (Norse) and Bronx Park on 25 May 1969 (Maguire et al.). The tiny Grassy Sprain population appears to have died out no later than 1954, so those in the Bronx/Inwood Parks beginning in 1963 may have been part of a new and unrelated recolonization process.

Elsewhere in New York City, singles had been recorded in Central Park only in 1935, 1953, 1954, 1963, and in Prospect Park only in 1944 (Carleton 1970). Away from the Bronx and Central and Prospect Parks, records were equally sparse. Most involved obvious spring migrants except for a handful of unsuccessful territorial ♂♂ on extreme eastern Long Island from 1959–69.

This sparse scenario changed abruptly in May–Jun 1970 when migrant Acadians appeared on Long Island's South Shore and widely throughout northern NJ and southeastern New York—and then began to nest in scattered locations. Wholly unexpectedly, they often chose dark Eastern Hemlock woods at higher elevations in NJ, especially near streams—strikingly different from their typical moist lowland deciduous habitat. Perversely, this occurred just when hemlocks began to be decimated by Asiatic Hemlock Woolly Adelgids and Elongate Hemlock Scale insects, in the process taking with them many only recently breeding Acadian Flycatchers. They had all but destroyed the entire New York Botanical Garden's famed Hemlock Grove old-growth trees before any Acadians attempted to breed there.

The early 1970s invasion and potential range expansion/consolidation notwithstanding, there were few Van Cortlandt Acadian records after 1964: on 25 May 1990 (Young), 17 May 1991 (Young), 13 May 1998 (Garcia, Lyons), and 10 May 2009 (Young). More recently, singing and potentially territorial ♂♂ were seen in 3 consecutive recent years in suitable habitat: on 9 Jun 2012 (Baksh), 20 Jun 2013 (Pehek et al.), and 19 May 2014 (Buckley, Young). Singles were also at Inwood Park on 3 Jun 2003 (DiCostanzo), in the Bronx Zoo on 9 May 2016 (Olson), and in Bronx Park on 5 May 2008 (Neville) and 14 (E. Olsen) and 25 May 2016 (B. Purcell). A migrant *Empidonax* believed to have been an Acadian was in the Northwest Forest on 24 Aug 2015 (Rosenberg), but a few should occur in Aug–Sep.

The nearest current breeders may be on the Palisades, elsewhere in Bergen County, NJ, or in Westchester County, and they did breed in Alley Pond Park between 2000 and 2005. Singing Acadians are being seen in May in New York City parks and in 2007 nested in Central, Prospect, and Forest Parks, at Jamaica Bay Wildlife Refuge, and on Staten Island, although in each case for only 1–2 years

(Lindsay 2007). Territorial ♂♂ appeared and disappeared at various Long Island locations during this period, but such erratic breeding behavior is not unexpected in a (re)colonizing species at the edge of its range. Neither NYSBBA I in 1980–85 nor II in 2000–2005 recorded breeders in study area Blocks 5852B and 5952A, and NYSBBA II also noted fewer occupied blocks in Westchester and Rockland Counties than did NYSBBA I. On Long Island and Staten Island, NYSBBA I found them in 5 confirmed and 6 probable blocks, but NYSBBA II in only 1 confirmed and 8 probable blocks (McGowan and Corwin 2008).

Specimens
Griscom (1923) ascribed 2 of his Bronx Region specimens to Dwight (19 Sep 1885, 13 May 1887), and the AMNH database lists another 5 between 1885 and 1890, but all 7 were only near New York City without more precise locations. Bicknell recorded egg dates in Riverdale from 10–25 Jun, and he collected 4 there in May, Jun, Jul 1876–78 (NYSM).

Alder Flycatcher *Empidonax alnorum*

NYC area, current
Throughout the New York City area, a regular Neotropical migrant in small numbers in spring (mid-May–mid-Jun) and in fall (late Aug–late Sep), and a very local and disappearing breeder only in Westchester County and northward. Alders are easily overlooked unless captured for banding or their diagnostic songs or calls are heard, given that otherwise they are all but inseparable from Willow Flycatcher.

Bronx region, historical
Bicknell's *Riverdale* 1872–1901: uncommon spring migrant

Eaton's *Bronx + Manhattan* 1910–14: map on p. 40, vol. 1 showed breeding no closer than Orange–Rockland County border and not in Westchester County

Griscom's *Bronx Region* as of 1922: single singing bird on 30 May 1915 Van Cortlandt; also breeding in nearby NJ in Great Swamp

Kuerzi's *Bronx Region* as of 1926: uncommon local breeder [locations not given] and presumably also migrant (15 May–20 Aug)

NYC area, historical
Cruickshank's *NYC Region* as of 1941: uncommon spring/fall migrant (11 May–26 Sep), with 6–12 pairs breeding in Westchester County, most north of White Plains

Bull's *NYC Area* as of 1963: Willows [still unnamed] have emerged as increasing New York City area breeders while true Alders vanished northward; naming silent migrants impossible

Bull's *New York State* as of 1975: status unchanged

Levine's *New York State* as of 1996: status unchanged

Until Stein's work (1958, 1963) elucidated their true status, Alder and Willow had not been recognized as separate biological species. Their striking song differences (*fee-bee-o* and *fitz-bew*) had been remarked on for decades but were assumed to represent only regional, dialectal, or subspecific differences within the old Alder. Only Alder migrated through and bred in the New York City area, and it sang only *fee-bee-o*. Because Willow did not invade the New York City area until the late 1930s–early 1940s, it can be safely concluded that Alder/

Willow breeders and spring/fall migrants before then were true Alders. (See "Alder/Willow Flycatchers Unidentified to Species" and also "Willow Flycatcher").

Study area, historical and current
A regular Neotropical spring/fall migrant in small numbers, certainly much more numerous than the few records suggest.

In nearby Riverdale, Bicknell considered Alder a regular spring migrant (but not a breeder), with extremes of 17 May 1896–15 Jun 1876, 1879. If it occurred in fall, he was unaware of that. However, he did find Alders breeding at 4 locations (the number of pairs was unstated) within Jerome Meadows in 1888 and 1893. As far as we have been able to determine, these remain the southeasternmost breeders ever found in New York State. This raises the possibility that Alder was a vestigial breeder in the New York City area late in the 19th century only to be replaced 30–40 years later by more widespread Willow. Alder's departure was not necessarily from competition with Willow, as climatic warming may have been sufficient cause and could be somewhat analogous to the situation with Blue-winged and Golden-winged Warblers, in which the former has been advancing steadily northward as a breeder in tandem with the latter's retreat. Despite scattered areas of current Alder/Willow sympatry as close as NJ's Troy Meadows and Great Swamp, we are aware of neither mixed pairs nor hybrids there or anywhere else—while acknowledging how difficult their detection would be. We are also unaware of any recent nuclear + mitochondrial DNA work on Alder and Willow Flycatchers that would permit recognition of hybrids and introgressed genes in reciprocal species, as has been done in Pacific-slope and Cordilleran Flycatchers.

Both Kuerzi (1927) and Cruickshank (1942) reported a few breeding pairs in extreme southern Westchester County, perhaps meaning Grassy Sprain Reservoir, but none were breeding within New York City at those times. The very last for the Bronx and New York City, save one (see below), were Bicknell's at Jerome Meadows; an adult with a fledged young at the Baychester Marshes on 24 Jul 1923 (L. N. Nichols); and, presumably from the same pair, 1 singing there on 18 Jun 1925 (Hickey et al.).

As of 2016, there are only 9 Van Cortlandt records of singing spring migrant Alders, all but 3 fairly recent: 30 May 1915 (Rogers), 27 May 1934 (Weber, Jove), 22 Jun 1969 (Sedwitz), 24 May 1985 (Buckley), 20 May 1989 (Young), 18 May 1990 (Young), 25 May 1996 (Lyons, Garcia), 25 May 2008 (Lyons, Garcia), and 22 May 2010 (Baksh). So few singing Alders in Van Cortlandt only mirror its relatively low intensity of coverage. The sole fall *fee-bee-o*–singing Alder was examined closely on 21 Sep 1958 (Steineck). We are unaware of any study area migrants having been heard giving their diagnostically sharp *pik* call, which is quite unlike Willow's and Least's soft *quip*, but this distinction is recent and not widely appreciated.

Young saw silent Alder/Willow migrants in Riverdale on 23 May 2005 and in fall in Van Cortlandt on 27 Sep 1999, 25 Sep 2000, 12 Sep 2003, 7 Aug 2004, 8 Aug 2004, 4 on 18 Aug 2005, 29 Aug 2007, 3 Sep 2009. Other observers have occasionally reported them in the same Aug–Sep period when both species are known from banding studies to be migrating southbound through the New York City area. Many of these fall migrants, maybe even most, would have been Alders, not Willows, given that its breeding range catchment north and west of the New York City area is far larger than Willow's.

The only other recent Bronx Alder migrants, all singing, were in Bronx Park on 16 May 2010 (Finger) and in the Bronx Zoo on 25 May 2013 (Bochnik). Almost as close were singles at Inwood Park on 3 Jun 2003 (DiCostanzo) and 15 May 2016 (Souirgi), and 1–2 there from 23–29 May 2014 (Souirgi, DiCostanzo et al.).

With this history, most unexpected was Farrand's careful report of a singing ♂ Alder in Van Cortlandt Swamp in Jun 1977 attending a nest whose construction clearly matched Alder's and differed strikingly from that of adjacent Willows. Given that ♀ Alders alone build their nests (Gorski 1969), this indicates a genuine breeding pair. It is unknown whether they were successful and young were fledged, because Farrand was unable to follow up and word was not spread until too late. This is the sole *modern* Alder Flycatcher breeding event in the study area and in New York City.

The nearest currently breeding Alders may be somewhere in northern Westchester County or in NJ at Troy Meadows and the Great Swamp. During NYSBBA I in 1980–85, Alders were reported in 5 possible or probable blocks in eastern/southeastern Westchester County and 4 possible or probable blocks on extreme eastern Long Island, but all 9 blocks were vacant by NYSBBA II in 2000–2005. The 4 during NYSBBA I and a different possible during NYSBBA II are the only potentially breeding Alders ever reported on Long Island, but none were ever verified. It seems more likely they were late migrants or nonbreeders. It is worth noting, though, that in 1965 2 pairs of Alders did breed at Westchester County Airport (Zupan), but singers at Rye, Purchase, and New Castle on 13–14 Jun 1998 (Greenwich-Stamford Summer Bird Count) may have been late migrants, nonbreeders, or potential breeders.

Alas, DOIBBS data combined Alder and Willow, so their respective New York State trends are not demonstrable (Sauer et al. 2007; McGowan and Corwin 2008). But NYSBBA II found Alder to have increased its number of occupied blocks by 45% since NYSBBA I, even as it retracted its distribution away from the immediate New York City area (McGowan and Corwin 2008).

Specimens
Bicknell collected 1 in Riverdale on 25 May 1876 (NYSM).

Alder/Willow Flycatchers Unidentified to Species

Until the mid-1980s conventional wisdom held resolutely that all silent *Empidonax* were unidentifiable (some still believe this). Then in a series of landmark papers Whitney and Kaufmann (1985a, 1985b, 1986a, 1986b, 1987) demonstrated this to be false, and astute observers began scrutinizing and then identifying many/most of their silent *Empidonax*, in the process frequently experiencing great satisfaction when one vocalized and proved its morphology-only identification. The most refractory to field separation have consistently been silent Pacific-slope/Cordilleran and silent Alder/Willow Flycatchers. But it was a revelation to most observers that the latter two species, which helpfully call often in migration and winter, are no more difficult to separate by voice than are Short- and Long-billed Dowitchers. Willow's soft *quip* stands apart from Alder's sharp *pik* or *peek* without the need for sonograms.

Often if silent Alder/Willows are examined closely, tall-crested darker heads, generally darker green dorsal color, and a more obvious eye ring, will point to Alder, while flatter, paler (often even grayish) heads, grayish-olive dorsal color with hints of brown, and an incomplete, fainter eye ring will point to Willow. Most such individuals are probably the species we think they are, but it is always rewarding when in spring they sing after having been scrutinized or in fall when they call fairly frequently—and proclaim their identities. These days with smart phones and excellent apps, one can query almost any suspected Alder/Willow with song or call at any season. Many will respond, an easy way to learn the subtleties of posture and plumage that characterize the two species.

Confounding the problem, the well-known English name Alder Flycatcher (*sensu lato*) changed to Traill's Flycatcher between 1957 and 1973—ending in the latter year when Willow and Alder were finally split in the AOU Checklist, the more northerly *fee-bee-o* singers becoming Alder (*sensu stricto*) and the newly colonizing, more southerly *fitz-bew* singers becoming Willow. Both names indicate well their respective preferred breeding vegetation. Unhelpfully, even after the AOU split the Bird Banding Laboratory maintained the invalid name Traill's Flycatcher for all banded birds unable to be assigned to Alder or to Willow. Regrettably, this spurious name (and not Alder/Willow) was then kleptoparasitized by writers, leading to an enduring literature morass.

Willow Flycatcher   *Empidonax traillii*

NYC area, current
A regular Neotropical migrant in small numbers in spring (mid-May–mid-Jun) and in fall (late Aug–late Sep), but unless diagnostic songs or diagnostic calls are heard, most are only tentatively identifiable. It is also a thinly but widely distributed breeder in all 5 New York City boroughs, which colonized the area in the late 1930s–mid-1940s and then spread quickly through it. Included within Traill's Flycatcher (formerly Alder Flycatcher) by the AOU from 1957 until 1973, when Willow and Alder were finally spilt as separate biological species.

NYC and study areas, historical
Historically, from Bicknell's Riverdale in 1872–1901 to Kuerzi's Bronx Region in 1926, *fitz-bew* singers (now known to be Willow Flycatchers) were never detected. Griscom's New York City Region (1922) reported Alders (without question *fee-bee-o* singers) breeding in NJ's Great Swamp and at Troy Meadows, so they were indeed modern Alders. Yet by 1941 Fischer found only Willows there, so sometime between 1923 and 1941 they moved in and most (but not all) Alders moved out. Cruickshank's NYC region as of 1941 did not mention any *fitz-bew* singers even though he reported Alder Flycatchers breeding in Queens—which we know today to have been the first breeding Willows in the New York City area (see below). Bull (1964, 1970) offered a good analysis of what was understood at the time, with information about migrants and breeders of undefined song-type as well as extensive discussion of the 2 song-types and their New York City area breeding distributions. (See additional discussion in the preceding taxon entry.)

Little appreciated is that the first unequivocal *fitz-bew*–singing *Empidonax* flycatchers ever breeding in the New York City area were detected by Fischer (1941, 1950) at Kissena Park, Queens, from 1937–41 (not 1939 of earlier authors). He also offered the useful comment (Fischer 1950) that every Alder Flycatcher he had ever heard in the New York City area was a *fitz-bew*–singer, but regrettably, beyond mentioning that as the only song type heard at Troy Meadows, NJ, during the 1940s, he gave readers no clue when and where else he had heard them. He did, however, carefully comment that their call note—an emphatic *whit* indistinguishable from that of Least Flycatchers—differed clearly from the sharp *pik* or *peek* call of *fee-bee-o* singers. In the Connecticut River valley of MA, Bagg and Eliot (1937) found MA's first-ever Willow in

Longmeadow on 6–12 Jun 1932 and noted that its song was "manifestly the same as the *fitz-bew* that Peterson (1934: 98) attributes to Ohio [Alders]." The cusp of the vanguard, perhaps.

The Bronx and study area occurrence of only Alder in the Alder/Willow duo ended when the first *fitz-bew* Willows—assuredly not present during the extensive breeding bird censuses of Van Cortlandt Swamp in 1937–38 (Sialis Bird Club)—were firmly first identified in Van Cortlandt Swamp on 26 May 1946 (Kieran, Eisenmann), a year when 6 pairs nested, hinting at a somewhat earlier arrival. They have bred there annually ever since, usually 4–5 pairs (Norse, Young), with 4–6 pairs in 2014.

The earliest study area arrival is 3 May 2011 (Young), and while most formerly arrived >15 May, since the 1980s this date has been slowly advancing in the New York City region to the point where 1–10 May arrivals no longer invite comment. Migration peak varies annually but is in late May–early Jun, continuing through mid-Jun. Daily maxima are all from breeders. Late spring and early fall migrants are clearly confounded with breeders, but the first in fall arrive in mid–late Aug, peak in the first half of Sep, and depart by the end of that month. The latest singing dates for presumed local breeders are 3 on 25 Jul 2004 (Klein) and 2 on 25 Aug 1955 (Buckley). There is only a single report of migrating fall Willow(s), on 10 and 17 Sep 2011 (Baksh), but it lacks identification details.

Unfortunately DOIBBS data combined Alder and Willow, so their respective New York State trends are not demonstrable (Sauer et al. 2007). However, NYSBBA II in 2000–2005 found Willow to have increased its number of occupied blocks by 36% since NYSBBA I in 1980–85, as it enlarged its northward range while maintaining or even expanding its distribution in the southern part of the state (McGowan and Corwin 2008).

Comments

What follows must be treated with some circumspection, given that external morphology-only separation of migrant Alder and Willow Flycatchers remains exceedingly difficult (see Pyle 1997).

Two AMNH specimens of migrating Willow Flycatcher taken by Dwight on 26 Sep 1889 (AMNH #369807) and on 8 Sep 1890 (AMNH #369806) "near New York City" merit discussion. They seem to have no more exact locations than that, but many of Dwight's New York City/Westchester specimens taken at that time were so marked. Both of these are labeled *Empidonax traillii brewsteri*, whose breeding range is west of the Cascades and in the Sierra Nevada from southwestern British Columbia south to southwestern California (Sedgwick 2000). This subspecies was described by Oberholser from Nevada in 1918 as the western North American subspecies of "Alder" long before Willow was separated from Alder by Stein (1958, 1963). (Monotypic Alder does not breed in the contiguous United States west of northeastern ND.) Then in 1932 *adastus*, breeding east of the Cascades from MT to NM, was further separated by Oberholser from *brewsteri*. These 2 New York Dwight Willow specimens may have been specifically and subspecifically identified by Allan R. Phillips on one of his AMNH trips, and if correct would be among the few (only?) examples of this taxon from eastern North America and genuine vagrants. (Phillips normally initialed and dated his identifications in pencil on original labels.) But even if they eventually proved to be *adastus* or the eastern subspecies of Willow (nominate *traillii*), they would still be the first Willows recorded in the New York City area, 50 years before the first known breeders began to arrive in the late 1930s.

Least Flycatcher *Empidonax minimus*

NYC area, current
A regular but uncommon spring/fall Neotropical migrant and a vanishing breeder that may no longer nest within city limits that was not detected anywhere on Long Island during NYSBBA II in 2000–2005. It has bred in Central Park but not in Prospect. This is another Neotropical migrant that has been seen with greatly diminished frequency and in much smaller numbers since the 1960s, but if the breeding grounds and not the neotropics are responsible, forest maturation could be the culprit.

Bronx region, historical
Bicknell's *Riverdale* 1872–1901: common breeder, spring/fall migrant

Hix's *New York City* as of 1904: common migrant [breeding unmentioned]

Eaton's *Bronx + Manhattan* 1910–14: common breeder

Griscom's *Bronx Region* as of 1922: regular migrant; ex-common breeder now extirpated

Griscom's *Riverdale* as of 1926: nearly extirpated breeder

Kuerzi's *Bronx Region* as of 1926: fairly common breeder [locations not given] 21 Apr–4 Oct

NYC area, historical
Cruickshank's *NYC Region* as of 1941: common breeder, spring (maximum several dozen)/fall migrant (21 Apr–4 Oct); steadily decreased as breeder around New York City, suburbs [locations not given]

Bull's *NYC Area* as of 1968: 20–30 seen in May and Aug–Sep but counts may include other *Empidonax*; early spring date 20 Apr, late fall date 9 Oct

Bull's *New York State* as of 1975: late fall date 8 Nov; fall maximum 13

Study area, historical and current
An inconspicuous spring/fall migrant and extirpated breeder. It was not until the 1980s that observers were able to confidently identify many silent *Empidonax* (Whitney and Kaufmann 1985a, 1985b, 1986a, 1986b, 1987), so there are no firm identifications of silent migrant *Empidonax* other than Yellow-bellied before then. And while the majority almost certainly were and are Leasts, few observers have taken the trouble to record them.

In spring, first migrants now arrive in early May, peak in mid-May, and continue well into Jun, with extreme dates of 4 May 1955 (Buckley) and 2 Jun 1953 (Buckley) and a maximum of 2 on 22 May 1996 (Young).

It is odd to us in 2016, but in CT Leasts were reportedly common town, village, and even city (New Haven, Stamford) breeders from the mid-1880s through the 1930s (Zeranski and Baptist 1990), and when they also began to decline, as they did in MA, it was attributed to increased numbers of House Sparrows and loss of apple orchards (Griscom and Snyder 1955). In New York, Bicknell called it a common breeder in Riverdale from 1872–1901, and Eaton (1910, 1914) considered it equally at home in forests and in settled districts. But by 1923, Griscom pointedly remarked that it did not tolerate more civilized conditions, and in 1942 Cruickshank observed that while it "does not object to human proximity and occasionally nests in village parks and streets, it evidently does object to too much construction and confusion and has steadily decreased around New York City and its crowded suburbs."

From 1938–52, 2 pairs bred in Van Cortlandt on the Ridge (Norse), where the last nest was found on 8 Jun 1952 (Buckley, Penberthy). In addition, a pair bred in Van Cortlandt Swamp in 1961 and 1962 (Heath et al.), and breeding also occurred elsewhere in Van Cortlandt until 1964. There was also a small breeding cluster at

Grassy Sprain Reservoir through the mid-1950s (Weber, Buckley et al.). While it is possible that a pair may breed in Van Cortlandt occasionally, it must be regarded as extirpated. Until the mid-1980s the nearest breeders may have been at Pelham Bay Park (Künstler), but now they are no closer than central Westchester County.

Neither NYSBBA I in 1980–85 nor NYSBBA II in 2000–2005 recorded any breeders in New York City or Nassau County, and about 25 occupied blocks in Westchester County had dropped to 5 by NYSBBA II (McGowan and Corwin 2008), although singing breeders are often dismissed as migrants. DOIBBS data showed a 2.5% annual decrease statewide from 1980–2005 (Sauer et al. 2005). However, given that the New York City area is at or near Least's southern limit east of the Appalachians, retreat of its breeding range northward may only be tracking climate warming.

In fall, judging from Fire Island banding results (Mitra et al. MS), first migrants arrive in late Jul, peak in late Aug–early Sep, and depart by mid-Sep; in the study area the few extreme dates range between 23 Jul 1972 (Young) and 17 Sep 2001 (Young), all singles. Bicknell collected 1 in Riverdale as late as 4 Oct 1881 (Griscom 1927). A lingering blowback drift-migrant was in Bronx Park from 25 Nov–9 Dec 2017 (Evans et al.; photos).

Specimens
Bicknell collected 3 in Riverdale in May, Aug, Oct 1876–80 (NYSM).

Comments
Two very closely related blowback drift-migrant *Empidonax* have been vagrants to the New York City area. **Cordilleran Flycatcher**, *E. occidentalis*, was mist-netted at Fire Island Lighthouse 14–16 Sep 1995 (Buckley and Mitra 2003), and **Pacific-slope Flycatcher**, *E. difficilis*, was in Central Park 18–24 Nov 2015 (Post et al.) and possibly in Inwood Park 8–10 Dec 2016 (Keane et al.; photos). A third blowback drift-migrant vagrant *Empidonax*—**Hammond's Flycatcher,** *E. hammondii*—was in Central Park from 26 Nov–12 Dec 2017 (LaBella et al.; photos).

Eastern Phoebe   *Sayornis phoebe*

NYC area, current
A widespread and numerous spring/fall migrant but a very local and declining breeder in each New York City borough except Manhattan; rarely occurs in winter.

Bronx region, historical
Bicknell's *Riverdale* 1872–1901: common breeder, spring/fall migrant

Hix's *NYC* in 1904: common breeder Bronx Park

Eaton's *Bronx + Manhattan* 1910–14: common breeder

Griscom's *Bronx Region* as of 1922: common breeder and migrant

Griscom's *Riverdale* as of 1926: now extirpated breeder

Kuerzi's *Bronx Region* as of 1926: common breeder [locations not given], spring/fall migrant (5 Mar–26 Nov); 2 recent Jan records

NYC area, historical
Cruickshank's *NYC Region* as of 1941: common breeder and spring/fall (maximum 30+) migrant, although rare breeder near New York City [locations not given]; very rare in winter, especially inland

Bull's *NYC Area* as of 1968: now reported every winter

Bull's *New York State* as of 1975: maxima spring 35, fall 85

Study area, historical and current

A conspicuous spring/fall migrant, infrequent winter resident, and scarce or irregular breeder in Van Cortlandt, Woodlawn, and probably also at Jerome and Hillview.

In spring, first migrants arrive in late Mar, peak in early Apr, and depart by early May, with extreme dates of 28 Feb 2007 (Young) and 18 May 1957 (Weintraub), and a maximum of 15 on 27 Mar 2007 (Klein). Until Jun 1967, the spring maximum in Central Park was 15 on 15 Apr 1943 (Aronoff).

In the late 1800s, it was a common breeder in Riverdale (Bicknell) and in Van Cortlandt (Dwight), continuing through the first half of the 20th century but in declining numbers. Total breeding numbers were generally unrecorded, but in 1955, 2 pairs bred in Van Cortlandt plus 3 pairs in Riverdale (Kane, Buckley). Both NYSBBAs found breeders in study area Block 5852B (McGowan and Corwin 2008).

There may be almost as many study area breeding pairs as there are bridges over water or structures near water bodies, but probably <7–8 pairs. Most are dismissed as migrants or overlooked when they become almost completely silent following nest construction as early as Apr. As of 2016 in the study area, single pairs may be breeding most years at each of the 3 golf course ponds, around Van Cortlandt Lake, and at the ponds between the lanes of the Henry Hudson Parkway where there was a territorially behaving pair on 7 May 2002 (Jaslowitz). Recent late singles in Van Cortlandt on 13–16 Jun 1980 (Sedwitz, Ephraim), 4 Aug 1991 (Collerton), 1 Jun 1995 (Garcia, Lyons), 21 Jun 2004 (Klein), 15 Jun 2006 (Klein), 25 Jun 2011 (Baksh), 9 Aug 2014 (Souirgi), 27 Jun 2015 (Willow), 2–3 from 12–27 Aug 2015 (Aracil, Hudda), and Woodlawn on 31 May 2014 (Winston), were probably all local breeders. Single pairs may also be breeding, if intermittently, near the nursery above the Sycamore Swamp and at Jerome and Hillview. Nearest other Bronx breeders are singles or a few pairs each in Riverdale, Bronx Park, Bronx Zoo, and Pelham Bay Park.

NYSBBA I in 1980–85 and II in 2000–2005 had breeders in all Westchester County blocks almost to the Bronx line, but within New York City there were few occupied blocks: in the Bronx (2), Staten Island (2), Brooklyn (1), Queens (2). Southern Nassau County lost blocks in II, but Suffolk County's losses were offset by gains (McGowan and Corwin 2008). DOIBBS data showed no consistent statewide pattern, just fluctuating increases and decreases from 1966–2005 (Sauer et al. 2005).

In fall, first migrants arrive in early Sep, peak in early–mid-Oct, and depart by late Oct, with extreme dates of 8 Sep 2005 (Young) and 26 Nov 1954 (Buckley), and a maximum of 12 on 26 Sep 1946 (Komorowski). Until Jun 1967, the fall maximum in Prospect Park was 24 on 9 Oct 1944 (Soll).

The first study area winter records were on 27 Dec 1936 (Norse et al.) and then on 7 Feb 1937 (Peterson)—overwintering?—but not until the 1950s did they became regular through late Nov, and every 5–8 years into late Dec. Phoebes may have successfully overwintered 2–3 times, but data are sparse. Recently seen in Woodlawn on 28 Dec 1997 (Lyons et al.), and in Van Cortlandt from 21 Dec 2002–9 Jan 2003 (Lyons et al.), from mid-Jan–11 Feb 2012 (Fiore et al.), and 2 (the maximum) through Jan 2013 (Young et al.). One in Riverdale 3 Feb 2003 may also have been attempting to overwinter (unknown observer, *Kingbird*). In winter they often take up residence near open water.

Specimens

Foster collected 1 in Woodlawn on 5 Apr 1890 (AMNH) and Bicknell 2 in Riverdale in Mar 1877–79 (NYSM).

Comments

Blowback drift-migrant **Say's Phoebe**, *S. saya*, is an annual fall vagrant to the Northeast with most locally on coastal Long Island, so one in Valhalla on 16 Sep 2007 (Panko), Westchester County's first, was odd only as to location. Wholly unexpected as to both location and date, one in Prospect Park on 27 Apr 2013 (Bass et al.) was the first in spring in the New York City area and probably overwintered somewhere in the Southeast.

Great Crested Flycatcher   *Myiarchus crinitus*

NYC area, current
A regular spring/fall Neotropical migrant and a widespread but thinly distributed breeder in all 5 New York City boroughs and surrounding counties.

Bronx region, historical
Bicknell's *Riverdale* 1872–1901: common breeder, spring/fall migrant

Hix's *NYC* in 1904: common breeder Bronx Park

Eaton's *Bronx + Manhattan* 1910–14: fairly common breeder

Griscom's *Bronx Region* as of 1922: uncommon breeder, common migrant

Kuerzi's *Bronx Region* as of 1926: common breeder, spring/fall migrant (28 Apr–14 Oct)

NYC area, historical
Cruickshank's *NYC Region* as of 1941: common breeder, spring (maximum 32)/fall migrant, 17 Apr–25 Oct; disappearing as breeder on outskirts of New York City [locations not given]

Bull's *NYC Area* as of 1968: late fall date 8 Nov, possibly later but cf. below [blowback drift-migrants]

Bull's *New York State* as of 1975: status unchanged

Study area, historical and current
A regular spring/fall Neotropical migrant and widespread breeder in Van Cortlandt and Woodlawn.

In spring, first migrants arrive in early May, peak in mid-May, and depart by late May, with extreme dates of 25 Apr 1961 (Horowitz) and 7 (the maximum) on 21 May 1958 (Horowitz). Until Jun 1967, the spring maximum in Central Park was 12 on 7 May 1964 (Post).

In the late 1800s it was a common breeder in Riverdale (Bicknell) and in Van Cortlandt (Dwight), a status that continues unchanged. While total breeding numbers are unknown, in 1953 3 pairs bred in just a small strip of Van Cortlandt woods along Central Ave. (Komorowski), so there could easily be 10–20 pairs throughout the study area, with <15 in Van Cortlandt.

Crested's New York City distribution was quite similar to Eastern Phoebe's, as were its gains and losses there and on Long Island, but both remained quite widely distributed (McGowan and Corwin 2008). DOIBBS data showed little statewide change from 1966–2005, although there was a 1.3% annual decrease after 1980 (Sauer et al. 2005).

In fall, first migrants arrive in early Aug, peak in late Aug, and depart by late Sep, with extreme dates of 2 Aug 1959 (Horowitz) and 5 Oct 1934 (Cantor) and a maximum of 5 on 24 Aug 1996 (Young).

Specimens
Foster collected 1 in Woodlawn on 17 May 1890 (AMNH) and Bicknell 2 in Riverdale in May, Aug 1876–79 (NYSM).

Comments
Any mid-Oct or later Great Crested Flycatcher should be scrutinized and photographed

whenever possible, as the vast majority of such late *Myiarchus* flycatchers in the Northeast—(lingering) blowback drift-migrants—have since the mid-1980s proved to be not Great Crested but vagrant western **Ash-throated Flycatchers**, *M. cinerascens*. Singles were in Larchmont on 22–24 Nov 1971 (Bahrt et al.), and at nearby Wave Hill in Riverdale on 24 Nov 1993 (Tudor, McGuinness et al.). They have also been found in Central and Prospect Parks.

Western Kingbird *Tyrannus verticalis*

NYC area, current
A regular but rare fall Neotropical migrant and blowback drift-migrant into Nov that occasionally lingers into winter and now sometimes occurs in spring, often into Jun. Most are on or near Long Island beaches, but a few have been inland. Recorded in Central and Prospect Parks.

NYC and study areas, historical and current
On 19 Oct 1875 Bicknell collected an HY ♂ in Riverdale a short distance west of Van Cortlandt, the first New York State record of this species; the second was not detected—on Long Island—until 37 years later. During the 1930s it was seen with increasing frequency following observers' greater time afield on Long Island barrier beaches, where it is most frequent. By 1942, Cruickshank was calling it a rare but annual fall migrant (maximum of 6) on Long Island's South Shore (19 Aug–1 Jan); Riverdale's remained the sole inland record. By 1964, Bull noted its increasing records, spread of dates (14 Aug–14 Jan), inland reports (2 more), number of early winter/Christmas Bird Count occurrences, and the first 2 in spring. Bull's (1974) status was unchanged.

It remains a rarity away from Long Island, and there are only 2 study area records: an HY came down the Ridge with migrating raptors in Van Cortlandt on 18 Sep 1956 (Post), and another HY was photographed in Woodlawn on 24 Sep 1960 (Horowitz, Buckley et al.).

In the Bronx it must occur more frequently than the few records would suggest. It has been at Baxter Creek in Sep/Oct 1954 (Buckley 1958), in Bronx Park on 17 Sep 1971 (Hackett) and 11 Nov–16 Dec 2006 (Matsushito et al.) and also in the Bronx Zoo on 16 Dec 2006 (McGann)—presumably the same bird, and in Pelham Bay Park on 14 Sep 1988 (DeCandido et al.) and 12 Nov 1989 (Schulze, Iula). Another was at Fort Tryon Park on 3 Nov 1967 (Gracies).

Specimens
Bicknell's 1875 Riverdale specimen's disposition was unmentioned by all authors from Eaton (1910, 1914) to Levine (1998), and it did not appear in VertNet output from any of the museums with significant Bicknell holdings (AMNH, ANSP, NYSM, USNM). It may be lost, undigitized, or an uncatalogued display mount.

Comments
Southwestern—and recently southeastern—**Scissor-tailed Flycatcher**, *T. forficatus*, now occurs almost annually in the New York City area as a vagrant with complex origins. It is known in the Bronx from only 2 records: in Riverdale on 21 May 1957 (Kallmann) and at Pelham Bay Park on 5 May 2008 (fide DeCandido; photos). It has been found in Prospect but not Central Park.

Eastern Kingbird *Tyrannus tyrannus*

NYC area, current
A common spring/fall Neotropical migrant and localized breeder that is unrecorded in winter.

Bronx region, historical
Bicknell's *Riverdale* 1872–1901: common breeder, spring/fall migrant

Hix's *New York City* as of 1904: common breeder

Eaton's *Bronx + Manhattan* 1910–14: common breeder

Griscom's *Bronx Region* as of 1922: rare breeder but common migrant

Kuerzi's *Bronx Region* as of 1926: common breeder, spring/fall migrant (28 Apr–14 Oct)

NYC area, historical
Cruickshank's *NYC Region* as of 1941: common breeder, spring (maximum 50)/fall (maximum several hundred) migrant, 4 Apr–24 Oct; even though few pairs still breed within New York City is disappearing from ever-expanding suburbs

Bull's *NYC Area* as of 1968: late fall date 11 Nov [blowback drift-migrant]; fall maximum 900

Bull's *New York State* as of 1975: status unchanged

Study area, historical and current
A common spring/fall Neotropical migrant and uncommon local breeder.

In spring, first migrants arrive in late Apr, peak in mid-May, and depart by late May, with extreme dates of 6 Apr 1952 (Buckley)—a trans-Gulf overshoot as was a Yellow-throated Vireo the same day in Roslyn Park, Long Island (P. Gillen)—and 28 May in 1933 (Weber, Swift) and 1944 (S. Horowitz), with a maximum of 10 on 8 May 1955 (Kane).

Total breeding numbers are uncertain, but there are <12 pairs breeding annually in all 4 extant subareas (e.g., 4 on 12 Jun 2010: Baksh), with <5 in Van Cortlandt. Both NYSBBA I in 1980–85 and NYSBBA II in 2000–2005 affirmed this species' widespread distribution in and around New York City, with only 4 unoccupied blocks in the densest urban areas (McGowan and Corwin 2008). DOIBBS data showed a 2.5% annual statewide decrease from 1980–2005 (Sauer et al. 2005).

In fall, first migrants arrive in early Aug, peak in late Aug, and depart by mid-Sep, with extreme dates of 21 Jul 1966 (Kane) and 14 Oct 1922 (Hix), and a maximum of 50 on 30 Aug 1952 (Kane). There is a single study area reference—among very few for the entire New York City area—to the southbound premigratory flocks of hundreds that gather to fatten up in some northeastern states. Hix in Dec 1916 remarked that the number of Eastern Kingbirds gathering in Van Cortlandt Park during fall migration was remarkable, at times reaching 500 chiefly in the bushy area above the lake (reported verbally, as noted in the 1917 *Proceedings of the Linnaean Society of New York*, nos. 26–27). This would have been in the general area now occupied by the Van Cortlandt golf course and the Major Deegan Expressway, but flocks might still gather elsewhere in the park. These multiple-day staging flocks are very different from the actively migrating diurnal flocks of kingbirds sometimes seen along the immediate coast. Until Jun 1967, the fall maxima in Central and Prospect Parks were, respectively, 817 migrating diurnally on 29 Aug 1956 (Post) and 25 on 19 Aug 1944 (Soll).

Comments

The most enigmatic bird ever found in the Bronx was the **Great Kiskadee**, *Pitangus sulphuratus*, photographed on a Spuyten Duyvil apartment house patio on 11 Sep 2011 (Brickner, Brickner). Unexpectedly, one had also been photographed 8 smi (14 km) south on a midtown Hudson River pier on 31 Aug 2011 (fide Votta), 3 days after the eye of Hurricane Irene had passed directly overhead. The existing images of each are adequate to confirm that only a single SY individual (P. Pyle, pers. comm.) was involved.

Species identity was never at issue, but what of origins? As insectivores, kiskadees are difficult to keep in captivity, so an escape or release was unlikely, and occurrence at the time of a proven bird-storm like Irene intimated storm transport. Great Kiskadee is not a prominent vagrant but belongs to a family with exceptionally high vagrancy rates; kiskadees have reached South Dakota, Colorado, Kansas, and Louisiana (multiples), great distances from South Texas. Some thought the large population of introduced Great Kiskadees on Bermuda may have been this bird's source, and even if so, others dismissed it as probably ship-assisted because it was discovered near where Bermuda-bound vessels dock. But assisted passage (not to mention confinement or food provisioning) can rarely be refuted *or* demonstrated, and owing to its ubiquity, many regard it as little different from winter feeding. In the final analysis, all questions about this bird's origin and status remain open.

Shrikes: Laniidae

Loggerhead Shrike   *Lanius ludovicianus*

NYC area, current
Especially on Long Island barrier beaches, until the 1960s this species was an annual but rare Apr–May and Aug–Sep migrant (1–3 per day) and an occasional winter visitor/winter resident. It has always been exceedingly scarce in Central and Prospect Parks. It is now rarely seen anywhere in the area and is a New York State Endangered Species.

Bronx region, historical
Bicknell's *Riverdale* 1872–1901: once fall [location not given]

Eaton's *Bronx + Manhattan* 1910–14: rare migrant

Griscom's *Bronx Region* as of 1922: exceedingly rare fall migrant, 2 records

Kuerzi's *Bronx Region* as of 1926: irregular, uncommon fall migrant (4 Aug–16 Dec), once 24 Apr [locations not given]

NYC area, historical
Cruickshank's *NYC Region* as of 1941: very rare migrant in spring (22 Mar–3 May)/ uncommon fall (4 Aug–late Oct; maximum 4); very rare but increasing on Long Island in winter

Bull's *NYC Area* as of 1968: late spring date 28 May

Bull's *New York State* as of 1975: early fall date 2 Aug

Study area, historical and current
A formerly rare but annual spring/fall migrant and scarce winter visitor that all but disappeared in the 1960s and is known only from Van Cortlandt and Jerome.

In the late 1960s the Loggerhead population crashed in the Northeast, and since then it has been unrecorded in the study area, although there are hints of a slow northeastern recovery, locally evidenced by a few recent coastal Long Island fall migrants. NYSBBA II in 2000–2005 found this disappearing breeder to have decreased its number of occupied blocks since 1980–85 by a further 83% to only 4 far upstate possibles (McGowan and Corwin 2008).

In spring, first migrants, all singles, formerly arrived and peaked in Apr, with extreme dates of 12 Apr 1930 (Cruickshank) and 3 May 1930 (Sedwitz). Until Jun 1967, the spring maximum in Central Park was 3 on 8 Apr 1954 (Messing).

In fall, first migrants arrived in mid-Aug and peaked in Sep, with extreme dates of 6 Aug 1966 (Kane) and 18 Oct 1944 (Komorowski), all singles.

There are 3 winter records of singles: 7 Jan 1932 (Cruickshank); 8 Sep–25 Dec 1951 (Komorowski, Kane et al.); 24–27 Dec 1955 (Boyajian, Sedwitz et al.). The very last study area records were in Van Cortlandt on 6 Aug 1966 (Kane) and at Jerome on 19 Aug 1967 (Sedwitz)—now a half century ago.

Northern Shrike   *Lanius excubitor*

NYC area, current
A formerly annual winter visitor/resident in small numbers that were occasionally greatly increased during major irruptions, as in 1921–22, 1926–27, 1930–31, 1949–50, 1978–79, 1995–96 (considered the largest ever), 1999–2000, and 2007–8. Always far more common on Long Island than inland, and in the Bronx on Long Island Sound shores.

Bronx region, historical
Bicknell's *Riverdale* 1872–1901: recorded 4 of every 5 winters, twice late Oct

Eaton's *Bronx + Manhattan* 1910–14: uncommon winter visitor

Griscom's *Bronx Region* as of 1922: occasional winter visitor

Kuerzi's *Bronx Region* as of 1926: irregular winter visitor, occasionally numerous in irruptions (30 Oct–22 Apr) [locations not given]

NYC area, historical
Cruickshank's *NYC Region* as of 1941: erratic, uncommon to rare winter visitor (12 Oct–25 Apr), periodically numerous during irruptions (maximum 6)

Bull's *NYC Area* as of 1968: Long Island winter maximum 12, late spring date 28 Apr

Bull's *New York State* as of 1975: status unchanged

Study area, historical and current
A formerly annual winter visitor/resident in small numbers, which since the 1960s has become an erratic, irruptive winter visitor often unrecorded for decades.

In Riverdale between 1872 and 1901, Bicknell recorded it 4 out 5 winters, the earliest appearing on the last few days of Oct, and collected 2 at Kingsbridge Meadows. It was obviously far more regular then as there are only 14 subsequent study area records (all but 1 in Van Cortlandt): 1 Mar 1927 (Kieran); 22 Feb–6 Mar 1928 (Cruickshank, L. N. Nichols); 30 Dec 1928 (Kieran); 2 Mar 1930 (Sedwitz); 9 Nov 1930 (Kieran); 9 Jan–12 Feb 1931 (Cruickshank, Kieran); 14 Mar 1931 (Sedwitz); Hillview on 19 Jan 1937 (Norse); 1 Dec 1938 (Kieran); 27 Feb 1943 (Kieran); 28 Dec 1949 (Kieran); 14 Feb 1950 (Kieran); adult and immature 2 Jan–27 Mar 1953, and on the last date the adult was in Brown Thrasher-like song (Kane, Buckley et al.); 29 Jan–1 Feb 1996 (Lyons, Garcia). One was barely outside the study area in Tibbett's Brook Park on 23 Dec 1995 on the Bronx-Westchester Christmas Bird Count.

Specimens
Bicknell collected 2 at Kingsbridge Meadows on 31 Mar 1877, 25 Nov 1875 (NYSM).

Vireos: Vireonidae

White-eyed Vireo   *Vireo griseus*

NYC area, current
An annual but uncommon spring/fall Neotropical migrant and very local breeder most numerous on eastern Long Island and Staten Island.

Bronx region, historical
Bicknell's *Riverdale* 1872–1901: common spring/fall migrant, breeder; earliest arrival 29 Apr 1880

Hix's *New York City* as of 1904: bred Bronx Park

Eaton's *Bronx + Manhattan* 1910–14: common breeder

Griscom's *Bronx Region* as of 1922: formerly common breeder but rare last 10 years; irregular migrant

Griscom's *Riverdale* as of 1926: formerly common breeder, but now extirpated; arrived as early as late Apr 3 times

Kuerzi's *Bronx Region* as of 1926: fairly common breeder, 29 Apr–2 Oct [locations not given]

NYC area, historical
Cruickshank's *NYC Region* as of 1941: very rare breeder near New York City yet a few pairs still nest [locations not given]; uncommon to rare spring (maximum 24)/fall migrant (21 Apr–17 Oct)

Bull's *NYC Area* as of 1968: fall maximum 20 dead Fire Island Lighthouse 1883; early spring date 15 Apr, late fall date 28 Oct

Bull's *New York State* as of 1975: early spring date 29 Mar, late fall date 22 Nov [blowback drift-migrant]

Study area, historical and current
Now an extirpated breeder and scarce spring/fall migrant.

In spring, first migrants arrive in early May, peak in mid-May, and depart by late May, with extreme dates of 30 Apr in 1959 (Horowitz) and 2015 (Furgoch), and 15 May 1958 (Heath), all singles save 2 on 9 May 1961 (Tudor).

From 1952–54, 3 pairs nested on the Ridge in Van Cortlandt (Buckley, Komorowski et al.)

but not certainly since then. It may have increased slightly in the New York City area as a breeder, but its disappearance from Van Cortlandt was due to successional changes, disturbance, or both. A single pair bred in Riverdale in 1954 (Kieran). One in the Northwest Forest on 27 Jun 2015 (Hudda) may have been a local breeder or a postbreeding disperser from southern Westchester County.

NYSBBA I in 1980–85 found this species quite widespread within and near New York City, including about 36 blocks in Westchester County, 9 on Staten Island, 4 in Queens, and 2 in Brooklyn, but by NYSBBA II in 2000–2005 Westchester County was down to 10 blocks, although New York City was holding its own. Neither NYSBBA I nor II recorded breeders in study area Blocks 5852B and 5952A, but 2 different Northeast Bronx blocks were occupied during the 2 atlases (McGowan and Corwin 2008). DOIBBS statewide data were too sparse for analysis (Sauer et al. 2005). The nearest current breeders would be in the Pelham Bay Park area, Grassy Sprain Reservoir, or elsewhere in southern Westchester County.

In the New York City area, first fall migrants arrive in mid-Aug, peak in late Aug–early Sep, and depart by mid-Sep, but there are only 4 fall study area migrants, all quite late singles: 8 Oct 1950 (Komorowski), 9 Oct 1931 (Cruickshank), and 23 Oct 2008 (Young). One singing on a warm 18 Nov 1954 (Buckley, Phelan) would have been a blowback drift-migrant, as indeed all Oct individuals may have been.

Specimens
Dwight collected 10 in Van Cortlandt in May, Jun, Jul, Aug, Sep 1887–90 (AMNH) and Bicknell 2 in Riverdale in May, Sep 1876–80 (NYSM).

Bell's Vireo   *Vireo bellii*

NYC area, current
An increasingly detected midwestern/southwestern blowback drift-migrant in fall and a northward-returning blowback drift-migrant in spring to the New York City area and the Northeast. Until very recently deprecated as unidentifiable locally.

Bronx region, historical
Cruickshank's *NYC Region* as of 1941: May, Aug Long Island records in 1930s

Bull's *NYC Area* as of 1968: Sep 1959 Long Island banding, 4 more spring observations; all deprecated as unidentifiable

Bull's *New York State* as of 1975: Sep 1970 Long Island specimen; all observations deprecated as unidentifiable

Levine's *New York State* as of 1996: status unchanged

NYC area, historical
In the mid-1960s, following a single small Blue-headed Vireo's misidentification as a Bell's until AMNH skins of several species of vireos were compared, all Bell's Vireo field identifications, past and future, were embargoed in Bull (1964). Immediately lost were a number of then (but no longer) exceptional northeastern spring birds in NY, NJ, and CT, many singing their unique songs and seen by multiple observers, several in Central and Prospect Parks. (NB: yet another singing ♂ at close range for an extended period in South Ozone Park, Queens on 19 May 1962: J. Mayer, Rose, was overlooked by subsequent authors.) Several occurrences were further compromised by incorrectly republished observation details or factually incorrect statements about Bell's plumages and soft parts. This morass has never been redressed, given that Bull (1974) and Levine (1998) only echoed Bull (1964). Eventually, someone will do

a thorough correction and reanalysis of *all* New York City area Bell's Vireo records, informed by objective information on its identification and on its status in spring and in fall along the East Coast, where it now occurs annually. For additional information, cf. Cruickshank (1942), Buckley and Post (1970), Boyle (2011), and McLaren (2012).

Study area, historical and current
A single study area record: one on 14 Apr 1954 in Van Cortlandt feeding with 2 Ruby-crowned Kinglets and scrutinized as close as 10 ft (3 m) for 30 minutes (Buckley, Scully). It was a smallish, dull vireo with wingbars and was not remotely suggestive of White-eyed, Blue-headed, or Yellow-throated. The bill was darkish, straight, and thin compared with the bills of those vireos. The eye was dark brown or black and surrounded by a distinct but not immediately noticeable white eye ring, connected to a thin white loral line from the eye to the upper mandible. The head was dull gray with a green tinge, fading into the typical greenish back of a vireo. It had two white wingbars, not broad but present. The grayish-green of its head did not extend more than 2–3 mm below the eye and blended, not contrasted, into the dull dirty white color of the throat, breast, and belly. The sides of the breast and belly, downward from about the shoulders, were yellowish-green, obvious but not in a heavy swath, and extended to and became more profuse at the undertail coverts, which were clear yellow. The tail was noticeably short and slightly notched, with some white visible from underneath. It was silent the entire time and so may have been a ♀.

Yellow-throated Vireo *Vireo flavifrons*

NYC area, current
An annual but uncommon spring/fall trans-Gulf Neotropical migrant and very local and disappearing breeder that is now extirpated from New York City itself.

Bronx region, historical
Bicknell's *Riverdale* 1872–1901: common spring/fall migrant, breeder

Hix's *New York City* as of 1904: uncommon Central Park breeder

Eaton's *Bronx + Manhattan* 1910–14: common breeder

Griscom's *Bronx Region* as of 1922: formerly common migrant, breeder but now uncommon, rapidly decreasing

Griscom's *Riverdale* as of 1926: unrecorded since 1917 but increasing in Bronx so may recolonize Riverdale

Kuerzi's *Bronx Region* as of 1926: fairly common breeder (29 Apr–1 Oct), perhaps increasing [locations not given]

NYC area, historical
Cruickshank's *NYC Region* as of 1941: increasing breeder, few pairs still nest Bronx [locations not given]; fairly common spring (maximum 14)/fall migrant (19 Apr–12 Oct)

Bull's *NYC Area* as of 1968: early spring date 21 Apr, late fall date 18 Oct [blowback drift-migrant]

Bull's *New York State* as of 1975: early spring date 19 Apr, late fall date 12 Nov [blowback drift-migrant]

Study area, historical and current
A spring/fall migrant and extirpated breeder along much of Tibbett's Brook and perhaps elsewhere in Van Cortlandt.

In spring, first migrants arrive in early May, peak in mid-May, and depart by late May, with the earliest on 15 Apr 1952 (Kane, Phelan, Buckley)—a trans-Gulf overshoot—and a maximum of 10 on 16 May 1936 (Imhof). Until Jun 1967, the spring maximum in Central Park was 6 on 11 May 1913 (Helmuth).

In the late 1800s, it was a common breeder in Riverdale (Bicknell) and in Van Cortlandt (Dwight), continuing intermittently until the mid-1950s. Certainly 1–3 pairs did breed in Van Cortlandt Swamp from 1926 (Holgate)–1939 (Norse) and then again from 1951 (Komorowski, Buckley)–1956 (Kieran, Buckley). Local breeders disappear or stop singing in late Jul, only to resume in Sep before departure (Kane).

Neither NYSBBA I in 1980–85 nor II in 2000–2005 found this species within New York City, with 2 northern Nassau County blocks and the small, long-standing eastern Suffolk County cluster the only Long Island sites. NYSBBA I and II did find them as close as Grassy Sprain Reservoir (McGowan and Corwin 2008), so it and the Palisades hold the closest breeders. DOIBBS data showed a slight but nonsignificant statewide decrease from 1966–2005 (Sauer et al. 2005; McGowan and Corwin 2008).

In fall, first migrants arrive in mid-Aug, peak in early Sep, and depart by mid-Sep, with extreme dates of 1 Aug 1953 (Buckley) and 9 Oct 1954 (Scully, Buckley), singles only.

Specimens
Dwight collected 5 in Van Cortlandt in Apr, May, Jun 1886–89 (AMNH) and Bicknell 4 in Riverdale in May, Jun, Aug 1876–79 (NYSM).

Blue-headed Vireo *Vireo solitarius*

NYC area, current
A moderately common spring/fall Neotropical migrant.

Bronx region, historical
Bicknell's *Riverdale* 1872–1901: regular spring migrant, only twice in Sep–Oct

Eaton's *Bronx + Manhattan* 1910–14: fairly common migrant

Griscom's *Bronx Region* as of 1922: fairly common migrant

Kuerzi's *Bronx Region* as of 1926: fairly common migrant in spring (20 Apr–28 May)/fall (9 Sep–27 Oct)

NYC area, historical
Cruickshank's *NYC Region* as of 1941: fairly common spring/fall (maximum 11) migrant (9 Apr–1 Dec)

Bull's *NYC Area* as of 1968: early spring date 8 Apr, late fall date 10 Dec; maxima spring 12, fall 20

Bull's *New York State* as of 1975: spring maximum 13; late fall date 16 Dec [blowback drift-migrant]

Study area, historical and current
A moderately common spring/fall migrant.

In spring, first migrants arrive in mid–late Apr, peak in and depart shortly after early May, with extreme dates of 12 Apr 1930 (Sedwitz) and 2 Jun 1952 (Kane) and a maximum of 7 on 30 Apr 1959 (Horowitz). A singing ♂ on the Ridge on 15 Jun 1952 (Buckley) was a nonbreeder or very late migrant. Until Jun 1967, the spring maximum in Prospect Park was 12 on 6 May 1950 (Kreissman).

In fall, first migrants arrive in early Sep, peak in mid-Oct, and depart by late Oct, with extreme dates of 5 Aug 1952 (Buckley) and 1 Nov 2016 at Hillview (Camillieri) and a maximum of 6 on 3 Oct 2005 (Young). Until Jun 1967, the fall maximum in Prospect Park was 20 on 15 Oct 1947 (Alperin, Jacobson).

Singles in Van Cortlandt on 16 Nov in 2005 (Young) and 2013 (Souirgi et al.) would have been lingering blowback drift-migrants, just like nearby singles in Bronx Park on 10 Nov 1957 (Maguire et al.), in the Bronx Zoo on 10 Nov 2014 and 29–30 Nov 2017 (Olson), and at Pelham Bay Park on 6 Dec 1991 (unknown observer, *Kingbird*) and from 16 Nov–4 Dec 2014 (Aracil).

DOIBBS data showed a slow but steady statewide increase from 1966–2005 (Sauer et al. 2007). Nearest breeders are in northwestern Rockland County and central and northeastern Westchester County, where they are holding their own.

Specimens
Dwight collected 2 in Van Cortlandt in Apr 1885–86 (AMNH) and Bicknell 2 in Riverdale in May, Oct 1876 (NYSM).

Warbling Vireo   *Vireo gilvus*

NYC area, current
An uncommon but annual spring/fall Neotropical migrant and local but increasing breeder.

Bronx region, historical
Bicknell's *Riverdale* 1872–1901: common spring/fall migrant, breeder

Eaton's *Bronx + Manhattan* 1910–14: common breeder

Griscom's *Bronx Region* as of 1922: almost unknown migrant, last bred 1918

Griscom's *Riverdale* as of 1926: extirpated breeder

Kuerzi's *Bronx Region* as of 1926: uncommon breeder, 28 Apr–20 Sep [locations not given]

NYC area, historical
Cruickshank's *NYC Region* as of 1941: widespread breeder decreasing toward New York City, but 4 pairs still breed Bronx [locations not given]; uncommon to rare spring (maximum 7)/fall migrant (28 Apr–13 Oct)

Bull's *NYC Area* as of 1968: late fall date 30 Oct, maximum 3; most reliable location is Van Cortlandt Park, where has bred for decades

Bull's *New York State* as of 1975: status unchanged

Study area, historical and current
A spring/fall migrant and breeder in Van Cortlandt and Woodlawn.

In spring, first migrants arrive in mid-Apr, peak in late Apr–early May, and depart by mid-May, with the earliest date 14 Apr 1993 (Young) and a maximum of 25 on 8 May 1955 (Kane et al.), some or many of which may have been local breeders. Until Jun 1967, the spring maximum in Prospect Park was 3 on 12 May 1943 (Soll, Whelen).

It is unexpected but informative that there are no specimens of Van Cortlandt breeders from the late 1800s/early 1900s in the extensive Dwight collection. Perhaps they were not breeding there, even though regular in Riverdale at the same time. Bicknell reported it a common breeder in Riverdale, now [1901] extirpated, Griscom (1923) formerly a breeder, last breeding in 1918, and Kuerzi (1927) called it an uncommon breeder. But in 1942, Cruickshank stated that at least 4 pairs bred in Bronx County without giving any further details, and Kieran confirmed that they had been breeding in Van Cortlandt

Swamp since the late 1920s. From the 1930s to the 1950s, 2–6 pairs bred there annually, increasing in 1956 to 18 pairs. Then sometime <2013, their numbers exploded. In 2013–14, 33 pairs were holding territories there, with another 5–6 pairs on Vault Hill, 2 pairs in the Northeast Forest, 4 pairs in the Shandler Area, and uncertain numbers on the 5 golf course sections. As of 2014, they also were breeding in Woodlawn (5–6 pairs) and around Jerome (2–3 pairs). This may now be the largest aggregation of breeding Warbling Vireos in the New York City area: ±50 pairs.

Both NYSBBA I in 1980–85 and II in 2000–2005 noted a broad New York City area breeding distribution, where they were understandably absent only from the most urbanized portions of all New York City boroughs (McGowan and Corwin 2008). DOIBBS data showed a consistent statewide increase from 1966–2005 (Sauer et al. 2005).

In fall migrants are all but undetectable, and there is no pronounced migratory period. Local breeders usually stop singing and vanish early, but late singers included 3 on 31 Jul 2010 (Willow) and 4 Sep 2014 (Young). Fall maxima are 15 on 9 Aug 2010 (Young)—mostly local breeders—and 6 on 5 Sep 1955 (Kane), possibly all migrants; the latest in the study area was on 20 Sep 2014 (Young), also singing.

Specimens
Bicknell collected 2 in Riverdale in May, Sep 1876–77 (NYSM).

Philadelphia Vireo   *Vireo philadelphicus*

NYC area, current
A Neotropical migrant, extremely scarce in spring but regular in fall in very small numbers. Increased spring records in Central Park may only indicate more observers.

Bronx region, historical
Bicknell's *Riverdale* 1872–1901: found dead 17 Sep 1885 (NYSM), only record

Eaton's *Bronx + Manhattan* 1910–14: rare migrant

Griscom's *Bronx Region* as of 1922: single Bicknell specimen

Kuerzi's *Bronx Region* as of 1926: very rare spring (7, 21 May)/rare fall (17 Sep–3 Oct) migrant

NYC area, historical
Cruickshank's *NYC Region* as of 1941: very rare spring/uncommon to rare fall migrant (4 May–20 Oct) [apparently only singles]

Bull's *NYC Area* as of 1968: 11 May–1 Jun, 20 Aug–23 Oct; maxima fall 5, spring 3; late fall date 23 Oct [blowback drift-migrant]

Bull's *New York State* as of 1975: late fall date 7 Nov [blowback drift-migrant]; fall maximum 10

Study area, historical and current
An uncommon migrant, extremely scarce in spring but regular in fall in very small numbers.

In spring, there are 10 records of singles, only a few of which were singing: 11 May 1934 (Cantor); 12 May 1942 (S. Horowitz); 19 May 1953 (Buckley); 23 May 1954 (Norse); 19 May 1955 (Kane, Buckley); 10 May 1970 (Ephraim); 22 May 1997 (Young); 22 May 1999 (Young); 8 May 2007 (Young); 30 Apr 2017 (Ravitts). Seen more frequently in spring in Bronx Park.

In fall, first migrants arrive in late Aug, peak in early Sep, and depart by late Sep, with extreme dates of 18 Aug 2012 (Young) and 1 Oct 1964 (Sedwitz), and a maximum of 3 on 5 Sep

1952 (Buckley). A very early fall migrant was at nearby Inwood Park on 26 Jul 1956 (Norse).

Specimens
Bicknell collected 1 in Riverdale on 17 Sep 1885 (NYSM).

Red-eyed Vireo   *Vireo olivaceus*

NYC area, current
A common spring/fall Neotropical migrant that also breeds widely throughout.

Bronx region, historical
Bicknell's *Riverdale* 1872–1901: regular spring/fall migrant, breeder

Hix's *New York City* as of 1904: abundant breeder

Eaton's *Bronx + Manhattan* 1910–14: abundant breeder

Griscom's *Bronx Region* as of 1922: common migrant, less common breeder

Kuerzi's *Bronx Region* as of 1926: very common breeder (24 Apr–14 Nov)

NYC area, historical
Cruickshank's *NYC Region* as of 1941: common, widespread breeder still on New York City outskirts [locations not given]; common to abundant spring (maximum 60)/fall migrant (18 Apr–14 Nov)

Bull's *NYC Area* as of 1968: fall maximum 91 dead Fire Island Lighthouse 1887

Bull's *New York State* as of 1975: late fall date 17 Nov [blowback drift-migrant]

Study area, historical and current
A common spring/fall migrant and widespread breeder.

In spring, first migrants arrive in early May, peak in mid-May, and depart by late May, with extreme dates of 28 Apr 2012 (Baksh) and 21 May 1953 (Buckley), with a maximum of 20 on 8 May 1955 (Kane). Until Jun 1967, the spring maximum in Prospect Park was 50 on 17 May 1945 (Soll).

In the late 1800s it was a common breeder in Riverdale (Bicknell) and in Van Cortlandt (Dwight), and as of 2016 breeds in Van Cortlandt (<10 pairs) and Woodlawn (8–10 pairs), maybe also at Jerome. Nearby historical breeding densities include 11 pairs at Inwood Park in 1937 (Karsch) and 18 pairs at Grassy Sprain in 1940 (Hickey).

NYSBBA I in 1980–85 and II in 2000–2005 both emphasized widespread breeding within and around New York City including the study area, but with loss of a few urbanized blocks in Brooklyn, Queens, and Staten Island (McGowan and Corwin 2008). DOIBBS data showed a 2.3% annual statewide increase from 1966–2005 (Sauer et al. 2005).

In fall, first migrants arrive in mid-Aug, peak in late Sep, and depart by early Oct, with extreme dates of 31 Jul 2010 (Willow) and 17 Oct 2008 (Young) and a maximum of 6 on 26 Sep 1989 (Young). However, one on 14 Nov 1926 (Cruickshank) was a blowback drift-migrant, as Bicknell's latest in Riverdale on 24 Oct 1880 may also have been.

Specimens
Dwight collected 6 in Van Cortlandt in May, Jun, Sep 1885–96; Foster 1 in Woodlawn on 9 Oct 1887 (all AMNH); and Bicknell 3 in Riverdale in May, Sep 1978–79 (NYSM).

Crows, Jays: Corvidae

Blue Jay *Cyanocitta cristata*

NYC area, current
A resident breeder and common migrant throughout.

Bronx region, historical
Bicknell's *Riverdale* 1872–1901: common spring/fall migrant, resident

Eaton's *Bronx + Manhattan* 1910–14: fairly common resident

Griscom's *Bronx Region* as of 1922: common breeder, resident, migrant

Kuerzi's *Bronx Region* as of 1926: common resident, abundant fall (late Sep–late Oct)/spring (Apr–May) migrant

NYC area, historical
Cruickshank's *NYC Region* as of 1941: common resident, spring/fall (maximum 400) migrant

Bull's *NYC Area* as of 1968: Long Island maxima 1000 spring, 3000 fall

Bull's *New York State* as of 1975: winter maximum 500 Bronx-Westchester Christmas Bird Count

Levine's *New York State* as of 1996: winter maximum 1300 on 1965, 1972 Bronx-Westchester Christmas Bird Counts

Study area, historical and current
A resident breeder and common spring/fall migrant.

In spring, the first noticeable migrants arrive in loose flocks in Mar–May, often well after local residents have eggs and are often moving into Jun, with a maximum of 29 on 15 Mar 2007 (Klein). Extreme spring/fall dates are meaningless.

Breeds in all 4 extant subareas including Jerome, with <15 pairs in Van Cortlandt, 3–4 in Woodlawn, and 2–3 at Jerome.

Jays were affected by West Nile Virus starting in the 1990s, which may have accounted for the loss of some highly urbanized New York City and Nassau County occupied blocks on NYSBBA II in 2000–2005 (McGowan and Corwin 2008). Nonetheless, DOIBBS data showed no statewide change from 1966–2005 (Sauer et al. 2007).

In fall, migration is much more conspicuous, often tracking hawks on the same ridges southbound and frequently accompanying them. The first migrants arrive in large numbers and peak in late Sep–early Oct and depart by mid-Nov, with a maximum of 1000 each on 30 Sep 1959 (Buckley) and 21 Sep 1960 (Steineck, Friedman). However, these numbers are only hints of how many may actually pass over the study area, as Boyajian (1969a) considered the 3000–6000 per day crossing the Hudson River from Yonkers to Alpine, NJ, during the peak of migration from 20 Sep–10 Oct to be normal numbers.

Largest fall and winter numbers occur when mast crops fail to the north and whole populations move south, many overwintering in the study area when they are quantified on Bronx-Westchester Christmas Bird Counts (Fig. 46). The recent study area maximum is 267 on 22 Dec 2013 between Van Cortlandt and Woodlawn (Lyons, Gotlib et al.).

Specimens
Foster collected 3 in Woodlawn in Oct 1883–89 (AMNH) and Bicknell 2 in Riverdale in Feb, Nov 1878–79 (NYSM).

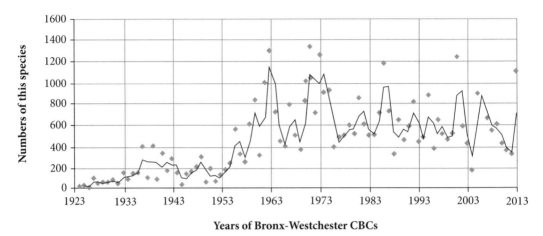

Figure 46

Black-billed Magpie   *Pica hudsonia*

NYC area, current
In the New York City area, a western vagrant that periodically irrupts eastward in fall, sometimes overwinters, and then appears in a light spring return flight. Unrecorded in New York State until 1935 but surprisingly frequent since then, especially on Long Island. Recorded in both Central and Prospect Parks.

NYC area, historical
Cruickshank's *NYC Region* as of 1941: vagrant fall/winter 1935, when 4 occurred

Bull's *NYC Area* as of 1968: very rare spring/fall migrant, mostly on Long Island Apr–May; 27+ reports, all singles save 4 doubles

Bull's *New York State* as of 1975: status unchanged

Study area, historical and current
In the study area there are 2 records. The first, in Van Cortlandt from 31 Oct 1935 (Sialis Bird Club)–22 Dec 1935 (Norse, Cantor, Kieran et al.), appeared during an incursion year and within the Oct–Nov window when most fall migrants have occurred. What may have been the same bird was found on the Palisades on 12 Feb 1936 (Weber et al.), but in the fall of 1935 there was a widespread incursion of magpies to the East, so it was equally likely a different individual. The second was in Woodlawn on 8–9 May 1961 (Horowitz, Heath, Honig), in the Apr–May window when most New York City area spring migrants have been detected. Boyajian watched another northbound migrant crossing the Hudson River from Yonkers to Alpine, NJ, on 8 May 1970, the same date as the Woodlawn bird 9 years prior.

American Crow   *Corvus brachyrhynchos*

NYC area, current
A spring/fall migrant, widespread breeder, and regular winter visitor and resident often collecting in enormous winter roosts draining a vast area. In the study area it is a resident breeder and spring/fall migrant, augmented by winter

arrivals. New York City and study area breeders have not recovered from recent decimation by West Nile Virus.

Bronx region, historical

Bicknell's *Riverdale* 1872–1901: common resident

Eaton's *Bronx + Manhattan* 1910–14: common resident, breeds

Griscom's *Bronx Region* as of 1922: common permanent resident

Kuerzi's *Bronx Region* as of 1926: common permanent resident

NYC area, historical

Cruickshank's *NYC Region* as of 1941: common permanent resident

Bull's *NYC Area* as of 1968: common resident, widespread breeder; abundant migrant inland, locally abundant at winter roosts

Bull's *New York State* as of 1975: status unchanged

Levine's *New York State* as of 1996: status unchanged

Study area, historical and current

A widespread and common breeder, migrant, and permanent resident throughout Riverdale and the study area until decimated by West Nile Virus, from which it may be slowly recovering.

In the study area there is a light spring movement in Mar–Apr, but extreme dates are pointless. In fall, they move in large numbers in Oct–Nov, but again extreme dates are meaningless. Nonbreeding maxima: *spring*, 14 on 17 Mar 2012 (Baksh); *fall*, 105 on 31 Oct 1965 (Norse) and 90 on 29 Nov 2008 (Young); *winter*, 114 between Van Cortlandt and Woodlawn on 27 Dec 2009 (Lyons et al.).

Crows are most evident in winter, when they typically develop the enormous (to 30,000) winter roosts that doubtless facilitated the rapid spread of West Nile Virus, but none have ever been established within or near the study area. Cruickshank (1942) noted that southern Westchester County and Bronx crows moved east across the Sound nightly to a huge roost (30,000–50,000) at Melville in western Suffolk County. In 2006, crows were pouring across Long Island Sound each evening near Pelham Bay Park to roost somewhere on western Long Island, and many also regularly cross the Hudson River at Riverdale to NJ winter roosts (Buckley). Recently, a 10,000-bird roost has been active near Battle Hill in White Plains (Askildsen) and may now be servicing the study area.

On the Bronx-Westchester Christmas Bird Count (Fig. 47), crow numbers barely exceeded

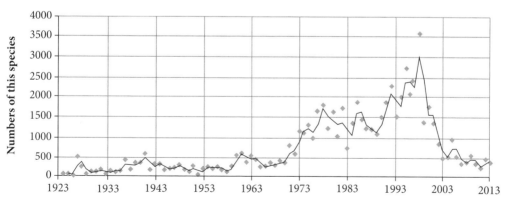

Figure 47

500 until the 1970s, when they began to increase rapidly. This continued until peaking at 3547 in 1998, after which they were decimated by West Nile Virus, whose epicenter was the Bronx Zoo. First recognized in 1999 (Lanciotti et al. 1999), it hit American Crows uniquely hard: on the Bronx-Westchester Christmas Bird Count maxima dropped 90% in 13 years (Fig. 47) with barely a hint of recovery by 2013. In the study area, overwintering crows all but vanished, so the Dec 2009 count above was reassuring.

Breeding American Crows were affected by West Nile Virus starting in the 1990s, which may have accounted for the loss of some highly urbanized New York City and Nassau County occupied blocks on NYSBBA II in 2000–2005 (McGowan and Corwin 2008). DOIBBS data supported a statewide decline since the 1990s (Sauer et al. 2007). How fast the New York State breeding population will recover remains to be seen. They breed in all 4 extant subareas, with a probable total of 10–20 pairs, <5 of them in Van Cortlandt. This is fewer than before West Nile Virus, but earlier quantitative breeding data are nearly nil.

Specimens
Foster collected 3 in Jun, Oct 1889–90 in Woodlawn (AMNH).

Fish Crow   *Corvus ossifragus*

NYC area, current
A widespread but usually scarce breeder, spring/fall migrant, and irregular winter resident/visitor. Most local breeders migrate southward, but variable numbers do overwinter at regional crow roosts.

Bronx region, historical
Bicknell's *Riverdale* 1872–1901: regular breeder, spring/fall migrant; does not overwinter

Hix's *NYC* in 1904: breeds Bronx Park

Eaton's *Bronx + Manhattan* 1910–14: fairly common resident, breeds

Griscom's *Bronx Region* as of 1922: common breeder, single winter record

Kuerzi's *Bronx Region* as of 1926: fairly common resident, in winter near Long Island Sound

NYC area, historical
Cruickshank's *NYC Region* as of 1941: common spring/fall migrant, widespread breeder; generally rare winter resident away from Long Island, but 12–48 usually on Bronx Long Island Sound shores between Pelham Bay Park and Hunt's Point; also on Staten Island

Bull's *NYC Area* as of 1968: Long Island maxima fall 200, spring 175, Pelham Bay Park winter maximum 77

Bull's *New York State* as of 1975: status unchanged

Levine's *New York State* as of 1996: winter maximum 8063 16 Dec 1995 at enormous Fresh Kills, Staten Island dump

Study area, historical and current
A widespread but usually scarce breeder, spring/fall migrant, and irregular winter resident/visitor.

There is a light spring movement of flocks of 5–10 from Mar–May, with extremes of 12 Feb 2011 (Baksh) and 3 on 11 May 1958 (Mayer), and maxima of 25 at Jerome on 18 Mar 1962 (Sedwitz) and 30 at Hillview on 23 Mar 2016 (Camillieri). Until Jun 1967, the spring maxima in Central and Prospect Parks were, respectively, 19 on 10 May 1929 (Helmuth) and 7 on 15 Mar 1953 (Restivo).

Fish Crows breed solitarily in all 4 extant subareas, but no more than 10 pairs throughout, with 1–2 pairs in Van Cortlandt.

Breeding Fish Crows were also affected by West Nile Virus starting in the 1990s, which may have accounted for the loss of some highly urbanized New York City, Nassau County, and Westchester County occupied blocks on NYSBBA II in 2000–2005 (McGowan and Corwin 2008). Sparse DOIBBS data showed a slight statewide increase in the 1970s and a stronger one in the 1990s, followed by a dip from West Nile Virus. However, the increases were largely in upstate New York, where this species was previously very thinly and locally distributed (Sauer et al. 2007).

In fall they move from Aug–Oct with the earliest (and the maximum) 13 on 19 Aug 1959 (Horowitz) and the latest on 2 Dec 2012 (Stuart).

It is assumed that most local breeders depart in winter and are replaced by immigrants. Scattered individuals or groups may occur anywhere in the study area in winter, with a maximum of 17 in Woodlawn on 26 Dec 1960 (Horowitz), but from the 1980s until about 2005 they were most reliable at the south end of Jerome, where up to 10 roosted. Along Broadway in at least the West 230s, Fish Crows now move in winter to feed in parking lots near restaurants and supermarkets, just as they do on Long Island. As many as 77 were at a Pelham Bay Park roost on 27 Dec 1964 (Russak, Stepinoff), where they were mobbing a Snowy Owl. Since about 2010, a Fish Crow–only roost of 500–1000 birds has been active in downtown White Plains (Askildsen), and it may now be partly servicing the study area. Curiously, their generally very low numbers on the Bronx-Westchester Christmas Bird Count (Fig. 48) took a sharp turn upward in the early 2000s at the same time that American Crow numbers were being decimated by West Nile Virus.

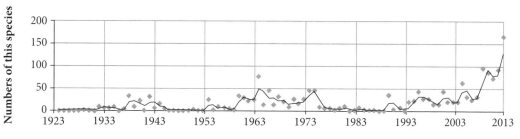

Figure 48

Common Raven   *Corvus corax*

NYC area, current

A very recent colonizer of the urban New York City area that first moved down the Hudson River to the Palisades in the early 2000s. It is now breeding in all 5 New York City boroughs as well as in Nassau and Suffolk Counties and is rapidly increasing. Prior to the first Bronx-Westchester Christmas Bird Count record since 1941 in 2003, it had been almost unrecorded in the immediate New York City/Long Island area since the mid-1800s, with <12 records since 1848.

NYC area, historical

Cruickshank's *NYC Region* as of 1941: last Long Island specimen 1848; next 2 New York City area records 28 Dec 1941 in New Rochelle, Easthampton

Bull's *NYC Area* as of 1968: 2 migrating Bear Mountain Nov 1946; 9+ downstate in Dec–May 1941–65 [details not given]

Bull's *New York State* as of 1975: first modern New York State breeding in Adirondacks 1968, 1974–75

Levine's *New York State* as of 1996: great upstate explosion as breeder, migrant, winter resident; closest breeding now Ulster, Dutchess Counties (1992), on Putnam/Westchester County line (1993); Long Island Apr record of 2

Study area, historical and current

A long-extirpated resident now actively recolonizing the study area and Riverdale.

Unrecorded in the New York City area for a century after the last 2 on Long Island—in 1836 and 1848—until singles in both New Rochelle and East Hampton on 28 Dec 1941 (Cruickshank 1942). Bull (1964) reported that 8 more had been seen by competent and experienced observers between Dec and May 1941–63 but omitted all details.

Ravens first nested in northwestern NJ in 1991, and then without warning a nesting pair appeared on a microwave relay tower in western Nassau County in 2009, on a water tower in the middle of urban Queens in 2010, and then on an apartment house water tower at Pelham Bay Park in 2012.

In the Bronx, future status changes may have been presaged by singles crossing the Hudson from Yonkers to Alpine on 8 Nov 1969 (Boyajian), at Baxter Creek on 6 Nov 1971 (Rafferty), at Pelham Bay Park on 23 Nov 1971 (Rafferty)—the latter 2 maybe the same individual—and on 21 Oct 1984 when was one being mobbed by crows along Broadway at W. 246th St. (Sedwitz). It was another 25 years until the next, in Van Cortlandt on 20 Jan 2009 (Young), probably 1 of the 2 in Riverdale on 26 Mar 2009 (Drucker), possibly associated with another at Pelham Bay Park on 18 Apr 2009 (Jett).

From that point forward everything changed, and it has now been seen on every Bronx-Westchester Christmas Bird Count since 2009, with numbers increasing each year. Pairs wander through Van Cortlandt and Woodlawn at any time of the year, and up to 3 are being seen at Hillview. They may be nesting along the Hudson River on the Henry Hudson Bridge linking Inwood Park to Spuyten Duyvil, in Riverdale on apartment house water towers—perhaps Skytop in the vicinity of W. 259th St. and Riverdale Ave.—and on apartments along the south side of Van Cortlandt near W. 240th St. or around Jerome Reservoir. It is a potential breeder anywhere on the periphery of the study area and at Woodlawn and both reservoirs.

During NYSBBA I in 1980–85 ravens were breeding no closer to the study area than in 10 Catskill blocks, but by NYSBBA II in 2000–2005 they had saturated New York State in a 500% increase in occupied blocks. Still, none were closer to New York City than the Piermont Pier area (McGowan and Corwin 2008). DOIBBS data showed a consistent statewide increase from 1966–2005 that began accelerating to almost geometric levels in the 1990s (Sauer et al. 2005).

Larks: Alaudidae

Horned Lark   *Eremophila alpestris*

UNIDENTIFIED TO SUBSPECIES

NYC area, current
Over the last 100 years, the two regular subspecies (*alpestris*, Northern Horned Lark, and *praticola*, Prairie Horned Lark) collectively have gone from being a regular, sometimes abundant fall migrant and local winter visitor/resident largely restricted to Long Island outer beaches and marshes, to an abundant, widespread migrant and burgeoning breeder, then to a thinly distributed breeder, and now back to an uncommon migrant and winter visitor/resident with a rapidly disappearing breeding population. Horned Lark is now a New York State Species of Special Concern.

Study area, current
A spring/fall migrant, erratic overwinterer formerly (only?) in large flocks on Van Cortlandt Parade Ground and at Hillview and Jerome. In days gone by, migrants and overwinterers were routinely quantified by subspecies; this all but ceased after the 1970s.

The earliest arrival date is 13 on 17 Sep 1955 (Kane), with migrants generally beginning to return in Oct, peaking in Nov, the maximum fall count (flocks with both subspecies) on the Parade Ground being 800 on 6 Nov 1951 and 1955 (Kane, Kieran, Buckley). Since then, we know of no 3-digit counts until a flock of 100+ unidentified to subspecies on 27 Nov 2014 (Drogin), and the largest Nov flocks have been 20–25, with only a handful >10 since about 2000 (Young). This trend is reflected clearly in Bronx-Westchester Christmas Bird Count data (unsorted by subspecies) for that same period (Fig. 49).

Spring migrants arrive in late Feb, peak in Mar, and most are gone by Apr; no maxima or extreme dates are at hand.

NORTHERN HORNED LARK

Bronx region, historical
Bicknell's *Riverdale* 1872–1901: unmentioned

Eaton's *Bronx + Manhattan* 1910–14: common winter visitor

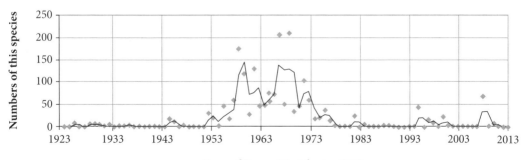

Figure 49

Griscom's *Bronx Region* as of 1922: rare winter visitor/resident

Kuerzi's *Bronx Region* as of 1926: common fall/less common spring migrant, uncommon in winter (9 Oct–13 Apr)

NYC area, historical
Cruickshank's *NYC Region* as of 1941: common fall, spring migrant, winter resident (Long Island maximum several hundred) from 21 Sep–3 May; rare winter visitor Van Cortlandt Parade Ground

Bull's *NYC Area* as of 1968: coastal winter maximum 1500

Bull's *New York State* as of 1975: status unchanged

Study area, historical and current
A fall migrant in greatly reduced numbers on the Parade Ground and the reservoirs; occasional in winter and spring.

Fall migrants begin to return in early Oct, peak in Nov, and depart by late Nov, with the earliest 2 on 28 Sep 2013 (Souirgi; photos). The maximum reported fall count of *alpestris* on the Parade Ground is 100 on 28 Oct 1955 (Birnbaum), but about 80% of the 2 largest mixed flocks—800 in Nov 1951 and 1955—were also *alpestris*. Until Jun 1967, the fall maximum in Central Park was 29 on 8 Nov 1927 (Staloff), and in Prospect Park 25 on 4 Oct 1950 (Whelen).

Snow cover allowing, small flocks will attempt to overwinter, and unless identified as *praticola* winter larks are assumed to be *alpestris*. Recent Parade Ground maxima are 30 on 30 Dec 2017 (Allen) and 60 on 3 Jan 2018 (Karlson), with 37 at Hillview on 6 Jan 2017 (Adams), although successful overwintering may not have occurred since the 1960s.

Spring migrants arrive in late Feb and peak in Mar, and most are gone by Apr. The latest date is 15 Apr 1937 (Norse), and the maximum is 30 at Hillview on 22 Mar 1925 (Kuerzi).

PRAIRIE HORNED LARK

Bronx region, historical
Bicknell's *Riverdale* 1872–1901: unmentioned

Eaton's *Bronx + Manhattan* 1910–14: occasional winter visitor, migrant

Griscom's *Bronx Region* as of 1922: single Jul 1916 Parade Ground record

Kuerzi's *Bronx Region* as of 1926: rare migrant in tiny groups; 5 records most likely this subspecies in Nov (twice), Mar, and May

NYC area, historical
Cruickshank's *NYC Region* as of 1941: increasing breeder in New York City region; in Bronx rare Nov, Feb migrant

Bull's *NYC Area* as of 1968: coastal breeders tripled since Cruickshank (1942), 5 pairs bred Baxter Creek 1955

Bull's *New York State* as of 1975: status unchanged

Study area, historical and current
An extirpated breeder and a fall migrant in greatly reduced numbers on the Parade Ground and the reservoirs; occasional in winter and spring.

Fall migrants begin to arrive in late Oct, peak in Nov, and depart by late Nov, with the earliest 2 on 25 Oct 1934 (Kuerzi, Kuerzi). We can find no explicit fall maxima for *praticola* on the Parade Ground beyond 30 on 2 Nov 1954 (Scully), but about 20% of the 2 largest mixed flocks—800 on both 6 Nov 1951 and on 6 Nov 1955—were *praticola*. Most disappear quickly once snow cover appears, and only flocks <6 will occasionally attempt to overwinter (as in 1952–53: Kane, Buckley), usually vanishing by early Jan, with 2 as late as 20 Jan 1956 (Scully) and 22 Feb 1952 (Norse).

The few spring migrants pass through all but unnoticed in Mar—e.g., 17 Mar 1951 (Kane)—and by early Apr only breeders are left. Local breeders were singing and nest building by mid-Mar with fledged young in Apr–early Jun (Kieran, Komorowski, Buckley).

Major expansion of *praticola*'s breeding range into the New York City area began in the late 1930s, although one on the Parade Ground on 29 Jul 1916 (Griscom 1923) could indicate earlier (intermittent?) nesting, and Eaton (1910, 1914) did list it as a breeder in his Bronx + Manhattan county but without any details. Several on the Parade Ground on 2 May 1926 (Kessler, Kuerzi) support this idea, and even though there were no other breeding season records in this period, they could have been overlooked.

At least 2 pairs were nesting in Dutchess County by 1930 (Crosby) and 7 pairs at Perth Amboy, NJ, on 12 Jun 1930 (Urner). Sedwitz (1940a) noted that by 1936 Horned Larks were nesting on Long Island and in NJ with greater frequency than ever before, and Sedwitz (1940b) recorded their breeding in 1937 at Idlewild (Queens), Canarsie (Brooklyn), and Morris County, NJ. But as of 1942 and 1944, they were not yet breeding in Van Cortlandt or elsewhere in the Bronx (Cruickshank 1942; Bull 1946).

Nonetheless, it is clear that by 1948 2–3 pairs were nesting on the Parade Ground and adjacent Vault Hill edges with prefledged young in late Apr–early May, continuing until 1962 (Kieran, Norse, Komorowski, Buckley). At that time they were also breeding at Baxter Creek (5+ pairs: Pappalardi) and at Pelham Bay Park. Throughout the New York City area their 1950s–60s range expansion and consolidation reached a peak in the 1970s, when Bull (1975) reported them breeding in every single New York State county, although they quickly began to abandon most breeding sites, and as of 2016 they are largely restricted to barrier beaches and overgrown landfills in the New York City area. The last time they bred in Van Cortlandt was 1962, although their exceedingly early nesting (they often began singing on warm days in mid-Feb, and 28 Feb is the earliest local egg date in Cruickshank 1942) easily leads to their being overlooked. A few pairs were also breeding at Hillview and Jerome at the same time.

NYSBBA I in 1980–85 found breeding larks throughout Suffolk County and along the shore in Nassau County, Queens, and Staten Island but otherwise no closer to the study area than 8 blocks in Orange County. By NYSBBA II in 2000–2005 breeders were no longer found in any New York City blocks and at only a handful of Nassau County barrier beaches, although locally numerous in eastern Suffolk County, and Orange County hosted 4 blocks (McGowan and Corwin 2008). DOIBBS data showed a consistent statewide decline from 1966–2005 that leveled off slightly after 1980 (Sauer et al. 2005). It is unclear where the nearest breeders are as of 2016—perhaps on the Pelham Bay Park landfill, the new Ferry Point Park golf course, Hillview, the Rockaway Peninsula, or Staten Island's Fresh Kills landfill.

Swallows: Hirundinidae

Purple Martin *Progne subis*

NYC area, current
This colonial breeder, the least common breeding swallow, has always been strikingly local, absent over vast areas despite nominally appropriate habitat and available nest boxes. It is an annual but uncommon

early spring and rare early fall Neotropical migrant.

Bronx region, historical

Bicknell's *Riverdale* 1872–1901: rare transient recorded only 6 times spring, 4 fall

Eaton's *Bronx + Manhattan* 1910–14: uncommon breeder

Griscom's *Bronx Region* as of 1922: no record since 30 Apr 1886

Kuerzi's *Bronx Region* as of 1926: rare spring/fall migrant (13 Apr–24 Sep), breeds Rye

NYC area, historical

Cruickshank's *NYC Region* as of 1941: rare migrant in spring/uncommon to rare (maximum 100+) in fall (1 Apr–6 Oct); probably overlooked migrating high; no longer nests nearer New York City than Seaford and Rye

Bull's *NYC Area* as of 1968: fall maximum 2000+ grounded by fog; early spring date 9 Mar, late fall date 9 Oct; new Staten Island colony 1951–61

Bull's *New York State* as of 1975: Staten Island hosts sole New York City colony (maximum 75 pairs)

Study area, historical and current

An infrequent migrant or visitor, usually spring singles, and has always been so. It is conceivable that martins were breeding somewhere in the study area in the late 1890s as Dwight's 4 taken on the same day suggest, but we have located no contemporaneous support for breeding. Bicknell found it in Riverdale only 6 times in spring (22 Apr 1893–21 May 1885), once in midsummer (23 Jun 1888), and thrice in fall (17 Jul 1885–28 Aug 1880) but did not mention any breeding sites.

There are only 9 spring records, all singles except as noted: Dwight collected 4 on 30 Apr 1886 (AMNH); Jerome on 11 May 1927 (Cruickshank); 11 May 1952 (Kane, Post, Komorowski); 4 May 1953 (Kane); 3 at Jerome on 22 May 1953 (Buckley); 15 May 1954 (Kane, Buckley); 21 Apr 1956 (Phelan, Buckley); Jerome on 20 Apr 1992 (Meyer); 19–25 May 2012 (Russ et al.), and only 3 in fall: 30 Aug 1952 (Kane) and 18 Sep 1954 (Kane, Buckley) in Van Cortlandt, and 24 Jul 2017 at Hillview (Camillieri). One each in Van Cortlandt on 21 Jun 1952 (Buckley) and 26 Jun 1999 (Young) were nonbreeders or wanderers from some nearby colony. An extremely late blowback drift-migrant adult ♂ was with 2 Cave Swallows on the Hudson River in Riverdale on 30 Oct 2010 (Buckley). It and Bicknell's 3 in Jul–Aug are the only area fall records. Singles in Bronx Park on 27 Apr 2013 (Datlen) and 4 Jul 2014 (Aldana) and the Bronx Zoo on 20 Aug 14 (Olson) were among the very few there in recent decades.

Many areas in the northeastern United States reported precipitous declines in breeders following House Sparrow and Starling introductions and their ensuing usurpation of martin nesting structures. Only tediously slowly were the swallows able to reestablish a few beachheads within their former breeding range, in the New York City area no closer than Rye, Westchester County (after 1912), and Princess Bay, Staten Island (after 1950). These remain the closest colonies to the study area; within New York State martins are most numerous—even abundant—only within 50–100 smi (80–160 km) of Lakes Erie and Ontario.

DOIBBS data showed no significant statewide trends from 1966–2005 (Sauer et al. 2005). However, the number of occupied blocks in New York State dropped by 39% between NYSBBA I in 1980–85 and II in 2000–2005, although its already-low New York City area breeding distribution did not change much, its epicenter remaining eastern Suffolk County. (McGowan and Corwin 2008).

Specimens

Dwight collected 4 in Van Cortlandt on 30 Apr 1886.

Tree Swallow *Tachycineta bicolor*

NYC area, current
A common, locally abundant spring/fall Neotropical migrant, a widespread but local breeder in all 5 New York City boroughs, and an erratic winter resident on Long Island, when it feeds on bayberries.

Bronx region, historical
Bicknell's *Riverdale* 1872–1901: spring/fall migrant, especially at Kingsbridge Meadows

Eaton's *Bronx + Manhattan* 1910–14: occasional breeder, abundant migrant

Griscom's *Bronx Region* as of 1922: common migrant, unknown to breed

Kuerzi's *Bronx Region* as of 1926: common, frequently abundant spring (20 Mar–13 Jun)/fall (1 Jul–6 Dec) migrant; no breeding sites

NYC area, historical
Cruickshank's *NYC Region* as of 1941: common spring/abundant (thousands coastally) fall migrant, occasional in winter on Long Island South Shore; rare local breeder western Long Island, northern Westchester County; single Bronx breeding pair 1938 [location not given]

Bull's *NYC Area* as of 1968: fall coastal estimates 50,000–100,000; coastal winter maximum 100+, overwintering now regular except in years with poor Bayberry crops

Bull's *New York State* as of 1975: status unchanged

Study area, historical and current
A breeder and common spring/fall migrant.

In spring, first migrants arrive in late Mar, peak in early May, and depart by late May, with the earliest 6 on 3 Mar 2012 (Baksh), and a maximum of 50+ on 3 Apr 1956 (Kane). Until Jun 1967, the spring maximum in Central Park was 50 on 19 May 1956 (Carleton).

From 1938 (Norse) to the 1970s–80s (Buckley, Young) and onward, 6–8 pairs bred continually (if not continuously) in Van Cortlandt in natural tree cavities (more recently in nest boxes) and intermittently at Woodlawn (Horowitz) and at Jerome (Sedwitz, Meyer, C. Young), and Hillview (Camillieri). As many as 10 nest boxes were occupied at various places on the Van Cortlandt golf courses between 1998 and 2012 and some produced young swallows each year (Künstler). In 2014, 20 pairs were breeding in Van Cortlandt alone (Young, Buckley), several using natural tree cavities. They also bred in 1955 in Pelham Bay Park, where a few pairs still nest (Künstler), and on NYSBBA I in 1980–85 and II in 2000–2005 they were in almost every Bronx block.

As both NYSBBA I in 1980–85 and II in 2000–2005 showed, a widespread breeder in all 5 New York City boroughs (although overlooked in many highly urbanized sections), on Long Island, in Westchester County, and in study area Blocks 5852B and 5952A (McGowan and Corwin 2008). DOIBBS data quantified a 2.5% annual statewide increase from 1980–2005 (Sauer et al. 2005).

In fall, first migrants arrive in late Jul, peak in late Sep, and depart by early Oct, with extreme dates of 31 Jul 1959 (Horowitz) and 15 on 30 Oct 2010 (Baksh), and a maximum of 500 at Jerome on 21 Sep 1962 (Sedwitz). Until Jun 1967, the fall maxima in Central and Prospect Parks, respectively, were 1200 on 29 Aug 1956 (Post) and 10,000 overhead on 13 Oct 1950 (Whelen).

Though some may be migrating coastwise through Nov, there are no such study area records and only 3 later in Van Cortlandt: 23 Dec 1956 (Kane, Buckley), 22 Dec 1957 (Scully,

Buckley), and 2 on 23 Feb 1955 (Buckley)—the latter perhaps having overwintered not far away.

Specimens
Bicknell collected 3 at Kingsbridge Meadows in Apr, Sep 1876–79 (NYSM).

Northern Rough-winged Swallow *Stelgidopteryx serripennis*

NYC area, current
A loosely colonial breeder and common spring Neotropical migrant.

Bronx region, historical
Bicknell's *Riverdale* 1872–1901: common spring/fall migrant, breeder especially at Kingsbridge Meadows

Eaton's *Bronx + Manhattan* 1910–14: fairly common but local breeder

Griscom's *Bronx Region* as of 1922: uncommon migrant; extirpated from Riverdale, still breeds near Van Cortlandt, probably Jerome

Kuerzi's *Bronx Region* as of 1926: fairly common breeder, 10 Apr–9 Sep [locations not given]

NYC area, historical
Cruickshank's *NYC Region* as of 1941: common spring (maximum 33)/rarely seen very early fall migrant, from 31 Mar–2 Oct; fairly common Bronx breeder, along Hudson, around larger reservoirs [locations not given]

Bull's *NYC Area* as of 1968: early spring date 24 Mar

Bull's *New York State* as of 1975: spring maximum 40

Study area, historical and current
A breeder and common spring Neotropical migrant.

In spring, first migrants arrive in early Apr, peak in early May, and depart by mid-May, with extreme dates of 25 Mar 2009 (Young) and 28 May 2011 (Baksh), and a maximum of 43 on 27 Apr 1954 (Buckley). Until Jun 1967, the spring maximum in Prospect Park was 5 on 2 May 1943 (Soll).

They have been emblematic Van Cortlandt and Kingsbridge Meadows breeders since Bicknell's day, and Kieran noted them in Van Cortlandt and Jerome continuously from 1914–56. Since the 1920s, at least 30 pairs have bred at Van Cortlandt Swamp and Lake (10), Woodlawn (2), Jerome (8), and Hillview (10), where they nest in cracks in concrete, holes in banks, and pipes, etc. Sedwitz located 18 pairs in the Jerome–Riverdale area in 1981. In the 1980s they were nesting in artificial drainpipes along Van Cortlandt golf courses (Turner). All swallow numbers in Van Cortlandt became much more difficult to track with removal of the overhead wires along the Putnam Division track bed near Van Cortlandt Lake in the 1970s, so study area breeders may now be routinely underestimated.

Both NYSBBA I in 1980–85 and II in 2000–2005 found them widespread within and around New York City (although not in Brooklyn), as well as in both study area blocks (McGowan and Corwin 2008). DOIBBS data showed a slight but significant statewide increase from 1966–2005 (Sauer et al. 2005).

In fall, first migrants arrive in early Jul and peak in late Jul–early Aug, after which they vanish. The latest single was on 29 Aug 2015 (Hudda), and the maximum of 22 on 19 Jul 1994 (Young) may have contained locals, migrants, or both. Reports after Aug usually pertain to immature Tree Swallows.

Specimens
Dwight collected 3 in Van Cortlandt in Apr, May 1886 (AMNH), Bicknell 5 at Kingsbridge Meadows in May, Jul, 1877–79 (NYSM, MCZ).

Bank Swallow   *Riparia riparia*

NYC area, current
A spring/fall Neotropical migrant, highly local colonial breeder ostensibly in reduced numbers since the 1960s, and now nesting regularly within New York City limits only on Staten Island and Hart's Island (east of Pelham Bay Park).

Bronx region, historical
Bicknell's *Riverdale* 1872–1901: regular spring/fall migrant, especially at Kingsbridge Meadows

Hix's *NYC* in 1904: common Van Cortlandt breeder, but migrant elsewhere

Eaton's *Bronx + Manhattan* 1910–14: common breeder

Griscom's *Bronx Region* as of 1922: uncommon migrant; does not breed

Kuerzi's *Bronx Region* as of 1926: common spring/fall migrant (2 Apr–11 Oct), breeds intermittently in favorable sites [locations not given]

NYC area, historical
Cruickshank's *NYC Region* as of 1941: common spring/fall (maximum 100+) migrant (2 Apr–11 Oct); closest breeders few pairs in Westchester County [locations not given]

Bull's *NYC Area* as of 1968: fall maximum 400

Bull's *New York State* as of 1975: status unchanged

Study area, historical and current
Now an inexplicably uncommon migrant but formerly a regular spring/fall migrant and breeder in Van Cortlandt. It may have also bred in Riverdale and Kingsbridge Meadows, where specimens were collected by Bicknell in 1876 and 1879.

In spring, first migrants arrive in late Apr, peak in early May, and depart by early Jun, with extreme dates of 17 Apr 1952 (Buckley) and 27 May 1934 (Norse), and a maximum of 15+ on 16 May 1936 (Weber et al.). Until Jun 1967, the spring maximum in Central Park was 5 on 19 May 1966 (Carleton).

Bank Swallows seen in midsummer may be occasional breeders or postbreeding wanderers—like those in Van Cortlandt on 2 Jul 1933 (Norse) and on 19 Jun 2010 (Baksh) or at Jerome on 8 Jul 1987 (Meyer)—but any later would be southbound migrants like those at Hillview on 28 Jul 1938 (Norse, Cantor) and 8 Aug 1936 (Norse), the 3 in a mixed swallow flock working over the Parade Ground on 18 Jul 1994 (Lyons). It is not widely appreciated that Bank Swallows will use man-made structures just as Rough-wingeds normally do, and both species sometimes nest together. It is uncertain whether this happens only when bankside habitat is lacking, but Bank Swallows should be sought from May–Jul at both Jerome and Hillview and on Van Cortlandt golf courses. The presence of up to 6 from late May–late Jul 2017 at Hillview in 2017 (Camillieri) suggests they may already be nesting there with Rough-wings, at least occasionally. In 1955, 8 (= 4 pairs) throughout May at Baxter Creek were probably breeding (C. Young, Kane et al.). Nearest current breeders are on Hart's Island off Pelham Bay Park (Künstler); others may be at Grassy Sprain Reservoir, along the Bronx and Hudson Rivers, and in the Hackensack Meadowlands.

In the New York City area, NYSBBA I in 1980–85 found possible breeders in the Bronx in the block below study area Block 5852B and from central Westchester County north and west. NYSBBA II in 2000–2005 added Hart's Island and another block on Staten Island. Otherwise the Long Island Sound shore in Nassau and Suffolk Counties supported the bulk of the species' New York City area population (McGowan and Corwin 2008). DOIBBS data showed a strong statewide decline from 1966–2005 that may only be sampling artifacts (Sauer et al. 2007).

In fall, first migrants arrive in early Aug, peak in late Aug, and depart by mid-Sep, with extreme dates of 3 at Hillview on 8 Aug 2017 (Camillieri) and 21 Oct 1934 (Weber, Jove), and a maximum of 5 at Jerome on 21 Sep 1962 (Sedwitz).

Specimens
Bicknell collected 2 at Kingsbridge Meadows in Apr 1877, 1879 (NYSM) and another there on 11 Oct 1891 (Griscom 1927) that may have been lost.

Cliff Swallow *Petrochelidon pyrrhonota*

NYC area, current
After Purple Martin, this colonial breeder and Neotropical migrant is the least frequently recorded eastern swallow and a very local breeder. As migrants they are generally seen in large mixed swallow flocks but usually only in single-digit numbers in spring. In adjacent NJ it appears to be in the process of slowly reclaiming its former range by breeding under bridges; only very recently has it begun rebreeding within New York City.

Bronx region, historical
Bicknell's *Riverdale* 1872–1901: until 1886 scarce breeder, common spring migrant, decreasing rapidly; summer southbound migrant, earliest 2 Jul 1883

Eaton's *Bronx + Manhattan* 1910–14: common migrant, fairly common breeder

Griscom's *Bronx Region* as of 1922: long-extirpated breeder; uncommon to rare migrant

Kuerzi's *Bronx Region* as of 1926: uncommon but regular migrant in spring (19 Apr–29 May)/fall (2 Jul–6 Oct)

NYC area, historical
Cruickshank's *NYC Region* as of 1941: uncommon to rare but regular spring/fall migrant, maximum 30 [in spring?]; breeds no closer than extreme northeastern Westchester County

Bull's *NYC Area* as of 1968: 7 Apr–11 Oct; fall maximum 500 near Peekskill; bred Pound Ridge Reservation 1947–51, otherwise not east of Hudson River

Bull's *New York State* as of 1975: status unchanged

Study area, historical and current
A scarce and infrequent spring migrant rarely detected in fall. It was far more frequent with regular double-digit counts <1950 but is now slowly recolonizing southern Westchester County and the Bronx, including the study area.

In spring, first migrants arrive in late Apr, peak in mid-May, and depart by late May, with extreme dates of 14 Apr 1940 (Norse, Cantor) and 27 May 1934 (Norse); usually singles or twos but a maximum of 40 on 16 May 1938 (Cantor). Until Jun 1967, the corresponding maxima in Central and Prospect Parks, respectively, were 18 on 11 May 1945 (Mackenzie) and 35 on 20 May 1937 (Jacobson et al.).

Three at Van Cortlandt Lake on 30 Jun 1952 (Buckley) may have been breeding nearby, as were 2 on 23–24 May 1953 (Buckley). Several pairs did breed on the Van Cortlandt Lake boathouse in 1956, when up to 12, including just-fledged juveniles, were present from 17 May–30 Jun (Cantor, Kieran, Buckley). This appears to have been the first New York City breeding since the late 1800s, when Bicknell noted that 1–2 pairs bred in Riverdale until 1881, and by extension in Van Cortlandt. One in Riverdale on 2 Jul 1883 (Bicknell) may also have been a local breeder. Eaton (1910, 1914) considered Cliff Swallow a fairly common breeder in his Bronx + Manhattan county, although no elucidation was forthcoming in its species account. No subsequent authors even hinted at New York City breeding.

The last previously known Long Island breeding was in 1924, although possible breeders were seen on Plum Island during NYSBBA II in 2000–2005. In 2012 a pair did breed successfully at Alley Pond Park (Perrault) and although they returned in 2014, breeding was unproved.

NYSBBA I in 1980–85 found no breeders in the New York City area closer than a single Block in northwestern Westchester County, although a single at Van Cortlandt Lake on 18 Jun 1980 (Sedwitz) was suspicious. But by NYSBBA II in 2000–2005, the New York State colonies nearest the study area were in 4 blocks in the general vicinity of SUNY at Purchase, 15 smi (24 km) northeast of Van Cortlandt Lake. We do not know how many colonies or pairs were nesting there or how long they had been active. Probably part of this local population are the current colonies on SUNY–Purchase and Westchester County airport buildings, which first appeared at SUNY in 2012 with 5–8 nests that grew to 10–15 by 2015 (Schlesinger). There were no other southern Westchester County sites we know of, and this remains so—with one exception discussed next. DOIBBS data showed no statewide change from 1966–2005, although data are sparse (Sauer et al. 2005).

To great surprise, by 2011—and perhaps earlier, judging by 1 calling persistently there on 30 May 2001 (Nadareski)—Cliff Swallows had established new colonies on buildings at the north and south ends of Hillview, the first study area breeding since 1956. But as part of the NYCDEP Bureau of Water Supply Waterfowl Management Program's for reducing waterbird contamination of New York City drinking water (much of which transits Hillview), the birds were harassed and their nests removed under USFWS permits over the winters of 2011–12 and 2012–13 (no later data are available). Additional information is vague, other than that during the winter of 2012–13, 13 Cliff Swallow nests were removed (NYCDEP 2011, 2012). However, Cliffs *were* present and breeding there during the summer of 2017 when 18 individuals were counted on 7 Jun (Camillieri) attending an unstated number of nests.

Around this same time, Cliff Swallows were also detected nesting in Pelham Bay Park and environs, but given typical levels of coverage levels there after May, it is possible breeders had been overlooked previously. The first 2 active nests were reported on the southernmost arch of the Pelham Bridge over the Hutchinson River (sometimes called the Shore Road Bridge) in Jun 2010 (Cole, Aracil), but not again until 7 Jun 2015 by Benoit, who noted the presence of other unoccupied nests in the same location. It is likely they had been nesting there during the intervening years.

There were no reports from any other sites until 7 Jun 2015, when Benoit did a kayak trip along the Pelham Bay Park shore. He saw 2 Cliff Swallows feeding along the north shore of Hunter's Island, 2 pairs below the City Island Bridge flying in and out of 2 locations he was unable to investigate, and 3 pairs on the northeastern corner of the Pelham Bridge: 1 pair in a completed nest and 2 more that seemed to be building new nests next to the completed one. Benoit was unable to kayak near Orchard Beach itself, but by 19 Jul, 9 swallows were attending 3 nests on buildings there (Keogh, Conte). Satisfying indeed was Conte's observation of a Starling being actively repulsed by several swallows. As of mid-Jul 2015 there were 10+

pairs nesting at 3 sites in the Pelham Bay Park area, present again in early Jul 2016 (Benoit).

It appears that Cliff Swallows are finally clawing their way back as breeders in the Bronx area, first at Hillview, then at Pelham Bay Park (only 5.3 smi[8.3 km east]). It is highly unlikely these are the only breeders, now that we know they are finally using buildings and bridges near/over water in New York City. There are many such Bronx locations that would repay rigorous searching. Two adults and a recently fledged juvenile actively feeding in a mixed swallow flock over the southeast corner of the Parade Ground from 18 Jun–4 Jul 2016 (Souirgi, Young et al.) may have been Hillview breeders or even breeders from the Van Cortlandt Lake golf course building or nearby structures.

In fall, the few New York City area migrants arrive in early Aug, then peak and depart in mid–late Aug, the same period as the only the study area fall records: Hillview on 8–12 Aug 1936 (Norse, Cantor), Van Cortlandt on 14 Aug 2004 (Young) and 23 Aug 2007 (Young), and Hillview on 30 Aug 2017 (Camillieri). The 2–3 in a large mixed swallow flock working over the Parade Ground on 18 Jul 1994 (Lyons, Garcia), another in Van Cortlandt on 20 Jul 2003 (unknown observer, *Kingbird*), and the 25 at Hillview on 26 Jul 2017 (Camillieri) would have been early migrants or postbreeding dispersers from some Westchester colony.

Specimens
Bicknell collected 1 in Riverdale on 2 May 1876 (NYSM).

Comments
Observers now must look carefully at any Oct–Nov Cliff Swallows, given that southwestern vagrant **Cave Swallows**, *P. fulva*, are now occurring almost annually as blowback drift-migrants. None have yet been found in the study area or in Central or Prospect Park, but 2 were along the Hudson River in Riverdale with a very late ♂ Purple Martin on 30 Oct 2010 (Buckley). Others were seen that day along the south shore of Long Island and on 23 Oct 2010 along the Hudson at Croton Point (Johansson). The only other Bronx records are a single at Pelham Bay Park on 12 Nov 2008 (Fiore), 5 on 14 Nov 2015 (Rothman, Benoit et al.), and another the next day (Aracil, Benoit). They can be expected along the Hudson and Long Island Sound in Oct–Nov after strong west or northwest winds with cold front passage following a sustained warm-sector southwesterly flow.

Barn Swallow   *Hirundo rustica*

NYC area, current
A spring/fall Neotropical migrant and a widespread breeder even in urban areas.

Bronx region, historical
Bicknell's *Riverdale* 1872–1901: common breeder, migrant

Hix's *New York City* as of 1904: common Van Cortlandt breeder

Eaton's *Bronx + Manhattan* 1910–14: common breeder

Griscom's *Riverdale* as of 1926: extirpated breeder

Kuerzi's *Bronx Region* as of 1926: common breeder, 7 Apr–9 Nov

NYC area, historical
Cruickshank's *NYC Region* as of 1941: abundant migrant in spring (maximum 200+)/fall (maximum several hundred), from 15 Mar–19 Dec; common breeder; few pairs still nest within New York City but are decreasing annually [locations not given]

Bull's *NYC Area* as of 1968: maxima spring 250, fall 25,000

Bull's *New York State* as of 1975: status unchanged

Study area, historical and current
A breeder and common spring/fall migrant. Curiously, spring maxima come from Van Cortlandt Lake, but fall maxima are all Parade Ground feeders—postbreeders or migrants?

In spring, first migrants arrive in early Apr, peak in late Apr, and depart by late May, with extreme dates of 30 Mar 2016 at Hillview (Camillieri) and 50 (the maximum) on 29 May 1937 (Imhof et al.) with another of 40 on 1 May 1955 (Buckley et al.). Until Jun 1967, the spring maximum in Central Park was 75 on 19 May 1966 (Carleton) and in Prospect Park 28 on 28 Apr 1945 (Soll).

As many as 40–50 pairs have been breeding annually between Van Cortlandt Swamp and Lake (10–12 pairs), Woodlawn (3–4 pairs), Jerome (12–14 pairs), and Hillview (15–20 pairs). At Jerome 12 pairs bred in Jun 1955 (C. Young) and 6–8 pairs annually from 1945–51 (Brigham).

Both NYSBBA I in 1980–85 and II in 2000–2005 indicated the near ubiquity of breeders in the New York City area (McGowan and Corwin 2008), although DOIBBS data showed a 2% annual statewide decrease from 1980–2005 but not from 1966–80 (Sauer et al. 2005).

In fall, first migrants arrive in late Jul, peak in late Aug, and depart by late Sep, with extreme dates of 7 Aug 1959 (Heath) and 15 Sep 1937 (Norse) and a maximum of 35 on 23 Aug 2008 (Young), some of which may have been local breeders. Until Jun 1967, the fall maximum in Central Park was 3600 on 29 Aug 1956 (Post). Barn Swallows are regular blowback drift-migrants into Nov on Long Island, but none have yet been found in the Bronx.

Specimens
Bicknell collected 1 in Riverdale on 4 Jul 1879 (NYSM).

Tits: Paridae

Black-capped Chickadee *Poecile atricapillus*

NYC area, current
A spring/fall migrant, a thinly distributed but widespread breeder even in New York City parks, and a regular winter visitor and resident.

Bronx region, historical
Bicknell's *Riverdale* 1872–1901: common breeder, migrant, winter resident

Eaton's *Bronx + Manhattan* 1910–14: common resident

Griscom's *Bronx Region* as of 1922: resident; common to abundant in fall migrant, winter resident

Kuerzi's *Bronx Region* as of 1926: resident; fairly common summer, usually abundant in spring/fall migration, winter

NYC area, historical
Cruickshank's *NYC Region* as of 1941: common summer, winter (maximum several

dozen) residents but disappearing as breeders around New York City [locations not given]; common fall migrant, usually overlooked in spring

Bull's *NYC Area* as of 1968: winter maximum 640 Bronx-Westchester Christmas Bird Count

Bull's *New York State* as of 1975: status unchanged

Study area, historical and current

A resident breeder and common spring/fall migrant, irregularly irrupting southward in large numbers, some of which then overwinter.

In nonirruption falls, first migrants arrive in early Sep, peak in Oct–Nov, and depart by late Nov, with a maximum of 25 on 11 Nov 2012 (Stuart).

By early Dec winter residents are left, and the maximum we know of in a flight year is 100 between Van Cortlandt and Woodlawn on 26 Dec 2005 (Lyons et al.). We suspect considerably higher daily totals in other incursions were lost when they were merged with those from other count areas.

Minor Black-capped Chickadee irruptions occur about every 5 years, but enormous irruptions (those sometimes containing Boreals) have reached the study area much less frequently: 1918–19, 1941–42, 1951–52, 1954–55, 1959–60, 1960–61, 1961–62, 1962–63, 1964–65, 1965–66, 1975–76, 1979–80, 1980–81, 1981–82, 1982–83, 1983–84, 1985–86, 1990–91, 1993–94, 1995–96, 2001–2, and 2005–6 (note adjacent-year clusters). In such irruptions the vanguard may arrive in Aug (e.g., 40 on 30 Aug 1961: Buckley), and hundreds might pass by daily in early Sep. Most continue south but enough overwinter in the study area to be detected on Bronx-Westchester Christmas Bird Counts. In the largest incursions, 1279 were found in 2001 and 1223 in 2005, in contrast with nonflight years when 167 were found in 2011 and 270 in 2013, an order of magnitude difference.

In spring, migrants are barely noticed even in irruption years, most passing through in Mar–Apr as local winter residents slip away, with a maximum of 20 on 15 Mar 2010 (Higgins).

Chickadees breed in all 4 extant subareas except Jerome, with a study area–wide total of <10 pairs in Van Cortlandt and another 1–2 in Woodlawn. The 7 in Van Cortlandt on 27 May 1934 (Weber, Jove) were assuredly breeding. As many as 10 Tree Swallow nest boxes were emplaced at various locations on the Van Cortlandt golf course between 1998 and 2012, and a few were occasionally used by chickadees (Künstler).

Both NYSBBA I in 1980–85 and II in 2000–2005 elucidated a broad New York City area breeding distribution, where they were understandably absent from the most urbanized portions of all New York City boroughs except Staten Island (McGowan and Corwin 2008). DOIBBS data showed a consistent statewide increase from 1966–2005 (Sauer et al. 2007).

Specimens

Dwight collected 6 in Van Cortlandt in Apr, Oct, Nov, Dec 1885–87 (AMNH) and Bicknell 3 in Riverdale in Feb, Oct 1878–79 (NYSM).

Comments

On 24 May 1967 a chickadee singing a nominal **Carolina Chickadee,** *P. carolinensis*, song was watched closely in nearby Inwood Park (Norse) at a time when field separation of the 2 species by plumage was not understood (see Sibley 2017). Given that chickadee songs are learned, it was most likely a Black-capped raised in the nearby NJ area, where both species breed and not infrequently hybridize. But the Black-capped/Carolina contact zone is slowly advancing northward (in NJ: L. Larson, pers. comm.; and in eastern North America: Taylor et al.

Boreal Chickadee   *Poecile hudsonicus*

NYC area, current
A rarely detected irruptive invader from boreal forests that sometimes accompanies Black-capped Chickadees during major population explosions. It has been recorded in Central Park but not Prospect.

Bronx region, historical
Bicknell's *Riverdale* 1872–1901: unrecorded

Eaton's *Bronx + Manhattan* 1910–14: no mention of irruptions south of northern New York State breeding range

Griscom's *Bronx Region* as of 1922: Van Cortlandt 29 Oct 1916, an irruption year

Kuerzi's *Bronx Region* as of 1926: as above, plus 5 New Rochelle Nov–Dec 1916

NYC area, historical
Cruickshank's *NYC Region* as of 1941: 1941–42 irruption brought another to southern Westchester County

Bull's *NYC Area* as of 1968: irruptions in 1951–52, 1954–55, 1959–60, 1961–62—the latter largest ever in New York City area, with 35–40 Boreals (including northern NJ); 29 Oct–18 Apr

Bull's *New York State* as of 1975: another irruption 1965–66

Levine's *New York State* as of 1996: Long Island maximum 5 in Norway Spruce grove in winter 1954–55

Study area, historical and current
There are 6 study area records, 5 in Van Cortlandt: 29 Oct 1916 (C. Lewis), 2 on 12 Nov 1916 (fide Rogers), 27 Nov 1916 (Kieran)—all during the 1917–1918 irruption; 31 Oct 1954 (Carleton), 31 Oct 1965 (Norse); and once at Hillview: 12 Dec 1954 (Sedwitz, Buckley et al.—during the 1954–55 and 1964–65 irruptions. As of 2016, its preferred dense conifer habitat in the study area occurs only in Woodlawn, where access has long been limited and coverage is nearly nil.

In several other irruptions they were found close to but not within the study area, including 5 in New Rochelle from 25 Nov–30 Dec 1916 (Coles); 2 at Grassy Sprain 23 Dec 1951 (Bull); in Riverdale on 12 Nov 1954 (Buckley); in Bronx Park 12–13 Oct 1957 (C. Young et al.); at Hawthorne on 10 Nov 1961, 2 from 2 Nov 1963–28 Feb 1964, 20 Dec 1965–mid-Feb 1996 (all Augustine); 2 at Grassy Sprain 10 Nov 1963 (Friedman); New Rochelle 8 Nov 1965 (Stepinoff); 3 in Bronx Park on 26 Oct 1969 (Maguire et al.); Port Washington 4–16 Feb 1970 (Brigg et al.).

In the winter of 1975–76, 24 were in nearby Westchester County and the Bronx, including 2 in Bronx Park on 25 Oct (Peszel, Ozard), singles at Fort Tryon Park 7 Oct (Raices) and Scarsdale 26–31 Oct (Moyle), plus 6 at Kensico Reservoir on 28 Nov (Howe) and 3 in Tarrytown from mid-Nov–late Feb (Howe). It would seem they were never searched for at Woodlawn or Hillview that winter, but it is hard to imagine none were there. A single at Fort Tryon Park 12 Oct 1982 (Pohner) is the most recent in the vicinity of the study area.

As a New York State breeder it is restricted to Adirondack spruce-fir forests, where any population trends are unclear (McGowan and Corwin 2008).

Tufted Titmouse   *Baeolophus bicolor*

NYC area, current
Now a widespread resident breeder and erratic migrant throughout, quite a different story from its status in the late 1800s.

Bronx region, historical
Bicknell's *Riverdale* 1872–1901: once, 29 Nov 1874–28 Mar 1875

Eaton's *Bronx + Manhattan* 1910–14: rare; single New York City breeding on Staten Island

Griscom's *Bronx Region* as of 1922: 5 records 1874–1920

Kuerzi's *Bronx Region* as of 1926: occasional visitor, 7 records 1874–1924

NYC area, historical
Cruickshank's *NYC Region* as of 1941: extremely uncommon east of Hudson River where possible anytime; recently bred near Pelham Bay Park, New Rochelle

Bull's *NYC Area* as of 1968: great increase after early 1950s, including first Bronx breeding Pelham Bay Park 1935, Westchester County's New Rochelle 1936, Long Island's first in 100+ years Oyster Bay 1960, Central Park 1968

Bull's *New York State* as of 1975: Long Island breeders reach Peconic Bay

Study area, historical and current
A resident breeder that did not appear in the study area until the early 1940s. In the Bronx it was not found after Bicknell's unique 1874–75 overwintering until singles in Riverdale on 12 Feb 1911 (Griscom) and in Bronx Park on 26 Mar 1914 (Saunders) and from 19 Nov 1919 (L. N. Nichols)–20 May 1920 (L.N. Nichols et al.), but not in the study area until 1942, when 4 were in Woodlawn on 15 Nov (Marien). They were next seen in Van Cortlandt on 18 Oct 1950 (Komorowski) and 21 Mar 1951 (Kane), but not regularly until Apr–May 1954 (maximum 3: Buckley), by which time 1–2 pairs may have been breeding there and in Woodlawn. Numbers increased steadily since then in both locations, and they also breed at Jerome and Hillview. They first nested in the Bronx near Pelham Bay Park in 1935 but not in the study area until about 1954, although possibly in Woodlawn as early as 1942. In Riverdale, 2–3 pairs were breeding by 1953 (Kane, Buckley).

Seasonal daily maxima are: *spring* 20 on 15 Mar 2012 (Baksh); *fall* 16 on 26 Nov 2008 (Young); *winter* 86 across Van Cortlandt and Woodlawn on 23 Dec 2007 (Lyons et al.). While there is no evidence of migration into or out of the study area, it seems to occur; see "Comments" below.

An estimated 20–25 pairs breed throughout the study area, with <10 in Van Cortlandt, 5 in Woodlawn, and 3–4 each at Jerome and Hillview.

Both NYSBBA I in 1980–85 and II in 2000–2005 noted a broad New York City area breeding distribution, where they were understandably absent only from urban portions of all New York City boroughs except Staten Island, although less restricted than Black-capped Chickadee (McGowan and Corwin 2008). DOIBBS data showed a 9.8% annual statewide increase from 1980–2005 (Sauer et al. 2005). As with Fish Crow, most of the increase occurred north of the New York City area, where titmice have been well established since the 1970s.

Its history on the Bronx-Westchester Christmas Bird Count (Fig. 50) illustrates its colonization and rise in numbers east of the Hudson River. It was first recorded in 1929, but numbers never exceeded 12 until 1955, when they began to increase. They peaked in the 1990s and then dropped back slightly, where they remain as of 2016.

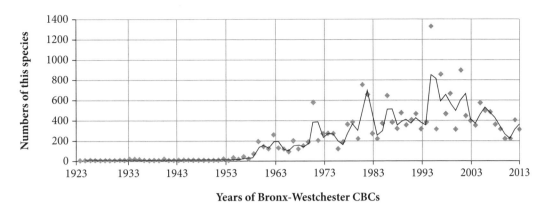

Figure 50

Comments

Titmice are not considered irruptive, but they certainly were in New York City in Oct–Nov 1978 (Post 1979) when they moved in flocks of 4–12 along city streets through midtown Manhattan. Many remained until Christmas Bird Count time, with a maximum of 125 in Central Park on 17 Dec, a tenfold increase over 1977. This event was detected as far away as Cape May and in southern VT and ME, where they moved in from the *south* in unprecedented numbers. This irruption's origins and triggers remain unexplained.

Nuthatches: Sittidae

Red-breasted Nuthatch *Sitta canadensis*

NYC area, current
A fall and spring migrant and winter visitor/resident in small numbers. It breeds sporadically in a few locations and fairly regularly irrupts southward from its preferred taiga breeding areas, occasionally remaining to breed.

Bronx region, historical
Bicknell's *Riverdale* 1872–1901: irregular, sometimes abundant fall migrant, only twice in spring; several early Jul records

Eaton's *Bronx + Manhattan* 1910–14: irregular migrant, rare winter visitor

Griscom's *Bronx Region* as of 1922: irregular fall migrant, rarely overwintering

Kuerzi's *Bronx Region* as of 1926: irregular spring/fall migrant, sometimes numerous, occasionally overwintering; recorded year-round

NYC area, historical
Cruickshank's *NYC Region* as of 1941: irregular, rare to abundant fall (maximum several dozen)/light spring migrant, erratically rare to abundant winter resident

Bull's *NYC Area* as of 1968: Long Island maxima fall (hundreds), winter 36

Bull's *New York State* as of 1975: spring maximum 28

Levine's *New York State* as of 1996: expanded breeding in Westchester County, Staten Island, Long Island

Study area, historical and current
A regular winter resident in Woodlawn conifers and an irregular fall migrant throughout. Given to periodic irruptions that often begin early (1–5 Jul 1886: Bicknell, 2 Aug 1952: Buckley, Kane), after which many may overwinter in study area conifer groves, with up to 20 each in Woodlawn and in the now-eradicated Hillview conifers. Unexpectedly, the widespread *Jun* 2016 appearances throughout the New York City area never led to an anticipated fall/winter mega-irruption.

First nonirruptive migrants normally arrive in late Sep, peak in late Oct, and depart by mid-Nov, with extremes of 12 Sep 1959 (Horowitz) and 13 Nov 2010 (Baksh) and a daily maximum of 2–3 on many occasions. But in irruption years, the first often appear in the first few days of Aug. Until Jun 1967, the fall maxima in Central and Prospect Parks were, respectively, 8 on 11 Sep 1943 (Bull, Eisenmann) and 13 on 16 Sep 1945 (Soll).

Daily winter maxima in appropriate habitat in nonirruption years are usually only 2–3 but 20+ in both Woodlawn and Hillview in irruption years (Kieran, Horowitz et al.). Even in nonirruption years one or more are usually findable in Woodlawn; on 22 Dec 2013 (a nonirruption year), 5 were in Woodlawn (Gotlib, McAlexander).

In spring, returnees are barely noticed, arriving in early Apr, peaking in late Apr, and departing by early May, with extremes of 8 Mar 1952 (Buckley) and 17 May 1958 (Mayer) and daily maxima of only 2 on many occasions.

After heavy irruption years they may breed in scattered coniferous locations in the New York City area (as do Pine Siskins and Red Crossbills) and are particularly frequent in Woodlawn, where, for example, one was calling persistently and behaving as if nesting on 16 Jun 1961 (Horowitz). Like many breeders, they can become quiet and very easy to overlook.

NYSBBA I in 1980–85 and II in 2000–2005 each showed a number of occupied blocks southeast of the closest breeding-block cluster in the Catskills, including singles on Staten Island and in Brooklyn, and about 40 in Nassau and Suffolk Counties. There were more blocks on NYSBBA I in 1980–85 than on NYSBBA II (McGowan and Corwin 2008), doubtless tracking irruption and nonirruption winters. DOIBBS data showed a consistent statewide increase from 1966–2005 (Sauer et al. 2007). But as with Fish Crow, most of the increase occurred well north of the New York City area, where its toehold remains tenuous.

Specimens
Bicknell collected 2 in Riverdale in Aug, Oct 1878 (NYSM).

White-breasted Nuthatch *Sitta carolinensis*

NYC area, current
An annual but uncommon spring/fall migrant, winter resident, and local breeder. It is least frequent on Long Island barrier beaches and most common on the North Shore and the mainland, and it is given to occasional low-number irruptions.

Bronx region, historical
Bicknell's *Riverdale* 1872–1901: common spring/fall migrant, resident

Eaton's *Bronx + Manhattan* 1910–14: common resident

Griscom's *Bronx Region* as of 1922: familiar resident, common in fall migration, winter

Kuerzi's *Bronx Region* as of 1926: fairly common resident

NYC area, historical
Cruickshank's *NYC Region* as of 1941: common fall, overlooked spring migrant; abundant, widespread winter resident (maximum 12); widespread but uncommon breeder

Bull's *NYC Area* as of 1968: maxima spring 10, fall 12, winter 76 on Southern Nassau Christmas Bird Count

Bull's *New York State* as of 1975: fall maximum 14

Study area, historical and current
A resident breeder in Van Cortlandt and Woodlawn and spring/fall migrant in all 4 extant subareas.

There is a very light and largely unnoticed spring movement from Mar–May that peaks in late Mar; the maximum is 9 on 28 Mar 1995 (Young).

Nuthatches breed in Van Cortlandt and Woodlawn, with an estimated population of <10 pairs, but they become silent and are easily overlooked.

NYSBBA I in 1980–85 and II in 2000–2005 showed a broad New York City area distribution. There were fewer blocks on NYSBBA II and only single blocks in Manhattan and Brooklyn (McGowan and Corwin 2008). DOIBBS data showed no statewide change from 1966–80, followed by a slight increase (Sauer et al. 2005).

Fall migration is more noticeable and occurs in Oct–Nov. There are occasional irruption years when they accompany Red-breasted Nuthatches and Hairy Woodpeckers. The daily maximum is only 6 on 14 Sep 1935 (Imhof). Until Jun 1967, the fall maximum in Central Park was 12 on 9 Oct 1943 (Aronoff).

In winter, migrants move in and swell local numbers; the study area maximum is 35 on the Bronx-Westchester Christmas Bird Count of 27 Dec 2015 (Lyons et al.).

Specimens
Dwight collected 2 in Van Cortlandt in Mar, Apr 1884–85 (AMNH) and Bicknell 2 in Riverdale on 25 Dec 1878 (NYSM).

Creepers: Certhiidae

Brown Creeper *Certhia americana*

NYC area, current
A regular spring/fall migrant, winter resident in small numbers, an occasional breeder on central and eastern Long Island and a declining breeder in Westchester County

Bronx region, historical
Bicknell's *Riverdale* 1872–1901: common spring/fall migrant, rare in winter

Eaton's *Bronx + Manhattan* 1910–14: fairly common winter visitor

Griscom's *Bronx Region* as of 1922: common spring/fall migrant, fairly common in winter

Kuerzi's *Bronx Region* as of 1926: common spring/fall migrant, winter resident (3 Sep–15 May); bred Van Cortlandt Swamp 1926

NYC area, historical
Cruickshank's *NYC Region* as of 1941: common spring/fall (maximum 20) migrant, winter resident (31 Aug–21 May)

Bull's *NYC Area* as of 1968: late spring date 5 Jun, early fall date 13 Aug; Long Island maxima spring 35, fall 75; Long Island breeding sites increasing

Bull's *New York State* as of 1975: 12 pairs breeding on Long Island at Connetquot State Park, also other new sites; fall maximum 150

Study area, historical and current
A regular spring/fall migrant and winter resident in small numbers.

In fall, most migrants move heavily in Sep–Oct, with extreme dates of 17 Sep 1959 (Horowitz) and 25 Nov 2012 (Baksh), and a maximum of 10 on 25 Oct 1952 (Buckley). Until Jun 1967, the fall maximum in Prospect Park was 9 on 10 Oct 1944 (Soll).

Overwinterers are in place by Nov, and typical counts for all of Van Cortlandt are 2–3, with a maximum of 10 on 27 Dec 1922 (Eisenstein).

In spring, migrants arrive with kinglets and Pine, Palm, and Myrtle Warblers in Apr, with extreme dates of 2 on 27 Mar 2007 (Klein) and 9 May 1961 (Horowitz), and a maximum of 3 on several dates. Until Jun 1967, the spring maximum in Central Park was 14 on 17 Apr 1963 (Post, Tudor).

One nest has been uncovered in the study area, in Van Cortlandt Swamp on 27 May 1926 (Cruickshank); creepers may have bred more often, but many observers are unfamiliar with their loud warbler-like songs.

NYSBBA I in 1980–85 found single occupied blocks in the New York City area in southern Westchester County and Queens, plus 3 in Nassau County and >30 in Suffolk County. By NYSBBA II in 2000–2005, these were down to none in southern Westchester County and Queens, 1 in Nassau County, and 10 in Suffolk County (McGowan and Corwin 2008). It is not clear to what extent these represent genuine declines or sampling issues. Sparse DOIBBS data showed no statewide changes from 1966–2005 (Sauer et al. 2007), and as Fish Crow and Red-breasted Nuthatch, most breed well north of the New York City area. The nearest regular breeders are in Bergen County, NJ, perhaps on the Palisades.

Specimens
Bicknell collected 2 in Riverdale in Nov, Dec 1876–81 (NYSM).

Wrens: Troglodytidae

House Wren   *Troglodytes aedon*

NYC area, current
A common spring/fall partial Neotropical migrant and a widespread breeder but exceedingly infrequent in winter.

Bronx region, historical
Bicknell's *Riverdale* 1872–1901: common spring/fall migrant, breeder

Eaton's *Bronx + Manhattan* 1910–1914: common [summer] resident

Griscom's *Bronx Region* as of 1922: uncommon spring, rare fall migrant; uncommon breeder, decreasing

Kuerzi's *Bronx Region* as of 1926: common breeder, spring (maximum 24)/fall migrant (12 Apr–23 Oct)

NYC area, historical

Cruickshank's *NYC Region* as of 1941: common breeder with many nesting in New York City [locations not given], common spring (maximum 24+)/fall migrant (12 Apr–late Oct); infrequent in winter

Bull's *NYC Area* as of 1968: fall maximum 12

Bull's *New York State* as of 1975: status unchanged

Study area, historical and current

A common spring/fall migrant and widespread breeder, exceedingly infrequent in winter.

In spring, first migrants arrive in late Apr, peak in early May, and depart by late May, with extreme dates of 14 Apr 2002 (Young) and 2 on 19 May 2012 (Baksh), and a maximum of 4 on 26 Apr 1953 (Buckley) and 3 May 1995 (Young). Until Jun 1967, the spring maximum in Central Park was 8 on 6 May 1964 (Post).

In the late 1800s, it was a common breeder in Riverdale (Bicknell) and in Van Cortlandt (Dwight) and it still is. As of 2016, 5–10 pairs breed in Van Cortlandt. But 10 Tree Swallow nest boxes were emplaced at various locations on the Van Cortlandt golf course between 1998 and 2012, and some were regularly used by House Wrens (Künstler), so the true total may be higher. It was a widespread study area and Riverdale breeder from the 1920s–50s (Kieran), and NYSBBA I in 1980–85 and II in 2000–2005 both found breeders in Blocks 5852B and 5952A, which include the study area and Riverdale (McGowan and Corwin 2008). Several pairs bred in Riverdale as late as 1955 (Kane, Buckley), and a few still do.

Both NYSBBA I in 1980–85 and II in 2000–2005 elucidated a broad New York City area breeding distribution, where they were understandably absent only from the most urban portions of all New York City boroughs except Staten Island (McGowan and Corwin 2008). DOIBBS data showed a slight statewide increase from 1980–2005 (Sauer et al. 2005).

In fall, first migrants arrive in late Aug, peak from mid-Sep–mid-Oct, and depart by early Nov, with extreme dates of 11 Aug 2008 (Young) and 26 Nov 1954 (Buckley), and a maximum of 4 on 28 Sep 2002 (Young). Until Jun 1967, the fall maximum in Central Park was 8 on 27 Sep 1956 (Post).

There are no Dec–Mar study area records, although singles were recorded several times in late Dec in Riverdale from the 1930s–50s (Cruickshank, Kieran, Buckley) and more recently on 29 Dec 1996 (Gibaldi et al.).

Specimens

Dwight collected 1 in Van Cortlandt on 3 May 1887 (AMNH) and Bicknell 4 in Riverdale in Sep, Oct 1876–81 (NYSM)

Winter Wren *Troglodytes hiemalis*

NYC area, current

A spring/fall migrant in small numbers, irregular and very local in winter.

Bronx region, historical

Bicknell's *Riverdale* 1872–1901: spring/fall migrant, uncommon in winter

Eaton's *Bronx + Manhattan* 1910–14: fairly common winter visitor

Griscom's *Bronx Region* as of 1922: rarely seen spring/fall migrant

Kuerzi's *Bronx Region* as of 1926: fairly common spring/fall migrant (17 Sep–13 May), frequently overwintering

NYC area, historical
Cruickshank's *NYC Region* as of 1941: regular but uncommon spring/fall (maximum 11) migrant, very uncommon winter resident; 11 Sep–31 May

Bull's *NYC Area* as of 1968: maxima fall 15, winter 12; bred northern Westchester County 1962

Bull's *New York State* as of 1975: status unchanged

Study area, historical and current
A spring/fall migrant in small numbers, irregular and very local in winter.

In fall, first migrants arrive in mid-Sep, peak in late Oct–early Nov, and depart by mid-Nov, with extreme dates of 9 Sep 1953 (Buckley, Phelan) and 20 Nov 1959 (Horowitz) and a maximum of 4 on several occasions. Until Jun 1967, the fall maximum in Prospect Park was 6 on 3 Oct 1944 (Soll).

Collectively, 6 or more may overwinter in Van Cortlandt and Woodlawn, and on 22 Dec 2013, 7 were there (Lyons, Gotlib et al.).

In spring, first migrants trickle in during late Mar, peak in late Apr–early May, and depart by mid-May, with extremes of 1 Apr 1960 (Horowitz, Heath) and 14 May 1951 (Kane), and a maximum of 3 on 23 Apr 2010 (Young).

Southeast of its occupied-block cluster in the Catskills, both NYSBBA I in 1980–85 and II in 2000–2005 found occupied blocks in Orange (4–5), Rockland (2–3), Putnam (4–5), and Westchester (7–9) Counties. NYSBBA II noted the number of occupied blocks to have increased by 49% statewide since NYSBBA I (McGowan and Corwin 2008). This mirrors a general southward expansion in the Northeast: they have established a growing breeding population on Martha's Vineyard, MA, and bred on the south shore of Long Island in 1976. Nearest breeders currently are on the Palisades and in Westchester County near Rye. DOIBBS data showed an annual statewide increase of 4.9% from 1980–2005 (Sauer et al. 2005).

Specimens
Dwight collected 1 in Van Cortlandt on 21 Dec 1885 (AMNH) and Bicknell 2 in Riverdale in Feb, Oct 1876–79 (NYSM).

Sedge Wren   *Cistothorus platensis*

NYC area, current
A formerly uncommon but now exceedingly rare spring/fall migrant, a former loosely colonial breeder, and a formerly occasional but now unknown winter resident. It is a New York State Threatened Species. Up to 1967, barely recorded in both Central (thrice, last in 1943) and Prospect (once, in 1952) Parks.

Bronx region, historical
Bicknell's *Riverdale* 1872–1901: rare Aug–Oct migrant or possible breeder at Kingsbridge Meadows

Eaton's *Bronx + Manhattan* 1910–14: rare local breeder

Griscom's *Bronx Region* as of 1922: 3 pairs nested Baychester Marshes Jun 1917; otherwise unrecorded

Kuerzi's *Bronx Region* as of 1926: rare spring/fall migrant (22 Apr–23 Oct) and rare breeder [locations not given]

NYC area, historical
Cruickshank's *NYC Region* as of 1941: very local, uncommon to rare breeder, until recently nesting at a few New York City sites [locations not given]; otherwise seldom recorded in spring or fall migration, exceedingly rare in winter coastally

Bull's *NYC Area* as of 1968: bred Queens to 1960, last on Long Island

Bull's *New York State* as of 1975: status unchanged

Study area, historical and current
Until the 1950s it was an exceedingly infrequent fall and spring migrant in Van Cortlandt Swamp and occasional in winter, although it never bred there. Throughout the Northeast it began to decline in the 1950s and since then has been unrecorded in the study area. The only 2 records for Inwood Park were on 19 Sep 1962 and 11 Oct 1966 (Norse).

There are 12 mid-20th-century Van Cortlandt records but none since 1956: 3 Sep 1933 (Norse); 4 (the maximum) on 14 Sep 1935 (Imhof, Norse); 3 on 29 Aug 1937 (Norse); 3 Oct 1937 (Norse); 3 Dec 1937 (Norse); 26 Dec 1948 (Norse, Cantor, Komorowski); 14 Sep 1952 (Norse); 3 Oct 1953 (Kane, Buckley); 13 May 1954 (Phelan, Buckley); singing on 27 Jul 1954 (Phelan); 17 Sep 1954 (Kane, Buckley); 6 May 1956 (Scully, Buckley).

Formerly a very local and rare New York City area breeder as close as Pelham Bay Park, where it nested continually through 1955 (Young 1958), and an exceedingly local breeder throughout the Northeast. During NYSBBA I in 1980–85 it was found near Rye, but during NYSBBA II in 2000–2005 the nearest probable breeders were in Ulster County (McGowan and Corwin 2008). It is unknown where the nearest regular breeders are in 2016.

In the late 1800s, Bicknell saw them only in fall at Kingsbridge Meadows in 1876, 1880, 1881, and 1895, with extreme dates of 12 Aug 1881–23 Oct 1880, collecting singles on 22 Sep 1876 and 13 Oct 1880 (NYSM). However, Sedge Wren is a normally a late Sep–Oct fall migrant, and it is now known that when they appear in appropriate habitat in Jul–Aug, as they frequently do, it is often for renesting after unsuccessful (or even successful) May–Jun breeding elsewhere. Bicknell would have been unaware of this pattern and so would not have followed up on it, and in any case they are easily overlooked by those not expecting them, missing their inconspicuous songs, or being infrequently afield in mid–late summer. In Van Cortlandt Swamp the 3 in Aug 1937 and the Jul 1954 single above may have also involved breeding attempts, but we'll never know. Eaton (1910, 1914) did list Sedge Wren as a breeder in his Bronx + Manhattan county with no additional details, but at the time Sedge Wren was not an uncommon breeder around New York City. Hence it possible that Sedge Wren may have been an occasional late summer breeder at Kingsbridge Meadows in the late 19th and very early 20th centuries.

Specimens
Bicknell collected singles at Kingsbridge Meadows in Sep, Oct 1876–80 (NYSM).

Marsh Wren *Cistothorus palustris*

NYC area, current
This loosely colonial breeder is widely but increasingly thinly distributed in cattail and *Phragmites* marshes (occasionally even in *Spartina* saltmarshes) and is a regular spring/fall migrant in small numbers even in urban parks.

Bronx region, historical
Bicknell's *Riverdale* 1872–1901: common spring/fall migrant, breeder in cattail marshes at Kingsbridge Meadows, Van Cortlandt

Eaton's *Bronx + Manhattan* 1910–14: abundant breeder

Griscom's *Bronx Region* as of 1922: uncommon breeder, formerly locally abundant

Griscom's *Riverdale* as of 1926: extirpated except Van Cortlandt

Kuerzi's *Bronx Region* as of 1926: common/abundant breeder (19 Apr–14 Nov), occasionally lingering into Jan

NYC area, historical
Cruickshank's *NYC Region* as of 1941: local, sporadic breeder in cattail marshes, otherwise rare spring/unknown fall migrant; rare, local winter visitor/resident

Bull's *NYC Area* as of 1968: winter maximum 22 Queens + Southern Nassau Christmas Bird Counts; 100 pairs Croton Point 1942

Bull's *New York State* as of 1975: status unchanged

Study area, historical and current
Formerly a spring/fall migrant throughout, and a breeder and winter resident in Van Cortlandt Swamp, Kingsbridge Meadows, Jerome Meadows, and Jerome Swamp. Now only an occasional spring or fall migrant in Van Cortlandt, but no longer a winter resident.

Since their extirpation as breeders/residents, first spring migrants (singles only) arrive in mid-Apr, peak in early May, and depart by mid-May, with extreme dates of 27 Apr 1994 (Young) and 20 May 2006 (Young), singing not in Van Cortlandt Swamp but in Tibbett's Marsh, as was another on 12 May 1995 (Garcia)—a potential new breeding site. The historical spring maximum would have been about 75 in late May 1938, when 48 occupied nests were in Van Cortlandt Swamp (Imhof et al.). Until Jun 1967, the spring maximum in Prospect Park was 3 on 23 May 1951 (Alperin, Whelen).

Until 1959 it was a resident breeder in the cattail portions of Van Cortlandt Swamp, then a migratory breeder until 1966, when there were 5 pairs, the last breeding we know of (Gera, Gera), although one on 11 Jun 1973 (McGuinness) might also have been. Since then it has only been a scarce migrant, although singing ♂♂ in Van Cortlandt Swamp on 28 May 1980 (Sedwitz) and 21 May 2006 (Delacretaz 2007) hold out hope for rebreeding in the future. Although not found in the study area on NYSBBA I in 1980–85 or II in 2000–2005, they were in the 4 blocks to the immediate east and southeast, including Pelham Bay Park, now the closest breeders (McGowan and Corwin 2008).

NYSBBA I in 1980–85 and II in 2000–2005 saw few occupied New York City area blocks. There were only 4 in Westchester County along the Hudson River, 3 in the Northeast Bronx, 8–10 in Queens, 2–4 in Brooklyn, and 8 on Staten Island. In Nassau and Suffolk Counties they were thinly distributed on Long Island Sound and ocean shores and at a handful of inland sites (McGowan and Corwin 2008). DOIBBS data showed a consistent statewide decrease from 1966–2005, but the data were too sparse for statistical analysis (Sauer et al. 2007). Presence of good-sized patches of cattails (optimal, but vanishing) or *Phragmites* remains the best nesting predictor.

In fall, first migrants are now arriving in Sep, peaking in Oct, and departing by late Oct, with extremes of 17 Sep in 1997 and 2001 (Young) and 2 Nov 2002 (Young), singles only. Bicknell collected migrants at Kingsbridge Meadows on 29 Sep 1880 and 14 Oct 1876 (NYSM).

In winter, up to 10 were routine until the 1940s, but once the Swamp had been decimated and *Phragmites* replaced cattails, in turn succeeded by woody shrubs, numbers gradually dropped until the last 2 overwintered in 1958–59 (Buckley et al.) even though breeding continued to 1966. One vocalizing in *Phragmites* on 28 Feb 2016 (Wilson) appears to be the first in winter since 1959.

Specimens
Bicknell collected singles at Kingsbridge Meadows in Sep, Oct 1876–80 (NYSM).

Carolina Wren   *Thryothorus ludovicianus*

NYC area, current
A widespread but very local resident and partial migrant. As a southern species that only began to colonize the New York City area in the 1800s, it is often extirpated locally during severe winters with heavy snow cover, but if so usually recovers and reoccupies favored habitat after a few years.

Bronx region, historical
Bicknell's *Riverdale* 1872–1901: recorded 10 of 15 years, bred 1879, 1881, 1888; migrants mostly in Mar–Apr, Sep–Nov

Eaton's *Bronx + Manhattan* 1910–14: rare resident, breeds

Griscom's *Bronx Region* as of 1922: several Riverdale records, bred once late 1880s; otherwise single 1909, 1911 records

Kuerzi's *Bronx Region* as of 1926: occasional visitor, rare resident

NYC area, historical
Cruickshank's *NYC Region* as of 1941: uncommon local resident, possible anywhere, anytime

Bull's *NYC Area* as of 1968: maximum 12–26 singing (♂♂?) Gardiner's Island

Bull's *New York State* as of 1975: winter maximum 70 Gardiner's Island

Study area, historical and current
A resident breeder in Van Cortlandt and Woodlawn. First recorded in Van Cortlandt in 1895 on 12 Apr (Chubb) and 4 May (Foster). Until the 1960s it seemed not to have begun breeding and had been recorded fewer than a dozen times throughout the years—almost always singles, sometimes singing (Kieran, Norse et al.). The early maximum was only 2 on 12 Sep 1925 (Kuerzi, Kuerzi) and on 27 May 1941 (E. G. Nichols).

Carolina Wrens are subject to drastic population diminution during severe winters with heavy snow, although less frequently than in the late 1800s–early 1900s. They did not become a regular study area bird until the 1960s–70s, and first breeding passed unrecorded, probably about 1975. They bred in Riverdale in 1879 (Bicknell), and again in 1944 (Kieran) but were not detected in the extensive 1955 Riverdale breeding census (Kane, Buckley). Because it is one of those species that may appear anywhere, anytime, observers seem not to pay much attention to it. Christmas Bird Count data (Fig. 51) do reflect the long-term study area trend well.

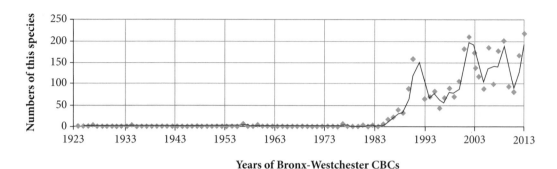

Figure 51

As of 2016 we estimate that 5–10 pairs breed in the study area, most of them in Van Cortlandt. Across the year daily maxima have usually not exceeded 3–4; the maximum of 16 was on 22 Dec 2013 between Van Cortlandt and Woodlawn (Lyons, Gotlib et al.). No study area migration information is at hand.

Both NYSBBA I in 1980–85 and II in 2000–2005 confirmed a broad New York City area breeding distribution, where they were understandably absent only from the most urbanized portions of all New York City boroughs except Staten Island (McGowan and Corwin 2008).

Sparse DOIBBS data showed a consistent statewide increase from 1980–2005 (Sauer et al. 2007). As with Fish Crow, most of the increase occurred well north of the New York City area, where Carolina Wren has been established for some time.

Specimens
Bicknell collected 1 in Riverdale on 2 May 1879 (NYSM).

Comments
Vanishing midwestern resident **Bewick's Wren**, *Thryomanes bewickii*, has always been the rarest of spring vagrants to the New York City area. It has been found only 4 times, all singing ♂♂ in spring: in Central Park from 10 Apr–8 May 1928 (Capen et al.); in Prospect Park on 13 May 1946 (Soll, Whelen) and from 15–18 Apr 1952 (Carleton, Restivo et al.); and at Croton Point on 8 May 1952 (Walsh et al.). Disappearing as a breeder from the northeastern portion of its range, it did breed as close as Fulton County, PA (130 mi [209 km] west of Philadelphia) until extirpated in the 1970s, and it has been unreported anywhere in PA since 1977. After the 2 in Apr–May 1952 there had been no southeastern New York State records and only a handful upstate, the last in 1975. To initial incredulity, a pair bred and fledged 3 young in the Catskills at Mohonk Lake, Ulster County, in Jul–Aug 1974 (Stapleton et al.), the sole New York State breeding event, but none of them ever returned. So it goes.

Gnatcatchers: Polioptilidae

Blue-gray Gnatcatcher   *Polioptila caerulea*

NYC area, current
A spring/fall Neotropical migrant and a highly local breeder that is slowly increasing its numbers and range. It is a southern United States species that has been slowly moving northward for 100+ years.

Bronx region, historical

Bicknell's *Riverdale* 1872–1901: unmentioned

Eaton's *Bronx + Manhattan* 1910–14: uncommon summer visitor

Griscom's *Bronx Region* as of 1922: 1 fall, 2 spring records 1895–1920

Kuerzi's *Bronx Region* as of 1926: rare spring (15 Apr–20 May)/fall (3 Sep–11 Oct) migrant, 12 records 1917–26

NYC area, historical

Cruickshank's *NYC Region* as of 1941: rare but regular spring (3–12 annually, maximum 3)/fall migrant (3 Apr–27 Oct); breeding unrecorded

Bull's *NYC Area* as of 1968: much commoner migrant since late 1940s, with spring maximum 9 and fall maximum 6; 3 Long Island breeding sites (1963, 1968) but none in Westchester County, Bronx; late fall date 27 Nov

Bull's *New York State* as of 1975: 5 more 1963–75 Long Island breeding sites; first in southern Westchester County

Levine's *New York State* as of 1996: spring maximum 20; late fall date 2 Dec

Study area, historical and current

A regular spring and scarce fall migrant and a recent breeder, not recorded until 30 Apr 1920 (Kieran), then on 25–28 Apr 1924 (Kuerzi, Kuerzi), 3 Sep 1925 (Cruickshank), 20 May 1926 (Cruickshank), and finally erratically in spring (almost never in fall) through the 1930s and '40s. Major spring flights in 1947 (when Russak found 5 on 13 Apr in Van Cortlandt and the first 3 northern NJ breedings occurred), in 1954, and in 1963 resulted in great increases in frequency and numbers seen around New York City, including the first breeding at Greenbrook Sanctuary on the Palisades opposite Riverdale (Roser). The first Westchester County breeding occurred sometime around 1963, but details are unavailable. With each succeeding year more and more arrived and consolidated their breeding range. They did not breed on Staten Island until 1982 (Siebenheller, Siebenheller) or on Long Island until 1968 at Lloyd's Neck (Connolly, Dove), so things were moving steadily, albeit slowly.

In spring, first migrants arrive in mid-Apr, peak in late Apr, and depart by early May, with extreme dates of 7 Apr 1952 (Buckley) and 31 May 1954 (Buckley), and a maximum of 9 on 3 May 2014 (Souirgi). Until Jun 1967, the spring maxima in Central and Prospect Parks were, respectively, 7 on 18 May 1958 (Post) and 8 on 26 Apr 1947 (Brooklyn Bird Club).

One of the enduring puzzles is the year when gnatcatchers first bred in the study area. They were seen annually around Van Cortlandt Swamp and the Sycamore Swamp in spring and later in increasing numbers through the 1960s, '70s, and '80s but seem not to have been reported breeding until the late '90s. Nonetheless, it is our belief that they first bred during the 1970s or even the late '60s; indeed, one on 7 Jul 1957 (Mayer) suggests first breeding may have been even earlier. Nest building begins by late Apr, when breeders would be quickly dismissed as migrants. During NYSBBA I in 1980–85 probable breeding was recorded in Block 5852B, and in NYSBBA II in 2000–2005 possible breeding was noted there, confirmed breeding in Block 5952A (McGowan and Corwin 2008). While these 2 blocks embrace the study area, exact locations were irretrievable. They have been breeding at Pelham Bay and Bronx Parks for years, but details are wanting. One in Riverdale 11 Jun 1955 (Buckley) was another potential breeder. As of 2016 they still breed in the study area but perhaps only 2–3 pairs in Van Cortlandt.

NYSBBA I in 1980–85 and II in 2000–2005 found occupied New York City area blocks in southern Westchester County (about 10), the Bronx (3), Brooklyn and Queens (1 each), Staten Island (2), plus about 50 in Nassau and Suffolk Counties. New York State DOIBBS data were too sparse for analysis, but statewide the number of occupied blocks increased by 19% by NYSBBA II (McGowan and Corwin 2008).

In fall they are easily overlooked unless calling, and early migrants are easily confounded with local breeders, so far fewer are recorded then. First migrants arrive in mid-Aug, peak in early Sep, and depart by late Sep, with extreme dates of 1 Aug 2017 at Hillview (Camillieri) and 3 Oct 1937 (Norse), plus a lingering blowback drift-migrant on 22 Dec 1987 (Naf). The current fall maximum, 4 on 14 Aug 2013 (Young), may involve local breeders; nearly all others were singles. In Bronx Park 2 on 31 Oct 1957 (Maguire et al.) and a single at Pelham Bay Park on 27 Nov 2011 (Rothman) were also blowback drift-migrants.

Kinglets: Regulidae

Golden-crowned Kinglet *Regulus satrapa*

NYC area, current
A spring/fall migrant and regular but thinly distributed winter resident.

Bronx region, historical
Bicknell's *Riverdale* 1872–1901: common spring/fall migrant, early as 10 Sep 1898

Eaton's *Bronx + Manhattan* 1910–14: common winter visitor

Griscom's *Bronx Region* as of 1922: formerly common, regular fall migrant, regular winter resident, now only occasionally overwintering

Kuerzi's *Bronx Region* as of 1926: common spring/fall migrant, fairly common in winter (10 Sep–7 May)

NYC area, historical
Cruickshank's *NYC Region* as of 1941: common spring/fall (maximum 100+) migrant, rare to fairly common winter resident (8 Sep–19 May)

Bull's *NYC Area* as of 1968: maxima spring 100, fall 150, winter 85

Bull's *New York State* as of 1975: early fall date 15 Aug; fall maximum 400; Dutchess, probable Westchester Counties breeding

Study area, historical and current
A spring/fall migrant and regular but thinly distributed winter resident.

In fall, first migrants arrive in late Sep, peak in mid-Oct, and depart by late Nov, with extreme dates of 24 Sep 1998 (Young) and 28 Nov 1995 (Young), and a maximum of 50+ on 12 Oct 1934 (Norse). Until Jun 1967, the fall maxima in Central and Prospect Parks were, respectively, 30 on 14 Oct 1945 (Aronoff) and 100 on 10 Nov 1956 (Soll, Whelen).

Depending on the year, from 2–50+ may overwinter in Woodlawn conifers, with other groups of 2–10 scattered throughout deciduous parts of the study area.

In spring, first migrants arrive in late Mar, peak in mid-Apr, and depart by late Apr, with extreme dates of 10 Mar 2012 (Higgins) and 11 May 1928 (Cruickshank), and a maximum of 21 on 27 Mar 2007 (Klein).

Southeast of its Catskills breeding-block cluster, during NYSBBA I in 1980–85 and II in 2000–2005 there were isolated (?) breeding events, almost always in dense Norway Spruce groves/plantations, in Orange (7), Rockland (1), Westchester (4), and Nassau Counties (1) (McGowan and Corwin 2008), and they have been reliable if local Norway

Spruce breeders in northern NJ since 1971 (Kane). DOIBBS data showed a consistent statewide increase from 1966–2005 (Sauer et al. 2007), but as with Fish Crow, most of the increase occurred well north of the New York City area. This species is a potential Woodlawn breeder in thick ornamental conifers.

Specimens
Dwight collected 8 in Van Cortlandt in Apr, Oct, Dec 1885–88 (AMNH).

Ruby-crowned Kinglet *Regulus calendula*

NYC area, current
A spring/fall (partially Neotropical) migrant and irregular, scarce winter resident.

Bronx region, historical
Bicknell's *Riverdale* 1872–1901: common spring/fall migrant, once as late as mid-Jan

Eaton's *Bronx + Manhattan* 1910–14: fairly common migrant

Griscom's *Bronx Region* as of 1922: common spring/fall migrant, once in winter

Kuerzi's *Bronx Region* as of 1926: common spring (20 Mar–31 May)/fall (14 Sep–22 Nov) migrant, occasional in winter

NYC area, historical
Cruickshank's *NYC Region* as of 1941: fairly common spring (maximum 60)/fall migrant, very rare winter resident (2 Sep–31 May)

Bull's *NYC Area* as of 1968: early fall date 22 Aug, spring maximum 100 Bronx Park, fall maximum 140

Bull's *New York State* as of 1975: status unchanged

Levine's *New York State* as of 1996: winter maximum 28 on 1984 Bronx-Westchester Christmas Bird Count

Study area, historical and current
A spring/fall migrant and scarce winter resident.

First migrants arrive in late Sep, peak in early Oct, and are usually gone in Nov, with extreme dates of 9 Sep 1953 (Phelan, Buckley) and 29 Nov 1953 (Norse), and a maximum of 20 on 18 Oct 1955 (Birnbaum).

Ruby-crowneds are completely absent in winter in some years, widespread (usually singles, occasionally small groups) in others; maximum 4 on 11 Dec 1931 (Cruickshank). Woodlawn is a favored winter site. Successful overwintering is uncertain, but 1 on 22 Feb 1951 (Kane) was well on its way. In Riverdale, Bicknell collected 1 on 11 Jan 1880 (NYSM).

In spring, first migrants arrive in late Mar, peak in mid-Apr, and depart by late Apr, with extreme dates of 15 Mar 1953 (Kane) and 18 May 1958 (Weintraub) and a maximum of 70 on 30 Apr 1965 (Norse). In nearby Bronx Park 100 were seen on 17 Apr 1944 (Komorowski). Until Jun 1967, the spring maximum in Central Park was 20 on 18 Apr 1965 (Carleton) and in Prospect Park 60 on 5 May 1950 (Jacobson, Whelen).

Nearest current breeders are scattered thinly through the Catskills. Sparse DOIBBS data showed a consistent but slight statewide decrease after 1966, the largest effect evident from 1980–2005 (Sauer et al. 2005; McGowan and Corwin 2008).

Specimens
Dwight collected 5 in Van Cortlandt in Apr 1885–88 (AMNH) and Bicknell 4 in Riverdale in Sep, Oct, Nov, Jan 1878–81 (NYSM).

Old-World Flycatchers: Muscicapidae

Northern Wheatear *Oenanthe oenanthe*

NYC area, current
A vagrant or very rare migrant from the eastern Canadian arctic whose entire population normally migrates back to Eurafrica for the winter—except for a handful that go south along the North American East Coast most years. There are 25 or so records from the New York City area, mostly from Long Island, where it is found once every 5–10 years. Recorded in Central but not Prospect Park.

NYC area, historical
Griscom's *NYC Region* as of 1922: 3 Long Island specimens from 1800s

Cruickshank's *NYC Region* as of 1941: singles in 1936, 1941

Bull's *NYC Area* as of 1968: singles in 1947, 1951, 1956, 1968

Bull's *New York State* as of 1975: single in 1970

Study area, current
Only 1 record and the sole wheatear in the Bronx: a breeding plumaged adult on 8 Jun 1968 on the Van Cortlandt Parade Ground (Goelet, Koelle). This carefully described bird falls into that narrow late May–early Jun interval for the very few East Coast wheatears detected returning in spring to arctic Canada, and it may have arrived with the one at Guilford, CT on 30–31 May 1968. The only other wheatear near the study area was at the expected fall time: at Inwood Park 22–28 Sep 1993 (Berkins et al.). Singles have been seen in Westchester County 4 times, including once in spring: at Peekskill on 15 Nov 1947 (Cruickshank), Croton Point on 6–8 Sep 2011 (Letts, Roberto), Rye on 17 Oct 1986 (unknown observer, *North American Birds*) and 7 May 1988 (unknown observer, *Kingbird*).

Thrushes: Turdidae

Eastern Bluebird *Sialia sialis*

NYC area, current
Formerly a numerous spring/fall migrant, breeder, and winter resident. Numbers crashed throughout New York State and elsewhere in the mid-1950s from biocides, competition with starlings and House Sparrows for nest holes, and loss of their grassland/farm-edge breeding habitat. With elimination of many of the most toxic pesticides used in agriculture and a widespread, aggressive nest box campaign, their numbers gradually rebounded, and by the 1980s–90s they were being seen in small flocks

in many locations. It is now back to its former migratory status, if not former numbers, and is increasing.

Bronx region, historical

Bicknell's *Riverdale* 1872–1901: common spring/fall migrant, breeder; irregular winter resident

Hix's *NYC* in 1904: breeds Bronx, Van Cortlandt Parks

Eaton's *Bronx + Manhattan* 1910–14: abundant migrant, common breeder, occasional resident

Griscom's *Bronx Region* as of 1922: decreasing, uncommon breeder, spring/fall migrant; rare in winter

Kuerzi's *Bronx Region* as of 1926: common spring/fall migrant, fairly common breeder [locations not given], occasionally overwintering

NYC area, historical

Cruickshank's *NYC Region* as of 1941: common breeder in more rural areas, decreasing slowly in immediate New York City vicinity [locations not given]; uncommon spring/common fall (maximum several hundred) migrant; widespread but very local in winter

Bull's *NYC Area* as of 1968: Long Island maxima fall 500–700, winter 200; greatly decreased as breeder

Bull's *New York State* as of 1975: status unchanged

Study area, historical and current

A former breeder, common spring/fall migrant, and regular winter resident. Now only an erratic spring/fall migrant, with a handful of very recent overwinterings.

Bluebird's study area status mirrored that throughout the New York City area, and by 1956 when they last bred in Van Cortlandt, they were becoming scare as migrants and winter residents. By the 1970s they were gone and went unrecorded for many years. The first few migrants began to appear in spring/fall by the late 1980s, but overwintering took another 30 years to recur.

In spring, first migrants arrive in mid-Mar, peak in late Mar–early Apr, and depart by late Apr, with extreme dates of 3 Mar 1920 (Gladden) and 27 Apr 1933 (Kieran), and a maximum of 10 on 3 Apr 1955 (Buckley).

In the late 1800s, it was a common breeder in Riverdale (Bicknell) and in Van Cortlandt (Dwight), continuing until the 1950s. As many as 6 pairs bred in Van Cortlandt Swamp, around golf course edges, and along the Putnam Division track bed annually through 1956 (Kieran, Buckley) but not since, although one on Vault Hill on 2 Jul 1957 (Mayer) was suspicious. With the resurgence of their numbers, judicious placement of predator-proofed, properly crafted nest boxes—especially inside the Swamp, along the Putnam Division track bed, adjacent to ponds on the golf courses, and in Woodlawn—would return this species as a study area breeder. Nearest breeders are those in nest boxes at Pelham Bay Park (Künstler) and in southern Westchester County.

NYSBBA I in 1980–85 found occupied blocks in southern Westchester County (6), on Staten Island (1), in northeastern Nassau County (5), and in Suffolk Co. (30). By II in 2000–2005, they were gone from Staten Island but increased in southern Westchester and were holding their own in Nassau and Suffolk Counties; a statewide increase of 54% was especially gratifying (McGowan and Corwin 2008). DOIBBS data showed a 3.2% annual statewide increase from 1980–2005 (Sauer et al. 2005). Well-thought-out nest box programs have played the major role in these increases.

In fall, first migrants arrive in early Oct, peak in early Nov when often in large calling flocks, and depart by late Nov with extreme dates of 8 on 14 Sep 1935 (Imhof) and 30 Nov 2012 (Young).

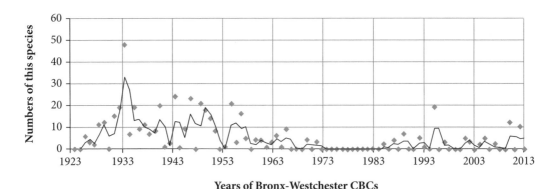

Figure 52

Before their decline, the maximum was 200 on 22 Oct 1946 (Komorowski); more recently, it was only 8 on 9 Nov 2006 (Young). A flock of 80 crossing the Hudson River from Riverdale on 30 Oct 2010 was an encouraging sign (Buckley). Until Jun 1967, the fall maximum in Prospect Park was 52 on 12 Nov 1944 (Soll, Whelen).

Bluebirds formerly overwintered in the study area annually in groups of 5–20, but since the mid-1950s only scattered singles have been occasionally reported on Bronx-Westchester Christmas Bird Counts, increasing slightly in the mid-1980s (Fig. 52). By the 1990s, 1–3 were lingering into Dec along the Putnam Division track bed or near the Sycamore Swamp (Garcia). Then, 3–5 were found on Vault Hill through 16 Feb 2013 (Fiore et al.), the remnants of a flock of 8 there on 3 Dec 2012 (Young)—the first successful overwintering in the study area since the 1960s.

Specimens
Dwight collected 3 in Van Cortlandt in Apr 1886–90 (AMNH) and Bicknell 2 in Riverdale in Mar, Nov 1886–87 (NYSM).

Comments
A western vagrant and blowback drift-migrant ♂ **Mountain Bluebird**, *S. currucoides*, was found in North Tarrytown, Westchester Co., on 20 Apr 1990 (R. Lewis) after having overwintered somewhere in the East/Southeast. Unrecorded in Central or Prospect Parks.

<u>Veery</u>   *Catharus fuscescens*

NYC area, current
This Neotropical migrant is a spring/fall migrant and nearly extirpated breeder.

Bronx region, historical
Bicknell's *Riverdale* 1872–1901: common spring/fall migrant, breeder

Hix's *New York City* as of 1904: breeds Bronx Park

Eaton's *Bronx + Manhattan* 1910–14: common breeder

Griscom's *Bronx Region* as of 1922: uncommon breeder, spring/fall migrant

Kuerzi's *Bronx Region* as of 1926: common breeder, 12 Apr–28 Sep

NYC area, historical
Cruickshank's *NYC Region* as of 1941: common breeder (but decreasing abruptly near New York City outskirts), abundant migrant spring (maximum 27)/fall (hundreds overhead at night); 12 Apr–5 Nov

Bull's *NYC Area* as of 1968: fall maximum 700 overhead at night, very few on ground next morning

Bull's *New York State* as of 1975: status unchanged

Study area, historical and current
A spring/fall migrant and formerly widespread breeder now almost extirpated.

In the study area in spring, first migrants arrive in early May, peak in mid-May, and depart by late May, with extreme dates of 29 Apr 1994 (Young) and 25 May 2006 (Young), and a maximum of 15 on 19 May 1955 (Kane et al.). Until Jun 1967, the spring maximum in Central Park was 25 on 5 May 1950 (Carleton).

In the late 1800s, it was a common breeder in Riverdale (Bicknell) and in Van Cortlandt (Dwight) continuing through the 1950s. It still breeds in Van Cortlandt but barely; there may be only 3–5 pairs left, most of which recent censusers may have dismissed as migrants. Otherwise there may no others breeding within New York City except a pair or 2 irregularly (?) at Pelham Bay Park (Künstler), and possibly in Bronx Park.

NYSBBA I in 1980–85 recorded New York City area occupied blocks in southern Westchester County, in the Bronx (1), Queens (1), and Staten Island (2), plus about 12 in mostly northern Nassau County and many throughout Suffolk County. By NYSBBA II in 2000–2005 in New York City there was only a single possible block on Staten Island and losses in Nassau County and Suffolk County (McGowan and Corwin 2008). DOIBBS data showed a consistent statewide decline from 1966–79, which accelerated to 2.1% per year from 1980–2005 (Sauer et al. 2005). However, the greatest decreases were in the northernmost parts of New York State.

In fall, when it is a distinct rarity on the ground, first migrants arrive in mid-Aug, peak in late Aug (when hundreds if not thousands can sometimes be heard passing overhead on favorable nights), and depart by late Sep, with extremes of 26 Aug 2001 (Young) and 15 Sep 2013 (Young), and a maximum of 4 on 3 Sep 1990 (Young).

Specimens
Dwight collected 14 in Van Cortlandt in May, Aug, Sep 1886–90 (AMNH) and Bicknell 2 in Riverdale in Aug, Sep 1878–79 (NYSM).

Gray-cheeked Thrush *Catharus minimus*

NYC area, current
An annual but increasingly infrequent spring/fall migrant. This is a Neotropical migrant that has been seen with conspicuously diminished frequency and in vastly smaller numbers since the 1960s.

Bronx region, historical
Bicknell's *Riverdale* 1872–1901: uncommon spring/fall migrant

Eaton's *Bronx + Manhattan* 1910–14: common migrant

Griscom's *Bronx Region* as of 1922: common spring/fall migrant

Kuerzi's *Bronx Region* as of 1926: common spring (5 May–6 Jun)/fall (9 Sep–21 Oct) migrant

NYC area, historical
Cruickshank's *NYC Region* as of 1941: fairly common spring (22 Apr–early Jun)/fall (31 Aug–10 Dec) migrant, spring maximum 14

Bull's *NYC Area* as of 1968: maxima spring 50, Long Island fall 300 overhead at night,

20 on ground next morning; late fall date 16 Dec [blowback drift-migrant]

Bull's *New York State* as of 1975: status unchanged

Study area, historical and current
An annual but increasingly uncommon spring/fall Neotropical migrant seen with greatly diminished frequency and in much smaller numbers since the 1960s.

In spring, first migrants arrive in early–mid-May, peak in and depart by late May, with extreme dates of one studied within 20 ft (6 m) alongside 6 Hermit Thrushes on 16 Apr 1952 (Kane, Phelan, Buckley)—a trans-Gulf overshoot—and 28 May 1996 (Young), and a maximum of 3 on 19 May 1955 (Kane, Buckley.) Bicknell collected 1 in Riverdale on the early date of 30 Apr 1878 (NYSM). Until Jun 1967, the spring maximum in Central Park was 50 on 20 May 1924 (Helmuth).

In fall, first migrants arrive in mid–late Sep, peak in early Oct, and normally depart by late Oct, with extreme dates of 27 Aug 1959 (Zupan) and 3 Oct 1937 (Norse)—we can find no later Oct dates—and a maximum of 2 on 11 Sep 2010 (Baksh). In the Bronx Gray-cheekeds have occurred with unanticipated frequency in Nov and extraordinarily twice into Dec: in the study area an injured bird appearing to be blind in 1 eye was studied at close range in Van Cortlandt Swamp on 8 Dec 1993 (Young) but was never refound, and in 1962 1 remained in Bronx Park until 23 Dec (C. Young et al.). These last 2 would have been lingering blowback drift-migrants.

Extrapolating from New York City area banding data and Central and Prospect Park singing Gray-cheekeds, as many as 5% of study area migrants may have always been Bicknell's Thrushes (see below), which when silent are separated only with great difficulty from Gray-cheekeds (Frey et al. 2008), although work is proceeding slowly on their field identification.

Specimens
Dwight collected 3 in Van Cortlandt in May 1890 (AMNH) and Bicknell 3 in Riverdale in Apr, May, Oct 1877–81 (NYSM, MCZ).

Bicknell's Thrush *Catharus bicknelli*

NYC area, current
An uncommon to rare but annual spring/fall Neotropical migrant in unknown numbers, given that some fraction of study area Gray-cheeked Thrushes in spring and in fall are actually Bicknell's Thrushes. Extrapolating from New York City area banding data and Central and Prospect Park singers, as many as 5% are Bicknell's, whose spring movement is mainly in May (peaking late) and in fall mainly in Oct (peaking early). It is a New York State Species of Special Concern.

Bronx region, historical
Bicknell's *Riverdale* 1872–1901: annual spring/fall migrant; specimens collected at Kingsbridge Meadows in fall 1875 (MCZ), in Riverdale May, Oct

Eaton's *Bronx + Manhattan* 1910–14: rare migrant

Griscom's *Bronx Region* as of 1922: Dwight Bronx region specimens 16 May 1887, 12 May 1890

Kuerzi's *Bronx Region* as of 1926: Dwight specimens 16 May 1887, 12 May 1890; Bicknell specimens 24 May 1877, 16 May 1883, 20 Sep 1881

NYC area, historical
Cruickshank's *NYC Region* as of 1941: uncertain migration abundance, frequency in

spring (11–27 May)/fall (7 Sep–8 Nov); no daily maxima

Bull's *NYC Area* as of 1968: from banding, specimen data: regular spring (12–22 May)/fall (17 Sep–15 Oct) migrant, with daily maxima 5 spring, 9 fall

Bull's *New York State* as of 1975: status unchanged

Study area, historical and current
Probably an annual spring/fall migrant whose true status is only glimpsed.

Apart from Dwight's specimen collected on 12 May 1890 in Van Cortlandt, there are only 2 other study area records: collected at Kingsbridge Meadows by Bicknell in fall 1875 (MCZ), and another enjoyed in Van Cortlandt at very close range on 7 May 2008 (Young). Bicknell collected others in Riverdale on 16 May 1883 (in AMNH ?), 24 May 1877 (MCZ), 27 Sep 1878 (MCZ), 20 Sep 1881 (in AMNH ?), and 8 Oct 1881 (USNM).

One in Inwood Park on 25 May 2014 (DiCostanzo) responded to a Bicknell's recording but ignored that of a Gray-cheeked. Another was banded in the Bronx Zoo on 10 Oct 2005 (Seewagen and Slayton 2006).

Earlier Bicknell's Thrush records of specimens and of banded migrants measured in the hand may have included some Gray-cheeks, once separation was first corrected by Wallace (1939) and then refined by Phillips (1991), Pyle (1997), Marshall (2001), Rimmer et al. (2001), and most recently Frey et al. (2008). Regrettably, key papers reporting New York City area Bicknell's specimens and especially bandings did not discuss the morphological criteria used for identification or proffer any measurements obtained. For example, in Queens for 8 years in the 1930s, 37% of Gray-cheekeds were identified as Bicknell's by Beals and Nichols (1940), and in Huntington, LI for 10 years in the 1960s, 30% of Gray-cheekeds were identified as Bicknell's by Lanyon et al. (1970). These percentages need to be verified because while they doubtless did handle *some* Bicknell's Thrushes, their true migration periods, numbers, and percentages compared with those of Gray-cheeked will remain uncertain unless supporting unpublished data are unearthed and meet current criteria. The gold standard for Bicknell's Thrush identification (Frey et al. 2008) is a wing chord (never a flattened wing) <93 mm in conjunction with a wing measurement (p8 – p1), whereas Bicknell's averaged 24.7 mm, Gray-cheeked 29.1 mm. The frustration of those most familiar with in-hand birds of both species has been summarized (with tongue in cheek) thus: a small, brownish thrush with a wing chord ⩽85 mm [*sic*] is almost certainly Bicknell's, and a large olive-gray bird with wing chord >100 mm is almost certainly Gray-cheeked, but individual and geographic variation, confounded by age and sex, complicate the situation. Strict application of this standard would upset more than a few status applecarts.

Bicknell's was not split from Gray-cheeked Thrush until 1995, and they are currently separated in the field only with great difficulty, although work is proceeding on the identification of silent individuals: see Lane and Jaramillo 2000, and McLaren 1995, 2012. With the ready availability of high-quality vocalizations on smart phones, suspected Bicknell's are increasingly being identified on spring migration. To experienced ears, both species have recognizably different songs, on-ground calls, and night-flocking calls, and they frequently vocalize in May as they near their breeding range, so vocal approaches are increasing our understanding of their respective New York City area statuses. Their external morphology (bill color and pattern, primary formula) also differs, and while the identification of silent individuals remains difficult, one hopes that with new information, identification of silent Bicknell's will eventually be possible by careful, experienced observers just as it is with silent *Empidonax*.

NYSBBA II in 2000–2005 found this species to have increased its number of occupied blocks by 46% since NYSBBA I in 1980–85, but the 2 surveys were not truly comparable (McGowan and Corwin 2008). The nearest current breeders are atop the tallest Catskill peaks.

Comments
Given the shifting criteria even for separating specimens, the identities of all Bicknell's Thrushes in older accounts as well as those of all extant specimens should be periodically reexamined.

Swainson's Thrush *Catharus ustulatus*

NYC area, current
An annual but increasingly uncommon spring/fall trans-Gulf Neotropical migrant that is occasionally found in winter. This species has also been seen with greatly diminished frequency and in much smaller numbers since the 1960s.

Bronx region, historical
Bicknell's *Riverdale* 1872–1901: regular spring/fall migrant; abundant 11 May 1876

Eaton's *Bronx + Manhattan* 1910–14: common migrant

Griscom's *Bronx Region* as of 1922: common spring/fall migrant

Kuerzi's *Bronx Region* as of 1926: very common spring (28 Apr–5 Jun)/fall (3 Sep–3 Nov) migrant

NYC area, historical
Cruickshank's *NYC Region* as of 1941: common spring (28 Apr–11 Jun)/fall (22 Aug–6 Nov) migrant, maximum 39 fall

Bull's *NYC Area* as of 1968: early spring date 24 Apr; maxima spring 200, Long Island fall 2000 overhead at night, 150 on ground next morning; first winter records, now increasing

Bull's *New York State* as of 1975: late fall date 10 Nov; additional winter records

Study area, historical and current
An annual but increasingly uncommon spring/fall Neotropical migrant seen with greatly diminished frequency and in much smaller numbers since the 1960s.

In spring, first migrants arrive in early May, peak in mid-May, and depart by late May, with extreme dates of 28 Apr 2009 (Young) and 31 May 1996 (Young), with an earlier maximum of 60 on 19 May 1955 (Kane, Buckley) but a more recent one of only 14 on 20 May 2005 (Young). Bicknell collected 1 in Riverdale on the very early date of 24 Apr 1877 (NYSM). Somewhat early, hundreds were passing overhead just before dawn on 8 May 1964 (Horowitz, Tudor). One on 17 Jun 1955 (Post) was a late migrant or nonbreeder. Until Jun 1967, the spring maxima in Central and Prospect Parks were, respectively, 200 on 11 May 1914 (Helmuth) and 20 on 5 May 1950 (Whelen).

NYSBBA I in 1980–85 and II in 2000–2005 identified two breeding-block clusters, in the Adirondacks and in the Catskills—the nearest regular breeders—with single southeastward outliers in Orange and Putnam Counties (McGowan and Corwin 2008). DOIBBS data hinted at a statewide decline from 1980–2005, but it was statistically nonsignificant (Sauer et al. 2005).

In fall, first migrants arrive in mid-Sep, peak in late Sep–early Oct, and depart by late Oct, with extreme dates of 3 Sep 1996 (Young) and 5 Nov 1933 (Norse), and maxima of 8 on 2 Oct 1987 (Young) and 14 on 7 Oct 1935 (Weber et al.). A number have been carefully studied in the Bronx in early Nov, plus singles in Bronx

Park from 28 Nov–26 Dec 1954 (C. Young, Buckley et al.) and Riverdale on 26 Dec 1977 (Sedwitz, Staloff), all of which would have been lingering blowback drift-migrants.

Specimens
Dwight collected 9 in Van Cortlandt in May, Sep 1886–90 (AMNH) and Bicknell 4 in Riverdale in Apr, May, Sep 1877–81 (NYSM).

Hermit Thrush   *Catharus guttatus*

NYC area, current
A common spring/fall migrant, very localized breeder, irregular and scarce in winter.

Bronx region, historical
Bicknell's *Riverdale* 1872–1901: common spring/fall migrant, occasional in winter

Eaton's *Bronx + Manhattan* 1910–14: abundant migrant, occasional winter visitor

Griscom's *Bronx Region* as of 1922: common spring/fall migrant, occasional in winter

Kuerzi's *Bronx Region* as of 1926: common spring (23 Mar–17 May)/fall (11 Aug–late Nov) migrant, occasionally overwintering

NYC area, historical
Cruickshank's *NYC Region* as of 1941: common to abundant spring (maximum 48)/fall migrant, very rare winter resident; Long Island Pine Barrens breeder

Bull's *NYC Area* as of 1968: Long Island maxima spring 250, fall 300

Bull's *New York State* as of 1975: shrinking Suffolk County breeding population

Study area, historical and current
A common spring/fall migrant, irregular and scarce in winter.

In fall, first migrants arrive in early Sep, peak in late Oct–early Nov, and depart by late Nov, with extreme dates of 5 Aug 1952 (Kane)—plus 11 Aug 1926 in adjacent Riverdale (Griscom)— and 29 Nov 1959 (Horowitz), and a maximum of 13 on 10 Oct 2005 (Young). Until Jun 1967, the fall maxima in Central and Prospect Parks were, respectively, 100 on 13 Oct 1953 (Messing) and 172 on 30 Oct 1944 (Grant, Soll).

Singles and twos regularly linger into late Dec, and while only a few have overwintered at Woodlawn, more may be doing so in recent warmer winters.

In spring, first migrants arrive in early Apr, peak in late Apr, and depart by early May, with extreme dates of 7 Mar 1994 (Young) and 14 May 1959 (Zupan), and a maximum of 36 on 18 Apr 1965 (Horowitz), the same day as 50 in nearby Inwood Park (Norse).

NYSBBA I in 1980–85 and II in 2000–2005 demonstrated that east of the Catskill breeding-block cluster, there is another small cluster of occupied blocks in Sullivan and Rockland Counties crossing the Hudson River into Putnam and Dutchess Counties. Southeast of that there are only 4–5 in eastern Westchester County, then none until a cluster in the eastern Long Island Pine Barrens in Suffolk County. By NYSBBA II Nassau County had lost 4 blocks and western Suffolk County 10, but possible singles persist in northeastern Nassau County and on Staten Island (McGowan and Corwin 2008). Nearest breeders to the study area may be on the Palisades. DOIBBS data showed a 4.4% annual statewide increase from 1980–2000 (Sauer et al. 2005).

Specimens
Dwight collected 3 in Van Cortlandt in Apr, May 1886–90 and Foster 4 from Woodlawn in Apr, Oct 1887–90, and Meyer found one dead at Jerome on 21 Nov 1983 (all AMNH). Bicknell collected 4 in Riverdale in Apr, May, Oct 1876–78 (NYSM).

Wood Thrush   *Hylocichla mustelina*

NYC area, current
A relatively common spring/fall Neotropical migrant and widespread breeder in intact forest patches with a good herbaceous layer. May breed in all 5 New York City boroughs, though possibly declining there.

Bronx region, historical
Bicknell's *Riverdale* 1872–1901: common breeder, spring/fall migrant as late as 12 Nov 1888

Hix's *New York City* as of 1904: uncommon breeder Bronx, Central Parks

Eaton's *Bronx + Manhattan* 1910–14: abundant breeder, including Central Park

Griscom's *Bronx Region* as of 1922: common spring/fall migrant, breeder

Kuerzi's *Bronx Region* as of 1926: common breeder (23 Apr–23 Oct)

NYC area, historical
Cruickshank's *NYC Region* as of 1941: common breeder (13 Apr–10 Nov), spring (maximum 36)/fall migrant; decreasing as breeder at New York City edges [locations not given]

Bull's *NYC Area* as of 1968: spring maximum 40+; early spring date 17 Apr, late fall date 25 Nov; <10 winter records but increasing

Bull's *New York State* as of 1975: status unchanged

Levine's *New York State* as of 1996: additional winter records, overwintering twice

Study area, historical and current
A relatively common spring/fall migrant and widespread breeder in Van Cortlandt and occasionally Woodlawn.

In spring, first migrants arrive in late Apr, peak in mid-May, and depart by late May, with extreme dates of 22 Apr 1961 (Horowitz) and 31 May 1955 (Buckley), and a maximum of 40 on 8 and 19 May 1955 (Kane et al.). Until Jun 1967, the spring maxima in Central and Prospect Parks were 40 (Aronoff) and 35 (Jacobson, Whelen), respectively, both on 5 May 1950.

Wood Thrush has long bred in Van Cortlandt and may do so intermittently in Woodlawn. While the exact Van Cortlandt breeding population is uncertain, based on territorial ♂♂ it is >25 pairs. Observers' impressions are that this species has declined as a study area breeder, but supporting data are sparse. For example, 9 on 15 Jun 2006 (Klein) in just the Northwest Forest indicates a healthy population.

Both NYSBBA I in 1980–85 and II in 2000–2005 exposed a broad New York City area breeding distribution, where they were understandably absent only from the most urbanized portions of all New York City boroughs except Staten Island. By NYSBBA II in 2000–2005 some attrition had occurred in both Staten Island and Long Island occupied blocks (McGowan and Corwin 2008). DOIBBS data showed a 3.1% annual statewide decline from 1980–2005, even greater in the Adirondacks (Sauer et al. 2005).

In fall, migrants are found only infrequently, the first arriving in early Sep, peaking in late Sep, and departing by early Oct, with extreme dates of 4 Aug 1991 (Collerton) and 20 Oct 1961 (injured: Horowitz), and a maximum of 4 on 22 Aug 1997 (Young). A few blowback drift-migrants have lingered into Dec, including one in Bronx Park to 27 Dec 1981 (Maguire et al.).

Specimens
Dwight collected 8 in Van Cortlandt in May, Jul, Aug, Sep 1886–90 (AMNH), and Bicknell 2 in Riverdale in Apr, Sep 1875–79 (NYSM).

American Robin   *Turdus migratorius*

NYC area, current
A spring/fall migrant, ubiquitous and rural breeder, and regular winter visitor and resident often in large flocks at roosts.

Bronx region, historical
Bicknell's *Riverdale* 1872–1901: common breeder, migrant; uncommon winter

Eaton's *Bronx + Manhattan* 1910–14: abundant breeder, occasional resident

Griscom's *Bronx Region* as of 1922: abundant spring/fall migrant, breeder; overwinters almost annually

Kuerzi's *Bronx Region* as of 1926: abundant breeder, frequently overwintering

NYC area, historical
Cruickshank's *NYC Region* as of 1941: abundant breeder, spring/fall migrant, uncommon but sometimes locally abundant winter visitor/resident

Bull's *NYC Area* as of 1968: maxima spring 300, fall 3000

Bull's *New York State* as of 1975: fall maxima 4800, 5000 Long Island; winter maximum 500; Ulster County winter roost 30,000

Study area, historical and current
A widespread, conspicuous resident breeder and numerous spring/fall migrant.

In spring, first migrants can arrive anytime from mid-Mar–May, with maxima of 100 on the Parade Ground on 22 Mar 1984 (Sedwitz), 200+ there in mid-Apr 1946 (Kieran) and on 13 Apr 2007 (Klein), 250+ on 12 Apr 2010 (Young), but movements and aggregations are highly weather dependent.

Breeds in all 4 extant subareas, with 100+ pairs in Van Cortlandt and another 25 in Woodlawn. A flock of 40 on 25 Jul 2004 (Klein) would have been local breeders. DOIBBS data showed essentially no statewide change from 1966–2005 (Sauer et al. 2007).

In fall, the first genuine migrants don't arrive until mid-Oct and peak in late Oct–Nov, often in large flocks of thousands migrating diurnally, especially after passage of very strong cold fronts. These can be easily missed by those not doing systematic migration counts, and most are gone by late Nov. Smaller flocks seen from Jul–early Oct comprise area breeders, and many later flocks also overwinter, so extreme dates are trivial.

Robins overwinter in varying numbers depending on the season, often supplemented by midwinter hard-weather arrivals from farther north. Variably large flocks track local winter berry crops, with counts ranging from 10 to the low hundreds, depending on snow cover and weather severity.

Specimens
Dwight collected 3 in Apr, Jun 1886–90 in Van Cortlandt (AMNH) and Bicknell 5 in Riverdale in May, Aug, Oct, Nov 1877–83 (NYSM).

Comments
American Robin ♂♂ with strongly black backs suggestive of Newfoundland's *nigrideus* (Black-backed Robin) can often be seen if searched for in large Nov and Apr robin flocks (e.g., 2 on 6 Apr 1957: Buckley).

Far western vagrant **Varied Thrush**, *Ixoreus naevius*, is an erratic, probably blowback drift-migrant, to the New York City area that has been found in Central and Prospect Parks, as well as Battery Park and Stuyvesant Town. So far unrecorded in the Bronx, it has occurred as close as Tarrytown, where one was present from 1–26 Jan 1963 (L. Brown et al.).

Mockingbirds, Thrashers: Mimidae

Gray Catbird   *Dumetella carolinensis*

NYC area, current
A common and widespread spring/fall Neotropical migrant and ubiquitous breeder, some of which attempt to overwinter annually.

Bronx region, historical
Bicknell's *Riverdale* 1872–1901: common breeder, spring/fall migrant; often into Nov, overwintering twice

Eaton's *Bronx + Manhattan* 1910–14: abundant breeder

Griscom's *Bronx Region* as of 1922: common breeder, spring/fall migrant

Kuerzi's *Bronx Region* as of 1926: very common breeder (16 Apr–30 Nov), several winter records, once overwintering Bronx Park

NYC area, historical
Cruickshank's *NYC Region* as of 1941: widespread common breeder, spring (maximum 60)/fall migrant, rare winter resident coastally

Bull's *NYC Area* as of 1968: spring maxima 100 Bronx Park, fall 150 Long Island

Bull's *New York State* as of 1975: status unchanged

Study area, historical and current
A common, widespread spring/fall migrant and breeder, a few of which occasionally attempt to overwinter.

In spring, first migrants arrive in mid-Apr, peak in early May, and depart by late May, with extreme dates of 16 Apr 1925 (Hickey) and 33 on 21 May 2011 (Bochnik), and maxima of 40 on 1 May 1955 (Buckley et al.) and 8 May 2011 (Aracil), and 44 on 14 May 2006 (Klein). Until Jun 1967, spring maxima in Central and Prospect Parks were 18 on 14 May 1933 (Helmuth) and 60 on 14 May 1950 (Whelen).

Breeds in all 4 extant subareas in large numbers. In late May 2014, an estimated 17 pairs were in Woodlawn and 100+ pairs throughout Van Cortlandt (Buckley, Young). DOIBBS data showed no statewide changes from 1966–2005 (Sauer et al. 2007).

In fall, first genuine migrants don't arrive until early Sep, peak in mid–late Sep, and depart by early Oct, with extreme dates of 15 on 13 Aug 2006 (Klein) and 16 Nov 2013 (Souirgi), and a maximum of 34 on 26 Aug 2015 (Aracil). It is likely that some of these and of the 30 on 25 Jul 2004 (Klein) were local breeders.

In winter, Catbirds are infrequent in the study area, mirroring the relative scarcity of berry-laden thickets. One is found in Dec every 5–6 years, but successful overwintering has never been proved, although those on 13 Mar 2008 (Klein) and 3 on 17 Feb 2013 (Nachtmann) were good candidates. One was collected in Van Cortlandt on 21 Dec 1885 (Dwight: AMNH). Three remained through mid-Jan 1957 in a Riverdale thicket (Scully).

Specimens
Dwight collected 7 in Van Cortlandt in May, Aug, Sep, Dec 1886–96 (AMNH), and Bicknell 3 in Riverdale in May, Oct 1876–86 (NYSM, AMNH).

Brown Thrasher *Toxostoma rufum*

NYC area, current
A spring/fall migrant, rare but regular winter resident, and formerly widespread breeder whose local numbers and range have greatly decreased since the 1970s, possibly following loss of second-growth habitat to succession.

Bronx region, historical
Bicknell's *Riverdale* 1872–1901: common breeder, spring/fall migrant; rare in winter, twice to Feb

Hix's *New York City* as of 1904: breeder Bronx Park

Eaton's *Bronx + Manhattan* 1910–14: common breeder; occasional winter visitor

Griscom's *Bronx Region* as of 1922: common breeder, spring/fall migrant

Kuerzi's *Bronx Region* as of 1926: common breeder (10 Apr–22 Nov), once in winter

NYC area, historical
Cruickshank's *NYC Region* as of 1941: common breeder slowly disappearing within New York City [excepting Central Park, locations not given]; common spring (maximum 40)/fall migrant; rare in winter, less common than Catbird

Bull's *NYC Area* as of 1968: fall maximum 50 on Long Island

Bull's *New York State* as of 1975: status unchanged

Study area, historical and current
Now an uncommon but annual migrant and possibly extirpated breeder, infrequent in winter.

In spring, first migrants arrive in early Apr, peak in late Apr–early May, and depart by late May, with extreme dates of 31 Mar 2008 (Young) and 17 May 1999 (Young) and a maximum of 50 on 8 May 1955 (Kane). Until Jun 1967, the spring maximum in Prospect Park was 23 on 12 May 1945 (Soll).

Thrashers were conspicuous breeders throughout the study area and Riverdale from the late 1870s (Bicknell, Dwight) through the mid- and later 1900s (Kieran, Cruickshank, Norse, Buckley). Until the 1970s they were easily seen, with 10–15 or more pairs breeding parkwide. Through the 1990s they were breeding in all 4 extant subareas, with 1 pair noted in a 24 ac (10 ha) plot in the Northwest Forest in 1991–92 (Jaslowitz and Künstler 1992, Jaslowitz 1993). But in a comprehensive breeding bird census of the entire Northwest Forest in 2006, only 1 was heard singing (Delacretaz 2007), and by 2008 none were located in any Van Cortlandt forested areas (Künstler and Young 2010). Even the traditional 1–3 pairs in Woodlawn (Horowitz) had also vanished by 2014 (Buckley). To our knowledge the only possible study area breeders since 2006 were on 19 Jun 2010 (Baksh) and on 21 Jun 2015 (Hudda), but numerous open areas with highly suitable thrasher breeding habitat, especially on the restricted-access golf courses, have not been searched in many years. Until that has occurred, we are reluctant to write Brown Thrasher's study area breeding obituary.

On NYSBBA I in 1980–85 they were found in all Bronx and Queens blocks but one, but by NYSBBA II in 2000–2005 they had disappeared from nearly all Queens and Bronx sites, although reported in study area Blocks 5852B and 5952A. NYSBBA II in 2000–2005 recorded the loss of dozens of occupied blocks since NYSBBA I in 1980–85, within New York City, Nassau, and western Suffolk Counties. Nearest breeders would now be in Riverdale or on the Palisades. Statewide, the loss was 30% of blocks (McGowan and Corwin 2008), and DOIBBS data quantified a statewide decline beginning in 1970 and continuing at 5% per year (Sauer et al. 2007). This has been attributed to loss of habitat, plant succession, biocide contamination, and vehicle strikes.

In fall, first migrants arrive in early Sep, peak in late Sep, and depart by early Oct, with extreme dates of 6 (the maximum) on 8 Sep 1972 (DiCostanzo) and 10 Nov 1963 (Horowitz), and another maximum of 6 on 2 Oct 1959 (Horowitz).

Normally far less frequently encountered in winter in the New York City area than Gray Catbird, there are a few more thrasher than catbird study area winter records, all singles: 30 Nov–21 Dec 1913 (Rogers), 16 Dec 1956 (Sedwitz et al.), 10 Dec 1960 (Tudor, Zupan), 10 Nov–22 Dec 1963 (Horowitz, Tudor), 1 Dec 2005 (Young), 24 Jan 2006 (Klein), 18 Jan 2007 (Klein), although it has never overwintered successfully.

Specimens
Dwight collected 5 in Van Cortlandt in May, Sep 1886–90 (AMNH) and Bicknell 2 in Riverdale in Apr, Sep 1876–79 (NYSM).

Northern Mockingbird   *Mimus polyglottos*

NYC area, current
A relatively recent range consolidator that has been occurring erratically year-round for almost 200 years but that did not breed until 1956. It is now a ubiquitous but not especially numerous resident and well-known coastwise migrant.

Bronx region, historical
Bicknell's *Riverdale* 1872–1901: 4 records

Eaton's *Bronx + Manhattan* 1910–14: rare summer resident [no locations given]

Griscom's *Bronx Region* as of 1922: 4 records 1877–1920

Kuerzi's *Bronx Region* as of 1926: recorded 9 times 1877–1922, often protracted stays

NYC area, historical
Cruickshank's *NYC Region* as of 1941: rare/very rare, occurs annually almost anywhere, recent records 1 Aug–10 May, least often in spring

Bull's *NYC Area* as of 1968: detailed discussion of recent New York City area increases, including numerous New York City breeding area locations; fall maximum 4

Bull's *New York State* as of 1975: increased breeding numbers, sites

Study area, historical and current
A widespread but not especially numerous resident and vague migrant.

Until the 1960s, it had always been an exceedingly rare and unpredictable visitor to the Bronx. In the study area the first certain Van Cortlandt record was 17 Feb 1912 (Griscom, Hix), followed by 9 Oct 1922 (R. Kuerzi), 4 Dec 1938 (Kramer), 25 Apr 1952 (Kieran), 24 Apr 1954 (Buckley), 23 Sep 1954 (Buckley), 8 May 1955 (Kane et al.), and 22 Oct 1955 (Sutton).

Then in May 1959 a pair bred for the first time along the Putnam Division track bed (Zupan, Temple, Mayer). Since then it has become a resident in all 4 extant subareas, with a current total of 15+ breeding pairs, <5 of them in Van Cortlandt. First bred in Central Park in 1963, but the first Prospect Park breeding went unrecorded.

First found on a Bronx-Westchester Christmas Bird Count in 1935 but not again until

Figure 53

1960, after which it increased rapidly, peaking at about 250 between 1980 and 2000, then dropping back to about 100 (Fig. 53).

DOIBBS data indicated an increasing statewide population after 1965 that suffered a slight decline from 1980–2005 (Sauer et al. 2005).

There is no evidence of any study area migration periods beyond slight movements in Apr–May (e.g., 26 Apr 1962: Tudor) and Sep–Oct (e.g., 3 on 17 Sep 1964: Steineck).

Maxima by seasons: *spring* 4 on several occasions; *summer* 15+ pairs, with 9–10 in Woodlawn alone in 2014 (Buckley); *fall* 6 on 28 Aug 2010 (Baksh); *winter* 18 Van Cortlandt and Woodlawn on 22 Dec 2013 (Lyons, Gotlib et al.).

Comments
Close reading of Giraud (1844), Chapman (1906), Braislin (1907), Eaton (1914), Griscom (1923), and Cruickshank (1942) reveals only a handful of vague suspected or rumored Mockingbird breedings, so the 3 pairs in 1956 in Queens and Suffolk Counties were actually the first proved in New York City and on Long Island.

Specimens
Bicknell collected 1 in Riverdale in Nov 1877 (NYSM).

Starlings: Sturnidae

European Starling   *Sturnus vulgaris*

NYC area, current
A ubiquitous, abundant resident, migrant, and winter resident that may have been at least partly responsible for declines in some native hole-nesters. Forms large winter roosts and often joins those of native blackbirds.

Bronx region, historical
Bicknell's *Riverdale* 1872–1901: first found 9 May 1891, next 16 Oct 1893

Dwight's *Bronx* specimens 1880s–1910: none

Eaton's *Bronx* + *Manhattan* 1910–14: abundant resident by 1900, hundreds overwintering near Kingsbridge; undesirable because they usurp nesting sites of Purple Martins, Tree Swallows, Eastern Bluebirds, White-breasted Nuthatches, woodpeckers

Griscom's *Bronx Region* as of 1922: common to abundant throughout New York City region

Kuerzi's *Bronx Region* as of 1926: abundant, ever-increasing resident

NYC area, historical
Cruickshank's *NYC Region* as of 1941: first in Bronx 1899; abundant resident

Bull's *NYC Area* as of 1968: maximum 150,000 at Manhattan winter roost

Bull's *New York State* as of 1975: maximum 200,000 at Manhattan winter roost

Study area, historical and current
A resident breeder introduced in Central Park in 1890–91 (80 on 6 Mar 1890, 40 on 25 Apr 1891: Chapman 1906), first recorded only a month later in Riverdale on 9 May 1891 (Bicknell) and seen elsewhere in the Bronx by Dwight in 1899. Starlings' aggressive cavity usurpation is regarded as responsible (in part) for the New York City area diminution or disappearance of breeding Red-headed Woodpeckers, Common Flickers, Purple Martins, and Eastern Bluebirds. Breeds in all 4 extant subareas and joins the episodic winter blackbird roosts in Van Cortlandt Swamp.

In the study area <30 pairs nest within Van Cortlandt, with about 12–15 in Woodlawn and similar numbers in and around Jerome and Hillview. DOIBBS data showed a 1.2% statewide decline from 1980–2005 (Sauer et al. 2005).

Seasonal maxima include *spring* 200, *summer* 500, *fall* 400, and *winter* 350, but few observers bother to record large flocks, and while there are no study area data on migration, starlings are migratory and many depart in winter.

Pipits, Wagtails: Motacillidae

American Pipit   *Anthus rubescens*

NYC area, current
A spring/fall migrant, regular winter visitor, and winter resident in small numbers, especially coastally.

Bronx region, historical
Bicknell's *Riverdale* 1872–1901: once in spring, common fall migrant at Kingsbridge Meadows

Eaton's *Bronx + Manhattan* 1910–14: common migrant

Griscom's *Bronx Region* as of 1922: rare spring, uncommon fall migrant

Kuerzi's *Bronx Region* as of 1926: uncommon spring (22 Feb–15 May)/common fall (28 Aug–1 Dec) migrant

NYC area, historical
Cruickshank's *NYC Region* as of 1941: uncommon spring/common to abundant fall (maximum 500) migrant (19 Aug–30 May) on coastal plain, Hudson Valley; very rare in winter inland but more regular coastally

Bull's *NYC Area* as of 1968: Long Island maxima fall 600, winter 125, spring 50; early fall date 13 Aug

Bull's *New York State* as of 1975: status unchanged

Study area, historical and current
A variably abundant fall migrant, almost unnoticed spring migrant, and erratic winter visitor in all 4 extant subareas but especially on the Van Cortlandt Parade Ground where flocks of 100+ were formerly normal in Oct–Nov, and at Hillview.

In fall the first migrants arrive in late Sep, peak in late Oct–early Nov, and depart by mid-Nov, with extreme dates of 13 Sep 2017 at Hillview (Camillieri) and 28 Nov 1937 (Norse), and Parade Ground maxima of 120 on 3 Oct 1953 (Kane, Buckley) and 300 from 25 Oct–6 Nov 1952 (Kane, Kieran et al.). But since the 1960s, maxima have usually been only in the low to middle double digits, although the 105 on 30 Oct 2010 (Baksh) was a hopeful sign. Massive Parade Ground use by various sporting activities and replacement of rough native grasses and flowers by a hybrid turfgrass monoculture are partly to blame. Until Jun 1967, the fall maxima in Central and Prospect Parks were 30 on 6 Oct 1956 (Messing, Post) and 24 on 8 Nov 1959 (Raymond), respectively.

If pipits are present in winter, they occur in tiny numbers and are most easily detected when snow reduces bare ground to patches. Consequently, even though more numerous in mild, snow-free winters, they are recorded far less often then. The winter maximum is only 6 on 8 Dec 2012 (Baksh).

In spring the few migrants first arrive in early Mar, peak in late Mar, and depart by early Apr, with extreme dates of 11 Mar 1997 (Young) and 13 May 2000 (Young), and older maxima of 50 on 18 Mar 1933 (Cruickshank) and 25 at Hillview on 27 Mar 1955 (Kane), now reduced to 2–5 anywhere in the study area.

Specimens
Foster collected 1 on 26 Oct 1889 at Woodlawn (AMNH) and Bicknell another on 18 Oct 1876 at Kingsbridge Meadows (NYSM).

Waxwings: Bombycillidae

Cedar Waxwing *Bombycilla cedrorum*

NYC area, current
A scarce but widespread resident breeder and spring/fall migrant most commonly seen in winter, usually in large wandering flocks.

Bronx region, historical
Bicknell's *Riverdale* 1872–1901: common breeder, erratically abundant

Eaton's *Bronx + Manhattan* 1910–14: fairly common breeder

Griscom's *Bronx Region* as of 1922: uncommon spring, abundant fall migrant, erratic in winter; bred Van Cortlandt 1919

Griscom's *Riverdale* as of 1926: extirpated breeder

Kuerzi's *Bronx Region* as of 1926: common spring/fall migrant, uncommon breeder, occasional in winter

NYC area, historical
Cruickshank's *NYC Region* as of 1941: common Westchester County breeder, decreasing close to New York City [locations not given]; uncommon spring/common fall (maximum 100+) migrant; variable in winter but usually extremely rare

Bull's *NYC Area* as of 1968: maxima spring 350, fall 1000+, winter 300

Bull's *New York State* as of 1975: status unchanged

Study area, historical and current
A scarce but annual resident breeder and spring/fall migrant that is most common in winter.

In fall, first migrants arrive in late Sep, peak in Oct, and depart by late Nov, with extreme dates of 5 Sep 1955 (Kane) and 25 Nov 2012 (Baksh), and a maximum of 120 on 19 Nov 2013 (Young). Numbers of fall migrants and winter residents can vary by an order of magnitude year to year: sometimes absent, sometimes in flocks of several hundred. Until Jun 1967, the fall maximum in Prospect Park was 50 on 2 Nov 1950 (Whelen).

If they are present at all, it is in winter that waxwings are most conspicuous, when they roam the study area in noisy flocks seeking out fruiting winterberry and the like but rarely staying in one location very long. Most winter flocks are 10–30, with maximum of 200+ on 6 Jan 1949 (Kieran). This variability comes across clearly in the Christmas Bird Count data in Figure 54.

In spring, first migrants arrive in late Apr, peak in mid-May, and depart by late May, with extreme dates of 29 Mar 1959 (Horowitz) and 25 May 1957 (Horowitz), and a maximum of 100+ on 24 May 1994 (Young).

Probably >20 pairs breed in Van Cortlandt with a few more in Woodlawn, occasionally also at either reservoir, and they can appear in the study area at any time. A late breeder (egg dates 12 Jun–15 Sep: Cruickshank 1942) whose study area young have fledged as late as 7 Sep 1961 (Enders); a pair at a nest in Van Cortlandt Swamp on 16 Jun 2001 (Garcia) and 10 on 18 Jul 1954 (Buckley) were also local breeders.

Both NYSBBA I in 1980–85 and II in 2000–2005 affirmed a broad New York City area breeding distribution, where they were understandably absent only from the most urbanized portions of all New York City boroughs except Staten Island (McGowan and Corwin 2008). DOIBBS data showed a statewide decline in the 1980s followed by stabilization beginning around 1993 (Sauer et al. 2007).

Specimens
Dwight collected 5 in Van Cortlandt in Mar, Apr 1884–86 (AMNH) and Bicknell 2 in Riverdale in Apr, May 1876–78 (NYSM).

Comments
Holarctic taiga breeder and sometime irruptive **Bohemian Waxwing**, *B. garrulus*, has always been excessively rare in the New York City area. It has been seen once in Prospect Park on 22 Dec 2007 (D. Gochfeld), never in the Bronx, and since about 2000 a few times on Long Island, the closest recent one to the study area having been at Sands Point on 25 Jan 2015 (Perrault).

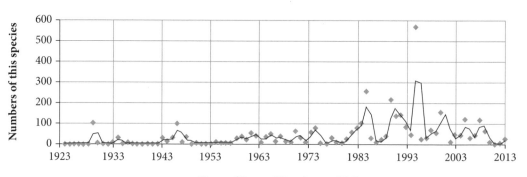

Figure 54

Longspurs: Calcariidae

Lapland Longspur *Calcarius lapponicus*

NYC area, current
An annual but uncommon fall migrant and winter visitor/resident in small flocks in open grassy areas, especially coastally. Frequently found with Horned Larks and recorded in Central and Prospect Parks.

Bronx region, historical
Bicknell's *Riverdale* 1872–1901: unmentioned

Eaton's *Bronx + Manhattan* 1910–14: rare winter visitor

Griscom's *Bronx Region* as of 1922: unrecorded

Kuerzi's *Bronx Region* as of 1926: rare and irregular winter visitor 22 Oct–13 Mar [locations not given]

NYC area, historical
Cruickshank's *NYC Region* as of 1941: uncommon fall migrant, less common winter visitor/winter resident, rare spring migrant (12 Aug–18 Apr); extremely rare Van Cortlandt Parade Ground, usually only Nov

Bull's *NYC Area* as of 1968: Long Island maxima fall 75, winter 125; late spring date 9 May

Bull's *New York State* as of 1975: Long Island winter maximum 200

Study area, historical and current
An irregular fall migrant and winter visitor on the Van Cortlandt Parade Ground, and also at Hillview and Jerome. It might even occur annually, but the requisite searching is now infrequently attempted in season on days when the Parade Ground is empty of people.

There are 16 records, mostly singles with Horned Larks, occasionally with Snow Buntings, from Oct–Feb but most often in Nov: 4 (the maximum) from 3–14 Nov 1925 (Kuerzi, Cruickshank, Kieran); 7 Nov 1933 (Kuerzi, Kuerzi); 25 Oct 1934 (Kuerzi, Kuerzi); 6 Nov 1935 (Weber); at Hillview on 15 Feb 1936 (Norse et al.); 2 on 20 Nov 1943 (Kieran); 11 Nov 1952 (Kieran); 2 on 26 Oct 1993 (Lyons); 2 on 26 Oct 1994 (Garcia); 20–21 Nov 2003 (Lyons); 25 Nov 2005 (Garcia); 2 on 20 Jan 2003 (Fiore); 2 on 23 Dec 2007 (Lyons, Garcia); 1 Jan 2013 (Baksh); 28–30 Nov 2013 (Drogin et al.); and at Hillview on 27 Dec 2017 (Camillieri). Most seen in the Bronx have been singles, twos, or very small flocks; the largest flock was 50 at Baxter Creek on 13 Dec 1970 (Kane), a fine count anywhere in the New York City area.

Snow Bunting *Plectrophenax nivalis*

NYC area, current
A common fall migrant and winter resident on Long Island and Long Island Sound shores, and regular but less numerous inland in large open areas.

Bronx region, historical
Bicknell's *Riverdale* 1872–1901: irregular at Kingsbridge Meadows, recorded only after major midwinter snowstorms; common 31 Jan–14 Mar 1875; maximum 20 on 16 Jan 1877

Eaton's *Bronx + Manhattan* 1910–14: common winter visitor

Griscom's *Bronx Region* as of 1922: unrecorded

Kuerzi's *Bronx Region* as of 1926: irregular uncommon migrant, winter visitor (1 Nov–18 Mar)

NYC area, historical
Cruickshank's *NYC Region* as of 1941: common coastal Long Island winter resident (maximum 500+), rare but regular fall/winter on Long Island Sound shores in Bronx, Westchester Co.; 3 Oct–14 Apr

Bull's *NYC Area* as of 1968: late spring date 9 May; early fall date 1 Oct; maxima fall 400, winter 1500

Bull's *New York State* as of 1975: status unchanged

Study area, historical and current
Since the early 1900s, it has been a regular fall migrant and winter visitor on the Van Cortlandt Parade Ground, Woodlawn, Jerome, and Hillview. Parade Ground flocks rarely linger because of increasing disturbance from sports activities and dogs but do feed actively when present. Recorded in Central and Prospect Parks but usually only singles or 2s, rarely remaining long.

A predictable Nov–Dec migrant, occasional in winter. Bicknell recorded it 4 times in midwinter on Kingsbridge Meadows after heavy snows: common from 31 Jan–14 Mar 1875, 20 on 16 Jan 1877, a single from 8–13 Feb 1882, and 2 on 13 Feb 1883. On 24 Jan 1881, there was a flock of 200 in nearby Yonkers (A. K. Fisher). Most study area records involve singles or small groups up to a dozen. Extreme dates are Woodlawn on 30 Oct 1960 (Horowitz, Heath) and the Parade Ground on 21 Mar 1963 (Norse), with maxima of 50 across Van Cortlandt and Woodlawn on 23 Dec 2007 (Lyons et al.) and 60 on 13 Nov 1959 (Rafferty). Daily Parade Ground and Hillview coverage from 1 Oct–31 Dec should prove this species to be annual. On the Bronx-Westchester Christmas Bird Count they have always been scarce, and only 5 times have they totaled >20. The largest flock ever in the Bronx contained 150 at Baxter Creek on 14 Nov 1970 (Kane).

New-World Warblers: Parulidae

Ovenbird   *Seiurus aurocapilla*

NYC area, current
A common spring/fall Neotropical migrant but a greatly reduced urban breeder. This is another Neotropical migrant that has been seen in smaller numbers since the 1960s.

Bronx region, historical
Bicknell's *Riverdale* 1872–1901: abundant breeder, regular spring/fall migrant

Hix's *New York City* as of 1904: breeds Bronx Park

Eaton's *Bronx + Manhattan* 1910–14: abundant migrant, breeds

Griscom's *Bronx Region* as of 1922: common breeder, spring/fall migrant

Kuerzi's *Bronx Region* as of 1926: common, frequently abundant spring/fall migrant, breeder [locations not given]; 15 Apr–6 Nov [blowback drift-migrant]

NYC area, historical
Cruickshank's *NYC Region* as of 1941: common spring (maximum 36+)/fall migrant, 15 Apr–23 Nov, plus Dec 1935 at Inwood [blowback drift-migrants]

Bull's *NYC Area* as of 1968: maxima spring 200, fall 78 dead Empire State Building; 10 Apr–24 Nov, plus 3 in Dec, 1 in Jan–Feb [blowback drift-migrants]

Bull's *New York State* as of 1975: fall maximum 85 dead Empire State Building; 10 Dec records [blowback drift-migrants]

Study area, historical and current
A common spring/fall Neotropical migrant; also a breeder now in greatly reduced numbers.

In spring, first migrants arrive in late Apr, peak in early–mid-May, and depart by late May, with extreme dates of 24 Apr 1961 (Horowitz) and 28 May 1961 (Horowitz) and maxima of 50 on 19 May 1955 (Kane, Buckley) and 55 on 13 May 1963 (Horowitz). Until Jun 1967, the spring maximum in Central Park was 200 on 11 May 1914 (Helmuth) and in Prospect Park 160 on 5 May 1950 (Jacobson, Whelen).

In the late 1800s, it was a common breeder in Riverdale (Bicknell) and in Van Cortlandt (Dwight), continuing but in rapidly declining numbers into the 1930s, '40, and '50s. Breeding in Van Cortlandt in 1935 (Weber), 1937 (Norse) and 1951 (Komorowski) was not regarded as particularly noteworthy. It breeds thinly throughout Van Cortlandt forests as of 2016 but with an estimated population of no more than 3–5 pairs, although on 19–20 May 2014, some 14 territorially behaving ♂♂ were located (Buckley, Young). One in Van Cortlandt on 1 Jun 1995 (Garcia) may also have been a breeder. Even though NYSBBA I in 1980–85 did, and NYSBBA II in 2000–2005 did not, record breeders in Blocks 5852B and 5952A, which include the study area (McGowan and Corwin 2008), this discrepancy may have been due to atlas fieldwork timing and dismissal of breeders as migrants. They breed in good numbers on the Palisades but may now be gone from Riverdale along the Hudson River. In the Bronx they may now breed only in Van Cortlandt.

Both NYSBBA I in 1980–85 and II in 2000–2005 supported a greatly reduced breeding distribution in New York City and Nassau County, where they are now extirpated except in the Northwest Bronx, on Staten Island, in about 9 clustered blocks in northeastern Nassau County, and throughout most of Suffolk County (McGowan and Corwin 2008). On Staten Island most recent breeding attempts have been unsuccessful (Veit et al.). DOIBBS data showed a 1.5% annual statewide increase from 1980–2005 (Sauer et al. 2005).

In fall, first migrants arrive in early Aug, peak near mid-Sep, and depart by early Oct, with extreme dates 5 Aug 1952 (Buckley) and 27 Oct 1994 (Young) and a maximum of 4 on 5 Sep 1955 (Kane) and on 14 Sep 1999 (Young).

Lingering blowback drift-migrant Ovenbirds have occurred very late nearby with unexpected frequency: Inwood Park on 17–21 Dec 1935 (Karsch, Norse), late Nov–21 Dec 1969 (Norse et al.), and 18 Dec 2016 (O'Reilly); Bronx Park 5 Dec 2001–26 Jan 2002 (unknown observer, *North American Birds*); Bronx Zoo 26 Dec 2004 (Krauss) and through mid-Dec 2006 (Olson); Pelham Bay Park 18 Dec 1978 (H. Martin), 20 Dec 1984 (unknown observer, NYRBA), and 20 Jan 1985 (Morris).

Specimens
Dwight collected 5 in Van Cortlandt in Apr, May, Jun, Jul 1886–89 (AMNH), Bicknell 3 in Riverdale in May, Sep 1877–80 (NYSM).

Worm-eating Warbler *Helmitheros vermivorum*

NYC area, current
A probably extirpated New York City breeder and an uncommon spring/fall Neotropical migrant that has been seen with greatly diminished frequency and in smaller numbers since the 1960s.

Bronx region, historical
Bicknell's *Riverdale* 1872–1901: common breeder, spring/fall migrant

Eaton's *Bronx + Manhattan* 1910–14: rare breeder

Griscom's *Bronx Region* as of 1922: uncommon migrant, last bred 1895 (Bicknell, Dwight)

Griscom's *Riverdale* as of 1926: rare migrant, extirpated breeder

Kuerzi's *Bronx Region* as of 1926: fairly common spring/fall migrant (30 Apr–2 Oct), breeder [locations not given]

NYC area, historical
Cruickshank's *NYC Region* as of 1941: decreases as breeder south of White Plains, almost gone in Bronx [locations not given], still breeds Inwood Park; rare spring (maximum 50 [includes breeders])/very rare fall migrant; 19 Apr–11 Oct

Bull's *NYC Area* as of 1968: variably rare spring/fall migrant; maxima spring 12, fall 7; early spring date 10 Apr, late fall date 17 Oct

Bull's *New York State* as of 1975: status unchanged

Study area, historical and current
An extirpated breeder and an uncommon spring/fall Neotropical migrant.

In spring, first migrants arrive late Apr–early May, peak in mid-May, and depart by late May, with extreme dates of 23 Apr 1952 (Buckley) and 27 May 1934 (Weber, Jove), and a maximum of 10 on 6 May 1950 (Komorowski). Until Jun 1967, the spring maxima in Central and Prospect Parks were, respectively, 5 on 27 Apr 1925 (Griscom) and 12 on 5 May 1950 (Jacobson, Whelen).

Worm-eating's New York State center of distribution has always been in the Hudson Valley. In the late 1800s, it was a common breeder in Riverdale (Bicknell) and in Van Cortlandt (Dwight), but it has not bred certainly in Van Cortlandt since then, a time when it was strikingly more common in the Bronx than in the 1950s. Judging from Cruickshank's counts, they were quite numerous as close as central Westchester County through the early 1940s, when a pair may have been in Van Cortlandt, given that they also were at Inwood then, and at Grassy Sprain into the mid-1950s (Buckley et al.). Their weak song can easily be overlooked or passed off as a Chipping Sparrow's, so they might still be breeding in very small numbers as close to Van Cortlandt as Inwood Park, along the Hudson in Riverdale, and at Grassy Sprain. They should be carefully listened for in Van Cortlandt on dry hillsides with tall oaks, though none were heard in a careful search in late May 2014 (Young, Buckley).

NYSBBA I in 1980–85 and II in 2000–2005 found them throughout Westchester County, NYSBBA I south to Grassy Sprain Reservoir but by NYSBBA II no closer than the White Plains area, in a small remnant population along the north shore of Long Island in eastern Nassau and western Suffolk Counties, and in 4 more isolated blocks farther east (McGowan and Corwin 2008). Thus, nearest breeders to the study area would be on the Palisades. Sparse DOIBBS data showed an apparent but nonsignificant statewide decline from 1966–2005 (Sauer et al. 2005).

In fall, first migrants arrive in late Jul, peak in mid-Aug, and depart by early Sep, with extreme dates of 23 Jul 2015 (Hudda; photos) and 27 Sep 1994 (Young), singles all.

Specimens
Dwight collected 13 in Van Cortlandt in May, Jun, Jul, Aug 1885–95 (AMNH), and Bicknell 5 in Riverdale in May, Jun, Jul 1876–79 (NYSM, ANSP, USNM).

Louisiana Waterthrush   *Parkesia motacilla*

NYC area, current
An annual but uncommon spring/fall Neotropical migrant and an extirpated breeder within New York City.

Bronx region, historical
Bicknell's *Riverdale* 1872–1901: spring/fall migrant, breeds at single site

Eaton's *Bronx + Manhattan* 1910–14: common breeder

Griscom's *Bronx Region* as of 1922: rare spring/fall migrant; bred Van Cortlandt 1917, still breeds northeast Yonkers

Griscom's *Riverdale* as of 1926: extirpated breeder

Kuerzi's *Bronx Region* as of 1926: fairly common breeder (7 Apr–4 Oct) [locations not given]

NYC area, historical
Cruickshank's *NYC Region* as of 1941: fairly common spring (maximum 23)/uncommon fall migrant, 28 Mar–12 Oct; breeds Westchester County but not immediate New York City vicinity

Bull's *NYC Area* as of 1968: early spring date 25 Mar; spring maximum 4

Bull's *New York State* as of 1975: status unchanged

Study area, historical and current
An annual but uncommon spring/fall Neotropical migrant and an extirpated breeder.

In spring, first migrants arrive in mid-Apr, peak in early May, and depart by mid-May, with extreme dates of 27 Mar 2007 (Klein) and 27 May 1934 (Weber, Jove) and a maximum of 4 on 19 May 1955 (Kane, Buckley). Until Jun 1967, the spring maximum in Prospect Park was 5 on 15 May 1949 (Alperin et al.). In the AMNH collection there are 4 Apr, May 1886–95 Van Cortlandt specimens that could have been migrants or local breeders.

Bred in Van Cortlandt until 1917 (Griscom 1923) and probably later; also at Grassy Sprain Reservoir through the 1930s–50s (Norse, Buckley et al.), possibly later. As of 2016, the nearest breeders would be on the Palisades.

As demonstrated on both NYSBBA I in 1980–85 and II in 2000–2005, this species' breeding center is the southeastern one-third of mainland New York State, when the closest Westchester County breeders were in the Ardsley-Scarsdale area. Alarmingly, NYSBBA II showed a statewide decrease of 21% in occupied blocks since NYSBBA I (McGowan and Corwin 2008) and DOIBBS data affirmed a strong statewide decline from 1980–2005 (Sauer et al. 2005).

In fall, first migrants arrive (often unseen) in early Jul, peak in Aug, and depart by early Sep, with extreme dates of 29 Jul in 1962 (Heath, Zupan) and 2014 (Young), and 17 Sep 1964 (Steineck), all singles.

Specimens
Dwight collected 5 in Van Cortlandt in Apr, May 1886–95 (AMNH) and Bicknell 1 in Riverdale on 13 May 1879 (NYSM). At the time it bred in both locations.

Northern Waterthrush *Parkesia noveboracensis*

NYC area, current
A common spring/fall migrant.

Bronx region, historical
Bicknell's *Riverdale* 1872–1901: common spring/fall migrant

Eaton's *Bronx + Manhattan* 1910–14: common migrant

Griscom's *Bronx Region* as of 1922: common spring/fall migrant

Kuerzi's *Bronx Region* as of 1926: common migrant in spring (23 Apr–8 Jun)/fall (30 Jul–20 Oct)

NYC area, historical
Cruickshank's *NYC Region* as of 1941: common spring (maximum 27)/fall migrant, 14 Apr–30 Oct, plus late Nov

Bull's *NYC Area* as of 1968: fall maximum 55; 30 Nov, late Dec records [blowback drift-migrants]

Bull's *New York State* as of 1975: spring maximum 21; now 4 Dec–Jan [blowback drift-migrants]

Study area, historical and current
A common spring/fall migrant.

In spring, first migrants arrive in late Apr, peak in early May, and depart by late May, with extreme dates of 10 Apr 2005 (Klein)—a trans-Gulf overshoot—and on 28 May in 1997 (Young) and in 2006 (Garcia, Lyons), with maxima of 25 on 19 May 1955 (Kane, Buckley) and 30 on 29 Apr 1956 (Kane). Until Jun 1967, the spring maximum in Central was 25 on 15 May 1957 (Post). A nonbreeder oversummered in Van Cortlandt from 30 May–15 Aug 1934 (Cantor) with no evidence of breeding.

NYSBBA I in 1980–85 found this species widely if thinly distributed in Rockland, Sullivan, Orange, Ulster, Putnam, and Dutchess Counties, with 5–7 more blocks in Westchester County. But by NYSBBA II in 2000–2005, half of the occupied blocks in those 6 counties were gone and only 4 were left in Westchester County. The nearest breeders east of the Hudson River were near Pound Ridge Reservation in northeastern Westchester County and west of the Hudson River near Hook Mountain in Rockland County (McGowan and Corwin 2008). Sparse DOIBBS data showed no consistent statewide trend from 1966–2005 (Sauer et al. 2005).

In fall, first migrants arrive in early Aug, peak in early Sep, and depart by late Sep, with extreme dates of 31 Jul 1966 (Kane) and 20 Oct 1995 (Young; a blowback drift-migrant?), and a maximum of 4 on 14 Sep 1955 (Kane). Until Jun 1967, the fall maximum in Prospect Park was 55 on 13 Sep 1964 (Yrizarry).

Specimens
Dwight collected 7 in Van Cortlandt in May, Aug, Sep 1885–90 (AMNH), and Bicknell 3 in Riverdale in May, Sep 1876–86 (NYSM, AMNH).

Golden-winged Warbler *Vermivora chrysoptera*

NYC area, current
One of the scarcest spring/fall Neotropical migrants. It has been declining throughout its breeding range for 100 years as it ceded breeding sites to its close relative Blue-winged Warbler while its own range inexorably collapsed northward. It is a New York State Species of Special Concern.

Bronx region, historical

Bicknell's *Riverdale* 1872–1901: rare spring/fall migrant

Eaton's *Bronx + Manhattan* 1910–14: uncommon migrant

Griscom's *Bronx Region* as of 1922: rare spring/fall migrant

Griscom's *Riverdale* as of 1926: rare spring migrant

Kuerzi's *Bronx Region* as of 1926: uncommon but regular migrant in spring (4–21 May)/fall (20 Jul–15 Sep)

NYC area, historical

Cruickshank's *NYC Region* as of 1941: astonishingly rare spring/very rare fall migrant (29 Apr–6 Oct)

Bull's *NYC Area* as of 1968: annual but rare spring (maximum 12)/fall (maximum 2) migrant, 26 Apr–10 Oct

Bull's *New York State* as of 1975: status unchanged

Study area, historical and current

A scarce but annual migrant that has never bred in the Bronx; see comments about hybrids under Blue-winged Warbler. There have been very few records since the 1960s, a mirror of this species' general decline and northward shift as a breeder as it is relentlessly supplanted by Blue-wingeds.

In spring, first migrants arrive in early May, peak in mid-May, and depart by late May, with extreme dates of 26 Apr 1953 (Buckley et al.) and 3 on 26 May 1947 (Ryan), and normally only singles but maxima of 4 on 13 May 1964 (M. Gochfeld) and 8 (plus another 4 in Bronx Park) on 8 May 1943 (Komorowski). The most recent study area occurrence was in Woodlawn on 10 May 2003 (Gambino). Until Jun 1967, the spring maximum in Prospect Park was 7 on 10 May 1946 (Jacobson et al.).

There are fewer than 15 fall records of singles of this very early fall migrant, most of them in Aug, with extremes of 11 Aug 1959 (Zupan) and 11 Sep 1996 (Young).

This species has greatly decreased throughout the New York City area as a migrant since the 1960s, as its range has been pushed ever northward by climate warming, loss of tropical overwintering habitat, oldfield succession, and perhaps breeding pressure from Blue-wingeds. While it has never bred in the study area (not even in mixed pairs with Blue-wingeds), both Brewster's- and Lawrence's-type hybrids and intergrades have bred in Van Cortlandt, last in 1963. Notwithstanding, even in Bicknell's day Golden-wing singles or pairs never bred anywhere in New York City or on Long Island back to DeKay (1844), Giraud (1844), Chapman (1906), and Braislin (1907). Yet Eaton (1914: 386) called it a rare summer resident on Long Island even though in his massive Local List section the column entries under Breeds are blank for Kings, Queens, and Nassau Counties and for Suffolk County. Accordingly, we treat his Long Island breeding statement as a *lapsus calami*.

In 1928, the last, very small Golden-wing breeding population in Westchester County was near Peekskill (Kuerzi), perhaps the location of the sole Westchester County confirmed breeding on NYSBBA I in 1980–85, at which time in the Rockland, Sullivan, Orange, Ulster, Putnam, and Dutchess Counties area there were >50 blocks reporting them. But by NYSBBA II in 2000–2005 this was down to 29, all but 9 in Orange County (McGowan and Corwin 2008). DOIBBS data showed a steep annual statewide decline of 5.6% from 1966–2005 (Sauer et al. 2005). As of 2016 the nearest breeding population is in the Sterling Forest area of Orange County.

Specimens

Dwight collected 1 in Van Cortlandt on 22 Aug 1890 (AMNH) and Bicknell another in Riverdale on 11 Aug 1881 (NYSM).

Blue-winged Warbler   *Vermivora cyanoptera*

NYC area, current
An uncommon but annual spring/fall Neotropical migrant and a nearly extirpated breeder within New York City but locally numerous on Long Island and in Westchester County.

Bronx region, historical
Bicknell's *Riverdale* 1872–1901: common spring/fall migrant, breeder

Hix's *New York City* as of 1904: common breeder northern Bronx

Eaton's *Bronx + Manhattan* 1910–14: common breeder

Griscom's *Bronx Region* as of 1922: common migrant, breeder

Kuerzi's *Bronx Region* as of 1926: common spring/fall migrant (26 Apr–25 Sep), breeder [locations not given]; 2 Dec–Jan Bronx Park records [blowback drift-migrants]

NYC area, historical
Cruickshank's *NYC Region* as of 1941: disappearing as breeder in densely populated areas, few still remain within New York City [locations not given]; common to abundant spring (maximum 100)/fall migrant; 25 Apr–6 Oct

Bull's *NYC Area* as of 1968: early spring 15 Apr date, late fall 29 Nov date; fall maximum 12

Bull's *New York State* as of 1975: status unchanged

Study area, historical and current
An uncommon but annual spring/fall migrant and former breeder.

In spring, first migrants arrive in late Apr, peak in early May, and depart by late May, with extreme dates of 16 Apr 1886 (Dwight: AMNH) and 23 Apr 1953 (Buckley), to 21 May 1961 (Horowitz), with a maximum of 25 on 8 May 1955 (Kane et al.). Until Jun 1967, the spring maximum in Prospect Park was 25 on 4 May 1953 (Usin).

A widespread breeder in Van Cortlandt, Woodlawn, and Riverdale in the late 1800s, evidenced by AMNH specimens (Dwight, Foster) and Bicknell's comments and specimens. Kieran noted 6–8 breeding pairs in the 1920s–30s divided between Lincoln Marsh and the Sycamore Swamp, with ♂♂ routinely singing into Jul. One in the AMNH found dead in Van Cortlandt on 16 Jun 1925 would also have been a local breeder. From the 1930s through the last pair in 1964 (Russak), 3–5 pairs also nested in open areas on the Ridge (Norse, Komorowski, Buckley). With increased disturbance and loss of much shrub habitat following succession, Blue-winged no longer breeds in Van Cortlandt. One on Vault Hill in Van Cortlandt on 6 Jun 2006 (Künstler) was a late migrant or nonbreeder.

Both NYSBBA I in 1980–85 and II in 2000–2005 noted its strong Hudson Valley and Long Island breeding base, even though as of 2016 it is absent from New York City except for 1–2 Staten Island blocks, and from the lower half of Nassau County and southwestern Suffolk County The nearest NYSBBA II block was at Grassy Sprain Reservoir (McGowan and Corwin 2008), suggesting it would not be difficult for it to breed in Van Cortlandt again should proper habitat appear. DOIBBS data showed it more or less constant statewide from 1966–2005 (Sauer et al. 2007).

In Van Cortlandt, a ♂ of the Brewster's flavor raised young with a ♀ Blue-winged in Jun 1953 (Buckley, Bauer, Bauer), and a ♂ of the Lawrence's persuasion was feeding young with a ♀ Blue-winged in 1923 (Griscom) and in 1963 (Schmidt, Simon). They and various other hybrid phenotype combinations have been seen as migrants on several occasions from the 1920s to the 1970s.

In fall, first migrants arrive in early Aug, peak in mid-Aug, and depart by late Aug, with extreme dates of 4 Aug 1947 (Norse) and 27 Sep 1952 (Kane), and a maximum of 2 on 11 Aug 2007 (Young). One in Bronx Park from 10 Dec 1899–6 Jan 1900 (Britton) would have been a lingering blowback drift-migrant.

Specimens
Dwight collected 23 in Van Cortlandt in Apr, May, Jun, Jul, Aug in 1885–92, another was taken there on 16 Jun 1925, Foster took 4 in Woodlawn in May–Jul 1890 (all AMNH), and Bicknell took 6 in Riverdale in May, Jun, Aug 1876–81 (NYSM, ANSP, USNM). Clearly, it was a widespread and numerous breeder and migrant in the area at that time.

Black-and-white Warbler   *Mniotilta varia*

NYC area, current
A common spring/fall Neotropical migrant, formerly a thinly but highly local breeder now almost extirpated from New York City although widespread in mature woods elsewhere.

Bronx region, historical
Bicknell's *Riverdale* 1872–1901: breeder, spring/fall migrant

Hix's *NYC* in 1904: breeds Bronx Park

Eaton's *Bronx + Manhattan* 1910–14: common migrant, fairly common breeder

Griscom's *Bronx Region* as of 1922: common migrant, uncommon breeder

Griscom's *Riverdale* as of 1926: now extirpated breeder

Kuerzi's *Bronx Region* as of 1926: common breeder (18 Apr–1 Nov)

NYC area, historical
Cruickshank's *NYC Region* as of 1941: common breeder but rare in New York City [locations not given]; very common to abundant spring (maximum 67)/fall migrant; 13 Apr–14 Nov

Bull's *NYC Area* as of 1968: early spring 3 Apr date, late fall dates 2 in Dec [blowback drift-migrants]; maxima spring 100, fall 27

Bull's *New York State* as of 1975: fall maximum 66 dead Empire State Building

Study area, historical and current
A common spring/fall migrant and possibly an occasional breeder.

In spring, first migrants arrive in mid–late Apr, peak in early May, and depart by late May, with extreme dates of 5 Apr 1956 (Young)—a trans-Gulf overshoot—and 25 May 1961 (Horowitz), and maxima of 100 on 8 May 1955 (Kane et al.) and 200 on 4 May 1962 (Friedman, Steineck). Until Jun 1967, the spring maxima in Central and Prospect Parks were, respectively, 20 on 13 May 1910 (Helmuth) and 62 on 12 May 1945 (Soll).

In the late 1800s, it was a common breeder in Riverdale (Bicknell) and in Van Cortlandt (Dwight). In the study area Black-and-whites bred in large tracts of mature woods with a good herbaceous layer in Van Cortlandt (Kieran et al.). Since the late 1800s this small population has waxed and waned, but it may have been extirpated by c. 1948 (Norse, Komorowski). It is possible that a pair may attempt to breed there occasionally, as it may also do along the Hudson River in Riverdale, still does on the Palisades and perhaps also at Grassy Sprain Reservoir.

On both NYSBBA I in 1980–85 and II in 2000–2005, they were widely distributed in, although slowly disappearing from, mainland southeastern New York counties, and scattered in Nassau County but widespread in Suffolk County On NYSBBA II they were gone from New York City with the sole exception of a possible (migrant?) somewhere in study area Block 5852B, but more precise information is lacking (McGowan and Corwin 2008). DOIBBS data showed a consistent statewide decline from 1966–2005 (Sauer et al. 2007).

In fall, first migrants arrive in early Aug, peak in late Aug–early Sep, and depart by early Oct, with extreme dates of 3 on 21 Jul 1953 (Buckley) that may have been local breeders and 15 Oct 2011 (Baksh), and a maximum of 30 on 3 Aug 1953 (Phelan, Buckley et al.). Lingering blowback drift-migrants were in Van Cortlandt Swamp on 25 Nov 2017 (Cole et al.) and probably the same bird on 23 Dec 2017 (Dolan et al.), at Inwood Park from 10–17 Dec 2017 (Barry et al.), at Pelham Bay Park on 20 Nov 2002 (unknown observer, *Kingbird*), and another in adjacent New Rochelle on 27 Dec 2015 (Tozer et al.).

Specimens
Dwight collected 9 in Van Cortlandt in Apr, May, Jun, Aug, Sep 1886–90 (AMNH), and Bicknell 3 in Riverdale in Apr, Sep 1875–81 (NYSM).

Prothonotary Warbler   *Protonotaria citrea*

NYC area, current
A scarce but annual and increasing spring Neotropical migrant and trans-Gulf overshoot near the northern edge of its range that is attempting to establish breeding sites that are often quickly abandoned. Its early fall migration is infrequently detected.

Bronx region, historical
Bicknell's *Riverdale* 1872–1901: single 1895 Yonkers record

Eaton's *Bronx + Manhattan* 1910–14: unrecorded

Griscom's *Bronx Region* as of 1922: single 1895 record

Kuerzi's *Bronx Region* as of 1926: 1895 record, plus 2 Bronx Park

NYC area, historical
Cruickshank's *NYC Region* as of 1941: extremely rare spring/almost unrecorded fall migrant, 42 New York City area records, including Bronx, all but 4 spring [locations not given]

Bull's *NYC Area* as of 1968: 19 Apr–15 Oct; rare but annual spring, very rare fall migrant; spring/fall maxima 2

Bull's *New York State* as of 1975: early spring date 27 Mar [trans-Gulf overshoot]

Study area, historical and current
A spring-only infrequent migrant in the study area that with increased coverage would be detected more frequently, as it is in the Bronx Zoo and Bronx, Central, and Prospect Parks.

There are 17 records, all >1952 and from Van Cortlandt Swamp/Lake except as noted: in the Sycamore Swamp on 27–28 Apr 1952 (Komorowski et al.); 10 May 1955 (Ephraim et al.); 27 May 1956 (Steineck, Yeganian); 15 May 1957 (Mayer); 4 May 1962 (Friedman, Steineck); 4 May 1964 (Stepinoff, C. Young); in the Sycamore Swamp on 2 May 1974 (Meyer); 24–25 Apr 1977 (Oswald et al.); 29 Apr 1991 (Young); 23 May 1991 (Young); 24 Apr 1992 (Young); 26 Apr 1994 (Young); 6 May 1995 (Jaslowitz); 16 May 1995 (Garcia); 26 Apr 2005 (Young); 27 Apr 2009 (Young); singing ♂ on 23 May 2015 (Furgoch et al.). Kieran found 1 in Riverdale from 28 Apr–6 May 1952, and 1 was at Tibbett's Brook Park on 25 Apr 2009 (Bochnik), 2 days before 1 in Van Cortlandt (above). Central Park recorded its first in 1926, but Prospect Park not until 1953.

Bicknell located a singing ♂ on 2 Jun 1895 more than a mile east of the Hudson River and half that distance north of Van Cortlandt Park, which would place it along Tibbett's Brook close to McLean Ave., Yonkers. At the time it was the first in New York State away from Long Island, where there were only 2 records. Given its flooded woodland habitat, future breeding in Van Cortlandt Swamp would not be unexpected. The first Long Island breeding was 1–2 pairs at Belmont Lake SP in 1983 (File), but as of 2016 the nearest breeders are no closer than NJ's Great Swamp. Prothonotary is at the northeastern edge of its range in southeastern New York State, so fluctuations in breeding numbers and sites are normal. Thus the 50% statewide decline in blocks occupied during NYSBBA I in 1980–85 (22) and NYSBBA II in 2000–2005 (11) is neither surprising nor worrisome (McGowan and Corwin 2008).

Swainson's Warbler *Limnothlypis swainsonii*

NYC area, current
An extremely infrequent Neotropical spring range-prospector first recorded in 1950 and most often seen in New York City parks, where there are <20 records. Found on multiple occasions in Central and Prospect Parks, the most recent singing in Central on 28 Apr 2016.

NYC area, historical
Bull's *NYC Area* as of 1968: 2 May records

Bull's *New York State* as of 1975: 5 Apr–May records

Study area
A southern spring range-prospector vagrant with a single record: 12 May 1964 in the Sycamore Swamp (Steineck, Friedman), described in great detail from a 45-minute study. Nearby, it has occurred twice in Bronx Park—on 6 May 1963 (Carleton, Horowitz et al.) and on 30 Apr 1977 (Rafferty, Maguire et al.). These 3 are the only Bronx records.

Tennessee Warbler *Oreothlypis peregrina*

NYC area, current
An annual but uncommon spring/fall Neotropical migrant whose numbers vary greatly year to year. Tennessee is a Spruce Budworm warbler whose breeding population size and success are tied to worm outbreaks and whose numbers in New York City parks in spring are also related to periodic outbreaks of small moth larvae, locally called green inchworms. Warblers of many species are attracted to them on oaks, gorging for days and sometimes weeks in the same trees. New York State's small Tennessee breeding population is stable (McGowan and Corwin 2008).

Bronx region, historical
Bicknell's *Riverdale* 1872–1901: very rare spring/fall migrant

Eaton's *Bronx + Manhattan* 1910–14: uncommon migrant

Griscom's *Bronx Region* as of 1922: variably common spring/fall migrant

Kuerzi's *Bronx Region* as of 1926: regular, fairly common spring (6–30 May)/sometimes abundant fall (9 Aug–17 Oct) migrant

NYC area, historical
Cruickshank's *NYC Region* as of 1941: uncommon spring (maximum 4)/common fall (maximum 40) migrant, 1 May–20 Oct

Bull's *NYC Area* as of 1968: 29 Apr–15 Nov, plus 1 in Jan; maxima spring 30, fall 42 dead Empire State Building

Bull's *New York State* as of 1975: spring maximum 40; late fall date 24 Nov, second Jan record [blowback drift-migrants]

Levine's *New York State* as of 1996: spring maximum 250

Study area, historical and current
An annual but uncommon spring/fall Neotropical migrant whose numbers vary greatly year to year.

In spring, first migrants arrive in early May, peak in mid-May, and depart by late May, with extreme dates of 25 Apr 1953 (Buckley et al.) and 25 May 1961 (Horowitz), and a maximum of 10 on 20 May 1954 (Buckley).

In fall, first migrants arrive in mid-Aug, peak in early Sep, and depart by late Sep, with extreme dates of 3 (the maximum) on 9 Aug 1953 (Buckley) and 5 Oct 1954 (Scully)—although Bicknell collected specimens in Riverdale on 16 Aug 1880 and 9 Oct 1876 (NYSM). Until Jun 1967, the fall maxima in Central and Prospect Parks, respectively, were 35 on 20 Sep 1958 (Messing) and 12 on 19 Aug 1944 (Soll). One at Inwood Park on 18 Dec 2000 (Lyons) was a lingering blowback drift-migrant.

Specimens
Bicknell collected 2 in Riverdale in Aug, Oct 1876–80 (NYSM).

Orange-crowned Warbler *Oreothlypis celata*

NYC area, current
A very scarce spring and annual but uncommon late fall Neotropical migrant that sometimes lingers into early winter.

Bronx region, historical
Bicknell's *Riverdale* 1872–1901: very rare: 9, 29 Oct 1876

Eaton's *Bronx + Manhattan* 1910–14: rare migrant

Griscom's *Bronx Region* as of 1922: 2 Bicknell Riverdale records

Kuerzi's *Bronx Region* as of 1926: very rare fall migrant, once May, once Jan

NYC area, historical
Cruickshank's *NYC Region* as of 1941: very rare spring/rare (usually singles) fall migrant, occasional in winter [locations not given], 10 Aug–22 May

Bull's *NYC Area* as of 1968: overwintered 3 times recently, maximum 3 at feeder

Bull's *New York State* as of 1975: status unchanged

Levine's *New York State* as of 1996: fall maximum 5

Study area, historical and current
Probably an annual migrant that is rarely noticed; it is almost unrecorded in spring and

winter, and scarce in fall. Its reported occurrence in Van Cortlandt reflects relatively low coverage.

There are only 4 spring records: 17–19 May 1931 (Cruickshank), 10–11 May 1952 (Kane, Buckley et al.); 26 Apr 1953 (Buckley)—all 3 singing ♂♂—and 27 Apr 1989 (unknown observer, *Kingbird*). Another was at adjacent Tibbett's Brook Park in Yonkers on 17 May 1931 (Cruickshank).

They occur far more frequently in fall, when they are particularly fond of feeding on flowering goldenrods. There are almost 2 dozen fall records of singles between 1876 and 2013, mostly in Oct with extremes of 22 Sep 2005 (Young) and 11 Nov 1959 (Young), plus Kingsbridge Meadows on 9 and 29 Oct 1876 (Bicknell).

Singles have also been seen twice in Van Cortlandt in early winter: 22 Dec 1919 (Pell, Pell) and 22 Dec 2002 (Garcia, Lyons). It has also occurred in Riverdale on 14 Dec 1953 (Kieran), 28 Dec 2003 (Fiore et al.), and 27 Dec 2015 (Fiore et al.).

Nashville Warbler   *Oreothlypis ruficapilla*

NYC area, current
An annual but uncommon spring/fall Neotropical migrant occasionally found in early winter.

Bronx region, historical
Bicknell's *Riverdale* 1872–1901: regular spring/fall migrant

Eaton's *Bronx + Manhattan* 1910–14: fairly common migrant

Griscom's *Bronx Region* as of 1922: common spring/fall migrant

Kuerzi's *Bronx Region* as of 1926: common spring (26 Apr–30 May)/fall (16 Aug–18 Oct) migrant, once Dec–Jan

NYC area, historical
Cruickshank's *NYC Region* as of 1941: fairly common spring (maximum 12+)/fall migrant, 21 Apr–30 Nov, plus 2 Dec/Jan records [blowback drift-migrants]

Bull's *NYC Area* as of 1968: migrant in spring (20 Apr–16 Jun)/fall (24 Jul–4 Dec; maximum 5); overwintering to 7 Feb

Bull's *New York State* as of 1975: maxima spring 25, fall 18 dead Empire State Building

Study area, historical and current
An annual but uncommon spring/fall Neotropical migrant occasional in early winter, with a unique breeding event.

In spring, first migrants arrive in late Apr, peak in early May, and depart by late May, with extreme dates of 21 Apr 2012 (Baksh) and 25 May 2006 (Young), and a maximum of 75 on 7 May 1955 (Kane). Until Jun 1967, the spring maxima in Central and Prospect Parks, respectively, were 5 on 28 Apr 1924 (Helmuth) and 105 on 17 May 1945 (Soll).

Unexpected was Eames's (1893) report that a pair spent the summer of 1892 in Woodlawn—given that neither Bicknell nor Eaton spoke of their breeding anywhere near New York City. However, Griscom (1923) did note its closest breeding in Passaic County, NJ, and that the ♀ collected in his Englewood Region on 12 Jun 1887 (Chapman) may have been breeding. In nearby CT, Bevier (1994) noted that it was reported breeding sparingly throughout the state by 1877 and by 1913 still was, although more frequently in the northern part. Its CT range began shrinking in the late 1800s, so the 1892 breeding would have occurred near its peak abundance in CT. This remains the sole Bronx and New York City breeding by this species.

NYSBBA I in 1980–85 and II in 2000–2005 found the southeastern edge of its New York State breeding range no closer than southeastern Sullivan and Ulster Counties (McGowan and Corwin 2008), so nearest breeders would be in northwestern Passaic County, NJ. DOIBBS data showed a flat statewide line from 1966–2005 (Sauer et al. 2005).

In fall, first migrants arrive in mid-Aug, peak in late Sep–early Oct, and depart by mid-Oct, with extreme dates of 1 Aug 1953 (Buckley) and 22 Oct 2012 (Young), singles only.

Recorded a number of times in very late fall/winter in Van Cortlandt and nearby, all lingering blowback drift-migrants. In the study area: 8–12 Nov 2005 (Young); 16 Dec 1918–9 Jan 1919 (Chubb, W. Miller)—not 1917–18 as in Griscom (1923) et seq.; 23 Nov–4 Dec 1925 (Chubb); 23 Dec 2001 (unknown observer, *Kingbird*); and in Woodlawn 27–30 Dec 2015 (Lyons, Gotlib et al.).

In Riverdale were the following additional very late occurrences: 27 Dec 1987 (Antenen), 26 Dec 1994 (Gibaldi et al.), 28 Dec 1996 (Cantor), found dead in late Feb 1997 (Wallstrom) but not saved, 27 Dec 1998 (Lyons et al.), and 23 Dec 2001 (Fiore et al.). Nearby, there were 2 in Yonkers on 23 Dec 2001 (unknown observer, Bronx-Westchester Christmas Bird Count), Fort Tryon Park 18 Dec 2011 (Fiore et al.), Inwood Park 6–7 Jan 2012 (Wilkinson et al.), and Bronx Zoo 17 Jan 2012 (Olson). Some or even many of these individuals in the eastern United States and Canada are now suspected of being lingering blowback drift-migrant western *ridgwayi* (see following).

Specimens
Bicknell collected 6 in Riverdale in May, Sep, Oct 1875–81 (NYSM).

Comments
On 1 May 2009 in Woodlawn, an unfamiliar warbler song was heard on a moderate flight day with 2 dozen or more Nashvilles feeding and singing overhead in catkins at the tops of mature oaks. Tracked down, this singer nonetheless proved to be a ♂ Nashville Warbler but one whose song was loud, melodious, and reminiscent of an off Chestnut-sided Warbler and lacking typical Nashvilles' slow terminal *ti-ti-ti-ti*. It was also vigorously tail wagging, behavior that accentuated its long-tailed look and bright yellow rump, grayish, barely olive back, and whitish belly as it worked over blossoms in a low flowering ornamental fruit tree within 15 ft (5 m) of the ground. The observer has had considerable experience with **Calaveras Nashville Warbler**, *O. (ruficapilla) ridgwayi*, the western North America Nashville subspecies (or species: Calaveras Warbler), including several times in winter in the northeastern United States, and is confident it was a ♂ *ridgwayi* (Buckley). Others at Jones Beach 31 Dec 1954 (DeAngelo, Buckley et al.), Staten Island (with a MacGillivray's) from 12–28 Dec 1999 (Bernick, Veit et al.), and Point Lookout 18 Nov 2006 (Fritz et al.) were also believed to have been blowback drift-migrant *ridgwayi*.

Connecticut Warbler   *Oporornis agilis*

NYC area, current
Now a scarce but formerly not uncommon fall Neotropical migrant that has always been all but unknown in spring. This is another Neotropical migrant whose post-1960 fall numbers are but a shadow of their levels in the 1920s–50s, to say nothing of the late 1800s when they were frequently found in double-digit lighthouse kills.

Bronx region, historical
Bicknell's *Riverdale* 1872–1901: tolerably common fall migrant, 4 Sep specimens 1875–81

Eaton's *Bronx + Manhattan* 1910–14: fairly common fall migrant

Griscom's *Bronx Region* as of 1922: irregular fall migrant, occasionally fairly common

Kuerzi's *Bronx Region* as of 1926: irregular fall migrant, occasionally numerous (20 Aug–12 Oct)

NYC area, historical
Cruickshank's *NYC Region* as of 1941: almost unknown spring (6–29 May)/rare to uncommon fall (17 Aug–26 Nov) migrant

Bull's *NYC Area* as of 1968: far more numerous <1910 than after; fall maxima 57 dead Fire Island Lighthouse 1877, but only 13 dead Gabreski Airport, Westhampton 1954; 5–6 seen alive few times; late spring date 3 Jun, early fall date 15 Aug

Bull's *New York State* as of 1975: status unchanged

Study area, historical and current
An annual but scarce fall migrant. In Bicknell's day, Connecticuts were far more abundant in the New York City area than in 2016, and he collected 4 in Riverdale in Sep 1875–81 (NYSM).

In fall, the first migrants arrive in early Sep, peak in mid-Oct, and depart by late Oct, with extreme dates of 11 Sep 1936 (Norse) and 17 Nov 1943 (Komorowski), usually only 1–2 but a maximum of 3 on 18 Sep 1962 (Friedman, Steineck). The Nov 1943 bird and 1 in Riverdale on 30 Oct 2010 (Buckley) may have been blowback drift-migrants. Until Jun 1967, the fall maximum in Prospect Park was 6 on 27 Sep 1950 (Whelen).

The only 3 in spring were carefully studied at close range by observers aware of this species' extreme rarity in that season and familiar with both Mourning and Connecticut Warblers: singing ♂♂ at the Sycamore Swamp on 25 May 1962 (Rafferty) and on 5 May 2018 (Lyons), and a silent ♂ on the Ridge on 16 May 1995 (Garcia). All appeared on days with large numbers of neotropical migrants.

Specimens
Bicknell collected 4 in Riverdale in Sep 1875–81 (NYSM).

Mourning Warbler *Geothlypis philadelphia*

NYC area, current
A spring/fall migrant in very small numbers. Until the 1960s this was strictly a Memorial Day weekend arriver in the New York City region, but it now appears as early as the first week in May. Although a Neotropical migrant, it is unclear whether it is being seen with lesser frequency or in smaller numbers in spring since the 1960s. In the past it was overlooked in fall because it is most frequent in Aug before many observers had become active again, but this has changed in the last 30 years.

Bronx region, historical
Bicknell's *Riverdale* 1872–1901: single fall, 3 spring records

Eaton's *Bronx + Manhattan* 1910–14: rare migrant

Griscom's *Bronx Region* as of 1922: very rare spring migrant, 3 recent records

Kuerzi's *Bronx Region* as of 1926: rare spring (18 May–5 Jun)/fall (5 Aug–24 Sep) migrant

NYC area, historical
Cruickshank's *NYC Region* as of 1941: rare spring (maximum 5)/very rare fall migrant, 5 May–12 Oct

Bull's *NYC Area* as of 1968: late fall date 19 Oct

Bull's *New York State* as of 1975: status unchanged

Study area, historical and current
A spring/fall migrant in very small numbers.

In spring, first migrants now appear in the first week of May, peak in mid–late May, and depart by early Jun, with extreme dates of 8 May 1943 (Komorowski) and 8 Jun 1956 (Kieran). Usually only singles, but Cruickshank (1942) noted that during exceptional flights he had seen 5 in a day in May in Van Cortlandt Swamp. Until Jun 1967, the spring maximum in Central Park was 7 on 21 May 1966 (Tozzi).

NYSBBA I in 1980–85 and II in 2000–2005 found the southeastern edge of its Catskill breeding-block cluster no closer than Delaware County (McGowan and Corwin 2008), the nearest breeders as of 2016. Even though suspected, northwestern NJ breeding remains unproved. DOIBBS data showed no significant statewide trend from 1966–2005 (Sauer et al. 2005).

In fall, first migrants arrive in early Aug, peak in late Aug–early Sep, and depart by late Sep, with extreme dates of 3 Aug 1953 (Phelan) and 7 Oct 1935 (Weber et al.), singles only. Both Cruickshank and Bull regarded them as rarely seen in fall, especially on Long Island, yet banding operations near Huntington through the 1960s and at Fire Island Lighthouse from 1969–1972 and 1995–1999 showed that they are regular then. Observers also began to report them increasingly in fall beginning in the 1970s, and we do not believe this indicates a change in status. Rather, we are confident they have long been overlooked as they are notorious skulkers, or have been misidentified as Connecticuts—believed to be more expected in fall than Mourning.

Kentucky Warbler   *Geothlypis formosa*

NYC area, current
A scarce spring and very scarce fall Neotropical migrant and a very rare and local breeder that formerly was more widespread, all but disappeared, and is now very slowly recolonizing the area.

Bronx region, historical
Bicknell's *Riverdale* 1872–1901: uncommon breeder in woods north of Van Cortlandt, at Grassy Sprain Reservoir until 1898

Eaton's *Bronx + Manhattan* 1910–14: common breeder

Griscom's *Bronx Region* as of 1922: formerly bred Riverdale, long since extirpated; once May 1917

Kuerzi's *Bronx Region* as of 1926: rare spring/fall migrant (12 May–11 Sep); now breeding no closer than Elmsford but formerly more widespread

NYC area, historical
Cruickshank's *NYC Region* as of 1941: rare spring/extremely rare fall migrant, 2 May–14 Sep; bred recent years Grassy Sprain Reservoir, Palisades

Bull's *NYC Area* as of 1968: spring maximum 3; 1 May–2 Oct; not breeding east of Hudson River

Bull's *New York State* as of 1975: early spring date 29 Apr, late fall date 12 Oct; bred central Long Island 1973

Study area, historical and current
A rare spring migrant rarely recorded in fall. Until the late 1940s it was an infrequent spring migrant, but it has been almost unrecorded since then, a function of reduced coverage. It is also a potential rebreeder.

Since 1953, 10 spring records: 16 May 1953 (Phelan); 21 May 1956 (Friedman, Steineck); 23 May 1959 (Zupan); 9 May 1961 (Tudor); 4 May 1962 (Steineck, Friedman); 19 May 1963 (Friedman, Steineck); 26 Apr 1995 (Young); 13 May 1997 (Young); 22 May 1997 (Suggs); 27 Apr 2002 (unknown observer, *Kingbird*); but only 3 in fall: 16–24 Sep 1951 (Kieran, Kane, Buckley)—note another in Prospect Park on 18 Sep 1951 (Russell); 27 Sep 1994 (Young); 19 Sep 1995 (Young). Kieran saw singles in Riverdale on 21 Aug 1930 and 23 May 1952 and another migrant was there on 26 May 2004 (Pehek, Love).

Through the late 1800s it was an uncommon breeder in Riverdale and in adjacent Van Cortlandt (Bicknell), and a few may have bred in Van Cortlandt Park until about 1910. At nearby Grassy Sprain Reservoir there were several pairs in 1927 (Kuerzi) and some persisted through the 1960s (Steineck et al.). During NYSBBA I in 1980–85 they were found 2 breeding blocks north of the study area in the Ardsley-Scarsdale area (McGowan and Corwin 2008).

On NYSBBA I in southeastern New York they were noted in single blocks in Dutchess, Putnam, Orange, 3 in Rockland, and 16 in Westchester Counties, plus 2 in Nassau and 5 in Suffolk Counties. By NYSBBA II in 2000–2005 these had fallen to singles in Dutchess and Orange Counties, plus 2 each in Westchester and Nassau Counties (McGowan and Corwin 2008). Kentucky is at the northern edge of its range, so fluctuations in breeding numbers and sites are normal. Nearest breeders, ever shifting, may at the moment be in central Westchester Co., northern Nassau County, or northeastern NJ. It is a potential study area recolonizer at any time, although there have been no recent persistently singing ♂♂.

Specimens
Bicknell collected a ♂ in Riverdale on 30 May 1876 (NYSM).

Common Yellowthroat *Geothlypis trichas*

NYC area, current
A common spring/fall Neotropical migrant and widespread breeder.

Bronx region, historical
Bicknell's *Riverdale* 1872–1901: common breeder, spring/fall migrant

Hix's *New York City* as of 1904: bred Bronx Park

Eaton's *Bronx + Manhattan* 1910–14: abundant breeder

Griscom's *Bronx Region* as of 1922: common breeder, spring/fall migrant

Griscom's *Riverdale* as of 1926: now extirpated breeder

Kuerzi's *Bronx Region* as of 1926: very common breeder (26 Apr–15 Nov)

NYC area, historical
Cruickshank's *NYC Region* as of 1941: common breeder, but decreasing in New York City; common spring (maximum 100+)/fall migrant; 10 Apr–late Dec, plus 7 mostly coastal records >Dec

Bull's *NYC Area* as of 1968: spring maximum 250, fall 100; rare in winter but many more records, overwintering several times

Bull's *New York State* as of 1975: status unchanged

Levine's *New York State* as of 1996: winter maximum 8

Study area, historical and current
A common spring/fall Neotropical migrant and widespread breeder.

In spring, first migrants arrive in late Apr, peak in mid-May, and depart by late May, with extreme dates of 26 Apr 1953 (Buckley et al.) and 6 on 21 May 2011 (Bochnik), and a maximum of 100+ on 11 May 1958 (Horowitz). Until Jun 1967, the spring maximum in Central Park was 250 on 11 May 1914 (Helmuth) and in Prospect Park 150 on 14 May 1950 (Kreissman, Whelen).

Breeds in all 4 study areas; the 1937 census located 4 pairs in Van Cortlandt Swamp, where Kieran estimated a dozen pairs by the early 1950s. As of 2016, the entire study area may support 10+ pairs, but <5 in Van Cortlandt.

Both NYSBBA I in 1980–85 and II in 2000–2005 identified its nearly ubiquitous southeastern New York breeding status and found it in all but 2–3 New York City blocks on NYSBBA I, up to 10 on NYSBBA II (McGowan and Corwin 2008). DOIBBS data showed a slight statewide increase from 1966–79, followed by a slight decrease to 2005 (Sauer et al. 2005).

In fall, first migrants arrive in late Jul, peak in late Aug, and depart by mid-Oct, with extreme dates of 7 Aug 1995 (Young) and 28 Oct 1955 (Birnbaum) and a maximum of 30 on 6 Sep 1964 (Horowitz). Until Jun 1967, the fall maximum in Prospect Park was 55 on 13 Sep 1964 (Yrizarry).

Probably lingering blowback drift-migrant singles were in Van Cortlandt Swamp 18–24 Dec 1955 (Scully et al.), on 10 Jan 1998 (unknown observer, *Kingbird*), on 23 Dec 2001 (Lyons et al.), and from 22 Dec 2002–9 Jan 2003 (Lyons et al.). Another remained in Riverdale until mid-Jan 1957 (Scully).

Specimens
Dwight collected 7 in Van Cortlandt in May, Jun 1885–89, and Foster took another in Woodlawn on 18 May 1889 (all AMNH), and Bicknell 6 in Riverdale in May, Jul, Aug 1876–81 (NYSM).

Hooded Warbler *Setophaga citrina*

NYC area, current
An annual but scarce spring/fall Neotropical migrant that has been steadily retreating from the immediate New York City area as a breeder since the 1960s.

Bronx region, historical
Bicknell's *Riverdale* 1872–1901: common spring/fall migrant, bred Riverdale, Van Cortlandt to 1893

Eaton's *Bronx + Manhattan* 1910–1914: common but local breeder

Griscom's *Bronx Region* as of 1922: formerly breeding Riverdale, West Farms but long extirpated; very rare spring migrant, only 2 records

Kuerzi's *Bronx Region* as of 1926: uncommon but annual spring/fall migrant (2 May–1 Oct); fairly common, increasing breeder [locations not given]

NYC area, historical
Cruickshank's *NYC Region* as of 1941: decreases as breeder south of White Plains (but still at Grassy Sprain Reservoir), now

extirpated from Bronx; rare spring/fall migrant, 28 Apr–late Sep

Bull's *NYC Area* as of 1968: does not breed Long Island, southern Westchester County; spring maximum 5; 12 Apr [trans-Gulf overshoot]–7 Dec [blowback drift-migrant]

Bull's *New York State* as of 1975: early spring date 4 Apr [trans-Gulf overshoot]

Study area, historical and current
An annual but scarce spring/fall migrant and extirpated breeder.

In spring, first migrants arrive in late Apr, peak in early May, and depart by late May, with extreme dates of 26 Apr in 1952 (Kieran, Kane, Buckley) and 2011 (Young), and 6 Jun 1962 (Zupan), with maxima of 4 on numerous occasions, 5 on 15 May 1955 (Kane, Buckley et al.) and a stunning 10 on 19 May 1961 that was erroneously published (Friedman and Steineck 1963) as 20 (P. Steineck, pers. comm.). Until Jun 1967, the spring maximum in Central Park was 4 on 11 May 1960 (Carleton) and in Prospect Park 3 on 12 May 1945 (Soll).

In the late 1800s, it was a common breeder in Van Cortlandt where Dwight collected 3 that would have been breeders: on 20 May 1885, 7 Jun 1889, and 3 Jun 1890 (AMNH). Bicknell considered it a common Riverdale breeder, collecting 7 between 1875 and 1879 (NYSM, MCZ, USNM), but they were gone before 1922 (Griscom). A singing ♂ there in Jun–Jul 1955 (Kane) was a nonbreeder. One along the edge of the Van Cortlandt golf course on 3 Jun 2003 (Jaslowitz) and another singing persistently on the Ridge on 12 May 2007 (Garcia) seem never to have found mates, but it is a potential Van Cortlandt rebreeder at any time.

As both NYSBBA I in 1980–85 and II in 2000–2005 showed clearly, its breeding epicenter is extreme southwestern New York near the PA border, with a small secondary breeding-block cluster in the Hudson Valley in Dutchess, Putnam, Westchester, Rockland, and Orange Counties. By NYSBBA II, 25 blocks in Westchester County had dropped to 8, and on Long Island 6 were down to 2 (McGowan and Corwin 2008). It remains to be seen if this decline is a secular trend or just normal range-edge variability. In any case, closest reliable breeders are on the Palisades. Somewhat sparse DOIBBS data showed a consistent statewide increase from 1980–2005, most of which occurred upstate (Sauer et al. 2005).

In fall, first migrants arrive in early Aug, peak in late Aug–early Sep, and depart by mid-Sep, with extreme dates of 1 Aug 1953 (Phelan et al.) and 9 Sep 1953 (Phelan), singles only.

Specimens
Dwight collected 3 in Van Cortlandt in May, Jun 1885–90 (AMNH), and Bicknell 7 in Riverdale in May, Jul, Aug 1875–79 (NYSM, MCZ, ANSP).

American Redstart *Setophaga ruticilla*

NYC area, current
A common spring/fall Neotropical migrant and a widespread but highly local breeder in areas outside New York City proper, where breeding occurs in only a few locations.

Bronx region, historical
Bicknell's *Riverdale* 1872–1901: abundant breeder, spring/fall migrant

Hix's *New York City* as of 1904: common breeder in city parks

Eaton's *Bronx + Manhattan* 1910–14: common breeder

Griscom's *Bronx Region* as of 1922: common spring/fall migrant, breeder

Griscom's *Riverdale* as of 1926: probably now extirpated breeder

Kuerzi's *Bronx Region* as of 1926: abundant spring/fall migrant (24 Apr–16 Oct), common breeder [locations not given]

NYC area, historical
Cruickshank's *NYC Region* as of 1941: common breeder, except rare around New York City outskirts [locations not given]; common to abundant migrant in spring (maximum 100+)/fall, 24 Apr–23 Nov, plus once in Dec

Bull's *NYC Area* as of 1968: fall maximum 180; 22 Apr–30 Nov, once in Dec [blowback drift-migrants]

Bull's *New York State* as of 1975: late fall date 30 Nov; fall maxima 200, 300

Study area, historical and current
A common spring/fall migrant with only a few breeders persisting. This is another Neotropical migrant that has been seen in smaller numbers since the 1960s.

In spring, first migrants arrive in late Apr, peak in early May, and depart by late May, with extreme dates of 28 Apr in 1924 (L. N. Nichols) and 2009 (Young) and 6 Jun 1961 (Horowitz), and a maximum of 100+ on 19 May 1955 (Kane, Buckley). Until Jun 1967, the spring maximum in Central Park was 60 on 23 May 1954 (Feinberg, Maumary, Post).

In the late 1800s, it was a common breeder in Riverdale (Bicknell) and in Van Cortlandt (Dwight), persisting in diminishing numbers through the 1920s, '30s, and '40s. They were breeding in Van Cortlandt Swamp during the 1950s (Kieran) and 2013–14 (Pehek et al., Young), and annually in the Northwest Forest, on the Ridge, and probably in the Northeast Forest (Kieran, Sedwitz et al.), and also in Riverdale from 1955 onward (Kieran, Kane, Sedwitz). They may breed in Woodlawn, and in Van Cortlandt, juveniles, some being fed, were seen on 12 Jun 2005 (Klein), 3 Jul 2014 (O'Reilly), and 7 Jul 2016 (Hudda). We estimate that <10 pairs breed, but with considerable interannual variation. NYSBBA I in 1980–85 did, but NYSBBA II in 2000–2005 did not, find breeders in study area Blocks 5852B and 5952A, but breeders are frequently confounded with migrants. Nearest other breeders are on the Palisades and in Pelham Bay Park (Künstler).

Both NYSBBA I in 1980–85 and II in 2000–2005 validated their broad New York City area breeding distribution including Long Island, and they were understandably absent only from the largest urban portions of all New York City boroughs except Staten Island (McGowan and Corwin 2008). DOIBBS data showed a 2% annual statewide decline from 1980–2005 (Sauer et al. 2007).

In fall, first migrants arrive in late Jul, peak in early Sep, and depart by mid-Oct, with extreme dates of 4 Aug 1947 (Norse) and 26 Oct 1960 (Buckley) and a maximum of 40 in Woodlawn on 6 Sep 1964 (Horowitz). Until Jun 1967, the fall maximum in Prospect Park was 100 on 13 Sep 1950 (Alperin et al.). One in Van Cortlandt on 11 Nov 2013 (Young) would have been a blowback drift-migrant, as was a lingerer taken there on 1 Dec 1886 (Dwight). Other blowback drift-migrants, some lingering, have been recorded in Bronx Park on 3 Nov 1957 (Maguire et al.), Inwood Park 10–27 Dec 1931 (Cruickshank), and at nearby Spuyten Duyvil on 23 Dec 2006 (Fiore et al.).

Specimens
Dwight collected 9 in May, Jul, Aug, Dec 1886–90 in Van Cortlandt (AMNH), and Bicknell 2 in Sep, Oct 1876–79 in Riverdale (NYSM).

Cape May Warbler   *Setophaga tigrina*

NYC area, current
A spring/fall Neotropical migrant, by years uncommon to exceedingly rare. This is another Neotropical migrant that has been seen with diminished frequency and in smaller numbers since the 1960s. Cape May is a Spruce Budworm warbler whose breeding population size and success are tied to worm outbreaks; warblers' numbers in New York City parks in spring are also related to periodic oak outbreaks of the small moth larvae locally called green inchworms. Warblers are attracted to them, gorging for days and sometimes weeks.

Bronx region, historical
Bicknell's *Riverdale* 1872–1901: very rare spring/fall migrant

Eaton's *Bronx + Manhattan* 1910–14: rare migrant

Griscom's *Bronx Region* as of 1922: uncommon spring/fall migrant

Kuerzi's *Bronx Region* as of 1926: fairly common spring (6 May–5 Jun)/fall (22 Aug–14 Oct) migrant

NYC area, historical
Cruickshank's *NYC Region* as of 1941: uncommon spring (maximum 21)/fairly common fall migrant with great interannual variation (29 Apr–28 Oct, once in Dec) [blowback drift-migrant]

Bull's *NYC Area* as of 1968: 23 Apr–5 Dec, plus 30, 31 Dec [blowback drift-migrants]; fall maximum 45

Bull's *New York State* as of 1975: fall maximum 112

Levine's *New York State* as of 1996: fall maximum 3000 dropped from the eye of Hurricane Gloria onto Jones Beach

Study area, historical and current
A spring/fall Neotropical migrant, by years uncommon to exceedingly rare. Neither Dwight in Van Cortlandt Park nor Bicknell in Riverdale collected any Cape May Warblers from 1872–1901, and Bicknell considered it very rare. Yet major Spruce Budworm outbreaks were occurring every 40 years or so at that time in northeastern US and Canada, one even beginning about 1878, and were mirrored in other Spruce Budworm warblers' numbers in Riverdale—a curious anomaly.

In spring, first migrants arrive in early May, peak in mid-May, and depart by late May, with extreme dates of 5 May 1962 (Horowitz) and 5 Jun 1926 (Kuerzi) and a maximum of 6 on 19 May 1955 (Kane). Until Jun 1967, the spring maximum in Prospect Park was 6 on 14 May 1950 (Kreissman, Whelen).

In fall, when it is more regular and slightly more numerous, first migrants arrive in late Aug, peak in mid-Sep, and depart by early Oct, with extreme dates of 10 Aug 1953 (Buckley) and 15 Oct 1931 (Cruickshank) and a maximum of 8 on 6 Sep 1955 (Kane). Until Jun 1967, the fall maximum in Central Park was 14 on 9 Sep 1953 (Post).

Cerulean Warbler   *Setophaga cerulea*

NYC area, current
A very scarce but annually recorded spring Neotropical migrant and an erratic breeder in a handful of locations on eastern Long Island. Most frequently detected in Central and Prospect Parks. Its numbers are declining

continentally, probably from fragmented breeding and Neotropical winter habitats augmented by cowbird brood parasitism. It is now a New York State Species of Special Concern.

Bronx region, historical

Bicknell's *Riverdale* 1872–1901: unrecorded

Eaton's *Bronx + Manhattan* 1910–14: rare

Griscom's *Bronx Region* as of 1922: single record 14 May 1921

Kuerzi's *Bronx Region* as of 1926: 2 records, including Van Cortlandt 27 May 1926

NYC area, historical

Cruickshank's *NYC Region* as of 1941: excessively rare, mostly spring migrant, 4 on Long Island, 4 in Central Park, 2 in Bronx (last in 1926), 2 in Westchester County

Bull's *NYC Area* as of 1968: very rare to rare spring (27 Apr–5 Jun)/extremely rare fall (24 Jul–5 Oct) migrant

Bull's *New York State* as of 1975: early spring date 23 Apr

Study area, historical and current

An infrequent spring migrant perhaps overlooked in its treetop feeding locations and by those unfamiliar with its song. It is a late Jul–Aug migrant that has only been detected once in fall.

Only 15 spring records, all but one of singing ♂♂, doubtless reflecting low coverage: 27 May 1926 (Cruickshank); 13 May 1954 (Buckley); 19 May 1954 (Kane, Buckley); 17 May 1956 (Buckley); 19 May 1961 (Steineck, Friedman); 6 May 1962 (Horowitz et al.); 4 May 1963 (Norse); 12 May 1963 (Steineck, Friedman); 12–13 May 1964 (M. Gochfeld et al.); 23 May 1964 (Norse); 29 May 1967 (Norse); 30 Apr 2007 (unknown observer, *Kingbird*); 1 May 2009 (Buckley); 10 May 2012 (Buckley); 3 May 2014 (Furgoch, Souirgi).

Ceruleans are very early (Jul–Aug) fall migrants and so are rarely seen then. The sole study area fall record was a ♀ with 4 Parulas in Van Cortlandt on 13 Oct 1981 (Sedwitz)—both a blowback drift-migrant and the latest Cerulean in the entire New York City area. (NB: this bird was *not* in Central Park as stated in *Kingbird* 22: 72.) A more typical fall migrant was at nearby Inwood Park on 24 Jul 1953 (Carleton), where others in fall were found on 16 Sep 1961 and 14 Aug 1963 (Norse).

Kieran saw singles in Riverdale in the same tree 2 years running: 17 May 1950 and 13 Jun 1951—the latter a nonbreeder or very late migrant—and there was another in Riverdale on 14 May 2001 (unknown observer, *Kingbird*). A singing ♂ at Grassy Sprain Reservoir on 27 May 1925 (Cruickshank) was also a nonbreeder or late migrant.

NYSBBA I in 1980–85 and II in 2000–2005 reaffirmed the 100-year-old upper Hudson Valley population, but by NYSBBA II occupied blocks in Westchester, Putnam, Dutchess, Ulster, Orange, and Rockland Counties had dropped from about 40 to about 30, and to none in Westchester and Rockland Counties (McGowan and Corwin 2008). This trend has been mirrored throughout its range, and much attention is now focusing on its Neotropical overwintering areas (Jones et al. 2008; Buehler et al. 2008). Nearest breeders from 1950 to the mid-1960s were on the Palisades opposite Riverdale (Boyajian 1966), but they are no longer there, so the closest would now be in southeastern Orange County. After a number of territorial ♂ false starts in eastern Suffolk County, a pair finally bred in 1981 at Sag Harbor (Salzman), and in 1992 a cluster of 2–3 pairs was located in northern Easthampton (Lindsay, Veso).

Northern Parula *Setophaga americana*

NYC area, current
A common spring/fall Neotropical migrant and a rare but possibly increasing returning breeder. Like many Neotropical migrants, it was far more numerous in spring before 1960 than after.

Bronx region, historical
Bicknell's *Riverdale* 1872–1901: common spring/fall migrant, earliest 26 Apr 1878, nonbreeder/late migrant 8 Jun 1879, latest 26 Oct 1879

Eaton's *Bronx + Manhattan* 1910–14: common migrant, local breeder

Griscom's *Bronx Region* as of 1922: common spring/fall migrant

Kuerzi's *Bronx Region* as of 1926: very common spring (23 Apr–8 Jun)/fall (14 Aug–28 Oct) migrant; nonbreeder Grassy Sprain 10 Jun–mid-Jul 1923

NYC area, historical
Cruickshank's *NYC Region* as of 1941: common to abundant spring (several hundred)/fall migrant, 14 Apr–18 Nov

Bull's *NYC Area* as of 1968: maxima spring 200, fall 34 dead Gabreski Airport, Westhampton; 1 Apr–9 Dec [blowback drift-migrant]

Bull's *New York State* as of 1975: late fall date 23 Dec [blowback drift-migrant]

Study area, historical and current
A common spring/fall Neotropical migrant.

In spring, first migrants arrive in late Apr, peak in early May, and depart by late May, with extreme dates of 16 Apr 1952 (Kane, Buckley et al.)—a trans-Gulf overshoot—and 25 May 1961 (Horowitz) and maxima of 200–300 on 7–8 May 1955 (Kane et al.). A territorial ♂ at nearby Grassy Sprain Reservoir in Jun–Jul 1923 (Kuerzi) was a nonbreeder. Until Jun 1967, the spring maxima in Central and Prospect Parks were, respectively, 20 on 23 May 1954 (Feinberg, Maumary, Post) and 200 on 14 May 1950 (Whelen).

Even though their main New York State breeding stronghold is the Adirondacks, Parulas have long bred at isolated and fluctuating locations throughout southeastern New York, from the Catskills to Montauk. They bred regularly on Long Island until the mysterious disappearance of hanging Old Man's Beard lichens in the late 1930s. It took area breeders many decades to accommodate to this habitat loss throughout the Northeast, but they eventually began to recolonize, starting along the Delaware River in NJ and eventually making tentative moves back to Long Island. The closest of these to the study area was in Alley Pond Park in Queens, and an extended discussion of this process is in Salzman (2001). In northern NJ they nest in Norway Spruces and in suspended flood debris in Delaware River American Sycamores (R. Kane, pers. comm.). Between NYSBBA I in 1980–85 and II in 2000–2005 they held their own on Long Island and at 8–10 sites in the lower Hudson Valley (McGowan and Corwin 2008). DOIBBS data, although sparse, showed a 5.9% statewide increase from 1980–2005 (Sauer et al. 2005).

In fall, first migrants arrive in late Aug, peak in mid-Sep, and depart by mid-Oct, with extreme dates of 21 Aug 1953 (Buckley) and 22 Oct 2011 (Baksh), and a maximum of 12+ on 1 Oct 2015 (Aracil).

Specimens
Dwight collected 4 in Van Cortlandt in May, Sep 1888–90 (AMNH) and Bicknell 6 in Riverdale in May, Sep, Oct 1876–81 (NYSM).

Magnolia Warbler   *Setophaga magnolia*

NYC area, current
A relatively common spring/fall migrant. This is another Neotropical migrant that has been seen in smaller numbers since the 1960s.

Bronx region, historical
Bicknell's *Riverdale* 1872–1901: common spring/fall migrant

Eaton's *Bronx + Manhattan* 1910–14: common migrant

Griscom's *Bronx Region* as of 1922: common spring/fall migrant

Kuerzi's *Bronx Region* as of 1926: very common spring (1 May–12 Jun)/fall (15 Aug–19 Oct) migrant

NYC area, historical
Cruickshank's *NYC Region* as of 1941: common spring (maximum 100)/fall migrant, 27 Apr–23 Oct

Bull's *NYC Area* as of 1968: late fall date 30 Nov, once 30 Dec [blowback drift-migrants]; maxima spring 250, fall 32 dead Empire State Building

Bull's *New York State* as of 1975: status unchanged

Study area, historical and current
A relatively common spring/fall migrant.

In the study area in spring, first migrants arrive in early May, peak in mid-May, and depart by late May, with extreme dates of 28 Apr 1954 (Buckley) and 2 Jun 1952 (Buckley) and maxima of 40 on 11 May 1958 (Horowitz) and 13 May 1964 (M. Gochfeld) and 100+ on 17 May 1962 (Friedman, Steineck). Until Jun 1967, the spring maxima in Central and Prospect Parks were 250 on 23 May 1954 (Feinberg, Maumary, Post) and 70 on 10 May 1948 (Whelen).

As NYSBBA I in 1980–85 and II in 2000–2005 showed clearly, the southeastern breeding-block cluster in its New York State range ends abruptly at the Sullivan-Orange County line, beyond which are only a half-dozen sites no closer than Orange and Putnam Counties (McGowan and Corwin 2008). DOIBBS data showed a 2.1% annual statewide increase from 1966–2005 (Sauer et al. 2005).

In fall, first migrants arrive in late Aug, peak in early Sep, and depart by late Sep, with extreme dates of 2 on 1 Aug 1953 (Buckley et al.) and 16 Oct 2005 (Young) and a maximum of 25 on 6 Sep 1955 (Kane). Until Jun 1967, the fall maximum in Prospect Park was 45 on 13 Sep 1964 (Yrizarry). One in Riverdale 23 Dec 2012 (Shen, Fiore, Wallstrom; photos) and another in Bronx Park 5 Nov 1957 (Maguire et al.) were lingering blowback drift-migrants.

Specimens
Dwight collected 3 in Van Cortlandt in May 1885–90 (AMNH) and Bicknell 8 in Riverdale in May, Aug, Sep, Oct 1876–81 (NYSM).

Bay-breasted Warbler   *Setophaga castanea*

NYC area, current
An uncommon but annual spring/fall Neotropical migrant. Bay-breasted is a Spruce Budworm warbler whose breeding population size and success are tied to worm outbreaks; warblers' numbers in New York City parks in spring are also related to periodic oak outbreaks of small moth larvae colloquially called green inchworms. Warblers are attracted to them, gorging on them for days and occasionally weeks.

Bronx region, historical
Bicknell's *Riverdale* 1872–1901: rare spring/fall migrant, early as 5 Jul 1886, 26 Jul 1875

Eaton's *Bronx + Manhattan* 1910–14: rare migrant

Griscom's *Bronx Region* as of 1922: variably common spring/fall migrant

Kuerzi's *Bronx Region* as of 1926: common spring (4 May–6 Jun)/fall (25 Jul–13 Oct) migrant

NYC area, historical
Cruickshank's *NYC Region* as of 1941: fairly common spring (maximum 32)/fall migrant, 1 May–13 Oct

Bull's *NYC Area* as of 1968: late spring date 23 Jun; late fall date 1 Dec [blowback drift-migrant]; maxima spring 42, fall 63 dead Empire State Building

Bull's *New York State* as of 1975: early spring date 29 Apr; spring maximum 50

Study area, historical and current
An uncommon spring/fall Neotropical migrant.

In spring, first migrants arrive in early May, peak in mid-May, and depart by late May, with extreme dates of 9 May (thrice) and 26 May 1961 (Friedman, Steineck), and a maximum of 15 on 19 May 1955 (Kane, Buckley) and 17 May 1962 (Friedman, Steineck). Until Jun 1967, the spring maxima in Central and Prospect Parks were, respectively, 15 on 23 May 1954 (Feinberg, Maumary, Post) and 12 on 16 May 1953 (Usin).

In fall, first migrants arrive in early Aug, peak in early Sep, and depart by late Sep, with extreme dates of 3 Aug 1953 (Phelan, Buckley) and 16 Oct 1946 (Komorowski) and a maximum of 6 on 19 Sep 2005 (Young). Until Jun 1967, the fall maximum in Central Park was 14 on 3 Sep 1944 (Bull, Eisenmann). Bicknell collected 1 in Riverdale as late as 13 Oct 1876.

Nearest breeders are no closer than the Adirondacks, where they are scarce and may be declining (McGowan and Corwin 2008).

Specimens
Bicknell collected 5 in Riverdale in May, Aug, Oct 1876–81 (NYSM).

Blackburnian Warbler *Setophaga fusca*

NYC area, current
A spring/fall migrant in small numbers. This is another Neotropical migrant that has been seen with greatly diminished frequency and in smaller numbers since the 1960s.

Bronx region, historical
Bicknell's *Riverdale* 1872–1901: as rare a spring/fall migrant as Bay-breast, not recorded annually

Eaton's *Bronx + Manhattan* 1910–14: fairly common migrant

Griscom's *Bronx Region* as of 1922: common spring but rare fall migrant

Kuerzi's *Bronx Region* as of 1926: common spring (1 May–6 Jun)/fall (12 Aug–4 Oct) migrant

NYC area, historical
Cruickshank's *NYC Region* as of 1941: fairly common spring (maximum 18)/fall migrant, 23 Apr–18 Oct

Bull's *NYC Area* as of 1968: spring (19 Apr–13 Jun; maximum 15)/fall (3 Aug–20 Oct)

migrant; fall maximum 17 dead Empire State Building

Bull's *New York State* as of 1975: late fall date 3 Nov [blowback drift-migrant]

Study area, historical and current
A spring/fall Neotropical migrant in small numbers.

First spring migrants arrive in early May, peak in mid-May, and depart by late May, with extreme dates of 26–27 Apr 1952 (Kane) and 27 May 1951 (Norse), and a maximum of 25 on 13 May 1933 (Weber et al.). Until Jun 1967, the spring maximum in Prospect Park was 10 on 23 May 1948 (Whelen).

As NYSBBA I in 1980–85 and II in 2000–2005 showed clearly, the southeastern breeding-block cluster in its New York State breeding range ends abruptly at the Sullivan-Orange County line, beyond which are <10 Blocks no closer than Orange and Putnam Counties (McGowan and Corwin 2008). DOIBBS data indicated a nonsignificant statewide 1.1% increase from 1980–2005 (Sauer et al. 2005). If persisting, the nearest breeders would be in hemlocks at the north end of the Palisades (Boyajian 1971).

In fall, first migrants arrive in late Aug, peak in early Sep, and depart by late Sep, with extreme dates of 5 Aug 1952 (Kane) and 23 Sep 1954 (Buckley), singles only.

Specimens
Dwight collected 2 in Van Cortlandt in May 1885, 1890 (AMNH) and Bicknell 5 in Riverdale in May, Aug, Sep 1876–79 (NYSM).

Yellow Warbler *Setophaga petechia*

NYC area, current
A widespread spring/fall trans-Gulf Neotropical migrant and breeder. It is a notably early fall migrant, so overlooked then.

Bronx region, historical
Bicknell's *Riverdale* 1872–1901: common spring/fall migrant, breeder

Eaton's *Bronx + Manhattan* 1910–14: common breeder

Griscom's *Bronx Region* as of 1922: uncommon and decreasing breeder

Kuerzi's *Bronx Region* as of 1926: common breeder, spring/fall migrant; 20 Apr–5 Oct

NYC area, historical
Cruickshank's *NYC Region* as of 1941: widespread breeder, many still breeding New York City [locations not given]; common spring (maximum 50+)/fall migrant, 18 Apr–12 Oct

Bull's *NYC Area* as of 1968: fall maximum 8

Bull's *New York State* as of 1975: late fall date 24 Oct

Study area, historical and current
A spring/fall migrant and study area breeder.

In spring, first migrants arrive in mid-Apr, peak in early May, and depart by late May, with extreme dates of 1–3 Apr 1956 (Buckley, Kane)—a trans-Gulf overshoot—and 25 May 1961 (Horowitz), with a maximum of 50 on 8 May 1955 (Kane et al.). Until Jun 1967, the spring maxima in Central and Prospect Parks were, respectively, 25 on 23 May 1954 (Feinberg, Maumary, Post) and 14 on 18 May 1945 (Soll).

Breeds in Van Cortlandt and intermittently in Woodlawn. In 1937, 27 pairs bred in just Van Cortlandt Swamp (Sialis Bird Club). A reasonable estimate of 2014–15 study area breeders would be 30–40 pairs in Van Cortlandt and another 10 in Woodlawn, all of which usually vanish by early Aug.

Both NYSBBA I in 1980–85 and II in 2000–2005 indicated a broad New York City area breeding distribution, where they were understandably absent only from the most urbanized portions of all New York City boroughs except the Bronx and Staten Island (McGowan and Corwin 2008). DOIBBS data showed no statewide changes until 1980 and then a slight but significant decline until 2005 (Sauer et al. 2005; McGowan and Corwin 2008).

In fall, first migrants arrive in early Jul, peak in early Aug, and depart by late Aug, with extreme dates of 5 at Hillview on 18 Jul 2017 (Camillieri) and 18 Oct 1955 (Birnbaum)—the latter probably a blowback drift-migrant—with a maximum of 7 on 16 Aug 1977 (Young).

Specimens
Dwight collected 2 in Van Cortlandt in May 1885–90 (AMNH) and Bicknell 3 in Riverdale in May, Jul, Aug 1876–77 (NYSM).

Chestnut-sided Warbler   *Setophaga pensylvanica*

NYC area, current
A moderately numerous spring/fall Neotropical migrant. As a breeder, this early succession species was formerly widespread following regeneration of oldfields and forests, but since the 1960s it has been disappearing as its habitat vanished.

Bronx region, historical
Bicknell's *Riverdale* 1872–1901: common spring/fall migrant, not breeding

Hix's *New York City* as of 1904: unmentioned as breeder

Eaton's *Bronx + Manhattan* 1910–14: common migrant, does not breed

Griscom's *Bronx Region* as of 1922: common spring migrant, unrecorded fall

Kuerzi's *Bronx Region* as of 1926: common spring/fall migrant (26 Apr–9 Oct); fairly common breeder [locations not given]

NYC area, historical
Cruickshank's *NYC Region* as of 1941: common spring (maximum 67)/fall migrant, 26 Apr–9 Oct; widespread local breeder, but very few pairs around New York City outskirts [locations not given]

Bull's *NYC Area* as of 1968: 16 Apr–13 Nov [blowback drift-migrant], fall maximum 16 dead Empire State Building

Bull's *New York State* as of 1975: status unchanged

Study area, historical and current
A moderately numerous spring/fall Neotropical migrant and breeder that seems to have been breeding only continually >1960s but that may have recently reestablished itself in Van Cortlandt.

In spring, first migrants arrive in early May, peak in mid-May, and depart by late May, with extreme dates of 26 Apr 1953 (Buckley et al.) and 25 May 1961 (Horowitz), and a maximum of 40 on 8 May 1955 (Kane). Until Jun 1967, the spring maxima in Central and Prospect Parks were, respectively, 45 on 23 May 1954 (Feinberg, Maumary, Post) and 30 on 10 May 1946 (Soll).

In the late 1800s, it seems not to have been breeding in Riverdale (Bicknell) or in Van Cortlandt (Dwight) or if so, only in very small numbers. It was not until about 1922 that a few pairs of this quintessentially second-growth, oldfield succession breeder nested on the Ridge in Van Cortlandt, with 3 pairs staying until 1959 (Zupan). There was also a breeding pair or 2 in Van Cortlandt Swamp during the 1940s–early 1950s (Kieran), and another pair was attending a nest on the Ridge on 1 Jun 1985 (Block). Their breeding departure in 1960 followed successional changes as park forests matured and filled in, but hurricanes in more recent years have opened up new areas. In 2014, there were 12 singing ♂♂ in late May in recently cleared forest openings from Hurricanes Irene (2011) and Sandy (2012), behaving as if territorial on the Ridge, Northeast Forest, Vault Hill, and Northwest Forest (Buckley, Young). A conservative parkwide breeder estimate would be 5–6 pairs.

NYSBBA I in 1980–85 identified widespread breeding that stopped abruptly as one neared New York City except for a single occupied block in Queens, plus 6 in Nassau County and 40 in Suffolk County. But by NYSBBA II in 2000–2005 none were in New York City, northern Nassau County was down to 2 blocks, and Suffolk County to 17. The southernmost Westchester County blocks had also retreated northward (McGowan and Corwin 2008). DOIBBS data showed a 0.8% annual statewide decline from 1966–2005 (Sauer et al. 2007). The nearest breeding locations are on the Palisades and at Sands Point, Long Island, although this species can move in quickly when appropriate habitat appears. A pair may have bred in a revegetating storm clearing in Pelham Bay Park in the mid-1980s (Künstler) but not since. The status of a singing ♂ in Riverdale on 5 Jul 2005 (Pehek, Stanley) was uncertain, but it was not a local breeder.

In fall, first migrants arrive in mid-Aug, peak in early Sep, and depart by late Sep, with extreme dates of 2 on 3 Aug 1952 (Buckley et al.) and 1 Nov 1954 (Kane)—the latter a blowback drift-migrant—and a maximum of 8 on 26 Aug 2011 (Young).

Specimens
Dwight collected 9 in Van Cortlandt in May 1885–90 (AMNH) and Bicknell 4 in Riverdale in May, Aug 1876–78 (NYSM).

Blackpoll Warbler   *Setophaga striata*

NYC area, current
A relatively common spring/fall Neotropical migrant that has been seen with diminished frequency and in smaller numbers since the 1960s. Blackpoll is a Spruce Budworm warbler whose breeding population size and success are tied to worm outbreaks; warblers' numbers in New York City parks in spring are also related to periodic oak outbreaks of small moth larvae, colloquially called green inchworms. Warblers are attracted to them, gorging for days and sometimes weeks.

Bronx region, historical
Bicknell's *Riverdale* 1872–1901: common spring/fall migrant; as late as 5 Jul in exceptionally cold 1886 spring

Eaton's *Bronx + Manhattan* 1910–14: abundant migrant

Griscom's *Bronx Region* as of 1922: very common spring/fall migrant

Kuerzi's *Bronx Region* as of 1926: abundant spring (3 May–17 Jun)/fall (2 Sep–1 Nov) migrant; nonbreeder to 5 Jul

NYC area, historical
Cruickshank's *NYC Region* as of 1941: common to abundant spring (maximum 100+)/fall migrant, 2 May–20 Nov

Bull's *NYC Area* as of 1968: common to abundant spring/fall migrant, 30 Mar–3 Dec [blowback drift-migrant]; maxima spring 130, fall 356 dead Fire Island Lighthouse 1887

Bull's *New York State* as of 1975: status unchanged

Study area, historical and current
A relatively common spring/fall Neotropical migrant.

In spring, first migrants arrive in late Apr–early May, peak in late May, and depart by early Jun, with extreme dates of 28 Apr 1962 (Heath, Zupan) and 6 Jun 1961 (Horowitz) plus a 10 Jun 1886 specimen (Dwight: AMNH), and a maximum of 50 on 19 May 1955 (Kane, Buckley). A late migrant or nonbreeder was in Riverdale from 18–26 Jun 1974 (Pasquier). Until Jun 1967, spring maxima in Central and Prospect Parks were, respectively, 35 on 23 May 1954 (Feinberg, Maumary, Post) and 130 on 17 May 1945 (Soll).

In fall, first migrants arrive in early Sep, peak in late Sep–early Oct, and depart by mid-Oct, with extreme dates of 30 Aug 1961 (Buckley) and 25 Oct in 1953 (Buckley, Harrison et al.) and 1955 (Birnbaum), and a maximum of 15 on 25 Sep 2012 (Young). Until Jun 1967, the fall maximum in Central Park was 200 on 24 Sep 1963 (Kleinbaum). Blowback drift-migrants frequently linger into Nov coastally but almost never inland.

As a New York State breeding species, Blackpoll is centered in the Adirondacks and Catskills (nearest breeders), but its long-term status is unclear (McGowan and Corwin 2008).

Specimens
Dwight collected 4 in Van Cortlandt in May, Jun 1886–90 (AMNH) and Bicknell 4 in Riverdale in May, Sep 1976–79 (NYSM).

Black-throated Blue Warbler *Setophaga caerulescens*

NYC area, current
A common spring Neotropical migrant, scarcer in fall.

Bronx region, historical
Bicknell's *Riverdale* 1872–1901: regular spring/fall migrant

Eaton's *Bronx + Manhattan* 1910–14: common migrant

Griscom's *Bronx Region* as of 1922: common spring/fall migrant

Kuerzi's *Bronx Region* as of 1926: common spring (26 Apr–28 May)/fall (16 Aug–29 Oct) migrant; once Nov–Dec Irvington [blowback drift-migrant]

NYC area, historical
Cruickshank's *NYC Region* as of 1941: common spring (maximum 46)/fall migrant, 25 Apr–30 Nov, 2 Dec records [blowback drift-migrants]

Bull's *NYC Area* as of 1968: 11 Apr–11 Dec; late dates 27 Nov, 1–31 Dec [blowback drift-migrants]; maxima spring 50, fall 48 dead Fire Island Lighthouse 1883

Bull's *New York State* as of 1975: 4 Dec–Jan records [blowback drift-migrants]

Study area, historical and current
A common spring Neotropical migrant, scarcer in fall.

In spring, first migrants arrive in early May, peak in mid-May, and depart by late May, with extreme dates of 15 Apr 1951 (Kane) and 25 May 1961 (Horowitz), and a maximum of 40 on 4 May 1962 (Friedman, Steineck). Until Jun 1967, the spring maxima in Central and Prospect Parks were, respectively, 40 on 23 May

1954 (Feinberg, Maumary, Post) and 30 on 10 May 1946 (P. Wells, Whelen).

In fall, first migrants arrive in late Aug, peak in late Sep, and depart by early Oct, with extreme dates of 2 on 1 Aug 1953 (Buckley et al.) and 19 Oct 1996 (Young), and a maximum of 7 on 6 Oct 2002 (Young). This species is another with lingering blowback drift-migrants but only 1 has been found in the study area—in Van Cortlandt on 21 Nov 2000 (unknown observer, *Kingbird*)—although one overwintered in Inwood Park in 1997–98 (DiCostanzo et al.), and another was in Riverdale on 9 Nov 2014 (Wright et al.).

As NYSBBA I in 1980–85 and II in 2000–2005 showed clearly, the southeastern breeding-block cluster in its New York State breeding range ends abruptly near the Sullivan-Orange County line, beyond which are only a dozen blocks in Orange and Putnam Counties plus one outlier along the Hudson River at Bear Mountain—the nearest breeders (McGowan and Corwin 2008). DOIBBS data showed a 2.3% annual statewide decline from 1980–2005 (Sauer et al. 2007).

Specimens
Dwight collected 3 in Van Cortlandt in Apr, May 1885–90 and Foster 1 in Woodlawn in May 1890 (all AMNH), and Bicknell 1 in Riverdale on 8 Sep 1881 (NYSM).

Palm Warbler *Setophaga palmarum*

UNIDENTIFIED TO SUBSPECIES

NYC area, current
A common spring/fall Neotropical migrant, occasional in winter. Two distinctive subspecies occur: Yellow Palm (*hypochrysea*) (YP), more common in spring, and Western Palm (*palmarum*) (WP), predominating in fall and winter. Throughout the last 150 years, few observers have bothered to identify and record them by subspecies, so except when stated otherwise, all spring records are assumed to apply to *hypochrysea* and all fall records to *palmarum*.

Yellow Palms breed in taiga west to western Quebec, migrate southwestward in fall to their Gulf Coast overwintering sites, then in spring northeastward mostly along the Atlantic coast east of the Appalachians. Western Palms breed in taiga east to western Quebec, migrate southeastward in fall to their Florida and West Indies overwintering areas, then in spring move northwestward largely west of the Appalachians and slightly later than Yellow Palms. In effect, they cross each others' main migration routes in spring/fall, but with some directional leakage by each during both seasons. Their breeding, migration, and overwintering areas are shown in Dunn and Alderfer (2017: 574).

Bronx region, historical
Bicknell's *Riverdale* 1872–1901: WP: never recorded (?); YP: common spring/uncommon fall migrant

Eaton's *Bronx + Manhattan* 1910–14: WP: rare migrant; YP: common migrant

Griscom's *Bronx Region* as of 1922: WP: rare spring, uncommon fall migrant; YP: common spring/fall migrant

Kuerzi's *Bronx Region* as of 1926: WP: uncommon spring (20 Apr–15 May)/common fall (5 Sep–26 Oct) migrant; YP: common spring (24 Mar–18 May)/fall (16 Sep–27 Nov) migrant

NYC area, historical
Cruickshank's *NYC Region* as of 1941: WP: rare spring/uncommon to locally common

fall (maximum 50+) migrant, very rare winter coastally; YP: common spring (maximum 100+) /fall migrant, very rare winter coastally

Bull's *NYC Area* as of 1968: WP: fall maximum 80; commoner than YP in winter, maximum 10; YP maxima: spring 300, fall 34

Bull's *New York State* as of 1975: WP: late spring date 3 Jun; winter maximum 19

YELLOW PALM WARBLER

In the study area the expected subspecies in spring migration, vastly outnumbering Western; regular although far less numerous in fall. Unrecorded in winter.

In spring, first migrants arrive in early Apr, peak in mid–late Apr, and depart by early May, with extreme dates of 2 on 17 Mar 1953 (Buckley) and 15 May 1953 (Kane) and maxima of 300+ on 1 May 1950 (Russak) and 200+ in nearby Bronx Park on 23 Apr 1955 (Solomon, Kane et al.). Until Jun 1967, the spring maximum in Central Park was 200 on 21 Apr 1929 (Watson et al.).

In fall, first migrants arrive in late Sep, peak in early Oct, and depart by late Oct, with extreme dates of 6 Sep 1964 (Horowitz) and 25 Nov 1938 (Norse, Cantor) and a maximum of 6 on 29 Oct 1960 (Horowitz, Buckley).

Specimens
Dwight collected 4 (all presumed *hypochrysea*) in Van Cortlandt in Apr 1885–90 (AMNH), and Bicknell 8 in Apr, May in Riverdale 1877–80—all presumed *hypochrysea* except possibly 1 on 12 May (NYSM).

WESTERN PALM WARBLER

In the study area the expected subspecies in fall migration and winter; infrequent in spring but probably overlooked.

In fall, first migrants arrive in late Sep, peak in early Oct, and most depart by late Oct, with extreme dates of 2 on 10 Sep 2011 (Baksh) and 27 Nov 1937 (Norse, Cantor), with a maximum of 26 on 27 Oct 2002 (Young). Some attempt to overwinter in the New York City area every year, usually coastally, but there are no study area winter records.

There are only 7 spring study area records because few observers bothered to routinely separate and record seasonal numbers for each subspecies: 23 Apr 1936 (Norse, Cantor); 29 Apr 1939 (Norse); 10 May 1953 (Norse); 12 Apr 1954 (Buckley); 3 on 30 Apr 1954 (Buckley); 2 on 18–30 Apr 1956 (Post, Scully et al.); 2 on 10 May 1956 (Buckley). There ought to be more spring records: 1 collected by Bicknell in Riverdale on 12 May 1877 has not been examined, but the late date suggests Western.

Pine Warbler   *Setophaga pinus*

NYC area, current
A widespread breeder in conifers in Suffolk County, northern Nassau County, and eastern Westchester County but undetected as a breeder within New York City until 2005, when several pairs bred in Bronx Park. It is a potential breeder in ornamental conifers in large cemeteries, where it has not been systematically sought.

Bronx region, historical
Bicknell's *Riverdale* 1872–1901: very rare spring/fall migrant, may have bred 1890

Eaton's *Bronx + Manhattan* 1910–14: uncommon migrant

Griscom's *Bronx Region* as of 1922: uncommon spring, rare fall migrant

Kuerzi's *Bronx Region* as of 1926: uncommon spring (24 Mar–10 May)/fall (23 Sep–7 Nov) migrant, once in winter; uncertain breeding Kensico Reservoir

NYC area, historical
Cruickshank's *NYC Region* as of 1941: uncommon spring/rare fall (maximum 20) migrant, late Mar–late Nov, with Dec and Feb records

Bull's *NYC Area* as of 1968: decreased markedly from 1948–63; rare to fairly common spring (maximum 15)/fall migrant; a dozen winter records

Bull's *New York State* as of 1975: status unchanged

Study area, historical and current
A moderately common spring/fall migrant and potential breeder that has occasionally been found in winter.

In spring, first migrants arrive in early Apr, peak in late Apr, and depart by early May, with extreme dates of 15 Mar 1953 (Kane) and at Hillview on 15 May 1953 (Kane; a breeder?), and a maximum of 10 on 5 Apr 2012 (Young). Until Jun 1967, the spring maximum from both Central and Prospect Parks was 6 on, respectively, 8 Apr 1954 (Feinberg, Post) and 23 Apr 1948 (Whelen).

Even though searched for unsuccessfully in Woodlawn in 2014, Pine Warbler may nonetheless breed there given that it does so as close as Bronx Park, the Bronx Zoo, and the Palisades. Both NYSBBA I in 1980–85 and II in 2000–2005 did record probable breeders in Block 5852D, which includes Bronx Park and the Bronx Zoo (McGowan and Corwin 2008). The first proven Bronx and New York City breeders since a possible pair in Riverdale in 1890 (Bicknell) were feeding young in mature Eastern White Pines in Bronx Park on 4 Jul 2005 (DeCandido and Allen 2005a), where presumed breeders had been seen since 2001 (Block). They also bred in the adjacent Bronx Zoo from 2006–10 (Olson).

As NYSBBA I in 1980–85 and II in 2000–2005 showed clearly, this species breeds in distinct habitat-bound clusters in New York State, the southeast having one of them. Most occupied blocks are in Suffolk County, a few in northern Nassau County, jumping then to Westchester, Putnam, and Rockland Counties—excepting the recent Bronx colonization. It appears that prior to NYSBBA II, all had been in the Suffolk County Pine Barrens, with the next cluster in southeastern Sullivan County, but some of this range expansion may have been due to better census coverage on NYSBBA II (McGowan and Corwin 2008). DOIBBS data showed no statewide change until the late 1980s, then a strong increase continuing to 2005 (Sauer et al. 2005).

In fall, first migrants arrive in early Sep, peak in mid-Oct, and depart by early Nov, with extreme dates of 21 Aug 1997 (Young) and 10 Nov 2008 (Young), and a maximum of 6 on 13 Oct 2007 (Young).

In winter, it has never been seen in Woodlawn or at Hillview on a Christmas Bird Count, although singles were in Van Cortlandt Swamp from 5–17 Jan 1952 (Kane, Buckley), in Woodlawn from 9 Nov–7 Dec 1957 (Buckley et al.), and on the Ridge in the second week of Jan 2013 (Murray). Another was at a Riverdale feeder on 23 Dec 2000 (Fiore, Wallstrom), and 2 were in Bronx Park on 23 Dec 1951 (Kane, Buckley), where there were also 6 on 26 Dec 2004 (Olson) and 2 on 27 Dec 2009 (Olson).

Specimens
Dwight collected 4 in Van Cortlandt in Apr 1885–88 (AMNH).

Myrtle Warbler   *Setophaga (coronata) coronata*

NYC area, current
A sometimes abundant spring/fall migrant that in winter is all but restricted to the immediate Long Island South Shore, where it migrates southbound by the thousands in Oct.

Bronx region, historical
Bicknell's *Riverdale* 1872–1901: common to abundant spring/fall migrant

Eaton's *Bronx + Manhattan* 1910–14: abundant migrant, uncommon winter visitor

Griscom's *Bronx Region* as of 1922: abundant spring/fall migrant, rare winter

Kuerzi's *Bronx Region* as of 1926: abundant spring (5 Apr–5 Jun)/fall (23 Aug–16 Nov) migrant, occasionally overwintering

NYC area, historical
Cruickshank's *NYC Region* as of 1941: abundant spring/fall (maximum many hundreds) migrant, rare and irregular in winter away from Long Island

Bull's *NYC Area* as of 1968: maxima spring 650 Bronx Park, fall 2500, winter 800 both Long Island

Bull's *New York State* as of 1975: 10,000+ normal Oct counts on barrier beaches

Study area, historical and current
A widespread, abundant spring migrant that is less numerous in fall and only occasional in winter.

In spring, first migrants arrive in early Apr, peak in late Apr, and depart by mid-May, with extreme dates of 5 Mar 1950 (Norse) and 25 May in 3 different years, and a maximum of 300 on 8 May 1955 (Kane et al.). One on 21 Jun 1955 (Post, Buckley) was a nonbreeder or very late migrant. Until Jun 1967, the spring maximum in Prospect Park was 325 on 30 Apr 1953 (Usin).

As NYSBBA I in 1980–85 and II in 2000–2005 showed clearly, the southeastern breeding-block cluster in its New York State breeding range ends abruptly at the Sullivan-Orange County line, beyond which are only a half-dozen sites no closer than southwestern Orange and southeastern Dutchess County except for 2 blocks near Bear Mountain that hold the closest breeders (McGowan and Corwin 2008). DOIBBS data showed an annual statewide increase of 4% from 1966–2005 (Sauer et al. 2005).

In fall, first migrants arrive in late Aug, peak in mid-Oct, and depart by early Nov, with extreme dates of 5 Aug 1952 (Buckley) and 26 Nov 1954 (Buckley), and a maximum of 120 on 24 Oct 1995 (Young). Until Jun 1967, the fall maximum in Prospect Park was 1000 on 15 Oct 1950 (Alperin, Jacobson)—the coastal effect in action.

Singles are occasional anytime between 1 Dec and 1 Mar—most recently on 22 Dec 2013 (Gotlib, McAlexander)—but overwintering is unknown.

Specimens
Dwight collected 3 in Van Cortlandt in Apr, May 1886–88 (AMNH) and Bicknell 4 in Riverdale in Apr, Oct 1876–78 (NYSM).

Comments
Myrtle is sometimes lumped in Yellow-rumped Warbler with **Audubon's Warbler**, *S. (coronata) auduboni*. This blowback drift-migrant is a vagrant to Long Island but has never been recorded in Central or Prospect Parks or the Bronx. It has appeared inland in southern New York twice: at a feeder near Red Oaks Mill,

Dutchess County, from 5–9 Feb 1976 (DeOrsey, DeOrsey) and in western Rockland County at the Mt. Peter hawkwatch on 17 Oct 1987 (Tramontano).

Yellow-throated Warbler *Setophaga dominica*

NYC area, current
A southern Neotropical migrant species that has been inching northward as a breeder from extreme southern NJ since the 1950s, finally breeding on Long Island since 2013. It is an annual but very scarce spring trans-Gulf overshoot or range-prospector most often found in New York City parks, and very rarely as a blowback drift-migrant in fall, notably on Long Island barrier beaches.

Bronx region, historical
Bicknell's *Riverdale* 1872–1901: unrecorded

Eaton's *Bronx + Manhattan* 1910–14: unrecorded

Griscom's *Bronx Region* as of 1922: unrecorded

Kuerzi's *Bronx Region* as of 1926: Westchester County May record

NYC area, historical
Cruickshank's *NYC Region* as of 1941: exceedingly rare spring range-prospector, 7 Long Island, 2 Central Park, 2 Westchester County, no Bronx records

Bull's *NYC Area* as of 1968: rare but annual in spring [trans-Gulf overshoot/range-prospector] (15 Apr–6 Jun), rarely detected in fall (8 Jul–6 Oct, all coastal); singles only

Bull's *New York State* as of 1975: early spring date 3 Apr

Study area, historical and current
A sporadic spring range-prospector, increasing in frequency as it slowly expands its breeding range northward. With increased coverage in Apr–May, this species would prove to be almost annual, as it now is in Central and Prospect Parks.

Only 8 records, all since 1952: Sycamore Swamp from 20–27 Apr 1952 (Komorowski, Norse et al.); 24 Apr 1954 (Norse, Buckley et al.); 2 May 1954 (Kieran); 7–10 May 1956 (Scully, Buckley, Sutton); 22 Apr 1957 (Buckley); 14 May 2001 (unknown observer, *Kingbird*); 20 Apr 2004 (Young); 23 Apr 2006 (Nulle). Recorded much more frequently in nearby Bronx Park and also at Pelham Bay Park on 28 Apr 1951 (Russak, Solomon), where it and a 17 Dec 1950 Pine Warbler had been feeding on Pine Needle Scale insects, and at Grassy Sprain Reservoir on 28 May 1933 (Sedwitz). A blowback drift-migrant at Pelham Bay Park on 17 Oct 2007 (Rothman) appears to be the sole fall Bronx record.

Comments
It is unclear whether any of the spring birds were trans-Gulf overshoots or if all were range-prospectors in a species that has been slowly expanding its range northward and eastward. Several showed the characteristics of the white-lored and -chinned, shorter-billed, Ohio-valley sycamore-loving subspecies *albilora* (colloquially called Sycamore Warbler), whose nearest breeders are now along the Delaware River in NJ and in extreme southwestern Orange County. Others resembled the yellow-lored and -chinned, longer-billed, coastal pine-loving nominate *dominica*, whose nearest breeders were in the NJ Pine Barrens until a pair established a breeding beachhead at Connetquot River State Park, Suffolk County, in Jun 2013, reaching 3 pairs in 2015.

Prairie Warbler *Setophaga discolor*

NYC area, current
A fairly common and widespread spring/fall Neotropical migrant, and a very local open area, second-growth breeder favoring pines, so all but restricted to Suffolk County on Long Island. It seems to be extirpated as a breeder from Nassau County and New York City and is almost gone from southern Westchester County. It does not become more widespread until northern Westchester and western Rockland Counties.

Bronx region, historical
Bicknell's *Riverdale* 1872–1901: rare spring migrant, once fall

Eaton's *Bronx + Manhattan* 1910–14: fairly common migrant

Griscom's *Bronx Region* as of 1922: rare spring/fall migrant

Kuerzi's *Bronx Region* as of 1926: common spring (26 Apr–28 May)/fairly common fall (16 Aug–30 Sep) migrant

NYC area, historical
Cruickshank's *NYC Region* as of 1941: fairly common spring (maximum 20 Westchester County)/fall (maximum 5) migrant, 16 Apr–24 Oct

Bull's *NYC Area* as of 1968: spring maximum 30; late fall date 29 Nov

Bull's *New York State* as of 1975: late fall dates of 3, 15, 26 Dec [blowback drift-migrants]

Study area, historical and current
A fairly common and widespread spring/fall Neotropical migrant.

In spring, first migrants arrive in late Apr–early May, peak in mid-May, and depart by late May, with extreme dates of 7 Apr 1951 (Kane) and 25 May 1940 (Norse), and a maximum of 15 on 8 May 1955 (Kane et al.). Until Jun 1967, the spring maximum in Prospect Park was 30 on 5 May 1946 (Soll).

One might imagine study area breeding at some time, given that the nearest breeders may be nesting on the Palisades. And though both NYSBBA I in 1980–85 and II in 2000–2005 did record breeders in the 2 blocks immediately north of study area Block 5952A (McGowan and Corwin 2008), we know of no historical breeding anywhere in the Bronx. Sparse DOIBBS data showed a 4% annual statewide increase from 1966–2005 (Sauer et al. 2005).

In fall, first migrants arrive in late Aug, peak in early Sep, and depart by late Sep, with extreme dates of 1 Aug 1953 (Buckley et al.) and 11 Oct 1994 (Young) and a maximum of 3 on 31 Aug 2012 (Young). In the Bronx, very late Prairies, lingering blowback drift-migrants, were at Ferry Point Park on 26 Dec 1994 (Walter, Jaslowitz) and at Pelham Bay Park from 23 Dec 2012–19 Jan 2013 (Rothman et al.).

Specimens
Dwight collected 2 in Van Cortlandt on 6 May 1885 (AMNH) and Bicknell 2 in Riverdale in May 1877 (NYSM).

Black-throated Green Warbler *Setophaga virens*

NYC area, current
A common spring/fall Neotropical migrant that has been seen with diminished frequency and in smaller numbers since the 1960s.

Bronx region, historical
Bicknell's *Riverdale* 1872–1901: common spring/fall migrant

Eaton's *Bronx + Manhattan* 1910–14: common migrant

Griscom's *Bronx Region* as of 1922: common spring/fall migrant

Kuerzi's *Bronx Region* as of 1926: common spring (21 Apr–9 Jun)/fall (21 Aug–7 Nov) migrant; several pairs breeding locally [locations not given]

NYC area, historical
Cruickshank's *NYC Region* as of 1941: common to abundant spring (maximum 100+)/fall migrant, 30 Mar–21 Nov, 2 Dec records [blowback drift-migrants]

Bull's *NYC Area* as of 1968: late fall date 9 Dec, single Jan record [blowback drift-migrants]; fall maximum 22

Bull's *New York State* as of 1975: status unchanged

Study area, historical and current
A common spring/fall Neotropical migrant.

In spring, first migrants arrive in mid–late Apr, peak in early May, and depart by mid-May, with extreme dates of 25 Mar 1982 (Sedwitz)—a trans-Gulf overshoot—and 2 Jun 1953 (Buckley), and a maximum of 50+ on 1 May 1955 (Kane et al.). Until Jun 1967, the spring maximum in Prospect Park was 75 on 10 May 1946 (Soll).

In fall, first migrants arrive in early Sep, peak in late Sep, and depart by early Oct, with extreme dates of 20 Aug 2008 (Young) and 18 Oct 1955 (Birnbaum), and a maximum of 20 on 6 Sep 1955 (Kane).

An adult ♂ in a White Pine grove near the Van Cortlandt Lake boathouse from 8 Nov 1943–1 Jan 1944 (Komorowski) and others on 20 Nov 1928 (Chubb) and from 19–25 Nov 2017 (Souirgi et al.) were lingering blowback drift-migrants, as were singles in Riverdale on 9 Nov 2014 (Wright et al.) and in Inwood Park 31 Dec 2011–6 Jan 2012 (Savaresse).

As NYSBBA I in 1980–85 and II in 2000–2005 showed clearly, the southeastern breeding-block cluster in its New York State range ends abruptly at the Sullivan-Orange County line, beyond which are several dozen sites in Orange, Rockland, Putnam, and central Westchester Counties (McGowan and Corwin 2008). Nearest breeders are those on the northern Palisades, although a pair may have nested near Little Neck Bay, Queens in 2001 (J. Miller), and it may breed intermittently in a few Suffolk County blocks. DOIBBS data showed an annual statewide increase of 4% from 1966–2005 (Sauer et al. 2007).

Specimens
Dwight collected 3 in Van Cortlandt in Apr 1885–87 (AMNH) and Bicknell 3 in Riverdale in Apr, Aug, Sep in 1876–78 (NYSM).

Comments
Three closely related western blowback drift-migrants that overwintered somewhere in the East/Southeast have occurred as vagrants in New York City parks in spring. **Townsend's Warbler**, *S. townsendi*, has been found 15–20 times, with multiple records in Central and Prospect Parks. A ♂ in Bronx Park on 9 May 1964 (C. Young, Stepinoff et al.) is the sole Bronx record. **Black-throated Gray Warbler**, *S. nigrescens*, has been seen in Central and other New York City parks in both spring and fall, but remains undetected in the Bronx or in Prospect Park. One was as close as Hawthorne, Westchester County, on 23 Apr 2015 (Seneca). The rarest of this group in the East, **Hermit Warbler**, *S. occidentalis*, has been found only once in New York City, in Central Park on 1 May 2016 (Lamek, Fung, Collerton; photos).

Canada Warbler *Cardellina canadensis*

NYC area, current
A reasonably common spring/fall migrant. This is another Neotropical migrant that has been seen in smaller numbers since the 1960s.

Bronx region, historical
Bicknell's *Riverdale* 1872–1901: common spring/fall migrant

Eaton's *Bronx + Manhattan* 1910–14: common migrant

Griscom's *Bronx Region* as of 1922: common spring/fall migrant

Kuerzi's *Bronx Region* as of 1926: very common spring (2 May–11 Jun)/fall (9 Aug–29 Oct) migrant

NYC area, historical
Cruickshank's *NYC Region* as of 1941: common spring (maximum 40+)/fall migrant, 2 May–13 Nov

Bull's *NYC Area* as of 1968: maxima spring 75, fall 35; 23 Apr–12 Jun, 26 Jul–13 Nov [blowback drift-migrant]

Bull's *New York State* as of 1975: fall maximum 74; bred Suffolk County (2 sites)

Study area, historical and current
A reasonably common spring/fall migrant.

In spring, first migrants arrive in early May, peak in mid-May, and depart by late May, with extreme dates of 2 May 1953 (Kieran) and 6 Jun 1962 (Zupan), and maxima of 55 on 13 May 1964 (M. Gochfeld) and an impressive 200 on 19 May 1961 (Friedman, Steineck). Until Jun 1967, the spring maximum in Central Park was 75 on 23 May 1954 (Feinberg, Maumary, Post).

In fall, first migrants arrive in mid-Aug, peak in late Aug, and depart by mid-Sep, with extreme dates of 5 Aug 1952 (Buckley) and 26 Sep 2006 (Young) and a maximum of 6 on 20 Aug 1999 (Young). Until Jun 1967, the fall maximum in Prospect Park was 25 on 12 Aug 1953 (Usin).

Between NYSBBA I in 1980–85 and II in 2000–2005, breeding sites southeast of the Catskills dropped from several dozen in Westchester, Rockland, Orange, and Putnam Counties to 2 in extreme southwestern Orange, 3 in Rockland, <10 in northern Putnam County, and none in Westchester County, and even their Catskill and Adirondack centers of abundance were drastically thinned (McGowan and Corwin 2008). DOIBBS data showed an annual statewide decrease of 5% from 1966–2005 (Sauer et al. 2007). Nearest breeders may be on the Palisades (Boyajian 1966) or in southwestern Fairfield County, CT (Bevier 1994). However, a pair did breed at Silver Lake, near White Plains, in 1964 (Bowen), and they can be easy to overlook when not singing.

Specimens
Dwight collected 3 in Van Cortlandt in May, Aug 1886–90 (AMNH) and Bicknell 6 in Riverdale in May, Aug, Sep 1876–81 (NYSM).

Wilson's Warbler *Cardellina pusilla*

NYC area, current
An annual but scarce spring/fall Neotropical migrant, with blowback drift-migrants lingering to late Dec.

Bronx region, historical
Bicknell's *Riverdale* 1872–1901: rare spring/fall migrant, as late as 22 Nov 1885 [blowback drift-migrant]

Eaton's *Bronx + Manhattan* 1910–14: fairly common migrant

Griscom's *Bronx Region* as of 1922: uncommon spring/fall migrant

Kuerzi's *Bronx Region* as of 1926: fairly common spring migrant (8 May–6 Jun)/uncommon in fall (14 Aug–22 Nov [blowback drift-migrant])

NYC area, historical
Cruickshank's *NYC Region* as of 1941: uncommon to fairly common spring (maximum 43)/fall migrant, 30 Apr–25 Nov [blowback drift-migrant]

Bull's *NYC Area* as of 1968: spring maximum 100; 29 Apr–7 Jun, 26 Jul–30 Nov, 2 in Dec [blowback drift-migrants]; fall maximum 10

Bull's *New York State* as of 1975: status unchanged

Study area, historical and current
An annual but scarce spring/fall migrant with late-lingering blowback drift-migrants.

In spring, first migrants arrive in early–mid-May, peak in late May when they depart, with extreme dates of 4 May 1965 (Horowitz) and 30 May 1954 (Buckley) and a maximum of 15 from 15–19 May 1955 (Kane, Buckley). Until Jun 1967, the spring maximum in Central Park was 100 on 23 May 1954 (Feinberg, Maumary, Post) and in Prospect Park 8 on 14 May 1950 (Whelen).

In fall, first migrants arrive in mid-Aug, peak in early Sep, and depart by late Sep, with extreme dates of 3 (the maximum) on 1 Aug 1953 (Phelan et al.) and 4 Oct 2015 (O'Reilly); 1 on 24 Oct 1995 (Young) may have been a blowback drift-migrant. Until Jun 1967, the fall maximum in Prospect Park was 10 on 16 Sep 1954 (Yrizarry).

Lingering blowback drift-migrants were in Van Cortlandt 28 Nov 2007 (Young), 3–10 Dec 1993 (Garcia, Lyons), and in Riverdale 22 Nov 1885 (Bicknell) and 27 Dec 1987 (Buckley, Antenen et al.). Another was in the Bronx Zoo 15 Dec 2015–22 Jan 2016 (Olson et al.; photos). As with winter Nashvilles, some of these were suspected of having been western in origin, in this case presumably belonging to *pileolata*, often called Pileolated Warbler.

Specimens
Dwight collected 2 in Van Cortlandt in May, Aug 1887–90 (AMNH) and Bicknell 3 in Riverdale in May, Aug, Sep 1876–81 (NYSM).

Yellow-breasted Chat *Icteria virens*

NYC area, current
An annual but very uncommon spring/fall Neotropical migrant. It is an erratic breeder in tangled second growth, clusters of pairs arriving and departing rapidly when conditions change, and it is also an emblematic blowback drift-migrant that frequently attempts to overwinter, especially along the coast. It is a New York State Species of Special Concern.

Bronx region, historical
Bicknell's *Riverdale* 1872–1901: common spring/fall migrant, breeder

Hix's *NYC* in 1904: breeds Bronx Park

Eaton's *Bronx + Manhattan* 1910–14: common breeder

Griscom's *Bronx Region* as of 1922: uncommon migrant, breeder, formerly more numerous

Griscom's *Riverdale* as of 1926: extirpated breeder

Kuerzi's *Bronx Region* as of 1926: fairly common breeder (1 May–7 Oct), locally abundant Grassy Sprain ridge

NYC area, historical
Cruickshank's *NYC Region* as of 1941: erratic New York City area breeder, all but extirpated from New York City outskirts [locations not given]; uncommon to rare spring (maximum 17 Westchester County)/rare fall migrant, 28 Apr–13 Nov; 2 in winter [blowback drift-migrants]

Bull's *NYC Area* as of 1968: early spring date 26 Apr; >1949 annual fall, maximum 6, greatly increased winter in thickets, feeders [blowback drift-migrants]

Bull's *New York State* as of 1975: status unchanged

Study area, historical and current
An annual but scarce spring/fall migrant and an extirpated breeder that is occasionally found in early winter.

In spring, first migrants arrive in early May, peak in mid-May, and depart by late May, with extreme dates of 30 Apr 1992 (Young) and 25 May 1961 (Horowitz), all singles apart from Van Cortlandt breeders.

In the late 1800s, it was a common breeder in Riverdale (Bicknell) and Van Cortlandt (Dwight). Since then chats have intermittently bred in ephemeral small clusters, one such in the study area being the 3–4 pairs that nested on the Ridge from 1934 (Weber) to 1952–54 (Komorowski, Buckley et al.). The nearest current breeders may be somewhere in southern Westchester County, where there were an estimated 20 pairs at Grassy Sprain Reservoir in 1961–63 (Friedman and Steineck 1963) and where some were found by NYSBBA I in 1980–85 (McGowan and Corwin 2008). From NYSBBA I to NYSBBA II in 2000–2005, chat-occupied blocks dropped statewide from 122 to 27. Much of the loss occurred in the New York City area, where the combined total of occupied blocks in Orange, Putnam, Rockland, and Westchester Counties; Staten Island; and Nassau and Suffolk Counties fell from 46 to 9.

In fall, first migrants arrive in early Aug, peak in late Aug–early Sep, and depart by mid-Sep, with extreme dates of 3 Aug 1953 (Phelan et al.) and 5 Oct 1961 (Horowitz), singles only.

Study area chats in Van Cortlandt on 6 Nov 1951 (Kane) and in Woodlawn in snow on 22 Dec 2002 (Futuyma, Buckley), as well as in Riverdale from 26–31 Dec 1957 (Scully, Buckley) and on 23 Dec 2005 (Fiore), were lingering blowback drift-migrants, as are all in late fall and winter. There are a number of other Bronx winter records, and it has long been an open question whether any Northeast winter chats belong to the western subspecies *auricollis*.

Specimens
Dwight collected 5 in Van Cortlandt in May, Jun, Aug 1887–90 (AMNH), and Bicknell 2 in Riverdale in May, Aug 1876–80 (NYSM).

New-World Sparrows: Emberizidae

Eastern Towhee *Pipilo erythrophthalmus*

NYC area, current
A common but declining breeder, spring/fall migrant, and irregular winter resident/visitor.

Bronx region, historical
Bicknell's *Riverdale* 1872–1901: common spring/fall migrant, breeder

Hix's *New York City* as of 1904: some breed Bronx Park

Eaton's *Bronx + Manhattan* 1910–14: abundant breeder

Griscom's *Bronx Region* as of 1922: common breeder, several winter records

Kuerzi's *Bronx Region* as of 1926: common breeder, few winter records

NYC area, historical
Cruickshank's *NYC Region* as of 1941: very common breeder, still nests within New York City [locations not given]; abundant spring (maximum 100+)/fall migrant, rare in winter

Bull's *NYC Area* as of 1968: spring maximum 200, winter 7–8 at single feeder

Bull's *New York State* as of 1975: status unchanged

Study area, historical and current
Formerly a common spring/fall migrant and irregular winter resident/visitor. But sometime shortly after 2008, after having been a widespread and numerous study area breeder, for unknown reasons they have all but disappeared in the breeding season.

In spring, first migrants arrive in mid-Apr, peak in late Apr–early May, and depart by late May, with extremes of 3 Mar 2012 (Baksh) and 3 on 21 May 2011 (Bochnik) and a maximum of 40 on 1 May 1955 (Kane, Phelan, Buckley). Until Jun 1967, the spring maxima in Central and Prospect Parks were, respectively, 110 on 30 Apr 1947 (Sedwitz) and 59 on 1 May 1944 (Soll).

Towhees have been conspicuous breeders throughout the study area from the late 1870s (Bicknell, Dwight) through the mid- and late 1900s (Kieran, Cruickshank, Norse, Buckley) to the early 2000s (Young). Through the 1990s, they were breeding in all 4 extant subareas, with 5 pairs breeding in a 24 ac (10 ha) plot in the Northwest Forest in 1991–92 (Jaslowitz and Künstler 1992; Jaslowitz 1993). But in 2006 only a single pair of towhees was in the Northwest Forest west of the study plot, and they were gone in 2012 (Young). In a breeding bird census of the entire Northwest Forest in 2006, 4 pairs were found (Delacretaz 2007), and in 2008 5 pairs were scattered throughout various Van Cortlandt forested areas (Künstler and Young 2010). Censuses of the Northwest and Northeast Forests in 2009 and 2010 (E. Pehek, pers. comm.) located in 2009 what we interpret to have been 4–9 and 4 pairs, respectively, and in 2010 what we interpret to have been 4 and 2 pairs, respectively. But in 2014 and 2015, only a single study area pair was located, in the Northwest Forest (Young, Buckley et al.). We are reluctant to read too much into these data beyond that towhee numbers in 2 of the 4 main forested areas of Van Cortlandt are noticeably lower than they were, say, 30 years ago. It appears that parkwide <5 pairs may still persist.

Even though almost ubiquitous in the New York City area on NYSBBA I in 1980–85, by NYSBBA II in 2000–2005 14 blocks had been lost in the most urbanized portions of the Bronx, Queens, and southern Nassau County, although still in study area Blocks 5852B and 5952A (McGowan and Corwin 2008). DOIBBS data identified a steepening statewide decline between 1970 and 1980 that continued at an increased annual rate of 3.5% to 2005 (Sauer et al. 2005).

No explanations for its reduced numbers have been proffered beyond that this is yet another forest interior ground-nester (like Eastern Whippoorwill, Black-and-white and Worm-eating Warblers, and Ovenbird) evidencing strong declines. Perhaps cowbirds play an important role. In the study area we are also inclined to blame domestic cats foraging from adjacent residences, feral cats, and possibly native quadruped predators like Eastern Coyotes, Raccoons, Striped Skunks, and even Virginia Opossums and Red and Gray Foxes.

In fall, first migrants arrive in late Aug, peak in mid–late Oct, and depart by mid-Nov, with extremes 3 on 28 Aug 2010 (Baksh) and 30 Nov 2013 (Baksh), with a maximum of 14 on 18 Oct 1959 (Young).

Irregular in winter in Van Cortlandt and Woodlawn, mostly singles but occasionally 2 into late Dec. Overwintering has never certainly occurred, but 2 on 11 Mar 1925 (Kuerzi) and singles there on 3 Mar 2012 (Baksh) and 6 Mar 2009 (Young) and in Woodlawn on 15 Feb 1958 (Horowitz) may have been successful.

Specimens
Dwight collected 3 in Van Cortlandt in Jun, Aug 1889–90 and Foster 1 in Woodlawn on 10 Oct 1887 (all AMNH). Bicknell collected 1 in Riverdale on 12 Oct 1878 (NYSM).

Comments
Lingering western blowback drift-migrant **Spotted Towhee**, *P. maculatus*, has occurred twice in the Bronx, both in the Bronx Zoo: an HY ♀ from 28 Nov–23 Dec 1958 (Peszell et al.; Buckley 1959) and an AHY ♂ on 1 Feb 1973 (Conway). The latter record has been overlooked by all subsequent writers. Unrecorded in Central or Prospect Park.

American Tree Sparrow *Spizelloides arborea*

NYC area, current
A spring/fall migrant and regular winter visitor and resident that has been found in reduced numbers throughout since the 1980s.

Bronx region, historical
Bicknell's *Riverdale* 1872–1901: abundant winter resident; latest 29 Apr

Eaton's *Bronx + Manhattan* 1910–14: abundant winter visitor

Griscom's *Bronx Region* as of 1922: common winter resident

Kuerzi's *Bronx Region* as of 1926: abundant winter resident (20 Oct–29 Apr)

NYC area, historical
Cruickshank's *NYC Region* as of 1941: common fall (maximum 300)/spring migrant, winter resident (16 Sep–8 May)

Bull's *NYC Area* as of 1968: late spring date 12 May; winter maximum 280

Bull's *New York State* as of 1975: status unchanged

Study area, historical and current
A spring/fall migrant and regular winter visitor and resident, which for reasons unknown has been found in greatly diminished numbers since the 1980s.

In fall, first migrants arrive in early Nov, peak in late Nov–early Dec, and depart by mid-Dec, with extremes of 16 Sep 1934 (Norse) and 30 Nov 1953 (Buckley) and 20–30 many times.

The winter population is usually established by mid-Dec, but there has always been noticeable interannual variation. An average study area midwinter daily count from the 1960s–80s would have ranged between 10 and 40 (Weber's 55 in Van Cortlandt on 11 Jan 1936 was considered high), with a maximum of 92 between Van Cortlandt and Woodlawn on 26 Dec 1960 (Tudor et al). Until 1967, the winter maximum in Prospect Park was 7 on 23 Feb 1945 (Soll). But since the 1980s the study area winter population has been steadily declining to the point at which by the 2010s it has come close to being unrecorded some years, the recent maximum being only 17 on 17 Feb 2008 (Klein). Indeed, 7 active observers working in Van Cortlandt

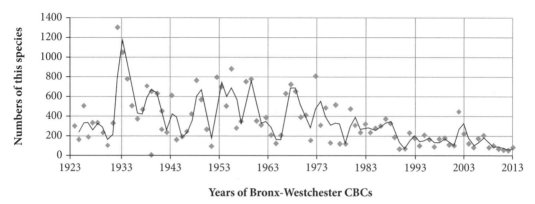

Figure 55

from 1 Dec 2012–28 Feb 2013 encountered only singles on 5 occasions and a flock of 3 but once (Young et al.) Suitable Tree Sparrow habitat appears largely unchanged apart from loss of favored cattails in Van Cortlandt Swamp, but the decline is also evident in the Bronx-Westchester Christmas Bird Count data (Fig. 55), which bottomed out at 26 in 2015 (that year not shown). It may be that this is only a regional example of global climate change manifested by milder winters and birds' shifting winter ranges.

In spring, most migrants pass through and residents depart during Mar, all but vanishing by mid-Apr, with extreme late dates of 6 in diminishing numbers from 16 Apr–1 May 1956 (Buckley et al.) and 1 with Chipping and Field Sparrows on 27 May 1934 (Weber, Jove)—very late, though there are Jun records from upstate New York. The spring maximum is 40 on 21 Mar 1954 (Phelan, Buckley).

Specimens
Dwight collected 8 in Feb, Apr, Dec 1884–89 in Van Cortlandt (AMNH), and Bicknell 5 in Mar, Apr, Nov 1876–79 in Riverdale (NYSM).

Chipping Sparrow   *Spizella passerina*

NYC area, current
A numerous spring/fall migrant, widespread breeder, and infrequent winter visitor.

Bronx region, historical
Bicknell's *Riverdale* 1872–1901: common breeder, spring/fall migrant

Eaton's *Bronx + Manhattan* 1910–14: abundant breeder

Griscom's *Bronx Region* as of 1922: common migrant, breeder, very rare in winter

Kuerzi's *Bronx Region* as of 1926: very common breeder, spring/fall migrant (21 Mar–15 Nov), 2 Dec–Jan records

NYC area, historical
Cruickshank's *NYC Region* as of 1941: common breeder, very common spring (maximum 84)/fall migrant, extremely rare in winter

Bull's *NYC Area* as of 1968: maxima spring 200, fall 250, winter 3

Bull's *New York State* as of 1975: status unchanged

Study area, historical and current
A numerous spring/fall migrant, breeder, and infrequent winter visitor.

In spring, first migrants arrive in late Mar, peak in late Apr, and depart by early May, with extremes of 3 Mar 1956 (singing: Buckley, Scully) and maxima of 25 in Woodlawn on 21 Apr 2012 (Rea) and 30+ on 29 May 1937 (Imhof et al.). Until Jun 1967, the spring maximum in Prospect Park was 200 on 2 May 1945 (Soll, Whelen).

The current study area breeding population is uncertain, but they are widespread in all 4 extant subareas. In late May 2014, 15 pairs were in Woodlawn alone, with another 10–15 in Van Cortlandt (Buckley).

Both NYSBBA I in 1980–85 and II in 2000–2005 showed a broad New York City area breeding distribution, where they were understandably absent only from the most urbanized portions of New York City boroughs except Staten Island (McGowan and Corwin 2008).

DOIBBS data showed a 0.8% annual statewide decline from 1966–2005 (Sauer et al. 2007).

In fall, the first migrants arrive in late Aug, peak in Oct, and most have departed by early Nov, with extremes of 3 on 3 Sep 1952 (Buckley) and 22 Nov 2014 (O'Reilly). Although no maxima higher than the 30 on 29 Sep 1935 (Weber, Jove) are at hand, daily counts of 40–50 have not been uncommon. Until Jun 1967, the fall maximum in Prospect Park was 125 in 9 Oct 1944 (Soll).

Chippings have always been singularly infrequent in winter, recorded in Van Cortlandt, Woodlawn, or Hillview once every 7–10 years with no proven overwinterings. One on 19 Jan 1929 (Cruickshank, Johnston) was the latest. Occasional small flocks occur, like the 21 on 23 Dec 2000 (Fiore, Veit, Buckley et al.).

Specimens
Dwight collected 5 in Van Cortlandt in Apr, May 1886–87 (AMNH) and Bicknell 2 in Riverdale in Oct 1876–78 (NYSM).

Clay-colored Sparrow *Spizella pallida*

NYC area, current
This Midwestern breeder was first recorded on Long Island in fall in the 1930s and was not found annually until the early 1950s. It has always been almost unrecorded away from the barrier beaches, but now that it has established a thin but increasing breeding population in upstate New York (first in 1971), it is being seen in New York City parks with greater frequency. Recorded in Central and Prospect Parks, but it has been largely overlooked in the Bronx, where it is known only from Van Cortlandt, Bronx, and Pelham Bay Parks.

NYC area, historical
Cruickshank's *NYC Region* as of 1941: Sep–Oct records 1933, 1934 first in New York City area

Bull's *NYC Area* as of 1968: rare but regular coastal fall migrant, 7 Sep–19 Nov; 2 spring records

Bull's *New York State* as of 1975: late fall date 30 Nov; another spring record; increasing as upstate breeder

Study area, historical and current
An increasing but exceedingly infrequent fall (only) migrant from the Midwest. There are only 3 Van Cortlandt records, mirroring the park's relatively low intensity of coverage: 9 Oct 1954 (Scully, Buckley); 30 Sep 1994 (Young); 16 Oct 2005 (Young, Garcia, J. Gillen), but if recent Bronx Park records are a guide, it should occur in the study area annually Sep–Oct.

Others were in Bronx Park on 4 Sep 1954 (Buckley) and Pelham Bay Park on 6 Oct 1956 (Scully, Buckley), 12 Oct 2014 (Benoit et al.), and 13 Sep 2015 (Benoit). One there from 21–27 Dec 2012 (Aracil et al.) is the sole Bronx winter record. In the last 10 years they have been seen regularly in Bronx Park on the following occasions: 15 Oct 2009, 17 Oct 2010 (Becker et al.); 28 Sep 2011 (Coco); 16–22 Oct 2011, 5–12 Oct 2012 (Coco, Becker); 29 Sep–15 Oct 2013 (Becker, Coco); 12 Oct 2014 (Becker). There may be additional Bronx Park records, but these are all the Bronx County records we know of, and there is but 1 Inwood Park record, 2 on 4 Nov 2012 (DiCostanzo). With this species' great expansion as a breeder in upstate New York, study area records will also only increase in proportion to observer searching—as has been happening in Bronx Park. No Bronx spring occurrences are known.

NYSBBA I in 1980–85 recorded breeding in 23 upstate blocks, increasing to 69 during NYSBBA II in 2000–2005. No sites have been closer than Sullivan and Ulster Counties (McGowan and Corwin 2008).

Field Sparrow   *Spizella pusilla*

NYC area, current
A formerly widespread spring/fall migrant, winter resident, and breeder, favoring shrubby oldfields. As this habitat vanished following succession and development, Field Sparrows have become scarcer and scarcer since the 1960s. Within New York City they now breed regularly only on Staten Island, at Floyd Bennett Field, and probably at JFK Airport, but they may also do so on former garbage dumps now closed to the public in the Bronx, Brooklyn, and Queens. They are no longer reliably located in winter flocks at accessible sites.

Bronx region, historical
Bicknell's *Riverdale* 1872–1901: common spring/fall migrant, uncommon winter; local breeder

Hix's *NYC* in 1904: common breeder Bronx, Van Cortlandt Parks

Eaton's *Bronx + Manhattan* 1910–14: abundant breeder

Griscom's *Bronx Region* as of 1922: common migrant and breeder, regular, sometimes common in winter

Kuerzi's *Bronx Region* as of 1926: common breeder, overwintering regularly, sometimes numerous

NYC area, historical
Cruickshank's *NYC Region* as of 1941: common breeder except in crowded metropolitan area [locations not given], common spring (maximum 50)/fall migrant, extremely rare to irregular in winter

Bull's *NYC Area* as of 1968: maxima fall 55, winter 115

Bull's *New York State* as of 1975: status unchanged

Levine's *New York State* as of 1996: winter maximum 150 on 1961 Bronx-Westchester Christmas Bird Count

Study area, historical and current
A spring/fall migrant, extirpated breeder, and perhaps decreasing winter resident/visitor.

In fall, first migrants arrive in late Sep, peak in Oct with frequent daily counts of 20–30, and depart by early Nov, with extremes of 12 Aug 1936 (Norse) and 22 Nov 1994 (Young), and a maximum of 38 on 18 Oct 1955 (Buckley). Until

Jun 1967, the fall maximum in Prospect Park was 15 on 12 Oct 1944 (W. Ferguson).

Erratic in winter—absent some years, numerous in others—but in general less abundant after the 1960s, as indicated in the Christmas Bird Count data (Fig. 56). Maxima include 41 across Van Cortlandt and Woodlawn on 26 Dec 1960 (Tudor et al.) and 50 on 15 Feb 1954 (Buckley).

In spring, first migrants arrive in mid-Mar, peak in late Apr, and depart by early May, with extremes of 20 Mar 2006 (Young) and 12 May

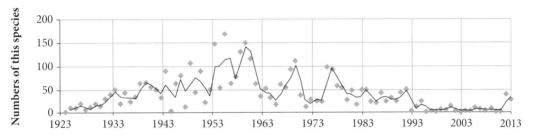

Figure 56

1983 (DiCostanzo) and a maximum of 50 on 28 Mar 1956 (Kane).

In the late 1800s, it was a common breeder in Riverdale and Jerome Meadows (Bicknell) and in Van Cortlandt (Dwight), occasionally at Woodlawn, Jerome, and Hillview, continuing to the mid-1960s (Kieran, Norse, Buckley, Heath). None have bred in the study area since about 1975, disappearing when succession and brush fire suppression finally eliminated their last patch of open, weedy habitat on Vault Hill. A singing ♂ for 3 weeks there in May–Jun 1981 (Turner) may have been the species' last breeding attempt. They were breeding at Pelham Bay Park in the 1980s (Künstler), but 1 along the Hudson Division track bed in Riverdale on 17 Jun 2002 (Stoechle) was unlikely to have been nesting.

NYSBBA I in 1980–85 indicated a broad New York City area breeding distribution, where they were understandably absent only from the most urbanized portions of all New York City boroughs except Staten Island and in southern Nassau County. By NYSBBA II in 2000–2005 they were reduced to single blocks in southern Westchester County, Brooklyn, and Queens and only 5 in Nassau County, and even Suffolk County lost almost 40 blocks (McGowan and Corwin 2008). DOIBBS data showed an annual statewide decline of 3.9% from 1966–2005 (Sauer et al. 2005).

Specimens
Dwight collected 12 in Van Cortlandt in Apr, Jun, Jul, Sep, Dec 1885–96 (AMNH), and Bicknell 4 in Riverdale in Jan, Oct, Nov 1878–79 (NYSM).

Vesper Sparrow *Pooecetes gramineus*

NYC area, current
Formerly a widespread and relatively common spring/fall migrant, local breeder, and uncommon winter visitor/resident. Now extirpated as a breeder except at a handful of outlying sites, all but unknown in winter, and an annual but

very scarce spring/fall migrant. It is a New York State Species of Special Concern.

Bronx region, historical
Bicknell's *Riverdale* 1872–1901: bred Jerome Meadows; common spring/fall migrant at Kingsbridge Meadows

Eaton's *Bronx + Manhattan* 1910–14: common breeder

Griscom's *Bronx Region* as of 1922: common migrant, now very rare breeder, winter

Kuerzi's *Bronx Region* as of 1926: common migrant, uncommon breeder [locations not given], rare winter (6 records)

NYC area, historical
Cruickshank's *NYC Region* as of 1941: diminishing breeder now found from Queens through Suffolk, Rockland, northern Westchester Counties but decreasing abruptly south of White Plains, long extirpated from Bronx; otherwise common spring (maximum 53)/fall migrant, very rare coastally in winter

Bull's *NYC Area* as of 1968: maxima fall 50, winter 35; now breeding only locally on extreme eastern Long Island

Bull's *New York State* as of 1975: status unchanged

Study area, historical and current
A spring/fall migrant in greatly reduced numbers since the 1960s and a formerly uncommon but now unrecorded winter visitor. It was most often seen on the Parade Ground (where it still occurs) but also at Hillview, Jerome, and Woodlawn. Bicknell recorded it annually in spring/fall at Kingsbridge Meadows, collecting 6 in Apr, Oct, Nov 1875–78 (NYSM).

In spring, the first migrants arrive in late Mar, peak in Apr, and depart by early May, with extreme dates of 30 (the maximum) on 1 Mar 1953 (Buckley) and 18 May 1957 (Weintraub). As of 2016 it is nearly unknown in spring, the only records >1960 being singles on 20 Apr 1994 and 18 Apr 2005 (both Young). Until Jun 1967, the spring maximum in Prospect Park was 7 on 18 Apr 1953 (Usin, Whelen).

Bicknell noted Vespers breeding (no years or numbers given) 2 smi (3 km) from the Hudson River and southeast of Van Cortlandt—that is, on Jerome Meadows. Eaton (1910, 1914) listed it as a breeder in his Bronx + Manhattan county with no additional details. It is possible they were breeding in the East Bronx and on Rikers Island through the 1950s, but the only firm territorial pair was in Pelham Bay Park in May–Jun 1955 (Young 1958).

NYSBBA I in 1980–85 and II in 2000–2005 found Vespers no closer to New York City than scattered single blocks in Orange and Dutchess Counties, except for a relict cluster of 19 in Suffolk County. NYSBBA I also noted breeding in 1097 upstate blocks, but by NYSBBA II in 2000–2005 occupied blocks upstate had dropped to 555, plus 7 in Suffolk County (McGowan and Corwin 2008). DOIBBS data quantified a strong statewide decline of 7.9% annually from 1966–2005, to the point where they have now almost disappeared from New York State survey routes (Sauer et al. 2005). As of 2016, the nearest breeders are no closer than Ulster, Dutchess, and eastern Suffolk Counties.

In fall, first migrants arrive in early Oct, peak in late Oct, and depart by early Nov, with extreme dates of 1 Sep 1937 (Norse) and 23 Nov 1930 (Cruickshank), and maxima of 24 on 27 Oct 1957 (Rafferty) and 30+ in Woodlawn from 9–16 Oct 1960 (Horowitz, Buckley). The highest recent fall counts are 7 on 11 Nov 1995 (Lyons, Garcia) and 6 on 26 Oct 2013 (Young et al.).

Vespers did not normally overwinter in the study area, but singles could appear for a few days at any time; the maximum was 2 at Hillview on 11 Jan 1936 (Weber). The last winter record was on the Parade Ground on 15 Jan 1962 (Norse). Always very rare on the Bronx-Westchester Christmas Bird Count (Fig. 57), numbers briefly peaked in the 1970s when winter conditions at Baxter Creek were optimal, supporting 62 on 27 Dec 1970 (Kane et al.). All too quickly numbers collapsed to zero.

Specimens
Bicknell collected 5 at Kingsbridge Meadows in Apr, Oct, Nov 1875–78 (USNM).

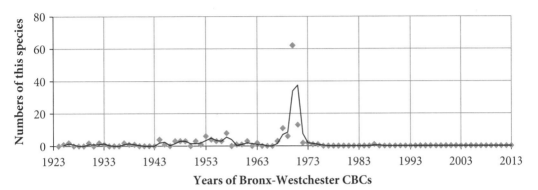

Figure 57

Lark Sparrow *Chondestes grammacus*

NYC area, current
Especially on Long Island barrier beaches, this species is an annual but rare Aug–Sep migrant (1–3 per day) but is almost unknown in winter and spring. Although not often, it has occurred in Central and Prospect Parks.

Bronx region, historical
Bicknell's *Riverdale* 1872–1901: unrecorded

Eaton's *Bronx + Manhattan* 1910–14: unrecorded

Griscom's *Bronx Region* as of 1922: unrecorded

Kuerzi's *Bronx Region* as of 1926: unrecorded

NYC area, historical
Cruickshank's *NYC Region* as of 1941: very rare on Long Island (32 mostly fall records), almost unknown in spring; the single Bronx spring record among only 4 in New York City area

Bull's *NYC Area* as of 1968: rare but annual fall migrant (maximum 3); 7 Jul–27 Nov, plus 11 Apr–12 Jun

Bull's *New York State* as of 1975: several more Dec–Jan

Study area, historical and current
In the Bronx a very rare early fall migrant, almost unrecorded in spring, when the only 2 occurrences include the study area's sole record, at the base of Vault Hill adjacent to the Van Cortlandt Parade Ground on the curious date of 15 Jun 1963 (Rafferty), with another singing in adjacent Riverdale on 11 Apr 1939 (Griscom). These are the only Bronx spring occurrences known. More regular coverage of the Parade Ground would add more of the expected Aug–Sep records, crowds permitting. Until Jun 1967, all in Central and Prospect Parks were singles save 2 on 21 Sep 1954 in Prospect (Carleton).

Scarce away from the South Shore of Long Island, in the Bronx it has been found in fall at Baxter Creek (5 records between 1 Sep and

12 Nov from 1955–72: Norse, Kane, Sedwitz, Buckley), Pelham Bay Park, the Bronx Zoo (3 Nov 1984: Krauss), and Bronx Park (21 Sep–6 Oct 2013: Becker et al.). Closer to the study area, another was at Inwood Park on 8 Nov 1963 (Norse).

Comments
High Plains vagrant **Lark Bunting**, *Calamospiza melanocorys*, in 2016 occurs as a blowback drift-migrant in the New York City area once a decade and has been recorded in the Bronx once: at Baxter Creek from 17–31 Oct 1971 (Kane, Sedwitz et al.). It has also occurred at Croton Point on 27 May 1973 (Howe) and in Rockland County at a New City feeder on 9 Oct 1977 (Conner) and at Piermont Pier on 11 Nov 1977 (Wilfred). Unknown in Central or Prospect Park.

Savannah Sparrow *Passerculus (sandwichensis) sandwichensis*

NYC area, current
A common spring/fall migrant, a very local winter resident, and a disappearing breeder. It may be reasonably common on airports and closed garbage dumps where access is normally interdicted.

Bronx region, historical
Bicknell's *Riverdale* 1872–1901: common spring/fall migrant, winter resident, especially at Kingsbridge Meadows; possible breeder

Eaton's *Bronx + Manhattan* 1910–14: abundant migrant, rare breeder

Griscom's *Bronx Region* as of 1922: common migrant, rare in winter on coast

Kuerzi's *Bronx Region* as of 1926: common migrant, several breeding pairs [locations not given], occasional in winter (6 Mar–mid-Nov)

NYC area, historical
Cruickshank's *NYC Region* as of 1941: common to abundant spring/fall (maximum several hundred) migrant; few pairs have nested on Bronx, Westchester County Long Island Sound shores [locations not given]; in winter locally common coastally

Bull's *NYC Area* as of 1968: Long Island maxima fall 1500–2000, winter 120–210; 100 pairs breeding on fill Jamaica Bay Wildlife Refuge 1953

Bull's *New York State* as of 1975: status unchanged

Study area, historical and current
A spring/fall migrant and occasional winter visitor, especially on the Parade Ground; a likely past and potential future breeder.

In spring, first migrants arrive in late Mar, peak in mid-Apr, and depart by mid-May, with extreme dates of 13 Mar 1954 (Norse) and 28 May 1997 (Young), and a maximum of 100+ on 8 May 1955 (Kane). Until Jun 1967, the spring maxima in Central and Prospect Parks were, respectively, 35 on 9 May 1958 (Post) and 6 on 27 Apr 1945 (Soll).

Bicknell described Kingsbridge Meadows as the best local site for this species (collecting 6 there in Apr–May and Sep–Oct 1876–80: NYSM) but did not specifically say whether they were breeding. Eaton (1910, 1914) listed it without details as a breeder in his Bronx + Manhattan county. However, several pairs were breeding on the Baychester Marshes in the mid-1920s (Kuerzi 1927), the same ones that Cruickshank (1942) reported breeding in the Bronx but only on Long Island Sound. Despite drastic declines in most grassland birds, Savannah Sparrow has increased as a breeder in the New York City area since the 1950s, particularly in the East Bronx. Until 1955, 5 territorial ♂♂ were at Baxter Creek and some remained there until the late 1990s, and 7, including fledglings, were on the new golf course there on 23 Jun 2016 (Sime). Advertising ♂♂ have also been on the Pelham Bay Park landfill in summer in recent years (Künstler), so these 2 sites would now hold the closest breeders to the study area. In the final analysis, we are uncertain whether Savannahs ever bred within the study area, but they probably did at Kingsbridge Meadows and Jerome Meadows in the late 1880s. Given that meadowlarks were nesting on Hillview in the late 1930s, Savannahs may also have been there. At the moment it is a potential breeder in Van Cortlandt, Woodlawn, and at Hillview, assuming reasonably undisturbed grassy areas throughout the breeding season.

NYSBBA I in 1980–85 found breeders widely but locally distributed across Orange and Putnam Counties, but southeastward in only 2 Rockland and 3 Westchester County blocks, 1 in the Bronx, 3 in Queens, 2 on Staten Island, none in Brooklyn, a dozen in Nassau County, and 3 dozen in Suffolk County. By NYSBBA II in 2000–2005, Rockland County, Westchester County, the Bronx, and Brooklyn were unchanged, but Staten Island was up to 4, Nassau County down to 4, and Suffolk County down to about 25. A probable pair was on the Pelham Bay Park landfill on 20 Jun 2001 (Künstler). DOIBBS data showed an annual statewide decrease of 2.6% from 1966–2005 (Sauer et al. 2005). The largest New York City area breeding population (200 pairs) is now on Staten Island's Fresh Kills landfill (Veit, Wollney et al.).

In fall, first migrants arrive in mid-Sep, peak in mid-Oct, and depart by mid-Nov, with extreme dates of 10 Aug 1953 (Buckley) and 4 on 19 Nov 2011 (Baksh), and a maximum of 50 on 23 Oct 2010 (Scully). The 500+ at Baxter Creek on 26 Sep 1954 (Buckley, Sedwitz) is the Bronx maximum.

Savannahs do not normally overwinter in the study area, but individuals and small flocks to a dozen or more may appear at any time on the Parade Ground and at Jerome and Hillview. The maximum is 24 on 11 Dec 1946 (Kieran).

Savannahs have had a checkered history on the Bronx-Westchester Christmas Bird Count (Fig. 58) that may have indexed availability of their preferred winter habitat: large open undisturbed weedy areas. This was far more prevalent through the 1930s–40s than in the 1980s–90s, and count numbers matched this trend. But their numerical peak from about 1954–84 did not necessarily track this habitat so may have been determined by breeding productivity elsewhere.

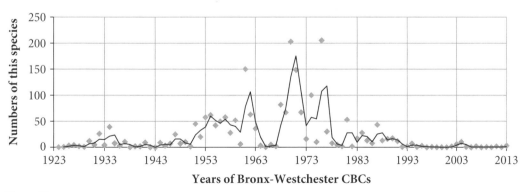

Figure 58

Specimens
Bicknell collected 6 at Kingsbridge Meadows in Apr, May, Sep, Oct 1876–80 (NYSM).

Comments
Closely related **Ipswich Sparrow**, *P. (sandwichensis) princeps*, a bird of seashore dunes, has not yet been recorded in Van Cortlandt, Central, or Prospect Parks. Formerly locally regular in the Bronx on Long Island Sound shores, especially Rodman's Neck in Pelham Bay Park, but may now occur only on Hart's Island and perhaps the new Ferry Point golf course. It has been reported along the Hudson at Croton Point on 12 Apr 1997 (unknown observer, *Kingbird*), from 26 Dec 2012–12 Jan 2013 (Bardwell et al.), and on 8 Mar 2017 (Heemstra; photos), as well as in *inland* Westchester County at Ward Pound Ridge Reservation on 2 Nov 2003 (Walter; photos). It thus becomes another Parade Ground migrant to search for in Nov.

Grasshopper Sparrow   *Ammodramus savannarum*

NYC area, current
Formerly a widespread clustered breeder now in severely reduced numbers at very few grassland sites. Now an annual but very scarce spring and uncommon fall migrant. It is a New York State Species of Special Concern.

Bronx region, historical
Bicknell's *Riverdale* 1872–1901: spring/fall migrant at Kingsbridge Meadows; numerous nesting colonies (see below)

Eaton's *Bronx + Manhattan* 1910–14: common breeder

Griscom's *Riverdale* as of 1926: all colonies long since extirpated

Kuerzi's *Bronx Region* as of 1926: formerly common breeder, few still breeding [locations not given]; 17 Apr–29 Oct

NYC area, historical
Cruickshank's *NYC Region* as of 1941: rare or overlooked spring/fall migrant (9 Apr–5 Nov); still breeds Queens, Staten Island, but gone from Bronx, southern Westchester County

Bull's *NYC Area* as of 1968: 23 Mar–23 Nov; maxima spring 8, fall 8; several recent Dec records; declining as breeder around New York City

Bull's *New York State* as of 1975: increasing Dec–Jan records

Study area, historical and current
An extirpated breeder and irregular, perhaps overlooked, spring/fall migrant. Nearly all records are of singles from Van Cortlandt. There are only 3 records after 1962, reflecting this species' breeding decline throughout the Northeast and New York State.

There are 15 study area spring records of migrants, all but 2 singles, a function of relatively low intensity of coverage: several seen on the Kingsbridge Meadows on 17 Apr 1886 (Bicknell); specimens there on 25 May 1876, 14 May 1877, and 10 May 1879 (Bicknell); specimens from Van Cortlandt on 13 May 1887 and 2 on 12 May 1890 (Dwight); seen on 10 May 1930 (Cruickshank); 11 May 1952 (Kane, Post, Komorowski); 15–19 May 1952 (Buckley et al.); 24–27 Apr 1953 (Cantor, Scully, Buckley); 2 May 1953 (Norse, Cantor); 2 on 14 May 1953 (Buckley, Phelan, Scully); 13–19 May 1954 (Kane, Buckley et al.); 15 May 1955 (Sedwitz); 2 on 10 May 2014 (Fiore et al.).

Grasshopper Sparrows bred in numerous nesting colonies near Van Cortlandt in 1887 (Dwight in Griscom 1923) and in other years unstated (Bicknell in Griscom 1927), the precise locations having been Jerome Meadows and Kingsbridge Meadows. Eaton (1910, 1914) considered it (vaguely) breeding in the West Bronx and on Staten Island. The last certain Bronx breeders were 3 pairs in Pelham Bay Park in 1955 (Young 1958) and at least 2 pairs at Baxter Creek in 1973 (Pappalardi). That area was finally destroyed in the 2000s for yet another New York City golf course, but future opportunistic nesting there and on the Pelham Bay Park landfill should be pursued.

NYSBBA I in 1980–85 noted probable/confirmed breeding in 495 blocks clustered along the PA border, down to 287 by NYSBBA II in 2000–2005, all upstate save 5 in Westchester and Orange Counties, plus a group of <20 blocks in eastern Suffolk County (McGowan and Corwin 2008). DOIBBS data determined a strong statewide decline of 9.4% annually from 1966–2005, to the point where they have almost disappeared from many New York State survey routes (Sauer et al. 2005). Today the nearest breeders are in north-central Westchester County and on the former Fresh Kills landfill on Staten Island, where there were 40 pairs in 2015 (Veit et al.).

There are only 8 fall records: Bicknell collected 1 at Kingsbridge Meadows on 12 Oct 1876; 2 were seen at Jerome on 22 Oct 1934 (Cantor); 2 on 3 Oct 1953 (Kane, Buckley); 23 Sep 1954 (Scully, Buckley); 26 Nov 1954 (Buckley, a blowback drift-migrant?); 2 Oct 1962 (Sedwitz); 15 Aug 1993 (Lyons)—the latter perhaps a postbreeding disperser from some nearby colony—and Hillview on 6 Oct 2015 (unknown observer, *Kingbird*). None have yet been detected in winter in the study area, but a likely blowback drift-migrant was at Pelham Bay Park on 26 Dec 1988 (Cech).

Specimens
Dwight collected 3 in Van Cortlandt in May 1887–90 (AMNH) and Bicknell 4 at Kingsbridge Meadows in May, Oct 1876–79 (NYSM).

Henslow's Sparrow *Ammodramus henslowii*

NYC area, current
Formerly an infrequently encountered, secretive spring/fall migrant and exceedingly local breeder but barely recorded since the early 1960s, coincident with its disappearance as a breeder in the Northeast for uncertain reasons. Recorded in Central and Prospect Parks but only twice since 1963 (in Central). It is a New York State Threatened Species.

Bronx region, historical
Bicknell's *Riverdale* 1872–1901: recorded thrice in Oct at Kingsbridge Meadows

Eaton's *Bronx + Manhattan* 1910–14: rare

Griscom's *Bronx Region* as of 1922: rare spring migrant, 2 recent records

Kuerzi's *Bronx Region* as of 1926: uncommon/rare spring (4 Apr–20 May)/fall (26 Sep–28 Oct) migrant

NYC area, historical
Cruickshank's *NYC Region* as of 1941: increasing breeder Rockland, Westchester Counties, rare but persisting on Long Island's South Shore from Jamaica Bay to Speonk; otherwise only rarest spring/fall migrant; 4 Apr–20 Nov

Bull's NYC Area as of 1968: > about 1952 breeders probably extirpated from Long Island, New York City, Westchester County; late fall date 27 Nov, plus 1 in Dec–Jan [blowback drift-migrants]

Bull's *New York State* as of 1975: late Long Island date 1 Dec [blowback drift-migrant]

Study area, historical and current
In the study area, this is now the rarest of all regularly occurring sparrows. It has been almost unrecorded since the late 1930s, with 15 records, all but 2 singles, and only 2 since 1955: in 1977 and 1990. It has never been recorded breeding in the Bronx.

Only 4 spring records: Jerome on 13 May 1926 (Kuerzi); Jerome on 24 Apr 1930 (Cruickshank); 19 May 1946 (Ephraim); 19 May 1953 (Buckley).

NYSBBA I in 1980–85 noted breeding in 344 upstate blocks, but by NYSBBA II in 2000–2005 occupied blocks had dropped to 70 upstate (McGowan and Corwin 2008). DOIBBS data affirmed a strong statewide decline from 1966–2005, to the point where they have now disappeared from survey routes (Sauer et al. 2005). The nearest breeders may now be in an isolated block in Ulster County or in northeastern PA.

Known fall records: specimens from Kingsbridge Meadows on 8 and 16 Oct 1880, 6 Oct 1881 (Bicknell); 2 seen in Van Cortlandt on 3 Oct 1926, another on 10 Oct (Kuerzi); at Jerome on 7 Oct 1935 (Weber); 3 in Van Cortlandt on 11–12 Oct 1935 (Weber, Norse, Cantor); 3 Oct 1936 (Norse); 3 Oct 1937 (Norse); 28 Sep 1954 (Buckley); 24 Oct 1954 (Buckley); at Jerome on 28 Oct 1955 (Buckley), 6 Nov 1977 (Sedwitz); in Van Cortlandt on 22 Sep 1990 (Young). The above multiple counts of migrants are among the very few known from the New York area, although they had to have been more frequent when the species was a regular breeder and far more numerous throughout the Northeast. One at nearby Inwood Park on 11 Oct 1965 (Norse) was also among the very last in the general area.

Through the 1930s, '40s, and early '50s, Henslow's Sparrows were being seen occasionally in the Bronx by active observers, but the last cluster of records involved the 3 singles seen in the spectacular sunflower field at Baxter Creek in Oct 1954 (Buckley 1958). Central Park has had its only 2 recently: on 4 Oct 1992 (Baumann, Baumann et al.) and on 31 Oct 2006 (Spitalnik et al.), and while Prospect Park has had 6, its last was on 10 Oct 1963 (Raymond). If as seems likely they were breeding, the 2 singing ♂♂ from May–Jul 1953 in weedy fields along Pennsylvania Ave. north of the Belt Parkway in Brooklyn (Kane, Buckley et al.) would have been among the very last on Long Island.

Specimens
Bicknell collected 3 at Kingsbridge Meadows in Oct 1880–81 (NYSM, MCZ).

Le Conte's Sparrow *Ammodramus leconteii*

NYC area, current
A very recent fall and occasional winter vagrant statewide, and a territorial ♂ oversummered along the St Lawrence River in 1995. Since about 2005 nearly annual around New York City but singles only. Most records are from coastal Long Island but also in Central and several other New York City Parks, including once in spring at Battery Park. Recorded in Dutchess and Sullivan Counties but not Rockland or Westchester County.

NYC area, historical
Bull's *New York State* as of 1975: except for 1895 Montezuma specimen, unrecorded until netted Tobay Pond in Oct 1970, followed by 2 more early 1970s upstate records

Levine's *New York State* as of 1996: increased since 1970 record, perhaps 20+ now

Study area, historical and current
Only a single record, on the Parade Ground on 13 Oct 1981 (Sedwitz). Typical of many Le Conte's Sparrows, this golden-headed gem allowed approach to within 10 ft (3 m), merely crouching when the accompanying Savannahs flushed at the observer's approach. This was the first Bronx record, with the second not until Pelham Bay Park on 28 Oct 2017 (Reisfeld et al.; photos). It may be regular there and at other East Bronx fresh- and saltmarshes during Oct, its peak fall migration period around New York City.

Nelson's Sparrow   *Ammodramus nelsoni*

NYC area, current
An uncommon and often overlooked spring/fall migrant in saltmarshes on Long Island but also inland in freshwater marshes, and which rarely overwinters (*nelsoni/alterus*). It is also a more numerous, regular spring/fall migrant in saltmarshes that sometimes, even regularly, overwinters, especially on Long Island (*subvirgatus*). When searched for at the right times of the year, both *nelsoni/alterus* and *subvirgatus* are regular in migration. Recorded in Central and Prospect Parks.

Bronx region, historical
Bicknell's *Riverdale* 1872–1901: unrecorded

Eaton's *Bronx + Manhattan* 1910–14: unrecorded; [nominate *nelsoni*] occurs every autumn in Hudson Valley, especially Croton River marshes 25 Sep–10 Oct; also specimens of *subvirgatus* there

Griscom's *Bronx Region* as of 1922: unrecorded except at Croton River marshes

Kuerzi's *Bronx Region* as of 1926: unmentioned

NYC area, historical
Cruickshank's *NYC Region* as of 1941: *alterus* probably regular fall migrant coastally, inland; *nelsoni* unknown from New York City area specimens

Bull's *NYC Area* as of 1968: from specimens, *nelsoni*, *alterus*, *subvirgatus* occur as migrants only, never in winter; all observations deprecated as unidentifiable

Bull's *New York State* as of 1975: status unchanged

Levine's *New York State* as of 1996: detailed update of all subspecies' statuses

Study area, historical and current
Almost certainly an overlooked, nearly annual spring/fall migrant. Only *nelsoni/alterus* (the prairie-breeding, so-called Freshmarsh Sparrow) is expected. It is a Sep–Oct and late May–early Jun migrant in small numbers (occasional in winter) in Pelham Bay Park, Randall's Island, and other East/South Bronx saltmarshes, and formerly at ephemeral sites like the Baxter Creek landfill.

There are 3 *nelsoni/alterus* Van Cortlandt Swamp records: on 3 Oct 1934 (Norse, Cantor), 19 Nov 1965 (Norse), and Tibbett's Marsh in May (date misplaced) 1998 (Lyons). The first 2 predate the split of Nelson's from Saltmarsh, but all 3 were clearly Nelson's. Judging from randomly collected museum specimens, the two subspecies may be almost equally likely as migrants. The 3 study area records fit well into the normal pattern of dates for *nelsoni/alterus*. Judging by recent records from saltmarshes fringing nearby Randall's Island, we conclude that Nelson's Sparrow should occur regularly in fall in Van Cortlandt Swamp. There are just too few observers there to see them these days. One in Bronx Park on 17 Oct 2015 (Becker) was right on time if not in quite the right place.

Maritime coast-breeding *subvirgatus* (the so-called Acadian Sparrow) also occurs with the same pattern in East Bronx saltmarshes but to our knowledge has never been detected out of that habitat in any *inland* New York City park. A. K. Fisher collected 3 *subvirgatus* at the base of Croton Point between 29 Sep–7 Oct 1880–86, perhaps southbound migrants from a small but now expanding breeding population along the St Lawrence River. See the map of all subspecies' ranges in Dunn and Alderfer 2017: 574.

Comments
For many decades the former Sharp-tailed Sparrow subspecies were carefully but routinely separated in the field around New York City, although observers sensibly pooled *alterus* and *nelsoni* (Peterson 1947: 271–72). *Subvirgatus*, on the other hand, was readily separated from them and from *caudacutus*, and the collective statuses of the 3 groups (*caudacutus*, *subvirgatus*, *nelsoni/alterus*) in the New York City area were reasonably well understood by the time of Cruickshank (1942). Attention to subspecies flagged after the deprecation of all as unidentifiable in the field in Bull (1964). But following the 1995 split of Sharp-tailed Sparrow into Saltmarsh (*caudacutus*) and Nelson's (*nelsoni, alterus, subvirgatus*), observers became reinterested in their field identification, and Sibley (2000, 2014) and Dunn and Alderfer (2011) took pains to reacquaint them with that process. Even field separation of *nelsoni* and *alterus* advanced, with the publication of a useful series of photographs of all Saltmarsh and Freshmarsh Sparrow subspecies by F. M. Smith (2011), although it is inadvisable to attempt it without a series of high-quality photographs of candidates. Nevertheless, separating the two main Nelson's groups from each other is usually relatively straightforward, and every time it is not done, important information on these quite different birds is lost—damaging, should *subvirgatus* be split from *nelsoni/alterus* in the future.

<u>Saltmarsh Sparrow</u>   *Ammodramus caudacutus*

NYC area, current
This nearly obligate saltmarsh bird is a relatively common spring/fall migrant, a local, loosely colonial breeder, and a rare winter resident. It has been recorded in Central and Prospect Parks.

Bronx region, historical
Bicknell's *Riverdale* 1872–1901: common breeder at Kingsbridge Meadows until 1882

Eaton's *Bronx + Manhattan* 1910–14: common breeder, also to Piermont Marshes, occasionally Newburgh

Griscom's *Bronx Region* as of 1922: occasionally breeding Baychester marshes, no winter records

Kuerzi's *Bronx Region* as of 1926: locally common breeder [locations not given], 10 Apr–14 Nov, once 26 Dec

NYC area, historical
Cruickshank's *NYC Region* as of 1941: common breeder on Long Island where saltmarshes remain, less commonly in East Bronx, Westchester County; few in winter, normally only Long Island

Bull's *NYC Area* as of 1968: formerly (only?) bred Piermont Marshes; maxima fall 300; 1000 after 1938 hurricane, including unknown but small numbers of Nelson's Sparrows

Bull's *New York State* as of 1975: status unchanged

Study area, historical and current
Bred through at least 1882 at Kingsbridge Meadows—where Bicknell collected 5 specimens between 1876 and 1879 (NYSM, MCZ)—as did Clapper Rail until 1924. It may have bred only until the 1930s in Piermont Marsh on the Hudson River, but careful breeding season work there seems not to have been done for many decades, so it may again do so. One was seen there on 24 May 1973 (Amos) but the subspecies of the then Sharp-tailed Sparrow was not given, so it may have been a migrant *nelsoni/alterus* Nelson's, particularly on that date. One at Croton Point on 29 Aug 1975 (Howe) was more likely to have been a locally breeding Saltmarsh Sparrow. It does nest as close to the study area as 3.7 smi (6 km) east-southeast in the Hutchinson River marshes at Pelham Bay Park, where an impressive 121+ pairs were breeding in 1955 before the Baychester marshes were destroyed (C. Young et al.). But apart from the extirpated Kingsbridge Meadows breeders, there are only 3 other study area records, singles in fall from Van Cortlandt Swamp on 20 Sep 1952 (Post, Messing) and 27 Sep 1997 (Lyons, Abramson), plus one in spring at Jerome on 11 May 1975 (Sedwitz).

Specimens
Bicknell collected 5 local breeders at Kingsbridge Meadows in Jul, Aug 1876–79 (NYSM, MCZ).

Comments
In 1995, Saltmarsh Sparrow was finally split from Nelson's Sparrow, but fortunately the above study area individuals were seen well and clearly identified as this species. This has not necessarily been true elsewhere, particularly at the Piermont Marshes. It is important to remember that through the late 1950s, the various migrant-only subspecies of then Sharp-tailed Sparrow were routinely studied and carefully identified in the field in the New York City area.

Seaside Sparrow   *Ammodramus maritimus*

NYC area, current
This obligate saltmarsh bird is a relatively common spring/fall migrant, a local, loosely colonial breeder, and a rare winter resident. It has been recorded in Central and Prospect Parks and is a New York State Species of Special Concern.

Bronx region, historical
Bicknell's *Riverdale* 1872–1901: nested Harlem River saltmarshes at Dyckman St. until [at least] 1881

Eaton's *Bronx + Manhattan* 1910–14: common Long Island breeder; also on lower Hudson to Piermont Marshes

Griscom's *Bronx Region* as of 1922: breeds Baychester marshes; no winter records

Kuerzi's *Bronx Region* as of 1926: fairly common breeder [locations not given]; 4 Apr–Dec

NYC area, historical
Cruickshank's *NYC Region* as of 1941: common breeder in Long Island saltmarshes, less commonly in East Bronx, Westchester

County; few in winter, normally only on Long Island

Bull's *NYC Area* as of 1968: status unchanged

Bull's *New York State* as of 1975: Jun 1973 at Piermont Marshes; breeding?

Study area, historical and current
Bred along the Harlem River in isolated patches of saltmarsh up to Dyckman St. until at least 1881 (Bicknell) and probably farther along Spuyten Duyvil Creek to the mouth of Tibbett's Brook—just as did Saltmarsh Sparrow and Clapper Rail. If Bicknell could overlook breeding Clapper Rails, he could have missed a few Seaside Sparrows.

They also bred in the Piermont Marshes until the late 1800s, but careful breeding season work there appears not to have been done for many decades, so they could still breed there. A pair seen there 15 Jun 1973 (Amos) argues that they do; a single at Croton Point 10 May 2015 (Roberto et al.) makes one wonder where else along the Hudson they might also be. They do nest as close as 3.7 smi (6 km) east-southeast of the study area in the Hutchinson River marshes (the remains of the former Baychester Marshes) in Pelham Bay Park, perhaps the source of some of Van Cortlandt Swamp's 4 birds: 30 Apr 1939 (Norse, Staloff); singing on 6 Aug 1954 (Phelan, Buckley); 3 Jun 1962 (Heath); 8 Oct 1997 (Garcia).

Red Fox Sparrow *Passerella iliaca iliaca*

NYC area, current
A fall and spring migrant from the Canadian taiga and a very local winter resident in small numbers.

Bronx region, historical
Bicknell's *Riverdale* 1872–1901: regular migrant in spring/fall (as early as 1 Oct 1880); winter resident

Eaton's *Bronx + Manhattan* 1910–14: common migrant, occasional winter visitor

Griscom's *Bronx Region* as of 1922: common migrant, occasional in winter

Kuerzi's *Bronx Region* as of 1926: common, occasionally abundant spring (6 Feb–30 Apr)/fall (1 Oct–2 Jan) migrant, frequently overwintering

NYC area, historical
Cruickshank's *NYC Region* as of 1941: common spring (maximum 200)/fall (maximum 100) migrant; 9 Aug–11 May; rare, local in winter

Bull's *NYC Area* as of 1968: winter maximum 300 on pooled Nassau/Suffolk County Christmas Bird Counts

Bull's *New York State* as of 1975: fall maximum 150

Study area, historical and current
An uncommon fall and spring migrant and very local winter resident. Most often seen in Woodlawn, especially in winter.

In fall, first migrants arrive in late Oct, peak in early and depart by late Nov, with extreme dates of 1 singing on 30 Sep 1928 (Kuerzi) and 30 Nov 1954 (Buckley), and a maximum of 7 on 23 Nov 2005 (Young).

Present in small numbers some winters, absent in others, and most reliable in Woodlawn, with usual daily counts there of 1–3 and maxima of 30 on 21 Dec and 50+ on 7 Dec 1963 (Horowitz, Tudor)—the latter possibly still migrating.

In spring, first migrants, often singing, arrive in early Mar, peak in late Mar, and depart by mid-Apr, with extremes of 3 Mar 2006 (Klein) and 27 Apr–9 May 1954 (Buckley et al.), and a maximum of 42 on 21 Mar 1954 (Phelan, Buckley). Until Jun 1967, spring maxima in Central and Prospect Parks were, respectively, 100 on 31 Mar 1933 (M. Rich) and 178 on 21 Mar 1944 (Soll).

Specimens
Dwight collected 4 in Van Cortlandt in Mar, Apr, Dec 1884–89 (AMNH), and Bicknell 3 in Riverdale in Apr, Nov 1877–79 (NYSM).

Comments
Closely related **Sooty Fox Sparrow**, *P. iliaca fuliginosa*, has occurred as a vagrant in Central Park on 13–14 May 2010 (Chang 2011; photos), as has **Slate-colored Fox Sparrow**, *P. iliaca shistacea*, at Fire Island Lighthouse on 12 May 1971 (Buckley 1974). Both are unique New York City area occurrences and were markedly later than spring migrating Red Fox Sparrows; each was a blowback drift-migrant that had overwintered somewhere in the East/Southeast.

Song Sparrow *Melospiza melodia*

NYC area, current
A common and widespread spring/fall migrant, winter resident, and breeder throughout.

Bronx region, historical
Bicknell's *Riverdale* 1872–1901: abundant spring/fall migrant, common resident

Eaton's *Bronx + Manhattan* 1910–14: abundant breeder, common resident

Griscom's *Bronx Region* as of 1922: common resident

Kuerzi's *Bronx Region* as of 1926: common resident

NYC area, historical
Cruickshank's *NYC Region* as of 1941: common resident, spring (maximum 200+)/fall migrant, winter resident

Bull's *NYC Area* as of 1968: maxima fall 400, winter 185

Bull's *New York State* as of 1975: status unchanged

Levine's *New York State* as of 1996: winter maximum 634 on 1990 Southern Nassau Christmas Bird Count

Study area, historical and current
A common and widespread spring/fall migrant, winter resident, and breeder.

In spring, first migrants arrive in early Mar, peak in late Mar, and depart by May, with no recorded extreme dates and frequent maxima of 30–50 from the 1930s–70s, although more recently only 20 on 3 Mar 2012 (Higgins). Until Jun 1967, the spring maximum in Prospect Park was 34 on 28 Mar 1945 (Soll).

Breeds in all 4 extant subareas, with an estimated 2013–14 total population of about 35 pairs: 15 in Van Cortlandt, 8–10 in Woodlawn, and 5–6 each at Jerome and at Hillview.

Both NYSBBA I in 1980–85 and II in 2000–2005 elucidated a broad New York City area breeding distribution, where they were understandably absent only from single blocks in urban Manhattan, Brooklyn, and Queens (McGowan and Corwin 2008). Notwithstanding its ubiquity, DOIBBS data for this species showed a 1% annual statewide decline from 1966–2005 (Sauer et al. 2005).

In fall, first migrants arrive in Sep, peak in Oct, and depart by late Nov with no available extreme dates, and frequent maxima of 75–100 through the 1970s, although recently only 18 on 9 Nov 2013 (Baksh). Until Jun 1967, the fall maximum in Central Park was 75 on 13 Oct 1953 (Carleton, Harrison).

Numbers in winter drop, but daily counts of 40–50 in warmer years are not uncommon. The study area maximum is 84 across Van Cortlandt and Woodlawn on 26 Dec 1960 (Tudor et al.).

Specimens
Dwight collected 9 in Van Cortlandt in Apr, May, Jun, Sep, Nov 1885–96, and Foster 1 in Woodlawn on 4 Jan 1890 (all AMNH); Bicknell collected 5 in Riverdale in Mar, Jun, Jul, Aug 1876–81 (NYSM, USNM).

Lincoln's Sparrow *Melospiza lincolnii*

NYC area, current
An annual but rare spring/fall Neotropical migrant that is exceedingly infrequent in early winter.

Bronx region, historical
Bicknell's *Riverdale* 1872–1901: once fall, never spring

Eaton's *Bronx + Manhattan* 1910–14: rare migrant

Griscom's *Bronx Region* as of 1922: uncommon but rare migrant, overlooked

Kuerzi's *Bronx Region* as of 1926: uncommon spring (25 Apr–26 May)/fall (12 Sep–12 Oct) migrant

NYC area, historical
Cruickshank's *NYC Region* as of 1941: uncommon spring (11 Apr–3 Jun)/fall (31 Aug–29 Nov) migrant

Bull's *NYC Area* as of 1968: late spring date 10 Jun; late fall date 4 Dec; maxima spring 7, fall 10; increasing Nov, Dec records

Bull's *New York State* as of 1975: status unchanged

Study area, historical and current
An annual but rare spring/fall Neotropical migrant that is exceedingly infrequent in early winter.

In fall, first migrants arrive in mid-Sep, peak in late Sep–early Oct, and depart by late Oct, with extreme dates of 10 Sep 2003 (Young) and 2 Nov 1980 at Jerome (Sedwitz) and a maximum of 4 on 4 Oct 1952 (Buckley). Until Jun 1967, the fall maximum in Prospect Park was 8 on 16 Oct 1966 (Yrizarry).

Exceedingly rare in winter, but throughout the New York City area lingering blowback drift-migrants have been very slowly increasing in frequency at this season since the late 1940s. There are 2 recent study area records from Van Cortlandt: 23 Dec 2000 (Veit, Buckley et al.) and 29 Dec 1996—the latter singing in Van Cortlandt Swamp as were Swamp and White-throated Sparrows on a still, sunny morning when the temperature reached 55° F (Veit, Buckley). Another was at Seton Falls Park, 2.7 smi (4.4 km) east of Woodlawn, on 22 Dec 1963 (Tudor, Horowitz).

In spring, first migrants arrive in early May, peak in mid-May, and depart by late May, with extreme dates of 27 Apr 1954 (Buckley) and 2 on 24 May 1956 (Buckley), and a maximum of 3 on 1 May 1955 (Kane). Until Jun 1967, the spring maximum in Central Park was 7 on 14 May 1956 (Post, Buckley).

DOIBBS data showed a slight but not significant statewide breeder decline from 1966–2005 (Sauer et al. 2005; McGowan and Corwin 2008).

Specimens
Bicknell collected 1 in Riverdale on 7 Oct 1881 (NYSM).

Swamp Sparrow *Melospiza georgiana*

NYC area, current
A common spring/fall migrant and regular but local winter visitor/resident, breeding extremely locally and continuing to disappear from many New York City and Long Island sites.

Bronx region, historical
Bicknell's *Riverdale* 1872–1901: common spring/fall migrant, breeder in all swamps from Riverdale to Kingsbridge Meadows and Dyckman St.; overwintered Riverdale 1875–76

Eaton's *Bronx + Manhattan* 1910–14: abundant breeder, occasional year-round

Griscom's *Bronx Region* as of 1922: resident, scarce in winter

Griscom's *Riverdale* as of 1926: extirpated breeder

Kuerzi's *Bronx Region* as of 1926: common resident, scarcer in winter [locations not given]

NYC area, historical
Cruickshank's *NYC Region* as of 1941: widespread, extremely local breeder [locations not given], common to abundant spring (maximum 70+)/fall migrant, locally uncommon in winter

Bull's *NYC Area* as of 1968: maxima fall 219, winter 65

Bull's *New York State* as of 1975: status unchanged

Study area, historical and current
A widespread spring/fall migrant, an expected winter resident and an extirpated breeder in Van Cortlandt Swamp, where it is no longer resident year-round.

In fall, first migrants arrive in mid-Sep, peak in Oct, and most depart by mid-Nov, with extreme dates of 12 Sep 2005 (Young) and 17 Nov 2012 (McGee), and a maximum of 40 on 3 Oct 1953 (Kane, Buckley).

In winter numerous in Van Cortlandt Swamp, with 1950s daily counts of 10+ and a maximum of 20 on 16 Jan 1954 (Buckley). However, since the 1990s only 1–2 usually overwintered, so 8 (some singing) on 29 Dec 2006 (Veit, Buckley) were surprising.

In spring, first migrants arrive in late Mar, peak in late Apr–early May, and depart by mid-May, with extremes of 2 Mar 2012 (Young) and 19 May 2012 (Baksh), and a maximum of 40 on 1 May 1955 (Kane et al.). Sadly, recent maxima have all been <10. Occurs annually in spring: one was even singing in Tibbett's Marsh on 12 May 2014 (Young). Until Jun 1967, spring maxima were 40 in Central Park on 14 May 1933 (Helmuth) and 13 in Prospect Park on 1 May 1944 (Soll).

Until the late 1950s, 5–6 pairs bred in Van Cortlandt Swamp, but except for the occasional singing ♂ in May, it no longer does even though much of the habitat appears suitable. It seems the last single pairs nested in 1962 and 1963 (Heath, Zupan). Neither NYSBBA I in 1980–85 nor II in 2000–2005 recorded breeders in study area Blocks 5852B and 5952A, but both did find them in adjacent blocks to the east and southeast.

Between NYSBBA I in 1980–85 and II in 2000–2005, in the New York City area many occupied blocks were lost on Staten Island and in Queens, Nassau County, Suffolk County, and Westchester County. NYSBBA II recorded a slight increase in occupied blocks upstate since NYSBBA I but a major loss on Long Island, in New York City (down from 20 to 13, 5 of which were on Staten Island), and in Westchester and Rockland Counties (McGowan and Corwin 2008). DOIBBS data showed no statewide trend from 1966–2005 (Sauer et al. 2005). The Pelham Bay Park area seems to be the nearest regular breeding location.

Specimens
Dwight collected 1 in Van Cortlandt Swamp on 12 May 1890 (AMNH) and Bicknell 5 in Riverdale in Apr, Jun, Aug, Oct, Nov 1876–80 (NYSM).

White-throated Sparrow *Zonotrichia albicollis*

NYC area, current
A common, sometimes abundant, spring/fall migrant and less common but widespread winter resident.

Bronx region, historical
Bicknell's *Riverdale* 1872–1901: abundant spring/fall migrant, regular winter resident

Eaton's *Bronx + Manhattan* 1910–14: abundant migrant, uncommon winter visitor

Griscom's *Bronx Region* as of 1922: common migrant, winter resident

Kuerzi's *Bronx Region* as of 1926: common migrant, winter resident

NYC area, historical
Cruickshank's *NYC Region* as of 1941: common spring/abundant fall (maximum several hundred) migrant, common in winter (15 Aug–2 Jul)

Bull's *NYC Area* as of 1968: daily maxima spring 700, fall 1200, winter 30

Bull's *New York State* as of 1975: spring maximum 2000

Study area, historical and current
A common, sometimes abundant, spring/fall migrant and less common winter resident.

In fall, first migrants arrive in late Sep, peak in Oct (daily counts of 50+ frequent), and depart by early Nov, with the earliest 12 Sep 1959 (Horowitz), and a maximum of 130 in Woodlawn on 12 Oct 1962 (Horowitz). Until Jun 1967, the fall maximum in Prospect Park was 1000 on 15 Oct 1950 (Alperin, Jacobson).

Winter numbers vary widely year to year, with daily counts of 10–20 and a maximum of 238 on 22 Dec 2013 between Van Cortlandt and Woodlawn (Lyons, Gotlib et al.).

In spring, first migrants arrive in mid-Mar, peak in Apr (daily counts of 50+ frequent), and depart by mid-May, with the latest 23 May 1954 (Norse) and a maximum of 125 on 1 May 1955 (Buckley et al.). Until Jun 1967, the spring maximum in Central Park was 700 on 30 Apr 1947 (Sedwitz).

NYSBBA II in 2000–2005 showed some loss of occupied Catskill blocks in Sullivan County since NYSBBA I in 1980–85, southeast of which there are only 6–8 isolated blocks in Rockland, Orange, Putnam, and Dutchess Counties (McGowan and Corwin 2008). DOIBBS data showed no statewide trend from 1966–2005 (Sauer et al. 2005). Nearest current breeders may be near Hook Mountain, Rockland County.

Specimens
Dwight collected 10 in Van Cortlandt in Apr, May, Dec 1885–90, and Foster 1 in Woodlawn on 26 Oct 1889 (all AMNH). Bicknell collected 4 in Riverdale in Jan, Apr, May, Sep 1877–81 (NYSM).

White-crowned Sparrow *Zonotrichia leucophrys*

NYC area, current
An uncommon but locally numerous spring/fall migrant that is increasingly being detected in winter, especially on eastern Long Island.

Bronx region, historical
Bicknell's *Riverdale* 1872–1901: 3 spring records (maximum 6), regular fall migrant

Eaton's *Bronx + Manhattan* 1910–14: uncommon migrant

Griscom's *Bronx Region* as of 1922: rare spring/fall migrant

Kuerzi's *Bronx Region* as of 1926: uncommon spring (2–29 May)/occasionally numerous fall (28 Sep–14 Nov) migrant

NYC area, historical
Cruickshank's *NYC Region* as of 1941: uncommon spring (maximum several dozen)/fairly common fall migrant, extremely rare in winter (23 Sep–14 Jun)

Bull's *NYC Area* as of 1968: early fall date 21 Sep; maxima spring several hundred, fall 118; winter increasing, maximum 5

Bull's *New York State* as of 1975: maxima fall 160, winter 20

Study area, historical and current
An uncommon but locally numerous spring/fall migrant, almost unrecorded in winter.

In fall, first migrants arrive in early Oct, peak in mid–late Oct, and depart by early Nov, with extreme dates of 1 Oct in 1938 (Norse) and 2007 (Young) and 30 Nov 2013 (Young), and a maximum of 20 on 7 Oct 1953 (Buckley).

There are only 4 winter records, all but 1 singles: 26 Dec 1951–17 Jan 1952 (Kane, Buckley); 2 from 23 Feb–14 Mar 1953 (Phelan, Buckley); 7 Dec 1963 (Horowitz, Tudor); Woodlawn 18–20 Jan 2014 (Stuart et al.). Although successful overwintering has not been proved, those in 1951 and 1953 may have been completed.

In spring when uncommon, first migrants arrive in late Apr, peak in early–mid-May, and depart by late May, with extreme dates of 17 Mar 1951 (Kane)—a local winter resident?—and 20 May 1950 (Cantor), and a maximum of 4 on 9 May 1961 (Horowitz). Until Jun 1967, the spring maximum in Central Park was 80 on 10 May 1956 (Post, Buckley).

Specimens
Bicknell collected 3 in Riverdale in Apr, May, Oct 1876–80 (NYSM).

Comments
Both adults and HYs of the central Canada taiga subspecies *gambelii* are not difficult to identify in the field if seen well, but it has yet to be detected in the study area. It is annual in the New York City area and is now being seen in winter, although it does exhibit strong interannual fluctuations in numbers.

Slate-colored Junco *Junco (hyemalis) hyemalis*

NYC area, current
An abundant, widespread spring/fall migrant and winter resident.

Bronx region, historical
Bicknell's *Riverdale* 1872–1901: abundant spring/fall migrant, winter resident

Eaton's *Bronx + Manhattan* 1910–14: abundant winter visitor

Griscom's *Bronx Region* as of 1922: abundant migrant, common winter resident

Kuerzi's *Bronx Region* as of 1926: abundant spring/fall migrant (10 Sep–2 Jun), common winter resident

NYC area, historical
Cruickshank's *NYC Region* as of 1941: common spring/fall (maximum several hundred) migrant, winter resident (15 Aug–5 Jun)

Bull's *NYC Area* as of 1968: fall maximum 1000

Bull's *New York State* as of 1975: status unchanged

Study area, historical and current
An abundant, widespread spring/fall migrant and winter resident.

In fall, first migrants arrive in late Sep, peak in mid–late Oct, and depart by mid-Nov, with extremes of 3 on 21 Sep 1960 (Steineck, Friedman) and 8 on 30 Nov 2013, and a maximum of 300+ in Woodlawn on 27 Oct 1963 (Horowitz). Until Jun 1967, the fall maximum in Prospect Park was 800 on 20 Oct 1944 (Soll).

In winter, numbers vary from year to year, but flocks of 20+ are not uncommon, with a maximum of 159 on 22 Dec 2013 between Van Cortlandt and Woodlawn (Lyons, Gotlib et al.).

In spring, first migrants arrive in mid-Mar, peak in late Mar–early Apr, and depart by the first week in May, with extremes of 3 on 3 Mar 2012 (Baksh) and 3 on 12 May 2012 (Baksh); no spring maxima are available, but daily counts of 50+, many singing, are not unusual. One on 23 Jun 1952 (Buckley et al.) was a nonbreeder or very late migrant.

Southeast of the Catskills in Sullivan County, there were only 4 occupied blocks in Rockland, Orange, and Putnam Counties during NYSBBA II in 2000–2005, down from 8 on NYSBBA I in 1980–85 (McGowan and Corwin 2008). Nearest breeders to the study area may be on the Palisades. DOIBBS data showed no statewide trends from 1966–2005 (Sauer et al. 2005).

Specimens
Dwight collected 8 in Van Cortlandt in Mar, Apr, May, Dec 1884–90 (AMNH), and Bicknell 2 in Riverdale in Sep, Oct 1878 (NYSM).

Comments
Two additional blowback drift-migrant juncos have occurred as vagrants nearby: an AHY ♂ **Pink-sided Junco**, *J. (hyemalis) mearnsi*, was at a Riverdale feeder from 26–30 Dec 2000 (Buckley, Veit et al.); an AHY ♂ **Oregon Junco**, *J. (hyemalis) montanus*, was at another Riverdale feeder on 16 Mar 1960 (Buckley); and another ♂ was at a Clason Pt. feeder 8 Jan–28 Feb 1979 (VanScoy). There are no other Bronx records, but a few Oregons occur almost annually in the New York City area, and it has also been in Central and Prospect Parks.

Tanagers, Grosbeaks, Buntings: Cardinalidae

Summer Tanager   *Piranga rubra*

NYC area, current
A rare but now annual trans-Gulf Neotropical migrant that is subject to overshooting in Apr when they can appear along the South Shore of Long Island and in New York City parks a month before normal migration dates. Very

rare but increasingly found in fall migration, and as of 2002 breeding in Long Island Pine Barrens.

Bronx region, historical

Bicknell's *Riverdale* 1872–1901: unrecorded

Eaton's *Bronx + Manhattan* 1910–14: rare summer visitor

Griscom's *Bronx Region* as of 1922: unrecorded; 13 Long Island records

Kuerzi's *Bronx Region* as of 1926: unrecorded

NYC area, historical

Cruickshank's *NYC Region* as of 1941: extremely rare but 1–2 almost every year, usually in spring [many are trans-Gulf overshoots]; only 2 Bronx records: 1927, 1939

Bull's *NYC Area* as of 1968: 6 Apr–19 Oct [blowback drift-migrant]

Bull's *New York State* as of 1975: early spring 25 Mar, late fall 2 Dec [blowback drift-migrants]

Study area, historical and current

An infrequent spring trans-Gulf overshoot and fall blowback drift-migrant.

In spring, 13 Van Cortlandt records: 21 Jun 1927 (Johnston); 22–23 Apr 1939 (Sialis Bird Club); 27 Apr 1956 (Kane, Buckley); 9–16 Apr 1958 (Weintraub et al., a trans-Gulf overshoot); 8 May 1960 (Young); 13 May 1964 (M. Gochfeld); 16 May 1970 (Stepinoff); 17 May 1972 (Young); 22 May 1997 (Young); 3 May 2003 (unknown observer, *Kingbird*); 28 May 2006 (Delacretaz 2007); 12 May 2007 (Young); 17 Apr 2018 (Waldron, Waldron, a trans-Gulf overshoot), plus 1 blowback drift-migrant 16–17 Sep 1954 (Scully, Buckley, Kane). A fully red ♂ was singing within sight of a ♂ Scarlet in Riverdale Park on 11 May 2012 (Buckley).

The few occurrences in Van Cortlandt are proportional to its relatively low intensity of coverage compared with the multiple annual reports in Central and Prospect Parks, and there are also many more records for Bronx Park. It is reported almost annually somewhere in the Bronx and is increasing in frequency in Inwood Park. As a breeder it is slowly creeping northward in NJ, having bred as close as Essex County, and a tiny population is now established at 3 sites in the eastern Long Island Pine Barrens (McGowan and Corwin 2008). A seemingly paired adult ♂ and ♀ in a Purchase, Westchester County, oak wood from 4–7 Jul 1955 (Schmidt et al.) but not relocated on 9 Jul, might have been the very first New York State breeders.

Scarlet Tanager *Piranga olivacea*

NYC area, current

A relatively common trans-Gulf spring/fall migrant and local breeder. This is another Neotropical migrant that has been seen with diminished frequency and in smaller numbers since the 1960s.

Bronx region, historical

Bicknell's *Riverdale* 1872–1901: common spring/fall migrant, breeder

Hix's *New York City* as of 1904: breeds Bronx Park

Eaton's *Bronx + Manhattan* 1910–14: common breeder

Griscom's *Bronx Region* as of 1922: common migrant, breeder

Kuerzi's *Bronx Region* as of 1926: common breeder, 29 Apr–23 Oct

NYC area, historical

Cruickshank's *NYC Region* as of 1941: common breeder, few pairs still nesting within New York City [locations not given]; common spring (maximum several dozen)/fall migrant, 12 Apr–21 Nov [blowback drift-migrant]

Bull's *NYC Area* as of 1968: early spring date 8 Apr; maxima spring 25, fall 25

Bull's *New York State* as of 1975: fall maximum 50

Study area, historical and current
A relatively common spring/fall migrant and local breeder.

In spring, first migrants arrive in early May, peak in mid-May, and depart by late May, with extreme dates of 28 Apr 1956 (Buckley) and 20 May 1958 (Horowitz) and a maximum of 18 on 14 May 1956 (Scully). Until Jun 1967, the spring maxima in Central and Prospect Parks were, respectively, 25 on 12 May 1958 (Post) and 11 on 11 May 1945 (Soll).

In the late 1800s, it was a common breeder in Riverdale (Bicknell) and in Van Cortlandt (Dwight). It still breeds in Van Cortlandt (and occasionally Woodlawn) with a 2014–15 total population of >10 pairs. NYSBBA I in 1980–85 found breeders in Blocks 5852B and 5952A, which include the study area, but none were detected during NYSBBA II in 2000–2005 (McGowan and Corwin 2008), when study area and Riverdale breeders may have been dismissed as migrants. However, it was in adjacent Yonkers blocks. In 1953 a pair nested in a Norway Maple 10 ft (3 m) over a driveway on Fieldston Rd. in Riverdale (Buckley), where it breeds along the Hudson River.

Both NYSBBA I in 1980–85 and II in 2000–2005 underscored a broad New York City area breeding distribution, where they were absent only from the most urbanized portions of Manhattan, Brooklyn, Queens, and southern Nassau County (McGowan and Corwin 2008). DOIBBS data showed an annual statewide decrease of 1.4% from 1966–2005 that accelerated to 2.4% after 1980 (Sauer et al. 2005).

In fall, first migrants arrive in mid-Aug, peak in early–mid-Sep, and depart by mid-Oct, with extreme dates of 15 Aug 1952 (Buckley) and 17 Oct 1948 (Norse), and a maximum of 4 on 11 Sep 2006 (Young). Bicknell collected 1 in Riverdale on 20 Oct 1876 (NYSM).

The following extremely late Bronx Scarlets would all have been lingering blowback drift-migrants: in Van Cortlandt on 30 Oct 1927 (Cruickshank), on 21 Nov 1927 (L.N. Nichols), ♂ found dead on 3 Dec 1973 (Gallagher), and in Pelham Bay Park on 27 Dec 1964 (Russak et al.).

Specimens
Dwight collected 4 in Van Cortlandt in May 1886–87, and another was found dead on 3 Dec 1973 (all AMNH). Bicknell collected 5 in Riverdale in May, Jun, Oct 1876–82 (NYSM, USNM).

Comments
Blowback drift-migrant **Western Tanager**, *P. ludoviciana*, has never been recorded in the Bronx but is known from Central and Prospect Parks. This western vagrant has also occurred twice as close as Inwood Park—on 2 Aug 1963 and 30 Sep 1965 (both Norse)—and should be kept in mind particularly when very late fall tanagers are encountered.

Northern Cardinal *Cardinalis cardinalis*

NYC area, current
A relatively widespread and common resident that shows only slight migratory tendencies.

Bronx region, historical
Bicknell's *Riverdale* 1872–1901: recorded Riverdale 1872–94, nesting 1889, 1890, 1894; several overwintered in various years

Hix's *New York City* as of 1904: several pairs resident Central Park

Eaton's *Bronx + Manhattan* 1910–14: fairly common but local resident

Griscom's *Bronx Region* as of 1922: very rare visitor, 6 records 1886–1920

Kuerzi's *Bronx Region* as of 1926: formerly bred, now occasional visitor; bred Bronx Park 1917, Scarsdale 1921–24; 7 other singles 1916–23.

NYC area, historical
Cruickshank's *NYC Region* as of 1941: by 1920 essentially extirpated from all New York State areas; recently slowly reestablishing itself northward, breeding regularly but sparsely in southern Westchester County and on Staten Island but elsewhere almost unknown as a breeder; little/no evidence of migration

Bull's *NYC Area* as of 1968: fairly common to locally abundant resident throughout lowland areas in suitable habitat; winter maximum 10

Bull's *New York State* as of 1975: still increasing

Study area, historical and current
A relatively widespread and not uncommon resident.

After an erratic presence in the late 1800s–early 1900s (Bicknell collected 1 in Riverdale on 17 Nov 1874, and Dwight 1 in Van Cortlandt on 14 Apr 1886), Cardinals all but vanished until they began reclaiming habitat in the early 1950s. Numbers built up fairly slowly, and they never became really numerous even though now resident in all 4 extant subareas.

Widespread throughout New York State, Cardinals were static between NYSBBA I in 1980–85 and II in 2000–2005 in the New York City area, and found in every block but 1 (McGowan and Corwin 2008). DOIBBS data showed an annual 1.9% statewide increase from 1966–2005 (Sauer et al. 2005).

Van Cortlandt maxima by seasons: *spring* 15 on 16 Mar 2013 (Perry); *summer* 10–15 pairs (2014: Young, Buckley); *fall* 12 on 26 Aug 2015 (Aracil); and *winter* 46 across Van Cortlandt and Woodlawn on 27 Dec 2009 (Lyons et al.). First recorded on the Bronx-Westchester Christmas Bird Count (Fig. 59) in 1927 but not annual and in only single digits until 1944. After peaking in the 1980s numbers more or less leveled off. Nothing is known of its study area migration patterns.

Specimens
Dwight collected 1 in Van Cortlandt on 14 Apr 1886 (AMNH) and Bicknell another in Riverdale on 17 Nov 1874 (NYSM).

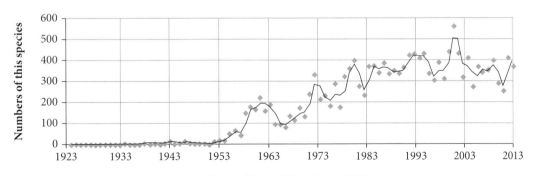

Figure 59

Rose-breasted Grosbeak   *Pheucticus ludovicianus*

NYC area, current
An annual but uncommon spring/fall Neotropical migrant and thinly distributed breeder, especially on Long Island.

Bronx region, historical
Bicknell's *Riverdale* 1872–1901: uncommon spring/fall migrant, first bred 1879, common breeder by 1882

Eaton's *Bronx + Manhattan* 1910–14: fairly common breeder

Griscom's *Bronx Region* as of 1922: uncommon breeder, migrant

Kuerzi's *Bronx Region* as of 1926: fairly common breeder; 26 Apr–16 Oct

NYC area, historical
Cruickshank's *NYC Region* as of 1941: widespread but very local breeder now all but extirpated from Bronx [locations not given]; common spring (maximum 27)/fall migrant, 20 Apr–mid-Oct

Bull's *NYC Area* as of 1968: maxima spring 35, fall 30; 16 Apr [trans-Gulf overshoot]–5 Dec, plus 2 in late Dec [blowback drift-migrants]

Bull's *New York State* as of 1975: spring maximum 111

Study area, historical and current
An annual but uncommon spring/fall Neotropical migrant and breeder.

In spring, first migrants arrive in early May, peak in mid-May, and depart by late May, with extremes of 2 on 30 Apr in 2006 (Klein) and 23 May 1959 (Horowitz), and a maximum of 6 on 9 May 1961 (Horowitz). Until Jun 1967, the spring maximum in Central Park was 50 on 18 May 1956 (Post).

In the late 1800s, it became a regular breeder in Riverdale (Bicknell) and in Van Cortlandt (Dwight). It has bred in Van Cortlandt ever since, with a maximum of 10+ pairs in 2014. It also breeds in Riverdale along the Hudson River and in Inwood Park. Both NYSBBA I in 1980–85 and II in 2000–2005 recorded breeders in Blocks 5852B and 5952A, which include the study area (McGowan and Corwin 2008).

Between NYSBBA I in 1980–85 and II in 2000–2005, occupied blocks were lost in southern Westchester County, Staten Island (all), Queens (all), Nassau County, and Suffolk County; the 2 study area blocks were the only ones left in the Bronx. However, it is known that New York City breeders were missed during NYSBBA II, so maybe others were as well (McGowan and Corwin 2008). DOIBBS data showed a slight statewide increase from 1966–79 followed by a decline through 2005 (Sauer et al. 2005).

In fall, first migrants arrive in early Sep, peak in mid-Sep, and depart by early Oct, with extreme dates of 26 Aug in 1996 (Young) and 2004 (Klein) and 19 Oct 1995 (Young), and a maximum of 6 on 14 Sep 2002 (Young). Until Jun 1967, the fall maximum in Prospect Park was 12 on 23 Sep 1956 (Carleton).

Lingering blowback drift-migrants have been in winter only once in the study area, an SY ♂ in Woodlawn on 6 Feb 2002 (Saphir, Poole), probably wandering from a nearby feeder. But they have occurred 7 times nearby in winter: in the Bronx Zoo from 1–24 Dec 1964 (Edler) and a ♀ on 28 Dec 2014 (Krauss, Olson) that moved between the zoo and Bronx Park through 14 Mar 2015 (O'Reilly, Olson, Crocker, Becker; photos); and in Bronx Park on 27 Dec 1959 (Maguire et al.), 21 Dec 1963 (Edler), and 26 Dec 1982 (DiCostanzo); at a Yonkers feeder from 16–22 Jan 1984 (Reynolds); and at Pelham Bay Park on 5 Jan 1987 (Stepinoff). A frazzled

AMNH specimen taken in the Bronx Zoo on 3 Jan 1907 clearly had been a cage bird.

Specimens
Dwight collected 5 in Van Cortlandt in May, Aug 1890 (AMNH), and Bicknell 1 in Riverdale on 17 Sep 1879 (NYSM).

Comments
A lingering western blowback drift-migrant and vagrant **Black-headed Grosbeak**, *P. melanocephalus*, overwintered within sight of Woodlawn Cemetery at a Williamsbridge feeder from 26 Jan–12 Apr 1976 (J. Gillen et al.). In addition, an adult ♂ Black-headed was studied at leisure in Spuyten Duyvil on 24 Sep 1980 (Sedwitz), the only other Bronx record. Others were at feeders in Englewood, NJ, from 24 Jan–28 Feb 1955 (Boyajian et al.); in extreme southeastern Rockland County in Palisades from 23–30 Apr 1955 (Stansbury); and in southern Westchester County in Hastings from 18–27 Dec 1971 (Moyle et al.), in Scarsdale on 8 Dec 1973 (Moyle), in Tuckahoe from 13 Feb–mid Apr 1976 (Forrest et al.), and Tarrytown from mid Dec 1978–31 Jan 1979 (Schaffner et al.). Note that the first ever in the New York City area, which overwintered in vacant lots in South Ozone Park, Queens in 1953–54, was not identified with certainty until a month after its discovery; the identity of winter *Pheucticus* grosbeaks should never be assumed. Unrecorded in Central or Prospect Park.

Blue Grosbeak   *Passerina caerulea*

NYC area, current
A southern breeder and Neotropical migrant that is slowly expanding northward and that increasingly occurs as a spring trans-Gulf overshoot/range-prospector and fall blowback drift-migrant, especially coastally. It is now recorded almost annually in spring in Central and Prospect Parks and along South Shore barrier beaches. The first New York City area breeding occurred on Staten Island in 1982 (only), followed by an unsuccessful attempt in Muttontown in 1995, and finally successfully by several pairs in the Riverhead-Eastport area of eastern Long Island from 1999–2016. In 2010 several, including singing adult ♂♂, were on Staten Island at Mt. Loretto, Cemetery of the Resurrection, and the Fresh Kills landfill, but breeding was not proved (Shanley 2013) until 2014 when 2 pairs were nesting at Fresh Kills (Fibikar et al. 2015). They are increasing in Bergen County NJ, in the Hackensack Meadowlands since 1973, and possibly also on the Palisades.

NYC area, historical
Cruickshank's *NYC Region* as of 1941: extremely rare spring/fall vagrant, 6 records, none from Bronx: 4 May, 2 Oct

Bull's *NYC Area* as of 1968: >1947 rare but annual spring migrant (31 Mar [trans-Gulf overshoot]–21 May), rare but annual fall migrant (27 Jul–24 Nov [blowback drift-migrant])

Bull's *New York State* as of 1975: bred Hackensack Meadowlands 1973–75

Levine's *New York State* as of 1996: spring maximum 13; once mid-Jan [blowback drift-migrant]

Study area, historical and current
The handful of records of singles reflects few observers. There have been 4 in spring: Vault Hill on 11 May 1992 (Young); Lincoln Marsh 31 May 1996 (Garcia); Parade Ground on 6 Apr 2007 (Young, a trans-Gulf overshoot); Parade

Ground on 6 May 2011 (Young); and 2 in fall: in Woodlawn on 29 Oct 1960 (Buckley; photos) and 5 Oct 2006 in Tibbett's Marsh (Garcia). Until Jun 1967, the fall maximum in Prospect Park was 2 on 23 Sep 1954 (Carleton, Restivo).

Near the study area it has occurred in Inwood Park on 13 May 1961, 8 Oct 1965, and 2 Oct 1966 (all Norse), 3 May 1981 (fide T. H. Davis), 11 Apr 1993 and 11 May 2002 (DiCostanzo), 30 Oct 2014 (Souirgi), 30 Oct 2016 (O'Reilly); in Bronx Park on 12 Oct 1954 (C. Young) and from 25 Oct–2 Nov 1970 (Maguire et al.). At Pelham Bay Park a ♂ oversummered from Jun–29 Aug 1990 (DeCandido), and others were seen on 2 Oct 1999 (Jaslowitz) and 9–10 May 2003 (Künstler). Shanley (2013) and Fibikar et al. (2015) summarized its breeding status in southeastern New York State through 2014.

Indigo Bunting *Passerina cyanea*

NYC area, current
A common spring/fall Neotropical migrant and a widespread but local breeder dependent on second-growth clearings that have been disappearing with development and succession. A frequent trans-Gulf overshoot in early Apr throughout the Northeast.

Bronx region, historical
Bicknell's *Riverdale* 1872–1901: spring/fall migrant, breeder

Hix's *New York City* as of 1904: breeds Bronx Park

Eaton's *Bronx + Manhattan* 1910–14: common breeder

Griscom's *Bronx Region* as of 1922: common breeder and migrant

Kuerzi's *Bronx Region* as of 1926: common breeder, 26 Apr–16 Oct

NYC area, historical
Cruickshank's *NYC Region* as of 1941: widespread but very local breeder, scarce to unknown in New York City [locations not given]; common spring (maximum 18)/uncommon fall migrant, 18 Apr–29 Oct

Bull's *NYC Area* as of 1968: fall maximum 15; 10 Apr [trans-Gulf overshoot]–10 Nov, plus Dec–Jan [blowback drift-migrants]

Bull's *New York State* as of 1975: late fall date 14 Nov; Feb–Mar overwintering [blowback drift-migrants]

Levine's *New York State* as of 1996: spring maximum 65

Study area, historical and current
A spring/fall migrant and scarce (irregular?) breeder.

In spring, first migrants arrive in early May, peak in mid-May, and depart by late May, with extreme dates of 10 Apr 2005 (Klein, a trans-Gulf overshoot) and 4 Jun 1960 (Horowitz), with a maximum of 18 on 19 May 1946 (Staloff). Until Jun 1967, the spring maxima in Central and Prospect Parks were, respectively, 5 on 13 May 1947 (Carleton) and 8 on 17 May 1943 (Soll).

Indigos have bred in Van Cortlandt since the early 1870s and may have peaked in the 1950s–60s at 20–30 pairs (Kieran), and they were described as very common along the Ridge during the 1960 breeding season (Horowitz), where in 1981 5 pairs were breeding (Sedwitz). But as forest succession proceeded, oldfield second growth rapidly disappeared there and elsewhere in Van Cortlandt through the 1980s. Breeding buntings did also, and by 1983 none were in Van Cortlandt. We are unsure when they began rebuilding their numbers, although NYSBBA I in 1980–85 and II in 2000–2005 both recorded breeders in Blocks 5852B and 5952A, which include the study area (McGowan and Corwin 2008), and several singing ♂♂ were in Van Cortlandt on 20 Jun 2010 (Lopes). Indigos left as forests matured, but Hurricanes Irene (2011) and Sandy (2012) opened up new forest clearings; in late May 2014, 24 ♂♂ singing persistently in new second-growth habitat were located on the Ridge, Vault Hill, and Northwest Forest (Buckley, Young). As of 2016 there are 5–10 pairs in Van Cortlandt, 2–3 in Woodlawn, with a pair or 2 at Hillview and Jerome.

On NYSBBA I in 1980–85 buntings were widely distributed in Westchester County, New York City, and (mostly northern) Nassau and Suffolk Counties, even though absent from Brooklyn and (except for 2 blocks) Queens. By NYSBBA II in 2000–2005 the 2 study area blocks were the only ones left in New York City, only 2 were left in southern Nassau County, and about 45 were gone from Suffolk County. (The stylized BBS New York City–Long Island map at the bottom of p. 583 in McGowan and Corwin 2008 made locating New York City blocks a challenge.) Statewide, DOIBBS data showed cyclic surges and retreats from 1966–2005 (Sauer et al. 2005).

In fall, first migrants arrive in mid-Sep, peak in late Sep, and depart by early Oct, with extreme dates of 5 (the maximum) on 20 Sep 1959 (Heath) and 21 Oct 2011 (Young). Until Jun 1967, the fall maximum in Prospect Park was 8 on 10 Oct 1947 (Jacobson). One in Riverdale on 23 Dec 2006 (Shen, Fiore, Wallstrom; photos) and another in Bronx Park on 21 Dec 1976 (Peszel) were blowback drift-migrants, as was one in Central Park that overwintered from Dec 1999–Mar 2000 (m. ob.), one of very few ever in the New York City area.

Specimens

Dwight collected 2 in Van Cortlandt in May, Sep 1886–88 (AMNH) and Bicknell 1 on 7 May 1886 (AMNH).

Painted Bunting   *Passerina ciris*

NYC and study areas, historical and current

A not quite annual spring trans-Gulf overshoot Neotropical migrant and a fall blowback drift-migrant to the Northeast that occasionally lingers into winter and has been found on several occasions in Central and Prospect Parks.

In the 1800s it was seen around New York City when cage-bird traffic—even of native species—was at its apex, and Griscom (1923) commented that no local specimens had been taken before cage-bird traffic had begun or after it had ended. But Cruickshank (1942) recorded 3 fall records of Sep–Nov blowback drift-migrants, and Bull (1964) added 3 more: 2 May trans-Gulf overshoots and another Oct blowback drift-migrant. Several others had been overlooked by previous authors.

The only study area record involved a blowback drift-migrant—an unsexable green individual so not an adult ♂—along the Putnam Division track bed in Van Cortlandt on 29 Sep 1937 (Petersen). The only other Bronx record, a worn adult ♂ collected by Bicknell in Riverdale on 13 Jul 1875 (NYSM 5598), could have been an escaped cage bird, as perhaps was another collected by A. K. Fisher in Croton Point area on 4 Jun 1874 (USNM 25659)—or both were wild. A ♀-type in New Rochelle 3 Jul 2013

(Block), an adult ♂ in Yonkers 8 Jan 2017 (Landdeck, Landdeck; photos) and a ♀-type in Elmsford 31 Mar–2 Apr 2018 (A. Rogers) were probably a postbreeding disperser, lingering blowback drift-migrant, and trans-Gulf overshoot, respectively.

Dickcissel *Spiza americana*

NYC area, current
An increasing migrant in spring (singles) and fall (annual in very small flocks), largely on coastal Long Island. The only nearby breeding attempts (3–4 pairs) in NJ since a pair in Burlington County in 1939 have been as close as the Hackensack Meadowlands. Most attempts have been unsuccessful, with the sites not being reoccupied in ensuing years.

Bronx region, historical
Bicknell's *Riverdale* 1872–1901: unrecorded

Eaton's *Bronx + Manhattan* 1910–14: collected Kingston NY Jun 1896, last New York State report

Griscom's *Bronx Region* as of 1922: unrecorded

Kuerzi's *Bronx Region* as of 1926: unrecorded

NYC area, historical
Cruickshank's *NYC Region* as of 1941: 6 Sep, Oct Long Island records 1927–41 plus Scarsdale

Bull's *NYC Area* as of 1968: beginning in 1946, accelerating in 1951, an expected coastal Long Island fall migrant 6 Aug–30 Nov in small groups to maximum 14; 53 recorded Fire Island Lighthouse 6 Sep–8 Nov 1969; increasing numbers (maximum 5 at feeder) attempting to overwinter, usually in House Sparrow flocks

Bull's *New York State* as of 1975: bred Hackensack Meadowlands 1974

Study area, historical and current
An irregular fall/winter migrant nearly unknown in spring. Only 8 records, proportional to Van Cortlandt's low intensity of coverage, all but 1 singles: 24 Sep 1951 (Kieran); 11 Oct 1953 (Cantor); singing ♂ 23–28 May 1954 (Buckley); 7 Oct 1954 (Kieran); 3 Jan 1955 (Kieran: unlikely the Oct individual); 3 in Woodlawn from 26 Oct–8 Nov 1960 (Horowitz, Buckley et al.); 15 Oct 2008 (Young). A calling flyover was at Hillview on 23 Aug 2017 (Camillieri).

Aug–Oct is the peak period of this species' movement in the New York City area, when others were in Riverdale on 2–7 Oct 1963 (Rosin), 22–23 Oct 1963 (Buckley), and 20 Oct 2001 (unknown observer, *Kingbird*). Some attempt to overwinter, there being 7 such records for Riverdale between 1961 and 1968 (Buckley, Heath et al.), plus one on 28 Dec 2008 (Fiore), as well as 3 others at Inwood Park: 29 Oct–12 Nov 1964 (Norse, Stepinoff), 18 Dec 2011–25 Jan 2012 (Knox et al.), and 7–10 Nov 2014 (Peltomaa et al.).

Elsewhere in the Bronx it has been recorded irregularly in fall and occasionally in winter in Bronx Park, the Bronx Zoo, and Pelham Bay Park, usually singles and often in House Sparrow flocks. In 1954, 4 were on the Baxter Creek landfill from 18 Sep–31 Oct (Buckley 1958), since the 1800s only the second occurrence of a flock in the Bronx.

Dickcissels were formerly localized grassland breeders and Neotropical migrants in the Northeast and until about 1880 bred throughout much of eastern North America north to New England and south to the Carolinas, where they also occurred during migration (Gross 1956). Explanations for their departure include changes in farming practices and increased brood parasitism by eastward-expanding cowbirds. Between 1880 and the late 1940s, there were fewer than a dozen records of migrants in the New York City area, mostly on Long Island. Then in the early 1950s they slowly began to increase as first fall and then occasionally spring migrants. Beginning in the late 1960s–early 1970s, groups of a dozen were appearing in the fall on Long Island barrier beaches, with 53 passing Fire Island Lighthouse from 6 Sep–8 Nov 1969 and a maximum of 15 on 29 Sep (Buckley, T. H. Davis). Around that time they also began appearing at feeders in flocks of House Sparrows and often attempted to overwinter with them. It is unclear whether those occurring in the fall are blowback drift-migrants intercepted en route to Venezuela or are merely moving eastward directly from the Midwest.

NYSBBA I in 1980–85 found this species in only 1 upstate block, which increased to 5 on NYSBBA II in 2000–2005 (McGowan and Corwin 2008). It has also bred at the Hackensack Meadowlands and intermittently at several south Jersey sites, as it claws its way back to a more widespread northeastern distribution.

Blackbirds, Orioles: Icteridae

Bobolink *Dolichonyx oryzivorus*

NYC area, current
An almost gone New York City and study area grassland breeder and Neotropical migrant that is regular in large flocks in fall and small groups in spring.

Bronx region, historical
Bicknell's *Riverdale* 1872–1901: regular spring, abundant fall migrant; common breeder at Kingsbridge Meadows, Jerome Meadows

Eaton's *Bronx + Manhattan* 1910–14: fairly common breeder

Griscom's *Bronx Region* as of 1922: extirpated breeder, last Throgg's Neck 1909; uncommon spring, very common fall migrant

Kuerzi's *Bronx Region* as of 1926: fairly common breeder, some still within New York City [locations not given]; uncommon spring/abundant fall migrant, 19 Apr–20 Oct

NYC area, historical
Cruickshank's *NYC Region* as of 1941: rare/unknown breeder near New York City; uncommon spring/abundant (maximum thousands) fall migrant, 19 Apr–2 Nov

Bull's *NYC Area* as of 1968: spring maximum 100; Long Island fall maxima (some at night) 3000–8000; late fall date 26 Nov [blowback drift-migrant]

Bull's *New York State* as of 1975: status unchanged

Study area, historical and current
An annual but uncommon spring and abundant fall migrant. Formerly bred but long ago extirpated.

In spring, first migrants arrive in early May, peak in mid-May, and depart by late May, with extreme dates of 19 Apr 1909 (Nichols, Nichols) and at Hillview on 27 May 1962 (Zupan), with a maximum of 26 on 12 May 1956 (Scully).

In the study area bred at Kingsbridge Meadows and Jerome Meadows through the late 1800s and early 1900s (Bicknell, Eaton) but not since. Bred at Purchase, Westchester County, in 1974 (unknown observer, *Kingbird*) and on Staten Island in 2003 (unknown observer, *Kingbird*). One in Van Cortlandt Swamp on 6 Jul 2005 (Garcia) was a nonbreeder, very late migrant, or failed breeder from not far away. A pair may have bred on the Edgemere, Queens landfill in Jun 2011 (Normandia et al.). Being secure and unmowed, vegetating landfills have high potential as breeding sites for several grassland species so are worth investigating whenever possible, although as of 2016 Bobolinks were not on Staten Island's Fresh Kills landfill. Hillview is another future site should it ever be left unmowed and the birds undisturbed.

During NYSBBA I in 1980–85, Putnam, Westchester, and Rockland Counties supported 11, 13, and 3 occupied blocks, respectively, down to 1, 5, and none by NYSBBA II in 2000–2005, while New York City and Nassau and Suffolk Counties hosted 1, 2, and 8 on NYSBBA I, down to 1, none, and 2 on NYSBBA II (McGowan and Corwin 2008). But upstate, which hosts the vast majority of occupied sites, DOIBBS data showed no significant statewide change from 1966–2005 (Sauer et al. 2007).

In fall, first migrants arrive in late Jul, peak in late Aug when flocks of 50+ are regular in Van Cortlandt Swamp, and depart by late Sep, with extreme dates of 2 on 20 Jul 2014 (Khalifa) and 2 on 23 Oct 1927 (Kessler). A flock of 22 at Pelham Bay Park on 14 Nov 2015 (Benoit) was remarkably late—blowback drift-migrants—and the latest in the Bronx. Van Cortlandt Swamp maxima of 300 on 1–2 Sep 1951 (Komorowski) and 250 on 31 Aug and 500+ on 1 Sep 1952 (Buckley et al.) have not been approached since, neither there nor anywhere else in the Bronx. Until Jun 1967, the fall maxima in Central and Prospect Parks were, respectively, 60 overhead on 27 Aug 1944 (Bull) and 30 on 2 Sep 1958 (Carleton).

Red-winged Blackbird   *Agelaius phoeniceus*

NYC area, current
A widespread, often abundant, spring/fall migrant, winter visitor/resident, and breeder throughout.

Bronx region, historical
Bicknell's *Riverdale* 1872–1901: common spring/fall migrant, breeder, occasional in winter

Eaton's *Bronx + Manhattan* 1910–14: common breeder, abundant migrant

Griscom's *Bronx Region* as of 1922: common migrant, breeder, occasional in winter

Griscom's *Riverdale* as of 1926: no longer breeds

Kuerzi's *Bronx Region* as of 1926: common breeder, uncommon in winter

NYC area, historical
Cruickshank's *NYC Region* as of 1941: common to abundant breeder, occurs year-round

Bull's *NYC Area* as of 1968: Long Island fall maximum 3000

Bull's *New York State* as of 1975: status unchanged

Study area, historical and current
A common spring/fall migrant, local breeder, and erratic winter visitor/resident in all 4 extant subareas but especially Van Cortlandt Swamp.

In spring, first migrant arrivals vary widely with temperature, but small flocks of ♂♂ can be expected in late Feb followed by ♀♀ in Mar, peaking in late Mar, and departing by mid-Apr. Counts are few, with a maximum of 500 on 20 Apr 1936 (Weber et al.).

Red-wings breed widely but are concentrated in Van Cortlandt Swamp, where in 1937, 50+ pairs bred (Imhof et al.), down in 2013–14 to 40 pairs. Throughout the study area, a total of 60+ pairs still nest.

Both NYSBBA I in 1980–85 and II in 2000–2005 documented a broad New York City area breeding distribution, where they were understandably absent only from the most urbanized portions of Manhattan, Brooklyn, and Queens (McGowan and Corwin 2008).

Notwithstanding Red-wing's near ubiquity, DOIBBS data showed a 1.6% annual decline statewide from 1980–2005 (Sauer et al. 2005).

In fall, the first large flocks of migrants appear in Aug, peak in Oct, and depart by Nov. Counts are few, but the maximum of 5000 on 2 Nov 1954 (Scully) is representative.

In some years, a winter blackbird roost develops in Van Cortlandt Swamp, with a Red-wing maximum of 303 across Van Cortlandt and Woodlawn on 27 Dec 2009 (Lyons et al.).

Specimens
Dwight collected 3 in Van Cortlandt in May, Jul 1889–95 (AMNH), and Bicknell 5 in Riverdale in Apr, Aug, Nov 1876–80 (NYSM).

Eastern Meadowlark   *Sturnella magna*

NYC area, current
A formerly widespread and numerous spring and fall migrant, localized breeder, and expected winter resident. With the loss of undisturbed agricultural areas, meadowlarks, like most other grassland species, began to decline rapidly in the 1960s, though by the 1980s the decline had abated. They have disappeared as New York City breeders except at JFK Airport and on closed landfills around Jamaica Bay and on Staten Island. Throughout the New York City area they are now so scarce in migration as to merit comment when found, and in winter often go unrecorded.

Bronx region, historical
Bicknell's *Riverdale* 1872–1901: common spring/fall migrant, breeder

Hix's *NYC* in 1904: common breeder Bronx, Van Cortlandt Parks

Eaton's *Bronx + Manhattan* 1910–14: common breeder, occasionally year-round

Griscom's *Bronx Region* as of 1922: unmentioned

Griscom's *Riverdale* as of 1926: extirpated breeder

Kuerzi's *Bronx Region* as of 1926: common resident [locations not given] migrants augmenting numbers; fewer in winter

NYC area, historical
Cruickshank's *NYC Region* as of 1941: fairly common breeder [locations not given]; common spring/fall (maximum 100+ coastally) migrant, uncommon in winter; almost unrecorded Central, Prospect Parks

Bull's *NYC Area* as of 1968: formerly much more numerous in winter; Long Island maxima fall 200, winter 375

Bull's *New York State* as of 1975: status unchanged

Study area, historical and current

Meadowlark's study area decline mirrored that of the rest of the New York City area, only more severely. A former breeder, it is now seen only occasionally in fall on the Parade Ground and is extremely scarce in spring.

In spring, first migrants formerly arrived in early Mar, peaked in early Apr, and departed by early May, with extreme dates of 6 Mar 1938 (Norse) and 14 May 1959 (Zupan), and a maximum of 11 in Woodlawn on 19 Apr 1959 (Horowitz). Until Jun 1967, the spring maximum in Prospect Park was 8 on 21 Apr 1943 (Soll).

A few mostly Parade Ground summer records—8 Jun 1935 (Weber), 20 Jun 1995 (Young), 25 Jul 2017 (Hillview: Sargent), 3 Aug 1953 (Buckley), 5 Aug 1952 (Buckley), 18 Aug 1951 (Kane)—are difficult to allocate given that there are no other certain summer records, with one glaring exception: 3–5 pairs bred at Hillview from 1934–40 (Norse), the only ones nesting in the study area since the early 1900s. Thus these other summer birds were postbreeding dispersers or failed/undetected local breeders, at that time a possibility in the Northeast Bronx and southern Westchester County. Note that 10 pairs were on territory at Baxter Creek in May–Jun 1955 (m. ob.).

Between NYSBBA I in 1980–85 and II in 2000–2005, the number of occupied blocks dropped statewide by 25%, but the heaviest losses were in the New York City area. There, the number of blocks with breeding Meadowlarks dropped from 18 to 4 in Westchester County, 3 to 2 on Staten Island, 3 to 1 in Queens, 8 to 2 in Nassau County, and about 40 to 13 in Suffolk County (McGowan and Corwin 2008). DOIBBS data showed a statewide decrease from 1966–2005, steeper between 1966 and 1979 than later (Sauer et al. 2005). Nearest breeders are on JFK Airport, Queens.

Recent "improvements" to the Parade Ground between 2000 and 2010 involved temporarily fencing off several areas to public access. Into these areas at different times of the year moved groups of Meadowlarks, Horned Larks, Killdeer (which bred), Vesper Sparrows, Snow Buntings, and Lapland Longspurs—clear evidence of human disturbance pressure and cursorial birds' immediate response to its lifting.

In fall, first migrants arrive in early Oct, peak in mid-Oct, and depart by early Nov, with extreme dates of 5 Oct 1995 (Garcia, Lyons) and 28 Nov 1995 (Lyons, Garcia), and a maximum of 20 on 17 Oct 1953 (Phelan, Buckley). Until Jun 1967, the fall maximum in Central Park was 22 on 20 Oct 1957 (Post).

Formerly a dozen meadowlarks were uncommon but more or less regular in winter on the Parade Ground plus Hillview, but there have been no Dec–Feb study area records since the last at Hillview on 17 Dec 1938 (Norse, Cantor). This area-wide winter decline is all too evident in Bronx-Westchester Christmas Bird Count data (Fig. 60).

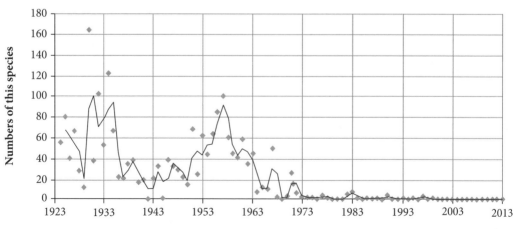

Figure 60

Yellow-headed Blackbird *Xanthocephalus xanthocephalus*

NYC area, current
An annual western visitor that is increasing in frequency of detection, often on Long Island in fall or winter at nighttime roosts in mixed blackbird flocks. Unappreciated by many is its preference for Brown-headed Cowbird flocks. Recorded in Central Park but not Prospect.

NYC area, historical
Cruickshank's *NYC Region* as of 1941: unrecorded until 1932, only 3 fall Long Island records

Bull's *NYC Area* as of 1968: very rare visitor, 14 records Mar–May, Aug–Oct, Jan

Bull's *New York State* as of 1975: 2 more Dec–Jan, Apr

Study area, historical and current
Recorded 4 times, thrice in Van Cortlandt Swamp—♀ on 9 Nov 1954 (Buckley); adult ♂ 22 Dec 1974–6 Jan 1975 (Meyer et al.); adult ♂ 9–16 Mar 2000 (Pirko et al.; photos); and once in Woodlawn: adult ♂ on 5 May 1977 (Teator). Otherwise known in the Bronx only from 2 at Baxter Creek in Sep–Oct 1954 (Buckley 1958).

Rusty Blackbird *Euphagus carolinus*

NYC area, current
A local and declining but annual spring/fall migrant and highly localized winter resident.

Bronx region, historical
Bicknell's *Riverdale* 1872–1901: spring/fall migrant, single winter record

Eaton's *Bronx + Manhattan* 1910–14: common migrant

Griscom's *Bronx Region* as of 1922: uncommon spring, common fall migrant; several midwinter records

Kuerzi's *Bronx Region* as of 1926: common migrant, fairly frequent in winter (18 Sep–12 May)

NYC area, historical
Cruickshank's *NYC Region* as of 1941: uncommon to locally common fall (maximum several hundred)/spring migrant, 1 Sep–3 Jun; small numbers overwinter extremely locally

Bull's *NYC Area* as of 1968: status unchanged

Bull's *New York State* as of 1975: status unchanged

Study area, historical and current
An annual but increasingly scarce spring/fall migrant and overwinterer. Van Cortlandt Swamp has been famous since the mid-1800s as the most reliable location to see this species in winter in the entire New York City area, but as of 2016 they have all but vanished.

In fall, first migrants arrive in early Oct, peak in mid–late Oct, and depart by early Nov, with extremes of 2 on 16 Sep 1954 (Buckley) and 30 Nov 1955 (Buckley), and maxima of 200+ on 25 Oct 1952 (Buckley et al.) and on 8 Oct 1954 (Scully). Until Jun 1967, the fall maximum in Prospect Park was 8 on 5 Nov 1944 (Soll).

Rusties have overwintered annually in Van Cortlandt Swamp since the 1870s, traditionally their most reliable location within New York City, with a winter maximum of 325 on 27 Dec 1970 (Kane, Sedwitz, Buckley). But winter numbers in Van Cortlandt have been declining steadily since their early 1970s peak and in recent years have been reduced to single digits, mirroring declines reported throughout eastern North America (Greenberg and Matsuoka 2010). Bronx-Westchester Christmas Bird Count data (Fig. 61) illustrate the generally low numbers prior to the early 1970s peak and the ensuing decline. NYSBBA II in 2000–2005 found this species to have decreased its number of occupied blocks by 23% since NYSBBA I in 1980–85, a trend appearing elsewhere in the Northeast (McGowan and Corwin 2008). Reasons for the nationwide decline are not understood.

In spring, first migrants arrive in early Mar, peak in late Mar–early Apr, and depart by late Apr, with extremes of 3 Mar 2012 (Rofe,

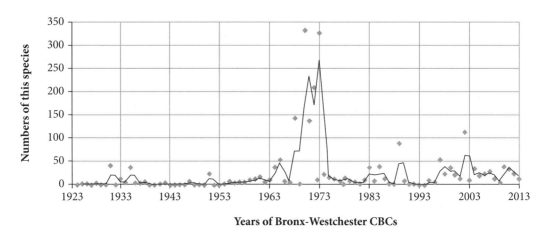

Figure 61

LaRosa) and 20 May 1932 (Ephraim), 27 May 1934 (Weber, Jove), and maxima of 200 on 14 Mar 1936 (Imhof) and 600 on 25 Mar 1998 (Künstler). The latter is by far the highest number ever recorded in 1 location in the New York City area or anywhere in NJ (Walsh et al. 1999), although a handful of 1935–67 4-figure counts and one of 10,000 (Nov 1990) are known from upstate New York (Bull 1974; Levine 1998).

Specimens
Dwight collected 1 in Van Cortlandt on 23 Dec 1885 (AMNH) and Bicknell 5 in Riverdale in Mar, May, Oct, Nov 1876–78 (NYSM).

Brewer's Blackbird   *Euphagus cyanocephalus*

NYC area, current
An increasingly reported blowback drift-migrant and vagrant from the west, in the New York City area almost exclusively from coastal Long Island in Oct–Dec, where it is now being found once every 2–3 years. Recorded in Central Park but not Prospect.

NYC area, historical
Cruickshank's *NYC Region* as of 1941: unrecorded

Bull's *NYC Area* as of 1968: all observations [no details given] deprecated as unidentifiable

Bull's *New York State* as of 1975: observations Oct, Dec 1970; others still deprecated

Levine's *New York State* as of 1996: status unchanged

Study area, historical and current
One study area record: a ♂ immediately picked out from a large blackbird flock feeding actively on the Parade Ground on 18 Oct 1981 (Sedwitz). After close examination over 2 hours, the careful description noted its short tail, bill smaller than a Red-wing's, yellow eyes, and the purple-blue sheen of its head against its greenish body sheen—as it fed alongside Rusty Blackbirds, Common Grackles, Red-winged Blackbirds, and Brown-headed Cowbirds.

While this is the first Bronx record, a ♂ was at a feeder in Pelham, less than 0.5 smi (1 km) from Pelham Bay Park, on 30–31 Dec 1970, and another on 1 Jan 1972 (Klots). Several other intriguing study area reports lacked sufficient information to eliminate aberrant Rusty Blackbirds, longtime fall and winter residents of Van Cortlandt Swamp. In the New York City area Brewer's has been seen 12 or more times in the last 25 years, mostly in Oct–Nov on Long Island. Identification is no longer the insoluble problem it was once believed to be.

Common Grackle   *Quiscalus quiscula*

NYC area, current
A widespread spring/fall migrant (Purple and Bronzed), breeder (Purple), and irregular but occasionally abundant winter visitor/resident (Bronzed).

Bronx region, historical
Bicknell's *Riverdale* 1872–1901: common spring (fall ?) migrant, breeder; rare in winter [when presumably only Bronzed]

Eaton's *Bronx + Manhattan* 1910–14: Purple common breeder; Bronzed fairly common migrant

Griscom's *Bronx Region* as of 1922: Purple common migrant, breeder; rare in winter; Bronzed status obscured by intergrades with Purple throughout New York City area

Griscom's *Riverdale* as of 1926: extirpated breeder

Kuerzi's *Bronx Region* as of 1926: Purple common breeder, replaced by Bronzed in winter

NYC area, historical
Cruickshank's *NYC Region* as of 1941: common to very common breeder; abundant (maximum thousands) spring/fall migrant; erratic in winter

Bull's *NYC Area* as of 1968: maxima fall 10,000, winter 15,000

Bull's *New York State* as of 1975: status unchanged

Study area, historical and current
A spring/fall migrant, breeder, and irregular but occasionally abundant winter visitor/resident.

First migrants arrive in late Feb, peak in mid-Mar, and depart by the end of May, but owing to the large number of overwinterers and breeders, extreme dates are pointless; the maximum is 150 on 5 Mar 1955 (Buckley). Until Jun 1967, the spring maximum in Prospect Park was 180 on 24 Mar 1951 (Kreissman).

Grackles breed in all 4 extant subareas, and while no recent hard census data are at hand, a low estimate would be on the order of 65 pairs: >40 in Van Cortlandt, 5 in Woodlawn, and another 10 pairs each at Jerome and Hillview. Some were feeding young as late as 24 Sep 1962 at Jerome (Sedwitz).

Both NYSBBA I in 1980–85 and II in 2000–2005 showed a broad New York City area breeding distribution, with only slight NYSBBA II losses in the most urbanized portions of Manhattan, Brooklyn, and Queens (McGowan and Corwin 2008). DOIBBS data showed a 2.7% annual statewide decrease from 1980–2005 (Sauer et al. 2005).

In fall, first migrants from outside the local breeding area arrive in early Oct, peak in late Oct–early Nov, and depart by mid-Nov. Flocks of hundreds are routine, and occasionally thousands are seen: 1100 on 4 Nov 2005 (Künstler), 2000+ on 6 Nov 1955 (Post, Buckley), 2300 at Jerome on 31 Oct 1983 (Sedwitz). On 7 Nov 1982, 8925 were moving south through Riverdale (Sedwitz). Until Jun 1967, the fall maximum in Central Park was 1350 on 1 Nov 1962 (Carleton).

The vast majority of grackles overwinter well to the south of the study area, and in most years they are scarce to absent, although in especially mild winters one or more flocks in the hundreds and occasionally thousands will wander throughout the Northwest Bronx, often roosting in Van Cortlandt Swamp nightly. The largest counts there are 4588 on 26 Dec 1982 (Sedwitz et al.) and 5030 on 27 Dec 2009 (Lyons et al.). Most flocks disappear with very low temperatures or major snowfalls, but a few sometimes linger, like the 1000+ in Van Cortlandt on 20 Jan 2008 (Garcia). On 26 Dec 1982, a flock of 4500+ was in Woodlawn (Sedwitz).

Specimens
Bicknell collected 5 in Riverdale in Mar, Apr, Nov 1874–80 (NYSM).

Comments
The New York City area is at or near the overlap zone for the 2 eastern Common Grackle subspecies. Most incoming breeders conform to what was formerly called Purple Grackle (*stonei*), with some morphologically intermediate between *stonei* and more northerly breeding Bronzed Grackle (*versicolor*). However, most migrants are Bronzed Grackles, and that is usually the taxon of overwintering flocks when they strike observers as noticeably different from local breeders.

Boat-tailed Grackle *Quiscalus major*

NYC area, current
A recent Long Island colonial breeder and colonizer from the south that first bred in 1979 in saltmarshes north of Long Beach, LI. Most or all are resident along the south shore from Jamaica Bay to Shinnecock Bay, with Jamaica Bay breeders moving to inland Queens during the winter, especially the Flushing Meadows Park area. Recorded in Central, Prospect, and Pelham Bay Parks (where not yet breeding), and an isolated colony established in Stratford CT is tediously attempting to expand northeastward.

NYC area, historical
Cruickshank's *NYC Region* as of 1941: unrecorded

Bull's *NYC Area* as of 1968: ♀ after Hurricane Carol in 1954 was first in New York State; then none until solo vanguard ♂ 1967–71

Bull's *New York State* as of 1975: 3 additional vanguard colonizers (2 ♂♂, single ♀)

Levine's *New York State* as of 1996: additional records of small groups, first breeding 1979; breeders now established east to Shinnecock Bay

Study area, current
There is but 1 record, the first in the Bronx: an adult ♀ on the Putnam Division track bed alongside Van Cortlandt Swamp on 25 Mar 1982 (Sedwitz). Nonbreeders from Jamaica Bay disperse to inland Queens in winter, so this Mar ♀ is not surprising. Presumed migrants or progeny from the Stratford, CT, colony have been seen in Mar–May, Nov at Pelham Bay Park (early 2000s–2017), including potential prospectors at Huckleberry Island on 27 May 2014 (Künstler) and North Brother Island on 22 May 2013 (NYCA surveys). An unstated number of pairs reportedly bred near Little Neck Bay, Queens, in 2011 (J. Miller), possibly the first on western Long Island away from the South Shore. In winter, birds from any of these locations could reach the study area. Another ♀ at Croton Point on 5 Apr 2015 (Trachtenberg) is the deepest inland penetration so far in the New York City area.

Brown-headed Cowbird *Molothrus ater*

NYC area, current
A common migrant, breeder, and erratic overwinterer, often in large flocks in migration and winter, when they roost with other blackbirds.

Bronx region, historical
Bicknell's *Riverdale* 1872–1901: common breeder, only occasional in winter

Eaton's *Bronx + Manhattan* 1910–14: common breeder

Griscom's *Bronx Region* as of 1922: uncommon breeder, decreasing; no midwinter records

Griscom's *Riverdale* as of 1926: no longer breeds

Kuerzi's *Bronx Region* as of 1926: common breeder, rare in midwinter

NYC area, historical
Cruickshank's *NYC Region* as of 1941: fairly common spring (maximum 100+)/fall (maximum hundreds) migrant; breeds widely but decreasing in New York City; rare but regular in winter, especially coastally

Bull's *NYC Area* as of 1968: maxima fall 1100, winter 1000

Bull's *New York State* as of 1975: winter roost maximum 5000

Study area, historical and current
A common migrant, breeder, and erratic overwinterer, often in large flocks in migration and winter, when they roost with other blackbirds.

In spring, first migrants arrive in early Mar, peak in early Apr, and depart by early May, but owing to the large number of overwinterers and breeders, extreme dates are pointless; usually only very small flocks are commonly seen (e.g., 30 on 20 Apr 2000: Young) and the maximum is 250 on 25 Mar 2007 (Klein). Until Jun 1967, the spring maximum in Prospect Park was 50 on 5 May 1950 (Jacobson, Whelen).

Cowbirds are brood-parasites of numerous species of local breeders in all 4 extant subareas, and although the mean number of pairs present each summer is uncertain, 20 were in Van Cortlandt Swamp alone on 29 May 1937 (Imhof et al.). As of 2016, probably 20–30 pairs breed in Van Cortlandt, with unknown numbers elsewhere in the study area.

Both NYSBBA I in 1980–85 and II in 2000–2005 emphasized its broad New York City area breeding distribution, where it was understandably absent only from the most urbanized portions of all New York City boroughs except the

Bronx (McGowan and Corwin 2008). DOIBBS data showed a statewide annual 2.7% decline from 1966–2005 (Sauer et al. 2007).

In fall, first migrants arrive in early Sep, peak in Oct, and depart by early Nov, with maxima of 1000 on 9 Oct 1957 (Rafferty), 1800 in Oct 2005 (Künstler), and a vast aggregation of whistling, wheeling clouds, thousands upon thousands on the Parade Ground on 23 Oct 1955 (Birnbaum). Until Jun 1967, the fall maximum in Central Park was 425 on 31 Oct 1957 (Carleton).

In winter there is usually a small population varying with winter severity. The largest winter count is 500+ on 5 Jan 1956 (Scully), but this is conservative, as 1670 were in Spuyten Duyvil on 28 Dec 1980 (Sedwitz).

Specimens
Dwight collected 6 in Van Cortlandt in Apr, May 1886–95 (AMNH), and Bicknell 5 in Riverdale in Apr, May, Sep, Oct 1877–80 (NYSM).

Orchard Oriole *Icterus spurius*

NYC area, current
This southern and midwestern species is near the northern edge of its range here. It is an increasing spring Neotropical migrant and sporadic breeder, often in small clusters that abandon breeding sites after only a few years. They are very early fall migrants only rarely detected southbound.

Bronx region, historical
Bicknell's *Riverdale* 1872–1902: common spring, very early fall migrant; breeder

Eaton's *Bronx + Manhattan* 1910–14: common breeder; commonest in New York City vicinity, lower Hudson Valley

Griscom's *Bronx Region* as of 1923: rare, irregular, almost unrecorded as migrant; bred Riverdale 1917, regularly near Baychester

Griscom's *Riverdale* as of 1926: extirpated breeder

Kuerzi's *Bronx Region* as of 1926: fairly common local breeder, especially at Long Island Sound; spring/fall migrant from 30 Apr–28 Aug

NYC area, historical
Cruickshank's *NYC Region* as of 1942: erratic, very local breeder, apparently no longer in Bronx; uncommon to rare spring/fall migrant, 28 Apr–27 Sep

Bull's *NYC Area* as of 1964: spring maximum 10; early spring date 20 Apr; late fall date 11 Dec [blowback drift-migrant]

Bull's *New York State* as of 1975: status unchanged

Study area, historical and current
An annual but scarce spring migrant and a rare but increasing breeder, usually unrecorded in fall.

In spring, first migrants arrive in early May, peak in mid-May, and depart by late May, with extreme dates of 27 Apr 2012 (Young) and 27 May 1984 (Sedwitz), and a maximum of 5 on several occasions in May since 2010. Until Jun 1967, the spring maximum from both Central and Prospect Parks was 3 on, respectively, 4 May 1956 (Messing, Post) and 25 May 1950 (W. Ferguson, Whelen).

Present in Riverdale and the study area in low and fluctuating numbers as breeder and spring migrant since the late 1800s (Bicknell, Dwight), it declined in the early 1900s, was not breeding again until 1924 (Cruickshank 1942), then not again until 1933 (Norse) and 1936 (Weber et al.), and annually only since

1951 (Komorowski). After 1962 it bred only intermittently until 1996, when a pair on the Ridge had a ♀ Baltimore Oriole as a nest helper (Garcia, Lyons), then from 2009–12 at Tibbett's Marsh (Young). In 2012, 5 widely spaced out and nominally territorial ♂♂ were in Van Cortlandt Swamp alone on 10 May (Buckley), and 7 (sexes unreported) were found on 12 and 19 Jun 2010 (Baksh), local breeders all. A singing ♂ in Woodlawn on 31 May 2014 (Winston) may also have been breeding. Orchard is reestablished as a breeder for the moment, although this could change abruptly. As many as 12 pairs may have been breeding in the study area from 2010–14.

From NYSBBA I in 1980–85 to NYSBBA II in 2000–2005, the number of occupied blocks within and near New York City increased greatly as this species seemed to be finally consolidating its glacially slow northeastward breeding range expansion. Statewide, the number of occupied blocks also jumped, from 223 to 389 (McGowan and Corwin 2008). Sparse DOIBBS data supported a steady statewide increase from 1966–2005 (Sauer et al.).

Orchard Orioles quietly disappear in Jul, when they begin heading south, and there are only 3 study area singles after Jun: 5 Sep 1952 (Buckley), 17 Sep 1954 (Kane, Buckley), and 27 Aug 2009 (Young).

Specimens
Dwight collected 1 in Van Cortlandt on 30 May 1887 (AMNH), Bicknell 2 in Riverdale in May, Jul 1879 (NYSM).

Baltimore Oriole   *Icterus galbula*

NYC area, current
A common spring/fall Neotropical migrant and widespread but highly local breeder even in New York City parks. Occasional in winter.

Bronx region, historical
Bicknell's *Riverdale* 1872–1901: common breeder, migrant

Hix's *New York City* as of 1904: common breeder in parks

Eaton's *Bronx + Manhattan* 1910–14: common breeder

Griscom's *Bronx Region* as of 1922: common migrant, breeder

Kuerzi's *Bronx Region* as of 1926: common breeder, 27 Apr–late Nov

NYC area, historical
Cruickshank's *NYC Region* as of 1941: common breeder but pairs few and diminishing in immediate New York City vicinity [locations not given]; common spring (maximum 50+)/fall migrant, 14 Apr–late Nov, increasing coastal winter records [blowback drift-migrants]

Bull's *NYC Area* as of 1968: maxima spring 38, fall 30; 14 Apr [trans-Gulf overshoot]–1 Dec; much more frequent in winter [blowback drift-migrants], with maxima of 7 twice

Bull's *New York State* as of 1975: fall maxima 200, 300; winter feeder maximum 9

Study area, historical and current
A common spring/fall migrant and regular breeder.

In spring, first migrants arrive in late Apr, peak in early May, and depart by mid-May, with extreme dates of 21 Apr 2012 (Baksh) and 20 on 21 May 2011 (Bochnik). On flight days 6–10 can usually be found, with a maximum of 27 on 14 May 2006 (Klein). Until Jun 1967, the corresponding maxima in Central and Prospect Parks were, respectively, 22 on 12 May 1933 (Helmuth) and 9 on 10 May 1953 (Usin).

Breeds in Van Cortlandt (>20 pairs) and Woodlawn (7 pairs), and sporadically (only?) at the 2 reservoirs, with a study area total of about 30 pairs. A count of 7 on 12 Jun 2005 (Klein) reflects the strong Van Cortlandt breeding population.

Both NYSBBA I in 1980–85 and II in 2000–2005 noted a broad New York City area breeding distribution, where they were understandably absent only from the most urbanized portions of all New York City boroughs except Staten Island (McGowan and Corwin 2008). DOIBBS data indicated an annual 1.2% statewide decline from 1980–2005 (Sauer et al. 2007).

In fall, first migrants arrive in mid-Aug, peak in early Sep, and depart by late Sep, with extreme dates of 22 Jul 1959 (Horowitz) and, surprisingly, 6 Sep 1955 (Kane), with a maximum of 5 on 12 Aug 2012 (Aracil). Singles on 8–11 Nov 1952 (Norse, Kieran et al.) and 13 Nov 2005 (Young, Klein) were blowback drift-migrants like those lingering to late Dec–Jan in several different years in Riverdale and in Bronx Park although not yet in the study area.

Specimens
Dwight collected 2 in Van Cortlandt in Jul 1889, Foster 1 in Woodlawn in May 1889 (all AMNH), and Bicknell 2 in Riverdale in May, Aug 1876–79 (NYSM).

Comments
Western blowback drift-migrant **Bullock's Oriole**, *I. bullockii*, has never been found as a vagrant in the study area or in Central or Prospect Park. The sole Bronx report is of a ♀ studied at length in Bronx Park on 23 Dec 1978 by 2 careful observers (Stepinoff, Hamers) although no description or photos survive. A well-described ♀ was in Larchmont on 1 Nov 1993 (Lehman); it is worth remembering that for every ♂ Bullock's seen in fall/winter in the New York City area, there are probably 30 ♀♀ claimed. This unbalanced sex ratio bolsters the widespread belief that the vast majority of these ♀ Bullock's are actually very pale HY ♀ Baltimores, so images or very detailed descriptions are essential.

Finches: Fringillidae

Pine Grosbeak   *Pinicola enucleator*

NYC area, current
An erratic winter irruptive in very low numbers and which may have the longest gaps between irruptions of any New York City area species. There has been not been 1 of any consequence for 50 years. Pine Grosbeaks are inordinately fond of crabapples and other red berries and fruits in winter, all of which happen to be singularly scarce within the study area.

Bronx region, historical
Bicknell's *Riverdale* 1872–1901: recorded 5 of 19 winters, abundant Jan–Mar 1884

Eaton's *Bronx + Manhattan* 1910–14: occasional winter visitor

Griscom's *Bronx Region* as of 1922: very rare irregular winter visitor, single recent record

Kuerzi's *Bronx Region* as of 1926: rare, irregular winter visitor, only 5 records 1916–23

NYC area, historical
Cruickshank's *NYC Region* as of 1941: rare, irregular winter visitor (25 Oct–12 Apr), last irruption 1929–30

Bull's *NYC Area* as of 1968: irruptions 1951–52, 1961–62; winter maximum 150; late spring date 20 Apr; early fall date 8 Oct

Bull's *New York State* as of 1975: status unchanged

Study area, historical and current
A winter irruptive, usually recorded only in major flights. In Riverdale Bicknell found it from 29 Jan–1 Apr 1875, on 7 Nov 1881, from 25 Dec 1882–10 Feb 1883, abundant (no counts offered) from 24 Jan–24 Mar 1884, and on 18 Jan 1890.

Nonetheless, there are only 6 study area records, all from Van Cortlandt (the most recent 40 years ago) and all but 1 singles: 8 on 5 Dec 1921 (F. Brown), 10 Feb 1944 (Komorowski), 22 Jan 1953 (Phelan), 27 Oct 1956 (Scully, Buckley), 31 Oct 1965 (Norse), and 26 Dec 1977 (Sedwitz). Nearby, there were singles in Riverdale on 23 Oct 1955 and 12 Nov 1957 (Buckley), and 2 in Inwood Park on 25 Oct 1965 (Norse). It has been on the entire Bronx-Westchester Christmas Bird Count only twice since 1962, in 1968 and 1977. Over the years there have been numerous records from Bronx Park, whose winter-fruiting ornamentals Pine Grosbeaks find irresistible. Until Jun 1967, the maximum in Prospect Park was 6 on 12 Nov 1951 (Whelen).

House Finch   *Haemorhous mexicanus*

NYC and study areas, historical and current
Introduced on western Long Island in the early 1940s as released cage birds (Elliott and Arbib 1953) and until the early 1950s largely restricted to a few locations in southwestern Nassau County, notably the vicinity of Hewlett and Lawrence. Even though the first Westchester County House Finch was at Tarrytown in May 1948, followed by others at Bedford and Armonk in 1951, and the first in CT was in Riverside in late 1951, it was not detected in Prospect Park until Oct 1963 or in Central Park until Apr 1966. It is now a widespread breeder as well as spring/fall migrant throughout the New York City area.

It was not certainly recorded in the Bronx until 2 in Bronx Park on 11 Feb 1951 (Maxwell, Maxwell). In the study area and nearby the first were a ♂ and ♀ in Van Cortlandt on 7 May 1953 (Buckley), then a singing SY ♂ in Riverdale on 4 May 1955 (Buckley) and an adult ♂ at a Riverdale feeder on 14 Oct 1957 (Buckley). Thereafter, there were none until a flock in Woodlawn 27 Oct–13 Nov 1960, with a maximum of 6 on 29 Oct (Horowitz, Buckley).

From that point on they increased steadily and were first detected in the study area on a Bronx-Westchester Christmas Bird Count in Woodlawn on 23 Dec 1963 (Tudor, Enders). The first study area breeding occurred on 4 Jul 1964, when Staloff found a pair nesting in English Ivy along Broadway at W. 251st St. They are now widespread breeders throughout, with a current population of >20 pairs, the majority nesting immediately outside the study area but feeding within it daily. Most unexpectedly given that House Finches in the West are considered nonmigratory, banding returns demonstrated development of a migratory pattern in northeastern House Finches within 10–20 years of their establishment in Westchester Co. and Fairfield Co. CT (Cant and Geis 1961).

In the mid-1990s an epidemic conjunctivitis infection caused by the bacterium *Mycoplasma*

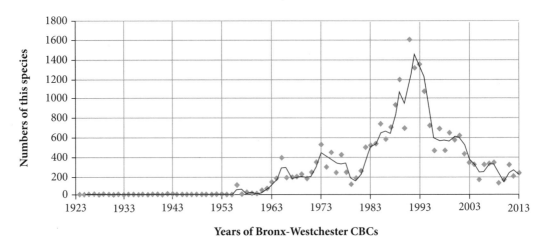

Figure 62

gallisepticum spread rapidly throughout eastern House Finches, facilitated by their strong migratory patterns. First detected in the winter of 1993–94 in the Washington, DC area, it raced through eastern breeders (Ley et al. 1996). House Finch's previously exponential population growth rate of 21% plummeted, and within 3 years numbers had dropped by 60% and more. Within New York City NYSBBA II in 2000–2005 detected loss of only a 4-contiguous-block portion of Manhattan-Brooklyn-Queens (McGowan and Corwin 2008). Their collapse in numbers on the Bronx-Westchester Christmas Bird Count finally seems to be bottoming out (Fig. 62).

Recent nonbreeding study area seasonal maxima are *spring* 12 on several occasions; *fall* 20 on 28 Sep 2013 (Souirgi), and *winter* 110 across Van Cortlandt and Woodlawn on 27 Dec 2015 (Lyons et al.)—evidence of some study area population recovery following the crash.

Purple Finch   *Haemorhous purpureus*

NYC area, current
A regular spring/fall migrant and winter visitor/resident, subject to irruptions and great swings in local populations. It had long bred in the New York City area, including Staten Island, Queens, and the Bronx, plus Nassau, Suffolk, and Westchester Counties, but thinly and inconsistently. Even so, since the 1980s it has all but vanished as a breeder; competition with House Finches has often been advanced as the cause, but that explanation may be facile.

Bronx region, historical
Bicknell's *Riverdale* 1872–1901: regular spring/fall migrant; abundant in winter 1877–78; bred 1876, 1881

Eaton's *Bronx + Manhattan* 1910–14: abundant migrant, rare breeder

Griscom's *Bronx Region* as of 1922: irregular spring, abundant fall migrant; rare in winter

Kuerzi's *Bronx Region* as of 1926: common migrant, uncommon in winter (14 Sep–26 May)

NYC area, historical
Cruickshank's *NYC Region* as of 1941: uncommon spring/common, occasionally abundant (maximum several hundred) fall migrant; variably numerous winter resident; few pairs breed Westchester Co. from Kensico Reservoir northward

Bull's *NYC Area* as of 1968: 18 Aug–26 May; irruption year maxima 400 fall, 250 winter, 330 spring; bred Grassy Sprain 1942

Bull's *New York State* as of 1975: fall maximum 600

Study area, historical and current
An erratically irruptive but otherwise annual migrant and overwinterer that has bred once.

In spring, first migrants arrive in late Mar, peak in mid-Apr, and depart by early May, with extreme dates of 8 Mar 2010 (Young) and 2 on 17 May 1953 (Buckley), and a maximum of 7 on 3 May 1959 (Horowitz). Until Jun 1967, the spring maxima in Central and Prospect Parks were, respectively, 70 on 14 Mar 1939 (Cantor) and 30 on 10 May 1946 (Soll).

Purples bred in Riverdale in 1876 and 1881 (Bicknell) and by extension in the Van Cortlandt area at the same time, and they bred in Central Park in 1888. In May 1945 Eisenmann watched a pair carrying nesting material into Van Cortlandt Swamp, the only modern study area breeding, although one on 8 Jun 1962 in Van Cortlandt may also have been a breeder (Horowitz). They have declined severely as breeders throughout the New York City area since the 1960s, which coincided (too?) nicely with the arrival and rapid spread of its congener, House Finch.

On NYSBBA I in 1980–85 occupied blocks were 9 in Westchester County; none in Rockland County, New York City, and Nassau County; and 30 in Suffolk County. But by NYSBBA II in 2000–2005, corresponding counts had dropped to 1 in Westchester County, 6 in Rockland County, none in New York City and Nassau County, and 1 in Suffolk County. Notwithstanding, DOIBBS data showed no statewide changes from 1966–2005, driven by their widespread upstate nesting in burgeoning conifer plantations (Sauer et al. 2007). The closest breeders now are in Bergen County, NJ, or in Rockland County.

In fall there tend to be 2 migration peaks: a small 1 in late Aug–early Sep that may presage irruptions, like the 3 on 5 Aug 1952 (Kane) and the 5 on 2 Sep 1936 (Norse), followed by a larger movement in Oct–Nov, especially in irruption years. Normal maxima are usually <10, but in irruption years are often in the 100s, like the 200 on 29 Nov 1953 (Buckley) and 200+ on 6 Oct 1957 and 13 Oct 1959 (Rafferty).

Winter numbers vary greatly. In nonirruption years they are absent or in scattered singles and flocks <6, whereas in irruption years larger numbers remain and they are widespread, with a study area maximum of 100+ on 19 Jan 1928 (Cruickshank). Their fluctuations over the last 90 years on the Bronx-Westchester Christmas Bird Count (Fig. 63) typify study area trends.

Specimens
Bicknell collected 4 in Riverdale in Apr, May, Oct 1877–86 (NYSM, AMNH).

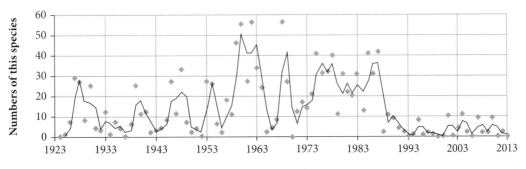

Figure 63

Red Crossbill   *Loxia curvirostra*

NYC area, current
An irregular fall/spring migrant and winter visitor/resident in very small numbers that periodically irrupts southward from its taiga breeding areas by the hundreds and thousands, largely coastwise, sometimes remaining to breed. But what is now called Red Crossbill may actually involve a dozen cryptic species differing in flight calls and preferring different conifer groups for food, and which are best identified by their calls; see Comments below. In the New York City area 3 or 4 call-types may be of regular occurrence.

Bronx region, historical
Bicknell's *Riverdale* 1872–1901: irregular winter 1874–94, abundant in winter 1874–75 with nest and eggs 22 Apr 1875; seen late as 10 May

Eaton's *Bronx + Manhattan* 1910–14: occasional winter visitor, rare breeder

Griscom's *Bronx Region* as of 1922: irregular winter visitor, last 1908

Kuerzi's *Bronx Region* as of 1926: rare, irregular in spring, fall, winter; most recent records Apr–May

NYC area, historical
Cruickshank's *NYC Region* as of 1941: irregular, exceedingly rare fall/winter migrant, 17 Sep–22 Apr; last irruption 1899–1900; spring return flight late, with 33% of all recent Bronx Park records in May

Bull's *NYC Area* as of 1968: irruption 1951–52 first since 1904–5, along Long Island South Shore in many flocks; Long Island fall maximum 200 in 1963

Bull's *New York State* as of 1975: fall 1973 irruption, with maximum 1800 at Riis Park; presumed breeding on Long Island 1972–76, occasionally other times

Study area, historical and current
Being dependent on conifers, it is almost exclusively found in Woodlawn (and formerly at Hillview) during major irruptions, but occasionally in the few remaining conifers in Van Cortlandt.

Generally unrecorded outside a few irruption winters (1952–53, 1963–64, 1969–70, 2012–13). Only 9 records: 8 in Van Cortlandt on 20 Nov 1952 (Kieran), 3 there on 11 Feb 1953 (Kieran); Hillview on 15 Feb 1953 (Norse); 20 in Van Cortlandt on 17 Feb 1953 (Phelan); 20 at Hillview from 14 Mar–11 Apr 1964 (Horowitz et al.); on the Ridge on 31 Oct 1965 (Norse); Hillview on 13 Dec 1969 (Kane, Buckley); 5 in Woodlawn on 21 Dec 1969 (Kane et al.); and 4 in Woodlawn on 23 Dec 2012 (Barry, Gotlib et al.). It must occur far more frequently than the few records imply, given the regularity with which it is seen as close as Bronx Park and Riverdale. Until Jun 1967, the fall maximum in Prospect Park was 40 on 23 Nov 1963 (Raymond) and in spring 7 on 16 Mar 1914 (Fleisher).

In Riverdale, Bicknell saw it numerous times, unfortunately never reporting numbers. In the winter of 1874–75, they were abundant in Riverdale from 3 Nov–10 May, and he located a nest with eggs on 22 Apr. Others were seen on 5–6 May 1877, 1 May 1879, 13 Jun–20 Jul 1878, 1 Apr–18 May 1884, 8 Nov 1887, 28 Jan 1888, 12 Jan–24 Apr 1890, and 22 Apr 1894. It is not unreasonable to conclude that it bred there more than once in this period and that others, breeding or not, were scattered throughout the study area at the same time.

NYSBBA II in 2000–2005 found this species to have decreased its number of occupied blocks by 64% since NYSBBA 1 in 1980–85, but this is related to irruptions—including the 4 Long Island blocks with 1980–85 breeders that were empty in 2000–2005 (McGowan and Corwin 2008). The nearest nonirruption regular breeders may be in the NJ or Long Island Pine Barrens or in conifer plantations west of the Catskills.

Following any irruption winter, this species should be considered a potential Woodlawn breeder, so beginning in Mar, late lingerers, especially any singing, should be watched carefully there.

Specimens
Bicknell collected 3 in Riverdale in May 1877–79 (NYSM, USNM).

Comments
Red Crossbills of several body and bill sizes occur and breed in New York State, some of them only (?) episodically. These had long been described as belonging to various traditional subspecies (see Dickerman 1986a, 1986b). But in a groundbreaking monograph, Groth (1993a, 1993b) described 8 different Red Crossbill call types throughout North America that were associated with conifers having small, medium, and large cones. Subsequently, 2 more call types were recognized, and there may be others; see Young 2012 and Young and Spahr 2017 for updated references showcasing all known to date. These call types may or may not be reproductively isolated from one another, and Groth did not designate any as new species. Moreover, call-type relationships to traditional morphology-only subspecies remain to be sorted out. Within New York State, Groth (1996) recorded a small form (Type 3) feeding on hemlocks and spruces; 2 medium forms (Types 1 and 4) feeding on White Pines and hemlocks that were common before 1900; and a large-billed form (Type 2) that prefers Red Pines and Pitch Pines. Little work has been done on which call types have occurred or are occurring in the New York City or study areas, but in addition to Types 1–4, Young (2011) has shown that Type 5 has also been confirmed in New York State. Because of more recent work by Dickerman, by Groth, and by M. Young, comments on New York City area subspecies of Red Crossbill in Bull (1964, 1974) should now be regarded as only work in progress.

White-winged Crossbill *Loxia leucoptera*

NYC area, current
An irregular fall migrant and winter visitor/resident subject to periodic irruptions when they are occasionally abundant, especially in Pitch Pines on Long Island barrier beaches.

Bronx region, historical
Bicknell's *Riverdale* 1872–1901: abundant in winter 1874–75 from 3 Nov–10 May, but no other records

Eaton's *Bronx + Manhattan* 1910–14: rare winter visitor

Griscom's *Bronx Region* as of 1922: irregular winter visitor, last recorded in winter 1919–20

Kuerzi's *Bronx Region* as of 1926: rare, irregular winter visitor, last found in winter 1922–23

NYC area, historical
Cruickshank's *NYC Region* as of 1941: very rare, irregular winter visitor; last flight 1899–1900, only 27 records of small flocks since, scattered throughout area; 25 Oct–29 May

Bull's *NYC Area* as of 1968: irruption 1952–53 first since 1899–1900, followed by another 1954–55; early fall date 6 Oct; Long Island fall maximum 200+

Bull's *New York State* as of 1975: Feb 1972 irruption with maximum 250 Fire Island

Study area, historical and current
Seen almost exclusively in the extensive conifers in Woodlawn and formerly at Hillview, occasionally in Van Cortlandt. Unrecorded in the study area except during a few irruption years (1952–53, 1953–54, 1956–57, 1963–64, 2012–13) but surely present during winter in 1899–1900, etc. even if unnoticed.

Only 12 study area records: 3 in Van Cortlandt on 26 Dec 1953 (Kane); 2 at Woodlawn 30 Oct 1955 (Buckley); 3 at Jerome on 23 Dec 1956 (Kane et al.); 3 at the Dutch Gardens on 26 Dec 1956 (Kane); Hillview on 21–22 Dec 1963 (Tudor, Horowitz); 3 at Hillview on 10 Apr 1964 (M. Gochfeld); Van Cortlandt 31 Oct 1965 (Norse); Woodlawn on 9 Jan 1978 (Teator) and 11 Dec 2008 (Fiore); 7 in Van Cortlandt on 25 Nov and 4 on 8 Dec 2012 (Baksh); 26 in Woodlawn on 23 Dec 2012 (Barry, Gotlib et al.). It has doubtless occurred far more frequently than these few records imply, given the regularity with which it occurs as close as Bronx Park, and the 300+ moving south along the Hudson River at Inwood Park on 25 Nov 1963 (Norse). Bicknell noted it only during 1 winter in Riverdale, when it was common from 3 Nov 1874–10 May 1875, but it has been seen there subsequently in small numbers during irruption years: e.g., 22 on 12 Dec 2001, 4 on 23 Dec 2001 (Lyons et al.). Until Jun 1967, the fall maximum in Prospect Park was 62 on 25 Nov 1963 (Raymond).

Common Redpoll *Acanthis (flammea) flammea*

NYC area, current
An irregularly irruptive winter visitor not recorded annually, but during exceptional irruptions thousands may suddenly appear, often not until Jan, Feb, or even Mar.

Bronx region, historical
Bicknell's *Riverdale* 1872–1901: irregular in winter, abundant Jan–May 1875, Nov 1878–Mar 1879

Eaton's *Bronx + Manhattan* 1910–14: irregular winter visitor

Griscom's *Bronx Region* as of 1922: irregular winter visitor, often abundant; recorded twice in 5 years

Kuerzi's *Bronx Region* as of 1926: irregular winter visitor, occasionally abundant 13 Nov–4 May

NYC area, historical
Cruickshank's *NYC Region* as of 1941: very rare, irregular winter visitor, maximum 100+, 18 Oct–4 May

Bull's *NYC Area* as of 1968: irruptions 1946–47, 1952–54, 1955–56, 1960–61—when 2000+ at Jones Beach late Jan

Bull's *New York State* as of 1975: Jan–Mar 1974 irruption; 2000+ Fire Island Lighthouse

Study area, historical and current
In the study area in nonirruption years, Redpolls may arrive in Nov and depart in Feb–Mar, with extremes of 25 (the maximum) on 3 Nov 1954 (Scully, Buckley) and 12 on 13 Feb 1944 (Komorowski). Irruptions, though, can arrive any time between Oct and Mar but often not until Jan or later, as on 16 Mar 1956, when 200+

appeared in Van Cortlandt during a snowstorm (Kane). When that happens redpolls also remain much later than in nonirruption years. Numbers in the study area in 1956, especially at the feeders then active in Woodlawn, were 40–50 daily with extensive turnover and built to a maximum of 300+ on 11 Apr, when they abruptly departed (M. Ferguson, Buckley). Bicknell's latest departure in Riverdale was 4 May 1875 after an irruption winter. Until Jun 1967, maxima in Central and Prospect Parks were, respectively, 40 on 24 Mar 1956 (Harrison et al.) and 60 on 11 Feb 1953 (Restivo, Usin), but these counts were surely exceeded in the 1959–60 winter.

Noteworthy New York City area irruptions since 1900 (the greatest are boldfaced) have occurred in the winters of **1919–20, 1935–36, 1946–47, 1952–53,** 1955–56, **1959–60, 1973–74,** 1975–76, **1977–78,** 1980–81, 1981–82, **1986–87, 1993–94,** 1995–96, **1997–98,** 2007–08, **2010–11.**

Specimens
Bicknell collected 2 in Riverdale on 14 Jan 1879 (NYSM, ANSP).

Comments
Owing to widespread ignorance about the field separation of vagrant **Hoary**, *Acanthis hornemanni*, and **Greater**, *A. (flammea) rostrata*, Redpolls during Common Redpoll irruptions, observers being discouraged from identifying them, and regional compilers' deprecation of all nonbanding or nonspecimen records as unidentifiable, over the years many of both taxa must have been overlooked, given that we now know they appear during major Common Redpoll irruptions. For years, the only New York City and study area Hoary Redpoll record was an AMNH SY ♂ *exilipes* Hoary taken in Van Cortlandt Park on 24 Mar 1888 by Dwight. But in the 1960s it was discovered that it had actually been taken not in Van Cortlandt Park but in Tuckahoe in nearby Westchester, so the study area lost an interesting species. Both Hoary and Greater are also unrecorded from Central and Prospect Parks.

During the huge 1960 Common Redpoll irruption, an SY ♂ *A. (hornemanni) exilipes* **Hoary** (the southernmost and expected Hoary subspecies) was collected as close as West Englewood NJ, on 1 Apr, and closer still an ASY ♂ was with Commons at a Riverdale feeder on 11–12 Mar (Buckley, Sedwitz). Other ♂ *exilipes* were photographed at a feeder in Kichawan, Westchester County, from 22 Feb (Weeks, Grierson) and on 6 Mar were studied with 2000 Commons at Jones Beach (Garland, Julig), the latter's identity further confirmed after examination of AMNH skins. But most interesting of all, on 13 Mar at Jacob Riis Park in Queens, a third adult ♂ Hoary was studied by Norse with another 2000 redpolls. But this particular bird was huge, pale with only the faintest pink blush on the chest, and so white at a distance that the observer at first suspected a Snow Bunting. Closer examination and study of AMNH specimens confirmed it to have been a classic ♂ *hornemanni* Hoary Redpoll, the very high arctic breeder sometimes treated as a species separate from *exilipes*, **Hornemann's Redpoll**, *A. (hornemanni) hornemanni*. To our knowledge this is the only time this taxon has been detected in the New York City area. But despite discouragement, *exilipes* Hoaries eventually began to be sought and found in New York City area Common flocks during irruption years, including at Montauk on 24 Nov 1975 (Richards), 2 ♂♂ at a Rye feeder from 24 Feb–31 Mar 1978 (Gee et al.), and another ♂ at Brookhaven on 13 Jan 1981 (Raynor, Larsen).

Greater Redpoll is an equally striking but very different bird (also sometimes treated as a separate species) that also typically appears in minuscule numbers during massive Common Redpoll irruptions. Its identification too was interdicted. But some Greaters were fortunately collected during major irruptions: several on Shelter Island in Feb 1879 and in the Croton Point area in Feb 1883. Then one was studied at close range at Tuckahoe on 28 Feb 1936 (Bull), another was examined through a telescope at Jones Beach on 14 Feb 1938 (Elliott), and 2 were in Scarsdale on 14 Mar 1939 (Lichten, Rosenheim). Two additional reports come from nearby NJ: on the Palisades on 12 Jan 1936 (Weber et al.) and at a Bloomfield feeder on 19 Mar 1960 (Buckley, Eisenmann, Grant et al.). In all cases observers were immediately struck by the Greaters' large size, black face, and very dark ventral streaking. Additionally, on 20 Mar 1958 in Westchester County observers independently reported 3 Greater Redpolls in Mt. Kisco (Braem) and a Hoary Redpoll in Katonah (W. Russell), but no details for either record seem to have been archived.

Great progress been made in redpoll identification in the last 25 years, and the imposed curtains of obscurity that clouded our understanding of the true status of all redpoll taxa in the New York City area are finally being parted. For thorough discussions of redpoll identification illustrated with splendid paintings and color photos of all 4 taxa, readers are referred to Beadle and Henshaw 1996 and Brinkley et al. 2011.

Pine Siskin *Spinus pinus*

NYC area, current
An irruptive fall migrant and winter resident completely absent most years, widespread in others. Typically a coastwise migrant, where daily peaks in the low thousands occur during major irruptions.

Bronx region, historical
Bicknell's *Riverdale* 1872–1901: overwintered twice, remained late as 15 May 1883, 10 Jun 1893

Eaton's *Bronx + Manhattan* 1910–14: fairly common to abundant winter visitor

Griscom's *Bronx Region* as of 1922: common fall migrant, rare in winter; big flights 4 of every 5 years

Kuerzi's *Bronx Region* as of 1926: common, frequently abundant migrant, occasional winter visitor; 27 Sep–10 Jun

NYC area, historical
Cruickshank's *NYC Region* as of 1941: irregularly common fall migrant (maximum several hundred), in irruption years remaining through the winter; 5 Sep–10 Jun

Bull's *NYC Area* as of 1968: Long Island maxima fall 1500–2000, winter 1500, spring 100

Bull's *New York State* as of 1975: Long Island maximum fall 5000

Study area, historical and current
An irruptive fall migrant and winter resident, completely absent in many years, widespread in others. Study area numbers are always orders of magnitude lower than coastal counts.

In fall, first migrants arrive in early–mid-Oct, peak in mid–late Oct, and diminish to a small overwintering population by early Nov, with extremes of 45 (the maximum) on 9 Oct 1946 (Komorowski) and 2 on 11 Nov 2012 (Stuart, Finger).

The winter population is normally very small, with daily counts rarely exceeding a dozen by early Dec but with a maximum of 250 on 12 Feb 1948 (Russak). Until Jun 1967, the winter maximum in Prospect Park was 200 on 11 Dec 1963 (Raymond).

Northbound returning migrants trickle through in Mar–Apr, with extreme late dates of 5 May 1952 (Kane) and 11 May 1928 (Cruickshank) and a maximum of 50 on 2 Apr 1953 (Kane, Buckley).

Siskins normally breed no closer to the New York City area than the Catskills, where their numbers fluctuate greatly during irruption years. Compare their statewide distributions during NYSBBA–I in 1980–85 and NYSBBA–II in 2000–05 for a striking example. Siskins are always potential study area breeders following irruption winters, when Woodlawn should be checked for possible breeding in Apr and May. It bred at Croton Point in 1883 (A.K. Fisher) and on Staten Island in 1982 (Siebenheller, Siebenheller). Given that the few New York City area egg dates for Pine Siskin are 12–25 May, it is possible that pairs in Riverdale on 10 Jun 1893 (Bicknell) and from 5–19 Jun 1955 (Buckley) were nesting there amidst the numerous ornamental conifers.

American Goldfinch *Spinus tristis*

NYC area, current
A common breeder, spring/fall migrant, and overwinterer throughout. Goldfinches, like siskins, are irruptive fall migrants, rare in some years and widespread in others.

Bronx region, historical
Bicknell's *Riverdale* 1872–1901: common resident

Eaton's *Bronx + Manhattan* 1910–1914: common resident, breeds

Griscom's *Bronx Region* as of 1922: common migrant, resident, irregular in winter

Kuerzi's *Bronx Region* as of 1926: common resident, usually scarce in winter

NYC area, historical
Cruickshank's *NYC Region* as of 1941: common resident, sometimes abundant fall, spring (maximum 600) migrant

Bull's *NYC Area* as of 1968: Long Island fall maximum 2000+

Bull's *New York State* as of 1975: status unchanged

Study area, historical and current
A common breeder, spring/fall migrant, and overwinterer throughout. Typically a coastwise migrant with daily peaks in the low hundreds during in major irruptions; study area numbers are normally orders of magnitude lower.

In fall, first migrants arrive in Sep–Oct, peak in Oct–Nov, and depart by early Dec, with extremes of 14 on 13 Aug 2006 (Klein) and 29 Nov 2012 (Finger), and a maximum of 550 in a single hour on 7 Nov 1946 (Komorowski). Until Jun 1967, the fall maximum in Central Park was 60 on 4 Oct 1952 (Aronoff).

In most years there is a small overwintering population (daily counts rarely exceed a dozen) scattered throughout the study area, with maxima of 75 on 25 Dec 1969 (Scully) and 109 between Van Cortlandt and Woodlawn on 26 Dec 2004 (Lyons et al.).

Northbound returning migrants move through in Apr–May, with extremes of 3 Mar 2012 (Baksh) and 28 May 2011 (Baksh) and a maximum of 60 on 5 and 8 May 1955 (Kane et al.). Until Jun 1967, the spring maximum in Prospect Park was 100 on 14 May 1946 (Soll).

Small numbers breed annually in Van Cortlandt (>20 pairs) and Woodlawn, maybe also at Jerome and Hillview, with a total breeding population of >30 pairs. They are late breeders (egg dates 26 Jun–24 Aug: Cruickshank 1942) that are often missed or underestimated by breeding bird surveys. A flock of 9 on 25 Jul 2004 (Klein) consisted of local breeders.

Both NYSBBA I in 1980–85 and II in 2000–2005 confirmed a broad New York City area breeding distribution, where they were understandably absent only from the most urbanized portions of all New York City boroughs (McGowan and Corwin 2008). DOIBBS data showed a statewide decline from 1966–79, followed by stability through 2005 (Sauer et al. 2007).

Specimens
Dwight collected 8 in Van Cortlandt in Apr, May, Sep 1886–96, Foster 2 in Woodlawn on 26 Oct 1889 (all AMNH), and Bicknell 5 in Riverdale in May, Jul, Aug, Nov, Dec 1875–80 (NYSM).

European Goldfinch — *Carduelis carduelis*

NYC area, historical
An introduced species whose final center of breeding distribution was in Nassau and extreme southwestern Suffolk Counties and whose numbers peaked in the late 1940s–early 1950s. It was first introduced in Hoboken in 1878, reached Central Park in 1879, and then spread out to Long Island—first reported at Massapequa, Nassau County, about 1910—but slowly disappeared, and was believed extirpated when Griscom (1923) was published.

But it had *not* been extirpated. Rather, it had quietly shifted its center of abundance to southern Nassau County and extreme southwestern Suffolk County, where a small population prospered through the 1930s, '40s, and late '50s. Following the postwar development of central and southern Nassau County, goldfinch numbers plummeted, and they were presumed to have been extirpated by the early 1960s. One of the very last of the Long Island breeder population was at Orient on 30 May 1961 (Latham). Elliott (1956) studied them on Long Island for years and believed they had finally been done in by Common Grackle egg predation. During their years of residence most disappeared from Long Island postbreeding, as noted by Cruickshank (1942): "Immediately after the nesting season the birds seem to vanish, and records until the following spring are few and far between. Where the birds go is a mystery." This question remains unanswered; only a very few were ever recorded throughout New York City and western Long Island, and only once (1941) on the Southern Nassau Christmas Bird Count, which included the bulk of its summer range. (See Cruickshank 1942 for an expanded analysis of its New York City area history.)

Bronx region, historical
Eaton (1914) described its early history in the Bronx:

> Introduced at Hoboken NJ in 1878. The following year it appeared in Central Park, New York City, and soon spread over the northern portions of Manhattan island and surrounding country. Locally it was not an uncommon resident. In the winter of 1891 many were noticed flocking with American Goldfinches at Dobbs Ferry [Westchester County], but several were found dead in the snow, evidently the severity of the winter proving too much for this species. In the spring of 1900 I noticed several pairs that were endeavoring to build their nests in Central Park, and in the country about Kings Bridge and Spuyten Duyvil, New York City.

Bicknell also recorded them in Riverdale around that time, on 12–24 May 1901 and 1 Jan 1902, when they were breeding there.

Study area, historical and current
In the study area, in addition to the Kingsbridge Meadows breeding, there is one Van Cortlandt record: 21 Oct 1956 (Petersen, Buckley et al.). Nearby, Kieran found it 3 times in Riverdale:

21 Aug 1930, 3 Oct 1943, and 3 Jan 1955, and Buckley saw it there at feeders on 24 Dec 1952 and 28 May 1955. Cruickshank found one at University Heights on 19 Oct 1930. There have been other study area records since 1960, including 2 at Tibbett's Marsh on 12 Oct 1995 (Jaslowitz), but they have not been collated.

Comments
New York City area observers have encountered 1–3 European Goldfinches on a frequent if unpredictable basis since the early 1960s (most were never formally reported), and more than a few people have had difficulty accepting that they are *all* the result of recurring releases, or exceedingly long-lived individuals, or continual midwestern cagebird catastrophes. A small persistent breeding population—perhaps one more dispersed than that which died out (?) in the late 1950s on central southern LI—would also explain them, but there has been no evidence of post-1960 breeding anywhere in the Northeast. Still, winter flocks of up to 8 in Brooklyn (2016) and in Hartford, CT (2015), and at least 12 on Governor's Island (2017), suggest that it will soon be time for a thorough reanalysis of its New York City area status.

Evening Grosbeak   *Coccothraustes vespertinus*

NYC area, current
An irruptive, erratic fall migrant, winter visitor/resident, and spring migrant, occurring singly, in small groups, or by the hundreds during major irruptions. There are often many decades between major irruptions, yet during them they have bred as close as northern NJ. Peak New York City area numbers occurred between the late 1950s and late 1980s, and since then they have been relatively scarce. Like certain warblers, their interannual variation in abundance is tied to Spruce Budworm cycles, but their New York State breeding population is believed to be moderated more by bird feeder maintenance.

Bronx region, historical
Bicknell's *Riverdale* 1872–1901: unrecorded

Eaton's *Bronx + Manhattan* 1910–14: unrecorded

Griscom's *Bronx Region* as of 1922: irregular winter visitor, 3 recent records

Kuerzi's *Bronx Region* as of 1926: irregular winter visitor, 13 Nov–6 May

NYC area, historical
Cruickshank's *NYC Region* as of 1941: irregular, rare winter visitor, 27 Oct–6 May

Bull's *NYC Area* as of 1968: greatly increased >1946; 6 Sep–7 Jun, once Jul; most at feeders, especially during major irruption winters like 1951–52, when 6000 estimated at all area feeders (including northern NJ); first spring return flights in Apr–May; maxima spring 100, fall 150

Bull's *New York State* as of 1975: second Jul record

Study area, historical and current
Not found in the Bronx until the 1915–16 winter, this erratic visitor occurs in irruptions large and small, during which they may pass by, remain for the winter, or some combination. Feeders sustain most flocks locally.

Beginning in the 1910s–20s, recorded in all 4 extant subareas where the absence of feeders limits occurrences to mostly vocalizing flyovers. Area winter residents typically (a term used with reservations) arrive in Oct, remain through the winter in varying numbers, then depart in Mar, while those overwintering farther south pass back through in Apr–May. Extreme dates are 3 in Van Cortlandt on 9 Sep 1958 (Kane) and 2 there on 9 May 1964 (Russak), with fall and winter maxima of 50 in early Nov 1943 (Komorowski et al.) and from 1–8 Jan 1944 (Soll et al.)—an irruption winter. Until Jun 1967, the fall maximum in Central Park was 38 on 15 Oct 1961 (Bloom), and in Prospect Park 74 in early Nov 1951 (Brooklyn Bird Club). It has been unrecorded in the study area since 4 on 24 Apr 1966 (Kane), on the entire Bronx-Westchester Christmas Bird Count since 1987 has been found only in 1995 (1) and 2003 (5).

As NYSBBA I in 1980–85 and II in 2000–2005 clearly demonstrated, outside its Adirondack breeding-block cluster, sporadic nesting may occur anywhere in New York State after irruption winters but even then normally no closer than the Catskills (McGowan and Corwin 2008).

Old-World Sparrows: Passeridae

House Sparrow *Passer domesticus*

NYC and study areas, historical and current
A widespread but diminishing introduced exotic (nonnative) that is resident throughout. Cruickshank (1942) summarized its history well: "Eight pairs were liberated in Brooklyn in 1850 but did not thrive. In 1852 a larger number was brought over and kept in confinement throughout the winter. Those that survived were liberated the following spring in Green-Wood Cemetery, Brooklyn. These then quickly multiplied and spread rapidly. The species was first recorded in Chatham NJ in 1868 and in Caldwell, NJ in 1870." It reached its maximum abundance in the New York City area in the late 1800s, declining rapidly thereafter as horses (whose droppings were a major food source) were replaced by motor vehicles.

In the study area, Bicknell first found it in Riverdale on 11 Apr 1879, and Dwight and Foster collected their first in Van Cortlandt and Woodlawn in 1886–87, although they may have been present earlier. In any case, they quickly increased in numbers throughout the Bronx, peaking around 1900 and then plummeting with the disappearance of horse-drawn vehicles. As of 2016 they occur and breed in all 4 extant subareas but are quite local; recent Van Cortlandt seasonal maxima include *spring* 45, *summer* 105, *fall* 175, and *winter* 30. Over the years few observers have bothered recording numbers, so most counts are recent and barely representative.

House Sparrows are quite philopatric and appear to have very small home ranges. They regularly collect in flocks to 200 outside Van Cortlandt and then come into the park to feed, warbler-like, on spring buds high in trees (Young, Buckley). Probably <40 pairs breed within Van Cortlandt itself, with just as many around Jerome and Hillview and in Woodlawn. Another 100+ pairs also breed in urban areas immediately adjacent to Van Cortlandt.

Hard study area data are unavailable, but it is the authors' and others' perception that study area numbers have dropped significantly since the late 1940s–early 1950s, and again after the early 1990s. DOIBBS data did show a 2.1% annual statewide decline from 1966–79 that accelerated to 2.9% annually from 1980–2005 (Sauer et al. 2005).

Specimens

Dwight collected 4 in Van Cortlandt in Apr, Jun 1886 and others on 13 May 1887 and 12 May 1890; Foster took one in Woodlawn on 5 Oct 1887 (all AMNH). Bicknell collected singles in Riverdale in Sep, Oct 1880–81 (NYSM).

Appendix 1

Table 1. Approximate sizes of important study area components

Component	Size (ac/ha)
Van Cortlandt Park	1146/463
Van Cortlandt Swamp (1900)	70/28
Van Cortlandt Lake (water surface)	19/7
Tibbett's Marsh (2015)	4/1.6
Parade Ground (1890)	150/60
Northwest Forest	200/80
Vault Hill	45/18
Croton Forest (The Ridge)	180/72
Northeast Forest	150/60
Shandler Area	50/20
Golf Courses (Mosholu + Van Cortlandt)	200/80
Woodlawn Cemetery	400/161
Ponds (water surface)	1/0.4
Jerome Reservoir (water surface)	94/38
Hillview Reservoir (water surface)	90/36
Jerome Meadows (1880)	150/60
Jerome Swamp (1925)	90/36
Kingsbridge Meadows (1875)	130/52
Tibbett's Brook (length)	
Yonkers line–Van Cortlandt Lake	1.3 smi/2.1 km
Van Cortlandt Lake–Spuyten Duyvil Creek (1900)	1.2 smi/1.9 km

Table 2. Approximate distances from Van Cortlandt Lake to locations mentioned in the text. smi = statute miles; km = kilometers; distances >5 smi rounded to nearest smi, km. Distances measured with Google Earth Pro's Ruler.

Distances ≤5 smi

Baychester Marshes, Bronx (3.7 smi/6 km ESE)
Bronx Park (1.5 smi/2.4 km SE)
Bronx River (1.2 smi/1.9 km E)
Bronxville, Westchester County (4.4 smi/7.1 km NE)
Bronx Zoo (2 smi/3.2 km SSE)
Fort Tryon Park, Manhattan (3 smi/4.8 km SW)
Grassy Sprain Reservoir, Yonkers (4.9 smi/8 km NNE)
Hudson River (1.6 smi/2.6 km W)
Hutchinson River marshes, Bronx (3.7 smi/5.9 km ESE)
Inwood Hill Park, Manhattan (2.1 smi/3.4 km SW)
NJ Palisades (2.6 smi/4.2 km W)
Pelham Bay Park [Eastchester Bay, Goose Island, Hunter's Island, Orchard Beach, Rodman's Neck] (5 smi/8 km ESE)
Riverdale Park, Bronx (1.4 smi/2.2 km W)
Seton Falls Park, Bronx (2.7 smi/4.4 km E)
Sherman Creek, Manhattan (2.7 smi/4.4 km SSW)
Spuyten Duyvil, Bronx (2 smi/3.2 km SW)
Williamsbridge, Bronx (1.9 smi/3 km ESE)

Distances >5 smi

Alpine NJ Stateline Lookout (7 smi/11 km NNW)
Baxter Creek, Bronx (6 smi/10 km SE)
Central Park, Manhattan (9 smi/14 km SSW)
Connetquot River State Park, Suffolk County, NY (40 smi/64 km ESE)
Cross River Reservoir, Westchester County, NY (28 smi/45 km NNE)
Croton Point, Westchester County, NY (20 smi/32 km N)
Dutchess County, NY (42 smi/66 km N)
Extreme northeastern Westchester County, NY (35 smi/56 km NE)
Ferry Point Park/Whitestone Bridge, Bronx (6 smi/10 km SE)
Great Swamp, Morris County, NJ (32 smi/51 km SW)
Greene County, NY (95 smi, 152 km NNW)
Hackensack Meadowlands, Bergen County, NJ (14 smi/22 km SW)
Huckleberry Island, Westchester County, NY (7 smi/11 km E)
Jamaica Bay, Brooklyn/Queens (20 smi/32 km SSE)
Kensico Dam, Westchester County, NY (14 smi/22 km NNE)
Larchmont, Westchester County, New York (8 smi/13 km (ENE)
Lenoir Preserve, Westchester County, NY (6 smi/9 km N)
Morris/Passaic Counties, NJ (21 smi/33 km W)
NJ Pine Barrens (95 smi/152 km SSW)
North/South Brother Islands, Bronx (7 smi/10 km S)
Northern Westchester County, NY (30 smi/48 km N)
Orange County, NY (25 smi/40 km NW)
Piermont Marsh, Rockland County, NY (9 smi/14 km NNW)
Piermont Pier, Rockland County, NY (11 smi/17 km NNW)
Prospect Park, Brooklyn (17 smi/28 km SSE)
Purchase (SUNY), Westchester County, NY (15 smi/24 km NE)
Rockland County, NY (12 smi/19 km NNW)
Rye, Westchester County, NY (13 smi/21 km NE)

Distances >5 smi	
Sands Point, Nassau County, NY (9 smi/14 km ESE)	
Sandy Hook, Monmouth County, NJ (29 smi, 48 km SSW)	
Scarsdale, Westchester County, NY (10 smi/16 km NNE)	
Sterling Forest (Tuxedo Park), NY (29 smi/48 km NW)	
Sullivan County, NY (60 smi/96 km NW)	
Tappan Zee Bridge, NY (12 smi/19 km N)	
Throgs Neck Bridge, Bronx (7 smi/11 km SE)	
Troy Meadows, Morris County, NJ (26 smi/42 km WSW)	
Tuckahoe, NY (5 smi/8 km NE)	
Ulster County, NY (48 smi/78 km NNW)	
Westchester County Airport, Harrison (15 smi/23 km NE)	
White Plains, NY (11 smi/18 km NNE	

Table 3. Years by decades when new species (n = 51) have been added to the study area cumulative list after 1941, the cutoff date for material in Cruickshank (1942). Ruffed Grouse, Wild Turkey, Pileated Woodpecker, and Common Raven were the first since their extirpation in the 1700s–1800s.

Species	Year	Species	Year
Tufted Titmouse	1942	Buff-breasted Sandpiper	1964
Red-necked Phalarope	1943	Lesser Black-backed Gull	1964
Short-billed Dowitcher	1944	Fulvous Whistling-Duck	1965
Willow Flycatcher	1946	Black-headed Gull	1965
Golden Eagle	1946	Northern Wheatear	1968
Sanderling	1951	Tricolored Heron	1972
Dunlin	1951	Monk Parakeet	1974
Stilt Sandpiper	1951		
Yellow-throated Warbler	1952	Brewer's Blackbird	1981
Prothonotary Warbler	1952	Boat-tailed Grackle	1982
House Finch	1953	Le Conte's Sparrow	1982
Dickcissel	1953	Common Raven	1984
Bell's Vireo	1954		
Summer Tanager	1954	Wild Turkey	1994
Seaside Sparrow	1954	Cackling Goose	1996
Clay-colored Sparrow	1954		
Yellow-headed Blackbird	1954	White-winged Dove	2000
Ruffed Grouse	1955	Wood Stork	2003
Least Tern	1956	American Golden-Plover	2005
Western Kingbird	1956	American White Pelican	2006
Tufted Duck	1956	Black Vulture	2008
Blue Grosbeak	1960	Great Cormorant	2010
Pileated Woodpecker	1962	Greater White-fronted Goose	2011
Cattle Egret	1962	Barnacle Goose	2112
Lark Sparrow	1963	Eurasian Collared-Dove	2014
Black Skimmer	1963	Pink-footed Goose	2016
Swainson's Warbler	1964	Sandhill Crane	2017

Table 4. New York non-Long Island Sound sites no farther than 3 smi (5 km) from Van Cortlandt Lake that have recorded 48 species unknown in the study area. * = probably regular, if not annual, in small numbers in that location; *italics* = post-hurricanes only. Eastern Willet and Ruddy Turnstone are regular or annual in the East Bronx and may already have been recorded at Spuyten Duyvil, Sherman Creek, or nearby locations.

Sherman Creek
 White-rumped Sandpiper *
 Western Sandpiper *

Fort Tryon Park
 Calliope Hummingbird

Inwood Hill Park
 Mississippi Kite
 American Avocet
 Thayer's Gull
 Pacific-slope Flycatcher
 Western Tanager

Spuyten Duyvil
 Tundra Swan
 White-rumped Sandpiper *
 Western Sandpiper *
 Long-billed Dowitcher
 Red Phalarope
 Jaeger sp.

Riverdale
 Rufous Hummingbird
 Scissor-tailed Flycatcher
 Ash-throated Flycatcher
 Great Kiskadee
 Cave Swallow
 Oregon Junco
 Pink-sided Junco
 Hoary Redpoll
 Black-headed Grosbeak

Hudson River
 Tundra Swan
 King Eider

 Common Eider
 Cory's Shearwater
 Audubon's Shearwater
 Wilson's Storm-Petrel *
 Leach's Storm-Petrel
 Northern Gannet
 Brown Pelican
 Eastern Willet
 Parasitic Jaeger
 Thick-billed Murre
 Black-legged Kittiwake
 Black Tern *
 Arctic Tern
 Forster's Tern *

Yonkers
 Purple Gallinule

Williamsbridge
 Black-headed Grosbeak

Bronx Park/Zoo
 Mississippi Kite
 Whimbrel
 Baird's Sandpiper
 White-rumped Sandpiper
 Wilson's Phalarope
 Thick-billed Murre
 Rufous Hummingbird
 Townsend's Warbler
 Spotted Towhee

Table 5. Species that have occurred in only 1 of the 6 non–Van Cortlandt Park study subareas. All but Dovekie are annual or regular in or over the Bronx and so might appear in Van Cortlandt Park proper any time.

Kingsbridge Meadows	Jerome Reservoir
Clapper Rail	Black Scoter
Western Willet	Buff-breasted Sandpiper
	Dovekie
	Bonaparte's Gull

Table 6. Study area breeders (and their locations) that have nested only outside but immediately adjacent to Van Cortlandt Park proper (n = 9); *italics* are extirpated breeders. Nighthawk is also likely to have nested within Van Cortlandt but details are lacking.

Species	Subarea
Clapper Rail	Kingsbridge Meadows
Common Nighthawk	Adjacent buildings; Hillview + Woodlawn + Jerome?
Nashville Warbler	Woodlawn Cemetery
Vesper Sparrow	Jerome Meadows
Grasshopper Sparrow	Kingsbridge Meadows, Jerome Meadows
Saltmarsh Sparrow	Kingsbridge Meadows
Bobolink	Kingsbridge Meadows, Jerome Meadows
Eastern Meadowlark	Hillview Reservoir
European Goldfinch	Kingsbridge Meadows

Table 7. Cumulative species lists after 1871 for Van Cortlandt, Central, and Prospect Parks. JR = Jerome Reservoir, KM = Kingsbridge Meadows.

Species	Van Cortlandt	Central	Prospect
Fulvous Whistling-Duck	x		
Pink-footed Goose	x		
Greater White-fronted Goose	x		x
Snow Goose	x	x	x
Brant	x	x	x
Barnacle Goose	x	x	x
Cackling Goose	x		x
Canada Goose	x	x	x
Mute Swan	x	x	x
Trumpeter Swan		x	
Tundra Swan		x	x
Wood Duck	x	x	x
Gadwall	x	x	x
Eurasian Wigeon	x	x	x
American Wigeon	x	x	x
American Black Duck	x	x	x
Mallard	x	x	x
Blue-winged Teal	x	x	x
Northern Shoveler	x	x	x
Northern Pintail	x	x	x
Eurasian Teal	x	x	
Green-winged Teal	x	x	x
Canvasback	x	x	x
Redhead	x	x	x
Ring-necked Duck	x	x	x
Tufted Duck	x	x	
Greater Scaup	x	x	x
Lesser Scaup	x	x	x
Common Eider		x	

(*Continued*)

Table 7. (Continued)

Species	Van Cortlandt	Central	Prospect
Surf Scoter	x	x	x
White-winged Scoter	x	x	x
Black Scoter	x/JR		x
Long-tailed Duck	x	x	x
Bufflehead	x	x	x
Common Goldeneye	x	x	x
Hooded Merganser	x	x	x
Common Merganser	x	x	x
Red-breasted Merganser	x	x	x
Ruddy Duck	x	x	x
Northern Bobwhite	x	x	x
Ring-necked Pheasant	x		
Ruffed Grouse	x	x	
Wild Turkey	x	x	x
Red-throated Loon	x	x	x
Common Loon	x	x	x
Pied-billed Grebe	x	x	x
Horned Grebe	x	x	x
Red-necked Grebe	x	x	x
Wood Stork	x		
Great Cormorant	x	x	x
Double-crested Cormorant	x	x	x
Anhinga		x	x
American White Pelican	x	x	
American Bittern	x	x	x
Least Bittern	x	x	x
Great Blue Heron	x	x	x
Great Egret	x	x	x
Snowy Egret	x	x	x
Little Blue Heron	x	x	x
Tricolored Heron	x	x	x
Cattle Egret	x	x	x
Green Heron	x	x	x
Black-crowned Night-Heron	x	x	x
Yellow-crowned Night-Heron	x	x	x
White Ibis			x
Glossy Ibis	x	x	x
Black Vulture	x	x	x
Turkey Vulture	x	x	x
Osprey	x	x	x
Swallow-tailed Kite	x		x
Mississippi Kite		x	x
Bald Eagle	x	x	x
Northern Harrier	x	x	x
Sharp-shinned Hawk	x	x	x
Cooper's Hawk	x	x	x
Northern Goshawk	x	x	x
Red-shouldered Hawk	x	x	x
Broad-winged Hawk	x	x	x

Species	Van Cortlandt	Central	Prospect
Red-tailed Hawk	x	x	x
Rough-legged Hawk	x	x	x
Golden Eagle	x	x	x
Yellow Rail	x		
Clapper Rail	x/KM	x	x
King Rail	x		
Virginia Rail	x	x	x
Sora	x	x	x
Purple Gallinule		x	x
Common Gallinule	x	x	x
American Coot	x	x	x
Sandhill Crane	x		
Black-necked Stilt			x
Black-bellied Plover	x	x	x
American Golden-Plover	x		x
Semipalmated Plover	x	x	
Killdeer	x	x	x
Spotted Sandpiper	x	x	x
Solitary Sandpiper	x	x	x
Greater Yellowlegs	x	x	x
Eastern Willet			x
Western Willet	x/KM		
Lesser Yellowlegs	x	x	x
Upland Sandpiper	x	x	x
Whimbrel		x	
Stilt Sandpiper	x		
Sanderling	x		x
Dunlin	x		
Purple Sandpiper			x
Least Sandpiper	x	x	x
White-rumped Sandpiper		x	
Buff-breasted Sandpiper	x/JR		
Pectoral Sandpiper	x	x	x
Semipalmated Sandpiper	x	x	x
Western Sandpiper		x	
Short-billed Dowitcher	x		x
Wilson's Snipe	x	x	x
American Woodcock	x	x	x
Red-necked Phalarope	x		x
Red Phalarope		x	
Dovekie	x/JR		x
Black-legged Kittiwake			x
Bonaparte's Gull	x/JR	x	x
Black-headed Gull	x	x	x
Laughing Gull	x	x	x
Ring-billed Gull	x	x	x
Herring Gull	x	x	x
Iceland Gull	x	x	x

(Continued)

Table 7. (Continued)

Species	Van Cortlandt	Central	Prospect
Lesser Black-backed Gull	x	x	x
Glaucous Gull	x	x	x
Great Black-backed Gull	x	x	x
Least Tern	x		x
Caspian Tern	x	x	x
Black Tern			x
Common Tern	x	x	x
Forster's Tern			x
Black Skimmer	x	x	x
Rock Pigeon	x	x	x
Eurasian Collared-Dove	x	x	
Passenger Pigeon	x	x	x
White-winged Dove	x		x
Mourning Dove	x	x	x
Yellow-billed Cuckoo	x	x	x
Black-billed Cuckoo	x	x	x
Barn Owl	x	x	x
Eastern Screech-Owl	x	x	x
Great Horned Owl	x	x	x
Snowy Owl	x	x	
Barred Owl	x	x	x
Long-eared Owl	x	x	x
Short-eared Owl	x		
Boreal Owl		x	
Northern Saw-whet Owl	x	x	x
Common Nighthawk	x	x	x
Chuck-will's-widow		x	x
Eastern Whip-poor-will	x	x	x
Chimney Swift	x	x	x
Ruby-throated Hummingbird	x	x	x
Rufous Hummingbird		x	
Belted Kingfisher	x	x	x
Red-headed Woodpecker	x	x	x
Red-bellied Woodpecker	x	x	x
Yellow-bellied Sapsucker	x	x	x
Downy Woodpecker	x	x	x
Hairy Woodpecker	x	x	x
Black-backed Woodpecker	x	x	
Northern Flicker	x	x	x
Pileated Woodpecker	x	x	x
American Kestrel	x	x	x
Merlin	x	x	x
Peregrine Falcon	x	x	x
Monk Parakeet	x	x	x
Olive-sided Flycatcher	x	x	x
Eastern Wood-Pewee	x	x	x
Yellow-bellied Flycatcher	x	x	x
Acadian Flycatcher	x	x	x

Species	Van Cortlandt	Central	Prospect
Alder Flycatcher	x	x	x
Willow Flycatcher	x	x	x
Least Flycatcher	x	x	x
Hammond's Flycatcher		x	
Pacific-slope Flycatcher		x	
Eastern Phoebe	x	x	x
Say's Phoebe			x
Ash-throated Flycatcher		x	x
Great Crested Flycatcher	x	x	x
Western Kingbird	x	x	x
Eastern Kingbird	x	x	x
Scissor-tailed Flycatcher			x
Loggerhead Shrike	x	x	x
Northern Shrike	x	x	x
White-eyed Vireo	x	x	x
Bell's Vireo	x	x	x
Yellow-throated Vireo	x	x	x
Blue-headed Vireo	x	x	x
Warbling Vireo	x	x	x
Philadelphia Vireo	x	x	x
Red-eyed Vireo	x	x	x
Blue Jay	x	x	x
Black-billed Magpie	x	x	x
American Crow	x	x	x
Fish Crow	x	x	x
Common Raven	x	x	x
Horned Lark	x	x	x
Purple Martin	x	x	x
Tree Swallow	x	x	x
N. Rough-winged Swallow	x	x	x
Bank Swallow	x	x	x
Cliff Swallow	x	x	x
Barn Swallow	x	x	x
Black-capped Chickadee	x	x	x
Boreal Chickadee	x	x	
Tufted Titmouse	x	x	x
Red-breasted Nuthatch	x	x	x
White-breasted Nuthatch	x	x	x
Brown Creeper	x	x	x
House Wren	x	x	x
Winter Wren	x	x	x
Sedge Wren	x	x	x
Marsh Wren	x	x	x
Carolina Wren	x	x	x
Bewick's Wren		x	x
Blue-gray Gnatcatcher	x	x	x
Golden-crowned Kinglet	x	x	x
Ruby-crowned Kinglet	x	x	x

(*Continued*)

Table 7. (Continued)

Species	Van Cortlandt	Central	Prospect
Northern Wheatear	x	x	
Eastern Bluebird	x	x	x
Veery	x	x	x
Gray-cheeked Thrush	x	x	x
Bicknell's Thrush	x	x	x
Swainson's Thrush	x	x	x
Hermit Thrush	x	x	x
Wood Thrush	x	x	x
American Robin	x	x	x
Varied Thrush		x	x
Gray Catbird	x	x	x
Northern Mockingbird	x	x	x
Brown Thrasher	x	x	x
European Starling	x	x	x
American Pipit	x	x	x
Bohemian Waxwing			x
Cedar Waxwing	x	x	x
Lapland Longspur	x	x	x
Snow Bunting	x	x	x
Ovenbird	x	x	x
Worm-eating Warbler	x	x	x
Louisiana Waterthrush	x	x	x
Northern Waterthrush	x	x	x
Golden-winged Warbler	x	x	x
Blue-winged Warbler	x	x	x
Black-and-white Warbler	x	x	x
Prothonotary Warbler	x	x	x
Swainson's Warbler	x	x	x
Tennessee Warbler	x	x	x
Orange-crowned Warbler	x	x	x
Nashville Warbler	x	x	x
Connecticut Warbler	x	x	x
Mourning Warbler	x	x	x
Kentucky Warbler	x	x	x
Common Yellowthroat	x	x	x
Hooded Warbler	x	x	x
American Redstart	x	x	x
Cape May Warbler	x	x	x
Cerulean Warbler	x	x	x
Northern Parula	x	x	x
Magnolia Warbler	x	x	x
Bay-breasted Warbler	x	x	x
Blackburnian Warbler	x	x	x
Yellow Warbler	x	x	x
Chestnut-sided Warbler	x	x	x
Blackpoll Warbler	x	x	x
Black-throated Blue Warbler	x	x	x
Palm Warbler	x	x	x

Species	Van Cortlandt	Central	Prospect
Pine Warbler	x	x	x
Myrtle Warbler	x	x	x
Yellow-throated Warbler	x	x	x
Prairie Warbler	x	x	x
Black-throated Gray Warbler		x	
Townsend's Warbler		x	x
Hermit Warbler		x	
Black-throated Green Warbler	x	x	x
Canada Warbler	x	x	x
Wilson's Warbler	x	x	x
Yellow-breasted Chat	x	x	x
Eastern Towhee	x	x	x
Bachman's Sparrow			x
American Tree Sparrow	x	x	x
Chipping Sparrow	x	x	x
Clay-colored Sparrow	x	x	x
Field Sparrow	x	x	x
Vesper Sparrow	x	x	x
Lark Sparrow	x	x	x
Savannah Sparrow	x	x	x
Grasshopper Sparrow	x	x	x
Baird's Sparrow		x	
Henslow's Sparrow	x	x	x
Le Conte's Sparrow	x	x	
Nelson's Sparrow (inland)	x	x	x
Saltmarsh Sparrow	x	x	x
Seaside Sparrow	x	x	x
Red Fox Sparrow	x	x	x
Sooty Fox Sparrow		x	
Song Sparrow	x	x	x
Lincoln's Sparrow	x	x	x
Swamp Sparrow	x	x	x
White-throated Sparrow	x	x	x
White-crowned Sparrow	x	x	x
Slate-colored Junco	x	x	x
Oregon Junco		x	x
Summer Tanager	x	x	x
Scarlet Tanager	x	x	x
Western Tanager		x	x
Northern Cardinal	x	x	x
Rose-breasted Grosbeak	x	x	x
Blue Grosbeak	x	x	x
Indigo Bunting	x	x	x
Painted Bunting	x	x	x
Dickcissel	x	x	x
Bobolink	x	x	x
Red-winged Blackbird	x	x	x
Eastern Meadowlark	x	x	x

(*Continued*)

Table 7. (Continued)

Species	Van Cortlandt	Central	Prospect
Yellow-headed Blackbird	x	x	
Rusty Blackbird	x	x	x
Brewer's Blackbird	x	x	
Common Grackle	x	x	x
Boat-tailed Grackle	x	x	x
Brown-headed Cowbird	x	x	x
Orchard Oriole	x	x	x
Baltimore Oriole	x	x	x
Pine Grosbeak	x	x	x
Purple Finch	x	x	x
House Finch	x	x	x
Red Crossbill	x	x	x
White-winged Crossbill	x	x	x
Common Redpoll	x	x	x
Pine Siskin	x	x	x
American Goldfinch	x	x	x
European Goldfinch	x	x	x
Evening Grosbeak	x	x	x
House Sparrow	x	x	x
TOTALS	**301**	**303**	**298**
337	89%	90%	88%
Unique to one park:	12	14	11
37/337 = 11% single-park only	4%	4%	3%

Table 8. Species unique to Van Cortlandt, Central, or Prospect Park. JR = Jerome Reservoir, KM = Kingsbridge Meadows.

Van Cortlandt Park (n = 12) Landbirds: 2/12
  Fulvous Whistling-Duck
  Pink-footed Goose
  Ring-necked Pheasant
  Wood Stork
  Yellow Rail
  Sandhill Crane
  King Rail
  Dunlin
  Western Willet (KM)
  Stilt Sandpiper
  Buff-breasted Sandpiper (JR)
  Short-eared Owl

Central Park (n = 14) Landbirds: 8/14
  Trumpeter Swan
  Common Eider
  Whimbrel
  Western Sandpiper
  White-rumped Sandpiper
  Red Phalarope
  Boreal Owl
  Rufous Hummingbird
  Hammond's Flycatcher
  Pacific-slope Flycatcher
  Black-throated Gray Warbler
  Hermit Warbler
  Baird's Sparrow
  Sooty Fox Sparrow

Prospect Park (n = 11) Landbirds: 4/11
  White Ibis
  Black-necked Stilt
  Eastern Willet
  Purple Sandpiper
  Black-legged Kittiwake
  Black Tern
  Forster's Tern
  Say's Phoebe
  Scissor-tailed Flycatcher
  Bohemian Waxwing
  Bachman's Sparrow

Table 9a. Species recorded from Central and Prospect Parks but not Van Cortlandt (n = 11)

Tundra Swan
Anhinga
Mississippi Kite
Purple Gallinule
Chuck-will's-widow
Ash-throated Flycatcher
Bewick's Wren
Varied Thrush
Townsend's Warbler
Western Tanager
Oregon Junco

Table 9b. Species recorded from Van Cortlandt and Prospect Parks but not Central (n = 11)

White-fronted Goose
Cackling Goose
Black Scoter
Swallow-tailed Kite
American Golden-Plover
Sanderling
Short-billed Dowitcher
Red-necked Phalarope
Dovekie
Least Tern
White-winged Dove

Table 9c. Species recorded from Van Cortlandt and Central Parks but not Prospect (n = 12)

Eurasian Teal
Tufted Duck
American White Pelican
Semipalmated Plover
Eurasian Collared-Dove
Snowy Owl
Black-backed Woodpecker
Boreal Chickadee
Northern Wheatear
Le Conte's Sparrow
Yellow-headed Blackbird
Brewer's Blackbird

Table 10. Reasonably anticipated species never recorded from Central, Prospect, or Van Cortlandt Park (n = 10)

Ross's Goose	Franklin's Gull
Ruddy Turnstone	Thayer's Gull
Baird's Sandpiper	Cave Swallow
Long-billed Dowitcher	Audubon's Warbler
Wilson's Phalarope	Hoary Redpoll

Table 11. Cumulative known and potential (*italics*) breeding species after 1871 for Van Cortlandt, Central, and Prospect Parks

	Van Cortlandt		Central		Prospect	
Species	Has Bred	Potential Breeder	Has Bred	Potential Breeder	Has Bred	Potential Breeder
Canada Goose	x		x		x	
Mute Swan	x		x		x	
Wood Duck	x		x		x	
Gadwall		x		x		
American Black Duck	x				x	
Mallard	x		x		x	
Northern Bobwhite	x		x		x	
Ring-necked Pheasant	x					
Wild Turkey	x					
Pied-billed Grebe		x			x	
American Bittern	x					
Least Bittern	x					
Great Blue Heron		x				
Green Heron	x		x		x	
Black-crowned Night-Heron	x				x	
Yellow-crowned Night-Heron		x				
Black Vulture (nearby)		x				x
Turkey Vulture	x					
Osprey	x					
Northern Harrier	x					
Sharp-shinned Hawk	x					
Cooper's Hawk	x					
Red-shouldered Hawk	x				x	
Broad-winged Hawk	x					
Red-tailed Hawk	x		x		x	
Clapper Rail	x					
King Rail	x					
Virginia Rail	x					
Sora	x					
Common Gallinule	x					
American Coot	x					
Killdeer	x		x			
Spotted Sandpiper	x					

Species	Van Cortlandt		Central		Prospect	
	Has Bred	Potential Breeder	Has Bred	Potential Breeder	Has Bred	Potential Breeder
American Woodcock	x					
Rock Pigeon	x		x		x	
Mourning Dove	x		x		x	
Yellow-billed Cuckoo	x			x	x	
Black-billed Cuckoo	x				x	
Barn Owl		x				
Eastern Screech-Owl	x		x		x	
Great Horned Owl	x				x	
Barred Owl	x					
Common Nighthawk (nearby)	x				x	
Eastern Whip-poor-will	x		x			
Chimney Swift	x		x		x	
Ruby-throated Hummingbird	x		x		x	
Belted Kingfisher	x			x		x
Red-headed Woodpecker	x			x		
Red-bellied Woodpecker	x		x		x	
Downy Woodpecker	x		x		x	
Hairy Woodpecker	x		x		x	
Northern Flicker	x		x		x	
Pileated Woodpecker		x				
American Kestrel	x		x		x	
Monk Parakeet	x				x	
Eastern Wood-Pewee	x		x		x	
Acadian Flycatcher	x		x		x	
Alder Flycatcher	x					
Willow Flycatcher	x					
Least Flycatcher	x		x			
Eastern Phoebe	x		x		x	
Great Crested Flycatcher	x		x		x	
Eastern Kingbird	x		x		x	
White-eyed Vireo	x				x	
Yellow-throated Vireo	x		x		x	
Warbling Vireo	x		x		x	
Red-eyed Vireo	x		x		x	
Blue Jay	x		x		x	
American Crow	x		x		x	
Fish Crow	x		x		x	
Common Raven (nearby)		x				x
Horned Lark	x					
Purple Martin		x		x		x
Tree Swallow	x		x		x	
N. Rough-winged Swallow	x		x			
Bank Swallow	x		x			
Cliff Swallow	x					
Barn Swallow	x		x		x	

(*Continued*)

Table 11. (Continued)

Species	Van Cortlandt		Central		Prospect	
	Has Bred	Potential Breeder	Has Bred	Potential Breeder	Has Bred	Potential Breeder
Black-capped Chickadee	x		x		x	
Tufted Titmouse	x		x		x	
White-breasted Nuthatch	x		x		x	
Brown Creeper	x					
House Wren	x		x		x	
Sedge Wren		x				
Marsh Wren	x					
Carolina Wren	x		x		x	
Blue-gray Gnatcatcher	x		x		x	
Golden-crowned Kinglet						
Eastern Bluebird	x		x			
Veery	x		x			
Wood Thrush	x		x		x	
American Robin	x		x		x	
Gray Catbird	x		x		x	
Brown Thrasher	x		x		x	
Northern Mockingbird	x		x		x	
European Starling	x		x		x	
Cedar Waxwing	x		x		x	
Ovenbird	x				x	
Worm-eating Warbler	x		x			
Louisiana Waterthrush	x					
Blue-winged Warbler	x					
Black-and-white Warbler	x		x			
Nashville Warbler	x					
Kentucky Warbler	x					
Common Yellowthroat	x		x		x	
Hooded Warbler	x					
American Redstart	x		x		x	
Yellow Warbler	x			x	x	
Chestnut-sided Warbler	x		x			
Pine Warbler		x				x
Yellow-breasted Chat	x					
Eastern Towhee	x		x		x	
Chipping Sparrow	x		x		x	
Field Sparrow	x					
Vesper Sparrow	x		x			
Savannah Sparrow		x				
Grasshopper Sparrow	x					
Saltmarsh Sparrow	x					
Seaside Sparrow		x				
Song Sparrow	x		x		x	
Swamp Sparrow	x					
Scarlet Tanager	x				x	
Northern Cardinal	x		x		x	
Rose-breasted Grosbeak	x		x		x	
Indigo Bunting	x		x			x

	Van Cortlandt		Central		Prospect	
Species	Has Bred	Potential Breeder	Has Bred	Potential Breeder	Has Bred	Potential Breeder
Bobolink	x		x			
Red-winged Blackbird	x		x		x	
Eastern Meadowlark	x		x			
Common Grackle	x		x		x	
Brown-headed Cowbird	x		x		x	
Orchard Oriole	x		x		x	
Baltimore Oriole	x		x		x	
House Finch	x		x		x	
Purple Finch	x		x			
Red Crossbill						
American Goldfinch	x			x	x	
European Goldfinch	x		x			
House Sparrow	x		x		x	
TOTALS	**123**	**13**	**73**	**7**	**71**	**6**

Table 12. Study area cumulative breeding species that have never bred in either Central or Prospect Park (n = 39). Pied-billed Grebe is the sole species that has bred in either Central or Prospect Park but never in Van Cortlandt. However, it was suspected of having done so at various times from 1930 to 1955 even though absolute proof was never forthcoming. It is also possible that some of the remaining species might have bred in Central or Prospect Park at one time or another but were overlooked or never formally recorded breeding.

Ring-necked Pheasant
Wild Turkey
American Bittern
Least Bittern
Turkey Vulture
Osprey
Northern Harrier
Sharp-shinned Hawk
Cooper's Hawk
Broad-winged Hawk
Clapper Rail
King Rail
Virginia Rail
Sora
Common Gallinule
American Coot
Spotted Sandpiper
American Woodcock
Barred Owl
Belted Kingfisher
Red-headed Woodpecker
Alder Flycatcher
Willow Flycatcher
Horned Lark
Bank Swallow
Cliff Swallow
Brown Creeper
Marsh Wren
Louisiana Waterthrush
Blue-winged Warbler
Nashville Warbler
Kentucky Warbler
Hooded Warbler
Yellow-breasted Chat
Field Sparrow
Grasshopper Sparrow
Saltmarsh Sparrow
Swamp Sparrow
Bobolink

Table 13. Synopsis of museum specimens from the study area and Riverdale in (largely) the late 1800s—767 specimens of 136 species. VCP = Van Cortlandt Park, KM = Kingsbridge Meadows, WC = Woodlawn Cemetery, JR = Jerome Reservoir, RDL = Riverdale. For individual specimen details, see Appendixes 6a, b, c, d.

Species	VCP	KM	WC	JR	RDL	Totals
American Bittern		1				1
Green Heron		1				1
Sharp-shinned Hawk	1				2	3
Red-shouldered Hawk	1		1			2
Yellow Rail					1	1
Clapper Rail		1				1
Virginia Rail		1			1	2
Sora		2				2
Spotted Sandpiper		2				2
Solitary Sandpiper	1					1
Western Willet		1				1
Least Sandpiper		1				1
American Woodcock				1		1
Passenger Pigeon					1	1
Yellow-billed Cuckoo	5					5
Black-billed Cuckoo	2					2
Eastern Screech Owl					1	1
Long-eared Owl			1			1
Common Nighthawk					1	1
Eastern Whip-poor-will					1	1
Chimney Swift					1	1
Ruby-throated Hummingbird	3				3	6
Yellow-bellied Sapsucker	1		2		2	5
Downy Woodpecker	3		1		3	7
Hairy Woodpecker					1	1
Northern Flicker			1		2	3
Olive-sided Flycatcher					1	1
Eastern Wood-Pewee			1		5	6
Yellow-bellied Flycatcher					4	4
Acadian Flycatcher					4	4
Alder Flycatcher					1	1
Least Flycatcher					3	3
Eastern Phoebe			1		2	3
Great Crested Flycatcher			1		2	3
Northern Shrike		2				2
White-eyed Vireo	10				2	12
Yellow-throated Vireo	5				4	9
Blue-headed Vireo	2				2	4
Warbling Vireo					2	2
Philadelphia Vireo					1	1
Red-eyed Vireo	6		1		3	10
Blue Jay			3		2	5
American Crow			3			3
Purple Martin	4					4
Tree Swallow		3				3
N. Rough-winged Swallow	3	5				8
Bank Swallow		2				2
Cliff Swallow					1	1
Barn Swallow					1	1

Species	VCP	KM	WC	JR	RDL	Totals
Black-capped Chickadee	6				3	9
Red-breasted Nuthatch					2	2
White-breasted Nuthatch	2				2	4
Brown Creeper					2	2
House Wren	1				4	5
Winter Wren	1				2	3
Sedge Wren		2				2
Marsh Wren		2				2
Carolina Wren					1	1
Golden-crowned Kinglet	8					8
Ruby-crowned Kinglet	5				4	9
Eastern Bluebird	3				2	5
Veery	14				2	16
Gray-cheeked Thrush	3				4	7
Bicknell's Thrush	1	1			3	5
Swainson's Thrush	9				4	13
Hermit Thrush	3		4	1	4	12
Wood Thrush	8				2	10
American Robin	3				5	8
Gray Catbird	7				3	10
Brown Thrasher	5				2	7
Northern Mockingbird					1	1
American Pipit		1	1			2
Cedar Waxwing	5				2	7
Snow Bunting		2				2
Ovenbird	5				3	8
Worm-eating Warbler	13				5	18
Louisiana Waterthrush	5				1	6
Northern Waterthrush	7				3	10
Golden-winged Warbler	1				1	2
Blue-winged Warbler	23		4		6	33
Black-and-white Warbler	9				3	12
Tennessee Warbler					2	2
Nashville Warbler					6	6
Connecticut Warbler					4	4
Kentucky Warbler					1	1
Common Yellowthroat	7		1		6	14
Hooded Warbler	3				7	10
American Redstart	9				2	11
Northern Parula	4				6	10
Magnolia Warbler	3				8	11
Bay-breasted Warbler					4	4
Blackburnian Warbler	2				5	7
Yellow Warbler	2				3	5
Chestnut-sided Warbler	9				4	13
Blackpoll Warbler	4				4	8
Black-throated Blue Warbler	3		1		1	5
Palm Warbler	4				8	12
Pine Warbler	4					4
Myrtle Warbler	3				4	7
Prairie Warbler	2				2	4

(Continued)

Table 13. (Continued)

Species	VCP	KM	WC	JR	RDL	Totals
Black-throated Green Warbler	3				3	6
Canada Warbler	3				6	9
Wilson's Warbler	2				3	5
Yellow-breasted Chat	5				2	7
Eastern Towhee	3		1		1	5
American Tree Sparrow	8				5	13
Chipping Sparrow	5				2	7
Field Sparrow	12				4	16
Vesper Sparrow		5				5
Savannah Sparrow		6				6
Grasshopper Sparrow	3	4				7
Henslow's Sparrow		3				3
Saltmarsh Sparrow		5				5
Red Fox Sparrow	4				3	7
Song Sparrow	9		1		5	15
Lincoln's Sparrow					1	1
Swamp Sparrow	1				5	6
White-throated Sparrow	10		1		4	15
White-crowned Sparrow					3	3
Slate-colored Junco	8				2	10
Scarlet Tanager	5				5	10
Northern Cardinal	1				1	2
Rose-breasted Grosbeak	5				1	6
Indigo Bunting	2				1	3
Painted Bunting					1	1
Red-winged Blackbird	3				5	8
Rusty Blackbird	1				5	6
Common Grackle					5	5
Brown-headed Cowbird	6				5	11
Orchard Oriole	1				2	3
Baltimore Oriole	2		1		2	5
Purple Finch					4	4
Red Crossbill					3	3
Common Redpoll					2	2
American Goldfinch	8		2		5	15
House Sparrow	4		1		2	7
Total specimens	**367**	**53**	**34**	**2**	**311**	**767**
Total species	**77**	**22**	**22**	**2**	**106**	**136**
Total specimens from the four subareas	**456**					
Total species from the four subareas	**105**					

Table 14. HY Herring Gulls seen in Van Cortlandt Park in 1937 that were ascribable to their natal colonies by unique colorband combinations.

Number	Date	Natal Colony	Observers
1	7 Oct 1937	Kent I., NB	Kieran
1	7 Oct 1937	Penikese I., MA	Kieran
1	8 Oct 1937	Muscongus Bay, ME	Kieran
1	8 Oct 1937	Isles of Shoals, NH	Kieran
1	9 Oct 1937	Isles of Shoals, NH	Kramer
1	11 Nov 1937	off Fisher's I., NY	Herbert, Hickey
5	11 Nov 1937	Kent I., NB	Herbert, Hickey
1	11 Nov 1937	Heron I., ME	Herbert, Hickey
3	20 Nov 1937	Penikese I., MA	Kraslow

Table 15. The species of birds known to have bred in the study area after 1871 (n = 123). Breeding Ruffed Grouse and Pileated Woodpecker were extirpated before 1872.

Canada Goose
Mute Swan
Wood Duck
American Black Duck
Mallard
Northern Bobwhite
Ring-necked Pheasant
Wild Turkey
American Bittern
Least Bittern
Green Heron
Black-crowned Night-Heron
Turkey Vulture
Osprey
Northern Harrier
Sharp-shinned Hawk
Cooper's Hawk
Red-shouldered Hawk
Broad-winged Hawk
Red-tailed Hawk
Clapper Rail
King Rail
Virginia Rail
Sora
Common Gallinule
American Coot
Killdeer
Spotted Sandpiper
American Woodcock
Rock Pigeon
Mourning Dove
Yellow-billed Cuckoo
Black-billed Cuckoo
Eastern Screech-Owl
Great Horned Owl
Barred Owl
Common Nighthawk
Eastern Whip-poor-will
Chimney Swift
Ruby-throated Hummingbird
Belted Kingfisher
Red-headed Woodpecker
Red-bellied Woodpecker
Downy Woodpecker
Hairy Woodpecker
Northern Flicker
American Kestrel
Monk Parakeet
Eastern Wood-Pewee
Acadian Flycatcher
Alder Flycatcher
Willow Flycatcher
Least Flycatcher
Eastern Phoebe
Great Crested Flycatcher
Eastern Kingbird
White-eyed Vireo
Yellow-throated Vireo
Warbling Vireo
Red-eyed Vireo
Blue Jay
American Crow
Fish Crow
Horned Lark
Tree Swallow
N. Rough-winged Swallow
Bank Swallow
Cliff Swallow
Barn Swallow
Black-capped Chickadee

(Continued)

Tufted Titmouse	Chestnut-sided Warbler
White-breasted Nuthatch	Yellow-breasted Chat
Brown Creeper	Eastern Towhee
House Wren	Chipping Sparrow
Marsh Wren	Field Sparrow
Carolina Wren	Vesper Sparrow
Blue-gray Gnatcatcher	Grasshopper Sparrow
Eastern Bluebird	Saltmarsh Sparrow
Veery	Song Sparrow
Wood Thrush	Swamp Sparrow
American Robin	Scarlet Tanager
Gray Catbird	Northern Cardinal
Brown Thrasher	Rose-breasted Grosbeak
Northern Mockingbird	Indigo Bunting
European Starling	Bobolink
Cedar Waxwing	Red-winged Blackbird
Ovenbird	Eastern Meadowlark
Worm-eating Warbler	Common Grackle
Louisiana Waterthrush	Brown-headed Cowbird
Blue-winged Warbler	Orchard Oriole
Black-and-white Warbler	Baltimore Oriole
Nashville Warbler	House Finch
Kentucky Warbler	Purple Finch
Common Yellowthroat	American Goldfinch
Hooded Warbler	European Goldfinch
American Redstart	House Sparrow
Yellow Warbler	

Table 16. Post-1871 unconfirmed or suspected *past* (p) (n = 7), and reasonably likely *future* (f) (n = 10), study area breeders with their probable subareas (Σ = 13)

Species	Past or Future	Subarea
Gadwall	f	Van Cortlandt
Pied-billed Grebe	p	Van Cortlandt Swamp/Lake
Great Blue Heron	f	Van Cortlandt
Yellow-crowned Night-Heron	p, f	Woodlawn, Van Cortlandt
Black Vulture	f	Woodlawn
Barn Owl	p, f	Woodlawn, Jerome, Hillview
Pileated Woodpecker	p, f	Van Cortlandt, Woodlawn
Common Raven	f	Woodlawn, Jerome, Hillview
Purple Martin	f	Van Cortlandt Lake/Swamp
Sedge Wren	p	Kingsbridge Meadows
Pine Warbler	f	Woodlawn
Savannah Sparrow	p, f	Kingsbridge Meadows, Hillview
Seaside Sparrow	p	Kingsbridge Meadows

Table 17. History of species breeding in the study area after 1871 (n = 123). Potential historical breeders (n = 5) are in italics, and the 7 eras coincide roughly with published activity. Cumulative breeding species richness before 1966 (114 species) was strikingly higher than after 1965 (83 species), as it was during individual eras in those 2 periods. A Kruskal-Wallis test of the totals (bottom) across the 7 eras rejects the null hypothesis of no difference (adjusted H = 4.582, df = 1, P = 0.032) and confirms the higher breeding species richness in the eras before 1966.

Unique breeding: **Year(s)**
start or end of breeding: Year [at start or end of line]
year only approximate: ( )
breeding uncertain: ?
not breeding: blank
sparse data but at least intermittent breeding: ....
extended breeding: ——————

Species	1872–1901 30 y	1902–27 26 y	1928–41 14 y	1942–65 24 y	1966–80 15 y	1981–99 19 y	2000–15 16 y
Canada Goose			(1932) ——————————————————————				
Mute Swan						1982 ————— 2009	
Wood Duck		(1915)					
American Black Duck		(1915) ————————————————————————————————————					
Mallard				—————— 1942			
Northern Bobwhite			(1932) ————————				
Ring-necked Pheasant						(1997) ———————— (2008)	
Wild Turkey							
Pied-billed Grebe			?	?			
American Bittern	————1879		?	?			
Least Bittern				———————— 1961			
Green Heron							
Black-crowned Night-Heron	?	?	?	?	?	**1983–4**	?
Turkey Vulture							**2009**
Osprey	————1878						2010–12
Northern Harrier	————1896						

(*Continued*)

Table 17. (Continued)

Species	1872–1901 30 y	1902–27 26 y	1928–41 14 y	1942–65 24 y	1966–80 15 y	1981–99 19 y	2000–15 16 y
Sharp-shinned Hawk	....	....	....	———— 1952			
Cooper's Hawk	....	....	....	———— 1958	....	(1980s)	
Red-shouldered Hawk	....	....	....	———— 1964			
Broad-winged Hawk			....				
Red-tailed Hawk				———— 1952			
Clapper Rail		———— 1924					
King Rail		———— 1925	....	———— 1956			
Virginia Rail				———— 1961			
Sora				———— 1963			
Common Gallinule			(1927)				
American Coot		?	**1936**				
Killdeer			———— 1932	———— 1942			
Spotted Sandpiper							
American Woodcock							
Rock Pigeon							
Mourning Dove							
Yellow-billed Cuckoo							
Black-billed Cuckoo							
Barn Owl				?			
Eastern Screech-Owl							
Great Horned Owl					(1970) ————		
Barred Owl		———— 1926					
Common Nighthawk		———— 1900					
Eastern Whip-poor-will	(1900) ————						
Chimney Swift					....		
Ruby-throated Hummingbird							
Belted Kingfisher							
Red-headed Woodpecker			———— 1938				
Red-bellied Woodpecker						(1987) ————	
Downy Woodpecker							
Hairy Woodpecker							
Northern Flicker							
American Kestrel							
Monk Parakeet					1974		

Species			
Eastern Wood-Pewee			
Acadian Flycatcher	····· 1888–93	(1907)	
Alder Flycatcher		1946	
Willow Flycatcher		1962	
Least Flycatcher			
Eastern Phoebe			
Great Crested Flycatcher			
Eastern Kingbird		1954	
White-eyed Vireo		1956	
Yellow-throated Vireo			
Warbling Vireo			
Red-eyed Vireo			
Blue Jay			
American Crow			
Fish Crow	*1916?*	1948–62	?
Horned Lark		1938	
Tree Swallow			
N. Rough-winged Swallow			
Bank Swallow	···· 1926	?	?
Cliff Swallow	(1910)	*1956*	
Barn Swallow			2010–12
Black-capped Chickadee			
Tufted Titmouse		(1954)	
White-breasted Nuthatch			
Brown Creeper	*1926*		
House Wren			
Sedge Wren ···· 1880s?			
Marsh Wren		1966	
Carolina Wren		?	(1975)
Blue-gray Gnatcatcher			(1970s)
Eastern Bluebird		1956	
Veery			
Wood Thrush			
American Robin			

(Continued)

Table 17. (Continued)

Species	1872–1901 30 y	1902–27 26 y	1928–41 14 y	1942–65 24 y	1966–80 15 y	1981–99 19 y	2000–15 16 y
Gray Catbird							
Brown Thrasher							(2008)
Northern Mockingbird				1959			
European Starling	(1899)						
Cedar Waxwing							
Ovenbird							
Worm-eating Warbler			(1941)				
Louisiana Waterthrush	1917						
Blue-winged Warbler				1964			
Black-and-white Warbler				(1948)	?		
Nashville Warbler	1892						
Kentucky Warbler		(1910)					
Common Yellowthroat							
Hooded Warbler		(1926)					
American Redstart							
Yellow Warbler		(1922)					
Chestnut-sided Warbler				1954			
Yellow-breasted Chat							
Eastern Towhee							
Chipping Sparrow							
Field Sparrow					(1975)		
Vesper Sparrow	1880s	(1910s)					
Savannah Sparrow	?		?				
Grasshopper Sparrow	1880s	(1910s)					
Saltmarsh Sparrow	1882						
Seaside Sparrow	1880s?						
Song Sparrow				1963			
Swamp Sparrow							
Scarlet Tanager				(1944)			
Northern Cardinal							
Rose-breasted Grosbeak		(1910)					
Indigo Bunting							
Bobolink							
Red-winged Blackbird							

				1934–40			
Eastern Meadowlark	....						
Common Grackle	————						
Brown-headed Cowbird	————						
Orchard Oriole	....						
Baltimore Oriole	————						
House Finch					1964 ————		
Purple Finch	1876–81				*1945* ————		
American Goldfinch		?					
European Goldfinch							
House Sparrow	*1900* ————						
	1886 ————						
Extended + intermittent breeding:	97	93	87	91	74	76	79
Breeding uncertain:	4	4	4	7	6	2	1
TOTALS	**101**	**97**	**91**	**98**	**80**	**78**	**80**
% of 123 cumulative species:	82%	79%	74%	80%	65%	63%	65%

Table 18. Study area breeders lost by decades after 1871 (n = 50), their probable/ approximate year of last breeding, and likely reasons for their loss. Mute Swan, Osprey, Cooper's Hawk, Alder Flycatcher, and Cliff Swallow exclude (possible/likely) recent rebreedings. (Near-) ground nesters (n = 31) are in *italics*.

Species	Last Bred	Cause of Loss
Osprey	1878	Disturbance + shooting ?
American Bittern	1879	Habitat destroyed
Vesper Sparrow	1880s	Habitat destroyed
Grasshopper Sparrow	1880s	Habitat destroyed
Saltmarsh Sparrow	1882	Habitat destroyed
Alder Flycatcher	1888	Range contraction northward
Nashville Warbler	1892 only	Adventitious breeding event
Northern Harrier	1896	Habitat destroyed + shooting ?
Cliff Swallow	Late 1800s	Area-wide population decline
Acadian Flycatcher	1890s–1910s	Area-wide population crash
European Goldfinch	1900	Post-introduction apex
Bobolink	Early 1900s	Habitat destroyed
Bank Swallow	Mid-1900s	Habitat destroyed
Hooded Warbler	1910s	Unknown
Louisiana Waterthrush	1917	Habitat destroyed
Kentucky Warbler	Late 1910s	Range retraction ?
Eastern Whip-poor-will	1920s	Urbanization, ground predation
Clapper Rail	1924	Habitat destroyed
Brown Creeper	1926 only	Adventitious breeding event
Barred Owl	1926	Urbanization ?
American Coot	1936 only	Adventitious breeding event
Red-headed Woodpecker	1938	Area-wide population decline
Eastern Meadowlark	1940	Habitat destroyed
Worm-eating Warbler	Early 1940s	Urbanization
Common Gallinule	1942	Unknown
Northern Bobwhite	1942	Urbanization, range-edge loss
Purple Finch	1945 only	Adventitious breeding event
Black-and-white Warbler	1948	Urbanization
Sharp-shinned Hawk	1952	Urbanization ?
White-eyed Vireo	1954	Oldfield habitat lost
Yellow-breasted Chat	1954	Oldfield habitat lost
Yellow-throated Vireo	1956	Urbanization ?
King Rail	1956	Habitat + range-edge loss
Eastern Bluebird	1956	Area-wide population crash
Cooper's Hawk	1958	Biocide population crash
Virginia Rail	1961	Habitat loss via succession
Sora	1963	Habitat loss via succession
Least Bittern	1961	Habitat loss via succession

Species	Last Bred	Cause of Loss
Horned Lark	1962	Disturbance + range-edge loss
Swamp Sparrow	1963	Habitat loss via succession
Red-shouldered Hawk	1964	Urbanization ?
Least Flycatcher	1964	Range contraction northward
Blue-winged Warbler	1964	Habitat loss via succession
Marsh Wren	1967	Habitat loss via succession
Field Sparrow	1970s	Habitat loss via succession
Black-crowned Night Heron	1983	Area-wide population decline
Ring-necked Pheasant	2008	Area-wide population decline
Brown Thrasher	2008	Area-wide population decline, habitat loss via succession
Mute Swan	2009	Disturbance
Turkey Vulture	2009 only	Disturbance

Table 19. New study area breeders gained by decades after 1871, probable year of first breeding, and likely provenance (n = 32). Those underlined may have been breeding earlier but unreported, so their first known years are shown.

Species	First Bred	Provenance
Nashville Warbler	1892 only	Relict or adventitious breeder
European Goldfinch	1900 only	Recent introduction
Common Nighthawk	1900	Breeding behavior changes
American Black Duck	1915	Area population increase
Mallard	1915	Recent introduction
<u>Chestnut-sided Warbler</u>	Mid-1920s	Unmentioned but present earlier ?
King Rail	1925	Erratic NYC area breeder
Brown Creeper	1926 only	Adventitious breeding event
Common Gallinule	1927	Erratic NYC area breeder
Canada Goose	1932	Recent introduction
Ring-necked Pheasant	1932	Recent introduction
Killdeer	1932	Area population increase
Sharp-shinned Hawk	Early 1930s	Area population increase
American Coot	1936 only	Adventitious breeding event
<u>Broad-winged Hawk</u>	Late 1930s	Unmentioned but present earlier ?
Tree Swallow	1938	Area population increase
Willow Flycatcher	1946	Range expansion
Horned Lark	1948	Range expansion

(*Continued*)

Table 19. (Continued)

Species	First Bred	Provenance
Red-tailed Hawk	1952	Breeding behavior changes
Tufted Titmouse	1954	Range expansion
Cliff Swallow	1956	erratic New York City breeder
Northern Mockingbird	1959	Range expansion
House Finch	1964	Recent introduction
Great Horned Owl	1970	Breeding behavior changes
Monk Parakeet	1974	Recent introduction
Carolina Wren	1975	Range expansion
Mute Swan	1982	Area population increase
Black-crowned Night-Heron	1983	Adventitious breeding event
Red-bellied Woodpecker	1987	Range expansion
Wild Turkey	Late 1990s	Range expansion
Blue-gray Gnatcatcher	Late 1990s	Range expansion
Turkey Vulture	2009 only	Adventitious breeding event

Table 20. Current, extirpated, or potential breeding species within the study area that regularly or occasionally nest on or nearly on the ground, or are oldfield succession breeders. Upper: current, extirpated * (n = 25); potential (*italics*); (n = 43). Lower: obligate or near-obligate second growth/oldfield succession breeders (n = 8).

Black Duck
Canada Goose
Mallard
Gadwall
Mute Swan *
Northern Bobwhite *
Wild Turkey
Ring-necked Pheasant *
Ruffed Grouse *
Pied-billed Grebe
American Bittern *
Least Bittern *
Black-crowned Night-Heron
Turkey Vulture *
Black Vulture
Northern Harrier *
King Rail *
Virginia Rail *
Sora *
Common Gallinule *
American Coot *
Killdeer
Spotted Sandpiper
American Woodcock
Common Nighthawk
Eastern Whip-poor-will *

Horned Lark *
Veery
Ovenbird
Louisiana Waterthrush *
Worm-eating Warbler *
Blue-winged Warbler *
Black-and-white Warbler *
Kentucky Warbler *
Common Yellowthroat
Eastern Towhee *
Field Sparrow *
Vesper Sparrow *
Savannah Sparrow
Grasshopper Sparrow *
Song Sparrow
Swamp Sparrow *
Eastern Meadowlark *

White-eyed Vireo *
Brown Thrasher *
Blue-winged Warbler *
Chestnut-sided Warbler
Common Yellowthroat
Yellow-breasted Chat *
Field Sparrow *
Indigo Bunting

Table 21. Species believed to have bred in the study area more or less annually after 1871—*primary breeders* (n = 55). NB: Eastern Towhee and Brown Thrasher are almost extirpated.

Wood Duck	Eastern Kingbird	Common Yellowthroat
Green Heron	Warbling Vireo	American Redstart
Spotted Sandpiper	Red-eyed Vireo	Yellow Warbler
American Woodcock	Blue Jay	Chestnut-sided Warbler
Rock Pigeon	American Crow	Eastern Towhee
Mourning Dove	Fish Crow	Chipping Sparrow
Yellow-billed Cuckoo	N. Rough-winged Swallow	Song Sparrow
Black-billed Cuckoo	Barn Swallow	Northern Cardinal
Eastern Screech-Owl	Black-capped Chickadee	Rose-breasted Grosbeak
Ruby-throated Hummingbird	White-breasted Nuthatch	Indigo Bunting
Chimney Swift	House Wren	Red-winged Blackbird
Belted Kingfisher	Veery	Common Grackle
Downy Woodpecker	Wood Thrush	Brown-headed Cowbird
Hairy Woodpecker	American Robin	Orchard Oriole
Northern Flicker	Gray Catbird	Baltimore Oriole
American Kestrel	Brown Thrasher	American Goldfinch
Eastern Wood-Pewee	European Starling	House Sparrow
Eastern Phoebe	Cedar Waxwing	
Great Crested Flycatcher	Ovenbird	

Table 22. Species believed to have been breeding in the study area more or less annually from the years indicated—*secondary breeders* (n = 18). Mute Swans abandoned nesting in 2009 and failed again in 2016; *italics* = interrupted breeding from mid-1950s to mid-1980s; * = approximate year of first breeding.

Species	Year
American Black Duck	1915*
Mallard	1915*
Mute Swan	1982
Ring-necked Pheasant	1932*
Wild Turkey	1997*
Cooper's Hawk	1872
Red-tailed Hawk	1952
Killdeer	1932*
Great Horned Owl	1970*
Red-bellied Woodpecker	1987*
Monk Parakeet	1974
Willow Flycatcher	1946
Tree Swallow	1938
Tufted Titmouse	1954*
Carolina Wren	1975*
Blue-gray Gnatcatcher	1970s*
Northern Mockingbird	1959
House Finch	1964

Table 23. Known or potential (*italics*) study area breeding species (2 of them* recently extirpated after recolonization) that regularly or occasionally nest in or on artificial structures, especially nest boxes (n = 30)

Wood Duck	*Common Raven*
Black Vulture	Purple Martin
Osprey*	Tree Swallow
Killdeer	N. Rough-winged Swallow
Rock Pigeon	Bank Swallow
Barn Owl	Cliff Swallow*
Eastern Screech Owl	Barn Swallow
Common Nighthawk	Black-capped Chickadee
Chimney Swift	Tufted Titmouse
Red-bellied Woodpecker	White-breasted Nuthatch
Northern Flicker	House Wren
American Kestrel	Carolina Wren
Monk Parakeet	Eastern Bluebird
Eastern Phoebe	European Starling
Great Crested Flycatcher	House Sparrow

Table 24. Proven study area breeders >2000 whose breeding populations appear to be increasing (n = 13) or decreasing (n = 8) as of 2016. On- or near-ground nesters are in *italics*; ? indicates some uncertainty as to trend.

Increasing	**Decreasing**
Wild Turkey	*Ring-necked Pheasant*
Cooper's Hawk	Hairy Woodpecker ?
Red-tailed Hawk	American Crow
Red-bellied Woodpecker	*Veery*
Warbling Vireo	*Brown Thrasher*
Tree Swallow	*Ovenbird*
Cliff Swallow	*Eastern Towhee*
Blue-gray Gnatcatcher ?	House Sparrow ?
Yellow Warbler	
Chestnut-sided Warbler ?	
Indigo Bunting	
Orchard Oriole	
House Finch	

Table 25. Breeding birds of Woodlawn Cemetery in 1960–61 (n = 33), plus species (n = 21) and numbers of likely pairs in late May 2014. A total of 41 breeders are noted here but others have bred in the past. Blank = unrecorded; x = breeding, but numbers unrecorded.

Species	1960–61	2014
Ring-necked Pheasant	x	
Canada Goose	x	1
Mallard	x	
Red-tailed Hawk		2
Killdeer		2
Mourning Dove	x	
Northern Flicker	x	4
Chimney Swift	x	14
Hairy Woodpecker	x	
Downy Woodpecker	x	
American Kestrel	x	
Great Crested Flycatcher	x	
Eastern Kingbird	x	1
Warbling Vireo		7
Red-eyed Vireo	x	8
Blue Jay	x	
American Crow	x	
Fish Crow	x	
Black-capped Chickadee	x	
White-breasted Nuthatch	x	
House Wren	x	7
Carolina Wren		1
Wood Thrush	x	
American Robin	x	50
Gray Catbird	x	17
Brown Thrasher	x	
Northern Mockingbird		14
European Starling	x	
Yellow Warbler		1
Common Yellowthroat	x	
Chipping Sparrow	x	15
Song Sparrow	x	6
Scarlet Tanager		1
Northern Cardinal	x	7
Indigo Bunting	x	
Red-winged Blackbird	x	2
Common Grackle	x	
Brown-headed Cowbird	x	
Baltimore Oriole	x	7
House Finch		6
House Sparrow	x	

Table 26. Estimated numbers of species (n = 73) and pairs breeding in 2013–14 within Van Cortlandt Park and *immediately adjacent* urban areas. Data are from censuses done between 30 May and 12 Aug 2013 by Ellen Pehek and members of the NYC Parks Dept. Natural Resources Group, and from 19 May to 5 Jul 2014 by John L. Young, and from other ad hoc sources. All data were merged by the present authors to derive a subjective best estimate of the *minimum* number of pairs probably breeding park-wide between 2013 and 2014. Some species may not breed annually or in constant numbers. Note, however, that the 5 extensive golf course subsections were *not* censused in either year.

Species	Pairs
Canada Goose	1
Wood Duck	3
American Black Duck	1
Mallard	3
Wild Turkey	no estimate available
Green Heron	2–3
Cooper's Hawk	1–2
Red-tailed Hawk	2–3
Killdeer	2
American Woodcock	no estimate available
Spotted Sandpiper	1–2
Rock Pigeon	<10
Mourning Dove	<30
Black-billed Cuckoo	<5
Yellow-billed Cuckoo	<5
Eastern Screech-Owl	no estimate available
Great Horned Owl	1–3
Common Nighthawk (in adjacent areas)	6–10
Chimney Swift	20–40
Ruby-throated Hummingbird	2–3
Belted Kingfisher	1–2
Red-bellied Woodpecker	>25
Downy Woodpecker	>10
Hairy Woodpecker	6–8
Northern Flicker	<10
American Kestrel (+ adjacent areas)	1–2
Monk Parakeet (in adjacent areas)	no estimate available
Eastern Wood-Pewee	8–10
Willow Flycatcher	4–6
Eastern Phoebe	<5
Great Crested Flycatcher	<15
Eastern Kingbird	<5
Warbling Vireo	>40
Red-eyed Vireo	<10
Blue Jay	<15
American Crow	<5
Fish Crow	1–2
Tree Swallow	>20
N. Rough-winged Swallow	10
Barn Swallow	10–12
Black-capped Chickadee	<10
Tufted Titmouse	<10
White-breasted Nuthatch	<10

Species	Pairs
House Wren	5–10
Carolina Wren	6–8
Blue-gray Gnatcatcher	2–3
Veery	3–5
Wood Thrush	>25
American Robin	>100
Gray Catbird	>100
Northern Mockingbird	<5
European Starling (nesting *in* Van Cortlandt Park)	<30
Cedar Waxwing	>20
Ovenbird	3–5
Common Yellowthroat	<5
American Redstart	<10
Yellow Warbler	30–40
Chestnut-sided Warbler	5–6
Eastern Towhee	<5
Chipping Sparrow	10–15
Song Sparrow	15
Scarlet Tanager	>10
Northern Cardinal	10–15
Rose-breasted Grosbeak	>10
Indigo Bunting	5–10
Red-winged Blackbird	<50
Common Grackle	>40
Brown-headed Cowbird	20–30
Orchard Oriole	6–8
Baltimore Oriole	>20
House Finch	>20
American Goldfinch	>20
House Sparrow	<40

Table 27. Estimated numbers of breeding pairs across 77 years on censuses of Van Cortlandt Swamp in the 1930s (16.2 ha [40 ac] before decimation in 1949–50); in the 1960s (7.7 ha [19 ac] after landfilling), as published in *Bird-Lore*, *Audubon Field Notes*, and the *Linnaean Newsletter*; and then in 2013–14 ("2013") from censuses done from 30 May to 12 Aug 2013 by Ellen Pehek and members of the NYC Parks Dept. Natural Resources Group and from 19 May to 5 Jul 2014 by John L. Young. These latter 2 data sets were merged by the present authors to derive a subjective best estimate of the minimum number of pairs breeding in 2013–14. Brown-headed Cowbirds were present (+) during all censuses, but counts were unreliable; a blank indicates a species with no breeders that year. Note that the tabular summary for the 1937–38 and 1962–63–64 censuses in Heath and Zupan (1971–2) contains errors that are corrected here, including deletion of their 1961 grackle count.

Species	1937	1938	1961	1962	1963	1964	1965	1966	2013
Canada Goose	1	1			1				1
Wood Duck	5		2	3	1	1			3
American Black Duck	1					2	1		
Mallard	3	2	3	3	2	1	2	2	3
Northern Bobwhite	1	1							
Ring-necked Pheasant	2	2	3	3	3				
Least Bittern	3	5	1						
Green Heron		1	1		1	1		1	2
Virginia Rail	9	7	4						
Sora	2	1	1		1				
Common Gallinule	3	6							
Mourning Dove									3
Yellow-billed Cuckoo	3	2	1	3	3				
Black-billed Cuckoo			1	1	2				
Red-bellied Woodpecker									3
Downy Woodpecker	3	1	2	3	2	1	1	1	1
Northern Flicker	3	4	4	3	4	2	1	1	4
Willow Flycatcher			4	5	6	5	6	6	2–4
Least Flycatcher			1	1					
Eastern Kingbird	3	2	2	2	2	1	1	1	2
Warbling Vireo	3	4	2	3	2	3	3	2	33
Red-eyed Vireo		1	1			1	1	1	
Blue Jay									1
Tree Swallow									10
Black-capped Chickadee									2
House Wren	3	2							1
Marsh Wren	21	48	7	9	8	5	4	5	
Carolina Wren									2
Blue-gray Gnatcatcher									1
Eastern Bluebird		2							
Veery	1	1							
American Robin	9	23	15	19	22	30	17	28	24
Gray Catbird	16	19	20	24	23	30	10	15	17
Brown Thrasher	8	8	11	7	7	10	6	5	
European Starling	9	8	15	10	9	10	12	12	1
Common Yellowthroat	20	13	3	5	6	4	3	3	1–3
American Redstart									1
Yellow Warbler	32	27	15	18	19	15	14	18	27
Eastern Towhee				1	1	1			
Song Sparrow	28	22	15	18	20	20	20	22	12
Swamp Sparrow	9	6		1	1				
Northern Cardinal			2	2	1	1	1	1	4
Rose-breasted Grosbeak				1	1				

Species	1937	1938	1961	1962	1963	1964	1965	1966	2013
Red-winged Blackbird	125	165	90	80	86	70	60	76	41
Baltimore Oriole	2	4		1	3	3	2	3	8
Orchard Oriole			1						1
Common Grackle	1		20	3	2	2			7–10
Brown-headed Cowbird	+	+	+	+	+	+	+	+	+
American Goldfinch	3	5	1	2	3	2	2	3	6
House Sparrow			1						
Species richness [∑ = 50]	**31**	**32**	**30**	**28**	**30**	**25**	**21**	**21**	**31**
Pairs	**332**	**394**	**248**	**231**	**242**	**221**	**167**	**206**	**224–31**

Table 28. Estimated numbers of breeding pairs (transformed to density per 100 acres [40 ha] and rounded to the nearest whole digit for clarity) found on censuses of Van Cortlandt Swamp in the 1930s (40 ac [16.2 ha]) before decimation in 1949–50; numbers of singing ♂♂ (*not* breeding pairs) in the 1960s (19 ac [7.7 ha] after landfilling); and then from censuses done from 30 May to 12 Aug 2013 by Ellen Pehek and members of the NYC Parks Dept. Natural Resources Group and from 19 May to 5 Jul 2014 by John L. Young. These latter 2 data sets were merged by the present authors to derive a subjective best estimate of the minimum number of pairs breeding in 2013–14. Brown-headed Cowbirds were recorded on all censuses, but counts were unreliable; a blank indicates a species with no breeders found that year. Note that the tabular summary for the 1937–38 and 1962–63–64 censuses in Heath and Zupan (1971–2) contains errors that are corrected here, including deletion of the 1961 grackle count. See "Breeding Species" in Avifaunal Overview for discussion and references.

Species	1937	1938	1961	1962	1963	1964	1965	1966	2013
Canada Goose	3	3			5				3
Wood Duck	13		11	16	5	5			8
American Black Duck	3					11	5		
Mallard	8	5	16	16	11	5	11	11	8
Northern Bobwhite	3	3							
Ring-necked Pheasant	5	5	16	16	16				
Least Bittern	8	13	5						
Green Heron		3	5		5	5		5	5
Virginia Rail	23	18	21						
Sora	5	3	5		5				
Common Gallinule	8	15							8
Mourning Dove									
Yellow-billed Cuckoo	8	5	5	16	16				
Black-billed Cuckoo			5	5	11				
Red-bellied Woodpecker									8
Downy Woodpecker	8	3	11	16	11	5	5	5	3
Northern Flicker	8	10	21	16	21	11	5	5	10
Willow Flycatcher			21	27	32	27	32	32	5
Least Flycatcher			5	5					
Eastern Kingbird	8	8	11	11	11	5	5	5	5
Warbling Vireo	8	10	11	16	11	16	16	11	83
Red-eyed Vireo		3	5			5	5	5	
Blue Jay									3
Tree Swallow									25
House Wren	8	5							3
Black-capped Chickadee									3
Marsh Wren	53	120	37	48	43	27	21	27	
Carolina Wren									5
Blue-gray Gnatcatcher									3

(*Continued*)

Table 28. (Continued)

Species	1937	1938	1961	1962	1963	1964	1965	1966	2013
Eastern Bluebird		5							
Veery	3	3							
American Robin	23	58	80	101	117	159	90	148	60
Gray Catbird	40	48	106	128	122	159	53	88	43
Brown Thrasher	20	20	58	37	37	53	32	27	
European Starling	23	20	40	53	48	53	64	60	3
Common Yellowthroat	50	33	16	27	32	21	16	16	3
American Redstart									3
Yellow Warbler	80	68	80	95	101	80	74	95	68
Eastern Towhee				5	5	5			
Song Sparrow	70	55	80	95	106	106	106	117	30
Swamp Sparrow	23	15		5	5				
Northern Cardinal			11	11	5	5	5	5	10
Rose-breasted Grosbeak				5	5				
Red-winged Blackbird	313	415	477	424	456	371	317	403	103
Baltimore Oriole	5	8		5	16	16	11	16	20
Orchard Oriole			5						3
Common Grackle	3		—	16	11	11			18
American Goldfinch	8	13	5	11	16	11	11	16	15
House Sparrow		3							
Species richness (∑ = 41)	30	31	29	27	29	24	20	20	30
Pairs or singing ♂♂	841	996	1169	1226	1285	1172	884	1097	569

Table 29. Predicted ranges in the 1960s of breeding pairs (rounded for clarity) of the 28 species that nested in Van Cortlandt Swamp in both the 1930s and 1960s if breeding densities of the 1930s censuses had remained stable after the swamp was reduced in size by 48%. These are contrasted with actual 1960s numbers to highlight any density changes.

Species	Actual 1930s	Predicted 1960s	Actual 1960s	Density Change
Canada Goose	1	1	1	Stable
Wood Duck	5	2	1–3	Stable
American Black Duck	1	1	1–2	**Increased**
Mallard	2–3	1	1–3	**Increased**
Ring-necked Pheasant	2	1	3	**Increased**
Least Bittern	3–5	1	1	Stable
Green Heron	1	1	1	Stable
Virginia Rail	7–9	3–4	4	Stable
Sora	1–2	1	1	Stable
Yellow-billed Cuckoo	2–3	1	1–3	**Increased**
Downy Woodpecker	1–3	1	1–3	**Increased**
Northern Flicker	3–4	1	1–4	**Increased**
Eastern Kingbird	3	2	1–2	Stable
Warbling Vireo	3–4	2	2–3	Stable
Red-eyed Vireo	1	1	1	Stable
Marsh Wren	21–48	10–23	4–9	*Decreased*
American Robin	9–23	4–11	10–30	**Increased**
Gray Catbird	16–19	8–9	10–30	**Increased**

Species	Actual 1930s	Predicted 1960s	Actual 1960s	Density Change
Brown Thrasher	8	4	5–11	**Increased**
European Starling	8–9	4	9–15	**Increased**
Common Yellowthroat	13–20	6–10	3–6	*Decreased*
Yellow Warbler	27–32	13–15	14–19	**Increased**
Song Sparrow	22–28	13–15	15–22	**Increased**
Swamp Sparrow	6–9	3–4	1	*Decreased*
Red-winged Blackbird	125–166	60–80	60–90	Stable
Baltimore Oriole	2–3	1	1–3	**Increased**
Common Grackle	1	1	2–3	**Increased**
American Goldfinch	3–5	1–2	1–3	**Increased**

Table 30. Estimated numbers of occupied territories found on breeding bird censuses of a 25 ac (10 ha) mature forest plot (approximately a square 1037 ft [316m] on a side) in Van Cortlandt Park's Northwest Forest in 1991–92 and published in *Journal of Field Ornithology*. 0.5 = half the territory inside the plot, + = <25% of a territory within the plot; — = no occupied territories in 1991. See "Breeding Species" in Avifaunal Overview for discussion and references.

Species	1991	1992
Ring-necked Pheasant	1.5	1.5
Red-bellied Woodpecker	3	3.5
Hairy Woodpecker	1.5	0.5
Downy Woodpecker	3	2
Northern Flicker	2	2.5
Eastern Wood-Pewee	0.5	2
Great Crested Flycatcher	1	2
Red-eyed Vireo	1	1
Blue Jay	4.5	5
American Crow	—	1
Black-capped Chickadee	3	1.5
Tufted Titmouse	5	5
White-breasted Nuthatch	1	1
Wood Thrush	4.5	4.5
American Robin	16	18
Gray Catbird	7	6.5
Brown Thrasher	1	1
European Starling	2	3
Common Yellowthroat	1	1
Eastern Towhee	5	4.5
Northern Cardinal	6	4
Rose-breasted Grosbeak	1	1
Baltimore Oriole	—	+
Common Grackle	+	+
Brown-headed Cowbird	3.5	3
House Sparrow	+	1
Species richness (∑ = 26)	**24**	**26**
Occupied territories	**75**	**74**

Table 31. Woodlawn Cemetery species (n = 40) found on the 27 December 2015 Bronx-Westchester Christmas Bird Count. A Merlin was seen 3 days later.

Species	Number	Species	Number
Snow Goose	20	Tufted Titmouse	17
Hooded Merganser	1	Red-breasted Nuthatch	1
Great Blue Heron	1	White-breasted Nuthatch	18
Sharp-shinned Hawk	1	Ruby-crowned Kinglet	2
Cooper's Hawk	2	American Robin	131
Red-tailed Hawk	2	Northern Mockingbird	5
Peregrine Falcon	1	European Starling	500
Rock Pigeon	78	Cedar Waxwing	5
Mourning Dove	74	Nashville Warbler	1
Great Horned Owl	1	Eastern Towhee	1
Belted Kingfisher	1	Song Sparrow	5
Red-bellied Woodpecker	11	White-throated Sparrow	36
Yellow-bellied Sapsucker	7	Dark-eyed Junco	61
Downy Woodpecker	2	Northern Cardinal	10
Hairy Woodpecker	1	Red-winged Blackbird	70
Northern Flicker	2	Common Grackle	1350
Blue Jay	39	Purple Finch	3
American Crow	3	House Finch	108
Common Raven	2	American Goldfinch	44
Black-capped Chickadee	52	House Sparrow	13

Table 32. Species believed to have occurred and probably overwintered in the study area annually after 1871—*core winter residents* (n = 32). To them may be added those below the line that have done so annually since (and many well before) 2000 (n = 38).

Sharp-shinned Hawk	Black-capped Chickadee
Cooper's Hawk	White-breasted Nuthatch
Red-tailed Hawk	Brown Creeper
Red-shouldered Hawk	Winter Wren
Wilson's Snipe	Golden-crowned Kinglet
Herring Gull	American Robin
Great Black-backed Gull	Cedar Waxwing
Rock Pigeon	American Tree Sparrow
Eastern Screech-Owl	Red Fox Sparrow
Downy Woodpecker	Song Sparrow
Hairy Woodpecker	Swamp Sparrow
Northern Flicker	White-throated Sparrow
American Kestrel	Slate-colored Junco
Blue Jay	Red-winged Blackbird
American Crow	Purple Finch
Fish Crow	American Goldfinch
Canada Goose	Lesser Scaup
Wood Duck	Hooded Merganser
American Wigeon	Ruddy Duck
American Black Duck	Wild Turkey
Mallard	Pied-billed Grebe
Northern Shoveler	Great Blue Heron
Green-winged Teal	Turkey Vulture

Canvasback	Bald Eagle
Ring-billed Gull	Red-breasted Nuthatch
Mourning Dove	Carolina Wren
Great Horned Owl	Hermit Thrush
Red-bellied Woodpecker	Northern Mockingbird
Yellow-bellied Sapsucker	European Starling
Merlin	Northern Cardinal
Peregrine Falcon	Red-winged Blackbird
Tufted Titmouse	Common Grackle
Red-tailed Hawk	Brown-headed Cowbird
American Coot	House Finch
Killdeer	House Sparrow

Table 33. Estimated mean number of individuals per species found during Winter Bird Population Studies of Van Cortlandt Swamp in 1962–63, 1963–64, and 1965–66. (That for 1964–65 was done but apparently lost and never published in *Audubon Field Notes*.) < = a mean <0.5; blanks = not resident that winter; # = occurred but numbers not given. See "Winter Species" in Avifaunal Overview for additional information.

Species	1962–63	1963–64	1965–66
Canada Goose	42	31	#
Mute Swan	1	#	#
Wood Duck	1	+	
American Wigeon	1	6	< + #
American Black Duck	24	48	2 + #
Mallard	8	21	2 + #
Green-winged Teal		4	#
Greater Scaup	#		<
Bufflehead			<
Ring-necked Pheasant	10	6	6
American Bittern	1		
Great Blue Heron		<	
Black-crowned Night-Heron	<		
Red-tailed Hawk	1		1
Rough-legged Hawk	<	<	
Virginia Rail	1	<	
Killdeer	1		
Greater Yellowlegs	<		
Wilson's Snipe	1	1	3
Great Black-backed Gull			#
Herring Gull	#	#	#
Ring-billed Gull	#	#	#
Rock Pigeon		#	#
Mourning Dove			1
Belted Kingfisher		<	<
Downy Woodpecker	3	3	2
Hairy Woodpecker	1	1	<
American Kestrel		<	
Blue Jay	4	4	4
American Crow	5	4	5
Fish Crow	<		
Horned Lark	#		#
Black-capped Chickadee	9	9	4
Tufted Titmouse		2	2

(*Continued*)

Table 33. (Continued)

Species	1962–63	1963–64	1965–66
White-breasted Nuthatch	<	1	1
Brown Creeper	<		
American Robin	<		<
European Starling	16	15	15
Cedar Waxwing		<	
American Tree Sparrow	5	9	3
Song Sparrow	5	6	6
Swamp Sparrow	5	2	1
White-throated Sparrow	2	2	2
Slate-colored Junco	<		1
Northern Cardinal	<	1	2
Red-winged Blackbird	3	1	1
Rusty Blackbird	1	1	10
Common Grackle		1	<
Brown-headed Cowbird	<	<	6
House Finch			<
Pine Siskin		#	#
American Goldfinch	2	4	2
House Sparrow	1		
Species richness [∑ = 53]	**40**	**37**	**40**

Table 34. Estimated mean number of individuals per waterbird species found during Winter Bird Population Studies of Jerome Reservoir in 1963–64 and 1965–66. (That for 1964–65 was done but apparently lost and never published in *Audubon Field Notes*.) < = <0.5; blanks = not resident that winter. See "Winter Species" in Avifaunal Overview for additional information.

Species	1963–64	1965–66
Snow Goose		<
Canada Goose		1
American Wigeon	1	1
American Black Duck	261	211
Mallard	71	167
Northern Shoveler		3
Northern Pintail		1
Green-winged Teal	1	
Canvasback	269	458
Redhead		<
Ring-necked Duck	1	1
Greater Scaup		<
Lesser Scaup	17	7
Common Goldeneye	<	
Common Merganser	20	9
Hooded Merganser	<	<
Ruddy Duck	51	3
Pied-billed Grebe		1
Horned Grebe	<	
American Coot	1	2
Great Black-backed Gull	18	25
Herring Gull	365	1044
Ring-billed Gull	68	126
Species [∑ = 23]	**16**	**20**

Table 35. Migrating shorebird species (n = 15) forced down to the recently drained Van Cortlandt Lake bed by severe weather conditions on 10 May 1951. Those in **boldface** are species unique in the study area; the Stilt Sandpiper was in high breeding plumage.

Species	Number
Black-bellied Plover	2
Semipalmated Plover	5
Killdeer	5
Spotted Sandpiper	7
Solitary Sandpiper	21
Greater Yellowlegs	21
Lesser Yellowlegs	16
Stilt Sandpiper	**1**
Sanderling	**2**
Dunlin	**2**
Least Sandpiper	34
Pectoral Sandpiper	9
Semipalmated Sandpiper	12
Short-billed Dowitcher	3
Wilson's Snipe	2

Table 36. Years with the highest study area daily spring counts (plus 2 in fall only) for insectivorous/frugivorous Neotropical migrants (n = 51). Species with maxima of only 1–2 have been omitted. **Boldface**: maxima after 1965. Totals: 46 before 1966, 5 after 1965. A binomial test (z = 5.6, P = 0.000001) decisively rejects the null hypothesis of no difference between the 2 groups, confirming that the highest spring counts occurred before 1966.

Species	Year	Species	Year
Common Nighthawk *(fall)*	1953	Black-and-white Warbler	1962
Chimney Swift	1955	Tennessee Warbler	1954
Ruby-throated Hummingbird *(fall)*	**1995**	Nashville Warbler	1955
Eastern Wood-Pewee	**1997**	Mourning Warbler	1934
Yellow-bellied Flycatcher	1953	Common Yellowthroat	1958
Great Crested Flycatcher	1958	Hooded Warbler	1961
Eastern Kingbird	1955	American Redstart	1955
Yellow-throated Vireo	1936	Cape May Warbler	1955
Blue-headed Vireo	1959	Northern Parula	1955
Warbling Vireo	1955	Magnolia Warbler	1962
Red-eyed Vireo	1955	Bay-breasted Warbler	1955
Tree Swallow	1956	Blackburnian Warbler	1933
N. Rough-winged Swallow	1954	Yellow Warbler	1955
Bank Swallow	1936	Chestnut-sided Warbler	1955
Cliff Swallow	1938	Blackpoll Warbler	1955
Barn Swallow	1937	Black-throated Blue Warbler	1962
Blue-gray Gnatcatcher	**2014**	Prairie Warbler	1955
Veery	1955	Black-throated Green Warbler	1955
Gray-cheeked Thrush	1955	Canada Warbler	1961
Swainson's Thrush	1955	Wilson's Warbler	1955
Wood Thrush	1955	Scarlet Tanager	1956
Ovenbird	1963	Rose-breasted Grosbeak	1961
Worm-eating Warbler	1950	Indigo Bunting	1946
Louisiana Waterthrush	1955	**Orchard Oriole**	**2010**
Northern Waterthrush	1956	**Baltimore Oriole**	**2006**
Blue-winged Warbler	1955		

Table 37. Earliest study area spring dates in chronological order for mostly Neotropical migrant landbirds (n = 83); italicized species also winter in the southern US. Excluded are those with only 1–2 spring records. Null hypothesis: there is no significant difference between the number of Neotropical migrant species whose earliest arrival dates are after 1965 (n = 37) and those whose earliest arrival dates are before 1966 (n = 46). A binomial test (z = 0.88, P = 0.19) fails to reject the null hypothesis, confirming that earliest arrival dates have continued advancing since 1965.

Species	Earliest Spring Date
Blue-winged Warbler	16 Apr 1886
Grasshopper Sparrow	17 Apr 1886
Alder Flycatcher	17 May 1896
Bobolink	19 Apr 1909
American Redstart	28 Apr 1924
Gray Catbird	*16 Apr 1925*
Blue-headed Vireo	*12 Apr 1930*
Eastern Whip-poor-will	26 Apr 1936
Cliff Swallow	14 Apr 1940
Mourning Warbler	8 May 1943
Prairie Warbler	7 Apr 1951
Black-throated Blue Warbler	15 Apr 1951
Eastern Kingbird	6 Apr 1952
Blue-gray Gnatcatcher	*7 Apr 1952*
Yellow-throated Vireo	15 Apr 1952
Gray-cheeked Thrush	16 Apr 1952
Northern Parula	16 Apr 1952
Bank Swallow	17 Apr 1952
Yellow-throated Warbler	20 Apr 1952
Worm-eating Warbler	23 Apr 1952
Blackburnian Warbler	26 Apr 1952
Hooded Warbler	26 Apr 1952
Ruby-crowned Kinglet	15 Mar 1953
Palm Warbler	17 Mar 1953
Tennessee Warbler	25 Apr 1953
Chestnut-sided Warbler	26 Apr 1953
Common Yellowthroat	*26 Apr 1953*
Golden-winged Warbler	26 Apr 1953
Orange-crowned Warbler	*26 Apr 1953*
Eastern Wood-Pewee	27 Apr 1953
Canada Warbler	2 May 1953
Lincoln's Sparrow	27 Apr 1954
Magnolia Warbler	28 Apr 1954
Least Flycatcher	4 May 1955
Yellow Warbler	1 Apr 1956
Black-and-white Warbler	5 Apr 1956
Scarlet Tanager	28 Apr 1956
Summer Tanager	9 Apr 1958
White-eyed Vireo	*30 Apr 1959*
Wood Thrush	22 Apr 1961
Ovenbird	24 Apr 1961

Species	Earliest Spring Date
Great Crested Flycatcher	25 Apr 1961
Blackpoll Warbler	28 Apr 1962
Cape May Warbler	5 May 1962
Acadian Flycatcher	2 May 1964
Wilson's Warbler	4 May 1965
Olive-sided Flycatcher	10 May 1979
Black-throated Green Warbler	25 Mar 1982
Purple Martin	20 Apr 1992
Prothonotary Warbler	24 Apr 1992
Yellow-breasted Chat	30 Apr 1992
Warbling Vireo	14 Apr 1993
Veery	29 Apr 1994
Kentucky Warbler	26 Apr 1995
Ruby-throated Hummingbird	29 Apr 1997
Black-billed Cuckoo	4 May 2000
Common Nighthawk	5 May 2000
Yellow-bellied Flycatcher	5 May 2000
Yellow-billed Cuckoo	6 May 2001
House Wren	14 Apr 2002
Indigo Bunting	10 Apr 2005
Northern Waterthrush	10 Apr 2005
Field Sparrow	*20 Mar 2006*
Rose-breasted Grosbeak	30 Apr 2006
Eastern Phoebe	*28 Feb 2007*
Louisiana Waterthrush	27 Mar 2007
Blue Grosbeak	6 Apr 2007
Cerulean Warbler	30 Apr 2007
Bay-breasted Warbler	9 May 2007
Brown Thrasher	*31 Mar 2008*
N. Rough-winged Swallow	25 Mar 2009
Barn Swallow	7 Apr 2009
Swainson's Thrush	28 Apr 2009
Willow Flycatcher	3 May 2011
Eastern Towhee	*3 Mar 2012*
Tree Swallow	*3 Mar 2012*
Golden-crowned Kinglet	*10 Mar 2012*
Baltimore Oriole	21 Apr 2012
Nashville Warbler	21 Apr 2012
Orchard Oriole	27 Apr 2012
Red-eyed Vireo	28 Apr 2012
Chimney Swift	15 Apr 2013
Philadelphia Vireo	30 Apr 2017

Table 38. New York State and federal at-risk species known to have occurred in the study area (n = 33); *italics* = study area breeding known only in the past (n = 14); **boldface** = currently breeding; * may have bred in late 1800s but unproved; ** may have bred in mid 1900s but unproved. Wood Stork (Endangered) plus Bald Eagle and Peregrine Falcon (both now delisted) are the only federal at-risk species known from the study area.

Endangered	Threatened	Special Concern
Golden Eagle	Pied-billed Grebe**	Common Loon
Short-eared Owl	*Least Bittern*	*American Bittern*
Peregrine Falcon *(nearby)*	Bald Eagle	*Osprey*
Loggerhead Shrike	*Northern Harrier*	*Sharp-shinned Hawk*
	King Rail	**Cooper's Hawk**
	Upland Sandpiper	Northern Goshawk
	Least Tern	*Red-shouldered Hawk*
	Common Tern	Black Skimmer
	Sedge Wren*	**Common Nighthawk**
	Henslow's Sparrow	*Eastern Whip-poor-will*
		Red-headed Woodpecker
		Horned Lark
		Bicknell's Thrush
		Golden-winged Warbler
		Cerulean Warbler
		Yellow-breasted Chat
		Vesper Sparrow
		Grasshopper Sparrow
		Seaside Sparrow

Appendix 2

Scientific Names of All Organisms Mentioned in This Book Other Than Birds in the Species Accounts

English Name	Scientific Name
Fungi	
Chestnut Blight	*Cryphonectria parasitica*
Dutch Elm Fungus	*Ophiostoma ulmi, O. novo-ulmi*
Eelgrass Blight	*Labyrinthula zosterae*
Lichens	
Old Man's Beard	*Usnea* spp.
Vascular plants	
American Beech	*Fagus grandifolia*
American Chestnut	*Castanea dentata*
American Elm	*Ulmus americanus*
American Hornbeam	*Carpinus caroliniana*
American Sycamore	*Platanus occidentalis*
American White Oak	*Quercus alba*
Austrian Pine	*Pinus nigra*
Bayberry	*Myrica pensylvanica*
birches	*Betula* spp.
Black Cherry	*Prunus serotina*
Black Locust	*Robinia pseudoacacia*
Broad-leaved Cattail	*Typha latifolia*
Buttonbush	*Cephalanthus occidentalis*
Chestnut Oak	*Quercus prinus*
Common Reed *(exotic)*	*Phragmites australis australis*
Eastern Black Oak	*Quercus velutina*
Eastern Hemlock	*Tsuga canadensis*
Eastern White Pine	*Pinus strobus*
Eelgrass	*Zostera marina*
English Ivy	*Hedera helix*
firs	*Abies* spp.
goldenrods	*Solidago* spp.
hickories	*Carya* spp.
Japanese Black Pine	*Pinus thunbergii*
Jewelweed	*Impatiens capensis*
maples	*Acer* spp.

English Name	Scientific Name
Marsh Spikegrass	*Distichlis spicata*
Mullein	*Verbascum thapsus*
Narrow-leaved Cattail	*Typha angustifolia*
Norway Maple	*Acer platanoides*
Norway Spruce	*Picea abies*
oaks	*Quercus* spp.
Pitch Pine	*Pinus rigida*
Red Maple	*Acer rubrum*
Red Oak	*Quercus ruber*
Saltmarsh Cordgrass	*Spartina patens*
Sassafras	*Sassafras albidum*
Scarlet Oak	*Quercus coccinea*
Scots Pine	*Pinus sylvestris*
Smooth Cordgrass	*Spartina alterniflora*
spruces	*Picea* spp.
Sugar Maple	*Acer saccharum*
Sweet Gum	*Liquidambar styraciflua*
Tuliptree	*Liriodendron tulipifera*
Tupelo	*Nyssa sylvatica*
Water Chestnut	*Trapa natans*
White Ash	*Fraxinus americana*
willows	*Salix* spp.

Insects

English Name	Scientific Name
Four-spotted Mosquito	*Anopheles quadrimaculta*
Elongate Hemlock Scale	*Fiorinia externa*
green oak inchworm (Oak Leafroller?)	*(Archips semiferanus?)*
Hemlock Woolly Adelgid	*Adelges tsugae*
Pine Needle Scale Insect	*Chionaspis pinifoliae*
Spruce Budworm	*Choristoneura fumiferana*

Fishes

English Name	Scientific Name
Alewife	*Alosa pseudoharengus*
European Carp	*Cyprinus carpio*

Mammals

English Name	Scientific Name
American Bison	*Bison bison*
American Mink	*Neovison vison*
American Red Fox	*Vulpes fulva*
Brown Rat	*Rattus norvegicus*
Eastern Cottontail	*Sylvilagus floridanus*
Eastern Coyote (Coywolf)	*Canis latrans* (?)
Gray Fox	*Urocyon cinereoargenteus*
Gray Squirrel	*Sciurus carolinensis*
Long-tailed Weasel	*Mustela frenata*
Muskrat	*Ondatra zibethicus*
North American Beaver	*Castor canadensis*
North American River Otter	*Lutra canadensis*
Raccoon	*Procyon lotor*
Red Squirrel	*Tamiasciurus hudsonicus*
Short-tailed Weasel	*Mustela erminea*
Southern Flying Squirrel	*Glaucomys volans*

English Name	Scientific Name
Striped Skunk	*Mephitis mephitis*
Virginia Opossum	*Didelphis virginiana*
White-tailed Deer	*Odocoileus virginianus*

Exotic (nonnative) birds

Belcher's Gull	*Larus belcheri*
Brown-hooded Gull	*Chroicocephalus maculipennis*
Budgerigar	*Melopsittacus undulatus*
Chukar Partridge	*Alectoris chukar*
Gray Gull	*Leucophaeus modestus*
Gray-hooded Gull	*Chroicocephalus cirrocephalus*
Helmeted Guineafowl	*Numida meleagris*
Inca Tern	*Larosterna inca*
Mandarin Duck	*Aix galericulata*
Muscovy Duck	*Cairina moschata*
Nanday Parakeet	*Aratinga nenday*
Red-vented Bulbul	*Pycnonotus cafer*
Red Bishop sp.	*Euplectes orix/E. franciscanus*
Scaly-breasted Munia	*Lonchura punctulata*
Silver Gull	*Chroicocephalus novaehollandiae*
Tricolored Munia	*Lonchura malacca*

Additional native birds

Arctic Tern	*Sterna paradisaea*
Audubon's Shearwater	*Puffinus lherminieri*
Bachman's Sparrow	*Peuaea aestivalis*
Baird's Sparrow	*Ammodramus bairdii*
Black Brant	*Branta (bernicla) nigricans*
Black Guillemot	*Cepphus grylle*
Brown Booby	*Sula leucogaster*
Cory's Shearwater	*Calonectris diomedea*
Great Shearwater	*Puffinus gravis*
Green-tailed Towhee	*Pipilo chlorura*
Leach's Storm-Petrel	*Oceanodroma leucorhoa*
Long-tailed Jaeger	*Stercorarius longicaudus*
Manx Shearwater	*Puffinus puffinus*
Northern Gannet	*Morus bassanus*
Parasitic Jaeger	*Stercorarius parasiticus*
Royal Tern	*Thalasseus maximus*
Ruddy Turnstone	*Arenaria interpres*
Sabine's Gull	*Xema sabini*
Sandwich Tern	*Thalasseus sandvicensis*
Sooty Shearwater	*Puffinus griseus*
Thick-billed Murre	*Uria lomvia*
Townsend's Solitaire	*Myadestes townsendi*
Wilson's Storm-Petrel	*Oceanites oceanicus*

Appendix 3

Glossary of Symbols, Abbreviations, and Terms Used in This Book

♀, ♀♀	female, females
♂, ♂♂	male, males
<	less than, fewer than, before, pre-
>	greater than, more than, after, post-
≤	equal to or less (fewer) than
≥	equal to or greater (more) than
±	plus or minus, more or less
ac	acre(s)
ha	hectare(s)
km	kilometer(s)
km²	square kilometer(s)
smi	statute mile(s)
AHY	any year *after hatching year*
AMNH	American Museum of Natural History
ANSP	Academy of Natural Sciences of Philadelphia
CE	Common Era
CT	Connecticut
DOIBBS	US Department of Interior Breeding Bird Surveys (since 1966)
HY	hatching year; *first* calendar *year* of life
LI	Long Island, New York
m. ob.	multiple observers
MCZ	Museum of Comparative Zoology, Harvard University
NJ	New Jersey
NWR	National Wildlife Refuge
NYC	New York City
NYCA	New York City Audubon Society
NYCDEP	New York City Department of Environmental Conservation
NYRBA	New York [City area] Rare Bird Alert (telephone recording)
NYSBBA	New York State Breeding Bird Atlas
NYSBBA I	New York State Breeding Bird Atlas I (1980–85)
NYSBBA II	New York State Breeding Bird Atlas II (2000–2005)
NYSM	New York State Museum
SY	*second* calendar *year* of life
TY	*third* calendar *year* of life
USNM	United States National Museum

blowback drift-migrants: Vagrant species not expected to occur in eastern and northeastern North America in fall that originate in the far Southwest or southern Great Plains and are moved northeastward by abnormally strong and sustained warm-air southwesterly flows in advance of cold fronts. Also, fall migrants of normally occurring eastern species that occur (especially coastally) 1, 2, or even 3 months later than their last normal inland migrants. Formerly these had been considered just eastern breeders that occasionally happened to linger later than normal. We now understand that they too are blowback drift-migrants that have been intercepted in their southbound migration somewhere to the south or southwest and then blown back north(east)ward—exactly as the more exotic western vagrants are, and frequently accompanying them. Study area examples: White-winged Dove, Painted Bunting.

ecological connectivity: The condition in which discrete, perhaps isolated, natural areas are nonetheless connected by one or more sufficiently wide corridors of natural habitat to allow free passage of outside individuals between them from another source. Connectivity is often essential in maintaining adequate genetic diversity within a geographically or habitat-isolated population in order to prevent or ameliorate inbreeding depression. This occurs when inherent genetic diversity drops in very small populations as a result of random fluctuations that allow otherwise recessive genetic traits to surface, reducing fecundity and viability.

escape: An individual of any species, native or not to an area under discussion, that has somehow escaped confinement and is now free-living, whether breeding or not. Escapes may or may not have established self-sustaining breeding populations. **Escapee** is a needless synonym best reserved for felons. Study area examples: Ring-necked Parakeet, Muscovy Duck.

exotic species: Any species not native to the area under discussion but more often applied to intercontinental imports, introductions, or escapes. Exotics may or may not have established self-sustaining breeding populations. **Nonnative species** is a common synonym. Study area examples: Rock Pigeon, Helmeted Guineafowl.

flight year: An **irruption** year

floaters: Any individuals in a largely territorial population that are not defending a territory and whose movements encompass an area substantially larger than those of their average territorial conspecifics. They often quickly move in to replace paired ♂♂ or ♀♀ that have disappeared, but they also exist in year-round and winter populations and may occur singly or in groups. For additional information see Winker 1998. Out-of-range territorial ♂♂ are not considered floaters; see **nonbreeders**.

grasspipers: A collective term for sandpipers that prefer to feed on or in grass rather than mud or sand. In eastern North America these are Upland, Pectoral, Least, Baird's, and Buff-breasted, which are frequently associated in fall with American Golden Plovers.

hacking, hacked: Raising raptors in captivity for gradual release into the wild in order to reestablish extirpated breeding populations; birds so raised and released. Examples from close to the study area: Bald Eagle, Peregrine Falcon.

incursions: Unpredictable, uncertainly bidirectional, short-term movements, often over wide areas, that are so infrequent as to be almost unique. Study area examples: Dovekie, Black-backed Woodpecker, and Black-billed Magpie. **Incur** is not the corresponding verb, but **incourse** might be, although no term seems to be in wide use. **Influx** is a synonym.

introduced species: Any taxon not native to a specific area but which has been released into that area in the hope that it will establish a self-sustaining breeding population. Study area examples: European Starling, House Sparrow.

invasion: Massive, irregular, unpredictable, usually bidirectional abrupt movements but sometimes unidirectional and longer-term that may presage major breeding or wintering range changes over broad areas. Study area examples: Cattle Egret, Red-bellied Woodpecker. The corresponding verb is **invade**.

invasive species: A term more often used with exotic plants and applied to those that spread aggressively, displace native species, and occur in large monocultures. In this sense, European Starlings are often described as invasive, and House Sparrows, now in decline in North America, formerly were. In fact, an invasive species can be native, introduced, reintroduced, or exotic.

irruptions: Periodic, bidirectional, and usually massive postbreeding movements out of normal breeding areas and into areas where species normally do not occur in winter or other nonbreeding season. Study area examples: Snowy Owl, Common Redpoll. The corresponding verb is **irrupt**.

keystone species: A plant or animal whose presence or absence has a disproportionately large effect on its environment, one that plays a major role in maintaining the composition of its ecological community. Study area examples: Eastern Coyote, Broad-leaved Cattail.

metapopulation: A discrete population of interacting subpopulations, some of which regularly die out only to be refounded later by individuals from other subpopulations in the same metapopulation.

mitochondrial DNA, mtDNA: The DNA located in mitochondria, organelles within avian cells. Nuclear and mitochondrial DNA have different origins, with mtDNA originating in the circular genomes of bacteria that were engulfed by vertebrates' earliest ancestors. It is inherited asexually only from the mother (matrilineally), and its high mutation rates render it useful for untangling evolutionary relationships and details of the process of speciation.

nonbreeders: Any individuals that are not breeding at a time when their conspecifics usually are. They may occur within a species' normal range or far from it. They may be prebreeders not quite ready for their first breeding or adults that for whatever reason do not breed in a given year.

nuclear DNA, nuDNA: The DNA located on chromosomes with the nuclei of avian cells. It is inherited sexually under Mendelian rules and encodes for the majority of the genome in birds. It evolves more slowly than mtDNA, and because DNA moves more slowly from other species during hybridization in nuDNA than in mtDNA, it is more conservative and thus likely to give a truer evolutionary history picture. Combining results from mtDNA and nuDNA analyses usually yields the most robust results.

philopatry, philopatric: The tendency of organisms to stay in or habitually return to a particular area, in migratory birds to breeding or overwintering areas. Sexual differences in philopatry are common in waterfowl and have had significant evolutionary consequences, while strong natal island philopatry has led to high speciation rates in oceanic seabirds. **Site fidelity** is a synonym.

potential breeders: Species for which there is good reason but no firm evidence to believe they might now be breeding (even if erratically), are reasonably likely to have done so in the past, or are likely to do so in the near future. Note that this term is wholly different from the levels of breeding confidence used in Breeding Bird Atlases: possible, probable, confirmed. Study area examples: Barn Owl, Pine Warbler.

precontact: An anthropological term for the time and conditions existing before contact of indigenous peoples with any outside culture.

species richness: A simple count of the number of species in a given area, irrespective of the numbers of each species. It is often confused with **species diversity**, but that is a complex mathematical function incorporating the relative abundance of each species found. **Species evenness** is another mathematical facet of numerical abundance.

spring range-prospectors: Individuals of species normally breeding to the south of the study area that nonetheless appear somewhat out of range in spring migration but at expected times. These are often SYs in species that are expanding their breeding ranges northward and so are bellwethers of impending or actual range expansions. Study area examples: Prothonotary Warbler, Yellow-throated Warbler.

taxon, taxa: Any unit used in taxonomy (biological classification). In avian taxonomy the most frequently encountered taxa (the plural) are order, family, genus, species, and subspecies, and in this book, the last two especially. Taxa may include only a single level (species) or several (species and subspecies). Study area examples: waterfowl (the family Anatidae), Yellow Palm Warbler (a subspecies).

trans-Gulf overshoots: Individuals of species that migrate across the Gulf of Mexico in spring but that are intercepted before landfall at times normal for their arrival in southeastern states. They are then impelled rapidly to the northeast, often far out of their normal range, by strategically positioned low-pressure systems' counterclockwise winds and arrive weeks or even months early. Sometimes also referred to as slingshot migrants or trans-Gulf slingshot migrants. Study area examples: Eastern Kingbird, Yellow Warbler, Summer Tanager, Indigo Bunting.

wreck: An originally British term for an extensive mortality in seabirds caused by food shortages, oil spills, etc. Such birds are then often blown ashore in huge numbers, sometimes far inland, by abnormally strong winds and storms. Study area or nearby examples: Red Phalarope, Dovekie.

Appendix 4

Names of All Observers Appearing Anywhere in the Body of This Book

Abbott, C. G.
Abramson, Leonard
Adams, Michael
Aldana, Daniel
Allaire, Ken
Allen, Deborah
Allen, Robert P.
Alomeri, Mohammed
Alperin, Irwin
Alvarez, Glenn
Amos, Anthony
Antenen, Susan
Antonio, John
Aracil, Richard
Aronoff, Arthur
Arrowsmith, Anne
Askildsen, John
Atherton, Frederic
Auerbach, Anya
Augustine, Bob
Bahrt, Sid
Baksh, Andrew
Banner, Gilbert
Bannerman, Isabella
Barber, Art
Bardwell, Kyle
Barrett, David
Barron, Catherine
Barry, Ann
Bass, Ryan
Batren, Zach
Bauer, Herman
Bauer, Vernon
Baumann, Bill
Baumann, Wilma
Becker, Debbie
Bennett, Chris

Benoit, Matthieu
Benson, G.
Berkins, Nathaniel
Bernick, Andrew
Berthoud, C.
Bickford, Larry
Bicknell, Eugene
Birnbaum, Leonard
Black, Irving
Block, Andrew
Bloom, Joseph
Bochnik, Michael
Bourque, Ron
Bowen, R.
Boyajian, Ned
Braem, Helen
Brash, Alex
Breslau, Leo
Brickner, Alice
Brickner, Philip
Brigg, M.
Brigham, H. S.
Brinker, Lysle
Britt, Mike
Britton, Nathaniel
Brown, F. M.
Brown, Leroy
Brown, Malcolm
Buckley, P. A.
Bull, John
Bustamante, Vicki
Camillieri, Sean
Cantor, Irving
Cap, Edward
Capen, E. A.
Capodilupo, Jeanne
Carleton, Geoffrey

Caspers, E.
Caspers, John
Cassidy, Edward
Cech, Rick
Chang, Stephen
Chapman, Roy
Chavez, Dawn
Chubb, Samuel
Ciganek, Drew
Clayton, M.
Clinton, Jim
Clugston, Robin
Coco, J.
Cole, Jared
Coles, Robert
Collerton, Anthony
Collier, Jan
Collin, Norma
Collins, John
Conner, Robert
Connolly, Barbara
Conte, Isabel
Conway, William
Cooper, Mike
Corrigan, Bernard
Crocker, Sharron
Crosby, Maunsell
Cruickshank, Allan
Dadone, George
Dancis, Dale
Darrow, Harry
Datlen, Tracy
Davis, F.
Davis, Thomas
Davis, W.
DeAngelo, Angelo
Deed, Robert

DeCandido, Robert
Demes, Jim
DeOrsey, Joan
DeOrsey, Stan
Deppe, Gloria
DiCostanzo, Joseph
Dolan, Debbi
Dove, Aline
Dow, Tom
Drake, Kathy
Drogin, Alan
Drucker, Jacob
Duran-Ruiz, Adalaida
Dwight, Jonathan
Edelbaum, Evan
Edler, Eric
Eisenmann, Eugene
Eisenstein, H. J.
Ellen, Arthur
Elliott, John
Enders, Frank
Ephraim, Bill
Evans, Karen
Factor, Tim
Farnesworth, Andrew
Farrand, John
Fazzino, Frank
Feigin, William
Feinberg, Ezra
Ferguson, H. L.
Ferguson, Mike
Ferguson, Walter
File, Fran
Finger, Corey
Fiore, Thomas
Fischer, Richard
Fisher, A. K.

Fleisher, Edward
Fogarty, Brendan
Forrest, Mrs. R.
Foster, L. S.
Fraza, Louise
Freedman, Sharon
Friedman, Bob
Friedman, Ralph
Friedman, Stuart
Frielich, Jeff
Friton, Walter
Fritz, John
Fung, Karen
Furgoch, Mira
Futuyma, Doug
Gaillard, Ed
Gallagher, Cass
Gambino, Jack
Garcia, Yolanda
Garland, Len
Gee, John
Gell-Mann, Benedict
Gell-Mann, Murray
Gera, Joseph
Gera, Robert
Gershon, Richard
Gibaldi, Steve
Gilbert, Ben
Gillen, J.
Gillen, Paul
Gillian, Henry
Gilliard, Thomas
Girone, Mike
Giunta, Joe
Gladden, George
Gochfeld, Doug
Gochfeld, Michael
Goelet, Ogden
Gomes, Kenton
Gotlib, Adele
Gracies, S.
Grant, Bob
Grierson, Stanley
Grinnell, George B.
Griscom, Ludlow
Haas, John
Hackett, Bill
Hait, Sol
Hake, Mary
Hall, Burt
Hall, David
Haluska, William
Hamers, Sam
Hannay, Dawn
Harrison, Ricky
Harten, Chuck
Hastings, G. T.
Hastings, Watson
Heath, Fred
Heck, Otto
Heemstra, Valerie
Helmuth, William
Herbert, Richard A.
Herrick, Francis
Herskovics, Gene
Hickey, Joseph
Higgins, Conor
Hirschberg, E.
Hix, George
Holgate, J.
Honig, Howard
Hopper, Marjorie
Horan, Patrick
Horowitz, Joseph
Horowitz, Sol A.
Howard, Elliotte
Howe, William
Hudda, Neelakshi
Imhof, Thomas
Iula, Vincent
Jacobson, Malcolm
Jamison, Elizabeth
Jaslowitz, Carl
Jett, Rob
Johansson, Tait
Johnson, Herbert
Johnston, Tom
Johnston, Charles
Jove, José
Julig, Bill
Kallmann, Klaus
Kane, Richard
Karlson, Danny
Karsch, Henry
Kassoy, Irving
Keane, John
Keogh, Brendan
Kessler, Philip
Khalifa, Amin
Kieran, John
Kilanowski, Vicki
King, Ben
Klein, Coby
Kleinbaum, Michel
Klose, Bob
Klots, Alexander
Knoecklein, Karl
Knox, James
Koelle, Susan
Komorowski, George
Kramer, Robert
Krauss, Dave
Kravatz, Mike
Kreissman, Dave
Kriegeskotte, Karl
Kuerzi, John
Kuerzi, Richard
Künstler, David
LaBella, Linda
Landdeck, Kevin
Landdeck, Paula
La Dow, Stanley
Lamek, Alex
Lanyon, Bud
LaRosa, Seth
Larsen, Dave
Latham, Roy
Leary, F.
Lee, Andrew
Lehman, Paul
Lehrman, Daniel
Lehrman, Lawrence
Le Strange, Philip
Letts, Christopher
Levine, Emanuel
Lewis, C. L.
Lewis, Robert
Lichten, Alfred
Lichtenauer, C. E.
Lindsay, Patricia
Loeb, Betty
Logan, Elwood
Lopes, Heidi
Love, Richard
Lunt, Helene
Lutter, Pat
Lyons, Christopher
Machover, Bob
Mackenzie, Locke
MacKinnon, Roderick
Magill, Julia
Maguire, Edward
Mako, David
Malina, Alan
Malloy, John
Marien, Daniel
Mark, Evan
Martin, Adam
Martin, Hugh
Martin, Steve
Matsushito, W.
Maumary, Max
Maxwell, C.
Maxwell, J.
Mayer, John
Mayer, Paul
McAlexander, Chuck
McGann, Kevin
McGee, Gerald
McGovern, Ed
McGuinness, Hugh
McKellar, Mark
Messer, Alan
Messing, Pauline
Meyer, Paul
Michael, Keith
Miller, Eric
Miller, John
Miller, Waldron
Mitra, Shai
Morales, Omar
Morgan, Amanda
Morris, Art
Morrison, Bruce
Moss, William
Moyle, John
Mudge, Eugene
Mueller, Keith
Murray, William
Nachtmann, Aiden
Nadareski, Chris
Naf, Ulrich
Nagy, Christopher
Nathan, Bernard
Neville, Bruce
Nichols, Edward G.
Nichols, L. Nelson
Normandia, Mary
Norse, William
Nulle, Jeff
Nutter, David
O'Brien, Catherine
O'Dell, L.
Olsen, Arthur
Olsen, Emma
Olson, Todd
O'Reilly, Nathan
Orth, John

NAMES OF ALL OBSERVERS

Oswald, John
Ott, Al
Ozard, Steve
Pangburn, Clifford
Panko, Drew
Pappalardi, Felix
Pasquier, Roger
Pehek, Ellen
Pell, Morris
Pell, Walden
Peltomaa, Anders
Penberthy, Alan
Pendergrass, Mara
Perrault, Stephane
Perry, D.
Peszel, Theodore
Petersen, Erik
Peterson, R. T.
Phelan, Joseph
Pink, Eleanor
Pirko, Alex
Plunkett, Dick
Pohner, L.
Polhemus, Alma
Poole, Dorothy
Post, Peter
Prelich, Gregory
Previdi, Jim
Purcell, Anne
Purcell, Bill
Quinn, Glenn
Rafferty, Daniel
Raices, Florence
Rakowski, Miriam
Randall, Keir
Ravitts, Ricki
Raymond, Olney
Raynor, Gil
Rea, A.
Reisfeld, Peter
Restivo, Ernest
Reynolds, W.
Rich, Eva
Rich, Marc
Richards, A.
Rigerman, Alan
Ringer, David
Ritter, Jeffrey
Robben, Tom
Roberto, Charlie
Roche, Dave
Roderick, J.
Rodewald, Paul
Rodriguez, Diego
Rofe, Joe
Rogers, Adrian
Rogers, Charles
Root, P.
Rose, George
Rosenberg, Kenneth
Rosenheim, Richard
Roser, Dan
Rosin, Axel
Rothman, Jack
Rucht, Bob
Ruff, Frederick
Ruscica, John
Russ, Hilary
Russak, Marshall
Russell, George
Russell, Will
Ryan, Richard
Sadock, Ben
Salzman, Eric
Sanford, Lloyd
Saphir, Starr
Sargent, Jesse
Saunders, A. A.
Savaresse, G.
Schaffner, M.
Schlesinger, Lee
Schmidt, Gus
Schulman, Donna
Schulze, Eric
Schutz, Randolph
Schwartz, Vic
Schwartz-Weinstein, Zach
Scully, Robert
Sedwitz, Walter
Seneca, Jeff
Shaw, Ann
Shen, Peter
Shore, Lore
Siebenheller, Bill
Siebenheller, Norma
Sime, Sean
Simon, David
Sisinni, Susan
Skelton, Kathleen
Skrentny, Jeff
Small, Arnold
Smart, Zachary
Smith, D. W.
Smith, N. G.
Soll, Jerry
Solomon, Bill
Souirgi, Nadir
Spitalnik, Lloyd
Staloff, Charles
Stanley, Susan
Stansbury, Archer
Stapleton, James
Steffens, Frank
Steineck, Paul
Stephenson, O. K.
Stepinoff, Si
Stetson, Jeff
Stoechle, Mark
Stout, Gardiner
Stuart, Sam
Sturm, Chris
Suggs, John
Sutton, Al
Swift, Otis
Teator, Michael
Temple, Patrick
Thomas, Allen
Tiffany, J.
Tozer, Matt
Tozzi, Peter
Trachtenberg, Larry
Tramontano, John
Treacy, Ed
Trimble, Jeremiah
Trombone, Thomas
Tudor, Guy
Turner, Scott
Urner, Charles
Usin, Victor
Van Doren, Ben
Van Duesen, Hobart
VanScoy, R.
van Wert, A.
vanZyl, Debbie
Veit, Richard
Vellozzi, James
Veso, Tom
Victor, Kai
Vietor, Edward
Votta, John
Waldron, Michael
Waldron, Paula
Wallman, Josh
Wallstrom, Kristine
Walsh, Lester
Walter, Steve
Ward, Jeffrey
Waterman, Otis
Watson, Frank
Weber, William
Weeks, Arthur
Weintraub, Joel
Weiss, Carol
Wells, Della
Wells, Philip
Whelen, Ed
Whitt, Ray
Wiegmann, William
Wilcox, Leroy
Wilfred, Tom
Wilkinson, Christina
Willow, Gabriel
Wilson, Angus
Winston, Tod
Wollney, Seth
Wright, James
Yeaton, Sam
Yeganian, Charles
Young, Charles
Young, John L.
Yrizarry, John
Zabouli, Ray
Zupan, Jeffrey

Appendix 5

Stranger in a Strange Land: An Andean Gull in the Bronx

Walter Sedwitz, updated by P. A. Buckley

On 17 May 1980, Sedwitz found an unbanded hooded gull in a flock of Ring-billed Gulls resting on the sloping retaining wall at the north end of Jerome Reservoir. It was obviously not one of the many Laughing Gulls that frequented the location, nor was it a Black-headed, which had also occurred there 3 times in the 1960s–70s. With a much paler mantle than the Ring-billeds, it was striking, with a full blackish–brown hood, narrow semicircular eye ring around the posterior of the eye, and slender wine-red bill and legs. At rest, the suggestion of a large white patch proximal to the primary tips hinted at Franklin's Gull, but the mantle was far too light, and its full hood eliminated Black-headed.

Its identification as an Andean Gull was easily confirmed in the excellent gull series in the American Museum of Natural History, where Brown-hooded, Gray-hooded, Silver, and other gull species were quickly eliminated. Then on 2 November 1980, Sedwitz and Buckley found what appeared to be the same individual in the same place, only now in winter plumage, with a darker bill and legs, the hood replaced by a small dark ear-spot/crescent behind the eye. When it flew, Andean Gull's diagnostic black-striped white primary patch was conspicuous. It was seen again on 14 May 1981 resting on a nearby grassy area, still in a flock of Ring-billeds and back in breeding plumage, and then on 11 July and 14 August 1981, when it had moved to the Van Cortlandt Park Parade Ground, still with Ring-billeds but now starting to slip out of breeding plumage. That was the last time it was seen, so it was assumed to have departed or died, when surprisingly on 17 March 1984 it was back where first seen on Jerome Reservoir, still with Ring-billeds and now beginning its molt into breeding plumage.

This really was its last time seen, and given its consistent occurrence in flocks of Ring-billeds, one can only wonder if it followed them to some Great Lakes breeding colony and attempted to pair with one. Errant Black-headed Gulls have done this on several occasions in eastern North America, but no gull suggesting an Andean × Ring-billed hybrid was ever seen in the Jerome Reservoir/Van Cortlandt area.

Its molt and plumage sequencing matched that of Ring-billed and Laughing Gulls, an initially puzzling fact. But three factors bear on this. (1) This individual was surely an escape from captivity, as natural vagrancy from the west coast of South America is not only unlikely, but Andean Gull has evinced little or no history of such vagrancy even within South America; (2) when in captivity it is common for austral gulls to shift their breeding cycle to a boreal regimen and vice versa; and (3) its

Andean breeding range extends for about 3100 smi (5000 km) from southwestern Colombia south to southern Chile and Argentina. In the northern part of its range they breed from May to November, and in the southern part from September to January, indicating their intrinsic breeding flexibility.

So where did this particular individual come from?

From 1960–95 the Bronx Zoo had the only Andean Gulls in captivity anywhere in North America, in a walk-through aviary inhabited by a wide variety of other waterbirds and where the Andean Gulls were breeding successfully. The total number of Andean Gulls there is not available before 1970, but from 1967 to 1992 at least 41 Bronx-bred Andean Gull chicks fledged (Lindholm and Svanberg 2015). Unfortunately, on 6 February 1995 the aviary collapsed from heavy snowdrifts, and at least the following birds escaped: 12 Andean Gulls, 8 Gray Gulls, 1 Belcher's Gull, and 12 Inca Terns. If any were ever seen subsequently, they did not come to the attention of the local regional editors for *American Birds* and *Kingbird*.

With the proximity of the Bronx Zoo to Jerome Reservoir (2 smi [3.2 km] apart), there can be no doubt that the 1980s Andean Gull had also escaped (undetected or unreported) from that same aviary but sometime before about 1978, given its age when first seen at Jerome Reservoir. In so doing, it may have dodged a different demise 17+ years later, not a terribly long survival time for a gull in captivity.

An intriguing postscript is that an Andean Gull in breeding plumage was photographed and seen by many at Great Swamp National Wildlife Refuge in Morris County, NJ (32 smi [51 km] southwest of Van Cortlandt Lake) from 9 to 16 April 1981 (Leuzarder et al.) In the same plumage as the study area bird, it too was unbanded and bore no distinctive markings, so we'll never know if it was the same or another individual.

Appendix 6

Specimens in Museum Collections from Van Cortlandt Park, Kingsbridge Meadows, Woodlawn Cemetery and Jerome Reservoir, and Riverdale

Table 6a. Van Cortlandt Park specimens that have been digitized (n = 367 of 77 species) in the AMNH and MCZ collections. Most were in Dwight's original late 1800s collection (6-digit AMNH accession numbers). A handful of others lacking dates have been omitted.

Species	Accession Number	Collection Date
Sharp-shinned Hawk	AMNH 352124	3 May 1887
Red-shouldered Hawk	AMNH 98111	24 Nov 1887
Solitary Sandpiper	AMNH 98762	30 Apr 1889
Yellow-billed Cuckoo	AMNH 360103	14 Aug 1890
Yellow-billed Cuckoo	AMNH 360101	22 Aug 1890
Yellow-billed Cuckoo	AMNH 360102	22 Aug 1890
Yellow-billed Cuckoo	AMNH 98134	3 Sep 1888
Yellow-billed Cuckoo	MCZ 362059	5 Jun 1933
Black-billed Cuckoo	AMNH 360207	14 May 1890
Black-billed Cuckoo	AMNH 360206	18 May 1886
Ruby-throated Hummingbird	AMNH 361469	18 May 1886
Ruby-throated Hummingbird	AMNH 361471	2 Sep 1890
Ruby-throated Hummingbird	AMNH 361470	22 Aug 1890
Yellow-bellied Sapsucker	AMNH 98143	14 Apr 1888
Downy Woodpecker	AMNH 363367	21 Apr 1886
Downy Woodpecker	AMNH 363366	26 Feb 1884
Downy Woodpecker	AMNH 363365	26 Feb 1884
White-eyed Vireo	AMNH 378795	13 May 1887
White-eyed Vireo	AMNH 378796	13 May 1887
White-eyed Vireo	AMNH 378800	13 May 1887
White-eyed Vireo	AMNH 378801	14 Aug 1890
White-eyed Vireo	AMNH 378797	17 Jun 1889
White-eyed Vireo	AMNH 378798	5 Jul 1889
White-eyed Vireo	AMNH 378799	5 Jul 1889
White-eyed Vireo	AMNH 98451	5 Sep 1887
White-eyed Vireo	AMNH 98450	5 Sep 1887
White-eyed Vireo	AMNH 98449	5 Sep 1887
Yellow-throated Vireo	AMNH 379039	18 May 1886
Yellow-throated Vireo	AMNH 98439	28 Jun 1889
Yellow-throated Vireo	AMNH 379040	3 May 1887
Yellow-throated Vireo	AMNH 379038	30 Apr 1886

Species	Accession Number	Collection Date
Yellow-throated Vireo	AMNH 379037	6 May 1885
Blue-headed Vireo	AMNH 379101	22 Apr 1885
Blue-headed Vireo	AMNH 379102	26 Apr 1886
Red-eyed Vireo	AMNH 379368	10 Sep 1896
Red-eyed Vireo	AMNH 379371	12 May 1890
Red-eyed Vireo	AMNH 379369	13 May 1887
Red-eyed Vireo	AMNH 379370	19 May 1887
Red-eyed Vireo	AMNH 379372	20 May 1885
Red-eyed Vireo	AMNH 98419	25 Jun 1889
Purple Martin	AMNH 371504	30 Apr 1886
Purple Martin	AMNH 371505	30 Apr 1886
Purple Martin	AMNH 371506	30 Apr 1886
Purple Martin	AMNH 371507	30 Apr 1886
N. Rough-winged Swallow	AMNH 371263	3 May 1886
N. Rough-winged Swallow	AMNH 371262	30 Apr 1886
N. Rough-winged Swallow	AMNH 371264	30 Apr 1886
Black-capped Chickadee	AMNH 372806	1 Apr 1886
Black-capped Chickadee	AMNH 372807	1 Apr 1886
Black-capped Chickadee	AMNH 372805	23 Dec 1885
Black-capped Chickadee	AMNH 372808	23 Dec 1885
Black-capped Chickadee	AMNH 98671	24 Nov 1887
Black-capped Chickadee	AMNH 98670	29 Oct 1887
White-breasted Nuthatch	AMNH 373731	12 Mar 1884
White-breasted Nuthatch	AMNH 373730	13 Apr 1885
House Wren	AMNH 374413	3 May 1887
Winter Wren	AMNH 374570	21 Dec 1885
Golden-crowned Kinglet	AMNH 377630	11 Dec 1885
Golden-crowned Kinglet	AMNH 377631	13 Apr 1885
Golden-crowned Kinglet	AMNH 98673	14 Apr 1888
Golden-crowned Kinglet	AMNH 98674	14 Apr 1888
Golden-crowned Kinglet	AMNH 98675	21 Apr 1888
Golden-crowned Kinglet	AMNH 377629	23 Dec 1885
Golden-crowned Kinglet	AMNH 98676	29 Oct 1887
Golden-crowned Kinglet	AMNH 98677	29 Oct 1887
Ruby-crowned Kinglet	AMNH 377847	13 Apr 1885
Ruby-crowned Kinglet	AMNH 377848	13 Apr 1885
Ruby-crowned Kinglet	AMNH 98680	21 Apr 1888
Ruby-crowned Kinglet	AMNH 98679	21 Apr 1888
Ruby-crowned Kinglet	AMNH 377849	22 Apr 1885
Eastern Bluebird	AMNH 377161	14 Apr 1886
Eastern Bluebird	AMNH 377159	18 Apr 1890
Eastern Bluebird	AMNH 377160	18 Apr 1890
Veery	AMNH 376998	12 May 1886
Veery	AMNH 377001	12 May 1890
Veery	AMNH 377000	13 May 1887
Veery	AMNH 377006	13 May 1887
Veery	AMNH 377007	13 May 1887
Veery	AMNH 377002	14 May 1890
Veery	AMNH 377004	2 Sep 1890
Veery	AMNH 377003	22 Aug 1890
Veery	AMNH 377005	22 May 1890

(Continued)

Table 6a. (Continued)

Species	Accession Number	Collection Date
Veery	AMNH 377999	3 May 1887
Veery	AMNH 98691	3 Sep 1888
Veery	AMNH 98690	3 Sep 1888
Veery	AMNH 98689	30 May 1887
Veery	AMNH 98688	5 May 1888
Bicknell's Thrush	AMNH 376961	12 May 1890
Gray-cheeked Thrush	AMNH 376961	12 May 1890
Gray-cheeked Thrush	AMNH 376936	14 May 1890
Gray-cheeked Thrush	AMNH 376937	14 May 1890
Swainson's Thrush	AMNH 376770	12 May 1886
Swainson's Thrush	AMNH 376774	12 May 1890
Swainson's Thrush	AMNH 376775	12 May 1890
Swainson's Thrush	AMNH 376776	12 May 1890
Swainson's Thrush	AMNH 376772	13 May 1887
Swainson's Thrush	AMNH 376773	13 May 1887
Swainson's Thrush	AMNH 376771	18 May 1886
Swainson's Thrush	AMNH 376777	22 May 1890
Swainson's Thrush	AMNH 98700	5 Sep 1887
Hermit Thrush	AMNH 376979	14 Apr 1890
Hermit Thrush	AMNH 376578	26 Apr 1886
Hermit Thrush	AMNH 376579	3 May 1887
Wood Thrush	AMNH 376307	14 Aug 1890
Wood Thrush	AMNH 376304	14 Aug 1890
Wood Thrush	AMNH 376306	14 Aug 1890
Wood Thrush	AMNH 376308	2 Sep 1890
Wood Thrush	AMNH 376305	22 Aug 1890
Wood Thrush	AMNH 376309	5 Jul 1889
Wood Thrush	AMNH 98683	5 May 1888
Wood Thrush	AMNH 376303	7 May 1886
American Robin	AMNH 375968	14 Apr 1890
American Robin	AMNH 375969	19 Apr 1886
American Robin	AMNH 375967	3 Jun 1890
Gray Catbird	AMNH 375478	10 Sep 1896
Gray Catbird	AMNH 98643	15 Sep 1888
Gray Catbird	AMNH 375474	21 Dec 1885
Gray Catbird	AMNH 375476	22 Aug 1890
Gray Catbird	AMNH 375477	22 Aug 1890
Gray Catbird	AMNH 11240	3 May 1892
Gray Catbird	AMNH 375475	7 May 1886
Brown Thrasher	AMNH 375602	2 Sep 1890
Brown Thrasher	AMNH 375601	22 May 1890
Brown Thrasher	AMNH 375603	3 May 1886
Brown Thrasher	AMNH 98647	3 Sep 1888
Brown Thrasher	AMNH 375600	7 May 1886
Cedar Waxwing	AMNH 378276	12 Mar 1884
Cedar Waxwing	AMNH 378277	12 Mar 1884
Cedar Waxwing	AMNH 378280	12 Mar 1884
Cedar Waxwing	AMNH 378278	14 Apr 1886
Cedar Waxwing	AMNH 378279	14 Apr 1886
Ovenbird	AMNH 383551	30 Apr 1886
Ovenbird	AMNH 383554	4 Jun 1886

Species	Accession Number	Collection Date
Ovenbird	AMNH 383552	5 Jul 1889
Ovenbird	AMNH 383553	5 Jul 1889
Ovenbird	AMNH 383550	7 May 1886
Worm-eating Warbler	AMNH 379992	10 Jun 1886
Worm-eating Warbler	AMNH 379993	13 Jun 1887
Worm-eating Warbler	AMNH 379994	17 Jun 1889
Worm-eating Warbler	AMNH 379987	22 Aug 1890
Worm-eating Warbler	AMNH 379995	22 Aug 1890
Worm-eating Warbler	AMNH 379986	22 May 1890
Worm-eating Warbler	AMNH 379984	3 May 1887
Worm-eating Warbler	AMNH 379985	5 Jul 1889
Worm-eating Warbler	AMNH 379983	6 May 1885
Worm-eating Warbler	AMNH 379991	6 May 1885
Worm-eating Warbler	AMNH 379988	9 May 1895
Worm-eating Warbler	AMNH 379989	9 May 1895
Worm-eating Warbler	AMNH 379990	9 May 1895
Louisiana Waterthrush	AMNH 11249	1 May 1892
Louisiana Waterthrush	AMNH 383800	14 Apr 1886
Louisiana Waterthrush	AMNH 383801	19 Apr 1886
Louisiana Waterthrush	AMNH 383802	3 May 1887
Louisiana Waterthrush	AMNH 383803	9 May 1895
Northern Waterthrush	AMNH 383688	12 May 1890
Northern Waterthrush	AMNH 383688	12 May 1890
Northern Waterthrush	AMNH 383689	14 Aug 1890
Northern Waterthrush	AMNH 383689	14 Aug 1890
Northern Waterthrush	AMNH 383687	19 May 1885
Northern Waterthrush	AMNH 383690	2 Sep 1890
Northern Waterthrush	AMNH 383690	2 Sep 1890
Golden-winged Warbler	AMNH 380025	22 Aug 1890
Blue-winged Warbler	AMNH 188697	16 Jun 1925
Blue-winged Warbler	AMNH 380089	12 May 1886
Blue-winged Warbler	AMNH 380092	12 May 1890
Blue-winged Warbler	AMNH 380093	12 May 1890
Blue-winged Warbler	AMNH 380094	12 May 1890
Blue-winged Warbler	AMNH 380095	12 May 1890
Blue-winged Warbler	AMNH 380101	12 May 1890
Blue-winged Warbler	AMNH 383153	14 Apr 1886
Blue-winged Warbler	AMNH 380097	14 Aug 1890
Blue-winged Warbler	AMNH 380098	14 Aug 1890
Blue-winged Warbler	AMNH 380096	14 May 1890
Blue-winged Warbler	AMNH 380099	22 Aug 1890
Blue-winged Warbler	AMNH 380100	22 Aug 1890
Blue-winged Warbler	AMNH 380102	3 Jun 1890
Blue-winged Warbler	AMNH 380087	3 May 1886
Blue-winged Warbler	AMNH 380090	3 May 1887
Blue-winged Warbler	AMNH 11274	3 May 1892
Blue-winged Warbler	AMNH 383151	30 Apr 1885
Blue-winged Warbler	AMNH 383152	30 Apr 1885
Blue-winged Warbler	AMNH 380104	5 Jul 1889
Blue-winged Warbler	AMNH 380103	5 May 1889
Blue-winged Warbler	AMNH 380091	7 Jun 1889

(Continued)

Table 6a. (Continued)

Species	Accession Number	Collection Date
Blue-winged Warbler	AMNH 380088	7 May 1886
Black-and-white Warbler	AMNH 379788	12 May 1886
Black-and-white Warbler	AMNH 379785	14 Aug 1890
Black-and-white Warbler	AMNH 379789	19 Jun 1886
Black-and-white Warbler	AMNH 379786	22 Aug 1890
Black-and-white Warbler	AMNH 379783	3 May 1886
Black-and-white Warbler	AMNH 379790	3 May 1887
Black-and-white Warbler	AMNH 379787	30 Apr 1886
Black-and-white Warbler	AMNH 98453	5 Sep 1887
Black-and-white Warbler	AMNH 379784	7 Jun 1889
Common Yellowthroat	AMNH 384116	17 Jun 1889
Common Yellowthroat	AMNH 384120	19 May 1885
Common Yellowthroat	AMNH 384121	19 May 1887
Common Yellowthroat	AMNH 384117	5 Jun 1889
Common Yellowthroat	AMNH 384118	5 Jun 1889
Common Yellowthroat	AMNH 384119	5 Jun 1889
Common Yellowthroat	AMNH 384115	7 Jun 1889
Hooded Warbler	AMNH 384602	20 May 1885
Hooded Warbler	AMNH 384601	3 Jun 1890
Hooded Warbler	AMNH 384600	7 Jun 1889
American Redstart	AMNH 385027	1 Dec 1886
American Redstart	AMNH 385026	13 May 1887
American Redstart	AMNH 385030	14 Aug 1890
American Redstart	AMNH 385028	18 May 1886
American Redstart	AMNH 385029	18 May 1886
American Redstart	AMNH 385025	3 May 1887
American Redstart	AMNH 385031	5 Jul 1889
American Redstart	AMNH 385032	5 Jul 1889
American Redstart	AMNH 98632	5 May 1888
Northern Parula	AMNH 380682	14 May 1890
Northern Parula	AMNH 380681	14 May 1890
Northern Parula	AMNH 380683	22 May 1890
Northern Parula	AMNH 98480	22 Sep 1888
Magnolia Warbler	AMNH 381142	14 May 1890
Magnolia Warbler	AMNH 98513	19 May 1888
Magnolia Warbler	AMNH 381141	20 May 1885
Blackburnian Warbler	AMNH 382521	14 May 1890
Blackburnian Warbler	AMNH 382522	19 May 1885
Yellow Warbler	AMNH 380880	12 May 1890
Yellow Warbler	AMNH 380879	20 May 1885
Chestnut-sided Warbler	AMNH 382725	12 May 1886
Chestnut-sided Warbler	AMNH 382727	12 May 1890
Chestnut-sided Warbler	AMNH 382728	12 May 1890
Chestnut-sided Warbler	AMNH 382723	13 May 1887
Chestnut-sided Warbler	AMNH 382729	14 May 1890
Chestnut-sided Warbler	AMNH 98523	19 May 1888
Chestnut-sided Warbler	AMNH 98524	19 May 1888
Chestnut-sided Warbler	AMNH 382726	22 May 1890
Chestnut-sided Warbler	AMNH 382724	6 May 1885
Blackpoll Warbler	AMNH 382967	10 Jun 1886
Blackpoll Warbler	AMNH 382967	10 Jun 1886

Species	Accession Number	Collection Date
Blackpoll Warbler	AMNH 382966	22 May 1890
Blackpoll Warbler	AMNH 382966	22 May 1890
Black-throated Blue Warbler	AMNH 381371	13 May 1887
Black-throated Blue Warbler	AMNH 381372	14 May 1890
Black-throated Blue Warbler	AMNH 382195	22 Apr 1885
Palm Warbler (Yellow)	AMNH 383427	14 Apr 1890
Palm Warbler (Yellow)	AMNH 383428	14 Apr 1890
Palm Warbler (Yellow)	AMNH 383429	21 Apr 1885
Palm Warbler (Yellow)	AMNH 383430	22 Apr 1885
Pine Warbler	AMNH 383153	14 Apr 1886
Pine Warbler	AMNH 98561	14 Apr 1888
Pine Warbler	AMNH 383151	30 Apr 1885
Pine Warbler	AMNH 383152	30 Apr 1885
Myrtle Warbler	AMNH 381557	21 Apr 1886
Myrtle Warbler	AMNH 98505	5 May 1888
Myrtle Warbler	AMNH 98506	5 May 1888
Prairie Warbler	AMNH 383256	6 May 1885
Prairie Warbler	AMNH 383257	6 May 1885
Black-throated Green Warbler	AMNH 98551	17 Apr 1887
Black-throated Green Warbler	AMNH 98552	17 Apr 1887
Black-throated Green Warbler	AMNH 382195	22 Apr 1885
Canada Warbler	AMNH 384860	14 Aug 1890
Canada Warbler	AMNH 384861	18 May 1886
Canada Warbler	AMNH 384859	19 May 1887
Wilson's Warbler	AMNH 384653	13 May 1887
Wilson's Warbler	AMNH 384654	14 Aug 1890
Yellow-breasted Chat	AMNH 384458	13 May 1887
Yellow-breasted Chat	AMNH 384459	14 Aug 1890
Yellow-breasted Chat	AMNH 384460	14 Aug 1890
Yellow-breasted Chat	AMNH 384457	17 Jun 1889
Yellow-breasted Chat	AMNH 98614	19 May 1888
Eastern Towhee	AMNH 367981	17 Jun 1889
Eastern Towhee	AMNH 367982	17 Jun 1889
Eastern Towhee	AMNH 367983	22 Aug 1890
American Tree Sparrow	AMNH 403172	11 Dec 1884
American Tree Sparrow	AMNH 403173	11 Dec 1884
American Tree Sparrow	AMNH 403169	11 Dec 1885
American Tree Sparrow	AMNH 403175	13 Apr 1885
American Tree Sparrow	AMNH 98335	2 Apr 1887
American Tree Sparrow	AMNH 403174	21 Dec 1885
American Tree Sparrow	AMNH 403171	26 Feb 1884
American Tree Sparrow	AMNH 403170	8 Apr 1889
Chipping Sparrow	AMNH 403300	13 May 1887
Chipping Sparrow	AMNH 403301	13 May 1887
Chipping Sparrow	AMNH 403297	26 Apr 1886
Chipping Sparrow	AMNH 403298	26 Apr 1886
Chipping Sparrow	AMNH 403299	3 May 1887
Field Sparrow	AMNH 403713	10 Sep 1896
Field Sparrow	AMNH 403704	14 Apr 1886
Field Sparrow	AMNH 403708	14 Apr 1886
Field Sparrow	AMNH 403705	17 Jun 1889

(Continued)

Table 6a. (Continued)

Species	Accession Number	Collection Date
Field Sparrow	AMNH 403711	17 Jun 1889
Field Sparrow	AMNH 403707	18 Apr 1890
Field Sparrow	AMNH 403709	21 Apr 1886
Field Sparrow	AMNH 403710	21 Apr 1886
Field Sparrow	AMNH 403703	21 Dec 1885
Field Sparrow	AMNH 403712	3 Jun 1890
Field Sparrow	AMNH 98341	30 Apr 1889
Field Sparrow	AMNH 403706	5 Jul 1889
Grasshopper Sparrow	AMNH 400434	12 May 1890
Grasshopper Sparrow	AMNH 400435	12 May 1890
Grasshopper Sparrow	AMNH 400433	13 May 1887
Red Fox Sparrow	AMNH 404690	12 Mar 1884
Red Fox Sparrow	AMNH 404691	21 Dec 1885
Red Fox Sparrow	AMNH 404689	8 Apr 1889
Red Fox Sparrow	AMNH 404692	8 Apr 1889
Song Sparrow	AMNH 405199	10 Sep 1896
Song Sparrow	AMNH 405203	10 Sep 1896
Song Sparrow	AMNH 405204	10 Sep 1896
Song Sparrow	AMNH 405200	12 May 1890
Song Sparrow	AMNH 405536	13 Apr 1885
Song Sparrow	AMNH 405201	14 May 1890
Song Sparrow	AMNH 98361	24 Nov 1887
Song Sparrow	AMNH 405202	3 Jun 1890
Song Sparrow	AMNH 98360	3 Sep 1888
Swamp Sparrow	AMNH 405028	12 May 1890
White-throated Sparrow	AMNH 404430	12 May 1890
White-throated Sparrow	AMNH 404423	19 Apr 1886
White-throated Sparrow	AMNH 404424	19 Apr 1886
White-throated Sparrow	AMNH 404422	21 Dec 1885
White-throated Sparrow	AMNH 404426	21 Dec 1885
White-throated Sparrow	AMNH 404421	22 Apr 1885
White-throated Sparrow	AMNH 404427	23 Dec 1885
White-throated Sparrow	AMNH 404425	26 Apr 1886
White-throated Sparrow	AMNH 404428	26 Apr 1886
White-throated Sparrow	AMNH 404429	7 May 1886
Slate-colored Junco	AMNH 402123	1 Apr 1886
Slate-colored Junco	AMNH 402126	11 Dec 1885
Slate-colored Junco	AMNH 402122	11 Dec 1885
Slate-colored Junco	AMNH 402120	12 Mar 1884
Slate-colored Junco	AMNH 402124	14 Apr 1890
Slate-colored Junco	AMNH 402125	14 Apr 1890
Slate-colored Junco	AMNH 402121	6 May 1885
Slate-colored Junco	AMNH 402127	8 Apr 1889
Scarlet Tanager	AMNH 819465	3 Dec 1973
Scarlet Tanager	AMNH 364093	12 May 1886
Scarlet Tanager	AMNH 364094	12 May 1886
Scarlet Tanager	AMNH 364092	13 May 1887
Scarlet Tanager	AMNH 364095	18 May 1886
Northern Cardinal	AMNH 364381	14 Apr 1886
Rose-breasted Grosbeak	AMNH 364715	12 May 1890
Rose-breasted Grosbeak	AMNH 364716	12 May 1890

Species	Accession Number	Collection Date
Rose-breasted Grosbeak	AMNH 364717	14 May 1890
Rose-breasted Grosbeak	AMNH 364718	22 Aug 1890
Rose-breasted Grosbeak	AMNH 364718	22 Aug 1890
Indigo Bunting	AMNH 365045	12 May 1886
Indigo Bunting	AMNH 98389	15 Sep 1888
Red-winged Blackbird	AMNH 385887	5 Jul 1889
Red-winged Blackbird	AMNH 385888	5 Jul 1889
Red-winged Blackbird	AMNH 385886	9 May 1895
Rusty Blackbird	AMNH 386635	23 Dec 1885
Brown-headed Cowbird	AMNH 387127	10 May 1886
Brown-headed Cowbird	AMNH 387124	19 Apr 1886
Brown-headed Cowbird	AMNH 387126	3 May 1887
Brown-headed Cowbird	AMNH 387123	3 May 1887
Brown-headed Cowbird	AMNH 387125	9 May 1895
Brown-headed Cowbird	AMNH 387122	9 May 1895
Orchard Oriole	AMNH 98216	30 May 1887
Baltimore Oriole	AMNH 386469	5 Jul 1889
Baltimore Oriole	AMNH 386501	5 Jul 1889
American Goldfinch	AMNH 367178	10 Sep 1896
American Goldfinch	AMNH 367179	10 Sep 1896
American Goldfinch	AMNH 367180	10 Sep 1896
American Goldfinch	AMNH 367181	10 Sep 1896
American Goldfinch	AMNH 367173	3 May 1886
American Goldfinch	AMNH 367176	8 Apr 1889
American Goldfinch	AMNH 367177	8 Apr 1889
American Goldfinch	AMNH 367174	8 May 1889
House Sparrow	AMNH 409894	10 Jun 1886
House Sparrow	AMNH 409891	21 Apr 1886
House Sparrow	AMNH 409892	21 Apr 1886
House Sparrow	AMNH 409893	21 Apr 1886

Table 6b. Kingsbridge Meadows specimens (n = 53 of 22 species) that have been digitized in the NYSM (n = 47) and MCZ collections (n = 6).

Species	Accession Number	Collection Date
American Bittern	NYSM 2806	21 Oct 1876
Green Heron	NYSM 2107	29 May 1877
Clapper Rail	MCZ 275696	4 May 1878
Virginia Rail	NYSM 2559	9 Jun 1876
Sora	NYSM 2574	6 Sep 1876
Sora	NYSM 2575	14 Oct 1876
Spotted Sandpiper	NYSM 2699	26 Apr 1877
Spotted Sandpiper	NYSM 2700	20 May 1876
Western Willet HY *(mount)*	NYSM 7271	7 Sep 1880
Least Sandpiper	NYSM 2799	13 May 1880
Northern Shrike	NYSM 3741	31 Mar 1877
Northern Shrike	NYSM 3742	25 Nov 1875
Tree Swallow	NYSM 3633	25 Apr 1879
Tree Swallow	NYSM 3635	26 Apr 1879

Table 6b. (Continued)

Species	Accession Number	Collection Date
Tree Swallow	NYSM 3636	7 Sep 1876
N. Rough-winged Swallow	NYSM 3672	4 Jul 1879
N. Rough-winged Swallow	NYSM 3673	4 Jul 1879
N. Rough-winged Swallow	NYSM 3674	6 May 1878
N. Rough-winged Swallow	MCZ 208335	4 Jul 1879
N. Rough-winged Swallow	MCZ 310980	21 Apr 1877
Bank Swallow	NYSM 3661	21 Apr 1879
Bank Swallow	NYSM 3663	26 Apr 1877
Sedge Wren	NYSM 3862	13 Oct 1880
Sedge Wren	NYSM 3863	22 Sep 1876
Marsh Wren	NYSM 3851	29 Sep 1880
Marsh Wren	NYSM 3852	14 Oct 1876
Bicknell's Thrush	MCZ 275783	"fall" 1875
American Pipit	NYSM 3733	18 Oct 1876
Snow Bunting	NYSM 4342	13 Feb 1883
Snow Bunting	NYSM 4343	13 Feb 1883
Vesper Sparrow	NYSM 4881	2 Nov 1875
Vesper Sparrow	NYSM 4882	10 Oct 1876
Vesper Sparrow	NYSM 4883	17 Apr 1877
Vesper Sparrow	NYSM 4884	21 Apr 1877
Vesper Sparrow	NYSM 4885	14 Oct 1878
Savannah Sparrow	NYSM 4666	17 Apr 1876
Savannah Sparrow	NYSM 4667	19 Sep 1876
Savannah Sparrow	NYSM 4668	11 Oct 1876
Savannah Sparrow	NYSM 4669	25 Apr 1877
Savannah Sparrow	NYSM 4670	5 May 1877
Savannah Sparrow	NYSM 4672	9 Oct 1880
Grasshopper Sparrow	NYSM 4767	25 May 1876
Grasshopper Sparrow	NYSM 4768	12 Oct 1876
Grasshopper Sparrow	NYSM 4769	14 May 1877
Grasshopper Sparrow	NYSM 4770	10 May 1879
Henslow's Sparrow	NYSM 4758	6 Oct 1881
Henslow's Sparrow	NYSM 4759	16 Oct 1880
Henslow's Sparrow	MCZ 306914	8 Oct 1880
Saltmarsh Sparrow	NYSM 4729	29 Jul 1876
Saltmarsh Sparrow	NYSM 4730	8 Aug 1876
Saltmarsh Sparrow	NYSM 4731	4 Jul 1879
Saltmarsh Sparrow	NYSM 4732	4 Jul 1879
Saltmarsh Sparrow	MCZ 306973	4 Jul 1879

Table 6c. Woodlawn Cemetery (n = 34 of 22 species) and Jerome Reservoir specimens (n = 2 of 2 species) that have been digitized in the AMNH collections. All but 3 were in L. S. Foster's original late 1800s collection. A few others lacking dates have been omitted.

Species	Accession Number	Collection Date
Woodlawn Cemetery		
Red-shouldered Hawk	AMNH 98110	1 Sep 1890
Long-eared Owl	AMNH 98770	2 Jan 1893
Yellow-bellied Sapsucker	AMNH 98144	25 Sep 1887

Species	Accession Number	Collection Date
Yellow-bellied Sapsucker	AMNH 98145	21 Oct 1889
Downy Woodpecker	AMNH 98155	4 Jan 1890
Northern Flicker	AMNH 98149	5 Apr 1896
Eastern Wood-Pewee	AMNH 98173	13 Jun 1885
Eastern Phoebe	AMNH 98172	5 Apr 1890
Great Crested Flycatcher	AMNH 8165	17 May 1890
Red-eyed Vireo	AMNH 98434	9 Oct 1887
Blue Jay	AMNH 815120	28 Oct 1983
Blue Jay	AMNH 98195	21 Oct 1889
Blue Jay	AMNH 98196	10 Oct 1887
American Crow	AMNH 98200	4 Oct 1890
American Crow	AMNH 98201	3 Jun 1889
American Crow	AMNH 98202	3 Jun 1889
Hermit Thrush	AMNH 8703	5 Apr 1890
Hermit Thrush	AMNH 98704	26 Oct 1889
Hermit Thrush	AMNH 8705	31 Oct 1887
Hermit Thrush	AMNH 98706	31 Oct 1889
American Pipit	AMNH 98637	26 Oct 1889
Blue-winged Warbler	AMNH 98466	7 May 1890
Blue-winged Warbler	AMNH 98467	17 May 1890
Blue-winged Warbler	AMNH 98468	4 Jul 1890
Blue-winged Warbler	AMNH 98469	4 Jul 1890
Common Yellowthroat	AMNH 98592	18 May 1889
Black-throated Blue Warbler	AMNH 98494	17 May 1890
Eastern Towhee	AMNH 98375	10 Oct 1887
Song Sparrow	AMNH 98355	4 Jan 1890
White-throated Sparrow	AMNH 98325	26 Oct 1889
Baltimore Oriole	AMNH 8217	18 May 1889
American Goldfinch	AMNH 98262	26 Oct 1889
American Goldfinch	AMNH 98263	26 Oct 1889
House Sparrow	AMNH 98239	5 Oct 1887
Jerome Reservoir		
American Woodcock	AMNH 824697	11 Nov 1980
Hermit Thrush	AMNH 815089	21 Nov 1983

Table 6d. Bicknell Riverdale specimens (n = 311 of 106 species) that have been digitized in collections of the NYSM (n = 290), MCZ (n = 6), USNM (n = 6), AMNH (n = 5) —all assumed to be Bicknell skins—and ANSP (n = 4). A handful of others lacking dates have been omitted.

Species	Accession Number	Collection Date
Sharp-shinned Hawk	NYSM 2341	12 Oct 1876
Sharp-shinned Hawk	NYSM 2342	5 Sep 1879
Yellow Rail	NYSM 2583	2 Oct 1881
Virginia Rail	AMNH 27434	18 Sep 1898
Passenger Pigeon	MCZ 75697	23 Sep 1878
Eastern Screech-Owl	NYSM 3080	24 Jan 1900
Common Nighthawk	NYSM 3176	20 Oct 1876

(Continued)

Table 6d. (Continued)

Species	Accession Number	Collection Date
Eastern Whip-poor-will	NYSM 3166	10 May 1877
Chimney Swift	NYSM 3187	30 Apr 1879
Ruby-throated Hummingbird	NYSM 3203	26 Aug 1876
Ruby-throated Hummingbird	NYSM 3205	10 Jul 1879
Ruby-throated Hummingbird	NYSM 3207	17 Sep 1879
Yellow-bellied Sapsucker	NYSM 3332	23 Oct 1875
Yellow-bellied Sapsucker	NYSM 3335	14 Oct 1876
Downy Woodpecker	NYSM 3374	16 Oct 1876
Downy Woodpecker	NYSM 3376	14 Jan 1879
Downy Woodpecker	MCZ 279696	14 Jan 1879
Hairy Woodpecker	NYSM 3356	22 Oct 1875
Northern Flicker	NYSM 3250	14 Oct 1875
Northern Flicker	NYSM 3246	15 Oct 1878
Olive-sided Flycatcher	NYSM 3568	11 Sep 1876
Eastern Wood-Pewee	NYSM 3544	26 Sep 1876
Eastern Wood-Pewee	NYSM 3545	25 May 1876
Eastern Wood-Pewee	NYSM 3546	21 May 1877
Eastern Wood-Pewee	NYSM 3547	6 Oct 1881
Eastern Wood-Pewee	NYSM 3555	21 Aug 1879
Yellow-bellied Flycatcher	NYSM 3483	6 Oct 1881
Yellow-bellied Flycatcher	NYSM 3486	14 Sep 1881
Yellow-bellied Flycatcher	NYSM 3487	29 May 1879
Yellow-bellied Flycatcher	NYSM 3488	20 Aug 1881
Acadian Flycatcher	NYSM 3494	19 Jun 1876
Acadian Flycatcher	NYSM 3496	17 Jun 1876
Acadian Flycatcher	NYSM 3497	4 Jul 1878
Acadian Flycatcher	NYSM 3498	26 May 1877
Alder Flycatcher	NYSM 3501	25 May 1876
Least Flycatcher	NYSM 3529	6 May 1879
Least Flycatcher	NYSM 3530	29 Aug 1876
Least Flycatcher	NYSM 3533	4 Oct 1880
Eastern Phoebe	NYSM 3474	19 Mar 1879
Eastern Phoebe	NYSM 3470	29 Mar 1877
Great Crested Flycatcher	NYSM 3440	6 May 1879
Great Crested Flycatcher	NYSM 3441	1 Aug 1876
White-eyed Vireo	NYSM 5872	16 May 1876
White-eyed Vireo	NYSM 5873	1 Sep 1880
Yellow-throated Vireo	NYSM 5903	22 Aug 1876
Yellow-throated Vireo	NYSM 5904	16 May 1877
Yellow-throated Vireo	NYSM 5905	9 May 1877
Yellow-throated Vireo	NYSM 5906	22 Jun 1879
Blue-headed Vireo	NYSM 5892	12 May 1876
Blue-headed Vireo	NYSM 5893	9 Oct 1876
Warbling Vireo	NYSM 5955	4 Sep 1876
Warbling Vireo	NYSM 5956	3 May 1877
Philadelphia Vireo	NYSM 5917	17 Sep 1885
Red-eyed Vireo	NYSM 5936	7 Sep 1878
Red-eyed Vireo	NYSM 5937	19 Sep 1878
Red-eyed Vireo	NYSM 5938	15 May 1879

Species	Accession Number	Collection Date
Blue Jay	NYSM 6666	10 Feb 1879
Blue Jay	NYSM 6672	5 Nov 1878
Cliff Swallow	NYSM 3723	2 May 1876
Barn Swallow	NYSM 3699	4 Jul 1879
Black-capped Chickadee	NYSM 4239	22 Feb 1879
Black-capped Chickadee	NYSM 4240	10 Oct 1879
Black-capped Chickadee	NYSM 4241	19 Oct 1878
Red-breasted Nuthatch	NYSM 4290	22 Aug 1878
Red-breasted Nuthatch	NYSM 4291	1 Oct 1878
White-breasted Nuthatch	NYSM 4268	25 Dec 1878
White-breasted Nuthatch	NYSM 4269	25 Dec 1878
Brown Creeper	NYSM 4308	23 Dec 1881
Brown Creeper	NYSM 4309	15 Nov 1876
House Wren	NYSM 3814	5 Oct 1881
House Wren	NYSM 3815	12 Oct 1876
House Wren	NYSM 3816	14 Sep 1881
House Wren	NYSM 3817	12 Oct 1876
Winter Wren	NYSM 3831	10 Feb 1879
Winter Wren	NYSM 3833	19 Oct 1876
Carolina Wren	NYSM 3842	2 May 1879
Ruby-crowned Kinglet	NYSM 4203	28 Nov 1878
Ruby-crowned Kinglet	NYSM 4204	11 Jan 1880
Ruby-crowned Kinglet	NYSM 4218	11 Oct 1878
Ruby-crowned Kinglet	NYSM 4219	20 Sep 1881
Eastern Bluebird	NYSM 3944	7 Nov 1876
Eastern Bluebird	NYSM 3945	30 Mar 1877
Veery	NYSM 3987	26 Aug 1878
Veery	NYSM 3989	2 Sep 1879
Gray-cheeked Thrush	NYSM 4006	8 Oct 1881
Gray-cheeked Thrush	NYSM 4007	4 Oct 1881
Gray-cheeked Thrush	NYSM 4008	30 Apr 1878
Gray-cheeked Thrush	MCZ 275710	21 May 1877
Bicknell's Thrush	MCZ 275706	24 May 1877
Bicknell's Thrush	MCZ 275707	27 Sep 1878
Bicknell's Thrush	USNM 95546	8 Oct 1881
Swainson's Thrush	NYSM 4033	25 May 1877
Swainson's Thrush	NYSM 4034	27 Sep 1875
Swainson's Thrush	NYSM 4035	21 Sep 1881
Swainson's Thrush	NYSM 4036	24 Apr 1877
Hermit Thrush	NYSM 4062	2 May 1877
Hermit Thrush	NYSM 4063	30 Apr 1877
Hermit Thrush	NYSM 4064	19 Oct 1876
Hermit Thrush	NYSM 4065	12 Oct 1878
Wood Thrush	NYSM 4101	30 Apr 1879
Wood Thrush	NYSM 4102	28 Sep 1875
American Robin	NYSM 4146	2 May 1877
American Robin	NYSM 4147	11 Oct 1878
American Robin	NYSM 4148	5 Nov 1878
American Robin	NYSM 4150	26 Aug 1882
American Robin	NYSM 4151	28 Aug 1883

(Continued)

Table 6d. (Continued)

Species	Accession Number	Collection Date
Gray Catbird	NYSM 3898	1 May 1879
Gray Catbird	NYSM 3900	19 Oct 1876
Gray Catbird	AMNH 375482	7 May 1886
Brown Thrasher	NYSM 3927	20 Sep 1876
Brown Thrasher	NYSM 3929	29 Apr 1879
Northern Mockingbird	NYSM 3868	— Nov 1877
Cedar Waxwing	NYSM 3786	22 May 1876
Cedar Waxwing	NYSM 3787	5 Apr 1878
Ovenbird	NYSM 5656	17 May 1877
Ovenbird	NYSM 5657	19 Sep 1879
Ovenbird	NYSM 5658	10 May 1880
Worm-eating Warbler	NYSM 5705	13 May 1876
Worm-eating Warbler	NYSM 5706	15 May 1877
Worm-eating Warbler	NYSM 5707	22 May 1879
Worm-eating Warbler	USNM 21683	— Jun 1879
Worm-eating Warbler	ANSP 86611	24 Jul 1876
Louisiana Waterthrush	NYSM 5699	13 May 1879
Northern Waterthrush	NYSM 5685	16 May 1876
Northern Waterthrush	NYSM 5686	24 Sep 1882
Northern Waterthrush	AMNH 383674	7 May 1886
Golden-winged Warbler	NYSM 5142	11 Aug 1881
Blue-winged Warbler	NYSM 5152	22 Aug 1878
Blue-winged Warbler	NYSM 5153	19 May 1879
Blue-winged Warbler	NYSM 5154	22 May 1879
Blue-winged Warbler	NYSM 5155	19 Aug 1881
Blue-winged Warbler	USNM 21682	3 Jun 1876
Blue-winged Warbler	ANSP 86615	5 May 1880
Black-and-white Warbler	NYSM 5121	9 Sep 1875
Black-and-white Warbler	NYSM 5122	30 Apr 1877
Black-and-white Warbler	NYSM 5123	13 Sep 1881
Tennessee Warbler	NYSM 5173	9 Oct 1876
Tennessee Warbler	NYSM 5174	16 Aug 1880
Nashville Warbler	NYSM 5191	20 Sep 1875
Nashville Warbler	NYSM 5193	17 May 1877
Nashville Warbler	NYSM 5194	2 May 1881
Nashville Warbler	NYSM 5195	8 Sep 1881
Nashville Warbler	NYSM 5196	13 Sep 1881
Nashville Warbler	NYSM 5197	7 Oct 1881
Connecticut Warbler	NYSM 5781	20 Sep 1875
Connecticut Warbler	NYSM 5782	22 Sep 1878
Connecticut Warbler	NYSM 5783	22 Sep 1878
Connecticut Warbler	NYSM 5784	8 Sep 1881
Kentucky Warbler	NYSM 5772	30 May 1876
Common Yellowthroat	NYSM 5745	4 Jul 1878
Common Yellowthroat	NYSM 5746	11 Aug 1876
Common Yellowthroat	NYSM 5747	8 May 1877
Common Yellowthroat	NYSM 5748	27 Aug 1878
Common Yellowthroat	NYSM 5750	29 Aug 1880
Common Yellowthroat	NYSM 5751	3 May 1881
Hooded Warbler	NYSM 5800	20 Aug 1875

Species	Accession Number	Collection Date
Hooded Warbler	NYSM 5801	4 Jul 1878
Hooded Warbler	NYSM 5802	30 May 1879
Hooded Warbler	NYSM 5804	22 May 1877
Hooded Warbler	NYSM 5841	30 May 1879
Hooded Warbler	MCZ 309970	22 May 1877
Hooded Warbler	ANSP 86687	18 May 1877
American Redstart	NYSM 5617	12 Oct 1876
American Redstart	NYSM 5618	22 Sep 1879
Northern Parula	NYSM 5211	11 May 1876
Northern Parula	NYSM 5212	17 May 1876
Northern Parula	NYSM 5213	18 May 1877
Northern Parula	NYSM 5214	26 May 1877
Northern Parula	NYSM 5215	13 Sep 1881
Northern Parula	NYSM 5216	6 Oct 1881
Magnolia Warbler	NYSM 5437	18 May 1876
Magnolia Warbler	NYSM 5438	19 May 1876
Magnolia Warbler	NYSM 5439	31 Aug 1876
Magnolia Warbler	NYSM 5440	11 Oct 1876
Magnolia Warbler	NYSM 5441	30 Aug 1879
Magnolia Warbler	NYSM 5442	3 Sep 1879
Magnolia Warbler	NYSM 5443	8 Sep 1881
Magnolia Warbler	NYSM 5444	8 Sep 1881
Bay-breasted Warbler	NYSM 5594	26 Aug 1878
Bay-breasted Warbler	NYSM 5595	22 May 1879
Bay-breasted Warbler	NYSM 5596	18 Aug 1880
Bay-breasted Warbler	NYSM 5597	23 Aug 1881
Blackburnian Warbler	NYSM 5417	11 May 1876
Blackburnian Warbler	NYSM 5418	18 May 1877
Blackburnian Warbler	NYSM 5419	30 Aug 1879
Blackburnian Warbler	NYSM 5420	1 Sep 1879
Blackburnian Warbler	NYSM 5423	3 Aug 1876
Yellow Warbler	NYSM 5243	24 Jul 1876
Yellow Warbler	NYSM 5244	5 Aug 1876
Yellow Warbler	NYSM 5245	14 May 1877
Chestnut-sided Warbler	NYSM 5286	29 Aug 1876
Chestnut-sided Warbler	NYSM 5287	24 May 1878
Chestnut-sided Warbler	NYSM 5288	19 Aug 1878
Chestnut-sided Warbler	NYSM 5289	19 Aug 1878
Blackpoll Warbler	NYSM 5565	13 May 1876
Blackpoll Warbler	NYSM 5566	17 May 1877
Blackpoll Warbler	NYSM 5567	6 Sep 1878
Blackpoll Warbler	NYSM 5568	22 Sep 1879
Black-throated Blue Warbler	NYSM 5326	8 Sep 1881
Palm Warbler	NYSM 5520	13 Apr 1877
Palm Warbler	NYSM 5521	21 Apr 1877
Palm Warbler (Western ?)	NYSM 5522	12 May 1877
Palm Warbler	NYSM 5523	25 Apr 1879
Palm Warbler	NYSM 5524	25 Apr 1879
Palm Warbler	NYSM 5525	21 Apr 1880
Palm Warbler	NYSM 5526	21 Apr 1880
Palm Warbler	NYSM 5527	27 Apr 1880

(Continued)

Table 6d. (Continued)

Species	Accession Number	Collection Date
Myrtle Warbler	NYSM 5483	10 Oct 1876
Myrtle Warbler	NYSM 5484	21 Apr 1877
Myrtle Warbler	NYSM 5485	27 Apr 1877
Myrtle Warbler	NYSM 5487	14 Oct 1878
Prairie Warbler	NYSM 5384	8 May 1877
Prairie Warbler	NYSM 5385	12 May 1877
Black-throated Green Warbler	NYSM 5362	12 Sep 1876
Black-throated Green Warbler	NYSM 5363	30 Apr 1877
Black-throated Green Warbler	NYSM 5364	26 Aug 1878
Canada Warbler	NYSM 5837	20 May 1876
Canada Warbler	NYSM 5838	30 Aug 1876
Canada Warbler	NYSM 5839	19 Aug 1878
Canada Warbler	NYSM 5840	Aug 1878
Canada Warbler	NYSM 5842	13 Aug 1881
Canada Warbler	NYSM 5843	3 Sep 1881
Wilson's Warbler	NYSM 5811	13 May 1876
Wilson's Warbler	NYSM 5812	3 Sep 1879
Wilson's Warbler	NYSM 5813	31 Aug 1881
Yellow-breasted Chat	NYSM 5857	23 Aug 1876
Yellow-breasted Chat	NYSM 5858	5 May 1880
Eastern Towhee	NYSM 4939	12 Oct 1878
American Tree Sparrow	NYSM 4789	30 Nov 1876
American Tree Sparrow	NYSM 4790	21 Mar 1877
American Tree Sparrow	NYSM 4791	6 Apr 1877
American Tree Sparrow	NYSM 4792	13 Apr 1877
American Tree Sparrow	NYSM 4793	27 Nov 1879
Chipping Sparrow	NYSM 4832	11 Oct 1878
Chipping Sparrow	NYSM 4836	16 Oct 1876
Field Sparrow	NYSM 4858	18 Oct 1878
Field Sparrow	NYSM 4859	19 Oct 1878
Field Sparrow	NYSM 4860	1 Jan 1879
Field Sparrow	NYSM 4861	3 Nov 1879
Red Fox Sparrow	NYSM 4372	5 Nov 1878
Red Fox Sparrow	NYSM 4373	4 Apr 1877
Red Fox Sparrow	NYSM 4374	3 Nov 1879
Song Sparrow	NYSM 4424	26 Jun 1876
Song Sparrow	NYSM 4426	20 Mar 1877
Song Sparrow	NYSM 4427	29 Mar 1877
Song Sparrow	NYSM 4428	18 Jul 1878
Song Sparrow	USNM 21685	21 Aug 1881
Lincoln's Sparrow	NYSM 6734	7 Oct 1881
Swamp Sparrow	NYSM 4425	4 Aug 1876
Swamp Sparrow	NYSM 4489	28 Jun 1876
Swamp Sparrow	NYSM 4490	29 Apr 1880
Swamp Sparrow	NYSM 4491	21 Oct 1880
Swamp Sparrow	NYSM 4492	2 Nov 1880
White-throated Sparrow	NYSM 4552	26 Apr 1877
White-throated Sparrow	NYSM 4553	1 May 1877
White-throated Sparrow	NYSM 4554	21 Jan 1879
White-throated Sparrow	NYSM 4555	29 Sep 1881

Species	Accession Number	Collection Date
White-crowned Sparrow	NYSM 4510	2 May 1876
White-crowned Sparrow	NYSM 4511	16 Oct 1876
White-crowned Sparrow	NYSM 4512	28 Apr 1880
Slate-colored Junco	NYSM 4609	24 Sep 1878
Slate-colored Junco	NYSM 4610	17 Oct 1878
Scarlet Tanager	NYSM 5075	20 Oct 1876
Scarlet Tanager	NYSM 5076	12 May 1877
Scarlet Tanager	NYSM 5077	16 May 1880
Scarlet Tanager	NYSM 5078	3 Jun 1882
Scarlet Tanager	USNM 235635	4 Oct 1878
Northern Cardinal	NYSM 4999	17 Nov 1874
Rose-breasted Grosbeak	NYSM 4985	17 Sep 1879
Indigo Bunting	AMNH 365044	7 May 1886
Painted Bunting	NYSM 5598	13 Jul 1875
Red-winged Blackbird	NYSM 6076	27 Apr 1876
Red-winged Blackbird	NYSM 6077	15 Aug 1876
Red-winged Blackbird	NYSM 6078	27 Nov 1876
Red-winged Blackbird	NYSM 6079	13 Apr 1877
Red-winged Blackbird	NYSM 6080	15 Aug 1880
Rusty Blackbird	NYSM 6227	7 Mar 1876
Rusty Blackbird	NYSM 6228	17 Oct 1876
Rusty Blackbird	NYSM 6229	2 May 1877
Rusty Blackbird	NYSM 6230	17 Oct 1878
Rusty Blackbird	NYSM 6231	15 Nov 1878
Common Grackle	NYSM 6185	11 Nov 1874
Common Grackle	NYSM 6186	1 Nov 1875
Common Grackle	NYSM 6187	7 Apr 1877
Common Grackle	NYSM 6188	27 Mar 1878
Common Grackle	NYSM 6189	25 Nov 1880
Brown-headed Cowbird	NYSM 6259	30 Apr 1877
Brown-headed Cowbird	NYSM 6260	3 Apr 1877
Brown-headed Cowbird	NYSM 6261	2 May 1877
Brown-headed Cowbird	NYSM 6262	17 Oct 1878
Brown-headed Cowbird	NYSM 6263	2 Sep 1880
Orchard Oriole	NYSM 6018	4 Jul 1879
Orchard Oriole	NYSM 6019	23 May 1879
Baltimore Oriole	NYSM 5988	10 Aug 1876
Baltimore Oriole	NYSM 5989	23 May 1879
Purple Finch	NYSM 6447	27 Apr 1877
Purple Finch	NYSM 6448	18 Oct 1878
Purple Finch	NYSM 6449	18 Oct 1878
Purple Finch	AMNH 365615	7 May 1886
Red Crossbill	NYSM 6506	5 May 1877
Red Crossbill	NYSM 6507	6 May 1877
Red Crossbill	USNM 21681	1 May 1879
Common Redpoll	NYSM 6397	14 Jan 1879
Common Redpoll	ANSP 86789	14 Jan 1879
American Goldfinch	NYSM 6351	28 Nov 1875
American Goldfinch	NYSM 6352	25 Dec 1878

(*Continued*)

Table 6d. (Continued)

Species	Accession Number	Collection Date
American Goldfinch	NYSM 6353	19 May 1879
American Goldfinch	NYSM 6354	5 Jul 1880
American Goldfinch	NYSM 6355	22 Aug 1880
House Sparrow	NYSM 6567	27 Oct 1880
House Sparrow	NYSM 6568	30 Sep 1881

Literature Cited

Alperin, I. 1951. Migration mystery of the Yellow-breasted Chat. Linnaean Newsletter 4 (9): 1–2.

AOU (American Ornithologists' Union). 1957. Checklist of North American birds, 5th ed. Baltimore: American Ornithologists' Union, 691 pp.

AOU (American Ornithologists' Union). 1998. Checklist of North American birds, 7th ed. Washington, DC: American Ornithologists' Union, 829 pp.

Bagg, A.C., and S.A. Eliot. 1937. Birds of the Connecticut River Valley in Massachusetts. Northampton, MA: The Hampshire Bookshop, 813 pp.

Bagg, A.M. 1956. The changing seasons: A summary of the [1956] spring migration. Audubon Field Notes 10: 308–314.

Bagg, A.M. 1957. The changing seasons: A summary of the [1957] spring migration. Audubon Field Notes 11: 312–324.

Bagg, A.M. 1958. The changing seasons: A summary of the [1958] spring migration. Audubon Field Notes 12: 320–333.

Baird, J., A.M. Bagg, I.C.T. Nisbet, and C.S. Robbins. 1959. Operation Recovery: Report on mist-netting along the Atlantic Coast in 1958. Bird-Banding 30: 143–171.

Beadle, D., and B. Henshaw. 1996. Identification of "Greenland" [=Greater] Common Redpoll *Carduelis flammea rostrata*. Birders Journal 5: 44–47.

Beals, M.V., and J.T. Nichols. 1940. Data from a bird-banding station at Elmhurst, Long Island. Birds of Long Island 3: 57–76.

Bergstrom, E.A. 1951. Notes on Yellow-breasted Chats. Linnaean Newsletter 5 (1): 1.

Bevier, L.R., ed. 1994. The atlas of breeding birds of Connecticut. Geological and Natural History Survey of Connecticut Bulletin 113: 1–461.

Bien, J.T., and C.C. Vermeule. 1891a. Atlas of the metropolitan district and adjacent country, comprising the counties of New York, Kings, Richmond, Westchester, and part of Queens in the state of New York, the county of Hudson and parts of the counties of Bergen, Passaic, Essex, and Union in the state of New Jersey, Map II: City and county of New York [including Bronx County]. New York: Bien & Co.

Bien, J.T., and C.C. Vermeule. 1891b. Atlas of the metropolitan district and adjacent country, comprising the counties of New York, Kings, Richmond, Westchester, and part of Queens in the state of New York, the county of Hudson and parts of the counties of Bergen, Passaic, Essex, and Union in the state of New Jersey, Map IX: Westchester County, Yonkers to Dobbs Ferry, east to the state line. New York: Bien & Co.

Bolton, Reginald P. 1922. Indian paths in the great metropolis. New York: Museum of the American Indian, 280 pp.

Bolton, Robert. 1848. A history of the county of Westchester from its first settlement to the present time. Vol. 2. New York: Gould, 582 pp.

Boyajian, N. 1966. Warbler [breeding] populations on the east slope of the [NJ] Palisades, 1965. Linnaean Newsletter 20 (5): 1–6.

Boyajian, N. 1968. Diurnal migrants crossing the Hudson River. Linnaean Newsletter 22 (6): 2–3.

Boyajian, N. 1969a. Wintering Lesser Scaup on the Hudson River. Linnaean Newsletter 23 (1): 1–2.

Boyajian, N. 1969b. Some notes on the fall migration of swifts and swallows in the Lower Hudson Valley. Linnaean Newsletter 23 (4–5): 2–3.

Boyajian, N. 1969c. A note on fall Blue Jay flights. Linnaean Newsletter 23 (5–6): 2–4.

Boyajian, N. 1970. Some notes from the New Jersey Palisades. Linnaean Newsletter 23 (10): 1–3.

Boyajian, N. 1971. Notes on the summer birds of the New Jersey Palisades. Linnaean Newsletter 25 (6): 1–2.

Boyajian, N. 1972. Some notes on the [NJ Palisades] flight of May 13, 1972. Linnaean Newsletter 26 (3): 1–2.

Boyle, W.J. 2011. The birds of New Jersey: Status and distribution. Princeton: Princeton University Press, 308 pp.

Braislin, W.C. 1907. A list of the birds of Long Island, New York. Proceedings of the Linnaean Society of New York, nos. 17–19: 31–136.

Brinkley, E.S. 1999. Changing seasons: Low pressure. Fall migration, August–November 1998. North American Birds 53: 12–19.

Brinkley, E.S., P.A. Buckley, L.R. Bevier, and A.M. Byrne. 2011. Photo essay: Redpolls from Nunavut and Greenland visit Ontario. North American Birds 65: 206–215.

Buckley, P.A. 1958. The birds of Baxter Creek—fall and winter of 1954. Proceedings of the Linnaean Society of New York, nos. 66–70: 77–83.

Buckley, P.A. 1959. Recent specimens from southern New York and New Jersey affecting A.O.U. Checklist status. Auk 76: 517–520.

Buckley, P.A. 1974. Recent specimens of western vagrants at Fire Island National Seashore, Long Island, New York. Auk 91: 181–185.

Buckley, P.A., and F.G. Buckley. 1980. Population and colony-site trends of Long Island waterbirds for five years in the mid 1970s. Transactions of the Linnaean Society of NY 9: 23–56.

Buckley, P.A., and F.G. Buckley. 1984. Expanding Double-crested Cormorant and Laughing Gull populations on Long Island, NY. Kingbird 34: 146–155.

Buckley, P.A., and S.S. Mitra. 2003. Williamson's Sapsucker, Cordilleran Flycatcher, and other long-distance vagrants at a Long Island, New York stopover site. North American Birds 57: 292–304.

Buckley, P.A., and P.W. Post. 1970. Photographs of New York State rarities, 21: Bell's Vireo. *Kingbird* 20: 57–60.

Buehler, D.A., J.J. Giocomo, J. Jones, P.B. Hamel, C.M. Rogers, T.A. Beachy, D.W. Varble, C.P. Nicholson, K.L. Roth, J. Barg, R.J. Robertson, J.R. Robb, and K. Islam. 2008. Cerulean Warbler reproduction, survival, and models of population decline. Journal of Wildlife Management 72: 646–653.

Bull, J. 1946. The ornithological year 1944 in the New York City region. Proceedings of the Linnaean Society of New York, nos. 54–57: 28–35.

Bull, J. 1964. Birds of the New York area. New York: Harper & Row, 540 pp.

Bull, J. 1970. Supplement to Birds of the New York area. Proceedings of the Linnaean Society of New York, no. 71: 1–54.

Bull, J. 1974. Birds of New York State. New York: Doubleday, 655 pp.

Bull, J. 1976. Supplement to Birds of New York State. Cortland, NY: Federation of New York State Bird Clubs, 52 pp. plus correction sheet.

Cant, G., and H. Geis. 1961. The House Finch: A new East Coast migrant? EBBA News 24: 102–107.

Carleton, G. 1958. The birds of Central and Prospect Parks. Proceedings of the Linnaean Society of New York, nos. 66–70: 1–60.

Carleton, G. 1970. Supplement to The birds of Central and Prospect Parks. Proceedings of the Linnaean Society of New York, no. 71: 133–154.

Chang, S. 2011. A Sooty Fox Sparrow in Central Park, New York City. Kingbird 61: 203–205.

Chapman, F.M. 1906. Birds of the vicinity of New York City. Guide Leaflet No. 22, American Museum of Natural History, New York, 96 pp.

Cruickshank, A.D. 1942. Birds around New York City. New York: American Museum of Natural History, 489 pp.

DeCandido, R. 1990. The Pelham Bay hawkwatch, 1988–1989. *Kingbird* 40: 149–53.

DeCandido, R. 1991a. Pelham Bay hawkwatch results, 1990. Kingbird 41: 12–16.

DeCandido, R. 1991b. Fulvous Whistling-Duck found dead in Bronx County. Kingbird 41: 17.

DeCandido, R. 2005. History of the Eastern Screech-Owl (*Megascops asio*) in New York City, 1867–2005. Urban Habitats 3: 117–133. http://urbanhabitats.org/v03n01/screech-owl_full.html.

DeCandido, R., and D. Allen. 2005a. First nesting of Cooper's Hawk (*Accipiter cooperii*) in New York City since ca. 1955. Kingbird 55: 236–241.

DeCandido, R., and D. Allen. 2005b. First confirmed nesting of the Pine Warbler (*Dendroica pinus*) in New York City. Kingbird 55: 328–334.

DeCandido, R., and D. Allen. 2014. First nesting of Turkey Vulture (*Cathartes aura*) in Bronx County—spring 2009. Kingbird 64: 305–306.

DeCandido, R., and D. Allen. 2015. First recorded nesting of Ruby-throated Hummingbird (*Archilochus colubris*) in Central Park and New York County—spring 2014. Kingbird 65: 10–14.

Deed, R.F., and A.W. Wells. 2010. Birds of Rockland County, NY and the Hudson Highlands, 1844–1976, 2010 ed. New City, NY: Rockland Audubon Society, 573 pp. http://www.rocklandaudubon.org/pdf/DEED'S%20BIRDS%20OF%20ROCKLAND%20COUNTY%20120626.pdf.

DeKay, J.E. 1844. Zoology of New York. Part 2: Birds. New York: Appleton and Wiley and Putnam, 380 pp.

Delacretaz, N. 2007. Breeding and migrating bird census project in Van Cortlandt Park: A study [in 2006] in an important bird area. Unpublished report by New York City Audubon Society, 27 pp.

DeOrsey, S., and B.A. Butler. 2014. The birds of Dutchess County, New York. Poughkeepsie: Ralph T. Waterman Bird Club, 282 pp. http://watermanbirdclub.org/wp-content/uploads/2013/05/Online-BoDC-2014.pdf.

Dickerman, R. 1986a. A review of the Red Crossbill in New York State. Part 1: Historical and nomenclatural background. Kingbird 36: 73–78.

Dickerman, R. 1986b. A review of the Red Crossbill in New York State. Part 2: Identification of specimens from New York. Kingbird 36: 127–134.

Duncan, C.D. 1996. Changes in the winter abundance of Sharp-shinned Hawks in New England. Journal of Field Ornithology 67: 254 –262.

Dunn, J., and J. Alderfer. 2017. National Geographic field guide to the birds of North America, 7th ed. Washington, DC: National Geographic Society, 591 pp.

Eames, E. 1893. Notes from Connecticut. Auk 10: 89–90.

Eaton, E.H. 1910. The birds of New York. Part 1. Albany: Memoir 12 of the New York State Museum, 501 pp.

Eaton, E.H. 1914. The birds of New York. Part 2. Albany: Memoir 12 of the New York State Museum, 719 pp.

Eaton, S. 1964. The Wild Turkey in New York State. Kingbird 14: 4–12.

Eaton, S. 1988. Wild Turkey (*Meleagris gallopavo*). *In* R. Andrle and J. Carroll, eds., The atlas of breeding birds in New York State, 130–131. Ithaca: Cornell University Press, 551 pp.

Elliott, J.J. 1956. British [=European] Goldfinch on Long Island. Long Island Naturalist 5: 3–13.

Elliott, J.J., and R.S. Arbib. 1953. Origin and status of the House Finch in the eastern United States. Auk 70: 31–37.

Enders, F. 1966. Winter of 1965–66 bird population study of a fresh-water marsh and woodland swamp. Audubon Field Notes 20: 476.

Enders, F. 1975. Jerome Reservoir and nomenclatural matters. Linnaean Newsletter 29 (1–2): 4.

Enders, F. 1976. Where do New York City Canvasbacks feed? Linnaean Newsletter 30 (1): 1–2.

Enders, F., P. Temple, D. Cooper, R. Faden, J. Zupan, and F. Heath. 1962. Breeding bird census in 1961 of a fresh-water marsh and woodland swamp. Audubon Field Notes 16: 535–36.

Farrand, J. 1990. The Red-legged Black Duck. American Birds 44: 202–203.

Feigin, W., B. Gell-Mann, H. Karsch, R.G. Kramer, D.S. Lehrman, W.J. Norse, and O.K. Stephenson. 1937. Breeding bird census in 1937 of a fresh-water swamp. Bird-Lore 39: 373–376.

Feigin, W., J. Philips, O.K. Stephenson, and R.G. Kramer. 1938. Breeding bird census in 1938 of a fresh-water swamp. Bird-Lore 40: 351–353.

Fibikar, D., P. Cales, E.T. Brown, and R.R. Veit. 2015. Nesting of Blue Grosbeaks (*Passerina caerulea*) at the former Fresh Kills landfill, Staten Island, New York in 2014. Kingbird 65: 7–9.

Fischer, R.B. 1941. Alder Flycatcher breeding on Long Island. Proceedings of the Linnaean Society of New York, nos. 52–53: 144–47.

Fischer, R.B. 1950. Notes on the Alder Flycatcher. Linnaean Newsletter 4 (3): 3.

Fox, A.D., D. Sinnett, J. Baroch, D.A. Stroud, K. Kampp, C. Egevang, and D. Boertmann. 2012. The status of Canada Goose *Branta canadensis* subspecies in Greenland. Dansk Ornitologisk Forenings Tidsskrift 106: 87–92.

Frey, S.J., C.C. Rimmer, K.P. McFarland, and S. Menu. 2008. Identification and sex determination of Bicknell's Thrushes using morphometric data. Journal of Field Ornithology 79: 408–420.

Friedman, S.E., and P.L. Steineck. 1963. The warbler migration in Van Cortlandt Park: 1961–1963. Linnaean Newsletter 17 (4): 2–4.

Gehlbach, F.R. 1995. Eastern Screech-Owl (*Megascops asio*). *In* A. Poole and F. Gills, eds., The birds of North America, no. 165, 24 pp. Philadelphia: The Birds of North America.

Gera, J., and R. Gera. 1966. Breeding bird census in 1965 of a fresh-water marsh and woodland swamp. Audubon Field Notes 20: 659.

Gera, J., and R. Gera. 1967. Breeding bird census in 1966 of a fresh-water marsh and woodland swamp. Audubon Field Notes 21: 662.

Giraud, J.P. 1844. Birds of Long Island. New York: Wiley & Putnam, 397 pp.

Goll, P., F. Heath,, and F. Enders. 1964. Winter of 1963–64 bird population study of a fresh-water marsh and woodland swamp. Audubon Field Notes 18: 398–399.

Goll, P., F. Heath,, J. Zupan, and F. Enders. 1963. Winter of 1962–63 bird population study of a fresh-water marsh and woodland swamp. Audubon Field Notes 17: 370.

Gorski, L.G. 1969. Systematics and ecology of sibling species of Traill's Flycatcher. PhD thesis, University of Connecticut, Storrs, 162 pp.

Greenberg, R., and S.M. Matsuoka. 2010. Rusty Blackbird: Mysteries of a species in decline. Condor 112: 770–77.

Griscom, L. 1923. Birds of the New York City region. New York: American Museum of Natural History, 400 pp.

Griscom, L. 1927. The observations of the late Eugene P. Bicknell at Riverdale, New York City, fifty years ago. Proceedings of the Linnaean Society of New York, nos. 37–38: 73–87.

Griscom, L., and D. Snyder. 1955. Birds of Massachusetts. Salem: Peabody Museum, 295 pp.

Gross, A.O. 1956. The recent reappearance of the Dickcissel *(Spiza americana)* in eastern North America. Auk 73: 66–70.

Groth, J. 1993a. Evolutionary differentiation in morphology, vocalizations, and allozymes among nomadic sibling species in the North American Red Crossbill *(Loxia curvirostra)* complex. University of California Publications in Zoology 127: 1–143.

Groth, J. 1993b. [Red] Crossbill diagnosis page. http://research.amnh.org/vz/ornithology/crossbills/diagnosis.html.

Groth, J. 1998. Red Crossbill *Loxia curvirostra*. In E. Levene, ed., Bull's birds of New York State, pp. 564–566. Ithaca: Cornell University Press, 622 pp.

Guidice, J.H., and J.T. Ratti. 2001. Ring-necked Pheasant *Phasianus colchicus*. In Poole and Gill, The birds of North America, no. 572, 32 pp.

Heath, F., and F. Enders. 1964. Winter of 1963–64 bird population study of Jerome Park Reservoir. Audubon Field Notes 18: 409.

Heath, F., and J. Zupan. 1971–72. The breeding bird censuses of the Van Cortlandt Park Swamp: An analysis. Linnaean Newsletter:
25 (2): 3–4, Introduction, description
25 (3): 3–5, Description, coverage, populations
25 (4): 4–6, Populations
25 (5): 3–4, Populations
25 (6): 2–3, Populations
25 (7): 1–2, Recording nesting data
26 (1): 2–5, Nesting data
26 (2): 1–2, Bird banding, nest finding
26 (4): 1–2, Estimating breeding pairs, nest defense
26 (5): 1–3, The Cowbird, final statement, Index

Henning, J. 2015. Floral interactions in Van Cortlandt Park, Bronx, New York. PhD dissertation, City University of New York, 240 pp.

Hix, G. 1905. A year with the birds in New York City. Wilson Bulletin 17: 34–43.

Horowitz, J.L. 1963. Hurricane? Linnaean Newsletter 17 (5): 1–2.

Howell, S. 2002. Hummingbirds of North America: The photographic guide. San Diego: Academic Press Natural World, 219 pp.

Howell, S., and J. Dunn. 2007. Gulls of the Americas. Boston: Houghton Mifflin, 516 pp.

Howell, S., I. Lewington, and W. Russell. 2014. Rare birds of North America. Princeton: Princeton University Press, 428 pp.

Jaslowitz, C.A. 1993. Breeding bird census in 1992 of a mature urban deciduous forest. Journal of Field Ornithology 64 (1), supplement: 48.

Jaslowitz, C.A., and D.S. Künstler. 1992. Breeding bird census in 1991 of a mature urban deciduous forest. Journal of Field Ornithology 63 (1), supplement: 50.

Jones, J., D.R. Norris, M.K. Girvan, J.J. Barg, T.K. Kyser, and R.J. Robertson. 2008. Migratory connectivity and rate of population decline in a vulnerable songbird. Condor 110: 538–544.

Kane, R.P. 1974. Ring-necked Parakeets in the Bronx. Linnaean Newsletter 28 (1): 4.

Keil, J. 1974. More Ring-necked Parakeets in the Bronx. Linnaean Newsletter 28 (3): 3.

Kieran, J. 1959. A Natural History of New York City. Boston: Houghton Mifflin, 428 pp.

Kiviat, E., R. Schmidt, and N. Zeising. 1985. Bank Swallow and Belted Kingfisher nest in dredge spoil on the tidal Hudson River. Kingbird 35: 3–6.

Kuerzi, J. 1927. A detailed report on the birdlife of the Greater Bronx Region. Proceedings of the Linnaean Society of New York, nos. 37–38: 88–111.

Künstler, D. 1989. Huckleberry Island: 1988. Kingbird 39: 25–26.

Künstler, D. 1994. Early dates for Great Horned Owl on nest. Kingbird 44: 288.

Künstler, D. 2000. The Wild Turkey in the Bronx and lower Westchester Co., New York. Kingbird 50: 118–130.

Künstler, D. 2007. The colonial waterbirds of Goose Island, Pelham Bay Park, Bronx, New York: 1996–2006. Transactions of the Linnaean Society of New York 10: 229–237.

Künstler, D., and J.L. Young. 2010. Breeding bird survey in 2008 in the Northeast Forest and Croton Woods, Van Cortlandt Park, Bronx, New York. Unpublished report by New York City Dept. of Parks & Recreation, 17 pp.

Lanciotti, R., J. Roehrig, V. Deubel, J. Smith, and M. Parker. 1999. Origin of the West Nile Virus responsible for an outbreak of encephalitis in the northeastern United States. Science 286: 2333–2337.

Lane, D., and A. Jaramillo. 2000. Identification of *Hylocichla/Catharus* thrushes. Part 3: Gray-cheeked and Bicknell's thrushes. Birding 32: 318–331.

Lanyon, W.R., R.G. Van Gelder, and R.G. Zweifel. 1970. The Vertebrate Fauna of the Kalbfleisch Field Research Station [Dix Hills, Suffolk Co., NY]. New York: American Museum of Natural History, 78 pp.

Latham, R. 1957. Breeding hawks on eastern Long Island. Kingbird 7: 77–79.

Levine, E., ed. 1998. Bull's Birds of New York State. Ithaca: Cornell University Press, 622 pp.

Ley, D.H., J.E. Berkhoff, and J.M. McLaren. 1996. *Mycoplasma gallisepticum* isolated from House Finches (*Carpodacus mexicanus*) with conjunctivitis. Avian Diseases 40: 480–483.

Lindholm, J.H., and I. Svanberg. 2015. History of gulls in European and American zoos. Zoologische Garten 84: 277–83.

Lindsay, P. 2007. Multiple Acadian Flycatchers (*Empidonax virescens*) breeding in New York City and Long Island during summer 2007. Kingbird 57: 298–299.

Lunt, H. 1933. Dovekies in the Bronx. Journal of the New York Botanical Garden 34: 5–9.

Lyons, C. 2000. Early nesting of Great Horned Owls in the Bronx, NYC. Kingbird 50: 336–343.

Malling Olsen, K., and H. Larsson. 2004. Gulls of North America, Europe, and Asia. Princeton: Princeton University Press, 608 pp.

Marra, P.P., and C. Santella. 2016. Cat wars: The devastating consequences of a cuddly killer. Princeton: Princeton University Press, 216 pp.

Marshall, J.T. 2001. The Gray-cheeked Thrush *Catharus minimus* and its New England subspecies Bicknell's Thrush *Catharus minimus bicknelli*. Nuttall Ornithological Monographs No. 28, 136 pp.

McGowan, K.J., and K. Corwin, eds. 2008. The second atlas of breeding birds in New York State. Ithaca: Cornell University Press, 688 pp.

McLaren, I.A. 1995. Field identification and taxonomy of Bicknell's Thrush. Birding 27: 358–366.

McLaren, I.A. 2012. All the birds of Nova Scotia: Status and critical identification. Kentville, NS: Gasperau Press, 247 pp.

McLaren, I.A., B. Maybank, K. Keddy, P.D. Taylor, and T. Fitzgerald. 2000. A notable autumn arrival of reverse-migrants in southern Nova Scotia. North American Birds 54: 4–10.

Mitra, S.S., P.A. Buckley, and F.G. Buckley. MS. Breeding and migratory landbirds of the Lighthouse Tract of Fire Island National Seashore, Long Island, New York: A quantitative analysis of 1995–1999 patterns and habitat preferences in deciduous and coniferous maritime woodlands. Unpublished manuscript (62 pp. and 7 figs).

Mlodinow, S., P.F. Springer, B. Deuel, L.S. Semo, T. Leukering, T.D. Schonewald, W. Tweit, and J.H. Barry. 2008. Distribution and identification of Cackling Goose (*Branta hutchinsii*) subspecies. North American Birds 62: 344–360.

Murphy, R.C., and W. Vogt. 1933. The Dovekie influx of 1932. Auk 50: 325–49.

Nagy, C.M. 2012. Population dynamics and occupancy patterns of Eastern Screech-Owls (*Megascops asio*) in New York City parks and adjacent suburbs. PhD dissertation, City University of New York, 157 pp.

Nagy, C., and R. Rockwell. 2013. Occupancy patterns of Eastern Screech Owls in urban parks of New York City and southern Westchester County, NY, USA. Journal of Natural History 47: 2135–2149.

Newton, I. 2008. The migration ecology of birds. London: Academic Press, 976 pp.

Nichols, J.T. 1935. The Dovekie incursion of 1932. Auk 52: 448–449.

NYCDEP (New York City Department of Environmental Protection). 2011. Bureau of Water Supply Waterfowl Management Program: Unpublished report for 1 Apr 2010–31 Mar 2011 by Division of Drinking Water Quality Control, Valhalla, NY, 62 pp.

NYCDEP (New York City Department of Environmental Protection). 2012. Bureau of Water Supply Waterfowl Management Program: Unpublished report for 1 Apr 2011–31 Mar 2012 by Division of Drinking Water Quality Control, Valhalla, NY, 62 pp.

Pasquier, R. 1974. Recent additions to the birds of Central Park. Proceedings of the Linnaean Society of New York, no. 72: 80–81.

Peterson, R.T. 1934. A field guide to the birds, 1st ed. Boston: Houghton Mifflin, 167 pp.

Peterson, R.T. 1947. A field guide to the birds, 3rd ed. Boston: Houghton Mifflin, 290 pp.

Phillips, A.R. 1991. The known birds of North and Middle America. Part 2. Denver: privately published, 249 pp.

Post, P.W. 1964. On the reluctance of Bonaparte's Gulls to fly under objects. Kingbird 14: 91.

Post, P.W. 1979. An irruption of Tufted Titmice in the Northeast. American Birds 33: 249–50.

Pough, R.H., and D.R. Eckelberry. 1951. Audubon water bird guide. Garden City, NY: Doubleday, 352 pp.

Pyle, P. 1997. Identification guide to North American birds. Part 1: Columbidae to Ploceidae. Bolinas, CA: Slate Creek Press, 732 pp.

Rimmer, C.C., K.P. McFarland, W.G. Ellison, and J.E. Goetz. 2001. Bicknell's Thrush (*Catharus bicknelli*). *In* Poole and Gill, The birds of North America, no. 592, 28 pp.

Robbins, C.S., B. Bruun, H.S. Zim, and A. Singer. 1966. Birds of North America. New York: Golden Press, 340 pp.

Robbins, C.S., D. Bystrak, and P.H. Geissler. 1986. The breeding bird survey: Its first fifteen years (1965–1979). Washington, DC: USFWS Resource Publication No. 157, 205 pp.

Salzman, E. 2001. Northern Parula returns [as a breeder] to Long Island. Kingbird 51: 751–52.

Sanderson, E. 2009. Mannahatta. New York: Abrams, 352 pp.

Sauer, J.R., J.E. Hines, and J. Fallon. 2005. The North American breeding bird survey: Results and analysis, 1966–2006. Version 6.2.2005. Laurel, MD: United States Geological Survey, Patuxent Wildlife Research Center. www.mbr-pwrc.usgs.gov/bbs/bbs2005.html.

Sauer, J.R., J.E. Hines, and J. Fallon. 2007. The North American breeding bird survey: Results and analysis, 1966–2006. Version 7.23.2007. Laurel, MD: United States Geological Survey, Patuxent Wildlife Research Center. www.mbr-pwrc.usgs.gov/bbs/bbs.html.

Sedgwick, J.A. 2000. Willow Flycatcher (*Empidonax traillii*). *In* Poole and Gill, The Birds of North America, no. 533, 32 pp.

Sedwitz, W. 1940a. The ornithological year 1937 [read: 1936] in the New York City region. Proceedings of the Linnaean Society of New York, nos. 50–51: 41–49.

Sedwitz, W. 1940b. The ornithological year 1937 in the New York City region. Proceedings of the Linnaean Society of New York, nos. 50–51: 50–59.

Sedwitz, W. 1974. Some gull movements. Linnaean Newsletter 28 (6): 1–3.

Sedwitz, W. 1975. Which scaup is that? Linnaean Newsletter 29 (4): 2–3.

Sedwitz, W. 1977. A waterbird study of a limited area: Jerome Park Reservoir. Proceedings of the Linnaean Society of New York, no. 73: 65–79.

Sedwitz, W. 1979. Ruddy Duck observations. Linnaean Newsletter 33 (7): 1–3.

Sedwitz, W. 1982. Chimney Swifts—Perimeters and preferences. Linnaean Newsletter 36 (4): 3–4.

Sedwitz, W. 1985. Swan's way. Linnaean Newsletter 38 (6): 2–3.

Sedwitz, W., I. Alperin, and M. Jacobson. 1948. Gadwall breeding on Long Island, New York. Auk 65: 610–12.

Seewagen, C., and E. Slayton. 2006. Historical accounts of Bicknell's Thrush in New York City and a new record for Bronx County. Kingbird 56: 210–215.

Shanley, M. 2013. Recent status of the Blue Grosbeak on Staten Island and in New York State. Kingbird 63: 286–288.

Sibley, D.A. 2000. The Sibley guide to birds. New York: Knopf, 545 pp.

Sibley, D.A. 2014. The Sibley guide to birds, 2nd ed. New York: Knopf, 599 pp.

Sibley, D.A. 2017. Distinguishing Black-capped and Carolina Chickadees. http://www.sibleyguides.com/bird-info/black-capped-chickadee/black-capped-carolina-chickadee.

Smith, F.M. 2011. Photo essay: Subspecies of Saltmarsh Sparrow and Nelson's Sparrow. North American Birds 65: 368–377.

Stein, R.C. 1958. The behavioral, ecological, and morphological characteristics of two populations of the Alder Flycatcher, *Empidonax traillii* (Audubon). New York State Museum Science Service Bulletin no. 371, 63 pp.

Stein, R.C. 1963. Isolating mechanisms between populations of Traill's Flycatchers. Proc. Am. Philosophical Soc. 107: 21–50.

Stone, W. 1937. Bird studies at Old Cape May. Vols. 1, 2. Philadelphia: Delaware Valley Ornithological Club at Academy of Natural Sciences, 940 pp.

Taylor, S.A., T.A. White, W.M. Hochachka, V. Ferretti, R.L. Curry, and I. Lovette. 2014. Climate mediated movement of an avian hybrid zone. Current Biology 24: 671–676.

Terborgh, J. 1989. Where have all the birds gone? Princeton: Princeton University Press, 207 pp.

Tieck, W.A. 1968. Riverdale, Kingsbridge, Spuyten Duyvil: A historical epitome of the northwest Bronx. Old Tappan, NJ: Revell, 230 pp.

Tieck, W.A. 1971. Schools and school days in Riverdale, Kingsbridge, Spuyten Duyvil, New York City: The history of public education in the northwest Bronx. Old Tappan, NJ: Revell, 129 pp.

Turner, S. 1984. New York City Audubon Society study of the breeding and transient birds of Van Cortlandt Park, Bronx, New York. Unpublished report, 18 pp.

USACE (US Army Corps of Engineers). 2000. Expedited reconnaissance study [for] Section 905(B) (WRDA 86): Preliminary analysis [of the proposed] Hudson–Raritan Estuary environmental restoration. New York: Corps of Engineers, New York District, 38 pp.

Wallace, G. 1939. Bicknell's Thrush, its taxonomy, distribution, and life history. Proceedings of the Boston Society of Natural History 41: 211–402.

Walsh, J., R. Kane, and T. Halliwell. 1999. Birds of New Jersey. Bernardsville: New Jersey Audubon Society, 704 pp.

Whitney, B., and K. Kaufman. 1985a. The *Empidonax* challenge. Part 1: Introduction. Birding 17: 151–158.

Whitney, B., and K. Kaufman. 1985b. The *Empidonax* challenge. Part 2: Least, Hammond's, and Dusky Flycatchers. Birding 17: 277–287.

Whitney, B., and K. Kaufman. 1986a. The *Empidonax* challenge. Part 3: Willow and Alder Flycatchers. Birding 18: 153–159.

Whitney, B., and K. Kaufman. 1986b. The *Empidonax* challenge. Part 4: Acadian, Yellow-bellied, and Western Flycatchers. Birding 18: 315–327.

Whitney, B., and K. Kaufman. 1987. The *Empidonax* challenge. Part 5: Buff-breasted and Gray Flycatchers. Birding 19: 7–15.

Williamson, S. 2001. Hummingbirds of North America. Boston: Houghton Mifflin, 263 pp.

Winker, K. 1998. The concept of floater. Neotropical Ornithology 9: 111–119.

Young, C.F. 1958. The 1955 breeding season in the Pelham-Baychester area, Bronx County. Proceedings of the Linnaean Society of New York, nos. 66–70: 84–85.

Young, M. 2011. Red Crossbill call-types of New York: Their taxonomy, flight call vocalizations, and ecology. Kingbird 61: 106–123.

Young, M. 2012. North American Red Crossbill types: Status and flight call identification. http://ebird.org/content/ebird/news/recrtype/.

Young, M., and Tim S. 2017. Crossbills of North America: Species and Red Crossbill call types. https://ebird.org/pnw/news/crossbills-of-north-america-species-and-red-crossbill-call-types/.

Zeranski, J.D., and T.R. Baptist. 1990. Connecticut birds. Hanover, NH: University Press of New England, 328 pp.

Zupan, J., and F. Enders. 1966. Winter of 1965–66 bird population study of Jerome Park Reservoir. Audubon Field Notes 20: 476–477.

Zupan, J., F. Enders, P. Temple, A. Loss, I. Weiss, and F. Heath. 1963. Breeding bird census in 1962 of a fresh-water marsh and woodland swamp. Audubon Field Notes 17: 506–507.

Zupan, J., F. Heath, J. Gera, and R. Gera. 1965. Breeding bird census in 1964 of a fresh-water marsh and woodland swamp. Audubon Field Notes 19: 625.

Zupan, J., P. Temple, D. Simon, F. Enders, A. Nelson, and F. Heath. 1964. Breeding bird census in 1963 of a fresh-water marsh and woodland swamp. Audubon Field Notes 18: 569–570.

About the Authors

P. A. Buckley grew up in Riverdale in the Northwest Bronx and was an undergraduate at Columbia and then a graduate student at Cornell. Always studying birds, he began with the behavioral genetics of *Agapornis* parrots but quickly shifted to colonially breeding waterbirds. With his wife and colleague, Francine G. Buckley, he examined the breeding ecology of Royal Terns, the distribution and populations of breeding waterbirds in and around New York City, and the role of migratory birds in the spread of tick-borne diseases. He was Professor of Ecology/Oceanography at Old Dominion, Hofstra, and Rutgers Universities and the University of Rhode Island; the first Chief Scientist for the National Park Service's North Atlantic Region; and finally Senior Scientist (now Emeritus) with the US Geological Survey's Patuxent Wildlife Research Center. A cofounder of the Waterbird Society and its second president, he was also president of the Nuttall Ornithological Club and is now editor of *Nuttall Ornithological Monographs*. He has long had an interest in distribution, migration, and the evolutionary importance of avian vagrancy. He was a regional editor for, and is now an associate editor of, *North American Birds*, and coauthored *The Birds of Barbados: An Annotated Checklist* (2009, British Ornithologists' Union). His other books include *Neotropical Ornithology* (1985, American Ornithologists' Union) and *Avian Genetics: A Population and Ecological Approach* (1987, Academic Press), but this one returns him to the scene of his first warblers so many springs ago.

Walter Sedwitz was a native New Yorker brought up in Manhattan. He was elected president of the biology club at Stuyvesant High School, but his father didn't believe in college, so immediately after graduation he began his business career. Living in Manhattan through the 1920s, '30s, and '40s, he became one of the most active New York City and Long Island observers, regularly ranging from Prospect Park to Montauk. Jones Beach was an especially favored haunt, and it was there in the late 1940s that he found New York's first (and the East Coast's third) breeding Gadwalls; this discovery was later published in *Auk*. After living in Fordham from 1950, he finally moved to Riverdale in 1971 and resumed intensive West Bronx activity. He was afield nearly daily along the Hudson River or in Van Cortlandt Park, but closest to his heart was Jerome Reservoir, where he diligently tracked the comings and goings of its rich waterbird population. When the first New York State Breeding Bird Atlas work began in 1980, he assumed leadership responsibilities for Block 5852B, covering Riverdale, Jerome Reservoir, Van Cortlandt Park, Hillview Reservoir, and Woodlawn Cemetery. He was indefatigable and made important discoveries in all those locations. He was a prolific and adept writer, documenting not only his own observations in the *Linnaean Newsletter* and *Kingbird* but summarizing several years' activity for the entire New York City area in the *Proceedings of the Linnaean Society of New York*. He was a longtime area Christmas Bird Count participant, and those with him in Riverdale on the 1984 Bronx-Westchester Count were devastated when he died less than a month later.

William J. Norse was a son of Inwood in northern Manhattan, where he spent most of his adult life. He graduated from City College of New York, after which his career in Wall Street brokerage firms kept him in New York City. Beginning in 1932 he was a near-daily observer in adjacent Inwood Park, one of the first to recognize its ecological importance and strategic location for migratory birds. He was in the field there many times a week throughout the 1930s, '40s, '50s, and most of the '60s. He also spent a great deal of time in Van Cortlandt Park, only a short bus or subway ride away. He also was afield frequently in Bronx Park and the Bronx Zoo and was a regular observer at Jerome and Hillview Reservoirs. A charter member of the Sialis Bird Club when it was founded in 1935, he was an eager participant in the Breeding Bird Census of Van Cortlandt Swamp in 1937, the first year of nationwide Breeding Bird Censuses. When he finally retired in the early 1960s, his bird activity resumed its earlier pace, and he was afield nearly daily. From 1960 to 1966 he was compiler/cocompiler of the Bronx-Westchester Christmas Bird Count, but in 1967 he left Inwood for family reasons and moved to an inherited home in southern Vermont. Within weeks of his arrival he was adding to the knowledge of bird distribution in his adopted state. He returned to Inwood and Van Cortlandt periodically, but eventually his travels became limited, and he died in 2006.

John Kieran was Bronx-born and -bred, growing up on Kingsbridge Terrace down the street from, but 16 years before, Allan Cruickshank. It was there in 1914 that he discovered newly opened Jerome Reservoir to be full of migratory waterfowl, some of them at the time most unusual within New York City. This started him on a lifelong bird and natural history odyssey in the northwest Bronx. After attending City College of New York, he graduated from Fordham University and eventually became a sports reporter at the *New York Times* and the *New York Herald Tribune*. He also was a panelist on NBC radio's long-running program *Information, Please!* But his main love was nature, not sports, and he spent many hours and days in the field along the Hudson in Riverdale and in Van Cortlandt Park, not far from both of his houses in Riverdale. (The first was condemned for Henry Hudson Parkway construction in the 1930s, but he lived in the second until moving to Cape Ann, Massachusetts in 1956.) He wrote several books on nature that drew heavily on his Bronx fieldwork, the most well-known of which were *Footnotes on Nature* (1947) and *A Natural History of New York City* (1959). He died in December 1981, and in 1987 in his honor Van Cortlandt Park created the John Kieran Nature Trail, looping around Van Cortlandt Lake and through the south end of Van Cortlandt Swamp—his most favored area in his favorite park.

Index of English Bird Names

Page numbers in **bold** indicate species accounts.

Anhinga 29, 136
Avocet, American 74, 179

Bishop, Red, sp. 60
Bittern, American 60, 74, **138–39**, 170
    Least 40, 45, 46, 60, 74, **139–40**, 148, 170
Blackbird, Brewer's 35, **392–93**
    Red-winged 46, **388–89**, 393
    Rusty 48, 74, **391–92**, 393
    Yellow-headed 35, 75, **391**
Bluebird, Eastern 37, 38, 45, 99, 61, 66, 71, 73, **300–302**, 314
    Mountain 58, 302
Bobolink 40, 52, 56, 59, 73, **387–88**
Bobwhite, Northern 45, 74, **123–24**
Booby, Brown 74
Brant 53, **85–86**
    Black 29
Budgerigar 60
Bufflehead **116–17**
Bulbul, Red-vented 60
Bunting, Indigo 56, 64, **384–85**
    Lark 364
    Painted 56, 58, 75, **385–86**
    Snow 22, 50, 59, 65, **317–18**, 390, 405

Canvasback 49, 74, **106–8**, 109
Cardinal, Northern 45, 46, 48, 74, **380–81**
Catbird, Gray 46, **310**, 311, 312
Chat, Yellow-breasted 49, 57, 67, **354–55**
Chickadee, Black-capped 46, 48, 51, 54, 59, **283–85**
    Boreal 35, 50, 76, **285**
    Carolina 284–85
Chuck-will's-widow 73, 224
Collared-Dove, African 60, 209
    Eurasian 60, 73, **208–9**
Conure, Nanday 60, 68
Coot, American 37, 41, **174–75**
Cormorant, Double-crested 47, 48, 74, **134–36**, 137
    Great 74, **136–37**
Cowbird, Brown-headed 338, 356, 386, 391, 393, **395–96**

Crane, Common 176
    Sandhill 34, 35, 73, **175–76**
Creeper, Brown 37, 38, 48, 51, **289–90**
Crossbill, Red 39, 50, 54, 68, 76, 288, **402–3**
    White-winged 50, 68, 76, **403–4**
Crow, American 48, 51, 58, 74, **268–70**, 271
    Fish 48, 58, 74, **270–71**, 286, 288, 290, 296, 299
Cuckoo, Black-billed 45, 46, **212–13**
    Yellow-billed 46, **211–12**
Curlew, Long-billed 185

Dickcissel 75, 76, **386–87**
Dove, Barbary 290
    Mourning 46, **210–11**
    Ring-necked 209
    White-winged 58, 75, **210**
Dovekie 1, 34, **194–95**
Dowitcher, Long-billed 74, 190, 249
    Short-billed 34, **190–91**, 249
Duck, American Black 74, **97–99**, 100, 101
    Harlequin 74, 114
    Long-tailed 76, **115–16**
    Mandarin 60
    Muscovy 60
    Red-legged Black 98
    Ring-necked **109**
    Ruddy 41, 49, 61, **121–23**
    Tufted 74, **109–10**
    Wood 48, 49, 61, 63, 71, **93–94**
Dunlin 53, **186–87**

Eagle, Bald 60, 73, **154–55**
    Golden 57, 74, **166**
Egret, Cattle 73, **144–45**
    Great 48, **141–42**, 143
    Snowy 76, **142–43**
Eider, Common 74, 114
    King 114

Falcon, Peregrine 60, 73, **240–41**
Finch, House 39, 48, 60, 74, **399–400**, 401
    Purple 37, 74, **400–401**

Flicker, Northern 58, 61, **235**, 314
Flycatcher, Acadian 37, 60, 65, 73, 76, **245–47**
  Alder 37, 38, 40, 68, 75, 76, **247–49**, 250–51
  Alder/Willow 68, 247–48, **249–50**, 251
  Ash-throated 57, 58, 75, 256
  Cordilleran 248, 249, 253
  Great Crested 61, **255–56**
  Hammond's 253
  Least 45, 46, 67, 248, 250, **252–53**
  Olive-sided **242–43**
  Pacific-slope 58, 248, 249, 253
  Scissor-tailed 34, 256
  Traill's 250
  Willow 45, 68, 74, 247–49, **250–51**
  Yellow-bellied 41, **244–45**, 252
Frigatebird, Magnificent 29

Gadwall 41, 47, 48, **94–95**
Gallinule, Common 37, 40, 45, 60, **173–74**
  Purple 29, 74, 174
Gannet, Northern 416
Gnatcatcher, Blue-gray 46, 52, 56, 66, 73, **296–98**
Godwit, Hudsonian 190–91
  Marbled 190–91
Golden-Plover, American 34, **177**
Goldeneye, Barrow's 74, 118
  Common 49, **117–18**
Goldfinch, American 42, 48, **407–8**
  European 37–38, 60, **408–9**
Goose, Barnacle 33, 49, 74, 83, **86–87**, 89, 91
  Cackling 33, 34, 49, 74, 76, **87–88**, 89, 91
  Canada 32–33, 49, 59, 63, 65, 74, 83, 84, 87, **88–91**, 92
  Canada (*interior*) 32–33, 89
  Greater White-fronted 33, 34, 49, 74, 83, **84**, 89, 91
  Pink-footed 33, 49, 74, **83–84**, 89
  Ross's 74, 85, 91
  Snow 53, 74, **84–85**
Goshawk, Northern 50, **160**
Grackle, Boat-tailed 73, **394–95**
  Bronzed 394–95
  Common **393–94**, 408
  Purple 394–95
Grebe, Eared 74, 132–33
  Horned 53, 54, 74, 76, **132–33**
  Pied-billed 35, **130–31**
  Red-necked 53, 54, 76, **133**
  Western 29, 74, 133
Grosbeak, Black-headed 58, 383
  Blue 56, 73, 76, **383–84**
  Evening 54, 74, **409–10**
  Pine 74, 76, 182, **398–99**
  Rose-breasted 45, 49, **382–83**

Grouse, Ruffed 35, 74, **126–27**
Guillemot, Black 30, 75
Guineafowl, Helmeted 60
Gull, Andean 60, 197, 468–69
  Belcher's 469
  Black-headed 59, 75, **196–97**
  Bonaparte's 57, 59, 74, **195–96**, 197
  Brown-hooded 468
  Franklin's 30, 58, 75, 198
  Glaucous 50, 201, **203**
  Gray 469
  Gray-hooded 468
  Great Black-backed 47, 50, 74, **204**
  Herring 33, 47, 50, 51, 74, 198, **200–201**, 202, 204
  Iceland 50, **201–2**, 203
  Laughing 50, 73, **197–98**
  Lesser Black-backed 50, 75, 201, **202–3**
  Little 75, 197
  Ring-billed 50, 59, 74, 197, **198–99**, 200
  Sabine's 30
  Silver 468
  Thayer's 30, 75, 202
Gyrfalcon 241

Harrier, Northern 49, 74, **155–56**
Hawk, Broad-winged 37, 41, 53, 57, 67, **162–63**, 206
  Cooper's 37, 41, 48, 60, 73, 157, **158–59**, 160
  Red-shouldered 41, 48, 67, 74, **160–62**
  Red-tailed 41, 74, 83, 124, 126, **163–65**, 216, 242
  Rough-legged 59, 124, **165**
  Sharp-shinned 37, 41, 48, 67, 74, **156–58**
  Swainson's 163
Heron, Great Blue 73, **140–41**
  Green **145–46**
  Little Blue 76, 142, **143**
  Tricolored **144**, 148
Hummingbird, Calliope 227
  Ruby-throated 47, 65, 67, 70, **226–27**
  Rufous 58, 75, 227

Ibis, Glossy 47, 73, 76, **149–50**
  White 29, 74, 150
  White-faced 73

Jaeger, Long-tailed 29
  Parasitic 416
Jay, Blue 39, 46, 48, 51, 58, **267–68**
Junco, Oregon 75, 77, 378
  Pink-sided 77, 378
  Slate-colored 48, 51, 77, **377–78**

Kestrel, American 38, 39, 41, 47, 48, 51, 61, 68, 71, 74, 124, **237–39**
Killdeer 22, 59, 65, 67, **178–79**, 390

Kingbird, Eastern 39, 52, 56, 58, **257–58**
    Western 57, 58, **256**
Kingfisher, Belted 47, 71, **227–28**
Kinglet, Golden-crowned 39, 48, 50, 73, **298–99**
    Ruby-crowned 49, 262, **299**
Kiskadee, Great 258
Kite, Mississippi 73, 154
    Swallow-tailed 34, 74, **153–54**
Kittiwake, Black-legged 196

Lark, Horned 22, 37, 39, 41, 50, 59, 65, 73, **273–75**, 317, 390
    Northern Horned **273–74**
    Prairie Horned 273, **274–75**
Longspur, Lapland 22, 59, 65, 76, **317**, 390
Loon, Common 53, 54, **130**
    Pacific 74
    Red-throated 76, **129**

Magpie, Black-billed **268**
Mallard 98, **99–101**
Mannikin, Nutmeg 60
    Tricolored 60
Martin, Purple 36, 38, 61, 71, 76, **275–76**, 280, 282, 314
Meadowlark Eastern 22, 39, 65, 73, 365, **389–90**
Merganser, Common 49, **119–20**
    Hooded 48, 74, **118–19**
    Red-breasted 74, 76, **120–21**
Merlin 48, 49, 74, **239**
Mockingbird, Northern 74, **312–13**
Munia, Scaly-breasted 60
    Tricolored 60
Murre, Thick-billed 416

Night-Heron, Black-crowned 37, 41, 48, 49, 51, **146–47**, 170
    Yellow-crowned 36, 39, 47, 76, **148**
Nighthawk, Common 39, 41, 47, 53, 57, 60, 67, 74, **222–23**
Nuthatch, Red-breasted 39, 49, 50, 54, 68, 73, **287–88**, 289, 290
    White-breasted 48, 51, 61, **288–89**, 314

Oriole, Baltimore 46, **397–98**
    Bullock's 30, 398
    Orchard 45, 56, 57, 73, **396–97**
Osprey 37, 38, 39, 47, 60, 61, 71, 73, **152–53**
Ovenbird 38, 55, **318–19**, 356
Owl, Barn 36, 38, 39, 47, 61, 67, 71, 76, **213–15**
    Barred 36, 76, 216, **218–19**
    Boreal 221
    Great Horned 36, 41, 42, 47, 62, 68, 74, 125, 126, **216–17**
    Long-eared 49, 59, 76, **219–20**
    Northern Saw-whet 49, 50, 59, 76, **221**
    Short-eared 59, 74, **220**
    Snowy 59, 165, **217–18**, 271

Parakeet, Black-hooded 60, 68
    Monk 47, 60, 61, 68, 73, **241–42**
    Nanday 60, 68
    Ring-necked 60
    Rose-ringed 60, 68, 241–42
Partridge, Chukar 60
Parula, Northern 55, 56, 338, **339**
Pelican, American White 74, **137–38**
    Brown 74, 137–38
Phalarope, Red 194, 197
    Red-necked 34, **193–94**
    Wilson's 75, 194
Pheasant, Ring-necked 36, 46, 48, 60, 74, 124, **125–26**
Phoebe, Eastern 39, 49, 58, 61, 67, 68, 71, **253–55**
    Say's 34, 58, 75, 254–55
Pigeon, Passenger **209–10**
    Rock 60, **208**, 216
Pintail, Northern **104–5**
Pipit, American 50, 59, 65, **314–15**
Plover, Black-bellied **176–77**
    Semipalmated **177–78**

Rail, Black 74, 168
    Clapper 39, 60, **168–69**, 170, 371, 372
    King 37, 40, 60, 63, 74, **169–70**
    Virginia 19, 65, 67, 167–68, **170–71**, 172, 240
    Yellow **166–68**
Raven, Common 36, 39, 68, 74, **271–72**
Redhead 76, **108–9**
Redpoll, Common 49, 54, 59, **404–6**
    Greater 405–6
    Hoary 405–6
    Hornemann's 405
Redstart, American 46, 55, **335–36**
Robin, American 39, 46, 48, 59, **309**
    Black-backed 309
Ruff 185–86

Sanderling 34, **186**
Sandpiper, Baird's 190
    Buff-breasted 74, **188**
    Least 53, 54, 56, **187–88**
    Pectoral **188–89**
    Purple 187
    Semipalmated 53, 54, **189–90**
    Solitary 54, **181**
    Spotted 39, 47, 48, 54, 67, 74, **179–81**
    Stilt 53, **185–86**

Sandpiper *(continued)*
   Upland 53, 73, **184–85**
   Western 189–90
   White-rumped 189–90
Sapsucker, Yellow-bellied 48, 49, 74, **231–32**
Scaup, Greater 49, 74, **110–12**
   Lesser 41, 49, 74, 108, **112–13**
Scoter, Black 53, 114, **115**
   Surf 53, **113–14**
   White-winged 53, 74, **114–15**
Screech-Owl, Eastern 47, 61, 68, **215–16**, 219
Shearwater, Audubon's 416
   Cory's 416
   Great 29
   Manx 29, 74
   Sooty 29
Shoveler, Northern 48, 74, **102–4**
Shrike, Loggerhead 74, **258–59**
   Northern 59, 76, **259–60**
Siskin, Pine 39, 49, 54, 68, 288, **406–7**
Skimmer, Black **207–8**
Snipe, Wilson's 48, 54, **191–92**
Solitaire, Townsend's 58
Sora 37, 40, 45, 46, 60, 67, 170, **172–73**
Sparrow, Acadian 370, 371
   American Tree 48, 51, 59, 74, **357–58**
   Bachman's 423, 425, 461
   Baird's 423, 425
   Chipping 49, 320, **358–59**
   Clay-colored 75, 76, **359–60**
   Field 67, 73, 74, 358, **360–61**
   Freshmarsh 369–70
   Grasshopper 40, 73, 76, **366–67**
   Henslow's 74, 76, **367–68**
   House 45, 60, 252, 276, 300, 386, 387, **410–11**
   Ipswich 77, 366
   Lark 75, **363–64**
   Le Conte's 58, 75, **368–69**
   Lincoln's 49, **374**
   Nelson's 76, **369–70**, 371
   Red Fox 49, 59, 77, **372–73**
   Saltmarsh 39, 60, 369, **370–71**, 372
   Savannah 39, 50, 73, 77, **364–66**, 369
   Seaside 36, 39, 60, **371–72**
   Sharp-tailed 36, 370–71
   Slate-colored Fox 77, 373
   Song 39, 46, 48, 51, **373–74**
   Sooty Fox 77, 373
   Swamp 45, 46, 60, 67, 374, **375–76**
   Vesper 40, 65, 73, **361–63**, 390
   White-crowned **377**
   White-throated 48, 51, 374, **376**

Starling, European 48, 51, 60, 124, 229, 230, 276, 281, 300, **313–14**
Stilt, Black-necked 29, 74, 179
Stork, Wood 60, **134**
Storm-petrel, Leach's 416
   Wilson's 416
Swallow, Bank 39, 47, 57, 68, 71, 74, 228, **279–80**
   Barn 39, 46, 61, **282–83**
   Cave 57, 58, 75, 276, 282
   Cliff 37, 38, 39, 47, 58, 61, 68, 73, **280–82**
   Northern Rough-winged 39, 46, 57, 68, **278**, 279
   Tree 46, 61, **277–78**, 284, 291, 314
   Violet-green 57
Swan, Mute 60, 63, **92–93**
   Trumpeter 73, 93
   Tundra 54, 93
Swift, Chimney 39, 53, 58, 68, **224–25**

Tanager, Scarlet **379–80**
   Summer 56, 73, 76, **378–79**
   Western 30, 58, 380
Teal, Blue-winged **101–2**
   Eurasian 77, **106**
   Green-winged 77, **105–6**
Tern, Arctic 75
   Black 34, 74, 207
   Bridled 75
   Caspian **205–6**
   Common 33, **206–7**
   Forster's 34, 73, 207
   Gull-billed 73, 205
   Inca 469
   Least 34, **205**
   Roseate 74
   Royal 30, 75
   Sandwich 30, 75
   Sooty 52, 197
Thrasher, Brown 38, 46, 49, 60, 67, 74, 260, **311–12**
Thrush, Bicknell's 32, 68, 76, **304–6**
   Gray-cheeked 32, 56, **303–4**, 305
   Hermit 48, 49, 304, **307**, 352
   Swainson's **306–7**
   Varied 30, 75, 309
   Wood 46, **308**
Titmouse, Tufted 39, 48, 73, **286–87**
Towhee, Eastern 38, 45, 46, 49, 60, 74, **355–57**
   Spotted 357
Turkey, Wild 36, 68, 69, 73, **127–29**
Turnstone, Ruddy 416, 426
Turtle-Dove, Ringed 60, 209

Veery 38, 45, 60, **302–3**
Vireo, Bell's 58, **261–62**
    Blue-headed 261, **263–64**
    Philadelphia **265–66**
    Red-eyed 39, 46, **266**
    Warbling 46, **264–65**
    White-eyed 67, **260–61**
    Yellow-throated 56, 67, 257, **262–63**
Vulture, Black 36, 39, 67, 68, 73, 76, **150–51**
    Turkey 37, 38, 41, **151–52**

Warbler, Audubon's 30, 34, 35, 75, 77, 349–50
    Bay-breasted 55, **340–41**
    Black-and-white 38, 55, 56, 60, 67, **325–26**, 356
    Blackburnian **341–42**
    Blackpoll 55, **344–45**
    Black-throated Blue 55, **345–46**
    Black-throated Gray 30, 352
    Black-throated Green 55, **351–52**
    Blue-winged 32, 57, 60, 67, 248, 322, 323, **324–25**
    Brewster's 323, 324
    Calaveras 330
    Canada 55, **353**
    Cape May 52, 55, 57, **337**
    Cerulean 76, **337–38**
    Chestnut-sided 55, 64, 74, 330, **343–44**
    Connecticut 52, 74, 227, **330–31**, 332
    Golden-winged 57, 65, 74, 242, 248, **322–23**
    Hermit 352
    Hooded 55, 56, 57, 67, **334–35**
    Kentucky 37, 38, 41, 56, 67, 73, 76, **332–33**
    Lawrence's 323, 324
    MacGillivray's 330
    Magnolia 55, **340**
    Mourning 57, 65, 227, **331–32**
    Myrtle 52, 55, 77, 290, **349–50**
    Nashville 37, 38, 41, 55, 58, **329–30**, 354
    Orange-crowned 58, **328–29**
    Palm 55, 68, 290, **346–47**
    Pileolated 354
    Pine 38–39, 49, 68, 290, **347–48**, 350
    Prairie 56, **351**
    Prothonotary 56, 73, 76, **326–27**
    Swainson's 56, 75, **327**
    Sycamore 350
    Tennessee **327–28**
    Townsend's 352
    Western Palm 68, **346–47**
    Wilson's 55, 58, **353–54**
    Worm-eating 32, 38, 57, 60, **320–21**, 356
    Yellow 46, 56, 57, **342–43**
    Yellow Palm 55, 290, **346–47**
    Yellow-rumped 349
    Yellow-throated 56, 73, 76, **350**
Waterthrush, Louisiana 37, 38, 57, 242, **321**
    Northern 41, 56, **322**
Waxwing, Bohemian 30, 316
    Cedar 42, 48, 59, **315–16**
Wheatear, Northern **300**
Whimbrel 53, 185
Whip-poor-will, Eastern 37, 76, **223–24**, 356
Whistling-Duck, Fulvous 74, **83**
Wigeon, American **96–97**
    Eurasian 76, **95–96**
Willet, Eastern 34, 73, 77, 182, 183
    Western 34, 74, 77, **182–83**
Wood-Pewee, Eastern 163, **243–44**
Woodcock, American 41, 42, 59, 67, **192–93**
Woodpecker, Black-backed **234**
    Downy 48, 51, 230, 231, **232–33**
    Hairy 48, 51, 61, 230, **233–34**, 289
    Pileated 35, 36, 73, **236–37**
    Red-bellied 46, 48, 56, 73, **230–31**, 233, 234
    Red-headed 76, **229–30**, 314
Wren, Bewick's 296
    Carolina 46, 48, **295–96**
    House 45, 61, **290–91**
    Marsh 46, 48, 60, 74, 170, **294–95**
    Sedge 39, 74, 76, **292–93**
    Winter 48, **291–92**

Yellowlegs, Greater 54, **181–82**
    Lesser **183–84**, 185
Yellowthroat, Common **333–34**

Index of Scientific Bird Names

Page numbers in **bold** indicate species accounts.

Acanthis (flammea) flammea 49, 54, 59, **404–6**
   (flammea) rostrata 405–6
   (hornemanni) exilipes 405–6
   (hornemanni) hornemanni 405
Accipiter cooperii 37, 41, 48, 60, 73, 157, **158–59**, 160
   gentilis 50, **160**
   striatus 37, 41, 67, 74, **156–58**
Actitis macularius 39, 47, 48, 54, 67, 74, **179–81**
Aechmophorus occidentalis 29, 74, 133
Aegolius acadicus 49, 50, 59, 76, **221**
   funereus 221
Agelaius phoeniceus 46, **388–89**, 393
Aix galericulata 60
   sponsa 48, 49, 61, 63, 71, **93–94**
Alectoris chukar 60
Alle alle 1, 34, **194–95**
Ammodramus bairdii 423, 425
   caudacutus 39, 60, 369, **370–71**, 372
   henslowii 74, 76, **367–68**
   leconteii 58, 75, **368–69**
   maritimus 36, 39, 60, **371–72**
   nelsoni alterus 369–70, 371
   nelsoni nelsoni 369–70, 371
   nelsoni subvirgatus 370, 371
   savannarum 40, 73, 76, **366–67**
Anas acuta **104–5**
   americana **96–97**
   clypeata 48, 74, **102–4**
   (crecca) carolinensis 77, **105–6**
   (crecca) crecca 77, **106**
   discors **101–2**
   penelope 76, **95–96**
   platyrhynchos 98, **99–101**
   rubripes rubripes 98
   rubripes tristis 74, **97–99**, 100, 101
   strepera 41, 47, 48, **94–95**
Anhinga anhinga 29, 136
Anser albifrons 33, 34, 49, 74, 83, **84**, 89, 91
   albifrons flavirostris 84
   brachyrhynchus 33, 49, 74, **83–84**, 89
Anthus rubescens 50, 59, 65, **314–15**
Antrostomus carolinensis 73, 224
   vociferus 37, 76, **223–24**, 356

Aquila chrysaetos 57, 74, **166**
Aratinga nenday 60, 68
Archilochus colubris 47, 65, 67, 70, **226–27**
Ardea alba 48, **141–42**, 143
   herodias 73, **140–41**
Arenaria interpres 416, 426
Asio flammeus 59, 74, **220**
   otus 49, 59, 76, **219–20**
Aythya affinis 41, 49, 74, 108, **112–13**
   americana 76, **108–9**
   collaris **109**
   fuligula 74, **109–10**
   marila 49, 74, **110–12**
   valisineria 49, 74, **106–8**, 109

Baeolophus bicolor 39, 48, 73, **286–87**
Bartramia longicauda 53, 73, **184–85**
Bombycilla cedrorum 42, 48, 59, **315–16**
   garrulus 30, 316
Bonasa umbellus 35, 74, **126–27**
Botaurus lentiginosus 60, 74, **138–39**, 170
Branta bernicla 53, **85–86**
   (bernicla) nigricans 29
   canadensis 32–33, 49, 59, 63, 65, 74, 83, 84, 87, **88–91**, 92
   canadensis canadensis 90
   canadensis interior 89
   canadensis maxima 90
   canadensis parvipes 88, 91
   hutchinsii 33, 34, 49, 74, 76, **87–88**, 89, 91
   hutchinsii hutchinsii 88
   hutchinsii taverneri 88
   leucopsis 33, 49, 74, 83, **86–87**, 89, 91
Bubo scandiacus 59, 165, **217–18**, 271
   virginianus 36, 41, 42, 47, 62, 68, 74, 125, 126, **216–17**
Bubulcus ibis 73, **144–45**
Bucephala albeola **116–17**
   clangula 49, **117–18**
   islandica 74, 118
Buteo jamaicensis 41, 74, 83, 124, 126, **163–65**, 216, 242
   lagopus 59, 124, **165**
   lineatus 41, 48, 67, 74, **160–62**

Buteo (continued)
    platypterus 37, 41, 53, 57, 67, **162–63**, 206
    swainsoni 163
Butorides virescens **145–46**

Cairina moschata 60
Calamospiza melanocorys 364
Calcarius lapponicus 22, 59, 65, 76, **317**, 390
Calidris alba 34, **186**
    alpina 53, **186–87**
    bairdii 190
    fuscicollis 189–90
    himantopus 53, **185–86**
    maritima 187
    mauri 189
    melanotos **188–89**
    minutilla 53, 54, 56, **187–88**
    pugnax 185–86
    pusilla 53, 54, **189–90**
    subruficollis 74, **188**
Calonectris diomedea 416
Cardellina canadensis 55, **353**
    pusilla pileolata 354
    pusilla pusilla 55, **353–54**
Cardinalis cardinalis 45, 46, 48, 74, **380–81**
Carduelis carduelis 37–38, 60, **408–9**
Cathartes aura 37, 38, 41, **151–52**
Catharus bicknelli 32, 68, 76, **304–6**
    fuscescens 38, 45, 60, **302–3**
    guttatus 48, 49, 304, **307**, 352
    minimus 32, 56, **303–4**, 305
    ustulatus **306–7**
Cepphus grylle 30, 75
Certhia americana 37, 38, 48, 51, **289–90**
Chaetura pelagica 39, 53, 58, 68, **224–25**
Charadrius semipalmatus **177–78**
    vociferus 22, 59, 65, 67, **178–79**, 390
Chen caerulescens atlantica 53, 74, **84–85**
    caerulescens caerulescens 85
    rossii 74, 85, 91
Chlidonias niger 34, 74, 207
Chondestes grammacus 75, **363–64**
Chordeiles minor 39, 41, 47, 53, 57, 60, 67, 74, **222–23**
Chroicocephalus cirrocephalus 468
    maculipennis 468
    novaehollandiae 468
    philadelphia 57, 59, 74, **195–96**, 197
    ridibundus 59, 75, **196–97**
    serranus 60, 197, 468–69
Circus cyaneus 49, 74, **155–56**
Cistothorus palustris 46, 48, 60, 74, 170, **294–95**
    platensis 39, 74, 76, **292–93**
Clangula hyemalis 76, **115–16**

Coccothraustes vespertinus 54, 74, **409–10**
Coccyzus americanus 46, **211–12**
    erythropthalmus 45, 46, **212–13**
Colaptes auratus 58, 61, **235**, 314
Colinus virginianus 45, 74, **123–24**
Columba livia 60, **208**, 216
Contopus cooperi **242–43**
    virens 163, **243–44**
Coragyps atratus 39, 67, 68, 73, 76, **150–51**
Corvus brachyrhynchos 48, 51, 58, 74, **268–70**, 271
    corax 36, 39, 68, 74, **271–72**
    ossifragus 48, 58, 74, **270–71**, 286, 288, 290, 296, 299
Coturnicops noveboracensis **166–68**
Cyanocitta cristata 39, 46, 48, 51, 58, **267–68**
Cygnus buccinator 73, 93
    colombianus 54, 93
    olor 60, 63, **92–93**

Dendrocygna bicolor 74, **83**
Dolichonyx oryzivorus 40, 52, 56, 59, 73, **387–88**
Dryocopus pileatus 35, 36, 73, **236–37**
Dumetella carolinensis 46, **310**, 311, 312

Ectopistes migratorius **209–10**
Egretta caerulea 76, 142, **143**
    thula 76, **142–43**
    tricolor **144**, 148
Elanoides forficatus 34, 74, **153–54**
Empidonax alnorum 37, 38, 40, 68, 75, 76, **247–49**, 250–251
    alnorum/traillii 68, 247–248, **249–50**, 251
    difficilis 58, 248, 249, 253
    flaviventris 41, **244–45**, 252
    hammondii 253
    minimus 45, 46, 67, 248, 250, **252–53**
    occidentalis 248, 249, 253
    traillii 45, 68, 74, 247–49, **250–51**
    traillii adastus 251
    traillii brewsteri 251
    traillii traillii 45, 68, 74, 247–249, **250–51**
    virescens 37, 60, 65, 73, 76, **245–47**
Eremophila alpestris 22, 37, 39, 41, 50, 59, 65, 73, **273–75**, 317, 390
    alpestris alpestris **273–74**
    alpestris praticola 273, **274–75**
Eudocimus albus 29, 74, 150
Euphagus carolinus 48, 74, **391–92**, 393
    cyanocephalus 35, **392–93**
Euplectes orix/franciscanus 60

Falco columbarius 48, 49, 74, **239**
    peregrinus anatum 60, 73, **240–41**
    peregrinus tundrius 60, 73, **240–41**

rusticolus, 241
sparverius 38, 39, 41, 47, 48, 51, 61, 68, 71, 74, 124, **237–39**
Fregata magnificens 29
Fulica americana 37, 41, **174–75**

Gallinago delicata 48, 54, **191–92**
Gallinula chloropus 37, 40, 45, 60, **173–74**
Gavia immer 53, 54, **130**
   pacifica 74
   stellata 76, **129**
Gelochelidon nilotica 73, 205
Geothlypis formosa 37, 38, 41, 56, 67, 73, 76, **332–33**
   philadelphia 57, 65, 227, **331–32**
   tolmiei 330
   trichas **333–34**
Grus canadensis 34, 35, 73, **175–76**
   grus 176

Haemorhous mexicanus 39, 48, 60, 74, **399–400**, 401
   purpureus 37, 74, **400–401**
Haliaeetus leucocephalus 60, 73, **154–55**
Helmitheros vermivorum 32, 38, 57, 60, **320–21**, 356
Himantopus mexicanus 29, 74, 179
Hirundo rustica 39, 46, 61, **282–83**
Histrionicus histrionicus 74, 114
Hydrocoloeus minutus 75, 197
Hydroprogne caspia **205–6**
Hylocichla mustelina 46, **308**

Icteria virens auricollis 355
   virens virens 49, 57, 67, **354–55**
Icterus bullockii 30, 398
   galbula 46, **397–98**
   spurius 45, 56, 57, 73, **396–97**
Ictinia mississippiensis 73, 154
Ixobrychus exilis 40, 45, 46, 60, 74, **139–40**, 148, 170
Ixoreus naevius 30, 75, 309

Junco (hyemalis) hyemalis 48, 51, 77, **377–78**
   (*hyemalis*) *mearnsi* 77, 378
   (*hyemalis*) *montanus* 75, 77, 378

Lanius excubitor 59, 76, **259–60**
   ludovicianus 74, **258–59**
Larosterna inca 469
Larus argentatus 33, 47, 50, 51, 74, 198, **200–201**, 202, 204
   belcheri 469
   delawarensis 50, 59, 74, 197, **198–99**, 200
   fuscus graellsii 50, 75, 201, **202–3**
   glaucoides glaucoides **201–2**
   glaucoides kumlieni 50, **201–2**, 203
   hyperboreus 50, 201, **203**
   marinus 47, 50, 74, **204**
   thayeri 30, 75, 202
Laterallus jamaicensis 74, 168
Leucophaeus atricilla 50, 73, **197–98**
   modestus 469
   pipixcan 30, 58, 75, 198
Limnodromus griseus 34, **190–91**, 249
   scolopaceus 74, 190, 249
Limnothlypis swainsonii 56, 75, **327**
Limosa fedoa 190–91
   haemastica 190–91
Lonchura malacca 60
   punctulata 60
Lophodytes cucullatus 48, 74, **118–19**
Loxia curvirostra 39, 50, 54, 68, 76, 288, **402–3**
   leucoptera 50, 68, 76, **403–4**

Megaceryle alcyon 47, 71, **227–28**
Megascops asio 47, 61, 68, **215–16**, 219
Melanerpes carolinus 46, 48, 56, 73, **230–31**, 233, 234
   erythrocephalus 76, **229–30**, 314
Melanitta americana 53, 114, **115**
   fusca 53, 74, **114–15**
   perspicillata 53, **113–14**
Meleagris gallopavo 36, 68, 69, 73, **127–29**
Melopsittacus undulatus 60
Melospiza georgiana 45, 46, 60, 67, 374, **375–76**
   lincolnii 49, **374**
   melodia 39, 46, 48, 51, **373–74**
Mergus merganser 49, **119–20**
   serrator 74, 76, **120–21**
Mimus polyglottos 74, **312–13**
Mniotilta varia 38, 55, 56, 60, 67, **325–26**, 356
Molothrus ater 338, 356, 386, 391, 393, **395–96**
Morus bassanus 416
Myadestes townsendi 58
Mycteria americana 60, **134**
Myiarchus cinerascens 57, 58, 75, 256
   crinitus 61, **255–56**
Myiopsitta monachus 47, 60, 61, 68, 73, **241–42**

Numenius americanus 185
   phaeopus 53, 185
Numida meleagris 60
Nyctanassa violacea 36, 39, 47, 76, **148**
Nycticorax nycticorax 37, 41, 48, 49, 51, **146–47**, 170

Oceanites oceanicus 416
Oceanodroma leucorhoa 416
Oenanthe oenanthe **300**
Onychoprion anaethetus 75
   fuscatus 52, 197

Oporornis agilis 52, 74, 227, **330–31**, 332
Oreothlypis celata 58, **328–29**
    peregrina **327–28**
    (ruficapilla) ridgwayi 330
    (ruficapilla) ruficapilla 37, 38, 41, 55, 58, **329–30**, 354
Oxyura jamaicensis 41, 49, 61, **121–23**

Pandion haliaetus 37, 38, 39, 47, 60, 61, 71, 73, **152–53**
Parkesia motacilla 37, 38, 57, 242, **321**
    noveboracensis 41, 56, **322**
Passer domesticus 45, 60, 252, 276, 300, 386, 387, **410–11**
Passerculus (sandwichensis) princeps 77, 366
    (sandwichensis) sandwichensis 39, 50, 73, 77, **364–66**, 369
Passerella (iliaca) fuliginosa 77, 373
    (iliaca) iliaca 49, 59, 77, **372–73**
    (iliaca) schistacea 77, 373
Passerina caerulea 56, 73, 76, **383–84**
    ciris 56, 58, 75, **385–86**
    cyanea 56, 64, **384–85**
Pelecanus erythrorhynchos 74, **137–38**
    occidentalis 74, 137–38
Petrochelidon fulva 57, 58, 75, 276, 282
    pyrrhonota 37, 38, 39, 47, 58, 61, 68, 73, **280–82**
Peuaea aestivalis 423, 425, 461
Phalacrocorax auritus 47, 48, 74, **134–36**, 137
    carbo 74, **136–37**
Phalaropus fulicarius 194, 197
    lobatus 34, **193–94**
    tricolor 75, 194
Phasianus colchicus 36, 46, 48, 60, 74, 124, **125–26**
Pheucticus ludovicianus 45, 49, **382–83**
    melanocephalus 58, 383
Pica hudsonia **268**
Picoides arcticus **234**
    pubescens 48, 51, 230, 231, **232–33**
    villosus 48, 51, 61, 230, **233–34**, 289
Pinicola enucleator 74, 76, 182, **398–99**
Pipilo erythrophthalmus 38, 45, 46, 49, 60, 74, **355–57**
    maculatus 357
Piranga ludoviciana 30, 58, 380
    olivacea **379–80**
    rubra 56, 73, 76, **378–79**
Pitangus sulphuratus 258
Plectrophenax nivalis 22, 50, 59, 65, **317–18**, 390, 405
Plegadis chihi 73
    falcinellus 47, 73, 76, **149–50**
Pluvialis dominica 34, **177**
    squatarola **176–77**
Podiceps auritus 53, 54, 74, 76, **132–33**
    grisegena 53, 54, 76, **133**
    nigricollis 74, 132–33

Podilymbus podiceps 35, **130–31**
Poecile atricapillus 46, 48, 51, 54, 59, **283–85**
    carolinensis 284–85
    hudsonicus 35, 50, 76, **285**
Polioptila caerulea 46, 52, 56, 66, 73, **296–98**
Pooecetes gramineus 40, 65, 73, **361–63**, 390
Porphyrio martinicus 29, 74, 174
Porzana carolina 37, 40, 45, 46, 60, 67, 170, **172–73**
Progne subis 36, 38, 61, 71, 76, **275–76**, 280, 282, 314
Protonotaria citrea 56, 73, 76, **326–27**
Psittacula krameri 60, 68, 241–42
Puffinus gravis 29
    griseus 29
    lherminieri 416
    puffinus 29, 74
Pycnonotus cafer 60

Quiscalus major 73, **394–95**
    quiscula **393–94**, 408
    quiscula stonei 394–95
    quiscula versicolor 394–95

Rallus crepitans 39, 60, **168–69**, 170, 371, 372
    elegans 37, 40, 60, 63, 74, **169–70**
    limicola 19, 65, 67, 167–68, **170–71**, 172, 240
Recurvirostra americana 74, 179
Regulus calendula 49, 262, **299**
    satrapa 39, 48, 50, 73, **298–99**
Riparia riparia 39, 47, 57, 68, 71, 74, 228, **279–80**
Rissa tridactyla 196
Rynchops niger **207–8**

Sayornis phoebe 39, 49, 58, 61, 67, 68, 71, **253–55**
    saya 34, 58, 75, 254–55
Scolopax minor 41, 42, 59, 67, **192–93**
Seiurus aurocapilla 38, 55, **318–19**, 356
Selasphorus calliope 227
    rufus 58, 75, 227
Setophaga americana 55, 56, 338, **339**
    caerulescens 55, **345–46**
    castanea 55, **340–41**
    cerulea 76, **337–38**
    citrina 55, 56, 57, 67, **334–35**
    coronata 349
    (coronata) auduboni 30, 34, 35, 75, 77, 349–50
    (coronata) coronata 52, 55, 77, 290, **349–50**
    discolor 56, **351**
    dominica 56, 73, 76, **350**
    dominica albilora 350
    dominica dominica 350
    fusca **341–42**
    magnolia 55, **340**
    nigrescens 30, 352
    occidentalis 352

palmarum hypochrysea 55, 290, **346–47**
palmarum palmarum 68, **346–47**
pensylvanica 55, 64, 74, 330, **343–44**
petechia 46, 56, 57, **342–43**
pinus 38–39, 49, 68, 290, **347–48**, 350
ruticilla 46, 55, **335–36**
striata 55, **344–45**
tigrina 52, 55, 57, **337**
townsendi 352
virens 55, **351–52**
Sialia currucoides 58, 302
 sialis 37, 38, 45, 99, 61, 66, 71, 73, **300–302**, 314
Sitta canadensis 39, 49, 50, 54, 68, 73, **287–88**, 289, 290
 carolinensis 48, 51, 61, **288–89**, 314
Somateria mollisima 74, 114
 spectabilis 114
Sphyrapicus varius 48, 49, 74, **231–32**
Spinus pinus 39, 49, 54, 68, 288, **406–7**
 tristis 42, 48, **407–8**
Spiza americana 75, 76, **386–87**
Spizella pallida 75, 76, **359–60**
 passerina 49, 320, **358–59**
 pusilla 67, 73, 74, 358, **360–61**
Spizelloides arborea 48, 51, 59, 74, **357–58**
Stelgidopteryx serripennis 39, 46, 57, 68, **278**, 279
Stercorarius longicaudus 29
 parasiticus 416
Sterna dougallii 74
 forsteri 34, 73, 207
 hirundo 33, **206–7**
 paradisaea 75
Sternula antillarum 34, **205**
Streptopelia decaocto 60, 73, **208–9**
 roseogrisea 60, 209
Strix varia 36, 76, 216, **218–19**
Sturnella magna 22, 39, 65, 73, 365, **389–90**
Sturnus vulgaris 48, 51, 60, 124, 229, 230, 276, 281, 300, **313–14**
Sula leucogaster 74

Tachycineta bicolor 46, 61, **277–78**, 284, 291, 314
 thalassina 57

Thalasseus maximus 30, 75
 sandvicensis 30, 75
Thryomanes bewickii 296
Thryothorus ludovicianus 46, 48, **295–96**
Toxostoma rufum 38, 46, 49, 60, 67, 74, 260, **311–12**
Tringa flavipes **183–84**, 185
 melanoleuca 54, **181–82**
 (semipalmata) inornata 34, 74, 77, **182–83**
 (semipalmata) semipalmata 34, 73, 77, 182, 183
 solitaria 54, **181**
Troglodytes aedon 45, 61, **290–91**
 hiemalis 48, **291–92**
Turdus migratorius migratorius 39, 46, 48, 59, **309**
 migratorius nigrideus 309
Tyrannus forficatus 34, 256
 tyrannus 39, 52, 56, 58, **257–58**
 verticalis 57, 58, **256**
Tyto alba 36, 38, 39, 47, 61, 67, 71, 76, **213–15**

Uria lomvia 416

Vermivora chrysoptera 57, 65, 74, 242, 248, **322–23**
 cyanoptera 32, 57, 60, 67, 248, 322, 323, **324–25**
Vireo bellii 58, **261–62**
 flavifrons 56, 67, 257, **262–63**
 gilvus 46, **264–65**
 griseus 67, **260–61**
 olivaceus 39, 46, **266**
 philadelphicus **265–66**
 solitarius 261, **263–64**

Xanthocephalus xanthocephalus 35, 75, **391**
Xema sabini 30

Zenaida asiatica 58, 75, **210**
 macroura 46, **210–11**
Zonotrichia albicollis 48, 51, 374, **376**
 leucophrys gambelii 377
 leucophrys leucophrys **377**

Index of Subjects

Page numbers followed by f refer to figures.

Adelgid, Woolly 246
alien species. *See* nonnative species
American Birds 29, 76
American Revolution, effect on study area 8, 24, 59
Audubon, John James 1, 34
Audubon Field Notes 29, 45, 50, 76
Audubon Magazine 29, 75
Auk 29

Bibby's Pond. *See* Van Cortlandt Lake
Bicknell, Eugene P. 1, 9, 27–28, 29, 31–32, 39–40, 75, 76
Biological Field Club (DeWitt Clinton High School) 51
bird banding 32–33
    Canada Goose recoveries 32–33, 89–90
    Herring Gull recoveries 33
Bird-Banding (journal) 29
Bird Banding Laboratory 32, 42, 207, 250
Bird-Lore 29, 45, 75, 76
Bison, American, in Van Cortlandt Park 24–25
blowback drift-migrants: origin and discussion 57–58
Bobwhite, Northern, origin of local populations 123–24
Breeding Bird Census (formal) 45, 50
breeding bird censuses (generic) 40
    issues confronting 40–42, 45
    Northwest Forest 40, 46
    study area 40
    Van Cortlandt Park 40, 47, 446–47, 451
    Van Cortlandt Swamp 40, 45, 47, 448–51
Breeding Bird Survey (US Dept. of Interior) 70
Broadway 4, 10, 19
Bronx, the
    areas relinquished by Westchester Co. 1, 9
    precontact forest 7
    species never recorded in 29–30
Bronx County Bird Club 28, 29, 40, 51, 75
Bronx Park 4, 9, 39, 51
Bronx Region (Griscom's) 29
Bronx Region (Kuerzi's) 29
Bronx River 4, 8, 47, 54
Bronx-Westchester Christmas Bird Count 33, 49, 51–52. *See also* Christmas Bird Counts
    long-term trends 51, 77

Bronx Zoo 25, 33, 39, 51, 79
Brooklyn Bird Club 34–35
Bull, John 30
Bulletin of the Nuttall Ornithological Club 29

Cape May, NJ 52, 57
cemeteries. *See names of individual cemeteries*
Central Park 9, 46, 53, 77
    avifauna compared to Prospect Park 34–36, 417–29
    avifauna compared to Van Cortlandt Park 34–36, 417–29
    cumulative breeding species 35–36, 426–29
    cumulative species list 34, 417–24
    winter species 48
Central (Park) Ave. 10, 79
Chestnut, American, loss to Chestnut Blight 24, 60, 245, 246
Christmas Bird Counts 28, 33, 48, 49, 50–52, 77. *See also* Bronx-Westchester Christmas Bird Count
Clason Point 10, 50, 51
Coastal Lowlands 4
Colen Donck 8
Concourse Subway Yards 25
conjunctivitis in House Finches 74, 399–400
continuing regional species disappearances
    American Tree Sparrow 357–58
    Black-crowned Night-Heron 147
    Brown Thrasher 311–12
    Eastern Towhee 356–57
Cortlandt's Wood in American Revolution 24
Crossbill, Red, call types 402–3
Cruickshank, Allan 28, 29, 31, 40, 63

data cutoff date 77
DeWitt Clinton High School 25
Dovekie, 1932 fallout of 1, 194–95
Dutch colonists 8, 22
Dutchess County, birds of 33
Dwight, Jonathan 1, 27, 31–32, 75

Eastern Deciduous Forest 7
eBird data 33, 76
ecological connectivity 36

ecological succession 24, 26, 38
Elm, American, loss to Dutch Elm Fungus 24
exotic species. *See* nonnative species

Flycatchers, Alder and Willow, convoluted regional history of 247–51
Fordham Ridge 4
Forest and Stream 29, 124
Fort Tryon Park 4
Foster, L. S. 31–32
Friends of Van Cortlandt Park 62

Goldfinch, European, persistence of 408–9
Great Depression 1, 63, 195
Griscom, Ludlow 29, 32, 75–76
Gull, Andean, in the Bronx 468–69

Harlem River 4, 10, 36, 38, 47, 49, 50, 67
Harlem River Ship Canal 12, 19
Hemlock, Eastern, loss to Woolly Adelgid 246
Hemlock Grove, Bronx Park, loss of 246
Henry Hudson Parkway 16
    destruction of study area habitat 22
Hillview Reservoir 1, 4, 10, 12, 25, 28, 31, 51, 61, 68
    breeding species 39
    conifer groves, loss of 25, 26, 39, 50
    construction of 25
    nonbreeding species 49–50
history and origin of introduced species
    American Black Duck (alleged) 97–98
    Canada Goose 90
    Eurasian Collared-Dove 208–9
    European Goldfinch 408–9
    European Starling 314
    House Finch 399
    House Sparrow 410
    Mallard 100
    Monk Parakeet 241
    Mute Swan 92
    Rock Pigeon 208
    Rose-ringed Parakeet 241
    Trumpeter Swan 73, 93
history and origins of New York City area
    Northern Bobwhite 123–24
    Ring-necked Pheasant 125
    Wild Turkey 128
Hornaday, William 24–25
Hudson, Henry 7
Hudson Highlands 4
Hudson River 4, 8, 10, 12, 26, 32, 36, 49, 50, 53, 54, 59
Hunter College Bronx Campus 25
hurricanes, notable bird-transporting
    Carol (1954) 394
    Connie (1955) 190, 205
    Donna (1960) 189, 190, 206
    Edna (1954) 212
    Emily (1987) 52
    Gloria (1985) 52, 57, 337
    Great Atlantic (1938) 24, 212, 371
    Hazel (1954) 212
    Irene (2011) 64, 137, 179, 194, 258, 344, 385
    Sandy (2012) 64, 194, 344, 385
    Wilma (2005) 52
Hutchinson River and Marshes 43
    breeding birds 43
Hutchinson River Parkway 43
hybridization between
    Alder and Willow Flycatchers 248
    Black-capped and Carolina Chickadees 284
    Blue-winged and Golden-winged Warblers 324–25
    Cackling and Canada Geese 91
    Mallard and American Black Duck 99, 100–101

identification of
    Alder and Willow Flycatchers 249–50
    Baltimore and Bullock's Oriole females 398
    Bicknell's Thrush 32
    Hoary and Greater Redpolls 405–6
    Nelson's Sparrow subspecies 370
    Yellow Rail by voice 167–68
invasive species 24, 64, 66, 79
Inwood (Hill) Park 53, 54

Jamaica Bay 47
James Bay 33
Jerome Ave. 10, 18
Jerome Meadows 4, 7, 12, 25, 28
    breeding species 40
Jerome Park Racetrack 12, 25
Jerome (Park) Reservoir 1, 10, 11–12, 25, 26, 28, 31, 34, 51, 68
    breeding species 39
    construction of 10, 25
    importance to waterbirds 49–50
    oiled birds at 107
Jerome Swamp 12, 25
    creation of 25
    possible breeding species 40
Jerome Subway Yards 25
Journal of Field Ornithology 29, 45

Kingbird 29, 76, 77
King's Bridge, The 19
Kingsbridge Heights 10
Kingsbridge Island 4

Kingsbridge Meadows 4, 7, 12, 18, 19, 28, 31
   breeding species 39
   destruction of 19, 39
   importance of 39
Kiskadee, Great, origin of vagrant 258
Kuerzi, John 29, 75

Lenni-Lenape Indians 7, 22
Linnaean Newsletter 29

Major Deegan Expressway 62
   construction of 18
malaria, *Plasmodium falciparum* 10
malaria, *Plasmodium vivax* 10
mammalian disrupters of study area birds 22, 38, 61–62, 125–26, 224, 356–57
mammals in the Bronx and study area 60, 68
   introduced 61–62
   native 62, 63
Manhattan Hills 4
Manhattan Prong 4
Marble Hill 19
migration
   age of vagrants 58
   blowback drift-migrants 57–58
   counts and maxima 52, 55, 56–57
   declines of spring Neotropical migrants 56, 455
   differential by age-groups 54, 56–57
   diurnal raptor 53, 54, 57, 68
   diurnal songbird 54
   earlier arrival of spring Neotropical migrants 56, 456–57
   fall 52–53, 54, 56–59
   fallouts 52–53, 55
   hurricanes' effect on 52
   irruptive species 54, 59
   leading lines 53–54
   nocturnal 53, 55
   numbers by night vs. by day 52
   routes of Palm Warbler subspecies 346
   shorebird 53, 54, 57
   spring 52, 54–56
   spring overshoots 55–56
   summer 56
   trans-Gulf overshoots 56
   vagrants 58
   weather and 52–53, 55, 57, 59
   winter 50–51, 59
Moses, Robert 10, 18, 19, 28, 34
Mosholu 7–8, 22
Mosholu Ave. 9, 22, 63
Mosholu Brook. *See* Tibbett's Brook
Mosholu Golf Course 24, 25
Mosholu Parkway Extension 16, 18

Mosholu Parkway Greenbelt 26
Mt. Auburn Cemetery (MA) 39
Mt. St. Vincent 26

nest box use by
   American Kestrel 61, 238
   Barn Owl 36, 213–15
   Black-capped Chickadee 284
   Eastern Bluebird 61, 300–301
   Eastern Phoebe 61
   Eastern Screech-Owl 216
   Hooded Merganser 119
   House Wren 291
   Northern Saw-whet Owl 220
   Purple Martin 36, 61, 275
   Tree Swallow 61, 277
New Amsterdam 8
New Croton Aqueduct 24
New England Geomorphic Province 4
New Jersey Palisades
   breeding Great Horned Owls 217
   breeding Peregrine Falcons 240
   fall migration across Hudson River 54, 58–59
   spring migration fallouts on 54
New Netherland 8
New Parks Act (1884) 9
New York Botanical Garden. *See* Bronx Park
New York City Area/Region 29
   avian status changes after 1964 73–75
   geographic limits for regional avifaunas 77
New York City Audubon Society 47–48
New York City Department of Environmental Conservation 28
New York City Department of Parks and Recreation 8, 19, 25, 26, 47, 66, 69
   Urban Park Ranger Program 26, 63
New York City maritime island breeding birds 47–48
New York State Breeding Bird Atlas I (1980–1985) 40, 42–43, 44f, 45
New York State Breeding Bird Atlas II (2000–2005) 40, 42–43, 44f, 45
New York State Department of Environmental Conservation 66
New York State Museum 31
New York Zoological Society 25
nonnative species 24, 60–62
North American Birds 29, 76
North Brother Island 47, 48
Northwest Bronx
   automobiles, increase in 10
   buses replacing trolleys in 10
   geology and topography 1, 4, 7
   historical forests 7, 24

Northwest Bronx *(continued)*
   historical surface waters 4
   horsecars in 10
   omnibuses in 10
   surface water and wetlands (fresh) 4
   surface water and wetlands (salt) 4
Northwest Forest (Van Cortlandt Park) 4, 9, 15, 22, 24, 28, 40, 45, 46, 53, 62, 63, 64, 65
   breeding species in 1991–92 46, 451

Old Croton Aqueduct 24
Olmstead, Frederick Law 26, 34

Paparinemin. *See* Kingsbridge Island
Parade Ground (Van Cortlandt Park) 7, 9, 22, 28, 33, 61, 63–64, 68, 69, 70–71
   history 7–8, 22
   importance to birds 22, 36, 49, 50
   military use of 22
   reseeding 22, 26, 65
Pelham Bay Park (area) 9, 24, 26, 35, 51, 53, 57
   Baychester Marshes 51
   City Island 10
   Eastchester Bay 50
   Huckleberry Island 47
   waterbird breeding colonies 47–48
Piermont Marsh 36
Piermont Pier 53
Pine Barrens (Long Island) 7
Pine Barrens (NJ) 7
predation 22, 38, 62, 68
predators, potential quadruped 38, 60–62
Proceedings of the Linnaean Society of New York 29
Prospect Park 9, 46, 53, 77
   avifauna compared to Central Park 34–36, 417–29
   avifauna compared to Van Cortlandt Park 34–36, 417–29
   cumulative breeding species 35–36, 426–29
   cumulative species list 34–35, 417–24
   winter species 48
Putnam Trail (Van Cortlandt Park) 25, 67
   plans for bike path 66

radar, in study of bird migration 52
raptors, migration of 53, 54, 57
reservoirs. *See names of individual reservoirs*
Riverdale 4, 7, 10, 12, 24, 29, 51, 54, 75
Riverdale Ridge 4, 53
Rockland County, birds of 33, 53

Sandy Hook, NJ 79
Scale Insect
   Elongate Hemlock 246
   Pine Needle 350

scaup species' habitat preferences 112
Screech-Owl, Eastern
   breeding densities 215–16
   breeding metapopulations 216
seasons, definition of 27
separate feral and wild populations of
   American Black Duck (alleged) 97–98
   Canada Goose 32–33, 89–91
   Mallard 99–101
sewage treatment plant (Yonkers) 26
Sherman Creek 4
Sialis Bird Club 29
South Brother Island 47, 48
species-level taxonomy used 77
species that have declined
   in New York City area 73–74
   in study area 36–37, 45–46, 60
   on Bronx-Westchester Christmas Bird Counts 77
species that have increased
   in New York City area 73–74
   in study area 36–37, 45–46, 48
   on Bronx-Westchester Christmas Bird Counts 77
specimens of birds 31–32
   taken in Kingsbridge Meadows 31
   taken in Riverdale 31–32
   taken in Van Cortlandt Park 31
Spruce Budworm warblers 327, 337, 340, 344
Spuyten Duyvil 1, 7
Spuyten Duyvil Creek 4, 7, 10, 12
   before Ship Canal creation 4, 7, 9–10
   filling of 19
Spuyten Duyvil Ridge. *See* Riverdale Ridge
streetcars. *See* trolleys
study area
   at-risk species 60, 64, 458
   bird control in 39, 61
   breeders gained since 1872 37, 441–42
   breeders increasing or decreasing after 2000 38, 444
   breeders lost since 1872 37, 440–41
   breeders that never bred in Central or Prospect Parks 35, 429
   breeders that use artificial structures 36, 38–39, 61, 444
   burning by Lenni-Lenape 7
   core winter residents 50, 452–53
   cumulative breeding species list 426–29
   cumulative species list 60, 417–24
   definition of 11–12
   egg and nest dates in 78
   future of 67–71
   historical spring earliest arrival dates by years 55–56, 455

historical spring migrant maxima by years 55–56, 456–57
information needs in 67–69
nearby species not found in 28, 416
nonnative birds in 60–61, 68
nonnative mammals in 61–62
nonnative plants in 24, 62, 64, 65
observer coverage, historical 27–28
on- or near-ground breeders 37, 38, 442
primary breeding species 38, 443
research needs for birds 69–70
resource concerns 59–67
resource management actions recommended 70–71
secondary breeding species 38, 443
species found only outside Van Cortlandt Park 28, 417
species gained since 1942 28, 415
species probably underreported 76
subareas of 10–26
unconfirmed or suspected breeders 36, 434
uniquely breeding Bronx species 37–38
uniquely occurring species 35, 425
subspecies in
   American Robin 309
   Cackling Goose 88
   Canada Goose 91
   Common Grackle 394
   Horned Lark 273–75
   Iceland Gull 201–2
   Lesser Black-backed Gull 203
   Nashville Warbler 354
   Nelson's Sparrow 370
   Palm Warbler 346–47
   Peregrine Falcon 240–41
   Snow Goose 85
   White-crowned Sparrow 377
   Willow Flycatcher 251
   Yellow-breasted Chat 355
   Yellow-throated Warbler 350
Swan Point Cemetery (RI) 39

Throgg's Neck 79
Throgs Neck Bridge 79
Tibbett's Brook 4, 10, 12, 14, 16, 18, 19
   daylighting of 67
   history 7–8
   water quality 62
Tibbett's Brook Park 14
Tippet's Wood in the American Revolution 24
Tippit, George 8
trains
   IND subway 25
   IRT subway 10, 25
   New York Central and Hudson River Railroad 9–10
   New York Central Putnam Division 4, 9–10, 15–16, 18, 25, 26, 27
   Yonkers Rapid Transit 9, 24, 26
Triboro Bridge 47
trolleys 10

US Army Corps of Engineers 18

vagrants
   blowback drift-migrants 57–58
   Greenland origin of geese unsupported 33, 89–90
   hummingbirds 227
Valentine's Hill 14
Van Cortlandt family and descendants 8–9, 14, 28
Van Cortlandt Golf Course 16, 18, 24, 25, 61
Van Cortlandt Lake 1, 9, 14, 19, 24, 26, 27, 28
   boathouse-breeding Chimney Swifts 225
   origin 8, 14, 28
   sedimentation and water quality 19, 26, 62
   spring shorebird fallout 53
Van Cortlandt Park 4, 10, 11, 12, 31, 51
   American Bison in 24–25
   avifauna compared to Central Park 34–36, 417–29
   avifauna compared to Prospect Park 34–36, 417–29
   Barn Owl hacking in 214
   breeding species 35–36, 426–29
   cumulative breeding species list 35–36, 426–29
   cumulative species list 34–36, 417–24
   devastation by Robert Moses 19, 28
   establishment of 9, 26
   forest stand age 22, 24, 64, 68
   forests 22, 24–25, 64–65
   hawkwatch 57
   illicit grazing and hunting 63
   Master Plan (2014) 65, 66–67, 69
   One Million Trees Program 64–65
   plant ecology 63–65, 68
   recent forest openings 64
   resource management actions taken 26, 61, 62, 64–66
   vandalism and fires 19, 62–63
   winter species 48–49
Van Cortlandt Park, locations in
   Allen Shandler Recreation Area 4, 25, 40, 47, 64
   Bibby's Pond. *See* Van Cortlandt Lake
   Birch Pond 18
   Cass Gallagher Trail 25
   Croton Forest. *See* Ridge, the
   Dutch Gardens 9, 19, 34, 59
   Dutch Gardens Marsh 12, 14, 18, 19
   Elm Pond 18
   Island, the 19
   John Kieran Trail 18, 25

Van Cortlandt Park, locations in *(continued)*
  John Muir Trail 25
  Lake Marsh 19, 26
  Lincoln Marsh 16, 26, 68
  Maple Pond 18
  Northeast Forest 4, 22, 28, 55, 62–63, 64, 67
  Northwest Forest. *See separate entry*
  Old Croton Aqueduct Trail 25, 26
  Parade Ground. *See separate entry*
  Putnam Trail. *See separate entry*
  Ridge, the 4, 15–16, 22, 24, 26, 28, 53, 55, 57, 64
  Southern Forest 25
  Sycamore Pond 18
  Sycamore Swamp 22, 28, 67, 68, 71
  Tibbett's Marsh (new) 19, 66, 68
  Tibbett's Valley 4
  Triangle, the. *See* Lake Marsh
  Van Cortlandt Mansion 8, 9, 18, 19, 22, 34, 61
  Vault Hill 4, 8, 9, 15, 18, 22, 24, 28, 36, 53, 64
Van Cortlandt Park Conservancy 25, 66, 69
Van Cortlandt Stadium 19
Van Cortlandt Swamp 4, 24, 25, 28, 55, 67, 71
  breeding bird censuses 1937–2014 40, 45–47, 448–51
  Causeway, The 18
  devastation by Robert Moses 18, 28
  history of 14–16, 18–19, 27–28, 63
  importance to waterbirds 15, 18–19, 28
  Phragmites replacing cattails in 19, 36, 45, 62, 65
  vegetational changes in 19, 36, 45–46, 60, 62, 64, 65–66
  winter species 48, 50, 51, 453–54
van der Donck, Adriaen 8
Van der Donck Meadows 4, 19, 22, 67
Vaux, Calvert 34
VertNet (Vertebrate Network) 31–32
Vireo, Bell's, New York City area historical status distortion 261–62

Warbler, Calaveras Nashville, in eastern North America 330
Weber, William 29
Weckquaesgeeks 7
West Nile Virus 74, 267, 269, 270, 271
Westchester County 1, 8, 9, 54, 75
  birds of 33, 36
Westchester Path 7–8
Wildlife Conservation Society. *See* Bronx Zoo
Winter Bird Population Studies 50
  Jerome Reservoir 50
  Van Cortlandt Swamp 50
Woodlawn Cemetery 4, 10, 11, 25, 28, 31, 34, 47, 51, 61, 68–69
  2015 Christmas Bird Count 49
  breeding species 38–39
  ponds 25, 48
  winter species 49
Works Progress Administration (WPA) 16

Yonkers Racetrack 14